MESOSCALE DYNAMICS

Mesoscale weather systems are responsible for numerous natural disasters, such as damaging winds, blizzards, and flash flooding. A fundamental understanding of the underlying dynamics involved in these weather systems is essential in forecasting their occurrence. This book provides a systematic approach to this subject, and covers a more complete spectrum of mesoscale dynamics than other texts.

The opening chapters introduce the basic equations governing mesoscale weather systems and their approximations. The subsequent chapters cover four major areas of mesoscale dynamics: wave dynamics, moist convection, front dynamics, and mesoscale modeling. Wave dynamics covers wave generation and maintenance, orographically forced flow, and thermally forced flow. The moist convection part covers mesoscale instabilities, isolated storms, mesoscale convective systems, orographic precipitation, and introduces tropical cyclone dynamics. The dynamics of synoptic-scale fronts, mesoscale fronts, and jet streaks are discussed in the front dynamics part. The last part of the book introduces basic numerical modeling techniques, parameterizations of major physical processes, and the foundation for mesoscale numerical weather prediction.

Mesoscale Dynamics is an ideal reference on this topic for researchers in meteorology and atmospheric science. This book could also serve as a textbook for graduate students, and it contains over 100 problems, with password-protected solutions available to instructors at www.cambridge.org/9780521808750. Modeling projects, providing hands-on practice for building simple models of stratified fluid flow from a one-dimensional advection equation, are also described.

YUH-LANG LIN's research in mesoscale dynamics and modeling includes moist convection, orographic effects on airflow and weather systems, gravity waves, tropical, lee and coastal cyclogeneses, storm dynamics, wake vortex, aviation turbulence, forest fire, and modeling of the Martian atmosphere.

MESOSCALE DYNAMICS

By Yuh-Lang Lin

CAMBRIDGE
UNIVERSITY PRESS

CAMBRIDGE UNIVERSITY PRESS
Cambridge, New York, Melbourne, Madrid, Cape Town, Singapore,
São Paulo, Delhi, Dubai, Tokyo, Mexico City

Cambridge University Press
The Edinburgh Building, Cambridge CB2 8RU, UK

Published in the United States of America by Cambridge University Press, New York

www.cambridge.org
Information on this title: www.cambridge.org/9780521004848

© Cambridge University Press 2007

First published 2007
First paperback edition 2010

A catalogue record for this publication is available from the British Library

ISBN 978-0-521-80875-0 Hardback
ISBN 978-0-521-00484-8 Paperback

Contents

Preface

Mesoscale weather systems, such as thunderstorms, mesoscale convective systems, supercells, fronts, jet streaks, gravity waves, severe downslope winds, low-level jets, sea breezes, heat island circulations, and clear air turbulence, are responsible for numerous natural disasters, such as blizzards, torrential rain, flash flooding, damaging winds, and aviation accidents. Thus, a fundamental understanding of their underlying dynamics, the mesoscale dynamics, is essential to help forecast their occurrence. Although textbooks are available in individual subdisciplines such as cloud dynamics, storm dynamics, convection, and synoptic-dynamic meteorology, there are no textbooks which take a systematic approach and cover a more complete spectrum of the mesoscale dynamics. In particular, due to the rapid advancements in research in the past three decades or so, there is a need for a mesoscale dynamics textbook.

The text is presented in four parts: wave dynamics, moist convection, front dynamics, and mesoscale modeling. There are no clear boundaries among these parts. In the opening chapters, the basic equations governing mesoscale weather systems and their approximations are introduced. The wave dynamics include wave generation and maintenance, orographically forced flow, and thermally forced flow. The moist convection part includes mesoscale instabilities, isolated storms, mesoscale convective systems, and orographic precipitation. Traditionally, tropical cyclones are not viewed as a mesoscale phenomenon due to the wide range of scales involved in their genesis, movement, circulations, and convective systems. However, we may also view a hurricane or typhoon as an intense, rotating convective system once it has formed. Thus, for completeness, tropical cyclone dynamics are briefly introduced in the moist convection part of the text. The front dynamics covers dynamics of large scale fronts, mesoscale fronts, and jet streaks.

The last part of the text is devoted to the introduction of mesoscale modeling and the foundation for mesoscale numerical weather prediction. Since the 1970s, numerical models have become an important tool for studying mesoscale weather systems, thus making it essential to understand the fundamental properties of numerical models. In this part of the text, we briefly introduce the basic knowledge on numerical modeling techniques and parameterizations of major physical processes such as planetary boundary layer, cumulus convection, microphysical processes, and radiative transfer. This part of the text is not intended to replace advanced textbooks in mesoscale meteorological modeling or numerical weather prediction, but will instead provide the basic knowledge and background so that the readers have a better understanding of numerical schemes when they choose to use models to investigate mesoscale weather phenomena, rather than using a numerical model as a black box. The modeling exercise provides a hands-on practice for building a simple model of stratified fluid flow from a one-dimensional advection equation, two-dimensional shallow-water model, and three-dimensional shallow-water model.

This textbook is based on two graduate courses, mesoscale dynamics and mesoscale modeling, taught by the author at the Department of Marine, Earth, and Atmospheric Sciences of the North Carolina State University since 1987. It is designed for a two-semester course in mesoscale dynamics at graduate level. It may also be used for a one-semester graduate course focused on mesoscale wave dynamics by using material from Chapters 1 through 7, on moist convection and front dynamics by using material from Chapters 1 through 3 and Chapters 8 through 11, or on mesoscale modeling by using material from Chapters 1 through 3 and Chapters 12 through 14. I have assumed that students should have a fundamental understanding of basic dynamic meteorology or geophysical fluid dynamics. Although I have attempted to provide as many references as possible, there are still many papers that have been left out of the text. The webpage at: http://www.cambridge.org/9780521808750 will be used by the author to communicate with the readers.

I would like to acknowledge Professor Ron Smith for introducing mountain meteorology and relevant mesoscale dynamics to me when I was a Ph.D. student at Yale University. Also, I would like to acknowledge Professor Harry Orville at the South Dakota School of Mines and Technology for teaching me cloud modeling and dynamics and getting me involved in the development of a microphysical parameterization scheme (LFO scheme) which was later adopted as a major scheme in cloud, mesoscale, and numerical weather prediction models. I also benefited from attending the 1982 summer school on mesoscale meteorology in France (sponsored by NATO), where I learned a wide range of mesoscale meteorology from lecturers and fellow attendees. I would also like to extend thanks to the students in my mesoscale dynamics and mesoscale modeling/numerical weather prediction classes in North Carolina State University, especially Hye-Yeong Chun, Ron Weglarz, Mike Kiefer, Heather Reeves, Paul Suffern, Chad Ringley, Chenjie Huang, Dave Vollmer, Kelly Mahoney, Karl Pfeiffer, and Sen Chiao, who have contributed useful suggestions in the text and exercises that have improved the quality of the book. I am indebted to all of my colleagues who have made thorough reviews and comments on the manuscript of the book, which has greatly improved the quality of the text and are highly appreciated. These include Shu-Hua Chen, Min-Dah Chou, Jay Charney, Hye-Yeong Chun, Ching-Yuang Huang, Jerry Janowitz, Mike Kaplan, Steve Koch, Gary Lackmann,

Hsin-Mu Lin, Matt Parker, Rich Rotunno, Dave Schultz, Roger Smith, Wen-Yih Sun, Miguel Teixeira, Si-Chee Tsay, Ken Waight, Ron Weglarz, Chun-Chieh Wu, Qin Xu, Ming Xue, Ming-Jen Yang, Sandra Yuter and Fuqing Zhang. I would also like to thank Robert Mera, Michelle Lin, and Ron Weglarz for their technical editing of the manuscript and Ya-Hui Lin for her skillful drafting of the figures and patience in doing so. Jessica Lin has also assisted the drafting of the figures. The detailed copy-editing by Jon Billam is appreciated. Parts of the book reflect my own research on mesoscale dynamics and modeling, which has been sponsored by the National Science Foundation, National Aeronautics and Space Administration, Office of Naval Research, Air Force Research Laboratory, Forest Service Office of Air Force Scientific Research, and National Oceanic and Atmospheric Administration.

Finally, the encouragement, support, and love from my wife, Emily, and my daughters, Michelle and Jessica, have made it easier to go through the long writing process. I would like to dedicate this book to them.

1
Overview

1.1 Introduction

Mesoscale dynamics uses a dynamical approach for the study of atmospheric phenomena with a horizontal scale ranging approximately from 2 to 2000 km. These mesoscale phenomena include, but are not limited to, thunderstorms, squall lines, supercells, mesoscale convective complexes, inertia-gravity waves, mountain waves, low-level jets, density currents, land/sea breezes, heat island circulations, clear air turbulence, jet streaks, and fronts. Mesoscale dynamics may be viewed as a combined discipline of dynamic meteorology and mesoscale meteorology. From the dynamical perspective, mesoscale concerns processes with timescales ranging from the buoyancy oscillation ($2\pi/N$, where N is the buoyancy (Brunt–Vaisala) frequency) to a pendulum day ($2\pi/f$, where f is the Coriolis parameter), encompassing deep moist convection and the full spectrum of inertia-gravity waves but stopping short of synoptic-scale phenomena which have Rossby numbers less than 1. The Rossby number is defined as U/fL, where U is the basic wind speed and L the horizontal scale of the disturbance associated with the phenomenon. The study concerned with the analysis and prediction of large-scale weather phenomena, based on the use of meteorological data obtained simultaneously over the standard observational network, is called synoptic meteorology. Synoptic-scale phenomena include, but are not limited to, extratropical and tropical cyclones, fronts, jet streams, and baroclinic waves. The synoptic scale is also referred to as the large scale, macroscale, or cyclone scale in the literature and in this textbook. Traditionally, these scales have been loosely used or defined. For example, tropical cyclones have been classified as synoptic-scale phenomena by some meteorologists but are classified as mesoscale phenomena by others (Table 1.1). The same is the case with the mesoscale. For example, a tornado has been classified as a mesoscale phenomenon by some meteorologists due to the scale of its environment for formation, while it is generally classified as a microscale phenomenon based on its scale of circulation. In the mean time, fronts have been classified as both large-scale and mesoscale phenomenon due to their different scales in the along-front and cross-front directions.

Before about 1980, due to the lack of observational data at the mesoscale, mesoscale meteorology had advanced at a slower rate than synoptic meteorology.

Table 1.1 *Atmospheric scale definitions. (Adapted after Thunis and Bornstein 1996.)*

Horizontal Scale	Lifetime	Stull (1988)	Pielke (2002)	Orlanski (1975)	Thunis and Bornstein (1996)	Atmospheric Phenomena
10 000 km	1 month		Synoptic Regional	Macro-α	Macro-α	General circulation, long waves
		Macro		Macro-β	Macro-β	Synoptic cyclones
2000 km	1 week					
200 km	1 day			Meso-α	Macro-γ	Fronts, hurricanes, tropical storms, short cyclone waves, mesoscale convective complexes
			Meso	Meso-β	Meso-β	Mesocyclones, mesohighs, supercells, squall lines, inertia-gravity waves, cloud clusters, low-level jets, thunderstorm groups, mountain waves, sea breezes
20 km		Meso				
	1 h			Meso-γ	Meso-γ	Thunderstorms, cumulonimbi, clear-air turbulence, heat island, macrobursts
2 km						
		Micro		Micro-α	Meso-δ	Cumulus, tornadoes, microbursts, hydraulic jumps
200 m	30 min		Micro			
				Micro-β	Micro-β	Plumes, wakes, waterspouts, dust devils
20 m	1 min					
					Micro-γ	Turbulence, sound waves
2 m	1 s	Micro δ		Micro-γ	Micro-δ	

For example, some observed isolated, unusual values of pressure and winds shown on synoptic charts were suspected to be observational errors. Even though this may be true in some cases, others are now thought to represent true signatures of subsynoptic disturbances having spatial and temporal scales too small to be properly analyzed and represented on standard synoptic charts. Since the advancement of observational techniques and an overall increase in the number of mesoscale observational networks after about 1980 and rapid advancement in numerical modeling techniques, more and more mesoscale phenomena, as well as their interactions with synoptic scale and microscale flows and weather systems, have been revealed and better understood. In order to improve mesoscale weather forecasting, it is essential to improve our understanding of the basic dynamics of mesoscale atmospheric phenomena through fundamental studies by utilizing observational, modeling, theoretical, and experimental approaches simultaneously. Since mesoscale spans a wide horizontal range approximately from 2 to 2000 km, there is no single theory (such as the quasi-geostrophic theory for the large scale), which provides a unique tool for studying the dynamical structure of the variety of mesoscale motions observed in the Earth's atmosphere. In fact, the dominant dynamical processes vary dramatically from system to system, depending on the type of mesoscale circulation involved.

1.2 Definitions of atmospheric scales

Due to different force balances, atmospheric motions in fluid systems with distinct temporal and spatial scales behave differently. In order to better understand the complex dynamical and physical processes associated with mesoscale phenomena, different approximations have been adopted to help resolve the problems. Therefore, a proper scaling facilitates the choice of appropriate approximations of the governing equations.

Scaling of atmospheric motions is normally based on observational and theoretical approaches. In the observational approach, atmospheric processes are categorized through direct empirical observations and the instruments used. Since observational data are recorded in discrete time intervals and the record of these data in the form of a standard surface or upper air weather map reveals a discrete set of phenomena, the phenomena are then also categorized into discrete scales. For example, sea breezes occur on a time scale of about 1 day and spatial scales of 10 to 100 km, while cumulus convection occurs on a time scale of about 30 min and encompasses a spatial scale of several kilometers. Figure 1.1 shows the atmospheric kinetic energy spectrum in the free atmosphere and near the ground for various time scales. In the free atmosphere, there are strong peaks at periods ranging from a few days (the synoptic scale) to a few weeks (the planetary scale – at which the β effect plays an important role). In addition, there are also peaks at 1 year and 1 day and a smaller peak at a few minutes, although this latter peak may be an artifact of the analysis. This energy spectrum therefore suggests a natural division of atmospheric phenomena into three distinct (but not wholly separable) scales: large scale, mesoscale, and microscale. From the kinetic energy spectrum, the mesoscale therefore appears as the scale on which energy is allowed

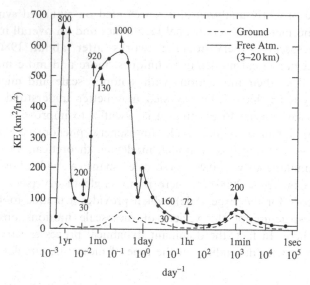

Fig. 1.1 Average kinetic energy of west–east wind component in the free atmosphere and near the ground. (Adapted after Vinnichenko 1970.)

to transfer from the large scale to the microscale and vice versa. Based on radar observations of storms, atmospheric motions can be categorized into the following three scales: (a) microscale: $L < 20$ km, (b) mesoscale: 20 km $< L < 1000$ km, and (c) synoptic (large) scale: $L > 1000$ km (Ligda 1951). The atmospheric motions have also been categorized into eight separate scales: macro-α ($L > 10\,000$ km), macro-β ($10\,000$ km $> L > 2000$ km), meso-α (2000 km $> L > 200$ km), meso-β (200 km $> L > 20$ km), meso-γ (20 km $> L > 2$ km), micro-α (2 km $> L > 200$ m), micro-β (200 m $> L > 20$ m), and micro-γ ($L < 20$ m) scales (Orlanski 1975; Table 1.1). Based on observations, atmospheric phenomena have also been categorized into masocale, mesoscale, misoscale, mososcale, and musoscale (Fujita 1986).

Atmospheric motions may also be categorized using a theoretical approach. For example, for airflow over a mountain, the scale of the mechanically induced quasi-steady waves corresponds roughly to the scale of the imposed forcing. For such problems, adoption of the *Eulerian* (fixed in space) *time scale* is reasonable. For example, for two steady cumulus clouds being advected by a steady basic wind, the time scale for a stationary observer located on the ground is approximately the horizontal scale of the mountain divided by the basic wind speed. However, the above time scale has little to do with the physical processes associated with the cloud development. Instead, it is more meaningful physically to use the *Lagrangian Rossby number* R_o, which is defined as the ratio of intrinsic frequency and the Coriolis parameter ($\omega/f = 2\pi/fT$, where T is the *Lagrangian time scale*), because the Lagrangian time

Table 1.2 *Lagrangian time scales and Rossby numbers for typical atmospheric systems. (Adapted after Emanuel and Raymond 1984.)*

Phenomenon	Time scale	Lagrangian R_o $(\approx \omega/f = 2\pi/fT)$
Tropical cyclone	$2\pi R/V_T$	V_T/fR
Inertia-gravity waves	$2\pi/N$ to $2\pi/f$	N/f to 1
Sea/land breezes	$2\pi/f$	1
Thunderstorms and cumulus clouds	$2\pi/N_w$	N_w/f
Kelvin–Helmholtz waves	$2\pi/N$	N/f
PBL turbulence	$2\pi h/U^*$	U^*/fh
Tornadoes	$2\pi R/V_T$	V_T/fR

where:
R = radius of maximum wind scale, ω = frequency, T = time scale, V_T = maximum tangential wind scale, f = Coriolis parameter, N = buoyancy (Brunt–Vaisala) frequency, N_w = moist buoyancy (Brunt-Vaisala) frequency, U^* = scale for friction velocity, h = scale for the depth of planetary boundary layer.

scale measures the time a fluid particle takes in following the motion. In the above example, the Lagrangian time scale is the time it takes an air parcel to rise to its maximum vertical displacement. Another example is the Lagrangian time scale for a cyclone, which is defined as $2\pi R/V_T$, where R is the radius of the circular motion and V_T is the tangential wind speed. The Lagrangian time scales and Rossby numbers for typical atmospheric systems are summarized in Table 1.2. Based on this type of theoretical considerations, the following different scales for atmospheric motions can be defined: (a) synoptic (large or macro) scale, for motions which are quasi-geostrophic and hydrostatic, (b) mesoscale, for motions which are non-quasi-geostrophic and hydrostatic, and (c) microscale, for motions which are non-geostrophic, non-hydrostatic, and turbulent (Emanuel and Raymond 1984). Based on this interpretation, the mesoscale may be defined as that scale which includes atmospheric circulations that are large enough in horizontal scale to be considered hydrostatic but too small to be described as quasi-geostrophically. Note that the hydrostatic assumption may not apply to some mesoscale weather systems, especially for those associated with convection.

Based on hydrostatic, convective, advective, compressible, and Boussinesq approximations of the governing equations – including temporal, horizontal and vertical spatial scales – in order to standardize existing nomenclature with regard to mesoscale phenomena – a more rigorous approach can be taken to define the atmospheric scales (e.g., Thunis and Bornstein 1996). This approach integrates existing concepts of atmospheric spatial scales, flow assumptions, governing equations, and resulting motions into a hierarchy which is useful in the classification of mesoscale motions. Horizontal and

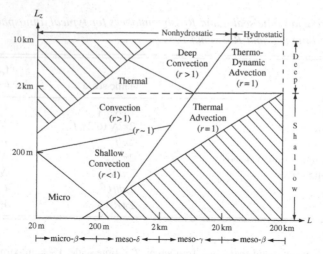

Fig. 1.2 Schematic of flow regimes under unstable stability conditions, where hatched zones indicate nonphysical flow regimes, the dotted line indicates merging of thermodynamic advection with macroscale, *r* represents scaled ratio of buoyancy and vertical pressure gradient forced perturbations, and the dashed line represents division of thermal convection into its deep and shallow regimes. (Adapted after Thunis and Bornstein 1996.)

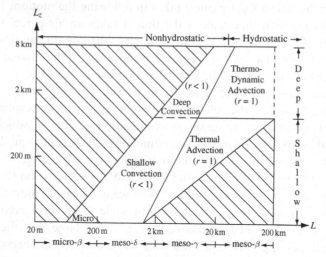

Fig. 1.3 As in Fig. 1.2 except for stable stability conditions. (Adapted after Thunis and Bornstein 1996.)

vertical scales of flow regimes under unstable and stable stability conditions for deep and shallow convection are shown in Figs. 1.2 and 1.3, respectively. Table 1.1 summarizes some examples of the horizontal and temporal scales for typical atmospheric phenomena as proposed by different authors. In this book, we will adopt Orlanski's scaling, except where otherwise specified.

1.3 Energy generation and scale interactions

Although many mesoscale circulations and weather systems are forced by large-scale or microscale flow, some circulations are locally forced at the mesoscale itself. Energy generation mechanisms for mesoscale circulations and weather systems may be classified into the following categories: (a) thermal or orographic surface inhomogeneities, (b) internal adjustment of larger-scale flow systems, (c) mesoscale instabilities, (d) energy transfer from either the large scale or microscale to the mesoscale, and (e) interaction of cloud physical and dynamical processes (Anthes 1986).

Examples of the first type of mesoscale weather systems are the land and sea breezes, mountain-valley winds, mountain waves, heat-island circulations, coastal fronts, dry lines, and moist convection. These mesoscale weather systems are more predictable than other types of systems that occur on the mesoscale. Examples of the second type of weather systems are fronts, cyclones, and jet streaks. These weather systems are less predictable since they are generated by transient forcing associated with larger-scale flows. Although instabilities associated with the mean wind or thermal structure of the atmosphere are rich energy sources of atmospheric disturbances, most atmospheric instabilities have their maximum growth rates either on the large scale through baroclinic, barotropic, and inertial instabilities or on the microscale through Kelvin–Helmholtz, conditional and potential (convective) instabilities. Symmetric instability appears to be intrinsically a mesoscale instability.

Energy transfer from small scales to the mesoscale also serves as a primary energy source for mesoscale convective systems. These mesoscale convective systems may start as individual convective cells that grow and combine to form thunderstorms and convective systems, such as squall lines, mesocyclones, and mesoscale convective complexes. On the other hand, energy transfer from the large scale to the mesoscale also serves as an energy source to induce mesoscale circulations or weather systems. For example, temperature and vorticity advection associated with large-scale flow systems may help the genesis of mesoscale frontal systems through scale contraction. Another possible energy source for producing mesoscale circulations or weather systems is the interaction of clouds' physical and dynamical processes. Mesoscale convective systems may be generated by this interaction process through scale expansion.

Scale interaction generally refers to the interactions between the temporally and spatially averaged zonal flow and a fairly limited set of waves that are quantized by the circumference of the earth, while it refers to multiple interactions among a continuous spectrum of eddies of all sizes in turbulence theory (Emanuel 1986). However, scale interaction should not be viewed as a limited set of interactions among discrete scales because, on average, the mesoscale is much more like a continuous spectrum of scales. Scale interaction depends on the degree of relative strength of fluid motions involved. For example, for a very weak disturbance embedded in a slowly varying mean flow, the interaction is mainly exerted from the mean flow to the weak disturbance. If this disturbance becomes stronger, then it may exert an increasing

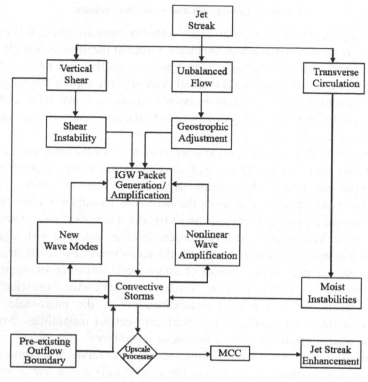

Fig. 1.4 Sketch of mutual interactions between the jet streak, inertial-gravity waves (IGW), and moist convection. (Adapted after Koch 1997.)

influence on the mean flow, and other scales of motion may develop. In this case, scale interactions become more and more numerous, and the general degree of disorder in the flow becomes greater. At the extreme, when the disturbance becomes highly nonlinear, such as in a fully developed turbulent flow, then the interactions become mutual and chaotic, and an explicit mathematical or analytic description of the interaction becomes problematic. Examples of scale-interactive processes which occur at mesoscale include: (i) synoptic forcing of mesoscale weather phenomena, (ii) generation of internal mesoscale instabilities, (iii) interactions of cloud and precipitation processes with mesoscale systems, (iv) influence of orography, boundary layer, and surface properties on mesoscale weather system development and evolution, (v) feedback contributions of mesoscale systems to larger-scale processes, (vi) energy transfer associated with mesoscale systems, and (vii) mechanisms and processes associated with stratosphere–troposphere exchange (Koch 1997). Figure 1.4 shows the mutual interactions between a jet streak, inertia-gravity waves, and strong mesoscale convection that can occur on the mesoscale.

Figure 1.5 shows the energy transfer process through geostrophic adjustment in the response of the free atmosphere to a cumulus cloud, which radiates gravity waves that

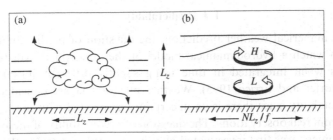

Fig. 1.5 Energy transfer process through geostrophic adjustment in the response of the free atmosphere to a cumulus cloud, which radiates gravity waves that lead to a lens of less stratified air whose width is the Rossby radius of deformation (NL_z/f). (a) The response of the free atmosphere to a cumulus cloud is the formation of gravity waves away from the cloud, which, in turn, lead to the formation of (b) a lens of less stratified air whose width is the Rossby radius of deformation. (Adapted after Emanuel and Raymond 1984.)

lead to a lens of less stratified air whose width is the Rossby radius of deformation. The process from state (a) to (b) in Fig. 1.5 represents a scale interactive process in which the system tends to reach geostrophic equilibrium. The above example of cumulus convection implies as least two distinct scales: (i) the cumulus scale $\sim L_z$, and (ii) the large scale $\sim NL_z/f$ (Rossby radius of deformation in a stratified fluid). The Rossby radius of deformation is the horizontal scale at which rotational effects become as important as buoyancy effects. The Rossby radius of deformation can be understood as the significant horizontal scale fluid parcels experience when a fluid undergoes geostrophic adjustment in a homogeneous fluid, such as water, to an initial condition such as

$$\eta = \eta_o \mathrm{sgn}(x), \tag{1.3.1}$$

where η is the vertical displacement of the fluid from the mean fluid depth, and η_o is the maximum η, x points eastward, and sgn is the sign function defined as $\mathrm{sgn}(x) = 1$ for $x > 0$ and -1 for $x < 0$. In the earlier stage, the motion is dominated by the pressure gradient force, and the fluid particles move toward the west ($-x$ direction). As time proceeds, the Coriolis force becomes more and more important and the fluid particles are deflected toward the right (north) in the Northern Hemisphere. The Coriolis force eventually reaches a geostrophic balance with the pressure gradient force. In this final stage, the basic flow is northward ($+y$ direction), and the Rossby radius of deformation is considered to be the horizontal distance from the location where the e-folding value of the vertical displacement is equal to the original average height of the homogeneous fluid.

In a shallow water fluid system, the Rossby radius of deformation is

$$\lambda_R = c_o/f, \tag{1.3.2}$$

where c_o is the shallow water wave speed ($\sqrt{gH}$, H is the fluid depth) induced by the gravitational (buoyancy) force.

1.4 Predictability

In mesoscale numerical weather prediction, the question of predictability concerns the degree to which a hydrodynamical model of the atmosphere will yield diverging solutions when integrated in time using slightly different initial conditions (e.g., Ehrendorfer and Errico 1995). Weather phenomena are considered to have limited predictability since there is an uncertainty associated with initial conditions determined from real observations. The question of predictability of mesoscale atmospheric phenomena was first investigated by using a simple model for the interaction of barotropic vorticity perturbations encompassing a number of diverse horizontal scales (Lorenz 1969). Those results suggested that the mesoscale may be less predictable, i.e. yielding perturbed solutions that diverge faster than the synoptic and planetary scales, essentially because the eddy timescale decreases on the horizontal scale. The predictability for synoptic scales is mainly limited by the nonlinear interactions between different waves with different wavelengths, i.e. different components of the wave spectrum. These interactions depend on the initial distribution of energy in the different wave numbers and on the number of waves the model can resolve. Errors and uncertainties in the resolvable-scale waves and errors introduced by neglecting unresolvable scales grow with time and spread throughout the spectrum, eventually contaminating all wavelengths and destroying the forecast (Anthes 1986). The predictability for mesoscale motions is mainly limited by the rapid transfer of energy between the large scale and the microscale. In addition, the predictability for small scales is mainly limited by three-dimensional turbulence. Inevitable errors or uncertainties in initial conditions in the small scale of motion will propagate toward larger scales and will reach the mesoscale sooner than the large scale, therefore rendering the mesoscale less predictable.

The response of a fluid system to a steady forcing tends to fall into one of the following four categories: (1) steady for a stable system, perfectly predictable, (2) periodic for a weakly unstable system, perfectly predictable, (3) aperiodic with a "lumpy" spectrum for a moderately unstable system, less predictable, and (4) aperiodic with a monotonic spectrum for a fully turbulent system, rather unpredictable (Emanuel and Raymond 1984). The atmospheric system falls into category (3). Monotonicity of the kinetic energy spectrum (Fig. 1.1) through the mesoscale implies that energy is mainly transferred from larger (large scale) to smaller (microscale) scales, although it can be generated intermittently at the mesoscale. This tends to limit the predictability at the mesoscale.

Besides the natural constraints imposed by forcing and physical processes, predictability of mesoscale phenomena is also affected by the initial conditions set up in a mesoscale numerical weather prediction model. If a mesoscale phenomenon does not exist at the beginning of the numerical simulation, then the predictability is less influenced by the accuracy of the initial conditions used in a mesoscale numerical weather prediction model. Under this situation, the mesoscale circulations are normally forced by surface inhomogeneities (thermal or orographic), internal adjustment of larger-scale flow systems, mesoscale instabilities, energy transfer from either the larger scale or the microscale, or the interaction of cloud physical and dynamical processes, as discussed earlier. For mesoscale circulations induced by larger-scale motion, the time scale for predictability of these types of mesoscale systems could exceed the actual time scale of the mesoscale systems themselves. On the other hand, if a mesoscale phenomenon exists at the beginning of the numerical prediction, then it is necessary to include the observed and analyzed motion and thermodynamic variables in the initial conditions in order to make an accurate numerical prediction. The accuracy of the numerical prediction relies more on observations in the beginning and less on the model because it takes time for the model to spin up. For example, if a numerical model starts its integration at a time where a tropical cyclone already exists, then the vortex and moisture fields associated with the tropical cyclone must be initialized in the model. Thus, observations and their implementation in the numerical model are very important at the beginning of the numerical prediction. Naturally, contributions of the model to numerical weather prediction become more and more important as time proceeds.

References

Anthes, R. A., 1986. The general question of predictability. In *Mesoscale Meteorology and Forecasting*, P. S. Ray (ed.), Amer. Meteor. Soc., 636–56.

Ehrendorfer, M. and R. M. Errico, 1995. Mesoscale predictability and the spectrum of optimal perturbations. *J. Atmos. Sci.*, **52**, 3475–500.

Emanuel, K., 1986. Overview and definition of mesoscale meteorology. In *Mesoscale Meteorology and Forecasting*, P. S. Ray (ed.), Amer. Meteor. Soc., 1–16.

Emanuel, K. and D. J. Raymond, 1984. *Dynamics of Mesoscale Weather Systems*, J. B. Klemp (ed.), NCAR.

Fujita, T. T., 1986. Mesoscale classifications: Their history and their application to forecasting. In *Mesoscale Meteorology and Forecasting*, P. S. Ray (ed.), Amer. Meteor. Soc., 18–35.

Koch, S. E., 1997. Atmospheric Convection. Lecture notes, North Carolina State University.

Ligda, M. G. H., 1951. Radar storm observations. In *Compendium of Meteorology*, Amer. Meteor. Soc. 1265–82.

Lorenz, E. N., 1969. The predictability of a flow which possesses many scales of motion. *Tellus*, **21**, 289–307.

Orlanski, I., 1975. A rational subdivision of scales for atmospheric processes. *Bull. Amer. Meteor. Soc.*, **56**, 527–30.

Thunis, P. and R. Bornstein, 1996. Hierarchy of mesoscale flow assumptions and equations. *J. Atmos. Sci.*, **53**, 380–97.

Vinnichenko, N. K., 1970. The kinetic energy spectrum in the free atmosphere – 1 second to 5 years. *Tellus*, **22**, 158–66.

2

Governing equations for mesoscale motions

2.1 Introduction

In this chapter, we will derive the governing equations for a stratified inviscid atmosphere on an f plane. The equations are based on the following physical laws: (a) Newton's second law of motion, (b) conservation of mass, and (c) the first law of thermodynamics. These laws are represented by the set of primitive equations that consist of the horizontal and vertical momentum equations, the continuity equation, and the thermodynamic energy equation. Several approximations used in simplifying the set of primitive equations will be discussed as well as appropriate upper and lower boundary conditions. It should be mentioned that wave motions behave completely differently from mass transport. Briefly speaking, fluid particles do not necessarily follow the disturbance in a wave motion, but they do always follow the mass transport process. For example, air parcels associated with gravity waves may oscillate in the vicinity of the source or forcing region, but the gravity waves themselves may propagate to great distances from their origin. On the other hand, air parcels within a cold pool generated by evaporative cooling associated with falling raindrops beneath a thunderstorm always move in concert with the density current.

2.2 Derivation of the governing equations

Considering an atmosphere on a planetary f plane, the momentum equations, continuity equation, and thermodynamic energy equation can be expressed in the following form:

$$\frac{Du}{Dt} - fv = -\frac{1}{\rho}\frac{\partial p}{\partial x} + F_{rx}, \qquad (2.2.1)$$

$$\frac{Dv}{Dt} + fu = -\frac{1}{\rho}\frac{\partial p}{\partial y} + F_{ry}, \qquad (2.2.2)$$

$$\frac{Dw}{Dt} = -\frac{1}{\rho}\frac{\partial p}{\partial z} - g + F_{rz}, \qquad (2.2.3)$$

$$\frac{D\rho}{Dt} + \rho\left(\frac{\partial u}{\partial x} + \frac{\partial v}{\partial y} + \frac{\partial w}{\partial z}\right) = 0, \tag{2.2.4}$$

$$\frac{D\theta}{Dt} = \frac{\theta}{c_p T}\dot{q}, \tag{2.2.5}$$

where $D/Dt = \partial/\partial t + u\partial/\partial x + v\partial/\partial y + w\partial/\partial z$ is the total (material) derivative, which represents the rate of change of a certain property within a fluid parcel following the motion, and F_{rx}, F_{ry}, and F_{rz} are the viscous terms or frictional forces per unit mass in the x, y, and z directions, respectively. The symbol c_p denotes the heat capacity of dry air at constant pressure, and $\dot{q}$ is the diabatic heating rate per unit mass in J kg^{-1} s^{-1}. Other symbols are defined as usual (also see Appendix A). In the *viscous sublayer*, which is a very thin layer near the Earth's surface, the viscous terms may be represented by molecular viscosity in the form $\nu\nabla^2 u$, $\nu\nabla^2 v$, and $\nu\nabla^2 w$, where ν is the kinematic viscosity coefficient associated with molecular viscosity. Note that ν is equal to μ/ρ, where μ is the dynamic viscosity coefficient and ρ is the air density. At sea level, ν has a value of about 1.46×10^{-6} m^2 s^{-1}. The molecular viscosity is almost completely negligible in the atmosphere above the viscous sublayer, where momentum and heat transfers are dominated by turbulent eddy motion. A number of parameterization schemes for turbulent eddy viscosity in the planetary boundary layer have been proposed in the literature and can be found in many numerical weather prediction textbooks. Some of these schemes will be discussed in Chapter 14.

The equation set (2.2.1)–(2.2.3) with no Coriolis term is often referred to as the *Navier–Stokes equations of motion*. The diabatic heating rate may be taken to represent, for example, surface sensible heating, elevated latent heating, or cloud-top radiative cooling. Note that the viscous terms on the right-hand side of (2.2.1) and (2.2.2) can be approximated by *Rayleigh friction* $(-\nu_o u, -\nu_o v)$, while the diabatic heating term of (2.2.5) can be approximated by the *Newtonian cooling* $(-\nu_o \theta)$, as is practiced in some theoretical studies to simplify the above system of governing equations. The coefficient ν_o is determined by the e-folding time scale of the disturbance. In the following, we will assume that the fluid is inviscid. In the above system (2.2.1)–(2.2.5), there are seven unknowns represented by five equations. In order to close the system, we need two additional equations, such as the equation of state for dry air (which is well represented by an ideal gas),

$$p = \rho R_d T, \tag{2.2.6}$$

and the Poisson's equation

$$\theta = T\left(\frac{p_s}{p}\right)^{R_d/c_p}, \tag{2.2.7}$$

where θ is the potential temperature, p_s is a constant reference pressure level (normally chosen as 1000 hPa) and R_d is the gas constant for dry air. For a moist atmosphere, the temperature in (2.2.6) is replaced by the virtual temperature, which takes into account the moist effects due to latent heat release, and the density is replaced by the total density, which is a sum of the dry air density and the total water density.

It should be noted that the continuity equation, (2.2.4), contains the implicit assumption that atmospheric mass is conserved. However, this assumption is violated when precipitation forms and falls to the surface. Similarly, evaporation from the surface adds atmospheric mass. Scale analysis indicates that the mass source/sink term, which should appear on the right side of (2.2.4) (e.g. Dutton 1986 – Section 8.1) is usually much smaller than other terms in the equation, but can be important in tropical cyclones or other heavily precipitating systems. The precipitation/evaporation mass sink/source term can play a non-negligible role in the dynamics of such systems, and should be included in numerical models used to simulate them (e.g., Lackmann and Yablonsky 2004).

The above equation set may be linearized by partitioning the field variables:

$$u(t, x, y, z) = U(z) + u'(t, x, y, z),$$

$$v(t, x, y, z) = V(z) + v'(t, x, y, z),$$

$$w(t, x, y, z) = w'(t, x, y, z),$$

$$\rho(t, x, y, z) = \bar{\rho}(x, y, z) + \rho'(t, x, y, z),$$

$$p(t, x, y, z) = \bar{p}(x, y, z) + p'(t, x, y, z),$$

$$\theta(t, x, y, z) = \bar{\theta}(x, y, z) + \theta'(t, x, y, z),$$

$$T(t, x, y, z) = \bar{T}(x, y, z) + T'(t, x, y, z),$$

$$\dot{q}(t, x, y, z) = q'(t, x, y, z),$$

(2.2.8)

where capital letters and overbars represent the basic state, such as synoptic scale flow in which the mesoscale disturbances evolve, and the primes indicate perturbations from the basic state, such as the mesoscale flow fields. The basic state is assumed to follow Newton's second law of motion, conservation of mass, and the first law of thermodynamics. The basic state is assumed to be in equilibrium. That is, it is in geostrophic wind balance,

$$U = -\frac{1}{f\bar{\rho}}\frac{\partial \bar{p}}{\partial y} \quad \text{and} \quad V = \frac{1}{f\bar{\rho}}\frac{\partial \bar{p}}{\partial x},$$

(2.2.9)

and in hydrostatic balance,

$$\frac{\partial \bar{p}}{\partial z} = -\bar{\rho}g, \tag{2.2.10}$$

where $\bar{p} = \bar{\rho} R_d \bar{T}$. Equations (2.2.9) and (2.2.10) automatically imply *thermal wind balance* for the basic state

$$U_z = -\frac{g}{f\bar{\theta}}\frac{\partial \bar{\theta}}{\partial y}; V_z = \frac{g}{f\bar{\theta}}\frac{\partial \bar{\theta}}{\partial x}, \tag{2.2.11}$$

where $\bar{\theta} - \bar{T}(p_s/\bar{p})^{R_d/c_p}$ and subscripts indicate partial differentiations.

Conservation of mass, (2.2.4), of the basic state leads to

$$\frac{D\bar{p}}{Dt} + \bar{p}\left(\frac{\partial U}{\partial x} + \frac{\partial V}{\partial y}\right) = \frac{1}{c_s^2}\frac{D\bar{p}}{Dt} = \frac{1}{c_s^2}\left(U\frac{\partial \bar{p}}{\partial x} + V\frac{\partial \bar{p}}{\partial y}\right) = 0, \tag{2.2.12}$$

where c_s is the sound wave speed defined as $c_s = \sqrt{\gamma R_d \bar{T}}$. Here $\gamma = c_p/c_v$. The last equality of (2.2.12) is consistent with the geostrophic wind relation. Conservation of the basic state thermal energy gives

$$U\frac{\partial \bar{\theta}}{\partial x} + V\frac{\partial \bar{\theta}}{\partial y} = 0, \tag{2.2.13}$$

which implies no basic state thermal advection by the basic wind, and will be assumed for deriving the perturbation thermodynamic equation. The left-hand side of (2.2.13) is required to satisfy the constraint that the vertical motion field vanishes at the surface and possibly at the upper boundary for some theoretical studies (Bannon 1986). In the *Eady* (1949) *model* of baroclinic instability, this term is assumed to be 0. In fact, if one assumes $V = 0$, then the above equation is automatically satisfied because $\partial \bar{\theta}/\partial x = (f\bar{\theta}/g)V_z = 0$, based on the basic-state thermal wind relations. Substituting (2.2.8) with (2.2.9)–(2.2.13) into (2.2.1)–(2.2.5) and neglecting the non-linear and viscous terms, the perturbation equations for mesoscale motions in the free atmosphere (i.e. above the planetary boundary layer) can be obtained,

$$\frac{\partial u'}{\partial t} + U\frac{\partial u'}{\partial x} + V\frac{\partial u'}{\partial y} + U_z w' - fv' + \frac{1}{\bar{\rho}}\frac{\partial p'}{\partial x} = 0, \tag{2.2.14}$$

$$\frac{\partial v'}{\partial t} + U\frac{\partial v'}{\partial x} + V\frac{\partial v'}{\partial y} + V_z w' + fu' + \frac{1}{\bar{\rho}}\frac{\partial p'}{\partial y} = 0, \tag{2.2.15}$$

$$\frac{\partial w'}{\partial t} + U\frac{\partial w'}{\partial x} + V\frac{\partial w'}{\partial y} - g\frac{\theta'}{\bar{\theta}} + \frac{p'}{\bar{\rho}H} + \frac{1}{\bar{\rho}}\frac{\partial p'}{\partial z} = 0, \tag{2.2.16}$$

$$\frac{1}{c_s^2}\left(\frac{\partial p'}{\partial t} + U\frac{\partial p'}{\partial x} + V\frac{\partial p'}{\partial y} + \bar{\rho}f(Vu' - Uv')\right) - \frac{\bar{\rho}}{H}w' + \bar{p}\nabla\cdot\mathbf{V}' = \frac{\bar{\rho}}{c_p\bar{T}}q', \quad (2.2.17)$$

$$\left(\frac{\partial\theta'}{\partial t} + U\frac{\partial\theta'}{\partial x} + V\frac{\partial\theta'}{\partial y}\right) + \frac{f\bar{\theta}}{g}(V_zu' - U_zv') + \frac{N^2\bar{\theta}}{g}w' = \frac{\bar{\theta}}{c_p\bar{T}}q', \quad (2.2.18)$$

where N is the buoyancy (Brunt–Vaisala) frequency and H is the scale height. The buoyancy frequency and scale height are defined, respectively, as

$$N^2 \equiv \frac{g}{\bar{\theta}}\frac{\partial\bar{\theta}}{\partial z}, H \equiv \frac{c_s^2}{g}. \quad (2.2.19)$$

Note that the *scale height* has also been defined in the literature as the height at which the basic density of the air at the surface (ρ_s) is reduced to its e-folding value, i.e. $\rho(z = H) = \rho_s e^{-1}$, assuming the air density decreases with height exponentially.

In deriving (2.2.14) and (2.2.15), we have assumed $|\rho'/\bar{\rho}| \ll |u'/U|$ and $|\rho/\bar{\rho}| \ll |v'/V|$, which is a good first approximation in the real atmosphere. The sum of the fourth and fifth terms of (2.2.16) represents the buoyancy force associated with the atmospheric motion, which is equal to $g\rho'/\bar{\rho} = -g\theta'/\bar{\theta} + p'/(\bar{\rho}H)$. This relation reduces to $\rho'/\bar{\rho} \approx -\theta'/\bar{\theta}$ for an incompressible or Boussinesq fluid. The incompressible and Boussinesq approximations will be discussed in Section 2.3. The buoyancy frequency represents the natural oscillation frequency of an air parcel displaced vertically from its equilibrium position by the buoyancy force in a stably stratified atmosphere (i.e. $N^2 > 0$). The vertical oscillation period for parcels in this type of atmosphere is $2\pi/N$. In deriving (2.2.17), we have substituted the equation of state and the first law of thermodynamics, $D(\ln\theta)/Dt = \dot{q}/(c_pT)$, into the total derivative of the air density ($D\rho/Dt$). This yields the diabatic term on the right-hand side of (2.2.17), which may be neglected for an incompressible or Boussinesq fluid since it is of the same order as other c_s^2 terms for most mesoscale flows. The continuity equation reduces to

$$\frac{\partial\rho'}{\partial t} + \mathbf{V}\cdot\nabla\rho' + w\frac{d\bar{\rho}}{dz} + (\bar{\rho} + \rho')\nabla\cdot\mathbf{V} = 0, \quad (2.2.20)$$

where $\mathbf{V} = (u, v, w)$. For small-amplitude perturbations, the above equation reduces to the following linear form:

$$\frac{1}{\bar{\rho}}\frac{\partial\rho'}{\partial t} + \frac{1}{\bar{\rho}}\mathbf{V}\cdot\nabla\rho' + \frac{w}{\bar{\rho}}\frac{d\bar{\rho}}{dz} + \nabla\cdot\mathbf{V}' = 0. \quad (2.2.21)$$

To formulate a more complete atmospheric system, we need to include nonlinear advective accelerations, viscosity and conservation equations for water substances (e.g., water vapor, cloud water, rain, ice, snow, and hail).

2.3 Approximations to the governing equations

Equations (2.2.14)–(2.2.18) form an elastic fluid system that may include the following types of waves:

(1) pure acoustic waves: c_s finite, $g = 0, f = 0$,
(2) acoustic-gravity waves: c_s/g finite, $f = 0$,
(3) pure gravity waves: $c_s \to \infty, g \neq 0, f = 0$, and
(4) inertia-gravity waves: $c_s \to \infty, g \neq 0, f \neq 0$.

In general, the system of (2.2.1)–(2.2.7) may include static (buoyant), shear (Kelvin–Helmholtz), symmetric, inertial, and baroclinic instabilities. Wave and instability dynamics will be discussed in later chapters. Our purpose here is to discuss some commonly used approximations and their limitations with regard to understanding the dynamics of real, observable mesoscale systems and circulations.

One of the simplest approximations is to assume that the scale height (H) and the adiabatic sound wave speed (c_s) are both independent of height. This corresponds to making the assumption of an isothermal atmosphere. It is a good first approximation, since the temperature in the troposphere, the layer in which most of the weather phenomena occur, varies vertically by about 20% (although somewhat more in the stratosphere). Another approximation that has often been adopted by meteorologists is the anelastic approximation. This approximation can be made by simply setting the terms involving c_s^2 equal to zero in the continuity equation, (2.2.17), while keeping H finite. The effect of this approximation is to eliminate all waves with very high propagation speeds associated with rapid (adiabatic) compression and expansion of the fluid. In this approximation, the continuity equation then becomes

$$\nabla \cdot \boldsymbol{V}' - \frac{w'}{H} = 0, \tag{2.3.1}$$

or

$$\nabla \cdot (\boldsymbol{V}' e^{-z/H}) = 0, \tag{2.3.2}$$

since the scale height is taken to be a constant. Equations (2.3.1) and (2.3.2) may also be expressed in an alternate form:

$$\nabla \cdot (\bar{\rho} \boldsymbol{V}') = 0. \tag{2.3.3}$$

Note that (2.3.3) is linked with (2.3.2) when the density decays exponentially with height. Equation (2.3.1), (2.3.2), or (2.3.3) is called the *anelastic* or *deep convection continuity equation*. Equation (2.3.3) was first proposed by Batchelor (1953), who defined $\bar{\rho}(z)$ to be the density in an adiabatic, stably stratified, horizontally uniform reference state. The name anelastic was coined by Ogura and Phillips (1962), who derived (2.3.3) through a rigorous scale analysis, along with approximate forms for the momentum and thermodynamic energy equations. Their scaling analysis assumes that: (a) all deviations of the potential temperature θ' from some constant mean value θ_o are small and (b) the time scale of the disturbance is comparable to the time scale for gravity wave oscillations. The terms that are neglected in the original anelastic equations are an order $\varepsilon (= \theta'/\theta_o)$ smaller than those that are retained. Thus, in the case of dry convection (where mixing will keep the environmental lapse rate close to the adiabatic lapse rate), ε will be small and the anelastic equations can be used to represent nonacoustic modes with complete confidence. For deep, moist convection or gravity wave propagation, however, the mean-state static stability can be sufficient to make ε rather large. For example, the θ' variations across a 10 km deep isothermal layer may reach as high as 40% of the mean θ_o.

Equation (2.3.1) may be further simplified, by assuming that the vertical scale (L_z) of the disturbance is much smaller than the scale height of the basic state atmosphere, $L_z/H \ll 1$,

$$\nabla \cdot V' = 0. \tag{2.3.4}$$

The above equation is called the *incompressible* or *shallow convection continuity equation*. This means that conservation of mass has become conservation of volume because density is treated as a constant. Thus, volume is a good proxy for mass under this approximation. It is important to distinguish the difference between the scale height (H) and the vertical scale of the disturbance or convection (L_z) because the scale height is controlled by the basic structure of the atmosphere, instead of by the fluid motion. Anelastic and incompressible approximations to the continuity equation can also be obtained by applying scale analysis to (2.2.17).

To improve the accuracy of the anelastic or incompressible hydrodynamic equations, the *pseudo-incompressible approximation* has been proposed (Nance and Durran 1994). Under the pseudo-incompressible approximation, the set of equations governing inviscid, rotating flow may be written:

$$\frac{Du}{Dt} - fv + (c_p\theta)\frac{\partial \pi'}{\partial x} = 0, \tag{2.3.5}$$

$$\frac{Dv}{Dt} + fu + (c_p\theta)\frac{\partial \pi'}{\partial y} = 0, \tag{2.3.6}$$

$$\frac{Dw}{Dt} + (c_p\theta)\frac{\partial \pi'}{\partial z} = g\frac{\theta'}{\bar{\theta}}, \tag{2.3.7}$$

$$\nabla \cdot (\bar{\rho}\bar{\theta}V) = \frac{\dot{Q}}{c_p\bar{\pi}}, \tag{2.3.8}$$

$$\frac{D\theta}{Dt} = \frac{\theta \dot{Q}}{c_p \bar{\pi} \, \overline{\rho \theta}},$$ (2.3.9)

where $\dot{Q}$ is the heating rate per unit volume $[= (\overline{\rho \theta}/\theta)\dot{q}]$ and π, the *Exner function*, is defined as

$$\pi = \left(\frac{p}{p_s}\right)^{R_d/c_p}.$$ (2.3.10)

The Exner function is partitioned into the basic state and the perturbation,

$$\pi = \bar{\pi}(z) + \pi'(t, x, y, z).$$ (2.3.11)

Note that the Exner function has been defined in different forms in the literature. In order to apply the pseudo-incompressible approximation in the investigation of mesoscale atmospheric motions, the following two criteria must be met: (1) $T \gg L/c_s$, i.e. the Lagrangian time scale associated with the disturbance must be much larger than the time scale associated with (adiabatic) sound wave propagation and (2) $\pi' \ll \bar{\pi}$.

A well-known approximation that has been widely used in theoretical studies is the *Boussinesq approximation*. For the fully nonlinear equations governing thermal convection in a compressible fluid, the conditions under which the Boussinesq approximation is applicable are: (a) the vertical dimension of the fluid motion is much less than any scale height and (b) the motion-induced fluctuations in density and pressure do not exceed, in order of magnitude, the total static variations of these quantities (Spiegel and Veronis 1960). In (2.2.14)–(2.2.18), the Boussinesq approximation is equivalent to assuming that (1) $L_z \ll H$, (2) density is treated as a constant except where it is coupled to gravity in the buoyancy term in the vertical momentum equation, and (3) $\bar{\rho}$ and $\bar{\theta}$ are replaced by ρ_o and θ_o, respectively, in all equations. Under the Boussinesq approximation, one may define the *perturbation buoyancy* ($b' = g\theta'/\theta_o$) and *kinematic pressure* (p'/ρ_o) to simplify the equation set. The Boussinesq approximation can be extended to a larger vertical scale of motion if the potential density and modified pressure are used in a low Rossby number flow (Janowitz 1977).

For a disturbance in which the horizontal scale is much larger than its vertical scale ($L_x \gg L_z$), the vertical acceleration generally becomes very small compared to the vertical pressure gradient force and buoyancy force and therefore may be neglected. This leads to the *hydrostatic approximation*. By eliminating the vertical accelerations, the perturbation pressure p' and density ρ' are in hydrostatic balance,

$$\frac{\partial p'}{\partial z} - \left(\frac{g\bar{\rho}}{\bar{\theta}}\right)\theta' = 0.$$ (2.3.12)

The above perturbation hydrostatic equation has a corresponding Boussinesq form,

$$\frac{\partial p'}{\partial z} - \left(\frac{g\rho_0}{\theta_0}\right)\theta' = 0. \tag{2.3.13}$$

Limitations of the hydrostatic approximation may be investigated by assuming a two-dimensional (in x and z), nonrotating, Boussinesq fluid with a uniform basic flow (i.e. $U = \text{constant}$; $V = 0$). Under these constraints, the linear set of perturbation equations (2.2.14)–(2.2.18) may be simplified and combined into a single equation for the vertical velocity w',

$$\left(\frac{\partial}{\partial t} + U\frac{\partial}{\partial x}\right)^2 \left(\frac{\partial^2 w'}{\partial x^2} + \frac{\partial^2 w'}{\partial z^2}\right) + N^2 \frac{\partial^2 w'}{\partial x^2} = 0. \tag{2.3.14}$$

It can be shown that the leftmost $\partial^2 w'/\partial x^2$ term is associated with the vertical acceleration term of the vertical momentum equation, i.e. the first two terms of (2.2.16). It can be shown by comparing the scales of $\partial^2 w'/\partial x^2$ and $\partial^2 w'/\partial z^2$ that in order to neglect the vertical acceleration term, $L_z/L_x \ll 1$ is required.

References

Bannon, P. R., 1986. Linear development of quasi-geostrophic baroclinic disturbances with condensational heating. *J. Atmos. Sci.*, **43**, 2261–74.

Batchelor, G. K., 1953. The condition for dynamical similarity of motions of a frictionless perfect-gas atmosphere. *Quart. J. Roy. Meteor. Soc.*, **79**, 224–35.

Dutton, J. A., 1986. *The Ceaseless Wind*. Dover.

Eady, 1949. Long wave and cyclone waves. *Tellus*, **1**, 33–52.

Janowitz, G. S., 1977. The effects of compressibility on the stably stratified flow over a shallow topography in the beta plane. *J. Atmos. Sci.*, **34**, 1707–14.

Lackmann, G. M. and R. M. Yablonsky, 2004. On the role of the precipitation mass sink in tropical cyclogenesis. *J. Atmos. Sci.*, **61**, 1674–92.

Nance, L. B. and D. R. Durran, 1994. A comparison of the accuracy of three anelastic systems and the pseudo-incompressible system. *J. Atmos. Sci.*, **51**, 3549–65.

Ogura, Y. and N. A. Phillips, 1962. Scale analysis of deep and shallow convection in the atmosphere. *J. Atmos. Sci.*, **19**, 173–9.

Spiegel, E. A. and G. Veronis, 1960. On the Boussinesq approximation for a compressible fluid. *Astrophys. J.*, **131**, 442–7.

Problems

2.1 Based on Poisson's equation and the equation of state, show that $g\rho'/\bar{\rho} \approx -g\theta'/\bar{\theta} + p'/\bar{\rho}H$, where $H \equiv \gamma R\bar{T}/g$ for a small-amplitude disturbance. You may assume that the basic state satisfies Poisson's equation and the equation of state.

2.2 Derive the linear set of equations (2.2.14)–(2.2.18) by substituting (2.2.8)–(2.2.13) into (2.2.1)–(2.2.5).

2.3 Obtain the analytical solution of the simplified governing equation of motion, $\partial u/\partial t = -\mu u$, with the initial value u_0, and determine the value of the coefficient of the Rayleigh friction if it takes 12 hours for u to reduce to its e-folding value, i.e. $u = u_0 e^{-1}$.

2.4 Perform scale analysis of (2.2.17) by (a) identifying scales of individual terms; (b) estimating magnitudes of individual terms by assuming the following fluid flow system: $U = 10 \text{ m s}^{-1}$, $W = 1 \text{ m s}^{-1}$, $L_x = 10$ km, $L_z = 10$ km, $H = 10$ km, $f = 10^{-4} \text{ s}^{-1}$, $\bar{p} = 1 \text{ kg m}^{-3}$, $\dot{q} = 1000$ W m $\text{kg}^{-1}/10$ km, $u' = 1 \text{ m s}^{-1}$, and $p' = 1$ mb; (c) finding the approximate form of the equation by keeping the highest order terms; and (d) identifying the approximation. What kind of weather systems does it describe?

2.5 (a) Let $\rho(t, x, y, z) = \bar{\rho}(z) + \rho'(t, x, y, z)$ in (2.2.4) and identify scales of individual terms. Make sure to differentiate the vertical scale of the basic state (H) and that of disturbance (L_z) and the scales of individual terms of the divergence term. (b) Consider the following scales of a weather system: $U = 10 \text{ m s}^{-1}$, $W = 1 \text{ m s}^{-1}$, $L_x = 10$ km, $L_z = 10$ km, $H = 10$ km, $\bar{\rho} = 1 \text{ kg m}^{-3}$, and $\rho' = 0.01 \text{ kg m}^{-3}$. Show that the approximate continuity equation is reduced to $(w'/\bar{\rho})(\partial\bar{\rho}/\partial z) + (\partial u'/\partial x + \partial v'/\partial y + \partial w'/\partial z) = 0$ by retaining only the highest two orders of magnitude. (c) Show that the approximate equation of (b) is identical to the anelastic equation (2.3.1) if the basic density decreases with height with an e-folding value of H. (d) Based on the scales identified in (a), derive the criterion for the shallow convection continuity equation (2.3.4).

2.6 Prove that (2.3.1), (2.3.2), and (2.3.3) are identical.

2.7 Apply the Boussinesq approximation to (2.2.14)–(2.2.18).

2.8 Derive (2.3.14) from the linear equations (2.2.14)–(2.2.18) by assuming a two-dimensional (in x and z), non-rotating, Boussinesq fluid with a uniform basic flow (i.e. $U = $ constant; $V = 0$). Show that: (a) the $\partial^2 w'/\partial x^2$ term is associated with the vertical acceleration term of the vertical momentum equation, i.e. the first two terms of (2.2.16), and (b) taking $L_z/L_x \ll 1$ is equivalent to making the hydrostatic approximation.

2.9 (a) Derive the *Reynolds* number (R_e) by taking the ratio of the scale of the *inertial term* (Du/Dt) and the molecular viscosity term $(\nu\nabla^2 u)$. Assume that the viscous term can be approximated by the horizontal part of the *Laplacian operator*. Estimate this number at the *viscous sublayer* very near the earth's surface with $U = 10 \text{ m s}^{-1}$, $L = 120$ km, and $\nu = 1.2 \times 10^{-5} \text{ m}^2 \text{ s}^{-1}$. What can you conclude?

(b) Assuming that the viscous term can be approximated by only the vertical derivative of the Laplacian operator, then use scale analysis to derive a new Reynolds number (R_{e2}). Assuming the new *eddy viscosity coefficient* $(\nu = K_m = 5 \text{ m}^2 \text{ s}^{-1})$ and $L_z = 1.2$ km, estimate R_{e2}. What can you conclude?

3

Basic wave dynamics

3.1 Introduction

When an air parcel is displaced from its initial position, a *restoring force* may cause it to return to its initial position. In doing so, inertia will cause the air parcel to overshoot and pass its initial equilibrium position moving in the opposite direction from that in which it is initially displaced, thereby creating an oscillation around the equilibrium position. Concurrently, a wave is produced that propagates from this source region to another part of the fluid system, which is the physical *medium of wave propagation*. A physical restoring force and a medium for propagation are the two fundamental elements of all wave motion in solids, liquids, and gases, including atmospheric waves, oceanic waves, sound (acoustic) waves, wind induced waves, seismic waves, or even traffic density waves. The ultimate behavior of the wave is dictated by the individual properties of the restoring force responsible for wave generation and the medium through and by which the wave propagates energy and momentum.

From scaling arguments of the so-called primitive equations based on horizontal scales of fluid motion, some of the more important categories of waves observed in the atmosphere may be classified as follows (Table 3.1): (a) sound (acoustic) waves, (b) mesoscale waves, and (c) planetary (Rossby) waves. In this chapter, mesoscale waves are defined in a more general manner, referring to waves that exist and propagate in the atmosphere with a mesoscale wavelength. Thus, mesoscale waves include pure gravity waves and inertia-gravity waves, instead of being limited to some long-lasting waves as are also defined in the literature. A more detailed classification of these waves and their probable restoring or wave generation forces are summarized in Table 3.1. Each group of waves exhibits multiple flow regimes. For example, pure gravity waves may be further categorized as either vertically propagating waves or evanescent waves, depending upon whether the wave energy is free to propagate vertically. The restoring forces for pure gravity waves and inertia oscillations are the buoyancy force ($-g\rho'/\bar{\rho}$ or $g\theta'/\bar{\theta}$) and Coriolis force, respectively. Sound waves derive their oscillations from longitudinal compression and expansion, while planetary waves derive their oscillations from the meridional variation of the Coriolis force or β effect. The compression force, buoyancy force, Coriolis force, and variation of Coriolis force are often represented by c_s, N, f, and β, respectively, in governing equations.

Table 3.1 *A summary of atmospheric waves.*

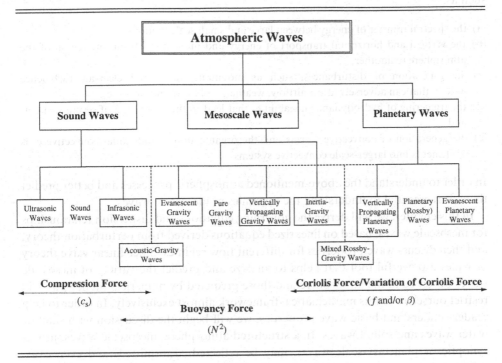

Restoring forces may also combine and work together to generate mixed waves, such as inertia-gravity waves, mixed acoustic-gravity waves, and mixed Rossby-gravity waves. Inertia-gravity waves are also known as Poincaré waves on the ocean surface and as boundary Kelvin waves along a rigid, lateral boundary such as a shoreline or coast. The oscillation period of the waves is determined by the strength of the restoring force and characteristics in the wave medium. The presence of mesoscale waves in the atmosphere and oceans can be inferred from (a) thermodynamic soundings, (b) microbarograph pressure traces, (c) visible and infrared satellite images, (d) radar echoes, and (e) vertical wind profilers. Data obtained from these sources and instruments may be used to help predict mesoscale wave generation and propagation, as well as to help explain the development and subsequent evolution of a variety of mesoscale scale weather phenomena associated with the passage of these waves.

In previous studies of large-scale dynamics and numerical weather prediction (NWP), mesoscale and sound waves were often regarded as undesirable "noise," since they often appear as small perturbations or disturbances embedded in the large-scale flow, and their presence can even trigger numerical instabilities in an operational forecast model if the grid interval of the NWP model is not sufficiently small. Therefore, they are normally filtered out from the primitive equations.

However, mesoscale waves do have major dynamical impacts on the atmosphere, such as:

(a) the spectral transfer of energy between large and small scale motions,
(b) the vertical and horizontal transport of energy and momentum from one region of the atmosphere to another,
(c) the generation of disturbances, such as mountain waves and clear-air turbulence (CAT), that can adversely affect airflow, weather, and aviation safety etc.,
(d) the triggering of hydrodynamic instabilities that lead to the generation of severe weather, and
(e) the generation of convective storms and the organization of individual convective cells or elements into larger-scale convective systems.

In order to understand the above-mentioned atmospheric processes and better predict them, it is essential to understand the dynamics of mesoscale waves.

In this chapter, we will derive the governing equations and dispersion relationships for mesoscale waves, based on linearized equations derived from perturbation theory, and then discuss wave properties for different flow regimes. Since linear wave theory provides a powerful tool that helps to analyze and predict the variety of mesoscale waves observed in the atmosphere and those predicted by numerical models, we will restrict ourselves to this mathematical framework almost exclusively. In order to help readers understand basic wave properties, we will begin the discussion with shallow water waves and sound waves. In a structured atmosphere, mesoscale waves, just as synoptic or planetary-scale waves, may be reflected, transmitted, and even over-reflected from certain internal boundaries. These properties will be discussed in the later part of the chapter. Wave generation, propagation, and maintenance mechanisms will also be presented in Chapter 4.

3.2 Basic wave properties

With the Boussinesq approximation, the linearized perturbation equations, (2.2.14)–(2.2.18), can be combined into a single equation for the vertical velocity w',

$$
\frac{D}{Dt}\left\{\frac{D^2}{Dt^2}\nabla^2 w' + f^2 w'_{zz} - \left(U_{zz}\frac{D}{Dt}+fV_{zz}\right)w'_x - \left(V_{zz}\frac{D}{Dt}-fU_{zz}\right)w'_y + N^2\nabla^2_H w' \right.
$$
$$
\left. +2f\left(U_z w'_{yz} - V_z w'_{xz}\right)\right\} + 2fU_z V_z\left(w'_{xx} - w'_{yy}\right)
$$
$$
+2f\left(V_z^2 - U_z^2\right)w'_{xy} - 2f^2\left(U_z w'_{xz} + V_z w'_{yz}\right) = \frac{g}{c_p T_o}\frac{D}{Dt}\nabla^2_H q'
$$

(3.2.1)

where $D/Dt = \partial/\partial t + U\partial/\partial x + V\partial/\partial y$. This equation governs the small-amplitude vertical velocity w' in a mesoscale system, which may contain the following mechanisms: (a) pure gravity and inertia-gravity wave generation, (b) static instability, (c) shear instability, (d) symmetric instability, and (e) baroclinic instability (Emanuel and Raymond, 1984). For different flow regimes, the Boussinesq form of (2.2.14)–(2.2.18) may reduce to different approximate equations, which will be further discussed in Sections 3.5 and 3.6.

Regardless of the restoring forces and propagation media, wave motions may be characterized by several fundamental properties, such as wave frequency, wave number, phase speed, group velocity, and dispersion relationship. Note that the dispersion relationship relates the wave frequency to the wave number. The period of oscillation (τ) of a wave determines the wave frequency ($\omega = 2\pi/\tau$), while the horizontal and vertical spatial scales (L_x, L_y, L_z) of a wave determine its horizontal and vertical wave numbers ($k = 2\pi/L_x$, $l = 2\pi/L_y$, $m = 2\pi/L_z$). A wave may be characterized by its amplitude and phase, $\varphi(x,y,z,t) = \text{Re}\{A\,\exp[i(kx + ly + mz - \omega t - \alpha)]\}$, where φ represents any of the dependent flow variables, Re means the real part, A the amplitude, $kx + ly + mz - \omega t - \alpha$ the phase, and α the phase angle. The phase angle is determined by the initial position of the wave. Lines of constant phase, such as wave crests, troughs, or any other particular part of the wave propagate through the fluid medium at a speed called the phase speed. Phase speeds in x, y, and z directions, are respectively given by

$$c_{px} = \omega/k; \quad c_{py} = \omega/l; \quad \text{and} \quad c_{pz} = \omega/m. \tag{3.2.2}$$

If an observed or experimental set of data is available, the phase speed of a wave may be estimated by determining and tracing a constant phase of the wave.

In a geophysical fluid system such as the atmosphere, waves normally are very complicated in form, due to the superposition of wave components with different wavelengths and their nonlinear interactions, and cannot be represented by a simple, single sinusoidal wave. However, a small-amplitude wave with any shape can be approximately represented by a linear superposition of wave trains of different wave numbers, i.e. a Fourier series of sinusoidal components. For example, a wave in the x direction may be decomposed into sinusoidal components of the form,

$$\varphi(x) = \sum_{n=1}^{\infty} (A_n \sin k_n x + B_n \cos k_n x), \tag{3.2.3}$$

where the Fourier coefficients A_n and B_n are determined by

$$A_n = \frac{2}{L} \int_0^L \varphi(x) \sin\frac{2\pi n x}{L}\,\mathrm{d}x, \tag{3.2.4}$$

$$B_n = \frac{2}{L} \int_0^L \varphi(x) \cos\frac{2\pi n x}{L}\,\mathrm{d}x. \tag{3.2.5}$$

The nth Fourier component or nth harmonic of the wave function φ_n is defined as $A_n \sin k_n x + B_n \cos k_n x$. In deriving (3.2.4) and (3.2.5), we have used the orthogonality relationships,

$$\int_0^L \sin \frac{2\pi nx}{L} \cos \frac{2\pi mx}{L} \, dx = 0, \quad \text{for all } n, m > 0,$$

$$\int_0^L \sin \frac{2\pi nx}{L} \sin \frac{2\pi mx}{L} \, dx = \left\{ \begin{array}{ll} 0, & n \neq m \\ L/2, & n = m \end{array} \right\}, \text{and}$$

$$\int_0^L \cos \frac{2\pi nx}{L} \cos \frac{2\pi mx}{L} \, dx = \left\{ \begin{array}{ll} 0, & n \neq m \\ L/2, & n = m \end{array} \right\}. \tag{3.2.6}$$

If a wave is composed of a series of Fourier components of different wavelengths, then the phase speed for each individual component may also be different, according to (3.2.2). If the phase speed is independent of wave number, then the wave will retain its initial shape and remain coherent as it propagates throughout the fluid medium. This type of wave is *nondispersive*. On the other hand, if the phase speed is a function of the wave number, then the wave will not be able to retain its initial shape and remain coherent as it propagates in the medium since each Fourier component is propagating at a different phase speed. In other words, the wave is *dispersive*. Thus, it becomes clear that the relationship between wave frequency and wave number determines whether or not the wave is dispersive.

Although visually, a dispersive wave may look like it is dissipative, dispersion and dissipation are completely different physical processes. In a dissipative, but nondispersive, wave, every Fourier component of the wave propagates at the same speed, while the wave amplitude decreases. Thus, individual wave groups preserve their phase (shape) during propagation. If the wave is nondispersive, then the wave pattern moves throughout the medium without any change in shape of the initial waveform. This means that the *phase velocity* of the individual wave crests (c_p) is equal to the *group velocity* of the slow-varying modulations or the envelope of Fourier wave components (c_g). The concept of group velocity is illustrated in Fig. 3.1, in which the simple group of two superimposed sinusoidal waves, represented by the wave function $\varphi(x,t) = \text{Re}\{A \exp[i((k + \Delta k)x - (\omega + \Delta\omega)t)] + \exp[i((k - \Delta k)x - (\omega - \Delta\omega)t)]\}$, propagates at the speed $\Delta\omega/\Delta k$. This velocity approaches $\partial\omega/\partial k$, which is defined as the group velocity in the x direction, as Δk approaches 0. Thus, the group velocity represents the velocity the slow-varying modulation of a wave propagates, which is given by the relation

$$c_g = c_{gx}\boldsymbol{i} + c_{gy}\boldsymbol{j} + c_{gz}\boldsymbol{k} = \frac{\partial\omega}{\partial k}\boldsymbol{i} + \frac{\partial\omega}{\partial l}\boldsymbol{j} + \frac{\partial\omega}{\partial m}\boldsymbol{k}. \tag{3.2.7}$$

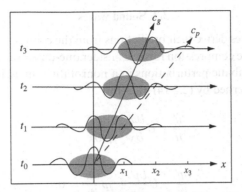

Fig. 3.1 Propagation of a wave group and an individual wave. The solid and dashed lines denote the group velocity (c_g) and phase velocity (c_p), respectively. Shaded oval denotes the concentration of wave energy which propagates with the group velocity. The phase speed c_p equals x_i/t_i, where $i = 1, 2$, or 3.

From the incompressible continuity equation, one can show that

$$\mathbf{k} \cdot \mathbf{V}' = 0, \tag{3.2.8}$$

where $\mathbf{k} = (k, l, m)$ is the *wave number vector*. For two-dimensional plane waves (e.g., $l = 0$), the above equation indicates that the wave motion or oscillation of fluid parcels is perpendicular to the wave number vector. At any instant t_0, a *wave front* is defined by setting the phase ($\mathbf{k} \cdot \mathbf{x} - \omega t - \alpha$) to be constant, which indicates that a family of parallel planes is normal to the wave number vector $\mathbf{k}$. As time proceeds, these planes move with the phase speed in the direction $\mathbf{k}$. Notice that the phase velocity can be derived as

$$\mathbf{c} = \omega \mathbf{k}/k^2. \tag{3.2.9}$$

That is, the phase velocity $\mathbf{c}$ is parallel to $\mathbf{k}$. As will be discussed in Section 3.6, inertia-gravity waves are an example of *transverse waves* because the fluid particle motion, $\mathbf{V}'$, is perpendicular to the phase velocity (Fig. 3.9 or Fig. 3.10a), as indicated by (3.2.8). Note that the phase speeds in the x and z directions do not comprise the phase velocity. That is,

$$\mathbf{c} \neq (\omega/k, \omega/l, \omega/m). \tag{3.2.10}$$

For sound waves (Section 3.3), it can be shown that

$$\mathbf{k} \cdot \mathbf{V}' \neq 0. \tag{3.2.11}$$

Thus, sound waves are an example of *longitudinal waves*.

3.3 Sound waves

Sound (acoustic) waves derive their oscillations from the compression and expansion of the medium due to the compression force. Consider one-dimensional ($\partial/\partial y = \partial/\partial z = 0$), small-amplitude, adiabatic perturbations in a nonrotating, inviscid, uniform (no basic wind shear) flow governed by (2.2.14)–(2.2.18),

$$\frac{\partial u'}{\partial t} + U\frac{\partial u'}{\partial x} + \frac{1}{\bar{\rho}}\frac{\partial p'}{\partial x} = 0, \tag{3.3.1}$$

$$\frac{\partial p'}{\partial t} + U\frac{\partial p'}{\partial x} + \gamma\bar{p}\frac{\partial u'}{\partial x} = 0. \tag{3.3.2}$$

The above two equations may be combined into a single equation for p',

$$\left(\frac{\partial}{\partial t} + U\frac{\partial}{\partial x}\right)^2 p' - c_{\rm s}^2\frac{\partial^2 p'}{\partial x^2} = 0. \tag{3.3.3}$$

Assuming a wave-like solution,

$$p(x,t) = p_0 e^{i(kx-\omega t)} = p_0[\cos(kx - \omega t) + i\sin(kx - \omega t)],$$

and substituting it into (3.3.3) leads to

$$\omega = (U \pm c_{\rm s})k. \tag{3.3.4}$$

For brevity, we omit the Re{ } notation in the wave-like solution, but it is to be understood that only the real part of the above solution has physical significance. The above method of the assumption of wave-like solutions for the small-amplitude perturbations is also referred to as the *method of normal modes*. The above equation is the dispersion relation for sound waves, which relates the wave frequency ω to the horizontal wave number k. From (3.3.4), we may obtain the horizontal phase speeds, which are given by

$$c_p = \frac{\omega}{k} = U \pm c_{\rm s}. \tag{3.3.5}$$

The above equation represents phase speeds of the downstream ($U + c_{\rm s}$) and upstream ($U - c_{\rm s}$) propagating sound waves, which are simultaneously being advected by the basic wind U. Sound waves are nondispersive since their phase speeds are independent of wave number. This nondispersive property of sound waves may also be verified by showing that the group velocities for these waves c_g are identical to their phase speeds c_p.

Consider a semi-infinite tube filled with gas whose right-hand side extends to infinity and whose left-hand side is confined by a piston. When the gas is alternatively compressed and expanded by oscillating the piston in and out of the left-hand side of the semi-infinite tube, an air parcel located adjacent to the piston will be forced to oscillate back and forth about its equilibrium position due to the oscillating horizontal pressure gradient and, concurrently, a sound wave will be excited that propagates toward the right at the speed $c_s = \sqrt{\gamma R \bar{T}}$. In a dry, isothermal atmosphere with a constant temperature of 300 K, a one-dimensional acoustic wave has a phase speed and group velocity of approximately 347 m s^{-1} ($\sim$ 776 mph).

In fact, the general solution of (3.3.3) may be written as,

$$p' = F[x + (U + c_s)t] + F[x + (U - c_s)t], \qquad (3.3.6)$$

where F is any arbitrary function whose amplitude is one-half that of the initial disturbance, and whose shape is identical to that of the initial disturbance. In a quiescent fluid, the first and second terms on the right-hand side of (3.3.6) represent the leftward and rightward propagating sound waves, respectively.

Since sound waves do not play significant dynamic roles in affecting most atmospheric motions, they are often eliminated from the primitive equations – in particular those that are commonly employed in most current operational NWP models. Although sound waves may have no particular relevance to atmospheric motions in the troposphere that are responsible for "weather," a special class of waves called *Lamb waves* has been observed. These waves can propagate horizontally in an isothermal atmosphere in the absence of vertical motion. Lamb waves, as well as sound waves and gravity waves, can be generated by latent heat release in a convective storm (e.g., Nicholls and Pielke 2000). In a two-dimensional ($\partial/\partial y = 0$), adiabatic, hydrostatic, nonrotating, inviscid, isothermal atmosphere with no vertical motion and basic wind shear, small-amplitude motions are governed by (2.2.14)–(2.2.18) reduced to (3.3.1), (3.3.2), and

$$\frac{1}{\bar{\rho}} \left(\frac{\partial p'}{\partial z} + \frac{p'}{H} \right) = g \frac{\theta'}{\bar{\theta}}, \qquad (3.3.7)$$

$$\frac{\partial \theta'}{\partial t} + U \frac{\partial \theta'}{\partial x} = 0. \qquad (3.3.8)$$

Note that in an isothermal atmosphere, the scale height, $H = c_s^2/g$, is a constant. Equations (3.3.7) and (3.3.8) may be combined to yield an equation that, when coupled with (3.3.3), forms the set of equations governing the evolution of Lamb waves.

3.4 Shallow water waves

Shallow water waves propagate horizontally along a water or ocean surface or along an interfacial boundary between two layers of fluid with distinctly different densities. Although this type of wave is observed more often in the oceans and lakes, they are also occasionally observed in the atmosphere. For example, the mesoscale waves

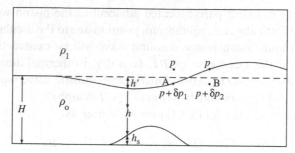

Fig. 3.2 A two-layer system of homogeneous fluids. Symbols H, h, h_s, and h' denote the undisturbed fluid depth, actual fluid depth, bottom topography, and perturbation (vertical displacement) from the undisturbed fluid depth, respectively. The densities of the upper and lower layers are ρ_1 and ρ_o, respectively. The pressure perturbations at A and B from p in the upper layer are denoted by $p + \delta p_1$ and $p + \delta p_2$, respectively.

generated below a thunderstorm that subsequently propagate along a temperature inversion capping a layer of very dense cold air produced by evaporative cooling, may be characterized and treated mathematically as shallow water waves. Another example is the internal hydraulic jump occurring over the lee slope of a mountain, as that shown in Fig. 3.5. Shallow water wave theory has also been applied to understand other weather systems, such as the propagation of tropical cyclones and tropical waves.

Consider a nonrotating, hydrostatic, two-layer fluid system with constant densities ρ_1 and ρ_o in the upper and lower layers, respectively, and assume that $\rho_1 < \rho_o$ (stably stratified) and that there exist no horizontal pressure gradients in the upper layer (Fig. 3.2). This two-layer shallow water system is governed by the following reduced forms of the horizontal momentum equations and the vertically integrated continuity equation,

$$\frac{\partial u}{\partial t} + u\frac{\partial u}{\partial x} + v\frac{\partial u}{\partial y} = -g'\frac{\partial(h + h_s)}{\partial x}, \tag{3.4.1}$$

$$\frac{\partial v}{\partial t} + u\frac{\partial v}{\partial x} + v\frac{\partial v}{\partial y} = -g'\frac{\partial(h + h_s)}{\partial y}, \tag{3.4.2}$$

$$\frac{\partial h}{\partial t} + u\frac{\partial h}{\partial x} + v\frac{\partial h}{\partial y} + h\left(\frac{\partial u}{\partial x} + \frac{\partial v}{\partial y}\right) = 0, \tag{3.4.3}$$

where h is the instantaneous depth of the fluid. The above three equations, corresponding to (A3.1.9), (A3.1.10), and (A3.1.14), respectively, are derived in Appendix 3.1. Decomposing u, v, and h into basic state and perturbation, $u = U + u'$, $v = V + v'$, and $h = H + h' - h_s$, and substituting them into (3.4.1)–(3.4.3) leads to the following perturbation equations:

$$\frac{\partial u'}{\partial t} + (U + u')\frac{\partial u'}{\partial x} + (V + v')\frac{\partial u'}{\partial y} + g'\frac{\partial h'}{\partial x} = 0, \tag{3.4.4}$$

$$\frac{\partial v'}{\partial t} + (U + u')\frac{\partial v'}{\partial x} + (V + v')\frac{\partial v'}{\partial y} + g'\frac{\partial h'}{\partial y} = 0, \tag{3.4.5}$$

$$\frac{\partial h'}{\partial t} + (U + u')\frac{\partial h'}{\partial x} + (V + v')\frac{\partial h'}{\partial y} + (H + h' - h_s)\left(\frac{\partial u'}{\partial x} + \frac{\partial v'}{\partial y}\right)$$
$$= (U + u')\frac{\partial h_s}{\partial x} + (V + v')\frac{\partial h_s}{\partial y}. \tag{3.4.6}$$

Note that H is the undisturbed upstream layer depth, h_s is the height of the bottom topography, h' is the vertical displacement from H. Thus, h_s is no longer included in the horizontal pressure gradient forces in the momentum equations (3.4.4) and (3.4.5). However, in the system of perturbation equations, (3.4.4)–(3.4.6), h_s serves as a forcing term for the vertical velocity $w' = Dh'/Dt$ in the vertically integrated mass continuity equation, (3.4.6).

For simplicity, consider the special case of small-amplitude (linear) perturbations in a one-layer fluid with a flat bottom and with $V = 0$. This system may be used as a first approximation of the air–water system since $\rho_{air} \ll \rho_{water}$. These assumptions yield $\Delta\rho = \rho_o - \rho_1 \approx \rho_o$, where $\rho_o = \rho_{water}$, $\rho_1 = \rho_{air}$, and $g' = g$. Under these assumptions, (3.4.4)–(3.4.6) become

$$\frac{\partial u'}{\partial t} + U\frac{\partial u'}{\partial x} + g\frac{\partial h'}{\partial x} = 0, \tag{3.4.7}$$

$$\frac{\partial v'}{\partial t} + U\frac{\partial v'}{\partial x} + g\frac{\partial h'}{\partial y} = 0, \tag{3.4.8}$$

$$\frac{\partial h'}{\partial t} + U\frac{\partial h'}{\partial x} + H\left(\frac{\partial u'}{\partial x} + \frac{\partial v'}{\partial y}\right) = 0. \tag{3.4.9}$$

The above set of equations can be combined into a wave equation governing the free surface displacement perturbation h'

$$\left(\frac{\partial}{\partial t} + U\frac{\partial}{\partial x}\right)^2 h' - (gH)\left(\frac{\partial^2 h'}{\partial x^2} + \frac{\partial^2 h'}{\partial y^2}\right) = 0. \tag{3.4.10}$$

Applying the method of normal modes with $h' = A\exp[i(kx - \omega t)]$ and substituting it into the two-dimensional form of (3.4.10) (i.e., with $\partial/\partial y = 0$), we obtain the dispersion relation for two-dimensional shallow water waves,

$$c = U \pm \sqrt{gH}. \tag{3.4.11}$$

With no basic flow, the above equation indicates that the solution to (3.4.10) consists of a leftward and a rightward propagating shallow-water wave along the free surface of the fluid with the shallow-water wave (phase) speeds of $-\sqrt{gH}$ and $+\sqrt{gH}$, respectively. With a basic flow, these two waves are advected by the basic flow speed (U). Since the phase speeds are independent of wave number, these two-dimensional shallow-water waves are nondispersive. It is straightforward to show that the group velocity is identical to the phase velocity in this system. In fact, we may obtain this conclusion directly from the general solution to the wave equation, namely,

$$h'(t, x) = \frac{1}{2}f(x + (U + \sqrt{gH})t) + \frac{1}{2}f(x + (U - \sqrt{gH})t), \qquad (3.4.12)$$

where $f(x)$ is the initial disturbance and f is any arbitrary function. The function represented by $(1/2)f$ will preserve the shape of the initial disturbance, but its amplitude will only be half that of the initial disturbance. An example for this is the shallow-water waves excited by the horizontal pressure gradient force along the free surface associated with the initial disturbance, which physically represents a localized, mesoscale high pressure system,

$$h'(x) = \frac{h_o b^2}{x^2 + b^2}, \qquad (3.4.13)$$

are represented mathematically as,

$$h'(x, t) = \frac{(h_o/2)b^2}{[x + (U + \sqrt{gH})t]^2 + b^2} + \frac{(h_o/2)b^2}{[x + (U - \sqrt{gH})t]^2 + b^2}. \qquad (3.4.14)$$

Equation (3.4.13) is called a *bell-shaped function* or the *Witch of Agnesi*, in which h_o and b are the maximum amplitude and half-width of the initial disturbance, respectively. Note that the waves generated have the same bell-shaped form but with their amplitude reduced to half that of the initial disturbance.

The equation for the total perturbation energy for a three-dimensional shallow water system can be derived,

$$\left(\frac{\partial}{\partial t} + U\frac{\partial}{\partial x}\right)\left[\frac{1}{2}\rho H(u'^2 + v'^2) + \frac{1}{2}\rho g h'^2\right] + \rho g H\left[\frac{\partial}{\partial x}(u'h') + \frac{\partial}{\partial y}(v'h')\right] = 0. \quad (3.4.15)$$

The first term inside the first square bracket is the perturbation kinetic energy, and the second term is the perturbation potential energy. The above equation indicates that the change in total perturbation energy per unit area is simply caused by the convergence or divergence of the mass flux.

Different flow regimes may result from different upstream flow conditions. For a two-dimensional, nonrotating, small-amplitude shallow water flow over an obstacle, the wave equation governing the evolution in vertical displacement can be derived

$$\left(\frac{\partial}{\partial t} + U\frac{\partial}{\partial x}\right)^2 h' - (gH)\frac{\partial^2 h'}{\partial x^2} = U\left(\frac{\partial}{\partial t} + U\frac{\partial}{\partial x}\right)\frac{\partial h_s}{\partial x}. \tag{3.4.16}$$

A steady state solution for (3.4.16) can be found

$$h'(x) = \left(\frac{F^2}{F^2 - 1}\right)h_s(x); \quad F \equiv \frac{U}{\sqrt{gH}}, \tag{3.4.17}$$

where F is the *shallow-water Froude number* that represents the ratio of the advection flow speed to the shallow water phase speed. The Froude number also represents the ratio of the square-root of the kinetic and potential energies of the undisturbed, upstream basic flow. Equation (3.4.17) implies that

$$h' \propto \begin{cases} h_s & \text{for } F > 1 \\ -h_s, & \text{for } F < 1 \end{cases}. \tag{3.4.18}$$

The above equation indicates that if $F > 1$ far upstream, $h'(x)$ follows the shape of $h_s(x)$. The interface or free surface will bow upwards over the obstacle as shown in Fig. 3.3a. That is, physically, the upstream flow has enough kinetic energy to overcome the potential energy barrier associated with the obstacle. This flow regime is called *supercritical flow*. On the other hand, when $F < 1$, $h'(x)$ decreases as $h_s(x)$ increases, as the perturbation flow converts its potential energy into enough kinetic energy to surmount the obstacle (Fig. 3.3e). Over the peak of the obstacle, the fluid reaches its maximum speed, as implied from (3.4.17). This flow regime is called

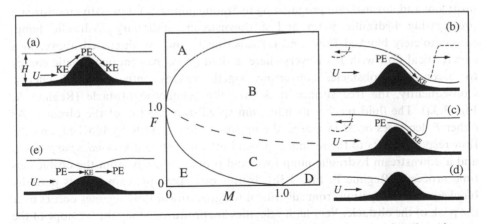

Fig. 3.3 Five flow regimes of the transient one-layer shallow water system, based on the two nondimensional control parameters ($F = U/\sqrt{gH}$, $M = h_m/H$). (a) Regime A: supercritical flow, (b) Regime B: flow with both upstream and downstream propagating hydraulic jump, (c) Regime C: flow with upstream propagating jump and downstream stationary jump, (d) Regime D: completely blocked flow, and (e) Regime E: subcritical flow. The dashed lines in (b) and (c) denote transient water surface. In regimes B and C, upstream flow is partially blocked. (Adapted after Baines 1995 and Durran 1990.)

subcritical flow. When the flow is supercritical ($F > 1$), small-amplitude disturbances cannot propagate upstream against the basic flow, and any obstacle along the bottom will tend to produce a purely local disturbance. When the flow is subcritical ($F < 1$), shallow water waves are able to propagate upstream since the shallow water wave speed is larger than the basic wind speed. The steady state effect of this response is to effectively increase the layer depth upstream, which increases the potential energy of the incoming flow. The potential energy is then converted into kinetic energy as the fluid surmounts the obstacle. Thus, the fluid reaches its maximum speed, and the water surface dips down over the peak of the obstacle creating a *Bernoulli* or *venturi* effect. The relationship between u' and h' can be derived,

$$\frac{u'}{U} = -\left(\frac{1}{F^2}\right)\frac{h'}{H}, \tag{3.4.19}$$

which indicates that the flow decelerates (accelerates) over a hump for a supercritical (subcritical) flow. In addition to F, (3.4.17) indicates that the *nondimensional mountain height* M ($\equiv h_m/H$, where h_m is the mountain height) also serves as a control parameter of the shallow-water flow over a bottom topography. In other words, the flow behavior of the shallow-water system is controlled by F and M.

The above discussion on wave characteristics is based on the steady-state solution (3.4.17). In transient flow, additional flow regimes may occur. Based on the non-dimensional control parameters F and M, five different flow regimes can be identified in transient, single-layer shallow water flow over an isolated obstacle (Fig. 3.3; Houghton and Kasahara 1968; Long 1970): (a) supercritical flow, (b) flow with both upstream and downstream propagating hydraulic jumps, (c) flow with an upstream propagating hydraulic jump and a downstream stationary hydraulic jump, (d) completely blocked flow, and (e) subcritical flow. As discussed above, in a supercritical flow with $F > 1$ everywhere, a fluid parcel has enough kinetic energy to ascend the obstacle, converting kinetic energy into potential energy. Consequently, the free surface rises over the symmetric obstacle (Regime A, Fig. 3.3a). The fluid reaches its minimum speed over the top of the obstacle. As either F decreases or M increases, the upstream flow is partially blocked, and the flow response shifts to the regime in which both an upstream *hydraulic jump* (*bore*) and a downstream hydraulic jump form and propagate away from the obstacle as time proceeds (Regime B, Fig. 3.3b). In this case, the upstream flow is partially blocked and a transition from subcritical to supercritical flow regimes occurs over the peak of the obstacle. Very high velocities are produced along the lee slope of the obstacle, which may help explain the occurrence of severe downslope winds for atmospheric flow over a mountain (Section 5.3), because the potential energy associated with the upstream flow is accumulated and converted into kinetic energy when the fluid passes over and descends the lee slope. Eventually, a steady state is established in the vicinity of the obstacle, and the shape of the free surface acquires a waterfall-like profile.

As *F* decreases further, the flow shifts to the regime in which an upstream hydraulic jump forms and propagates upstream, while the downstream hydraulic jump becomes stationary over the lee slope of the obstacle due to weaker advection associated with the incipient basic state flow (Regime C, Fig. 3.3c). Similar to Regime B, the upstream flow is also partially blocked. Both flow regimes B and C are characterized by high surface drag across the obstacle, as well as large flow velocities along the lee slope, and are referred to as the *transitional flow*. The *surface drag* is the pressure difference exerted by the shallow-water wave on the surface of the obstacle and is also called *mountain drag* or *form drag*.

For stationary mountain waves to exist in the atmosphere, the mountain must in some way be exerting a force against the flow so that the waves remain stationary. This force is called *mountain wave drag*, or *gravity wave drag*. Similar to Regime B, this transitional flow regime has also been used to help explain the formation of severe downslope windstorms. Figure 3.4a shows the severe downslope windstorm that

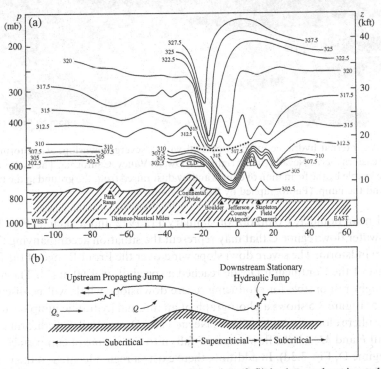

Fig. 3.4 (a) Analysis of potential temperature from aircraft flight data and rawinsondes for the January 11, 1972 Boulder windstorm. The bold dotted line separates data taken from the Queen Air aircraft (before 2200 UTC) and from the Saberliner aircraft (after 0000 UTC) (adapted after Klemp and Lilly 1975). The severe downslope wind reached a speed greater than 60 m s^{-1}. Data used to construct above and below the heavy dotted line are taken from different sources. (b) A sketch of flow Regime C of Fig. 3.3(c), which may be used to explain the phenomenon associated with (a). *Q* is the volume flux per unit width. (Adapted after Turner 1973.)

Fig. 3.5 An internal hydraulic jump associated with a severe downslope windstorm formed along the eastern Sierra Nevada (to the right) and Owens Valley, California. The hydraulic jump was made visible by the formation of clouds and by dust raised from the ground in the turbulent flow behind the jump. (Photographed by Robert Symons.)

occurred near Boulder, Colorado, on January 11, 1972 and a sketch (Fig. 3.4b) of shallow-water flow regime C that may represent the situation accompanying the 1972 Boulder windstorm. The severe downslope wind over the Front Range of the Rockies to the east of the Continental Divide reached a speed of over 60 m s^{-1}. The mechanisms thought responsible for producing severe downslope winds will be discussed in Chapter 5. Figure 3.5 shows a photograph of an internal hydraulic jump occurring in the atmosphere along the eastern Sierra Nevada and Owens Valley, California. With a very small F and $M > 1$, the flow response falls into the regime of completely blocked flow (Regime D, Fig. 3.3d). In addition, there exists a region located to the upper left corner of flow Regime B in Fig. 3.3b, in which the flow may be either supercritical or partially blocked (i.e. with hydraulic jumps present), depending upon the initial conditions (Long 1970). The flow characterized by $F < 1$ and $M < 1$ (Fig. 3.3e) is categorized as subcritical flow (Regime E), as discussed earlier.

If the nonlinear terms in the governing equations are included, then *wave steepening* may occur. The nonlinear effects on wave steepening may be illustrated

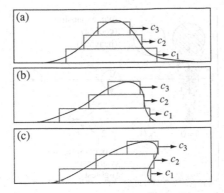

Fig. 3.6 The evolution of an initial symmetric wave, which is imagined to be composed of three rectangular blocks with shorter blocks on top of longer blocks. The wave speeds of these fluid blocks are approximately equal to $c_n = \sqrt{g(H + nh)}$, based on shallow-water theory, where $n = 1$, 2, and 3, H is the shallow-water layer depth, and h is the height of an individual fluid block. The wave steepening in (b) and wave overturning in (c) are interpreted by the different wave speeds of different fluid blocks because $c_3 > c_2 > c_1$.

by imagining an elevated wave that is composed of several rectangular blocks with shorter blocks of fluid on top of longer blocks (Fig. 3.6a). Since the shallow water wave speed is proportional to the mean layer depth H, the speed of fluid particles in the upper layer will be greater than that in the lower layer. Thus, the wave front will have a tendency to steepen (Fig. 3.6b) and possibly overturn (Fig. 3.6c). Once overturning occurs, the fluid becomes statically unstable and turbulence will be induced.

In a rotating shallow water system, the Coriolis force becomes more and more important when the Rossby number decreases. In this situation, the fluid may undergo geostrophic adjustment to an initial disturbance, as briefly discussed in Chapter 1, to a scale determined by the Rossby radius of deformation, $\lambda_R = c/f = \sqrt{gH}/f$.

3.5 Pure gravity waves

Consider small-amplitude (linear) perturbations in a two-dimensional ($\partial/\partial y = 0$), inviscid, nonrotating, adiabatic, Boussinesq, uniform basic state flow with uniform stratification, (2.2.14)–(2.2.18) reduce to

$$\frac{\partial u'}{\partial t} + U \frac{\partial u'}{\partial x} + \frac{1}{\rho_0} \frac{\partial p'}{\partial x} = 0, \tag{3.5.1}$$

$$\frac{\partial w'}{\partial t} + U \frac{\partial w'}{\partial x} - g \frac{\theta'}{\theta_0} + \frac{1}{\rho_0} \frac{\partial p'}{\partial z} = 0, \tag{3.5.2}$$

$$\frac{\partial u'}{\partial x} + \frac{\partial w'}{\partial z} = 0, \tag{3.5.3}$$

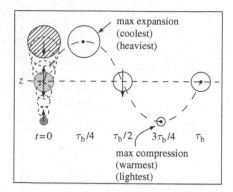

Fig. 3.7 Vertical oscillation of an air parcel in a stably stratified atmosphere when the Brunt–Vaisala frequency is N. The oscillation period of the air parcel is $\tau_b = 2\pi/N$ and the volume of the air parcel is proportional to the area of the circle. (Adapted after Hooke 1986.)

$$\frac{\partial \theta'}{\partial t} + U\frac{\partial \theta'}{\partial x} + \frac{N^2\theta_o}{g}w' = 0, \tag{3.5.4}$$

where θ_o is a constant reference potential temperature, and N^2 $(\equiv (g/\theta_o)\partial\bar{\theta}/\partial z)$ is the square of the Boussinesq Brunt–Vaisala (buoyancy) frequency. Figure 3.7 illustrates the vertical oscillation of an air parcel in a stratified atmosphere with a Brunt–Vaisala frequency N. The total oscillation period is $2\pi/N$ (τ_b in the figure). The air parcel expands and cools while it ascends, and reaches its maximum expansion and coolest state at $t = \tau_b/4$. At this level, the air parcel density perturbation is the largest. It then descends due to negative buoyancy and overshoots passing its original level at $t = \tau_b/2$. The air parcel compresses and warms adiabatically while it descends, and reaches its maximum compression and warmest state at $t = 3\tau_b/4$. At this level, the air parcel density perturbation is the lowest. It then ascends due to positive buoyancy, and returns to its original level at $t = \tau_b$. For a two-dimensional, nonrotating fluid flow, there is no need to retain the meridional (y-) momentum equation in our system of equations, because v' will keep its initial value for all time, as required by the reduced form of the y-momentum equation, namely, $\partial v'/\partial t + U\partial v'/\partial x = 0$. Note that the y-momentum equation needs to be kept if the fluid is two-dimensional and rotating since the initial v' will vary with time, although independent of y, due to the presence of the Coriolis force.

Equations (3.5.1)–(3.5.4) may be combined into a single equation for the vertical velocity w', which is a simplified form of the Taylor–Goldstein equation [(3.7.19)] in the absence of vertical wind shear,

$$\left(\frac{\partial}{\partial t} + U\frac{\partial}{\partial x}\right)^2 \left(\frac{\partial^2 w'}{\partial x^2} + \frac{\partial^2 w'}{\partial z^2}\right) + N^2\frac{\partial^2 w'}{\partial x^2} = 0. \tag{3.5.5}$$

Assuming a traveling sinusoidal plane wave solution of the form,

$$w' = \hat{w}(z)\, e^{i(kx-\omega t)}, \tag{3.5.6}$$

and substituting it into (3.5.5) yields the following linear partial differential equation with constant coefficients, which governs the vertical structure of w',

$$\frac{\partial^2 \hat{w}}{\partial z^2} + \left(\frac{N^2}{\Omega^2} - 1\right) k^2 \hat{w} = 0. \tag{3.5.7}$$

In the above equation, $\Omega \equiv \omega - kU$ is the *intrinsic (Doppler-shifted) frequency* of the wave relative to the uniform basic state flow. Equation (3.5.7) has the following two solutions:

$$\hat{w} = A\, e^{ik\sqrt{N^2/\Omega^2 - 1}\, z} + B\, e^{-ik\sqrt{N^2/\Omega^2 - 1}\, z}, \quad \text{for } N^2/\Omega^2 > 1, \tag{3.5.8}$$

and

$$\hat{w} = C\, e^{k\sqrt{1 - N^2/\Omega^2}\, z} + D\, e^{-k\sqrt{1 - N^2/\Omega^2}\, z}, \quad \text{for } N^2/\Omega^2 < 1. \tag{3.5.9}$$

Equation (3.5.8) represents a vertically propagating wave because it is sinusoidal with height. As will be discussed in Section 4.4, term A represents a wave with upward energy propagation, while term B represents a wave with downward energy propagation. Thus, for waves generated by orography, term B is unphysical and has to be removed because the wave energy source is located at the surface, as required by the *radiation boundary condition*. On the other hand, term C of (3.5.9) represents wave amplitude increasing exponentially with height, while term D represents a wave whose amplitude decreases exponentially from the level of wave generation. Thus, for waves or disturbances generated by orography, term C is unphysical. This is also called the *boundedness condition*. Under this situation, term D represents an evanescent wave or disturbance, whose wave amplitude decreases exponentially with height. In other words, there exist two distinct flow regimes for pure gravity waves (i.e. *vertically propagating waves* and *evanescent waves*) in the atmosphere, which are determined respectively by the following criteria:

$$N^2/\Omega^2 > 1 \quad \text{and} \quad N^2/\Omega^2 < 1. \tag{3.5.10}$$

The above two pure gravity wave flow regimes can be understood by considering steady-state responses of stably stratified airflow over a sinusoidal topography. In this particular case, $\Omega^2 = k^2 U^2$. When $N^2/\Omega^2 > 1$, we have $2\pi/N < L/U$, where $L = 2\pi/k$ is the dominant horizontal wavelength of the sinusoidal topography. Note that $2\pi/N$ is the buoyancy oscillation period and that L/U is the advection time an air parcel takes to cross over one wavelength of the mountain. Thus, fluid particles take less time to oscillate in the vertical, compared to the horizontal advection time required to pass over the mountain. This allows the wave energy to propagate vertically (Fig. 3.8a). On the other hand, when $N^2/\Omega^2 < 1$ or $2\pi/N > L/U$, fluid particles do not have enough time to

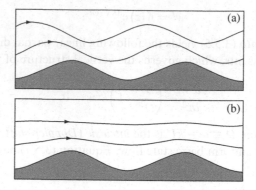

Fig. 3.8 (a) Vertically propagating waves and (b) evanescent waves for a linear, two-dimensional, inviscid flow over sinusoidal topography.

oscillate vertically because the time required for the particles to be advected over the mountain is shorter. Therefore, the wave energy cannot freely propagate vertically, and it is preferentially advected downstream, remaining near the Earth's surface (Fig. 3.8b). This type of wave or disturbance is also referred to as an evanescent wave or a *surface trapped wave.*

If the stratification of the fluid is uniform and the disturbance is sinusoidal in the vertical, then $\hat{w}$ may be written as $\hat{w} = w_o e^{imz}$, where w_o and m are the wave amplitude and vertical wave number, respectively. Substituting $\hat{w}$ into (3.5.7) yields the dispersion relation for pure gravity waves,

$$\Omega = \frac{\pm Nk}{\sqrt{k^2 + m^2}}. \tag{3.5.11}$$

For a quiescent fluid ($U = 0$), the above equation reduces to

$$\omega = \frac{\pm Nk}{\sqrt{k^2 + m^2}}, \tag{3.5.12}$$

or

$$\frac{\omega}{N} = \frac{\pm k}{\sqrt{k^2 + m^2}} = \pm \cos \alpha, \tag{3.5.13}$$

where α is the angle ($|\alpha| \le \pi/2$) between the wave number vector $\mathbf{k} = (k, m)$ and the x-axis. While the wave number vector is oriented in the same direction as the phase speed vector (c_p in Fig. 3.9), the wave front is oriented perpendicular. Fluid parcels oscillate in a direction perpendicular to the total wave number vector, as indicated by the incompressible continuity equation, $\mathbf{k} \cdot \mathbf{V}' = 0$. Therefore, the wave fronts or rays associated with particle oscillations tilt at an angle α with respect to the vertical. This characteristic behavior of nonrotating internal gravity waves has been verified in water tank experiments (Mowbray and Rarity 1967). For a given stratification,

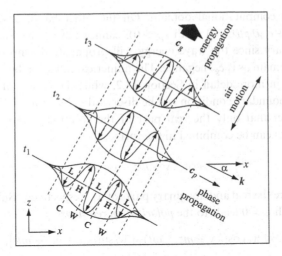

Fig. 3.9 Basic properties of a vertically propagating gravity wave with $k > 0$, $m < 0$, and $\omega > 0$. The energy of the wave group propagates with the group velocity (c_g; thick blunt arrow), while the phase of the wave propagates with the phase speed (c_p). Relations between w', u', p', and θ' as expressed by (3.5.16) and (3.5.17) are also sketched. Symbols H and L denote the perturbation high and low pressures, respectively, while W and C denote the warmest and coldest regions, respectively, for the wave at t_1. Symbol α defined in (3.5.13) represents the angle of the wave number vector k from the horizontal axis or the wave front (line of constant phase) from the vertical axis. (Adapted after Hooke 1986.)

waves with constant $\omega < N$ propagate at a fixed angle to the horizontal axis, which is independent of the wavelength.

From (3.5.12), we may obtain the horizontal and vertical phase velocities,

$$c_{px} = \frac{\omega}{k} = \frac{\pm N}{\sqrt{k^2 + m^2}}; \quad c_{pz} = \frac{\omega}{m} = \frac{\mp kN}{m\sqrt{k^2 + m^2}} \quad (3.5.14)$$

These expressions indicate that pure gravity waves are dispersive in both the x and z directions because both c_{px} and c_{pz} depend on wave number. The group velocities can be derived from (3.5.12),

$$c_{gx} = \frac{\partial \omega}{\partial k} = \frac{\pm m^2 N}{(k^2 + m^2)^{3/2}}; \quad c_{gz} = \frac{\partial \omega}{\partial m} = \frac{\mp kmN}{(k^2 + m^2)^{3/2}}. \quad (3.5.15)$$

Note that c_{px} and c_{gx} are directed in the same direction, while c_{pz} and c_{gz} are directed in opposite directions. This is also shown in Fig. 3.9. Due to these peculiar properties of internal gravity waves, the implementations for lateral and upper boundary conditions associated with mesoscale numerical models that resolve these waves must be carefully configured. Briefly speaking, a horizontal advection equation, $\partial \varphi / \partial t + c_{px} \partial \varphi / \partial x = 0$, where φ represents any prognostic dependent variable, can be applied at the lateral boundaries and can be implemented to help advect the wave energy out of the lateral boundary of the computational domain. On the other hand, a vertical advection equation, $\partial\varphi/\partial t + c_{pz}\partial\varphi/\partial z = 0$ (with $c_{pz} > 0$), cannot advect the wave energy out of the upper boundary since the wave energy will propagate downward back into the computational domain as is c_{gz} negative. The numerical radiation boundary conditions will be discussed in more detail in Section 13.2, while the details of the Sommerfeld (1949) radiation boundary condition will be discussed in Section 4.4.

Due to the fact that only the real part of the solution is physical, (3.5.6) and $\hat{w}(z) = w_o \exp(imz)$ can be combined in the form,

$$w' = \text{Re}\left(w_o e^{i(kx+mz-\omega t)}\right) = w_r \cos(kx + mz - \omega t) - w_i \sin(kx + mz - \omega t), \quad (3.5.16)$$

where w_r and w_i are the real and imaginary parts of w_o, respectively. Substituting w' into (3.5.1)–(3.5.4) with $U = 0$ leads to the *polarization relations*

$$u' = -(m/k)[w_r \cos(kx + mz - \omega t) - w_i \sin(kx + mz - \omega t)], \quad (3.5.17a)$$

$$p' = -(\rho_o \omega m/k^2)[w_r \cos(kx + mz - \omega t) - w_i \sin(kx + mz - \omega t)], \quad (3.5.17b)$$

$$\theta' = (\theta_o N^2/g\omega)[w_r \sin(kx + mz - \omega t) + w_i \cos(kx + mz - \omega t)]. \quad (3.5.17c)$$

The above relationships are also shown in Fig. 3.9 for the case where $k > 0$, $m < 0$, and $\omega > 0$. The wave frequency is assumed to be positive, in order to avoid redundant solutions. For $k > 0$, $m < 0$, and $\omega > 0$, (3.5.17a) indicates that u' is in phase with w', which is shown in Fig. 3.9 by fluid oscillating toward the right in regions of upward motion. Equation (3.5.17b) indicates that p' is also in phase with w'. Thus, high (low) pressure is produced in regions of upward (downward) motion. Equation (3.5.17c) indicates that θ' is out of phase with w' by $\pi/2$ (90°). Fluid particles lose (gain) buoyancy in regions of upward (downward) motion, according to (3.5.4) with $U = 0$. Therefore, the least buoyant (coldest) fluid parcels (denoted by C in t_1 of Fig. 3.9) will move toward regions of maximum upward motion. That is, internal gravity waves will move in the direction of phase propagation (toward the lower right corner of the figure), as denoted by c_p in the figure.

Returning to the vertical structure solutions, (3.5.8) and (3.5.9), there are two extreme cases that merit further discussion. When $N^2 \gg \Omega^2$, the buoyancy oscillation period $(2\pi/N)$ is much shorter than the oscillation period of the disturbance $(2\pi/\omega)$ or the advection time (L/U). Therefore, the wave energy will propagate purely in the vertical direction. In this situation, constant phase lines and group velocities are oriented vertically, while the total wave number vector is oriented horizontally. In this special flow regime, often referred to as the *hydrostatic gravity wave regime*, the vertical momentum equation (3.5.2) reduces to its hydrostatic form,

$$\frac{1}{\rho_o}\frac{\partial p'}{\partial z} = g\frac{\theta'}{\theta_o}. \quad (3.5.18)$$

This implies that the vertical pressure gradient force is in balance with the buoyancy force in the z direction. In other words, vertical acceleration Dw'/Dt plays an insignificant role in wave propagation. It can be shown from (3.5.8) that the waves repeat themselves in the vertical direction without losing their amplitude and have a wavelength of $2\pi\Omega/kN$ for a steady-state flow. For hydrostatic gravity waves, the wave equation (3.5.5) for the vertical velocity w' reduces to

$$\left(\frac{\partial}{\partial t} + U\frac{\partial}{\partial x}\right)^2 \frac{\partial^2 w'}{\partial z^2} + N^2 \frac{\partial^2 w'}{\partial x^2} = 0. \tag{3.5.19}$$

In the other limit, $N^2 \ll \Omega^2$, the buoyancy oscillation period is much greater than that of the disturbance $(2\pi/\omega)$ or advection time of the air parcel (L/U). Therefore, the buoyancy force plays an insignificant role in this flow regime. In this situation, the wave energy is not able to propagate vertically, and the wave disturbance will remain locally in the vicinity of the forcing. The vertical momentum equation, (3.5.2), reduces to

$$\frac{\partial w'}{\partial t} + U\frac{\partial w'}{\partial x} = -\frac{1}{\rho_0}\frac{\partial p'}{\partial z}. \tag{3.5.20}$$

Thus, only the vertical pressure gradient force contributes to the vertical acceleration. It can also be shown from (3.5.9) that the amplitude of the disturbance decreases exponentially with height. As discussed earlier, this special case is called the *evanescent flow regime*. The wave equation for w' reduces to

$$\left(\frac{\partial}{\partial t} + U\frac{\partial}{\partial x}\right)^2 \left(\frac{\partial^2 w'}{\partial x^2} + \frac{\partial^2 w'}{\partial z^2}\right) = 0. \tag{3.5.21}$$

If the flow starts with no relative vorticity in the y-direction (i.e. if $\partial u'/\partial z - \partial w'/\partial x = 0$ at $t = 0$), then the above equation reduces to a two-dimensional form of the Laplace's equation

$$\frac{\partial^2 w'}{\partial x^2} + \frac{\partial^2 w'}{\partial z^2} = 0. \tag{3.5.22}$$

Because this type of flow is everywhere vorticity-free, it is often referred to as *potential (irrotational) flow*.

3.6 Inertia-gravity waves

When the Rossby number $(R_o = U/fL)$ becomes smaller, rotational effects need to be considered. In this situation, buoyancy and Coriolis forces can act together as restoring forces and inertia-gravity waves can be generated. The governing equations are similar to (3.5.1)–(3.5.4), but with three-dimensional and rotational effects included,

$$\frac{\partial u'}{\partial t} + U\frac{\partial u'}{\partial x} - fv' + \frac{1}{\rho_0}\frac{\partial p'}{\partial x} = 0, \tag{3.6.1}$$

$$\frac{\partial v'}{\partial t} + U\frac{\partial v'}{\partial x} + fu' + \frac{1}{\rho_0}\frac{\partial p'}{\partial y} = 0, \tag{3.6.2}$$

$$\frac{\partial w'}{\partial t} + U\frac{\partial w'}{\partial x} - g\frac{\theta'}{\theta_0} + \frac{1}{\rho_0}\frac{\partial p'}{\partial z} = 0, \tag{3.6.3}$$

$$\frac{\partial u'}{\partial x} + \frac{\partial v'}{\partial y} + \frac{\partial w'}{\partial z} = 0, \tag{3.6.4}$$

$$\frac{\partial \theta'}{\partial t} + U\frac{\partial \theta'}{\partial x} + \frac{N^2\theta_0}{g}w' = 0. \tag{3.6.5}$$

The above set of equations can be combined into a single equation for w',

$$\left(\frac{\partial}{\partial t} + U\frac{\partial}{\partial x}\right)^2\left(\frac{\partial^2 w'}{\partial x^2} + \frac{\partial^2 w'}{\partial y^2} + \frac{\partial^2 w'}{\partial z^2}\right) + f^2\frac{\partial^2 w'}{\partial z^2} + N^2\left(\frac{\partial^2 w'}{\partial x^2} + \frac{\partial^2 w'}{\partial y^2}\right) = 0. \tag{3.6.6}$$

Again, applying the method of normal modes to w' in (x, y, t),

$$w' = \hat{w}(z)\exp[i(kx + ly - \omega t)], \tag{3.6.7}$$

and substituting it into (3.6.6) leads to

$$\frac{\partial^2 \hat{w}}{\partial z^2} + \frac{K^2(N^2 - \Omega^2)}{\Omega^2 - f^2}\hat{w} = 0, \tag{3.6.8}$$

where K is the horizontal wave number ($= \sqrt{k^2 + l^2}$). Similar to the pure gravity wave solutions, the above equation has solutions of the form,

$$\hat{w} = Ae^{imz} + Be^{-imz}, \tag{3.6.9}$$

where m, the vertical wave number, is defined as,

$$m^2 = \frac{K^2(N^2 - \Omega^2)}{\Omega^2 - f^2}. \tag{3.6.10}$$

It is clear from (3.6.9) that wave properties depend on the sign of m^2. Based on the signs of the numerator and denominator in (3.6.10) and on typical values of basic state flow parameters observed for waves in both the atmosphere and ocean (N is normally greater than f), three different flow regimes may be identified. The approximated governing equations and dispersion relations for the different flow regimes are summarized in Table 3.2. Their characteristics are described below.

The first flow regime occurs when $\Omega^2 > N^2 > f^2$. In this flow regime m is imaginary, so term A of (3.6.9) decays exponentially with height, while term B increases exponentially with height. The flow behavior is similar to the evanescent waves, as discussed in (3.6.10) for the case of $N^2/\Omega^2 < 1$, except that N^2 is required to be greater than f^2, and is referred to as the *high-frequency evanescent flow regime*.

When $\Omega^2 \gg N^2 > f^2$, (3.6.10) reduces to

$$m^2 \approx -K^2. \tag{3.6.11}$$

In this extreme case, both the buoyancy and Coriolis forces play insignificant roles in the process of wave generation and propagation. The governing equation for the vertical velocity w' reduces to what is essentially a three-dimensional version of (3.5.21),

$$\left(\frac{\partial}{\partial t} + U\frac{\partial}{\partial x}\right)^2 \left(\frac{\partial^2 w'}{\partial x^2} + \frac{\partial^2 w'}{\partial y^2} + \frac{\partial^2 w'}{\partial z^2}\right) = 0. \tag{3.6.12}$$

Thus, this extreme flow regime is characterized by *potential (irrotational) flow regime*, as discussed in the previous section. In this extreme case, for a steady state flow, the flow regime criterion becomes $L/U \ll 2\pi/N < 2\pi/f$. The second flow regime is when $N^2 > \Omega^2 > f^2$. In this flow regime, m is real and the waves are able to propagate freely in the vertical direction. Thus, the flow regime is referred to as the *vertically propagating inertia-gravity wave regime*. The two possible mathematical solutions of (3.6.9) represent either an upward or a downward propagation of energy. If the wave is generated by a low-level source such as stably stratified flow over a mountain, the radiation condition requires that the wave energy propagate away from the energy source, i.e., upward and away from the orographic forcing. This also applies to the boundary condition at $z = +\infty$ for elevated thermal forcing. However, both terms in (3.6.9) must be retained in the heating layer (forcing region) and in the layer between the heating base and the lower boundary (the Earth's surface). Above the forcing, either orographic forcing or elevated latent heating, the term Ae^{imz} should be retained to allow the energy to propagate upward, as required by the radiation boundary condition (Section 4.4).

Since the ratio N/f is typically large in both the atmosphere and the ocean, this flow regime is applicable to a wide range of intrinsic wave frequencies. When $N^2 > \Omega^2 \gg f^2$ and $O(N) = O(\Omega)$, (3.6.10) reduces to

$$m^2 \approx K^2 \left(\frac{N^2}{\Omega^2} - 1\right). \tag{3.6.13}$$

In this limit, rotational effects may be ignored, and the flow belongs to the *nonrotating* or *pure gravity wave regime*, as described in Section 3.5. Notice that for this extreme

Table 3.2 *Dispersion relations and approximated equations of w′ for mesoscale waves in different flow regimes.*

The governing equation for a linear, adiabatic, Boussinesq flow with a uniform basic state wind (*U*) and stratification (*N*) can be written:

$$\left(\frac{\partial}{\partial t} + U\frac{\partial}{\partial x}\right)^2 \left(\frac{\partial^2 w'}{\partial x^2} + \frac{\partial^2 w'}{\partial y^2} + \frac{\partial^2 w'}{\partial z^2}\right) + f^2\frac{\partial^2 w'}{\partial z^2} + N^2\left(\frac{\partial^2 w'}{\partial x^2} + \frac{\partial^2 w'}{\partial y^2}\right) = 0. \quad (3.6.6)$$

The dispersion relation is

$$m^2 = \frac{K^2(N^2 - \Omega^2)}{\Omega^2 - f^2}; \quad \Omega = \omega - kU, \quad (3.6.10)$$

and three major flow regimes are:

(I) High-frequency evanescent flow regime ($\Omega^2 > N^2 > f^2$; *m* imaginary)

   (i) Potential (irrotational) flow ($\Omega^2 \gg N^2 > f^2$)

$$m^2 \approx -K^2; \quad \left(\frac{\partial}{\partial t} + U\frac{\partial}{\partial x}\right)^2 \left(\frac{\partial^2 w'}{\partial x^2} + \frac{\partial^2 w'}{\partial y^2} + \frac{\partial^2 w'}{\partial z^2}\right) = 0, \quad (3.6.12)$$

   (ii) Nonrotating evanescent flow ($\Omega^2 > N^2 \gg f^2$)

$$m^2 \approx -K^2\left(1 - \frac{N^2}{\Omega^2}\right); \quad \left(\frac{\partial}{\partial t} + U\frac{\partial}{\partial x}\right)^2 \left(\frac{\partial^2 w'}{\partial x^2} + \frac{\partial^2 w'}{\partial y^2} + \frac{\partial^2 w'}{\partial z^2}\right) + N^2\left(\frac{\partial^2 w'}{\partial x^2} + \frac{\partial^2 w'}{\partial y^2}\right) = 0 \quad (3.6.14)$$

(II) Vertically propagating wave regime ($N^2 > \Omega^2 > f^2$; *m* real)

   (i) Pure gravity waves ($N^2 > \Omega^2 \gg f^2$ and $O(N) = O(\Omega)$)

$$m^2 \approx K^2\left(\frac{N^2}{\Omega^2} - 1\right); \quad \left(\frac{\partial}{\partial t} + U\frac{\partial}{\partial x}\right)^2 \left(\frac{\partial^2 w'}{\partial x^2} + \frac{\partial^2 w'}{\partial y^2} + \frac{\partial^2 w'}{\partial z^2}\right) + N^2\left(\frac{\partial^2 w'}{\partial x^2} + \frac{\partial^2 w'}{\partial y^2}\right) = 0 \quad (3.6.14)$$

   (ii) Hydrostatic gravity waves ($N^2 \gg \Omega^2 \gg f^2$)

$$m^2 = \left(\frac{KN}{\Omega}\right)^2; \quad \left(\frac{\partial}{\partial t} + U\frac{\partial}{\partial x}\right)^2\frac{\partial^2 w'}{\partial z^2} + N^2\left(\frac{\partial^2 w'}{\partial x^2} + \frac{\partial^2 w'}{\partial y^2}\right) = 0, \quad (3.6.16)$$

   (iii) Hydrostatic inertia-gravity waves ($N^2 \gg \Omega^2 > f^2$ and $O(\Omega) = O(f)$)

$$m^2 = \frac{K^2 N^2}{\Omega^2 - f^2}; \quad \left[\left(\frac{\partial}{\partial t} + U\frac{\partial}{\partial x}\right)^2 + f^2\right]\frac{\partial^2 w'}{\partial z^2} + N^2\left(\frac{\partial^2 w'}{\partial x^2} + \frac{\partial^2 w'}{\partial y^2}\right) = 0, \quad (3.6.18)$$

(III) Low-frequency evanescent flow regime ($N^2 > f^2 > \Omega^2$; *m* imaginary)

   (i) Quasi-geostrophic flow ($N^2 > f^2 \gg \Omega^2$)

$$m^2 \approx \frac{-K^2 N^2}{f^2}; \quad f^2\frac{\partial^2 w'}{\partial z^2} + N^2\left(\frac{\partial^2 w'}{\partial x^2} + \frac{\partial^2 w'}{\partial y^2}\right) = 0. \quad (3.6.20)$$

case, the flow regime criterion becomes $U/NL < 1/2\pi \ll R_o$ for a steady state flow. This implies that in order to generate pure gravity waves, the Rossby number of the basic state flow must be very large, normally much greater than 1. The governing equation for the vertical velocity w' in this extreme case becomes

$$\left(\frac{\partial}{\partial t} + U\frac{\partial}{\partial x}\right)^2 \left(\frac{\partial^2 w'}{\partial x^2} + \frac{\partial^2 w'}{\partial y^2} + \frac{\partial^2 w'}{\partial z^2}\right) + N^2\left(\frac{\partial^2 w'}{\partial x^2} + \frac{\partial^2 w'}{\partial y^2}\right) = 0. \tag{3.6.14}$$

When $N^2 \gg \Omega^2 \gg f^2$, (3.6.10) reduces to

$$m^2 \approx \left(\frac{KN}{\Omega}\right)^2. \tag{3.6.15}$$

This is identical to the *nonrotating hydrostatic gravity wave regime*, as discussed earlier. The governing equation for w' in this extreme case becomes

$$\left(\frac{\partial}{\partial t} + U\frac{\partial}{\partial x}\right)^2 \frac{\partial^2 w'}{\partial z^2} + N^2\left(\frac{\partial^2 w'}{\partial x^2} + \frac{\partial^2 w'}{\partial y^2}\right) = 0. \tag{3.6.16}$$

The flow regime criterion becomes $2\pi/N \ll L/U \ll 2\pi/f$ for a steady-state flow. This implies that the fluid parcel advection time is much longer than the period of buoyancy oscillation allowing the disturbance to propagate vertically, but much shorter than the inertial oscillation period. In this nonrotating hydrostatic wave regime, only upward propagating waves are allowed.

When $N^2 \gg \Omega^2 > f^2$ and $O(\Omega) = O(f)$, (3.6.10) reduces to

$$m^2 \approx \frac{K^2 N^2}{\Omega^2 - f^2}. \tag{3.6.17}$$

The flow response belongs to the *hydrostatic inertia-gravity wave regime*. Thus, the governing equation for w' becomes

$$\left[\left(\frac{\partial}{\partial t} + U\frac{\partial}{\partial x}\right)^2 + f^2\right]\frac{\partial^2 w'}{\partial z^2} + N^2\left(\frac{\partial^2 w'}{\partial x^2} + \frac{\partial^2 w'}{\partial y^2}\right) = 0. \tag{3.6.18}$$

For a basic flow with $N = 0.01$ s^{-1} and $f = 10^{-4}$ s^{-1}, the horizontal scale of typical hydrostatic inertia-gravity waves is on the order of 100 km.

The third flow regime is when $N^2 > f^2 > \Omega^2$. In this flow regime, m is imaginary. Similar to the first flow regime ($\Omega^2 > N^2 > f^2$), disturbances decay exponentially in the vertical away from the wave energy source. However, the wave frequency is low, thus the flow response is referred to as the *low-frequency evanescent flow regime*. When $N^2 > f^2 \gg \Omega^2$, inertial accelerations play an insignificant role in wave generation and propagation. The flow response is similar to a *quasi-geostrophic flow regime*. In this limiting case, (3.6.10) reduces to

$$m^2 \approx \frac{-K^2 N^2}{f^2}. \tag{3.6.19}$$

In this case, the fluid motion is quasi-horizontal and the governing equation for the vertical velocity w' becomes

$$f^2 \frac{\partial^2 w'}{\partial z^2} + N^2 \left(\frac{\partial^2 w'}{\partial x^2} + \frac{\partial^2 w'}{\partial y^2} \right) = 0. \tag{3.6.20}$$

The horizontal scale for this type of quasi-geostrophic flow is on the order of 1000 km for typical values of $N = 0.01$ s^{-1} and $f = 10^{-4}$ s^{-1}. The Rossby number of the basic state flow in this case is much smaller than 1.

In order to better understand the basic wave dynamics, we consider the case of hydrostatic inertia-gravity waves in a quiescent fluid $(U = 0)$. Substituting $\varphi' = \tilde{\varphi} \exp(kx + ly + mz - \omega t)$, where $\varphi = u$, v, w, p, or θ, into (3.6.1)–(3.6.5) and using the hydrostatic form of (3.6.3) lead to the following polarization relationships in wave number space,

$$\tilde{u} = \frac{1}{\rho_0} \left(\frac{\omega k + i l f}{\omega^2 - f^2} \right) \tilde{p}; \quad \tilde{v} = \frac{1}{\rho_0} \left(\frac{\omega l - i k f}{\omega^2 - f^2} \right) \tilde{p};$$

$$\tilde{w} = \frac{-\omega}{\rho_0 m} \left(\frac{k^2 + l^2}{\omega^2 - f^2} \right) \tilde{p}; \quad \tilde{\theta} = \frac{i N^2 \theta_0}{\rho_0 g m} \left(\frac{k^2 + l^2}{\omega^2 - f^2} \right) \tilde{p}. \tag{3.6.21}$$

The above relationships are depicted in Fig. 3.10. In the special case of two-dimensional flow $(\partial/\partial y = 0$ or $l = 0)$, it can be shown that the following solutions satisfy the two-dimensional form of (3.6.6),

$$u' = \tilde{u} \cos(kx + mz - \omega t); \quad v' = \left(\frac{f}{\omega} \right) \tilde{u} \sin(kx + mz - \omega t). \tag{3.6.22}$$

It can be easily shown from the above equations that the velocity vector associated with a plane inertia-gravity wave rotates anticyclonically with time in the Northern Hemisphere. The projection of the motion on the horizontal plane is an ellipse where ω/f is the ratio between the major and minor axes, as depicted in Fig. 3.10b. The velocity vector associated with an inertia-gravity wave rotates anticylonically with height for upward energy propagation. The particle motion and phase relationship for a Poincaré wave propagating on the ocean surface is similar to that of a hydrostatic inertia-gravity wave. It can also be shown that the ratio of the vertical to horizontal components of the group velocity vector for a two-dimensional, hydrostatic inertia-gravity wave is given by

$$|c_{gz}/c_{gx}| = |k/m| = \sqrt{\omega^2 - f^2}/N. \tag{3.6.23}$$

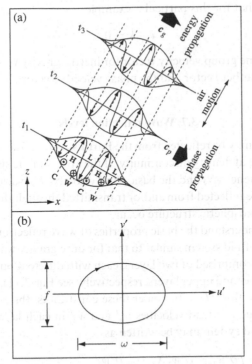

Fig. 3.10 (a) Similar to Fig. 3.9, except for a hydrostatic inertia-gravity wave with $m < 0$, $k > 0$, $l = 0$, $\omega > 0$, and $f > 0$. Meridional (i.e. north–south) perturbation wind velocities (v') are shown by arrows pointed into and out of the page. (b) The projection of fluid particle motion associated with a hydrostatic inertia-gravity wave onto the horizontal plane is an ellipse with ω/f as the ratio of major and minor axes. The velocity vector associated with a plane inertia-gravity wave rotates anticyclonically with height for upward energy propagation. (Adapted after Hooke 1986.)

Therefore, the wave energy of a hydrostatic inertia-gravity wave propagates more horizontally than that of a pure gravity wave with the same wave frequency.

Occasionally, it is observed that large-scale pressure gradients over the ocean are considerably smaller than those over the continents, which leads to a balance between Coriolis and centrifugal forces. Under this situation, fluid parcels follow circular paths, rotating in an anticyclonic sense in the horizontal plane, and that have an oscillation period of one pendulum day ($2\pi/f$). This type of flow is called an *inertial flow* or *inertial oscillation*. The radius (R) of curvature of the oscillation can be shown to be $R = -V/f$, where V is the non-negative horizontal wind speed along the direction tangential to the local velocity in the natural coordinates. The negative sign of R indicates the oscillation is anticyclonic (clockwise). The inertial oscillation has been used to explain the formation of low-level jets over the Great Plains to the east of the US Rocky Mountains (Section 10.6).

It can be shown that for this particular example

$$c_g \cdot k = 0. \tag{3.6.24}$$

This indicates that the group velocity vector for inertia-gravity waves is perpendicular to both the wave number vector and the phase velocity vector.

3.7 Wave reflection levels

Atmospheric waves may be reflected from the Earth's surface, the surface of a fluid, or the internal interface at density discontinuity. If the atmospheric structure, such as the Brunt–Vaisala frequency (N) and the basic wind velocity (U), varies with height, then gravity waves may be reflected from and/or transmitted through the interface at which rapid changes in atmospheric structure occur.

In order to help understand the basic properties of wave reflection and transmission, we consider a simple fluid system similar to that for pure gravity waves ((3.5.5)), except that the fluid is now comprised of two layers, each with different buoyancy frequencies (N_1 and N_2 in the lower and upper layers, respectively; see Fig. 3.11). We further assume that there is no basic flow ($U = 0$). Under these constraints, the governing equations for the small-amplitude vertical velocities w_1' and w_2' in each layer of this particular two-dimensional fluid system may be written as

$$\frac{\partial^2}{\partial t^2} \left(\frac{\partial^2 w_1'}{\partial x^2} + \frac{\partial^2 w_1'}{\partial z^2} \right) + N_1^2 \frac{\partial^2 w_1'}{\partial x^2} = 0, \quad 0 \leq z < H, \tag{3.7.1}$$

$$\frac{\partial^2}{\partial t^2} \left(\frac{\partial^2 w_2'}{\partial x^2} + \frac{\partial^2 w_2'}{\partial z^2} \right) + N_2^2 \frac{\partial^2 w_2'}{\partial x^2} = 0, \quad H \leq z. \tag{3.7.2}$$

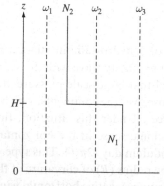

Fig. 3.11 Examples of wave reflection in a stratified flow with a piecewise constant profile in Brunt–Vaisala frequency (N). Case ω_1 has wave solutions in both layers; Case ω_2 (wave reflection case) has wave solutions in the lower layer, evanescent solutions in upper layer; and Case ω_3 has evanescent solutions in both layers, assuming the wave energy source is located at the surface ($z = 0$).

Applying the method of normal modes, $w' = \hat{w}(z) \exp[i(kx - \omega t)]$, to the above equations leads to the following equations for the vertical structure in each layer:

$$\frac{\partial^2 \hat{w}_1}{\partial z^2} + m_1^2 \hat{w}_1 = 0, \quad 0 \le z < H, \qquad (3.7.3)$$

$$\frac{\partial^2 \hat{w}_2}{\partial z^2} + m_2^2 \hat{w}_2 = 0, \quad H \le z, \qquad (3.7.4)$$

where

$$m_i^2 = k^2 \left(N_i^2 / \omega^2 - 1 \right), \quad i = 1, 2. \qquad (3.7.5)$$

In fact, (3.7.5) represents the dispersion relations in each layer, which are identical to (3.5.12) if m is replaced by m_i. Note that in the above equation (3.7.5), we have assumed that the wave frequencies and wavelength are identical in both layers. This simple type of piecewise layered model is able to provide the fundamental wave dynamics. It can be shown that the following solutions satisfy (3.7.3) and (3.7.4),

$$\hat{w}_1 = A\, e^{im_1(z-H)} + B\, e^{-im_1(z-H)}, \quad 0 \le z < H \qquad (3.7.6)$$

$$\hat{w}_2 = C\, e^{im_2(z-H)} + D\, e^{-im_2(z-H)}, \quad H \le z. \qquad (3.7.7)$$

Similar to the one-layer theory discussed in the previous section, the flow response is quite different depending on whether m_i (where $i = 1$ or 2) is real or imaginary (i.e. whether $N > \omega$ or $N < \omega$). If we assume a less stable layer sitting on top of a more stable layer ($N_2 < N_1$), then this leads to three possible cases: (a) $N_i > \omega$ for $i = 1, 2$; (b) $N_i < \omega$ for $i = 1, 2$; or (c) $N_2 < \omega < N_1$ (Fig. 3.11). This type of situation ($N_2 < N_1$) may occur in the vicinity of a thunderstorm in which the lower layer is more stable due to the evaporative cooling associated with rainfall. The opposite situation, ($N_2 > N_1$), often occurs when a mixed boundary layer is produced by surface sensible heating during the day, or when the stratosphere is considered in the problem, when N_2 is normally two to three times larger than N_1.

In the following, we will restrict our attention to the case with $N_2 < \omega < N_1$ (i.e. Case ω_2 of Fig. 3.11). Solutions for other cases may be obtained in a similar manner. The lower boundary condition for flow over a flat surface requires that the normal velocity component vanish, i.e. $w' = 0$ at $z = 0$. Applying this lower boundary condition to (3.7.6) yields

$$\hat{w}_1 = B\left[e^{-im_1(z-H)} - e^{2im_1 H} e^{im_1(z-H)} \right], \quad 0 \le z < H. \qquad (3.7.8)$$

In the above equation, the first term inside the square bracket represents waves with upward energy propagation, while the second term represents waves with downward energy propagation, since $z - H$ is negative. If the wave energy source is located at the surface, then the absolute value of $\exp(2im_1H)$ represents the reflection coefficient. In the upper layer, since m_2 is imaginary, (3.7.7) becomes

$$\hat{w}_2 = C\,e^{-n_2(z-H)} + D\,e^{n_2(z-H)}, \quad H \leq z, \tag{3.7.9}$$

where $n_2 = im_2 = k\sqrt{1 - N_2^2/\omega^2}$ is a real number. For a wave energy source located in the lower layer, the upper boundary condition requires a bounded solution, which, in turn, requires that $D = 0$. Thus, (3.7.7) reduces to

$$\hat{w}_2 = C\,e^{-n_2(z-H)}, \quad H \leq z. \tag{3.7.10}$$

The above equation represents evanescent waves, as described in Section 3.5. However, for cases with real values of m_2, the upper boundary condition is governed by the radiation boundary condition, which will be discussed in Section 4.4. Therefore, the radiation condition requires that $D = 0$ if the wave energy source is located in the lower layer. Coefficients B and C are determined by boundary conditions at the interface, i.e. by imposing the appropriate kinematic and dynamic boundary conditions at $z = H$. The kinematic boundary condition requires that

$$\boldsymbol{V}_1 \cdot \boldsymbol{n}_1 = \boldsymbol{V}_2 \cdot \boldsymbol{n}_2, \tag{3.7.11}$$

where the subscripts indicate the upper or lower layer; $\boldsymbol{V}_i$ is the total velocity in layer i; and $\boldsymbol{n}_i$ is the unit vector normal to the boundary in each layer. Note that $\boldsymbol{n}_i$ in (3.7.11) is a vector, which should not be confused with the n_i used in (3.7.9) and (3.7.10). For a small-amplitude (linear) perturbation, the unit normal vector is almost vertical since the wave amplitude is much smaller than the wavelength. This then implies that

$$w_1' = w_2'. \tag{3.7.12}$$

Another interface boundary condition is the dynamic boundary condition specifying continuity of pressure at $z = H$, i.e. $p_1' = p_2'$, which leads to

$$\frac{\partial w_1'}{\partial z} = \frac{\partial w_2'}{\partial z}. \tag{3.7.13}$$

Applying (3.7.12) to (3.7.8) and (3.7.10) at $z = H$ gives

$$\hat{w}_1 = \frac{C}{1 - e^{-2im_1H}}\left[e^{im_1(z-H)} - e^{-2im_1H}e^{-im_1(z-H)}\right], \quad 0 \leq z < H. \tag{3.7.14}$$

Applying the dynamic boundary condition, (3.7.13), to (3.7.14) and (3.7.10) yields the dispersion relation for this two-layer fluid system,

$$e^{-2im_1H} = \frac{n_2 + im_1}{n_2 - im_1}. \tag{3.7.15}$$

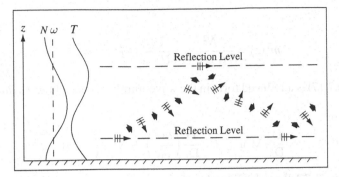

Fig. 3.12 Wave reflection in a continuously stratified fluid. N and T are the Brunt–Vaisala frequency and temperature of the sounding, respectively, and ω is the wave frequency. Ray paths are reflected at the reflection level at which $\omega = N$. A wave packet is also depicted in the figure. The short blunt arrows and long thin arrows denote the group and phase velocities, respectively, of the wave packet. Particle motions are parallel to the constant phase lines or wave fronts, which become vertically oriented at the reflection level since α, defined in (3.5.13) and also illustrated in Fig. 3.9, approaches 0. (Adapted after Hooke 1986.)

This expression may also be used to find the eigenvalues, ω and k, which are needed for substitution into (3.7.15) in order to obtain the eigenfunctions that describe the vertical structure of the wave,

$$\hat{w}_1 = C\left[\left(\frac{m_1 + in_2}{2m_1}\right)e^{im_1(z-H)} + \left(\frac{m_1 - in_2}{2m_1}\right)e^{-im_1(z-H)}\right]. \tag{3.7.16}$$

The reflection coefficient for a forcing located at the surface ($z = 0$) and the wave being reflected at $z = H$ may be obtained by taking the absolute value of the ratio of the first term (term A) to the second term (term B) inside the square bracket of the above equation, which gives the reflection coefficient $r = |(m_1 + in_2)/(m_1 - in_2)|$. Note that $z - H$ is negative in the lower layer.

In addition to interfacial discontinuities in stratification, wave reflection may also occur when the vertical profiles of the basic state wind and the stratification (Brunt–Vaisala frequency) vary continuously throughout the fluid (Fig. 3.12). To elucidate this further, we consider a two-dimensional, linear, nonrotating, inviscid, Boussinesq fluid system governed by (3.5.1)–(3.5.4), but whose basic state is generalized, to allow for both the background wind and Brunt–Vaisala frequency to vary with height. The equation governing this special type of fluid system may also be derived directly from the generalized linear equation set, (2.2.14)–(2.2.18) and is given by

$$\frac{\partial^2 \hat{w}}{\partial z^2} + m^2(z)\hat{w} = 0, \tag{3.7.17}$$

where

$$m^2(z) = \frac{N^2}{(U-c)^2} - \frac{U_{zz}}{U-c} - k^2, \quad c = \frac{\omega}{k}. \tag{3.7.18}$$

Equation (3.7.17) is a reduced form in the wave-number space of the *Taylor–Goldstein equation*,

$$\frac{D^2}{Dt^2} \nabla^2 w' - U_{zz} \frac{D}{Dt}\left(\frac{\partial w'}{\partial x}\right) + N^2 \frac{\partial^2 w'}{\partial x^2} = 0. \tag{3.7.19}$$

Note that m in (3.7.18) can be viewed as the vertical wave number, and the solutions of (3.7.17) can be written in the form of $\exp(\pm imz)$ when U and N are constant. If m^2 changes sign from positive to negative at a certain level, then m will change from a real to an imaginary number. A transition from the vertically propagating wave regime to the evanescent flow regime occurs based on an argument similar to that given for the solutions of (3.5.8) and (3.5.9). If a vertically propagating wave-like disturbance exists below that particular level, such as $z = H$ for ω_2 of Fig. 3.11, it will exponentially decay above that level. In this situation, the wave energy is not able to freely propagate vertically above $z = H$ and is therefore forced to reflect back. This level is called the *wave reflection level*. If such a reflection level exists above the lower, rigid, flat surface, this atmospheric layer then acts as a wave guide in trapping the wave energy between the reflection level and the surface and allows for the wave energy to effectively propagate far downstream horizontally. One well-known example of gravity wave reflection in the atmosphere is the lee waves (Section 5.2) generated by stratified airflow over a two-dimensional mountain ridge when the stratification (wind speed) is much stronger (smaller) in the lower layer compared to the upper layer. The idea of wave reflection in a slow-varying density gradient or stratification may also be traced by the paths of rays, whose directions (α) are defined locally by (3.5.13), as shown by the long thin arrows in Fig. 3.12. *Ray paths* are defined as paths where the tangent at any one point is in the direction of the group velocity (relative to the ground) of the waves. When a group of waves approaches the reflection level where $\omega = N$, α approaches 0, and the wave fronts are turned towards the vertical, reflecting the wave energy.

3.8 Critical levels

Another important phenomenon associated with gravity waves is the change in wave properties across a critical level. A *critical level* (z_c) is defined as the level at which the vertically sheared basic flow $U(z)$ is equal to the horizontal phase speed (c) of the wave or disturbance, i.e. $U(z_c) = c$. Note that the equation governing the vertical structure (3.7.17) has a singularity at the critical level. From an observational analysis of the Oklahoma squall line as depicted in Fig. 6.12 (Chapter 6), there exists a critical level near 6 km. Climatological studies indicate that most midlatitude squall lines exhibit a critical level in the mid-troposphere (Bluestein and Jain 1985; Wyss and Emanuel 1988).

Nonlinear numerical simulations indicate that a high-drag (severe-wind state) may be established after the upward propagating wave breaks above the mountain. The wave-breaking region is characterized by strong turbulent mixing with a local wind reversal on top of it. Note that the critical level coincides with the wind reversal level for a stationary mountain wave because the phase speed there is zero. The wave breaking region aloft might act as an internal boundary that reflects the upward propagating waves back to the ground and produces a high-drag state through partial resonance with the upward propagating mountain waves, as will be discussed in Chapter 5. One example of the wave-induced critical level simulated by a numerical model is given in Figs. 5.9 and 5.10 (Chapter 5).

Using an asymptotic method, such as the WKBJ method (e.g., Olver 1997), it is found that an upward propagating internal gravity wave packet in a nonrotating stably stratified fluid would approach the critical level for the dominant frequency and wave number of the packet, but it would not reach the critical level in any finite period of time since along a ray path, $dz/dt \propto (z - z_c)^2$ as $z \to z_c$, which gives $(t - t_o) \propto 1/(z - z_c)$ as $z \to z_c$ (Bretherton 1966). This means that it will take an infinite amount of time for a gravity wave packet to reach the critical level. Thus, the internal wave is physically absorbed (*wave absorption*) at the critical level, instead of being either transmitted or reflected, as discussed in the previous section. In the real atmosphere, a gravity wave is composed by a number of different wave modes which propagate at different phase speeds. This will form a layer of critical levels, i.e. a *critical layer*.

This particular type of internal wave behavior in a stratified fluid is illustrated in Fig. 3.13. When the wave energy associated with a wave packet propagates toward the critical level, the group velocity becomes more horizontal and eventually is oriented completely horizontally in the vicinity of the critical level. Near the critical level, the phase velocity is oriented downward since it is itself perpendicular to the group

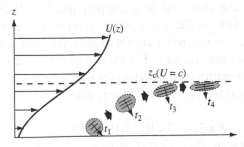

Fig. 3.13 The propagation of a wave packet upward toward a critical level located at $z = z_c$. The particle motions are parallel to the wave crests, which are denoted by straight lines. Note that the vertical wavelength decreases as the wave packet approaches the critical level. The phase lines are horizontally oriented at the critical level in this case. (Adapted after Bretherton 1966.)

velocity, and fluid parcel motions become horizontal. The vertical wavelength also decreases as the wave packet approaches the critical level. In the following discussion for obtaining the solution for (3.7.17), we will assume that the Richardson number, $R_i (= N^2/U_z^2)$ is always greater than $1/4$. R_i is also known as the *gradient Richardson number*.

Near $z = z_c$, $U(z)$ and $N(z)$ may be expanded in power series

$$U(z) = c + U_c'(z - z_c) + \cdots$$
$$N(z) = N_c + N_c'(z - z_c) + \cdots \tag{3.8.1}$$

where the prime denotes differentiation with respect to z and the subscript "c" denotes the function value at the critical level. We assume that z_c is a *regular singularity*, which requires $U'(z_c) \neq 0$. Hence, we seek a solution for (3.7.17) of the form,

$$\hat{w}(z) = \sum_{n=0}^{\infty} a_n (z - z_c)^{n+\alpha}, \quad a_0 \neq 0, \ n \text{ is an integer.} \tag{3.8.2}$$

Substituting the above expression into the Taylor–Goldstein equation, (3.7.17), leads to the indicial equation,

$$\alpha^2 - \alpha + R_{ic} = 0 \tag{3.8.3}$$

where $R_{ic} = \left(N_c/U_c'\right)^2$. The above indicial equation has the following solutions,

$$\alpha = 1/2 \pm i\mu; \quad \mu = \sqrt{R_{ic} - 1/4} \tag{3.8.4}$$

Thus, near $z = z_c$, a series solution may be found,

$$\hat{w}(z) \approx A(z - z_c)^{1/2+i\mu} + B(z - z_c)^{1/2-i\mu}$$
$$= A \exp[(1/2 + i\mu)(\ln|z - z_c| + i\arg(z - z_c))] \tag{3.8.5}$$
$$+ B \exp[(1/2 - i\mu)(\ln|z - z_c| + i\arg(z - z_c))]$$

where "arg" denotes the argument of a complex number. Both terms $\hat{w}_A$ and $\hat{w}_B$ (i.e., terms in (3.8.5) with coefficients A and B, respectively) have a branch point at $z = z_c$. In other words, both $\hat{w}_A$ and $\hat{w}_B$ are not single-valued functions when one winds a circle counterclockwise once around z_c. For the sake of definiteness, we may choose that branch of the natural logarithm (ln) function for which $\arg(z - z_c) = 0$ when $z > z_c$ and introduce the branch cut from $z = z_c$ along the negative x-axis. Therefore, we obtain

$$\hat{w}_A^+(z) = A\sqrt{|z - z_c|} \exp[i\mu \ln(z - z_c)], \quad \text{and}$$
$$\hat{w}_B^+(z) = B\sqrt{|z - z_c|} \exp[-i\mu \ln(z - z_c)], \quad \text{for } z > z_c. \tag{3.8.6}$$

In order to determine the appropriate argument, a small Rayleigh friction term can be added to the x-momentum equation (3.5.1) and a small Newtonian cooling term can be added to the thermodynamic energy equation (3.5.4). This gives when $\arg(z - z_c) = -\pi \operatorname{sgn}(U'_c)$ when $z < z_c$. Substituting the above expression into (3.8.5) yields

$$\hat{w}_A^- = A\sqrt{|z - z_c|} \exp\left[i\mu \ln|z - z_c| - (1/2)\pi i \operatorname{sgn} U'_c + \mu\pi \operatorname{sgn} U'_c\right],$$

$$\hat{w}_B^- = B\sqrt{|z - z_c|} \exp\left[-i\mu \ln|z - z_c| - (1/2)\pi i \operatorname{sgn} U'_c - \mu\pi \operatorname{sgn} U'_c\right], \text{ for } z < z_c. \tag{3.8.7}$$

Both solutions $\hat{w}_A$ and $\hat{w}_B$ in (3.8.6) and (3.8.7) satisfy (3.7.17) mathematically; however, they have different physical meanings and need to be properly determined. From (3.8.6) and (3.8.7), we have

$$\left|\frac{\hat{w}_A^+}{\hat{w}_A^-}\right| = \exp(-\mu\pi \operatorname{sgn} U'_c); \qquad \left|\frac{\hat{w}_B^+}{\hat{w}_B^-}\right| = \exp(\mu\pi \operatorname{sgn} U'_c). \tag{3.8.8}$$

For $U'_c > 0$ and low-level forcing, the amplitude of the disturbance generated in the lower layer should decrease as it passes across the critical level into the upper layer. Thus, we must choose $\hat{w}_A$. The proper solution can be found for other situations as well. Note that the above equation also indicates that the wave energy is exponentially attenuated through the critical level (Booker and Bretherton 1967). As mentioned earlier, the vertical wave number increases, and the perturbation velocity becomes increasingly horizontal as one approaches the critical level, because

$$m^2(z) \approx \frac{N^2}{(U - c)^2}. \tag{3.8.9}$$

This implies that $m \to \infty$ as $z \to z_c$. Thus, the vertical wavelength approaches zero near the critical level. This property is also depicted by Fig. 3.13. In the real atmosphere, a localized disturbance is often composed of many Fourier wave components, each with a different wavelength. Since each Fourier wave component has its own critical level, where $U = c = \omega/k$, an atmospheric layer of finite thickness composed of these critical levels, referred to as the critical layer, is often associated with a localized disturbance.

The flow near a critical level is highly nonlinear, since a small perturbation in the horizontal velocity field will necessarily exceed the basic horizontal flow velocity in the vicinity of the critical level, in a reference frame moving with the phase speed of the wave. Thus, linear theories are not accurate in the vicinity of a critical level. Based on numerical simulations of varying the Richardson number, three flow regimes have been found (Breeding 1971). For $R_i > 2.0$, the interaction between the incident wave and the mean flow is largely similar to that predicted by the linear theory as outlined above. That is, very little of the incident wave energy penetrates through the critical level. For $0.25 < R_i < 2.0$, a significant amount of the wave energy is reflected, part of which can be predicted by linear theory. This condition represents a balance between the outward diffusion of the added momentum from the incident wave and the rate at which it is absorbed at the critical level. When R_i falls into this range, some wave energy is also transmitted through the critical level. For $R_i < 0.25$, wave overreflection is predicted. Note that these regime boundaries could be more accurately defined with more recent and sophisticated numerical models. When wave overreflection occurs, more energy than that associated with the incident wave is reflected back from the critical level, because the flow possesses shear instability (Lindzen and Rosenthal 1983). In this situation, the wave is able to extract energy and momentum from the basic flow during the reflection process. If the overreflected waves are in phase with the incident waves, waves may grow exponentially with time by resonance, i.e. the normal mode instability exists. If the overreflected waves are partially in phase with the incident waves, waves may grow algebraically with time by partial resonance, i.e., the algebraic mode instability exists. Under certain conditions, such as an unstable layer containing a critical level and capping a stable layer, a wave duct can exist for mesoscale gravity waves, and wave absorption, transmission, and overreflection may occur at different ranges of R_i (see Fig. 4.13 and relevant discussions).

Figure 3.14 shows the linear, steady state responses to a prescribed heating in the layer below the critical level ($z_c = 0$) of a stably stratified flow with $R_i = 10$ and $R_i = 1$. Again, for steady state flow, the critical level coincides with the wind reversal level ($U = 0$). Upward motion is induced in the vicinity of the heat source, while two regions of very weak compensated downdrafts are produced upstream and downstream from the heat source. For the case with $R_i = 10$ (Fig. 3.14a), the flow above the critical level is almost undisturbed due to the exponential attenuation associated with critical level absorption. With $R_i = 1$ (Fig. 3.14b), upstream of the heat source is occupied by subsidence, while downstream of the heat source is occupied by ascending motion. This subsidence occurs due to strong advection (Section 6.2). The disturbance above the critical level is more pronounced because more energy is transmitted through the critical level into the upper layer.

The critical level dynamics can be extended to a rotating fluid flow. In a rotating fluid system, the governing equation for the small-amplitude vertical velocity $w'(t, x, z)$ for a two-dimensional, inviscid, Boussinesq flow on an f plane, can be derived (e.g., Smith 1986) and is given by

$$\frac{D}{Dt}\left(\frac{D^2}{Dt^2}\nabla^2 w' + f^2\frac{\partial^2 w'}{\partial z^2} - U_{zz}\frac{D}{Dt}\frac{\partial w'}{\partial x} + N^2\frac{\partial^2 w'}{\partial x^2}\right) - 2f^2 U_z\frac{\partial^2 w'}{\partial x \partial z} = 0. \qquad (3.8.10)$$

Taking a normal mode approach, substitution of $w' = \hat{w}(z)\exp[ik(x - ct)]$, leads to the following vertical structure equation:

$$\frac{\partial^2 \hat{w}}{\partial z^2} - \frac{2f^2 U_z}{(U - c)[k^2(U - c)^2 - f^2]}\frac{\partial \hat{w}}{\partial z} + \frac{k^2[N^2 - k^2(U - c)^2]}{k^2(U - c)^2 - f^2}\hat{w} = 0. \qquad (3.8.11)$$

The above equation indicates that in addition to the singularity at $U = c$, there are two additional singularities, $U = c \pm f/k$. These additional levels are called *inertial critical*

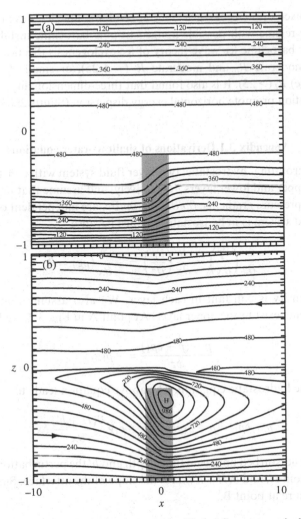

Fig. 3.14 (a) Streamlines for a linear, steady-state stratified airflow over an isolated heat source. The concentrated heating region is shaded. The basic flow has a linear shear ($U_z = $ constant) and its Richardson number (R_i) is 10. (b) Same as (a) except for $R_i = 1$. All contour values are nondimensionalized. The streamfunction (ψ) used for constructing the streamlines is defined as $u = \partial\psi/\partial z$ and $w = -\partial\psi/\partial x$. (After Lin 1987.)

levels (Jones 1967). For incident monochromatic (single wavelength) waves, the inertial critical levels can absorb wave energy in linear flow but tend to reflect wave energy in nonlinear flow (Wurtele *et al.* 1996). Based on the Richardson number ($R_i = N^2/U_z^2$) and the Rossby number ($R_o = U/fa$, where a is the forcing horizontal scale, such as the half-width of a bell-shaped mountain), four flow regimes for two-dimensional back-sheared flow over an isolated mountain ridge on an *f*-plane can be identified: (I) inertia-gravity wave regime, (II) mixed inertia-gravity wave and trapped baroclinic lee wave regime, (III) mixed evanescent wave and trapped baroclinic lee wave regime, and (IV) transient wave regime with possible nongeostrophic baroclinic instability (Shen and Lin 1999). The baroclinic lee wave theory of lee cyclogenesis (Section 5.5) belongs to regime II of moderate R_o and moderate R_i ($\sim$ 6.25), or small R_o ($\sim$0.4–0.8) and moderate (large) R_i ($\geq$ 25). It is also found that three-dimensionality, directional wind shear, and rotation promote horizontal energy dispersion (Shutts 2003).

Appendix 3.1 Derivations of shallow-water equations

Consider a nonrotating, hydrostatic, two-layer fluid system with constant densities ρ_1 and ρ_o in the upper and lower layers, respectively, and assume that $\rho_1 < \rho_o$. Note that the horizontal pressure gradients, $\partial p/\partial x$ and $\partial p/\partial y$, are independent of height in each layer if the fluid system is in hydrostatic balance because

$$\frac{\partial}{\partial z}\left(\frac{\partial p}{\partial x}\right) = \frac{\partial}{\partial x}\left(\frac{\partial p}{\partial z}\right) = -\frac{\partial}{\partial x}(\rho g) = 0, \tag{A3.1.1}$$

because the density is constant in each layer. We also assume that no horizontal pressure gradients exist in the upper layer. At point A in Fig. 3.2, we have

$$\frac{p - (p + \delta p_1)}{\delta z} = -\rho_1 g, \tag{A3.1.2}$$

according to the hydrostatic equation. The above equation leads to

$$p + \delta p_1 = p + \rho_1 g \delta z = p + \rho_1 g \frac{\partial(H + h')}{\partial x} \delta x, \tag{A3.1.3}$$

where H is the undisturbed upstream fluid depth and h' the perturbation or the vertical displacement from H, and $\partial(H + h')/\partial x$ is the slope of the interface. Similarly, we may derive the pressure at point B,

$$p + \delta p_2 = p + \rho_o g \delta z = p + \rho_o g \frac{\partial(H + h')}{\partial x} \delta x. \tag{A3.1.4}$$

Thus, the horizontal pressure gradient in the x-direction, $\partial p/\partial x$, at the interface can be approximated by

$$\frac{\partial p}{\partial x} = g \Delta \rho \frac{\partial(h + h_s)}{\partial x}, \tag{A3.1.5}$$

where $\Delta \rho = \rho_o - \rho_1$. In deriving (A3.1.5), we have used $h + h_s = H + h'$, where h is the instantaneous depth of the fluid, and h_s is the height of the bottom topography (Fig. 3.2). Similarly, we may derive the horizontal pressure gradient in the y-direction

$$\frac{\partial p}{\partial y} = g \Delta \rho \frac{\partial (h + h_s)}{\partial y}. \qquad (A3.1.6)$$

Therefore, the horizontal momentum equations become

$$\frac{\partial u}{\partial t} + u \frac{\partial u}{\partial x} + v \frac{\partial u}{\partial y} + w \frac{\partial u}{\partial z} = -g' \frac{\partial (h + h_s)}{\partial x}, \qquad (A3.1.7)$$

$$\frac{\partial v}{\partial t} + u \frac{\partial v}{\partial x} + v \frac{\partial v}{\partial y} + w \frac{\partial v}{\partial z} = -g' \frac{\partial (h + h_s)}{\partial y}, \qquad (A3.1.8)$$

where $g' = g \Delta \rho / \rho_0$ is called *reduced gravity*. Assuming that initially there is no vertical shear of the horizontal wind velocity ($\partial u / \partial z = \partial v / \partial z = 0$), it can be shown that u and v will be independent of z at any subsequent time. Under this constraint, (A3.1.7) and (A3.1.8) reduce to

$$\frac{\partial u}{\partial t} + u \frac{\partial u}{\partial x} + v \frac{\partial u}{\partial y} = -g' \frac{\partial (h + h_s)}{\partial x}, \qquad (A3.1.9)$$

$$\frac{\partial v}{\partial t} + u \frac{\partial v}{\partial x} + v \frac{\partial v}{\partial y} = -g' \frac{\partial (h + h_s)}{\partial y}. \qquad (A3.1.10)$$

The above equations give (3.4.1) and (3.4.2) respectively.

Integrating the continuity equation, (2.2.4), from the surface of the bottom topography ($z = h_s$) to the interface ($z = h + h_s = H + h'$) with respect to z leads to

$$\int_{h_s}^{H+h'} \left(\frac{\partial u}{\partial x} + \frac{\partial v}{\partial y} \right) dz + \int_{h_s}^{H+h'} \frac{\partial w}{\partial z} dz = 0, \qquad (A3.1.11)$$

because $D\rho / Dt = 0$ in each layer of the fluid. Since u and v are assumed to be independent of z initially, then both $\partial u / \partial x$ and $\partial v / \partial y$ will be independent of z for all time afterwards. Thus, the vertically integrated mass continuity equation, (A3.1.11), reduces to

$$h \left(\frac{\partial u}{\partial x} + \frac{\partial v}{\partial y} \right) + w(z = h + h_s) - w(z = h_s) = 0. \qquad (A3.1.12)$$

The vertical velocity is just the rate at which the height is changing: $w = Dz/Dt = \partial h / \partial t + u \partial h / \partial x + v \partial h / \partial y$. Substituting w into (A3.1.12) leads to

$$h \left(\frac{\partial u}{\partial x} + \frac{\partial v}{\partial y} \right) + \left(\frac{\partial (h + h_s)}{\partial t} + u \frac{\partial (h + h_s)}{\partial x} + v \frac{\partial (h + h_s)}{\partial y} \right) - \left(u \frac{\partial h_s}{\partial x} + v \frac{\partial h_s}{\partial y} \right) = 0.$$

$$(A3.1.13)$$

This can be further reduced to the expression

$$\frac{\partial h}{\partial t} + u\frac{\partial h}{\partial x} + v\frac{\partial h}{\partial y} + h\left(\frac{\partial u}{\partial x} + \frac{\partial v}{\partial y}\right) = 0 \qquad (A3.1.14)$$

because h_s, the height of the bottom topography, is generally assumed to be independent of time. Equation (A3.1.14) gives (3.4.3).

References

Baines, P. G., 1995. *Topographic Effects in Stratified Flows*. Cambridge University Press.

Bluestein, H. B. and M. H. Jain, 1985. Formation of mesoscale lines of precipitation: Severe squall lines in Oklahoma during the spring. *J. Atmos. Sci.*, **42**, 1711–32.

Booker, J. R. and F. P. Bretherton, 1967. The critical layer for internal gravity waves in a shear flow. *J. Fluid Mech.*, **27**, 513–39.

Breeding, R. J., 1971. A nonlinear investigation of critical levels for internal atmospheric gravity waves. *J. Fluid Mech.*, **50**, 545–63.

Bretherton, F. P., 1966. The propagation of groups of internal gravity waves in a shear flow. *Quart. J. Roy. Meteor. Soc.*, **92**, 466–80.

Durran, D. R., 1990. Mountain waves and downslope winds. *Atmospheric Processes over Complex Terrain*, Meteor. Monogr., No. 45, Amer. Meteor. Soc., 59–81.

Emanuel, K. and D. J. Raymond, 1984. Dynamics of Mesoscale Weather Systems. J. B. Klemp (ed.), NCAR.

Hooke, 1986. Gravity waves. In *Mesoscale Meteorology and Forecasting*, P. S. Ray (ed.), Amer. Meteor. Soc., 272–88.

Houghton, D. D. and A. Kasahara, 1968. Non-linear shallow fluid flow over an isolated ridge. *Comm. Pure Appl. Math.*, **21**, 1–23.

Jones, W. L., 1967. Propagation of internal gravity wave in fluids with shear and rotation. *J. Fluid Mech.*, **30**, 439–48.

Klemp, J. B. and D. K. Lilly, 1975. The dynamics of wave-induced downslope winds. *J. Atmos. Sci.*, **32**, 320–39.

Lin, Y.-L., 1987. Two-dimensional response of a stably stratified flow to diabatic heating. *J. Atmos. Sci.*, **44**, 1375–93.

Lindzen, R. S. and A. J. Rosenthal, 1983. Instabilities in a stratified fluid having one critical level. Part III: Kelvin-Helmholtz instabilities as overreflected waves. *J. Atmos. Sci.*, **40**, 530–42.

Long, R. R., 1970. Blocking effects in flow over obstacles. *Tellus*, **22**, 471–80.

Mowbray, D. E. and B. S. H. Rarity, 1967. A theoretical and experimental investigation of the phase configuration of internal waves of small amplitude in a density stratified liquid. *J. Fluid Mech.*, **28**, 1–16.

Nicholls, M. E. and R. A. Pielke Sr., 2000. Thermally induced compression waves and gravity waves generated by convective storms. *J. Atmos. Sci.*, **57**, 3251–71.

Olver, F. W. J., 1997. *Asymptotics and Special Functions*. A. K. Peters, Ltd.

Shen, B.-W. and Y.-L. Lin, 1999. Effects of critical levels on two-dimensional back-sheared flow over an isolated mountain ridge on an f plane. *J. Atmos. Sci.*, **56**, 3286–302.

Shutts, G., 2003. Inertia-gravity wave and neutral Eady wave trains forced by directionally sheared flow over isolated hills. *J. Atmos. Sci.*, **60**, 593–606.

Smith, R. B., 1979. The influence of mountains on the atmosphere. *Adv. in Geophys.*, **21**, 87–230.

Smith, R. B., 1986. Further development of a theory of lee cyclogenesis. *J. Atmos. Sci.*, **43**, 1582–1602.

Sommerfeld, A., 1949. *Partial Differential Equations in Physics*. Academic Press.

Turner, J. S., 1973. *Buoyancy Effects in Fluids*. Cambridge University Press.

Wurtele, M. G., A. Data, and R. D. Sharman, 1996. The propagation of gravity-inertia waves and lee waves under a critical level. *J. Atmos. Sci.*, **53**, 1505–23.

Wyss, J. and K. A. Emanuel, 1988. The pre-storm environment of midlatitude prefrontal squall lines. *Mon. Wea. Rev.*, **116**, 790–94.

Problems

3.1 Using the normal mode approach, derive the Lamb wave solution from (3.3.1), (3.3.2), (3.3.7) and (3.3.8).

3.2 (a) Derive (3.4.10) from (3.4.7)–(3.4.9). (b) Find the dispersion relation for linear, three-dimensional (x, y, t) shallow water waves. Are the waves dispersive or not? (c) From (b), find the group velocities, c_{gx} and c_{gy}.

3.3 Consider a two-dimensional, one-layer fluid flow over an obstacle. Estimate the Froude number (F) and nondimensional mountain height (M), assuming $H = 1000\,\mathrm{m}$, $U = 4\,\mathrm{m\,s^{-1}}$, $h_m = 200\,\mathrm{m}$. What is the flow regime, based on the flow regime diagram for one-layer shallow-water system (Fig. 3.3)? Change flow and/or orographic parameters to shift the above flow regime to the other four flow regimes.

3.4 Show that in the flow regime with $N^2 \gg \Omega^2$ the vertical momentum equation reduces to the hydrostatic equation, (3.5.18). Show that the vertical wavelength is $2\pi U/N$ for a steady state flow.

3.5 (a) Assuming $N^2 \ll \Omega^2$, prove that the vertical momentum equation, (3.5.2), reduces to (3.5.20).
(b) Show that (3.5.21) reduces to (3.5.22) for a flow starting with no vorticity at $t = 0$ (i.e. assume that the flow is initially irrotational).

3.6 Derive (3.6.6) from (3.6.1)–(3.6.5).

3.7 Consider a uniform basic flow, which has a uniform buoyancy frequency $N = 0.012\,\mathrm{s^{-1}}$, passing over a mountain with a horizontal scale of 20 km. What is the critical basic flow speed that separates the upward propagating waves and evanescent waves, assuming the Earth's rotation can be ignored?

3.8 (a) Derive the group velocities for inertia-gravity waves for $U = 0$.
(b) Derive (3.6.23) for the ratio of c_{gz}/c_{gx}.

3.9 Derive the dynamic interface boundary condition (3.7.13).

3.10 Derive (3.7.17) from (2.2.14)–(2.2.18) by assuming a two-dimensional $(\partial/\partial y = 0, V = 0)$, nonrotating, adiabatic, and Boussinesq flow.

4

Mesoscale wave generation and maintenance

4.1 Introduction

In this chapter, the generation mechanisms for mesoscale waves, which include pure gravity waves, inertial oscillations, and inertia-gravity waves, will be discussed. As mentioned in Chapter 3, due to their dispersive nature, most gravity waves generated by a variety of atmospheric sources tend to lose coherence after some time. Therefore, our discussion of wave maintenance will focus on waves, such as those defined in the literature as long-lasting mesoscale waves with perturbation pressure amplitudes of 0.2–7.0 hPa, horizontal wavelengths of 50–500 km, and periods of 1–4 h (Uccellini and Koch 1987).

As they propagate away from their source region, mesoscale waves in the atmosphere may also initiate severe convective storms, generate wind gusts, and create turbulence leading to aviation safety problems. The surface pressure perturbation associated with a large-amplitude mesoscale gravity wave may even be larger than that typically associated with midlatitude cyclogenesis and often creates some confusion in weather forecasts (Fig. 4.1). This type of event might be incorrectly interpreted as the onset of secondary cyclogenesis along a coastal front, such as the case which occurred on February 27, 1984, as pointed out by Bosart and Seimon (1988). Occasionally, mesoscale waves with horizontal wavelengths of 50–500 km may last for several hours and propagate distances of several hundred kilometers to more than 1000 km away from the wave source region. This type of long-lasting mesoscale wave may behave like an isolated wave of elevation or depression, or often as a localized wave packet. Therefore, understanding fundamental mesoscale wave generation, maintenance, and propagation mechanisms is important to help: (a) identify potential severe storm outbreak regions, (b) forecast gravity wave-generated wind gusts, (c) distinguish between the wave-generated low pressure and cyclone centers as well as accompanying precipitation patterns, (d) identify regions of clear air turbulence that may cause aviation safety problems, and (e) develop gravity wave parameterization schemes for global models.

4.2 Wave generation mechanisms

Mesoscale waves may be generated by, but are not limited to, any of the following sources: (a) orography, (b) surface heating or cooling, (c) density currents or density

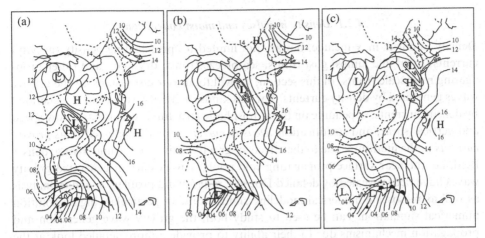

Fig. 4.1 An example of a mesoscale wave generated by moist convection. The wave is associated with the low (bold), starting over western Tennessee and Kentucky and traveling toward New England. Sea level isobaric analyses at (a) 1100, (b) 1300, and (c) 1500 UTC March 6, 1969 are shown. Isopleths are in hPa (e.g., 16 = 1016 hPa). The bold line in southern United States is a prefrontal squall line. (After Lin and Goff 1988.)

impulses, (d) moist convection, (e) mesoscale instabilities, (f) geostrophic adjustment, and (g) nonlinear interactions. It is well documented that orography can serve as a major mechanism in generating numerous mesoscale gravity waves. Orographically generated internal gravity waves are also known as mountain waves, which will be discussed, along with other orographically generated flows, in Chapter 5. Surface heating or cooling due to long-wave infrared radiation near the Earth's surface is also able to generate internal gravity waves. This type of gravity wave will be discussed, along with other thermally forced flows, in Chapter 6. In fact, surface cooling over a mountain slope can generate katabatic winds that behave similarly to density currents and are thus related to wave category (c). Moist convection can also serve as a source for internal gravity waves, such as those depicted in Fig. 4.1. Mesoscale instabilities in the interior of stratified flows may also generate mesoscale waves. For example, conditions for shear instability may be created by the presence of density currents associated with falling precipitation or an upper-tropospheric jet streak. This indicates that sources for mesoscale gravity wave generation may interact with each other. In addition to shear instability, other mesoscale instabilities and processes may also trigger inertia-gravity waves. Inertia-gravity waves might be excited in local regions of geostrophic imbalance, such as that resulting from an initial addition of momentum to a rotating fluid. Observations indicate that this often happens when a balanced jet streak impinges on an upper-tropospheric pressure ridge. In addition to the above mechanisms, nonlinear interactions may also serve as a generation mechanism for mesoscale atmospheric waves.

4.2.1 Density impulses and moist convection

Density impulses may include, but are not limited to, propagating density (gravity) currents, hydraulic jumps, gust fronts, sea-breeze fronts, dry lines, and other propagating mesoscale fronts. In this section, we will focus our discussions on the gravity waves generated by density currents. Various kinds of gravity waves can be generated by density currents, depending on the atmospheric structure, such as basic flow speed and vertical shear, and stratification. *Density currents* are also referred to as *gravity currents* in the literature, due to the effects of gravity acting on the underlying, denser fluid, i.e. the near surface propagating cold pool. Density current-generated gravity waves have been revealed by detailed Doppler radar, wind profiler, or sodar analysis as well as mesoscale numerical modeling simulations. In particular, high-resolution numerical simulations can be used to study both the gravity wave generation and propagation mechanisms due to their ability to provide a more detailed look at the horizontal and vertical structure of density currents and gravity waves than is typically resolved with field observations.

Figure 4.2 shows an example of gravity waves generated by a density current that occurred during the Mesogers field experiment in southwestern France. This event occurred in an environment containing a nocturnal inversion and an elevated neutral layer. The instrument systems recorded the passage of an atmospheric discontinuity that was marked by a distinct and abrupt shift in wind direction (southeasterly to north-westerly), large vertical velocities ($>1\,\mathrm{m\,s^{-1}}$), a sharp drop in temperature (3 to 4 K) up to a height of $z = 1.4\,\mathrm{km}$, and a sudden increase in surface pressure (2 hPa). The disturbance was subsequently identified as a density current moving to the east at a speed of $10\,\mathrm{m\,s^{-1}}$. Above the head of the density current (i.e. the gust front) and below a height of $z = 3.5\,\mathrm{km}$, there existed an updraft with vertical velocities greater than $1\,\mathrm{m\,s^{-1}}$.

The generation and propagation of the gravity waves shown in Fig. 4.2 are verified by numerical simulations from a nonhydrostatic, mesoscale numerical model (Fig. 4.3). The updraft in the layer above the gust front and below $z = 3.5\,\mathrm{km}$ is reasonably well simulated by the model. In addition, the vertically propagating gravity waves above $z = 3.5\,\mathrm{km}$ and the density current below $z = 1.25\,\mathrm{km}$ are clearly depicted. The overturning circulation within the head of the density current is also captured by the numerical simulation. The gust front updraft was generated by strong, near-surface convergence produced by the advancing density current. This robust gust front updraft forced the generation of gravity waves at higher levels aloft, above the propagating density current. These waves at higher tropospheric levels repeat themselves in the vertical, and therefore belong to the vertically propagating wave regime, as discussed in Chapter 3. The well-known "morning glory" phenomenon (Fig. 4.15) also belongs to this class of density-current generated waves. The mass-momentum imbalance during the collapse of a cold front may also generate inertia-gravity wave packets. The imbalance is associated with cross-frontal accelerations, $\partial(\partial u/\partial x)/\partial t$, and then acts as a source for the inertia-gravity waves (Ley and Peltier 1978).

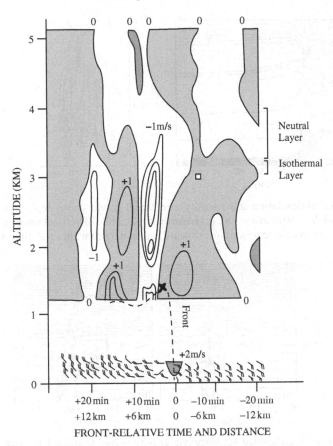

Fig. 4.2 An example of gravity waves generated by a density current, as revealed by a synthesis of sodar, rawinsonde, and wind profiler observations during the Mesogers field experiment in southwestern France. An observation gap exists between 350 and 1200 m; the sodar-observed vertical velocity is shown below the gap and the radar-observed vertical velocity above (contours at $1\ \mathrm{m\,s^{-1}}$ intervals with upward motions shaded). The rawinsonde-observed neutral and isothermal layers and depth of the cold air are also marked. The times shown on the abscissa are in reference to the time of frontal passage around 1900 UTC September 28, 1984. (From Ralph *et al.* 1993.)

Moist convection is another source of internal gravity waves. Occasionally, long-lasting, large-amplitude mesoscale gravity waves are generated by moist convection (e.g., Fig. 4.1). A convective storm, or even a cluster or complex of mesoscale convective systems, may act as a point source of gravity waves when it encounters stable overlying air, such as the base of the stratosphere or a temperature inversion at the top of the planetary boundary layer. Figure 4.4 shows an example of gravity wave generation in the stratosphere by penetrative moist convection associated with an idealized mesoscale convective storm, as simulated by a nonhydrostatic numerical model. In the developing stage of the storm (Fig. 4.4a), individual convective cells

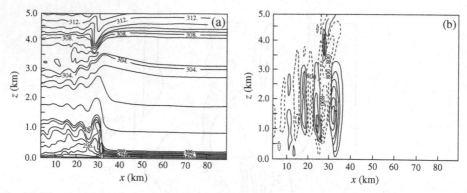

Fig. 4.3 Numerical simulation of the density current and gravity waves depicted in Fig. 4.2 with $\Delta x = 200$ m and $\Delta z = 50$ m at $t = 1$ h : (a) isentropes (K), and (b) vertical velocity (10^{-3} m s^{-1}). Solid (dotted) curves denote upward (downward) motion. (From Jin *et al.* 1996.)

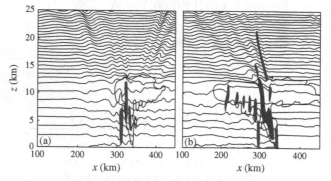

Fig. 4.4 Convectively generated stratospheric gravity waves simulated by a cloud model at (a) 2 h (developing stage) and (b) 4 h (mature stage). Vertical velocity (dark shaded for vertical velocity greater than 3 m s^{-1}) is superimposed on potential temperature contours (thin lines, 4 K interval) and cloud outline (solid curve; cloud water mixing ratio at 0.1 g kg^{-1}). (Adapted after Fovell *et al.* 1992.)

embedded within the storm have repeatedly overshot their levels of neutral buoyancy (LNB, see Chapter 7) and penetrated into the stratosphere. A series of gravity waves are generated by the buoyancy force associated with the overshooting convective cells. At the mature stage of the convective cloud, the waves are roughly symmetric about the main updraft (Fig. 4.4b). At this stage, the storm is composed of a series of weaker convective cells embedded in a rising, westward-tilted airflow. Accordingly, the generated gravity waves in the stratosphere are weaker. Due to the rearward (westward) propagation and development of these convective cells within the storm, the generated gravity waves in the stratosphere are biased to the western side. In general, gravity waves generated under these circumstances are highly dispersive. However, under certain circumstances, waves may sustain and are able to travel a significant distance. Various wave maintenance mechanisms will be discussed in the next section.

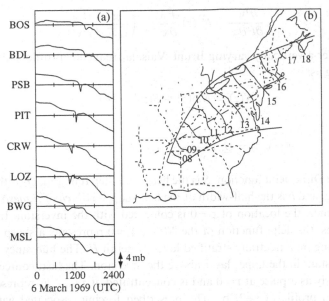

Fig. 4.5 (a) Surface pressure traces for selected stations on 6 March 1969: (1) BOS – Boston, MA; (2) BDL – Hartford, CT, (3) PSB – Philipsburg, PA; (4) PIT – Pittsburgh, PA; (5) CRW – Charleston, WV; (6) LOZ – London, KY; (7) BWG – Bowling Green, KY; and (8) MSL – Muscle Shoals, AL. (b) Isochrones (hours in UTC) of the minimum pressure indicating passage of the solitary wave. The heavy dashed line indicates the position of the sounding cross section shown in Fig. 4.17(b), as well as the primary direction of travel of the wave. (From Lin and Goff 1988.)

Figure 4.5a shows an isolated mesoscale wave of depression, which occurred on March 6, 1969, as revealed by surface pressure traces from several microbarographs. This mesoscale wave originated near a region of strong convective activity in Alabama and Mississippi early on the morning of March 6, 1969 (Fig. 4.1). It then propagated northeastward at a mean speed of $55 \, \mathrm{m \, s^{-1}}$ expanding within an arc of about 40° whose origin is located in central Mississippi (Fig. 4.5b). It traveled over a horizontal distance of more than 1000 km to Nova Scotia by 1800 UTC 6 March. This wave was generated by moist convection associated with a midlatitude squall line. Many other examples of moist convectively-generated mesoscale gravity waves have been observed, including the event that occurred on December 15, 1987, through which it was found that moist convection plays a significant role in wave generation (Powers and Reed 1993).

The mechanisms for internal gravity wave generation by density currents and moist convection may be understood by considering the transient or steady state dynamic response of a stratified airflow to a prescribed heat source. In the following, we present a linear theory, which includes a pronounced vertical temperature inversion, a quiescent basic-state wind ($U = 0 \, \mathrm{m \, s^{-1}}$), and a vertically varying buoyancy (Brunt–Vaisala) frequency. The theoretical model assumes a two-dimensional, inviscid, nonrotating, hydrostatic, Boussinesq flow. The small-amplitude vertical velocity is governed by

$$\frac{\partial^4 w'}{\partial t^2 \partial z^2} + N^2(z)\frac{\partial^2 w'}{\partial x^2} = \left(\frac{g}{c_p T_o}\right)\frac{\partial^2 \dot{q}}{\partial x^2}. \tag{4.2.1}$$

Now, consider a vertically varying Brunt–Vaisala frequency profile and an elevated, pulse heating as:

$$N(z) = N[\delta(z) + S(z)], -H \leq z, \tag{4.2.2a}$$

$$\dot{q}(t, x, z) = Q_o\left(\frac{b^2}{x^2 + b^2}\right)\delta(t)\,\delta(z - z_1),\ z_1 > 0, \tag{4.2.2b}$$

where δ is the Dirac delta function, S is the heavyside step function, Q_o is the amplitude of the heating, and b is the half-width of the bell-shaped (*Witch of Agnesi*) heat source. For convenience, the location of $z = 0$ is colocated with the inversion. In (4.2.2a), in physical terms, the delta function of the $N(z)$ at $z = 0$ represents a sharp temperature inversion riding on a neutrally stratified layer of depth H. The buoyancy frequency is assumed constant in the upper layer above the inversion. The heat source is activated instantaneously as a pulse at $t = 0$ and is concentrated at $z = z_1$, as represented by the delta function profile of (4.2.2b). The prescribed forcing associated with a density current may also physically represent evaporative cooling ($Q_o < 0$) associated with falling precipitation. The corresponding mathematical problem can be solved by determining the relevant Green's function solution. Taking Fourier transforms in x and t of (4.2.1), leads to the vertical structure equation:

$$\frac{\partial^2 \hat{w}}{\partial z^2} + \left(\frac{Nk}{\omega}\right)^2 \hat{w} = \frac{gQ_obk^2}{4\pi c_p T_o\omega^2}e^{-b|k|}\,\delta(z - z_1), \tag{4.2.3}$$

where the related Fourier transform pair (also see Appendix 5.1) is defined as

$$\hat{w}(k, z) = \frac{1}{2\pi}\int_{-\infty}^{\infty} w'(x, z)e^{-ikx}dx;\ w'(x, z) = \int_{-\infty}^{\infty} \hat{w}(k, z)e^{ikx}dk. \tag{4.2.4}$$

The lower boundary condition at the flat surface ($z = -H$) requires $\hat{w} = 0$, while the appropriate upper boundary condition is the Sommerfeld (1949) radiation boundary condition, i.e. $\hat{w} \sim \exp(iN|k|/\omega)$ as $z \to \infty$. There are four interface conditions required in this problem. The kinematic and dynamic conditions, as discussed in Chapter 3, require both $\hat{w}$ and $\hat{w}_z$ to be continuous across the interface at $z = z_1$. The dynamic condition is equivalent to integrating (4.2.3) across the interface. Likewise, at the inversion $z = 0$, $\hat{w}$ is continuous, but integrating (4.2.3) across $z = 0$ yields a condition that is related to the strength of the inversion; that is, $g' = g\Delta\theta/\theta_o$. The solution in the lower layer ($-H \leq z < 0$) is given by

$$\eta = \frac{-Q_o t^2(z + 1)e^{-Mt/x^2}\left[(x^2 - t^2)\sin\left(\frac{Mt}{|x|}\right) + Ft|x|\cos\left(\frac{Mt}{|x|}\right)\right]}{2|x|\left[(x^2 - t^2)^2 + (Ftx)^2\right]}, \tag{4.2.5}$$

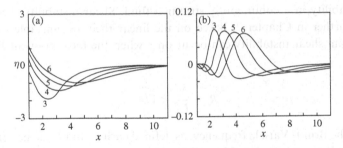

Fig. 4.6 (a) A wave of depression is generated at the inversion ($z = 0$) with $M = 0.1$, $F = 1.0$, and $Q_o = 0.5$, and the dimensional parameters $H = 5500$ m, $z_1 = 550$ m, $c_o = 55$ m s^{-1}, and $N = 0.01$ s^{-1} in the linear solution with a thermal forcing, (4.2.5). (b) A wave of elevation is generated at the inversion with $M = 2.0$, $F = 1.0$, and $Q_o = 0.5$, and the dimensional parameters $H = 1000$ m, $z_1 = 2000$ m, $c_o = 10$ m s^{-1}, and $N = 0.01$ s^{-1}. Waves at four nondimensional times ($\bar{t} = bt/c_o$) are shown. (From Lin and Goff 1988.)

where η is the vertical displacement defined by $\eta = \int w \, dt$. In the above equation, all variables have been nondimensionalized by scaling η and z by H, x by b, t by b/c_o, and Q_o by $c_p T_o c_o^2/g$, where $c_o = \sqrt{g'H}$ is the internal shallow water phase speed, $M = Nz_1/c_o$, and $F = NH/c_o$. The nondimensional number M represents an inverse internal Froude number associated with the lower layer of fluid, while F represents an inverse thermal Froude number. The asymptotic solution (4.2.5) was obtained through the method of stationary phase (Lin and Goff 1988).

Figure 4.6a shows the downstream (rightward) propagating *wave of depression* corresponding to (4.2.5). A dispersive wave of depression is clearly generated at the inversion by the pulse thermal forcing. As will be discussed in Subsection 4.3.2, the wave dispersion may be balanced by nonlinear wave steepening to produce a long-lasting, solitary wave. On the other hand, with a different set of flow parameters, a *wave of elevation* is generated at the inversion (Fig. 4.6b). The waves in the lower layer of the above fluid system are excited by the updraft impinging on the inversion, which vertically displaces the stable air in the upper layer. Gravity waves may also be generated in an opposite sense, i.e. being excited by downdrafts impinging from above on an underlying stable layer, thus displacing stable air vertically downward toward the surface.

4.2.2 Mesoscale instabilities

In addition to the above-mentioned wave generation mechanisms, mesoscale instabilities (Chapter 7), such as shear instability, symmetric instability, inertial instability, and even baroclinic instability, may also serve as generation mechanisms for mesoscale waves. Gravity waves that owe their existence to shear instability can impart energy from the mean flow into growing perturbations when a critical level exists (also see Chapters 3, 5 and 6). Shear instability is referred to as Kelvin–Helmholtz

(K–H) instability in a continuously stratified fluid. Shear instability theory will be discussed further in Chapter 7. Based on the linear theories, unstable wave modes resulting from shear instability can occur only when the local gradient Richardson number,

$$R_i = \frac{N^2}{U_z^2} < 1/4, \tag{4.2.6}$$

where N is the Brunt–Vaisala frequency, as defined earlier, and U_z is the vertical shear of the basic horizontal wind (Miles 1961; Howard 1961). As will be discussed in Chapter 7, the semicircle theorem implies that a critical level is required to exist somewhere within the flow domain for shear instability to occur. Further analysis of the Miles–Howard criterion (4.2.6) indicates that $R_i < 1/4$ should exist in the neighborhood of the critical level as a sufficient condition for unstable wave growth (Einaudi and Lalas 1973). A number of investigators have found that shear instability could have served as the generation mechanism for gravity waves observed in their cases. Figure 4.7 shows the vertical profiles of $U(z)$, $N(z)$, and $R_i(z)$ for a mesoscale gravity wave event observed to occur during the Cooperative Convective Precipitation Experiment (CCOPE). The U profile (Fig. 4.7a) indicates that a single critical level (z_c) exists at 5.5 km above the mean sea level, with positive intrinsic phase speeds ($c > U$) at all levels below z_c. Figure 4.7b indicates that the atmosphere is statically stable essentially everywhere above the planetary boundary layer (PBL). The R_i profile (Fig. 4.7c) indicates several levels where $R_i < 1/4$, but most importantly, one of them occurs in the vicinity of the critical level, $z_c = 5.5$ km. Observations also indicate that

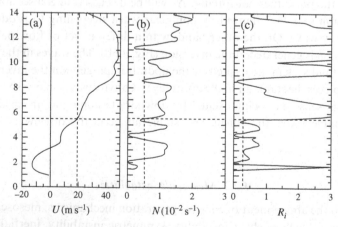

Fig. 4.7 An example of the environment in which the gravity waves were generated by shear instability. Shear instability analysis and wave vertical structure calculated from the Knowlton, Wyoming sounding at 0000 UTC July 12, 1981. Vertical profiles of U, N, and R_i are shown in (a), (b), and (c), respectively. A critical level is detected and denoted by dotted lines at 5.5 km. (Adapted after Koch and Dorian 1988.)

some mesoscale wave disturbances are generated near upper-tropospheric jet streaks, which indicates that the vertical wind shear associated with the jet could provide an energy source for the waves.

To elucidate the basic characteristics of internal gravity waves generated by shear instability, we consider a two-dimensional, small-amplitude, adiabatic, inviscid, Boussinesq flow governed by

$$\frac{D^2}{Dt^2}\nabla^2 w' - U_{zz}\frac{D}{Dt}\left(\frac{\partial w'}{\partial x}\right) + N^2\frac{\partial^2 w'}{\partial x^2} = 0. \tag{4.2.7}$$

The above equation is identical to the Taylor–Goldstein equation, (3.7.19). Taking the normal mode approach by assuming $w' = \hat{w}(z)\exp[i(kx - \omega t)]$, the above equation becomes

$$\frac{\partial^2 \hat{w}}{\partial z^2} + m^2(z)\hat{w} = 0, \tag{4.2.8}$$

where

$$m^2(z) = \frac{N^2}{(U-c)^2} - \frac{U_{zz}}{U-c} - k^2, c = \frac{\omega}{k}. \tag{4.2.9}$$

Over a flat surface, the lower boundary condition requires $\hat{w} = 0$ at $z = 0$ and the upper radiation boundary condition requires $\hat{w} \sim \exp(imz)$ in the stable layer above the shear layer. We assume a semi-infinite, vertically sheared, hyperbolic-tangent profile for the basic wind,

$$U(z) = U_o\tanh(z/h), \tag{4.2.10}$$

and a background density profile of the form,

$$\rho(z) = \rho_s e^{-z/h}. \tag{4.2.11}$$

In (4.2.10) and (4.2.11), 2h is the effective depth of the shear layer and ρ_s is the surface air density (Fig. 4.8a). By performing a linear stability analysis, the main characteristics of the most unstable modes can be determined. Figure 4.8b shows the nondimensional growth rate, $(hk/U_o)c_i$, and the nondimensional horizontal phase velocities, ω_r/kU_o, as a function of nondimensional wave number (hk) for different values of the *local Richardson number* R_i, i.e. $N^2/(U_o/h)^2$, at the basic wind inflection point for the most unstable (mode I) waves. If one assumes $2h = 5$ km and $R_i = 0.1$ at the inflection point, then the horizontal wavelength for the most unstable waves is about 39 km. The observed mesoscale gravity waves have horizontal wavelengths greater than those predicted by the theoretical model used in Fig. 4.8. This may be explained by the fact that the atmosphere acts to filter out small-scale waves, which decay in amplitude away from the shear zone more rapidly than mesoscale waves

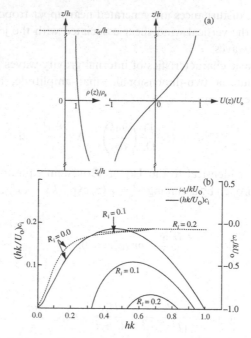

Fig. 4.8 (a) The normalized density and velocity profiles and the geometry of the basic flow (not in scale), (b) Growth rate $(hk/U_o)c_i$ (solid lines) and phase velocities ω_r/kU_o (dashed lines) as a function of hk (nondimensional wave number) for different values of local Richardson number R_i for the most unstable (mode I) waves. R_i is the Richardson number at the inflection point; $\rho(z)$ is the basic density; ρ_s is a scaling air density; U_o is a scaling horizontal velocity. (Adapted after Lalas and Einaudi 1976.)

(Uccellini and Koch 1987). Thus, only the waves with longer horizontal wavelengths would be detected at ground level.

Other mesoscale instabilities, such as symmetric instability and inertial instability, may also trigger mesoscale waves. As discussed above, the waves generated by shear instability often have a horizontal scale on the order of $O(10\,\text{km})$ and relatively high frequencies ($\omega \sim N$). However, inertia-gravity waves with horizontal wavelengths of 200 to 1000 km, vertical wavelengths of 1 to 5 km, and intrinsic frequencies near Coriolis frequency ($\omega \sim f$) are ubiquitous within the atmosphere (O'Sullivan and Dunkerton 1995). Apparently, these inertia-gravity waves are not often observed in the lower troposphere and cannot be easily explained by the mechanism of the generation of shear instability. This may be explained in Subsection 4.2.3.

4.2.3 Geostrophic adjustment

It has been documented that inertia-gravity waves are excited in local regions of geostrophic imbalance resulting from an initial impulsive addition of momentum to a rotating fluid (e.g., Uccellini and Koch 1987). The resulting wave train is highly dispersive and propagates perpendicular to the current in the two-dimensional flow regime; the wave eventually disperses and damps out as the current approaches a new geostrophically balanced state (Cahn 1945). One region favorable for inertia-gravity wave generation is in the vicinity of an unbalanced jet streak. It is found that the synoptic-scale flow becomes highly ageostrophic when an upper-level, midlatitude jet streak exists in an amplifying baroclinic wave and propagates downstream toward a quasi-stationary ridge in the geopotential height field (Kaplan and Paine 1977). Three-dimensional nonlinear modeling studies also indicate that inertia-gravity waves may also arise spontaneously as the tropospheric jet stream is distorted by baroclinic instability and strong parcel accelerations take place, primarily in the exit region of the upper tropospheric jet stream (O'Sullivan and Dunkerton 1995). Because of the large divergence tendency (4.2.30) of this type of flow, inertia-gravity waves are excited as the atmosphere attempts to reestablish a quasi-geostrophic balance. The divergence tendency physically accompanies local upper-tropospheric accelerations due to significant imbalances between the horizontal pressure gradient force, the Coriolis force, and the centripetal force, and can be diagnosed by the divergence tendency equation, which will be discussed later. The ageostrophic winds have the effect of transferring mass from the anticyclonic to the cyclonic side of the jet streak entrance region. This thermally direct circulation converts available potential energy into zonal kinetic energy, causing the flow to accelerate to the left of the stream. Such strong upper-level geostrophic adjustment processes are a common characteristic of synoptic-scale environments, where there are severe weather forms, and which are currently believed to be responsible for the generation of observed inertia-gravity waves.

A strong thermal inversion in the lower troposphere (north of a frontal boundary) and a jet streak propagating toward a ridge axis in the upper troposphere are commonly observed in all cases. In general, the area of wave activity is bounded by the 300-hPa jet axis to the west or northwest, a surface front to the southeast, a 500-hPa inflection axis (between the trough and ridge axes) to the southwest, and the 500-hPa ridge axis to the northeast (Fig. 4.9). The geostrophic adjustment process depends on the horizontal scale of the disturbance. The mass (pressure or temperature) field would adjust to the momentum (velocity) field when the horizontal scale of disturbance is small compared to the Rossby radius of deformation, $\lambda_R = c/f$, where c is the phase speed of the wave and f the Coriolis parameter. Otherwise, the momentum field will adjust to the mass field. As discussed in Chapter 3, the phase speed for a single-layer shallow-water system is $c = \sqrt{gH}$, where H is the undisturbed fluid layer depth. For a continuously stratified fluid, the phase speed may be approximately estimated as $N/m = NL_z/2\pi$, where m is the vertical wavelength and L_z is the vertical scale of motion. Thus, the Rossby radius of deformation for a stratified fluid becomes (Blumen 1972)

$$L_R \sim \frac{NL_z}{f}. \qquad (4.2.12)$$

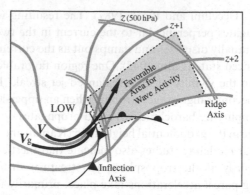

Fig. 4.9 Schematic of the environment conducive for mesoscale inertia-gravity wave generation by geostrophic adjustment. Depicted are the 300 hPa jet, 500 hPa height field (solid contours), and surface synoptic features. Positions of all atmospheric features shown on the schematic are approximate means during first half of wave episode, when wave generation mechanisms are assumed to operate most efficiently. The shaded region represents the area favorable for gravity wave activity during entire wave episode. Jet streak positions representing core of maximum wind speeds within the 300 hPa jet stream before (V_g) and after (V) the wave generation are denoted by thick arrows. (Adapted after Uccellini and Koch 1987.)

In the literature, L_R (λ_R) is also referred to as the *internal* (*external*) *Rossby radius of deformation*. One example of the geostrophic adjustment process is shown in Fig. 1.5, in which the energy is transferred from the disturbance associated with a cumulus cloud to a lens of less stratified air whose width is the Rossby radius of deformation. For typical midlatitude deep tropospheric motion with $N = 0.01 \text{ s}^{-1}$, $L_z = 10 \text{ km}$, and $f = 10^{-4} \text{ s}^{-1}$, the internal Rossby radius of deformation is $L_R = 1000$ km. This type of long-lived mesoscale wave has horizontal wavelengths of approximately 50 to 500 km, values that are contained within the range of internal Rossby radii of deformation specified by (4.2.12).

The geostrophic adjustment problem resulting from nonlinear inertial advective forcing accompanying a propagating mid-to-upper tropospheric jet streak, orographic forcing of an unusually intense summertime upper-tropospheric jet streak, or thermal forcing associated with low-to-midlevel latent heat release from nearby mesoscale convective systems, is mathematically equivalent to an initial value problem. The resulting evolution towards a new balanced state can be considered as a restructuring of the preexisting meso-α to synoptic-scale jet streak. Recent numerical simulations indicate that the subsynoptic jet streak geostrophic adjustment process is not well resolved by conventional upper-level rawindsonde data, yet it tends to play a major dynamical role in preconditioning the mesoscale environment for the formation of severe weather (Kaplan *et al.* 1998). The classical conceptual models of midlatitude synoptic-scale jet streaks and their associated ageostrophic circulations are based on either quasi- or semi-geostrophic dynamics that neglect the divergence tendencies necessary for the excitation of inertia-gravity waves. In reality, an unperturbed jet streak is in quasi-balance with the divergence associated with the zonal wind gradient and the ageostrophic entrance and exit region circulations responsible for its downstream propagation. Using a two-layer nonlinear primitive equation model, it has been shown that persistent unbalanced vertical motion, which is related to inertia-gravity wave modes, relative to the jet streak core, is generated, and its amplitude increases approximately as the square of the Rossby number (Van Tuyl and Young 1982). In addition, the "four-cell" pattern of vertical motion characterizing a quasi-geostrophically balanced, synoptic-scale jet streak (see Fig. 10.20) becomes more localized and subsequently evolves toward a "three-cell" and then "two-cell" pattern (with rising motion in the entrance region and sinking motion in the exit region) as the Rossby number ($R_o = U/fL$) increases from $\sim$0.29 to 0.52. These effects are attributed to nonlinear effects.

Most geostrophic adjustment studies have focused on the adjustment of initially unbalanced, impulsive, small-amplitude, ageostrophic perturbation introduced into a quiescent basic state. However, midlatitude jet streaks are never observed in quiescent basic-state flows. Therefore, advective effects should be taken into account, especially if jet streak adjustments or the dynamics of jet streak formation are formally considered. Some work has been done to investigate the response of a flow in gradient wind balance to mass and momentum forcing, and the effects of vertical wind shear on mass and momentum forcing in an idealized, two-layer model. Numerical simulations with a relatively finer horizontal grid resolution, such as 3.3 km, are able to reproduce observed mesoscale gravity waves with horizontal wavelengths $\sim$100–200 km , which are generated from the exit region of an upper-tropospheric jet streak (Zhang 2004). This implies that, through balance adjustment, the continuous generation of flow imbalance from the developing baroclinic wave leads to the continuous radiation of gravity waves. In the following, we will discuss some detailed dynamics of inertia-gravity wave generation through geostrophic adjustment and the diagnosis for unbalanced flow.

a. Inertia-gravity wave generation through geostrophic adjustment

The generation of inertia-gravity waves by geostrophic adjustment processes can be understood by imposing an unbalanced, small-amplitude zonal wind anomaly impulsively into a quiescent basic state or prescribing a zonal momentum forcing in the interior of the flow. The linearized equations governing small-amplitude baroclinic perturbations in a uniform, continuously stratified, hydrostatic, Boussinesq flow on an *f*-plane can be derived to have a form similar to (3.6.1)–(3.6.5),

$$\frac{Du'}{Dt} - fv' + \frac{1}{\rho_o}\frac{\partial p'}{\partial x} = F_x, \qquad (4.2.13)$$

$$\frac{Dv'}{Dt} + fu' + \frac{1}{\rho_o}\frac{\partial p'}{\partial y} = F_y, \qquad (4.2.14)$$

$$\frac{1}{\rho_o}\frac{\partial p'}{\partial z} - \frac{g}{\theta_o}\theta' = 0, \qquad (4.2.15)$$

$$\frac{\partial u'}{\partial x} + \frac{\partial v'}{\partial y} + \frac{\partial w'}{\partial z} = 0, \qquad (4.2.16)$$

$$\frac{D\theta'}{Dt} + \frac{N^2\theta_o}{g}w' = \frac{\theta_o}{c_p T_o}\dot{q}, \qquad (4.2.17)$$

where F_x and F_y are zonal and meridional momentum forcing, respectively, which are taken to represent the a priori assumed known parameterized or prescribed effects of nonlinear inertial advection ($F_x = -\boldsymbol{V}' \cdot \nabla u'$, $F_y = -\boldsymbol{V}' \cdot \nabla v'$, where $\boldsymbol{V}' = u'\boldsymbol{i} + v'\boldsymbol{j} + w'\boldsymbol{k}$) in the interior of the stratified flow, and $\dot{q}$ is the diabatic forcing (in $J\ kg^{-1}s^{-1}$). The presence of a nonzero basic-state flow incorporates the physical processes of perturbation geostrophic vorticity advection $-\boldsymbol{V} \cdot \nabla\zeta_g'$ and horizontal temperature advection $-\boldsymbol{V} \cdot \nabla\theta'$. These physical processes play an essential role in the development of midlatitude synoptic-scale systems when the zonal wind anomaly is viewed as an upper-level forcing interacting with low-level diabatically and orographically forced potential vorticity (PV) anomalies. Examples include, but are not necessarily limited to, (i) a preexisting PV anomaly that can interact with an isolated PV dipole through advection by the basic-state flow, (ii) the presence of external forcing in addition to an initial perturbation PV distribution, and (iii) self-advection of synoptic and mesoscale disturbances due to nonlinear advective processes. Understanding the wave generation mechanisms and the fundamental properties of this type of prescribed forcing and basic state atmosphere that control the wave characteristics are crucial steps toward improving the prediction of their formation, as well as developing a gravity wave parameterization for general circulation models or global climate models (e.g., Chun *et al.* 2004).

The equation governing the *perturbation PV* (q'_{pv}) may be derived from (4.2.13)–(4.2.17),

$$\frac{Dq'_{pv}}{Dt} = F_\zeta + \frac{fg}{N^2 c_p T_o}\frac{\partial\dot{q}}{\partial z}, \qquad (4.2.18)$$

where

$$q'_{pv} = \zeta' + \frac{f}{\rho_o N^2}\frac{\partial^2 p'}{\partial z^2}, \qquad (4.2.19)$$

and $F_\zeta = \partial F_y/\partial x - \partial F_x/\partial y$ is the external momentum forcing and $\zeta' = \partial v'/\partial x - \partial u'/\partial y$. Equations (4.2.13)–(4.2.17) may be combined into a single wave equation governing the pressure perturbation p',

$$\left[\left(\frac{\partial}{\partial t} + U\frac{\partial}{\partial x} + V\frac{\partial}{\partial y}\right)^2 + f^2\right]\frac{\partial^2 p'}{\partial z^2} + N^2\left(\frac{\partial^2 p'}{\partial x^2} + \frac{\partial^2 p'}{\partial y^2}\right) = S(x, y, z, t), \qquad (4.2.20)$$

where $S(x,y,z,t)$ is a forcing term that represents the initial perturbation PV distribution and forcing of momentum and diabatic sources,

$$S(x, y, z, t) = \rho_0 \left[N^2 \left(f q'_{pv} + F_\delta \right) + \frac{g}{c_p T_0} \frac{D}{Dt} \frac{\partial \dot{q}}{\partial z} \right], \qquad (4.2.21)$$

where $F_\delta = \partial F_x / \partial x + \partial F_y / \partial y$ is the divergence of the momentum forcing. Equations (4.2.18) and (4.2.20) form a mathematically closed system that requires specification of the three initial conditions p'_i, $\partial p'_i / \partial t$, and q'_{pvi}, in addition to the momentum and diabatic forcing terms $F(x, y, z, t)$ and $\dot{q}(r, t)$, if there are any. Linearized wave equations governing the other field variables (u', v', w', θ') can also be obtained, and are similar in form to (4.2.20).

In order to understand the fundamental response of the atmosphere to an initial PV perturbation, we may consider, in a first approximation, a vertically unbounded atmosphere. This approximation will be valid so long as the initial PV anomaly is not located too close to the Earth's surface or the tropopause. The corresponding initial value problem may then be solved by transforming the above equations into the Fourier space. The Fourier transform pair is defined by

$$\hat{p}(k, t) = \frac{1}{(2\pi)^3} \int_{-\infty}^{\infty} \int_{-\infty}^{\infty} \int_{-\infty}^{\infty} p'(r, t) e^{-ik \cdot r} dr, \qquad (4.2.22)$$

$$p'(r, t) = \int_{-\infty}^{\infty} \int_{-\infty}^{\infty} \int_{-\infty}^{\infty} \hat{p}(k, t) e^{ik \cdot r} dk, \qquad (4.2.23)$$

where $r = (x,y,z)$ is the position vector, and $k = (k, l, m)$ is the wave number vector. In the absence of both momentum and diabatic forcing, the solutions in Fourier space may be obtained from the following set of coupled equations:

$$\frac{\partial^2 \hat{p}}{\partial t^2} + 2i \Omega \frac{\partial \hat{p}}{\partial t} + \left(\frac{N^2 \kappa^2}{m^2} + f^2 - \Omega^2 \right) \hat{p} = -\frac{\rho_0 N^2 f}{m^2} \hat{q}_{pvi}(k), \qquad (4.2.24)$$

$$\hat{q}_{pvi}(k) = i[k \hat{v}_i(k) - l \hat{u}_i(k)] - \frac{m^2 f}{\rho_0 N^2} \hat{p}_i(k), \qquad (4.2.25)$$

where $\kappa^2 = k^2 + l^2$, $\Omega = kU + lV$ is the intrinsic wave frequency, and the subscript i denotes the initial disturbance. Substituting (4.2.25) into (4.2.24) and taking the inverse Fourier transform leads to

$$\hat{p}(k, t) = A(k) e^{i\omega_- t} + B(k) e^{i\omega_+ t} - \frac{\rho_0 N^2 f}{(N^2 \kappa^2 + m^2 f^2)} e^{-i\Omega t}, \qquad (4.2.26)$$

where

$$\omega_\pm = \Omega \pm f\sqrt{1 + \frac{N^2 \kappa^2}{m^2 f^2}}. \qquad (4.2.27)$$

The first two terms on the right-hand side of (4.2.26) represent the transient inertia-gravity waves excited by the non-zero divergence tendency of the ageostrophic wind field associated with the initial perturbation PV anomaly. The third term represents the linear, geostrophically balanced, steady-state equilibrium (which is advected downstream) that the initial ageostrophic potential vorticity state asymptotically approaches. The coefficients, A and B, in (4.2.26) are determined by specifying appropriate initial conditions. For example, if we are interested in the subsequent adjustment to geostrophic equilibrium arising from an ageostrophic initial state characterized by an unbalanced zonal wind anomaly, we can take our initial conditions to be

$$u_i'(\boldsymbol{r}) = u_{\mathrm{jet}}(\boldsymbol{r}) = u_{\mathrm{jo}}(x^2/a^2 + y^2/b^2 + 1)^{-3/2} e^{-z^2/d^2}, \qquad (4.2.28)$$

$$v_i'(\boldsymbol{r}) = p_i'(\boldsymbol{r}) = \theta_i'(\boldsymbol{r}) = 0,$$

$$q_{\mathrm{pvi}}'(\boldsymbol{r}) = -\frac{\partial u_{\mathrm{jet}}}{\partial y}(\boldsymbol{r}). \qquad (4.2.29)$$

Figure 4.10 shows the response in the wind fields predicted by the above linear theory during the first 36 h after the unbalanced zonal wind anomaly given by (4.2.28) is introduced into a quiescent ($U = V = 0$), vertically unbounded, stably stratified, Boussinesq atmosphere. The Rossby radius of deformation is about 2500 km, based on (4.2.12) and $L_z = 2d$. This value is much larger than typical tropospheric values, due to the large L_z value used in this unbounded atmosphere. Since the initial disturbance has a horizontal scale ($a = b = 500$ km) much smaller than the internal Rossby radius of deformation ($L_R = 2500$ km), we anticipate that the mass fields (p', θ') will adjust to the initial perturbation u_{jet} in the wind field (Blumen 1972). By $t = 3$ h (Fig. 4.10a), a zonally localized, northwest–southeast oriented jet streak exists at $z = -6$ km in the vicinity of the initial zonal wind anomaly and has a maximum magnitude of $10.5 \mathrm{m\,s^{-1}}$. Two isolated cells in w' characterize ascent/descent in the jet streak exit/entrance region and are flanked by radially propagating inertia-gravity waves to the east and west. These transient inertia-gravity waves are $180°$ out of phase. These wave modes may maintain their magnitude and propagate a relatively long distance away from the wave source region or perhaps even develop into a solitary wave in the presence of balanced nonlinear advection and wave dispersion effects, if there exists a wave duct in the atmosphere.

Other perturbation fields, such as p', θ', and u', at $t = 12$ h for the case of Fig. 4.10 are shown in Figs. 4.11a–c. A localized, positive wind anomaly is centered at the origin, which is supported by the evolving high–low pressure perturbation (Fig. 4.11a). The north–south wind perturbation v' is due to the meridional acceleration generated by the Coriolis force acting on the initial, unbalanced zonal wind anomaly. At $t = 12$ h (Fig. 4.10b), the jet streak is oriented in a southwest–northeast direction, and the total

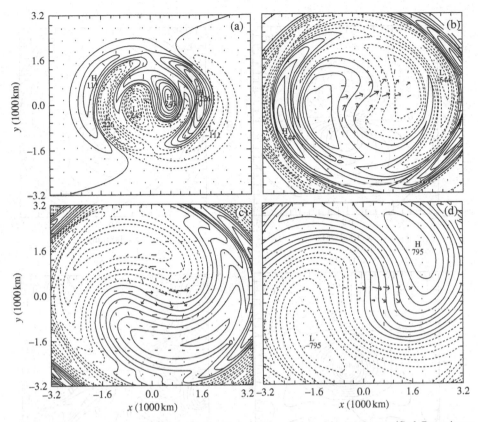

Fig. 4.10 Linear response of a quiescent, rotating, unbounded, continuously stratified, Boussinesq atmosphere to an impulsive addition of zonally localized momentum from an unbalanced jet given by (4.2.28). The perturbation horizontal wind vectors at $z = 0$ are superimposed on vertical motion w' (positive solid curves; negative dotted curves) at $z = -6$ km and at $t =$: (a) 3 h, (b) 12 h, (c) 24 h, and (d) 36 h. The jet and flow parameters are $N = 0.01$ s^{-1}, $f = 10^{-4}$ s^{-1}, $\rho_0 = 1$ kg m^{-3}, $\theta_0 = 273$ K, $u_{jo} = 20$ m s^{-1}, $a = b = 500$ km, and $d = 12.5$ km. (Adapted after Weglarz and Lin 1997.)

perturbation wind has a maximum magnitude of 9.76 m s^{-1}. Confluence/diffluence exists in the jet streak entrance/exit region. The anticyclonic rotation of the total perturbation vector wind v' from $t = 3$ h is due to the inertial oscillation, which has a period of $2\pi/f \sim 18$ h, experienced by the meridional wind perturbation v'. Because the group velocity associated with this component of the response to geostrophic equilibrium is nearly zero (e.g. Gill 1982), it will take several inertial periods before the fluid in the vicinity of the source region reaches balanced equilibrium. Dispersion will gradually take place as the longwave components are slowly radiated to the far field. At $t = 24$ h (Fig. 4.10c), the local jet streak is oriented more in a zonal direction, compared with its orientation at $t = 3$ h (Fig. 4.10a). This indicates that the amplitude of inertial oscillation in v' at the origin has been significantly reduced. Rather

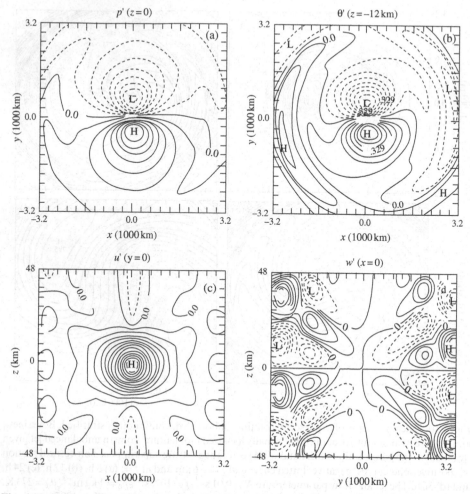

Fig. 4.11 Same as Fig. 4.10 except that (a) p' on $z = 0$ km, (b) θ' on $z = -12$ km, (c) u' on the east–west vertical cross section of $y = 0$, and (d) w' on the vertical cross section of $x = 0$. Solid and dashed contours denote positive and negative values, respectively. The time shown is $t = 12$ h. (Adapted after Weglarz and Lin 1997.)

significant cyclonic and anticyclonic circulations flank the core of the isolated jet streak at this time. By $t = 36$ h (Fig. 4.10d), v' indicates an isolated, primarily zonally oriented meso-α/synoptic scale jet streak of 10.8 m s^{-1}, whose length and width are approximately 1920 and 640 km, respectively. Also evident in the w' field at this time is the fact that longwave transient inertia-gravity waves continue to propagate out from the source region and slowly disperse into the surrounding fluid, allowing the source region to asymptotically approach geostrophic equilibrium.

Figures 4.11a–b indicate that the mass field quickly adjusts to the initial wind field perturbation, leaving a couplet of low-high perturbation pressure and cold (warm) air

north (south) of the origin at all levels below the level of maximum zonal perturbation wind. The θ' fields are reversed at all levels above $z=0$. Figure 4.11c shows
the east–west vertical cross section of u' through $y=0$ at $t=12$ h. At this time and
location, most of the significant, transient inertia-gravity wave activity has virtually
ceased, yielding a well-defined, vertically isolated zonal wind maximum. Figure 4.11d
shows the y–z cross section of w' at $x=0$ and $t=12$ h. Above and below the level
of the initial PV anomaly ($z=0$), transient, vertically propagating inertia-gravity
waves are clearly present and are shown dispersing ageostrophic wave energy into
the surrounding atmosphere as the fluid approaches a new balanced equilibrium in
the vicinity of $z=0$.

b. Diagnosis for unbalanced flow

Practically, one would be interested in how to detect the unbalanced states discussed in
Subsection 4.2.3a in the real atmosphere, in order to help better predict those regions
that have the greatest potential for generating inertia-gravity waves. It has been
hypothesized that the development of unbalanced motions is strongest at, and just
downwind of, the jet streak core (Van Tuyl and Young 1982). In the study of
geostrophic adjustment, various kinds of balances, such as geostrophic balance,
gradient wind balance, cyclostrophic balance, and the general state of the atmosphere
for which the nonlinear balance equation applies are all referred to as the balanced
states. Note that an atmosphere that obeys the quasi- or semi-geostrophic system of
equations can also be considered as being balanced even though parcel accelerations
do exist in these systems because the mass and momentum fields remains in a state of
balance with one another (Charney 1955).

It has been shown that as a jet streak approaches an upper-level ridge, an unbalanced state tends to develop, resulting in a significant increase in the divergence of the
upper-level winds (House 1961). This process may be most easily diagnosed with the
divergence tendency equation in isobaric coordinates (e.g., Haltiner and Williams 1980),

$$\frac{D\delta}{Dt} = -\delta^2 + f\zeta - \beta u - \nabla\omega \cdot \frac{\partial V}{\partial p} + 2J(u,v) - \nabla^2\phi, \qquad (4.2.30)$$

where δ is the horizontal divergence, $D\delta/Dt$ represents the divergence tendency
following a fluid parcel, ζ the vertical component of the relative vorticity, β the
latitudinal variation of the Coriolis parameter, ω the vertical motion in isobaric
coordinates, V the total wind velocity, $J(u,v)$ the Jacobian of the horizontal winds,
defined as $[(\partial u/\partial x)(\partial v/\partial y) - (\partial u/\partial y)(\partial v/\partial x)]$, and $\nabla^2\phi$ is the Laplacian of the geopotential height field. In most synoptic-scale situations, scale analysis of (4.2.30) shows
that the divergence tendency equation can be well approximated by the *nonlinear
balance equation*,

$$\nabla^2\phi - 2J(u,v) - f\zeta + \beta u = 0, \qquad (4.2.31)$$

which neglects the divergence tendency term and is composed of only the four largest terms. It has been shown that both the Laplacian and Jacobian terms in (4.2.31) are large and of the same sign in the diffluent exit region of the jet streak as it approaches the ridge axis in the geopotential height field, thereby invalidating the assumption of nonlinear balance (Kaplan and Paine 1977). The resulting imbalance among the four terms in (4.2.31) subsequently forces a large increase in the divergence tendency $D\delta/Dt$ for parcels exiting the jet streak and moving toward the ridge axis. If the advection of the ageostrophic wind is neglected (i.e., with the geostrophic momentum approximation; see Chapter 10), we have

$$V_a = \frac{1}{f}k \times \left[\frac{\partial V_g}{\partial t} + (V \cdot \nabla)V_g\right]. \tag{4.2.32}$$

In other words, we have assumed

$$\frac{DV}{Dt} = \frac{DV_g}{Dt} + \frac{DV_a}{Dt} \sim \frac{DV_g}{Dt} \tag{4.2.33}$$

in the equation of motion,

$$\frac{DV}{Dt} = fV_a \times k. \tag{4.2.34}$$

Equation (4.2.33) is equivalent to assuming that the *Lagrangian Rossby number* is much smaller than 1, i.e.

$$R_o = \frac{|DV/Dt|}{f|V|} \ll 1. \tag{4.2.35}$$

Substituting (4.2.34) into (4.2.35) leads to

$$R_o = \frac{|DV/Dt|}{f|V|} = \frac{|fV_a^\perp|}{f|V|} = \frac{|V_a^\perp|}{|V|} \ll 1, \tag{4.2.36}$$

where $V_a^\perp$ is the component of the ageostrophic wind normal to the flow. Thus, if R_o estimated by the last part of (4.2.36), i.e. $|V_a^\perp|/|V|$, is not much less than 1, then the flow is not in semi-geostrophic balance. In this case, inertia-gravity waves may be expected to be excited in regions of diagnosed imbalance.

Recent numerical simulation studies indicate that diabatic forcing plays an important role in unifying geostrophic adjustment and other types of atmospheric instability (e.g., shear instability) in a manner, which is not yet fully understood, that results in gravity wave genesis (e.g., Kaplan *et al.* 1997; Jewett *et al.* 2003). In addition, as noted in the detailed gravity wave climatology (e.g., Koppel *et al.* 2000), these waves typically occur in the presence of convection and just downstream from a major north–south oriented mountain range.

4.2.4 Nonlinear interactions

The concept of wave–wave interaction can be understood by considering the super-position of two plane waves of equal amplitude with wavenumbers k_1 and k_2, which correspond to frequencies ω_1 and ω_2, as given by the dispersion relation. The product of these two waves is

$$\varphi(x,t) = A_1 e^{i(k_1 x - \omega_1 t)} A_2 e^{i(k_2 x - \omega_2 t)} = (A_1 A_2) e^{i[(k_1+k_2)x - (\omega_1+\omega_2)t]}, \qquad (4.2.37)$$

where the multiplying coefficient $A_1 A_2$ is assumed to be small when the interaction is weak. In most cases, the above terms can be regarded as weak forcing for a new wave of wavenumber $k_1 + k_2$ and frequency $\omega_1 + \omega_2$, and the subsequent development of this new wave can be obtained from the governing equations. However, when the frequency $\omega_1 + \omega_2$ of the new wave equals the natural frequency ω for a wave with wavenumber $k_1 + k_2$, then wave resonance occurs and the new wave grows linearly with time (e.g., Gill 1982). This happens at least while the amplitude of the new wave remains small compared with the amplitude of the original waves. The time required for the new wave to reach amplitude comparable to the original waves is called the interaction time, which is inversely proportional to the amplitude of the original waves. Thus, the condition for wave resonance is

$$\omega = \omega_1 + \omega_2. \qquad (4.2.38)$$

A number of nonlinear interaction mechanisms have been proposed to explain the generation of mesoscale gravity waves, which include (a) nonlinear excitation of Kelvin–Helmholtz waves (Fritts 1982), and (b) nonlinear internal gravity waves generated by subharmonic excitation (Davis and Peltier 1979). Subharmonic excitation means that new waves are excited by existing waves with frequencies of ω/n, where n is an integer and ω is the dominant frequency.

4.3 Wave maintenance mechanisms

Occasionally, remarkably large-amplitude mesoscale waves have been recorded in regions where continuous forcing is absent or that are far removed from the wave genesis region. Under these situations, whatever the source of wave energy may be, long-lived, large-amplitude mesoscale waves can only exist if some mechanism exists either to trap the wave energy, to balance wave dispersion, or to supply energy, thus compensating the energy loss due to dissipation during wave propagation. Otherwise, these waves would lose their energy through horizontal or vertical radiation and/or dispersion, often before traveling the distance of one single horizontal wavelength. As mentioned earlier in this chapter, in this section we will focus on long-lived, large-amplitude mesoscale wave disturbances characterized by either an isolated wave of elevation (or depression) or localized wave packets with periods of 1–4 h, horizontal wavelengths of 50–500 km, and surface pressure perturbation amplitudes of 0.2–7.0 hPa.

The major mechanisms that have been proposed to explain the longevity of these long-lived, large-amplitude mesoscale waves are (a) linear wave ducting, (b) solitary waves, and (c) wave-CISK (Conditional Instability of the Second Kind). These mechanisms will be discussed next.

4.3.1 Linear wave ducting mechanism

Based on a two-dimensional, adiabatic, inviscid, nonrotating, Boussinesq flow, it has been proposed that linear mesoscale waves can be ducted in the lower atmosphere near the surface, under certain criteria (Lindzen and Tung 1976). The linear equation governing small-amplitude vertical velocities is the Taylor–Goldstein equation, (4.2.7). Applying the method of normal modes, with $w'(x, z, t) = \hat{w}(z) \exp[i(kx - \omega t)]$, to (4.2.7) yields the vertical structure equation

$$\frac{\partial^2 \hat{w}}{\partial z^2} + m^2(z)\hat{w} = 0, \tag{4.3.1}$$

where

$$m^2(z) = \frac{N^2}{(U - c)^2} - \frac{U_{zz}}{U - c} - k^2, \quad c = \frac{\omega}{k}. \tag{4.3.2}$$

In (4.3.2), m is the vertical wave number and c is the horizontal phase speed. A schematic diagram of a typical environment conducive to long-lived mesoscale waves is shown in Fig. 4.12a, which includes a cool dry layer near the surface, a warm moist stable layer, a convectively mixed layer, and an inversion with a critical level embedded within the convectively mixed layer. Figure 4.12a can be idealized as a three-layer atmosphere (Fig. 4.12b) in which the basic wind takes on constant values of $-U$ and U in layers 1 and 3, respectively, and varies linearly with height in layer 2. The shear layer is unstable (i.e., $R_i < 1/4$). The Richardson number is defined here as $R_i \equiv N_2^2/U_z^2 = N_2^2/\alpha^2$, where $\alpha = U_z$. The reflectivity of the unstable shear layer will be calculated to help determine the effectiveness of wave ducting in the lower layer. The Taylor–Goldstein equation (4.3.1) in the wavenumber or Fourier space applied to layer 2 can be written as

$$\frac{\partial^2 \hat{w}}{\partial z^2} + \left[\frac{R_i}{(z - z_c)^2} - k^2\right]\hat{w} = 0, \tag{4.3.3}$$

where $z_c = c/\alpha$ is the height of the critical level. Similarly, the vertical structure equations for layers 1 and 3 are

$$\frac{\partial^2 \hat{w}}{\partial z^2} + m_1^2 \hat{w} = 0, \quad m_1^2 = \frac{N_1^2}{(c + U)^2} - k^2 \quad \text{(layer 1)} \tag{4.3.4}$$

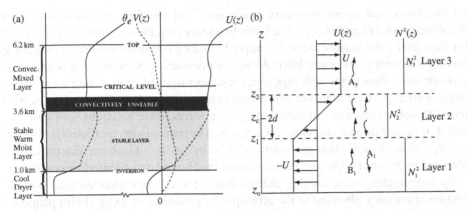

Fig. 4.12 (a) Schematic diagram of a typical environment conducive to long-lived mesoscale waves (adapted after Lindzen and Tung 1976). U and V are the mean wind speeds in x and y directions, respectively, and θ_e is the equivalent potential temperature; (b) Idealized three-layer atmosphere of (a). Displayed are the vertical profiles $U(z)$ of the basic wind and the Brunt–Vaisala frequency squared. The depth of the unstable sheared layer is $2d$. Arrows indicate the directions of wave energy propagation associated with different terms in (4.3.8).

and

$$\frac{\partial^2 \hat{w}}{\partial z^2} + m_3^2 \hat{w} = 0, \quad m_3^2 = \frac{N_3^2}{(c-U)^2} - k^2 \qquad \text{(layer 3)}. \qquad (4.3.5)$$

The kinematic and dynamic conditions require w' and p' to be continuous across the interfaces, respectively. The dynamic interface condition is equivalent to integrating the governing equation (4.3.1) across the interface from just below to just above with respect to z. This yields the interface conditions

$$\hat{w}_1 = \hat{w}_2, \quad \text{and} \quad \frac{\partial \hat{w}_1}{\partial z} = \frac{\partial \hat{w}_2}{\partial z} - \frac{\hat{w}_2}{z_1 - z_c}, \quad \text{at} \quad z = z_1, \qquad (4.3.6)$$

$$\hat{w}_3 = \hat{w}_2, \quad \text{and} \quad \frac{\partial \hat{w}_3}{\partial z} = \frac{\partial \hat{w}_2}{\partial z} - \frac{\hat{w}_2}{z_2 - z_c}, \quad \text{at} \quad z = z_2. \qquad (4.3.7)$$

The solutions in these three different layers are

$$\begin{aligned}
\hat{w}_1(z) &= A_1 e^{im_1(z-z_1)} + B_1 e^{-im_1(z-z_1)}, & 0 &\le z < z_1, \\
\hat{w}_2(z) &= A_2 (z-z_c)^{1/2+\mu} + B_2 (z-z_c)^{1/2-\mu}, & z_1 &\le z < z_2 \qquad (4.3.8) \\
\hat{w}_3(z) &= A_3 e^{im_3(z-z_2)} + B_3 e^{-im_3(z-z_2)}, & z_2 &\le z
\end{aligned}$$

where $\mu = (1/4 - R_i)^{1/2}$. There exists a branch point in the solution at $z = z_c$. We choose the branch $(z - z_c) = |z - z_c|$ for $z > z_c$ and $(z - z_c) = |z - z_c| e^{-i\pi}$ for $z < z_c$ from causality (Booker and Bretherton 1967). The six coefficients in (4.3.8) are then determined by the lower and upper boundary conditions, and the four interface conditions described by (4.3.6) and (4.3.7). The lower boundary condition requires $\hat{w} = 0$ at $z = 0$ for flow over a flat surface, while the upper boundary condition requires $B_3 = 0$ if the forcing is located in the lower layer. As will be discussed in Section 4.4, if both m and z are positive, then terms with $\exp(+imz)$ represent upward propagating waves while terms with $\exp(-imz)$ represent downward propagating waves. Thus, solutions with terms B_1 and A_3 represent upward propagating waves, while solutions with terms B_3 and A_1 represent downward propagating waves. The transmission coefficient is given by $|A_3/B_1|$, while the reflection coefficient is given by $|A_1/B_1|$. These coefficients can be obtained by solving (4.3.8) with the aforementioned boundary and interface conditions.

By investigating wave ducting characteristics for different characteristic vertical profiles commonly observed in the atmosphere, Lindzen and Tung (1976) proposed that the following four conditions are necessary for wave ducting: (1) a stable layer in which the wave can propagate (i.e., the *wave duct*) is adjacent to the ground; (2) the stable lower layer is capped by an unstable sheared layer with $R_i < 1/4$; (3) there is a critical level embedded in the unstable layer; and (4) the depth of the wave duct should be $(0.25 + n/2)\lambda$, where λ is the vertical wavelength and $n = 0,1,2\ldots$ Note that criterion (4) is occasionally misquoted as having a layer depth larger than $\lambda/4$. It was also found that the critical level embedded in the unstable layer acts as an almost perfect reflector when μ is near 0.4 or $R_i = 0.09$. For μ larger than 0.4 ($R_i < 0.09$), waves are "overreflected" from the unstable sheared layer, which means the waves are able to extract energy from the mean flow through the critical level. In theory, these waves should last indefinitely in an inviscid fluid. In reality, however, the wave duct conditions in the environment may not be satisfied along the path of the wave propagation, and the wave would be dissipated or dispersed when it moves into a region where it is no longer able to extract energy from the mean flow.

Characteristics of wave absorption, transmission, and overreflection are demonstrated in Fig. 4.13 by considering a three-layer flow, similar to that in Fig. 4.12b, over a small-amplitude mountain by varying the Richardson number. Some important flow features can be found in Fig. 4.13: (a) wave energy is absorbed at the critical level ($\tilde{z}_c = 0.605 + \sqrt{R_i}/2\pi$) for $R_i > 0.25$ (Figs. 4.13a–c), whereas it can be transmitted to the layer above the critical level for $R_i < 0.25$ (Figs. 4.13d–f); (b) the amplitude of the low-level disturbance increases as R_i decreases from 10 to 0.11, but decreases as R_i further decreases from 0.11 to 0.01; and (c) upward-propagating waves are predominant in the lowest layer when $R_i > 0.11$, whereas downward-propagating (reflected) waves dominate the lowest layer when $R_i < 0.11$, which can be seen from the phase reversal of the vertical velocity. There exist two wave modes of about equal strength when $R_i = 0.11$, as evidenced in the upright vertical tilt in constant phase lines (Fig. 4.13e). Note that the flow regime boundary values indicated in Fig. 4.13 are just for reference only. More accurate values may be obtained with more sophisticated, nonlinear numerical models. It has been proposed that mesoscale waves generated by geostrophic adjustment associated with a jet streak impinging on an upper-tropospheric ridge in the geopotential

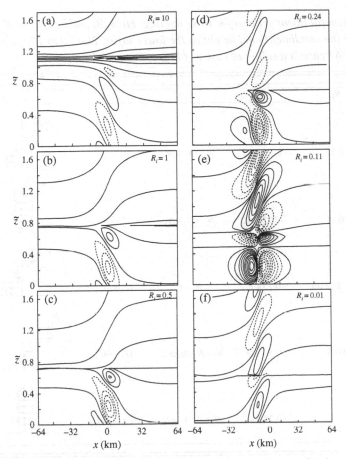

Fig. 4.13 Characteristics of wave absorption, transmission, and overreflection. The vertical velocity field for a two-dimensional, steady state, three-layer flow, similar to that in Fig. 4.12b, over a small-amplitude mountain with (a) $R_i = 10$, (b) 1.0, (c) 0.5, (d) 0.24, (e) 0.11, and (f) 0.01. The contour interval is 0.05 m s^{-1} (positive solid and negative dashed). The nondimensionalized $\tilde{z}_1$ is 0.605 and $\tilde{z}_c = \tilde{z}_1 + \sqrt{R_i}/2\pi$. (Adapted after Wang and Lin 1999.)

height field may maintain their coherent structure and last for a considerably long time when they exist in this type of ducting mechanism (Uccellini and Koch 1987).

In some cases (e.g. Ralph *et al.* 1993), some of the Lindzen and Tung (1976) ducting criteria do not seem to be satisfied. This may indicate that wave ducting criteria may be different for different basic state vertical profiles. In fact, the original linear wave ducting theory proposed by Lindzen and Tung (1976) has been generalized for a variety of different basic wind and Brunt–Vaisala frequency profiles, as shown in Table 4.1 (Wang and Lin 1999). Based on the table, Lindzen and Tung's (1976) case is a special case (Case 3) of those listed in Table 4.1. It was also found that linear wave ducting might occur over a relatively wide range of R_i in the presence of a critical level.

Table 4.1 *General linear criteria for wave ducting. Here R_i and $\tilde{z}_1$ are the Richardson number and the nondimensional height of the lower layer. Note that Lindzen and Tung's (1976) case is a subset of case 3. (From Wang and Lin 1999.)*

Case	R_i	$\tilde{z}_1$	$N(z)$	$U(z)$	Profile
1	≈ 0.11	$\approx 0.125 + n/2$	$N_3 = N_2 = N_1$	$U_3 = -U_1$	
2	< 0.11	$\approx 0.25 + n/2$	$N_3 = N_2 = N_1$	$U_3 \approx 0$	
3	$0.01 < R_i < 100$	$\approx 0.25 + n/2$	$N_2 \approx 0$	$U_3 = -U_1$	
4	< 0.03	$0.5 + n/2$	$N_2 = N_1, N_3 \gg N_1$	$U_3 = -U_1$	
5	∞	$\approx 0.25 + n/2$	$N_2 \approx 0, N_3 = N_1$	$U_3 = U_2 = U_1$	

For example, it is not necessary to require $R_i < 0.25$. Furthermore, it was found that the wave-ducting mechanism is applicable even in a nonlinear flow regime, although the intrinsic ducted waves may be strengthened over time, and that new-ducted wave modes may be generated when the lower layer depth in Fig. 4.12b is small.

The above wave ducting mechanism has been applied to explain the longevity of the mesoscale waves which occurred during the CCOPE field program. These waves were found to be ducted by a lower-tropospheric inversion or stable layer in the presence of a critical level (Koch and Dorian 1988). The wave ducting mechanism was found to play an important role in a gravity wave event on Mallorca (Balearic Islands, off the Mediterranean coast of Spain) in the summer of 1990. The sounding shown in Fig. 4.14a is composed by a shear layer in an unstable sheared layer and a lower stable layer (Monserrat and Thorpe 1996). A wave duct is found to exist in the lower shear layer adjacent to the ground, which trapped a neutral wave mode with a

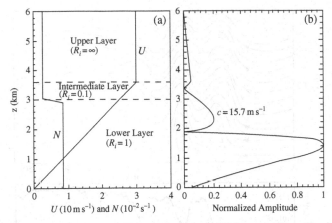

Fig. 4.14 (a) The wind (U) and Brunt–Vaisala frequency (N) for a profile with wind shear in the lower layer used to study wave duct problem, and (b) the modulus of the normalized vertical velocity for a neutral mode in the lower layer for profile in (a). The wavenumber is $k = 8 \times 10^{-4} \mathrm{m}^{-1}$. (Adapted after Monserrat and Thorpe 1996.)

horizontal phase speed of $15.7 \, \mathrm{m \, s^{-1}}$ (Fig. 4.14b). This indicates that there exists a critical level within the vertical shear layer at an elevation of about $z = 1.9 \, \mathrm{km}$. The wave energy associated with this wave mode will be mostly absorbed at the critical level.

4.3.2 Solitary wave mechanism

It has been proposed that the long-lived, large-amplitude mesoscale atmospheric waves propagated far away from their source regions are solitary waves (e.g., Lin and Goff 1988). A solitary wave is defined as a wave of finite amplitude that propagates without change of form. A solitary wave may form when there exists a balance between nonlinearity and dispersion. Physically, the nonlinear effect tends to steepen the waves (see Fig. 3.6), while the dispersive effect tends to flatten the waves by leaving shorter waves behind. In order to propagate for a considerable distance and avoid the vertical propagation of wave energy outside the waveguide through dispersion, these waves also require the right background conditions, such as a layered static stability profile and/or basic horizontal wind profile.

Solitary wave theories are well-developed disciplines in fluid dynamics. In this subsection, we will provide a brief review of these theories and link them to the maintenance of internal solitary waves, which have been observed in the atmosphere. Existing theories of solitary waves may be categorized into the following two classes: (a) the *classical (KdV) solitary waves* (Korteweg and de Vries 1895; Benjamin 1966), which are based on the shallow-water theory; and (b) the *algebraic (BDO) solitary waves* (Benjamin 1967; Davis and Acrivos 1967; Ono 1975), which are based on the

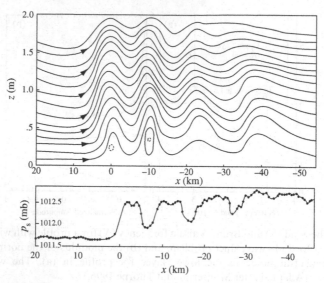

Fig. 4.15 Example of a morning glory as an undular bore observed on 4 October 1979 over the Gulf of Carpentaria, Australia: (a) cross section of the relative stream function; and (b) surface pressure. (Adapted after Clarke *et al.* 1981.)

deep-water theory. The KdV solitary waves may be represented mathematically by hyperbolic functions, such as the sech^2 function, and they are so stable that each wave component is able to preserve its own identity upon nonlinear interactions as if it were a single particle. Such individual wave component is called a *soliton*. On the other hand, the BDO solitary waves mathematically take the form of an algebraic function, such as the bell-shaped function, in the moving frame of reference of the wave itself, and are thus referred to as algebraic solitary waves, which are less stable than the KdV solitary waves. Both types of solitary waves have been observed in the atmosphere (e.g., Fig. 4.17 for KdV solitary waves and Fig. 4.15 for BDO solitary waves).

The KdV solitary waves typically have horizontal length scales on the order of 100 km, horizontal phase speeds on the order of $25-100 \text{ m s}^{-1}$, and may travel several hundred kilometers or even more than a thousand kilometers from their source region. Observations indicate that these waves may occur either as internal solitary waves of elevation or internal solitary waves of depression. On the other hand, algebraic solitary waves have horizontal length scales on the order of a few to 10 km and horizontal phase speeds on the order of 10 m s^{-1}. One well-known solitary wave of this type is the morning glory observed over the Gulf of Carpenteria in northern Australia and in the Southern Great Plains of the United States. The morning glory is a moving line of wind squalls, accompanied by a pressure jump, and often made visible by a long roll-cloud or series of such clouds (e.g., Fig. 4.15; Christie *et al.* 1978; Hasse and Smith 1984). It frequently occurs in the early morning, especially in October, in the vicinity of the Gulf of Carpenteria. The morning glory may form as an internal

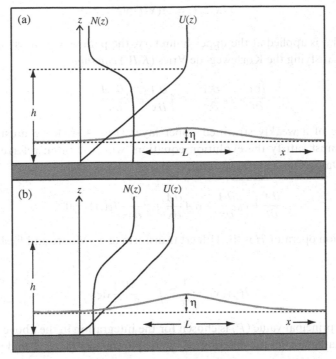

Fig. 4.16 Two possible waveguide environments in the atmosphere in which $N(z)$ is the basic state Brunt–Vaisala frequency, $U(z)$ the basic state horizontal velocity, and h is the height of the waveguide. The sketches show a typical streamline in the waveguide with η as its vertical displacement and L as its characteristic horizontal scale: (a) strong stability in the waveguide and weak stability above; (b) opposite to (a). (From Rottman and Einaudi 1993.)

undular bore, which has several smaller-scale waves (solitons) riding on a much broader elevated wave, propagating on the nocturnal and/or maritime inversion (Clarke *et al.* 1981). Its origin appears to lie frequently in the interaction of a deeply penetrating sea breeze front with a developing nocturnal inversion or occasionally from a katabatic flow.

Theories of solitary waves can be derived by considering the two-dimensional, nonlinear, nonrotating, nonhydrostatic, inviscid, adiabatic, Boussinesq fluid flow for the profiles sketched in Fig. 4.16. The governing equations may be simplified by defining two nondimensional parameters, $\delta = h/L$ and $\varepsilon = \eta/h$, where h is the depth of the wave guide, L is the characteristic horizontal scale of the wave, and η is the maximum vertical displacement of the wave. In order to obtain a balance between nonlinearity and linear dispersion, a particular relation must exist between ε and δ, and an amplitude function $A(x,t)$ must satisfy a particular evolution equation. For the first order of approximation in ε, the vertical displacement $\eta(x,z,t)$ can be expressed in terms of $A(x,t)$ and a vertical structure function $\phi(z)$ as

$$\eta(x, z, t) = A(x, t)\phi(z). \tag{4.3.9}$$

If a rigid lid is applied at the upper boundary, the parameter ε must satisfy $\varepsilon = \delta^2$ with $A(x,t)$ satisfying the Korteweg–de Vries (*KdV*) *equation*,

$$\frac{\partial A}{\partial t} + c_0 \frac{\partial A}{\partial x} + \alpha A \frac{\partial A}{\partial x} + \beta \frac{\partial^3 A}{\partial x^3} = 0. \tag{4.3.10}$$

For the case of a weakly stratified upper layer (Fig. 4.16a), we must have $\varepsilon = \delta$ and $A(x,t)$ must satisfy the evolution equation, which is often referred to as the Benjamin–Davis–Ono (BDO) equation,

$$\frac{\partial A}{\partial t} + c_0 \frac{\partial A}{\partial x} + \alpha A \frac{\partial A}{\partial x} + \gamma \frac{\partial^2}{\partial x^2} \mathcal{H}(A) = 0, \tag{4.3.11}$$

The integration operator $\mathcal{H}$ is the Hilbert transform (e.g., Moore and Feshbach 1978) defined as

$$\mathcal{H}[A(x)] = \frac{1}{\pi} \mathcal{P} \int_{-\infty}^{\infty} \frac{A(\xi)}{\xi - x} \, d\xi. \tag{4.3.12}$$

The Cauchy principal value ($\mathcal{P}$) is chosen for the integration in the above expression. The wave characteristics are determined by the parameters c_0, α, β, and γ in (4.3.10) and (4.3.11).

The third terms of both the KdV and BDO equations, (4.3.10) and (4.3.11) respectively, represent the effects of nonlinearity and the fourth terms represent the effects of dispersion. The dispersion relation for the linearized KdV and BDO equations are,

$$\omega = c_0 k - \beta k^3 \quad \text{(KdV)}, \tag{4.3.13}$$

$$\omega = c_0 k - \gamma k|k| \quad \text{(BDO)}. \tag{4.3.14}$$

The above equations indicate that short waves travel at slower speeds than longer waves in both fluid systems. In other words, short waves are left behind the long waves and the overall effect of the wave dispersion is analogous to an overall 'flattening' of the waveform. On the other hand, nonlinearity tends to steepen the wave front (e.g., Fig. 3.6), which may counteract the effects of dispersion under certain conditions, thereby allowing steady state solitary wave solutions of (4.3.10) and (4.3.11) to exist. The exact balance between nonlinearity and dispersion is regarded as the solitary wave mechanism. The amplitude function $A(x)$ for a steadily propagating solitary wave in a reference frame moving with the phase speed $c = c_0 + \varepsilon c_1$ given in the KdV theory can be obtained (e.g., Whitham 1974),

$$A(x) = \eta \, \text{sech}^2(x/\lambda), \tag{4.3.15}$$

where

$$\lambda^2 = \frac{12\beta}{\alpha\eta} \quad \text{and} \quad c_1 = \frac{1}{3}\alpha h. \tag{4.3.16}$$

In fact, the general solution of the KdV equation for all α and β and $0 \leq \alpha \leq \beta$ can be expressed in terms of the cnoidal function and is named aptly as the cnoidal wave (Korteweg and de Vries 1895). When $\alpha/\beta \to 1$, the cnoidal function can be approximated by the sech (hyperbolic secant) function, and the wave reduces to the classical (KdV) solitary wave, (4.3.15). Physically, this means that the nonlinearity, represented by α, is balanced by the dispersion, represented by β. When $\alpha/\beta \to 0$, the cnoidal function behaves like the cosine function. In this limit, the dispersive effect dominates and the nonlinear effect is negligible. When $\alpha \gg \beta$, the nonlinearity dominates the wave behavior, which leads to wave steepening and then eventually to wave breaking.

The amplitude function $A(x,t)$ for a steadily propagating solitary wave in a reference frame moving at the wave speed $c = c_o + \varepsilon c_1$ is given in the BDO theory as,

$$A(x) = \frac{\eta}{(x/\lambda)^2 + 1}, \tag{4.3.17}$$

where

$$|\lambda| = \frac{4\gamma}{\alpha\eta} \quad \text{and} \quad c_1 = \frac{1}{4}\alpha h, \tag{4.3.18}$$

and η is the maximum amplitude of the vertical displacement. In order to keep a positive wavelength λ, the amplitude η in (4.3.15) and (4.3.17) can be positive or negative. In other words, the background fluid determines whether the solitary wave is a wave of depression or elevation, both of which have been observed in the real atmosphere, as mentioned previously.

In order to explain the persistence of long-lived mesoscale waves that travel distances up to a few thousand kilometers, it was proposed that the KdV type of solitary wave mechanism might be responsible for the large-amplitude wave observed for the case of Fig. 4.1 (Lin and Goff 1988). The theory was improved by including critical layers embedded in regions of low Richardson number below and above the jet maximum, which typically occurs in the vicinity of the tropopause, where vertical shears are large (Rottman and Einaudi 1993). This background atmosphere supports wave trapping, which helps maintain the wave in addition to the solitary wave mechanism. In addition, regions of low Richardson number can act as a source of wave energy so that it is possible for waves to overreflect and actually gain energy from an encounter with such a layer. Figure 4.17a shows the streamline structure associated with the solitary wave of March 6, 1969, computed from KdV theory using an amplitude of 500 m propagating in a waveguide 7.6 km deep topped by a critical level. The streamline pattern matches reasonably well with the observations (Fig. 4.17b). The surface pressure tendency

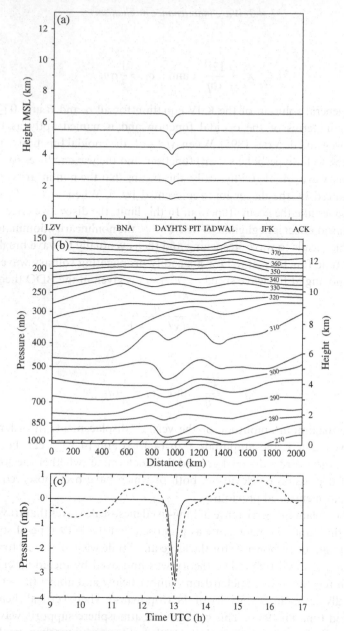

Fig. 4.17 A KdV solitary wave observed on March 6, 1969 (Lin and Goff 1988): (a) the streamlines computed from weakly nonlinear theory for a solitary wave with an amplitude of 500 m propagating in a waveguide 7.6 km deep; (b) as determined from an isentropic analysis (K) of the 1200 UTC sounding cross section; and (c) time series of the surface perturbation pressure at Pittsburgh: barograph trace (dashed line) and simulation from weakly nonlinear theory (solid line) as used in (a). The position of the cross section is indicated in Fig. 4.5 by a dashed line. The abbreviations of the sounding stations are: LZV, Little Rock, AK; BNA, Nashville, TN; DAY, Dayton, OH; HTS, Huntington, WV; PIT, Pittsburgh, PA; IAD, Washington, D.C.; WAL, Wallops Island, VA; JFK, New York City, NY; ACK, Nantucket, MA. (From Rottman and Einaudi 1993.)

computed using the KdV solitary wave theory for wave of depression also compares very well with that observed by the actual microbarographs (Fig. 4.17c).

In BDO theory, if the upper layer is weakly stratified, then the waves in the lower layer are no longer truly trapped, and some wave energy is radiated away through the upper layer (Maslowe and Redekopp 1980). For a weakly stratified upper layer, the ratio of the upper-layer buoyancy to that in the lower layer must be at least $O(\delta)$ for a solitary wave solution to exist (Rottman and Einaudi 1993). The atmosphere often does produce layers of very weak or nearly neutrally stratified air that are sufficiently deep to effectively trap solitary waves that propagate in the lower troposphere. Solitary waves can be maintained by different types of basic-state velocity profiles that provide sufficient trapping for substantial periods of time (Crook 1988). Explicit expressions for the minimum algebraic solitary wave amplitude necessary for long-lived propagation in waveguides bounded by weakly stratified shear flows can be derived (Doviak *et al.* 1991). For the weakly nonlinear KdV theory to be strictly valid, the ratio of the buoyancy frequency in the upper layer to that in the lower layer has to be infinite. This poses more of a problem in the atmosphere, because the increase in buoyancy frequency at the tropopause is typically only a factor of 3 to 4, but rarely as large as that required for it to be a good reflector.

Figures 4.18a and 4.18b show examples of algebraic (BDO) waves of elevation and depression, respectively. When the initial condition in the atmosphere matches the analytic solitary wave solutions, such as (4.3.15) and (4.3.17), then the wave will propagate as a steady state wave in a frame of reference moving with the horizontal phase speed $c = c_0 + \varepsilon c_1$. However, if the initial wave has a horizontal wavelength longer than that of the analytic solution, then it tends to break up into a number of smaller-scale waves. These smaller-scale waves then individually develop into solitary waves (i.e., solitons). During the transition period, one often observes a series of developing solitons riding over the initial long wave. This phenomenon is called the *undular bore*. Figure 4.15 shows an undular bore associated with the morning glory clouds observed on October 4, 1979 over the Gulf of Carpentaria, Australia. Note that this morning glory cloud consists of a family of four individual solitons or solitary waves and that these waves seem to be superimposed on a synoptic-scale positive pressure perturbation. Internal undular bores may also be generated by an accelerating hydraulic jump associated with a severe downslope windstorm over the lee slope of topography (e.g., Karyampudi *et al.*, 1995) or with nonclassical frontal propagation (e.g., Neiman *et al.* 2001).

4.3.3 Wave-CISK mechanism

The wave-CISK theory is based on the Conditional Instability of the Second Kind (CISK) theory. The CISK theory was originally developed to help explain tropical cyclone development (Section 9.3). The basic idea of CISK is that the latent heating released by moist convection produces a large-scale disturbance that, in turn,

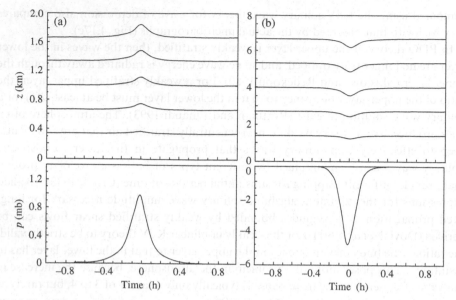

Fig. 4.18 (a) An example of an algebraic (BDO) solitary wave of elevation. Streamlines (upper panel) and surface pressure (lower panel) are shown for a steady state BDO type of solitary wave in an isothermal waveguide with constant shear when the layer above the waveguide is neutrally stratified. This wave has amplitude of 200 m and a wave speed of $27 \, \text{m s}^{-1}$. The waveguide has a depth of 2 km and a temperature of 250 K; the wind shear is $0.005 \, \text{s}^{-1}$. (b) An example of an algebraic (BDO) solitary wave of depression. Panel (b) is the same as (a) except with amplitude of 500 m and a wave speed of $70 \, \text{m s}^{-1}$. The waveguide has a depth of 10 km, a temperature of 250 K, and the wind shear is $0.005 \, \text{s}^{-1}$. (Adapted after Rottman and Einaudi 1993.)

generates low-level moisture convergence through boundary layer friction. The low-level convergence of the conditionally unstable air then produces latent heat to maintain the moist convection. This type of self-exciting convection mechanism has also been applied to help explain the coherent relationship between convection and waves associated with tropical cloud clusters and equatorial disturbances (Hayashi 1970; Lindzen 1974). The main difference between wave-CISK and conventional CISK is that the low-level convergence in wave-CISK is supported by internal gravity waves in an inviscid fluid, instead of PBL friction.

The wave-CISK theory can be understood through a simple mathematical formulation. Consider the governing equation, (4.2.1), for a two-dimensional, small-amplitude, hydrostatic, inviscid, non rotating, Boussinesq, quiescent fluid,

$$\frac{\partial^4 w'}{\partial t^2 \partial z^2} + N^2 \frac{\partial^2 w'}{\partial x^2} = \frac{\partial^2 \dot{Q}}{\partial x^2}, \qquad (4.3.19)$$

where $\dot{Q} = (g/c_p T_o)\dot{q}$ represents the thermal forcing that results from latent heating. The thermal forcing will eventually be described in terms of the wave fields themselves.

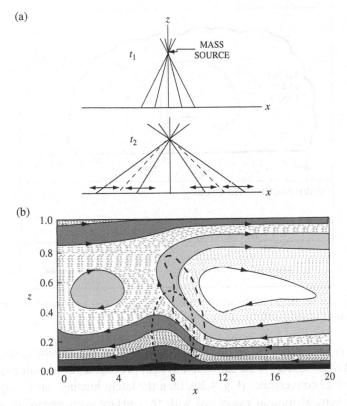

Fig. 4.19 (a) Constant phase lines of waves resulting from an impulsive mass source aloft. The two panels show how the waves evolve with time ($t_1 < t_2$). The arrows in the bottom panel show the direction of the horizontal wind at the surface. The solid (dashed) lines denote constant phases of divergence (convergence); (b) Streamlines for the rightward moving disturbance at $t = 100$, associated with the Oklahoma squall line analyzed by Ogura and Liou (1980) (Fig. 6.12), as simulated by a wave-CISK model. The streamfunction is calculated relative to the disturbance, which is assumed to move at a nondimensional speed of 0.1. The dashed and dotted lines, respectively, show the 30% contours of updraft and downdraft mass flux (corresponding to convective latent heating and evaporative cooling, respectively). (Adapted after Raymond 1983; 1984.)

The response of an unbounded stratified fluid with radiation boundary conditions to a point source, $\dot{Q} = \delta(x)\delta(z)\delta(t)$, can be described by the perturbation horizontal velocity as (Raymond 1983)

$$u'(x,t) = \frac{\cos(zt/x)}{2\pi x}. \qquad (4.3.20)$$

The evolution of constant phase lines of gravity waves emanating from an upper-level impulsive mass source, which is taken to represent the latent heat released in moist convection, is sketched in Fig. 4.19a. The dashed lines at $t = t_2$ denote constant phase

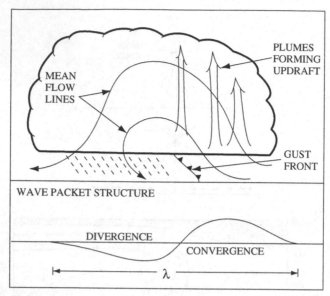

Fig. 4.20 Conceptual model of the wave-CISK mechanism. The horizontal wavelength of the wave-CISK mode is denoted by λ. (From Raymond 1975.)

lines of convergence produced by the inviscid wave motion, which propagate away from the center of mass source. If the convective system propagates at the same speed as the region of surface convergence (Fig. 4.20), then the latent heating can be supported by the upward vertical motion associated with the surface convergence in a coherent manner, thereby supporting the fundamental hypothesis of the wave-CISK mechanism. Figure 4.19b shows streamlines for the rightward moving disturbance associated with an observed squall line (Fig. 6.12), as simulated by a wave-CISK model. The simulated flow field compares reasonably well with the diagnosed streamlines. In particular, the level of zero relative wind (critical level) is well predicted, indicating an accurately predicted propagation speed by the wave-CISK model. In both the simulation and the observations, the downdraft is fed primarily by mid-level air entering the storm from the front (right), with only a slight contribution from inflow coming from the rear (left).

Note that the region of convective latent heating, which is approximately related to the cumulus mass flux as shown in Fig. 4.19b, is not completely in phase with the region of upward vertical motion. In conventional wave-CISK models, even for nonsheared mean flows, the lines of maximum w' normally slope with height (Lilly 1979). Diabatic heating proportional to w' or $\partial w'/\partial z$ at some given level, such as the level of free convection, is distributed uniformly in the heating layer as is normally assumed in wave-CISK heating profiles. Thus, the wave-CISK heating will not slope with height. This will cause the diabatic heating and vertical motion to be out of phase with one another. This particular property may therefore serve as a criterion for examining whether or not the wave-CISK mechanism is at work in mesoscale convection.

As also depicted in Fig. 4.20, the wave packet consists of one convergent and one divergent region with a dominant wavelength λ comparable to the scale of the squall line or mesoscale convective system. Convective plumes develop in the convergent region, passing into the divergent region as they decay and produce precipitation. When the squall line or the mesoscale convective system propagates at the same speed as the region of surface convergence or the forced gravity waves, then the latent heating can be supported by the upward vertical motion produced by the surface convergence in a coherent manner.

Based on the thermodynamic equation, (2.2.18), in a quiescent fluid, the release of latent heat (q') may result in two effects: (a) it increases θ' in the heating (cloud) layer; and (b) the heating-induced updraft may modify the w' term of (2.2.18) with a moist buoyancy frequency in the cloud region. In a wave-CISK model, the cloud is viewed as a plume or a hot tower, and the latent heating is not dependent on local vertical velocity. The latent heating is often represented as proportional to the vertical velocity at the level of free convection (z_b),

$$q'(x, z, t) = \alpha w'(x, z_b, t)g(z), \tag{4.3.21}$$

where α is a parameter to measure the degree of feedback between moisture convergence and condensation, and $g(z)$ is a vertical distribution function of heating. When $\alpha = 1$, converged moisture from the boundary layer is completely converted into condensation, while for $0 < \alpha < 1$, less water vapor is converted into condensation. The growth rates of unstable wave-CISK modes can then be found by substituting (4.3.21) into (4.3.19) and transforming (4.3.19) into the Fourier space, similar to the normal mode approach for instability problems.

Based on the fact that isolated moist convective cells generally appear to be short-lived in the absence of external forcing and that ensembles of convective cells persist much longer than isolated cells, wave-CISK may conceptually work in mesoscale convection (Emanuel 1982; Xu and Clark 1984). Some deficiencies of the wave-CISK mechanism have been raised, which include: (a) lack of scale selection due to the diabatic heating instantaneously related to the low-level moisture convergence of the large-scale wave flow; (b) sensitivity to the structure of their heating functions and the downdraft cooling and wind shear specifications; and (c) scale selection sensitive to the particular form of the phase-lagged updraft (Davies 1979) model (e.g. Xu and Clark 1984; Lim *et al.* 1990). Due to these deficiencies found in the wave-CISK mechanism and those of the CISK mechanism itself (Section 9.3; Raymond 1994), one may question the validity of this mechanism and use caution when it is applied.

4.4 Energy propagation and momentum flux

In the previous chapter, we discussed wave reflection from the interface boundary accompanying a discontinuity in stratification and the selection of the appropriate upper boundary condition. To better understand the wave reflection problem and the selection of appropriate upper and interfacial boundary conditions, we consider the vertical propagation of wave energy in a two-dimensional, steady-state, nonrotating, inviscid, adiabatic, and Boussinesq fluid system. The set of linear governing equations may be obtained from (2.2.14)–(2.2.18),

$$U\frac{\partial u'}{\partial x} + U_z w' + \frac{1}{\rho_0}\frac{\partial p'}{\partial x} = 0, \tag{4.4.1}$$

$$U\frac{\partial w'}{\partial x} + \frac{1}{\rho_0}\frac{\partial p'}{\partial z} - g\frac{\theta'}{\theta_0} = 0, \tag{4.4.2}$$

$$\frac{\partial u'}{\partial x} + \frac{\partial w'}{\partial z} = 0, \tag{4.4.3}$$

$$U\frac{\partial \theta'}{\partial x} + \frac{N^2\theta_0}{g}w' = 0. \tag{4.4.4}$$

The perturbation energy equation may be obtained from the above equations (Lin and Chun 1993),

$$\frac{\partial}{\partial x}(EU + p'u') + \frac{\partial}{\partial z}(p'w') = -\rho_0 U_z u'w', \tag{4.4.5}$$

where

$$E = \frac{1}{2}\rho_0\left[u'^2 + w'^2 + \left(\frac{g\theta'}{N\theta_0}\right)^2\right], \tag{4.4.6}$$

is the total perturbation energy. The first term on the left-hand side of (4.4.5) is the horizontal wave-energy flux through a unit vertical area, while the second term is the vertical wave-energy flux through a unit horizontal area. The term on the right-hand side of (4.4.5) is a secondary source of wave energy associated with the conversions between the kinetic energy of the basic wind shear and the perturbation wave energy. This term may be positive or negative depending upon whether the system is growing (unstable) or decaying (stable). If there exists no resonance, then the total perturbation energy approaches zero as $x \to \pm\infty$. Integration of (4.4.5) with respect to x from $-\infty$ to $+\infty$ yields

$$\frac{\partial}{\partial z}\int_{-\infty}^{\infty} p'w'dx = -\rho_0 U_z \int_{-\infty}^{\infty} u'w'dx = -U_z F, \tag{4.4.7}$$

where F is the vertical flux of horizontal momentum, which is also defined below in (4.4.9).

On the other hand, multiplying (4.4.1) by $Uu' + p'/\rho_0$ and integrating it from $x = -\infty$ to $+\infty$ yields

$$\int_{-\infty}^{\infty} p'w'\,dx = -\rho_0 U \int_{-\infty}^{\infty} u'w'\,dx. \tag{4.4.8}$$

In deriving the above equation, we have assumed $U_z \neq 0$. Taking the derivative of (4.4.8) with respect to z and comparing it with (4.4.7) leads to

$$F = \rho_0 \int_{-\infty}^{\infty} u'w'\,dx = constant, \quad \text{when} \quad U \neq 0. \tag{4.4.9}$$

This is the *Eliassen and Palm* (1960) *theorem*, which states that the vertical flux of horizontal momentum does not change with height except possibly at levels where $U = 0$ or in the layer of forcing. If the integration of (4.4.8) is taken over one horizontal wavelength, we have

$$\overline{p'w'} = -\rho_0 U \overline{u'w'}. \tag{4.4.10}$$

Thus, the vertical flux of wave energy is negatively proportional to the vertical flux of horizontal momentum if $U > 0$. For an energy source located in the lower troposphere, such as a mountain, the momentum flux is downward because the energy flux is upward.

Equations (4.4.1)–(4.4.4) may be reduced to

$$\frac{\partial^2 w'}{\partial x^2} + \frac{\partial^2 w'}{\partial z^2} + l^2(z)w' = 0, \tag{4.4.11}$$

where $l(z)$ is the Scorer parameter (Scorer 1949) and is defined as

$$l^2(z) \equiv \frac{N^2}{U^2} - \frac{U_{zz}}{U}. \tag{4.4.12}$$

For vertically propagating waves ($l^2 > k^2$), the solutions may be written as

$$w' = Ae^{i(kx+mz)} + Be^{i(kx-mz)}, \tag{4.4.13}$$

when l is constant and m is defined as

$$m^2 = l^2 - k^2. \tag{4.4.14}$$

Physically, the long wave assumption means that the basic flow has a stronger stratification or a weaker wind.

The vertical flux of perturbation wave energy may also be derived,

$$\overline{p'w'} = \frac{1}{2}\rho_0 U \frac{m}{k}\left(|A|^2 - |B|^2\right). \tag{4.4.15}$$

Therefore, term A represents internal waves in a stratified fluid with upward energy propagation. It is characterized by upstream tilted lines of constant phase, as indicated by (4.4.13), a result that has been observed in numerous studies of mountain waves

(Chapter 5). Conversely, term *B* represents downward energy propagation characterized by downstream tilted lines of constant phase. In other words, term *A* or *B* determines the direction of the energy propagation, which helps determine the appropriate radiation condition for any given problem. For flow over a mountain, term *B* is unphysical because the energy source is located at the Earth's surface. This is known as the *Sommerfeld* (1949) *radiation condition*. For an initial value problem, the boundary condition at infinity requires that the disturbance vanish there, which sometimes is referred to as the *boundedness condition*. In mesoscale numerical models, the radiation condition may be simulated by adding either a sponge layer on top of the upper boundary of the physical domain or by applying a numerical radiation condition applied in wavenumber (Fourier) space (see Chapter 13).

References

Benjamin, T. B., 1966. Internal waves of finite amplitude and permanent form. *J. Fluid Mech.*, **25**, 241–70.

Benjamin, T. B., 1967. Internal waves of permanent form in fluids of great depth. *J. Fluid Mech.*, **25**, 559–92.

Blumen, W., 1972. Geostrophic adjustment. *Rev. Geophys. Space Res.*, **10**, 485–528.

Booker, J. R. and F. P. Bretherton, 1967: The critical layer for internal gravity waves in a shear flow. *J. Fluid Mech.*, **27**, 513–559.

Bosart, L. F. and A. Seimon, 1988. A case study of an unusually intense atmospheric gravity wave. *Mon. Wea. Rev.*, **116**, 1857–86.

Cahn, A., 1945. An investigation of the free oscillations of a simple current system. *J. Meteor.*, **2**, 113–119.

Charney, J. G., 1955. The use of the primitive equations of motion in numerical weather prediction. *Tellus*, **7**, 22–6.

Christie, D. R., K. J. Muirhead, and A. L. Hales, 1978. On solitary waves in the atmosphere. *J. Atmos. Sci.*, **35**, 805–25.

Chun, H.-Y., I.-S. Song, J.-J. Baik, and Y.-J. Kim, 2004. Impact of a convectively forced gravity wave drag parameterization in NCAR CCM3. *J. Climate*, **17**, 3530–547.

Clarke, R. H., R. K. Smith, and D. G. Reid, 1981. The morning glory of the Gulf of Carpenteria: An atmospheric undular bore. *Mon. Wea. Rev.*, **109**, 1726–50.

Crook, N. A., 1988. Trapping of low-level internal gravity waves. *J. Atmos. Sci.*, **45**, 1533–41.

Davies, H. C., 1979. Phase-lagged wave-CISK. *Quart. J. Roy. Meteor. Soc.*, **105**, 325–53.

Davis, P. A. and W. R. Peltier, 1979. Some characteristics of Kelvin-Helmholtz and resonant overreflection modes of shear flow instability and their interaction through vortex pairing. *J. Atmos. Sci.*, **36**, 2394–412.

Davis, R. E. and A. Acrivos, 1967. Solitary internal waves in deep water. *J. Fluid Mech.*, **29**, 593–607.

Doviak, R. J., S. S. Chen, and D. R. Christie, 1991. A thunderstorm generated solitary wave observation compared with theory for nonlinear waves in a sheared atmosphere. *J. Atmos. Sci.*, **48**, 87–111.

Einaudi, F. and D. P. Lalas, 1973. On the growth rate of an unstable disturbance in a gravitationally stratified shear flow. *J. Atmos. Sci.*, **30**, 1707–10.

Eliassen, A. and E. Palm, 1960. On the transfer of energy in stationary mountain waves. *Geofys. Publ.*, Oslo, **22** (3), 1–23.

Emanuel, K. A., 1982. Inertial instability and mesoscale convective systems. Part II: Symmetric CISK in a baroclinic flow. *J. Atmos. Sci.*, **39**, 1080–97.

Fovell, R., D. Durran, and J. R. Holton, 1992. Numerical simulations of convectively generated stratospheric gravity waves. *J. Atmos. Sci.*, **49**, 1427–42.

Fritts, D. C., 1982. Shear excitation of atmospheric gravity waves. *J. Atmos. Sci.*, **39**, 1936–52.

Gill, A. E., 1982: *Atmosphere-Ocean Dynamics*. Academic Press.

Hasse, S. P. and R. K. Smith, 1984. Morning glory wave clouds in Oklahoma: A case study. *Mon. Wea. Rev.*, **112**, 2078–89.

Haltiner, G. J. and R. T. Williams, 1980. *Numerical Prediction and Dynamic Meteorology*. 2nd edn., John Wiley & Sons.

Hayashi, Y., 1970. A theory of large-scale equatorial waves generated by condensation heat and accelerating zonal wind. *J. Meteor. Soc. Japan*, **48**, 140–60.

House, D. C., 1961. The divergence equation as related to severe thunderstorm forecasting. *Bull. Amer. Meteor. Soc.*, **42**, 803–16.

Howard, L. N., 1961. Note on a paper by John Miles. *J. Fluid Mech.*, **10**, 509–12.

Jewett, B. F., M. K. Ramamurthy, and R. M. Rauber, 2003. Origin, evolution, and fine scale structure of the St. Valentine's Day mesoscale gravity wave observed during STORM-FEST. Part III: Gravity wave genesis and the role of evaporation. *Mon. Wea. Rev.*, **131**, 617–33.

Jin, Y., S. E. Koch, Y.-L. Lin, F. Martin Ralph, and C. Chen, 1996. Numerical simulations of an observed gravity current and gravity waves in an environment characterized by complex stratification and shear. *J. Atmos. Sci.*, **53**, 3570–88.

Kaplan, M. L. and D. A. Paine, 1977. The observed divergence of the horizontal velocity field and pressure gradient force at the mesoscale: Its implications for the parameterization of three-dimensional momentum transport in synoptic-scale numerical models. *Beitr. Phys. Atmos.*, **50**, 321–30.

Kaplan M. L., S. E. Koch, Y.-L. Lin, R. P. Weglarz, and R. A. Rozumalski, 1997. Numerical Simulations of a Gravity Wave Event over CCOPE. Part I: The Role of Geostrophic Adjustment in Mesoscale Jetlet Formation. *Mon. Wea. Rev.*, **125**, 1185–211.

Kaplan, M. L., Y.-L. Lin, D. W. Hamilton, and R. A. Rozumalski, 1998. The numerical simulation of an unbalanced jetlet and its role in the Palm Sunday 1994 tornado outbreak in Alabama and Georgia. *Mon. Wea. Rev.*, **126**, 2133–65.

Karyampudi, V. M., S. E. Koch, C. Chen, J. W. Rottman, and M. L. Kaplan, 1995. The influence of the Rocky Mountains on the 13–14 April 1986 severe weather outbreak. Part II: Evolution of a prefrontal bore and its role in triggering a squall line. *Mon. Wea. Rev.*, **123**, 1423–46.

Koch, S. E. and P. B. Dorian, 1988. A mesoscale gravity wave event observed during CCOPE. Part III: Wave environment and probable source mechanisms. *Mon. Wea. Rev.*, **116**, 2570–92.

Koppel, L. L., L. F. Bosart, and D. Keyser, 2000. A 25-yr climatology of large-amplitude hourly surface pressure changes over the conterminous United States. *Mon. Wea. Rev.*, **128**, 51–68.

Korteweg, D. J. and G. de Vries, 1895. On the change of form of long waves advancing in a rectangular canal, and on a new type of long stationary waves. *Phil. Mag.*, (**5**) **39**, 422–43.

Lalas, D. P. and F. Einaudi, 1976. On the characteristics of gravity waves generated by atmospheric shear layers. *J. Atmos. Sci.*, **33**, 1248–1259.

Ley, B. E. and W. R. Peltier, 1978. Wave generation and frontal collapse. *J. Atmos. Sci.*, **15**, 3–17.

Lilly, D. K., 1979. The dynamical structure and evolution of thunderstorms and squall lines. *Annu. Rev. Earth Planet. Sci.*, **7**, 117–61.

Lim, H., T.-K. Lim, and C.-P. Chang, 1990. Reexamination of wave-CISK theory: Existence and properties of nonlinear wave-CISK modes. *J. Atmos. Sci.*, **47**, 3078–91.

Lin, Y.-L. and H.-Y. Chun, 1993. Structures of dynamically unstable shear flows and their implications for shallow internal gravity waves. *Meteor. Atmos. Phys.*, **52**, 59–68.

Lin, Y.-L. and R. C. Goff, 1988. A study of mesoscale solitary wave in the atmosphere originating near a region of deep convection. *J. Atmos. Sci.*, **45**, 194–205.

Lindzen, R. S., 1974. Wave-CISK in the tropics. *J. Atmos. Sci.*, **31**, 156–79.

Lindzen, R. S. and K. K. Tung, 1976. Banded convective activity and ducted gravity waves. *Mon. Wea. Rev.*, **104**, 1602–17.

Maslowe, S. A. and L. G. Redekopp, 1980. Long nonlinear waves in stratified shear flows. *J. Fluid Mech.*, **101**, 321–48.

Miles, J. W., 1961. On the stability of heteorogeneous shear flows. *J. Fluid Mech.*, **10**, 496–508.

Monserrat, S. and A. J. Thorpe, 1996. Use of ducting theory in an observed case of gravity waves. *J. Atmos. Sci.*, **53**, 1724–36.

Moore, P. M. and H. Feshbach, 1978. *Methods of Theoretical Physics*. McGraw-Hill.

Neiman, P. J., F. M. Ralph, R. L. Weber, T. Uttal, L. B. Nance, and D. H. Levison, 2001. Observations of nonclassical frontal propagation and frontally forced gravity waves adjacent to steep topography. *Mon. Wea. Rev.*, **129**, 2633–59.

Ogura, Y. and M.-T. Liou, 1980. The structure of a midlatitude squall line: A case study. *J. Atmos. Sci.*, **37**, 553–67.

Ono, H., 1975. Algebraic solitary waves in stratified fluid. *J. Phys. Soc. Japan*, **39**, 1082–91.

O'Sullivan, D. and T. J. Dunkerton, 1995. Generation of inertia-gravity waves in a simulated life cycle of baroclinic instability. *J. Atmos. Sci.*, **52**, 3695–716.

Powers, J. G. and R. J. Reed, 1993. Numerical model simulation of the large-amplitude mesoscale gravity wave event of 15 December 1987 in the central United States. *Mon. Wea. Rev.*, **121**, 2285–308.

Ralph, F. M., C. Mazaudier, M. Crochet, and S. V. Venkateswaran, 1993. Doppler sodar and radar wind-profiler observations of gravity-wave activity associated with a gravity current. *Mon. Wea. Rev.*, **121**, 444–63.

Raymond, D. J., 1975. A model for predicting the movement of continuously propagating convective storms. *J. Atmos. Sci.*, **32**, 1308–17.

Raymond, D. J., 1983. Wave-CISK in mass flux form. *J. Atmos. Sci.*, **40**, 2561–72.

Raymond, D. J., 1984. A wave-CISK model of squall lines. *J. Atmos. Sci.*, **41**, 1946–58.

Raymond, D. J., 1994. Cumulus convection and the Madden-Julian oscillation of the tropical troposphere. *Physica D*, **77**, 1–22.

Rottman, J. W. and F. Einaudi, 1993. Solitary waves in the atmosphere. *J. Atmos. Sci.*, **50**, 2116–36.

Scorer, R. S., 1949. Theory of waves in the lee of mountains. *Quart. J. Roy. Meteor. Soc.*, **75**, 41–56.

Sommerfeld, A., 1949. *Partial Differential Equations in Physics*. Academic Press.

Uccellini, L. W. and S. E. Koch, 1987. The synoptic setting and possible energy sources for mesoscale wave disturbances. *Mon. Wea. Rev.*, **115**, 721–29.

Van Tuyl, A. H. and J. A. Young, 1982. Numerical simulation of nonlinear jet streak adjustment. *Mon. Wea. Rev.*, **110**, 2038–54.

Wang, T.-A. and Y.-L. Lin, 1999. Wave ducting in a stratified shear flow over a two-dimensional mountain. Part I: General linear criteria. *J. Atmos. Sci.*, **56**, 412–36.

Weglarz, R. P. and Y.-L. Lin, 1997. A linear theory for jet streak formation due to zonal momentum forcing in a stably stratified atmosphere. *J. Atmos. Sci.*, **54**, 908–32.

Whitham, G. B., 1974: *Linear and Nonlinear Waves*. John Wiley and Sons.

Xu, Q. and J. H. E. Clark, 1984. Wave CISK and mesocale convective systems. *J. Atmos. Sci.*, **41**, 2089–107.

Zhang, F. 2004. Generation of mesoscale gravity waves in upper-tropospheric jet-front systems. *J. Atmos. Sci.*, **61**, 440–57.

Problems

4.1 Derive (4.2.20) from (4.2.13)–(4.2.17).

4.2 Make the Fourier transform of $U^2(\partial^4 p'/\partial x^2 \partial z^2)$ (second term inside the square bracket) of (4.2.20).

4.3 (a) Derive (4.3.3)–(4.3.5) from (4.3.1) by assuming a three-layer atmosphere as shown in Fig. 4.12. (b) Derive the interface conditions for Lindzen and Tung's (1976) three-layer model, (4.3.6) and (4.3.7). [Hint for (b): Divide the integration across z_i into from z_i^- to z_i and from z_i to z_i^+, $i = 1, 2$.]

4.4 Show that the perturbation energy equation for a two-dimensional, uniform, time-dependent, stably stratified, inviscid, adiabatic flow is

$$\left(\frac{\partial}{\partial t} + U\frac{\partial}{\partial x}\right)E + \frac{\partial}{\partial x}(p'u') + \frac{\partial}{\partial z}(p'w') = 0,$$

where

$$E = \frac{1}{2}\rho_0\left[u'^2 + w'^2 + \left(\frac{g\theta'}{\theta_0 N}\right)^2\right].$$

4.5 Derive the vertical flux of perturbation energy (4.4.15). (Hint: Assume $w' = A\cos(kx + mz) + B\cos(kx - mz)$)

4.6 Derive the dispersion relation, (4.3.13), for the corresponding linear equation from the KdV equation. (a) Find the phase speed of the wave and show that shorter waves travel slower than longer waves; (b) Find the group velocity of the wave. What can you conclude about the propagation of an individual wave and a wave packet?

4.7 (a) Verify that $u(x,t) = -2k^2 \text{sech}^2[k(x - 4k^2t)]$ satisfies the KdV equation in the form $u_t - 6uu_x + u_{xxx} = 0$.

  (b) If $u(x,t) = \exp[2k(x - ct)]$ is a solution of the linearized KdV equation, $u_t + u_{xxx} = 0$. Determine c.

5

Orographically forced flows

Many well-known weather phenomena are directly related to flow over orography, such as mountain waves, lee waves and clouds, rotors and rotor clouds, severe downslope windstorms, lee vortices, lee cyclogenesis, frontal distortion across mountains, cold-air damming, track deflection of midlatitude and tropical cyclones, coastally trapped disturbances, orographically induced rain and flash flooding, and orographically influenced storm tracks. A majority of these phenomena are mesoscale and are induced by stably stratified flow over orography. Thus, understanding the dynamics associated with stably stratified flow over a mesoscale mountain is essential in improving the prediction of the above mentioned phenomena. In addition, understanding the dynamics of orographically forced flow will also help on different aspects of meteorology, such as turbulence which affects aviation safety, wind-damage risk assessment, pollution dispersion in complex terrain, and subgrid-scale parameterization of mountain wave drag in general circulation models.

5.1 Flows over two-dimensional sinusoidal mountains

Some fundamental properties of flow responses to orographic forcing can be understood by considering a two-dimensional, steady-state, adiabatic, inviscid, nonrotating, Boussinesq fluid flow over a small-amplitude mountain. The governing linear equations can be simplified from (2.2.14)–(2.2.18) to be

$$U\frac{\partial u'}{\partial x} + U_z w' + \frac{1}{\rho_0}\frac{\partial p'}{\partial x} = 0, \qquad (5.1.1)$$

$$U\frac{\partial w'}{\partial x} - g\frac{\theta'}{\theta_0} + \frac{1}{\rho_0}\frac{\partial p'}{\partial z} = 0, \qquad (5.1.2)$$

$$\frac{\partial u'}{\partial x} + \frac{\partial w'}{\partial z} = 0, \qquad (5.1.3)$$

$$U\frac{\partial \theta'}{\partial x} + \frac{N^2 \theta_0}{g} w' = 0. \tag{5.1.4}$$

The above set of equations can be further reduced to *Scorer's equation* (1954),

$$\nabla^2 w' + l^2(z)w' = 0, \tag{5.1.5}$$

where $\nabla^2 = \partial^2/\partial x^2 + \partial^2/\partial z^2$ is the two-dimensional Laplacian operator, and l is the Boussinesq form of the *Scorer parameter* (Scorer 1949), which is defined as:

$$l^2(z) = \frac{N^2}{U^2} - \frac{U_{zz}}{U}. \tag{5.1.6}$$

Equation (5.1.5) serves as a central tool for numerous theoretical studies of small-amplitude, two-dimensional mountain waves, which may also be interpreted as a vorticity equation upon being multiplied by U (Smith 1979). The first term, $U(w'_{xx} + w'_{zz})$, is the rate of change of vorticity following a fluid particle. The second term, $N^2 w'/U$, is the rate of vorticity production by buoyancy forces. The last term, $-U_{zz}w'$, represents the rate of vorticity production by the redistribution of the background vorticity (U_z). In the extreme case of very small Scorer parameter, (5.1.5) reduces to irrotational or potential flow,

$$\nabla^2 w' = 0. \tag{5.1.7}$$

As discussed in Chapter 3 [(3.5.22)], the buoyancy force is negligible in this extreme case. If the forcing is symmetric in the basic flow direction, such as a cylinder in an unbounded fluid or a bell-shaped mountain in a half-plane, then the flow is symmetric. For this particular case, there is no drag produced on the mountain since the fluid is inviscid.

In order to simplify the mathematics of the steady-state mountain wave problem, one may assume that $U(z)$ and $N(z)$ are independent of height, and a sinusoidal terrain

$$h(x) = h_m \sin kx, \tag{5.1.8}$$

where h_m is the mountain height and k is the wave number of the terrain. For an inviscid fluid flow, the lower boundary condition requires the fluid particles to follow the terrain, so that the streamline slope equals the terrain slope locally,

$$\frac{w}{u} = \frac{w'}{U+u'} = \frac{dh}{dx} \quad \text{at} \quad z = h(x). \tag{5.1.9}$$

For a small-amplitude mountain, this leads to the linear lower boundary condition

$$w' = U\frac{dh}{dx} \quad \text{at} \quad z = 0, \tag{5.1.10}$$

or

$$w'(x,0) = Uh_m k \cos kx \quad \text{at} \quad z = 0, \tag{5.1.11}$$

for flow over a sinusoidal mountain as described by (5.1.8). Due to the sinusoidal nature of the forcing, it is natural to look for solutions in terms of sinusoidal functions,

$$w'(x,z) = w_1(z) \cos kx + w_2(z) \sin kx. \tag{5.1.12}$$

Substituting the above solution into (5.1.5) with a constant Scorer parameter leads to

$$w_{izz} + (l^2 - k^2)w_i = 0, \quad i = 1, 2. \tag{5.1.13}$$

As discussed in Chapter 3 [(3.5.7)], two cases are possible: (a) $l^2 < k^2$ and (b) $l^2 > k^2$. The first case requires $N/U < k$ or $Na/U < 2\pi$, where a is the terrain wavelength. Physically, this means that the basic flow has relatively weaker stability and stronger wind, or that the mountain is narrower than a certain threshold. For example, to satisfy the criterion for a flow with $U = 10\,\mathrm{m\,s}^{-1}$ and $N = 0.01\,\mathrm{s}^{-1}$, the wavelength of the mountain should be smaller than 6.3 km. In fact, this criterion can be rewritten as $(a/U)/(2\pi/N) < 1$. The numerator, a/U, represents the advection time of an air parcel passing over one wavelength of the terrain, while the denominator, $2\pi/N$, represents the period of buoyancy oscillation due to stratification. This means that the time an air parcel takes to pass over the terrain is less than it takes for vertical oscillation due to buoyancy force. In other words, buoyancy force plays a smaller role than the horizontal advection. In this situation, (5.1.13) can be rewritten as

$$w_{izz} - (k^2 - l^2)w_i = 0, \quad i = 1, 2. \tag{5.1.14}$$

The solutions of the above second-order differential equation with constant coefficient may be obtained

$$w_i = A_i e^{\lambda z} + B_i e^{-\lambda z}, \quad i = 1, 2, \tag{5.1.15}$$

where

$$\lambda = \sqrt{k^2 - l^2}. \tag{5.1.16}$$

Similar to that described in Section 3.4, the upper boundedness condition requires $A_i = 0$ because the energy source is located at $z = 0$. Applying the lower boundary condition, (5.1.11), and the upper boundary condition ($A_i = 0$) to (5.1.15) yields

$$B_1 = Uh_m k; \quad B_2 = 0. \tag{5.1.17}$$

This gives the solution,

$$w'(x,z) = w_1(z) \cos kx = Uh_m k e^{-\sqrt{k^2 - l^2}\,z} \cos kx, \tag{5.1.18}$$

The vertical displacement (η) is defined as $w' = D\eta/Dt$ which reduces to

$$w' = \frac{D\eta}{Dt} = U\frac{\partial \eta}{\partial x} \qquad (5.1.19)$$

for a steady-state flow.

Equation (5.1.18) can then be expressed in terms of η,

$$\eta = \frac{1}{U}\int_0^x w'\,\mathrm{d}x = h_\mathrm{m}\sin kx\, \mathrm{e}^{-\sqrt{k^2-l^2}\,z}. \qquad (5.1.20)$$

The above solution is sketched in Fig. 5.1a. The disturbance is symmetric with respect to the vertical axis and decays exponentially with height. Thus, the flow belongs to the evanescent flow regime as discussed in Section 3.5. The buoyancy force plays a minor role compared to that of the advection effect. The other variables can also be obtained by using the governing equations and (5.1.18),

$$u' = U\, h_\mathrm{m}\sqrt{k^2-l^2}\,\sin kx\, \mathrm{e}^{-\sqrt{k^2-l^2}\,z}, \qquad (5.1.21)$$

$$p' = -\rho_0 U^2\, h_\mathrm{m}\sqrt{k^2-l^2}\,\sin kx\, \mathrm{e}^{-\sqrt{k^2-l^2}\,z}, \qquad (5.1.22)$$

$$\theta' = -(\theta_0 N^2/g)h_\mathrm{m}\sin kx\, \mathrm{e}^{-\sqrt{k^2-l^2}\,z}. \qquad (5.1.23)$$

The maxima and minima of u', p', and θ' are also denoted in Fig. 5.1a. The coldest (warmest) air is produced at the mountain peak (valley) due to adiabatic cooling (warming). The flow accelerates over the mountain peaks and decelerates over the

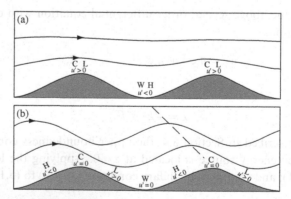

Fig. 5.1 The steady-state, inviscid flow over a two-dimensional sinusoidal mountain when (a) $l^2 < k^2$ (or $N < kU$), where k is the terrain wavenumber ($= 2\pi/a$, where a is the terrain wave length), or (b) $l^2 > k^2$ (or $N > kU$). The dashed line in (b) denotes the upstream tilt of the constant phase line. The maxima and minima of u', p' (H and L), and θ' (W and C) are denoted in the figures.

valleys. From the horizontal momentum equation, (5.1.1) with $U_z = 0$, or (5.1.22), a low (high) pressure is produced over the mountain peak (valley) where maximum (minimum) wind is produced. Note that (5.1.1) is also equivalent to the *Bernoulli equation*, which states that the pressure perturbation is out of phase with the horizontal velocity perturbation. Since no pressure difference exists between the upslope and downslope, this flow produces no net wave drag on the mountain (mountain drag). The *mountain drag* can be computed either from the horizontal pressure force on the mountain over a wavelength,

$$\mathcal{D} = \frac{k}{2\pi} \int_{-\pi/k}^{\pi/k} p'(x, z = 0) \left(\frac{dh}{dx}\right) dx, \qquad (5.1.24)$$

or equivalently, as the negative of the vertical flux of horizontal momentum (*momentum flux*) in the wave motion,

$$\mathcal{D} = -\frac{\rho_0 k}{2\pi} \int_{-\pi/k}^{\pi/k} u' w' dx. \qquad (5.1.25)$$

Note that *Eliassen and Palm's theorem*, (4.4.10), indicates that the vertical flux of horizontal momentum in a steady-state flow is negatively proportional to the vertical energy flux, $\overline{p'w'}$ (where the overbar denotes the average over a wavelength).

In the second case, $l^2 > k^2$, the flow response is completely different. This case requires $N/U > k$ or $Na/U > 2\pi$. As discussed in Section 3.5, this means that the basic flow has relatively stronger stability and weaker wind or that the mountain is wider. For example, and as mentioned earlier, to satisfy the criterion for a flow with $U = 10 \, \text{m s}^{-1}$ and $N = 0.01 \, \text{s}^{-1}$, the terrain wavelength should be larger than 6.3 km. Since $(a/U)/(2\pi/N) > 1$, the advection time is larger than the period of the vertical oscillation. In other words, buoyancy force plays a more dominant role than the horizontal advection. In this case, (5.1.13) can be written as

$$w_{izz} + m^2 w_i = 0, \quad m^2 = l^2 - k^2, \quad i = 1, 2. \qquad (5.1.26)$$

We look for solutions in the form

$$w_i(z) = A_i \sin mz + B_i \cos mz, \quad i = 1, 2. \qquad (5.1.27)$$

Substituting (5.1.27) into (5.1.12) leads to

$$w'(x, z) = C\cos(kx + mz) + D\sin(kx + mz) + E\cos(kx - mz) \\ + F\sin(kx - mz). \qquad (5.1.28)$$

In the above equation, terms of $(kx + mz)$ have an upstream phase tilt with height, while terms of $(kx - mz)$ have a downstream phase tilt. It can be shown that terms of $(kx + mz)$ have a positive vertical energy flux and should be retained since the energy source in this case is located at the mountain surface. This satisfies the Sommerfeld radiation boundary condition, as discussed in Section 4.4. Thus, the solution requires $E = F = 0$. This flow regime is characterized as the upward propagating wave regime, as discussed in Chapter 3. As in the first case, the lower boundary condition requires

$$C = Uh_mk, \ D = 0. \tag{5.1.29}$$

This leads to

$$w'(x,z) = Uh_mk\cos(kx+mz). \tag{5.1.30}$$

Other variables can be obtained through definitions or the governing equations,

$$\eta(x,z) = h_m\sin(kx+mz), \tag{5.1.31}$$

$$u'(x,z) = -Uh_mm\cos(kx+mz), \tag{5.1.32}$$

$$p'(x,z) = \rho_o U^2 h_m m\cos(kx+mz), \text{ and} \tag{5.1.33}$$

$$\theta'(x,z) = -\frac{N^2\theta_o h_m}{g}\sin(kx+mz). \tag{5.1.34}$$

The vertical displacement of the flow, and the maxima and minina of u', p', and θ' are depicted in Fig. 5.1b. Note that the flow pattern is no longer symmetric. The constant phase lines are tilted upstream (to the left) with height, thus producing a high pressure on the windward slope and a low pressure on the lee slope. Based on (5.1.32) or the Bernoulli equation (5.1.1), the flow decelerates over the windward slope and accelerates over the lee slope. The coldest and warmest spots are still located over the mountain peaks and valleys, respectively. The mountain drag can be calculated either from (5.1.24) or (5.1.25) to be

$$\mathcal{D} = \frac{1}{2}\rho_o U^2 h_m^2 k\sqrt{l^2 - k^2}. \tag{5.1.35}$$

The positive wave drag on the mountain is produced by the high pressure on the windward slope and the low pressure on the lee slope. This also can be understood through (5.1.25) and the out-of-phase relationship of u' and w' over the windward and lee slopes, as shown in Fig. 5.1b.

When $l^2 \gg k^2$, the flow approaches a limiting case in which the buoyancy effect dominates and the advection effect is totally negligible. In other words, the vertical pressure gradient force and the buoyancy force are roughly in balance and the vertical acceleration can be ignored. Thus, the mountain waves become hydrostatic. In this limiting case, the governing equation becomes

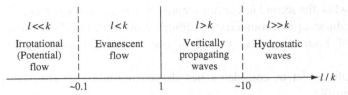

Fig. 5.2 Relations among different mountain wave regimes as determined by l/k, where l is the Scorer parameter and k is the wave number.

$$w'_{zz} + l^2 w' = 0. \tag{5.1.36}$$

The flow pattern repeats itself in the vertical with a wavelength of $\lambda_z = 2\pi/l = 2\pi U/N$, which is also referred to as the hydrostatic vertical wavelength. The regime boundary between the regimes of vertically propagating waves and evanescent waves can be found by letting $l = k$, which leads to $a = 2\pi U/N$. The relation among the mountain waves discussed in this subsection is sketched in Fig. 5.2.

5.2 Flows over two-dimensional isolated mountains

5.2.1 Uniform basic flow

The mountain wave problem in Section 5.1 may be extended to be more realistic by assuming an isolated mountain. Taking the one-sided Fourier transform (Appendix 5.1) of (5.1.5) yields

$$\hat{w}_{zz} + (l^2 - k^2)\hat{w} = 0. \tag{5.2.1}$$

The Fourier transform of the linear lower boundary condition, (5.1.10), is

$$\hat{w}(k, z = 0) = ikU\,\hat{h}(k). \tag{5.2.2}$$

For constant Scorer parameter, the solution of (5.2.1) can be written into two parts,

$$\hat{w}(k, z) = \hat{w}(k, 0)e^{i\sqrt{l^2 - k^2}z} \quad \text{for } l^2 > k^2 \text{ and} \tag{5.2.3a}$$

$$\hat{w}(k, z) = \hat{w}(k, 0)e^{-\sqrt{k^2 - l^2}z} \quad \text{for } l^2 < k^2. \tag{5.2.3b}$$

Taking the inverse one-sided Fourier transform of (5.2.3) yields the solution in the physical space,

$$w'(x, z) = 2\mathrm{Re}\left[\int_0^l ikU\,\hat{h}(k)e^{i\sqrt{l^2 - k^2}z}e^{ikx}dk + \int_l^\infty ikU\,\hat{h}(k)e^{-\sqrt{k^2 - l^2}z}e^{ikx}dk\right], \tag{5.2.4}$$

where Re represents the real part. The first integration on the right-hand side of (5.2.4) represents the upward propagating wave which satisfies the upper radiation boundary condition, while the second integration represents the evanescent wave which satisfies the boundedness upper boundary condition. Note that (5.2.4) is for a continuous spectrum of Fourier modes, instead of just one single mode as considered in Section 5.1.

For simplicity, let us consider a bell-shaped mountain or the *Witch of Agnesi* mountain profile,

$$h(x) = \frac{h_m a^2}{x^2 + a^2}, \qquad (5.2.5)$$

where h_m is the mountain height and a is the half-width where the mountain height is $h_m/2$. The advantage of using a bell-shaped mountain lies in that its one-sided Fourier transform (Appendix 5.1) is in a simple form,

$$\hat{h}(k) = \frac{h_m a}{2} e^{-ka}, \quad \text{for } k > 0. \qquad (5.2.6)$$

The Fourier transform for any k is $(h_m a/2) \exp(-|k|a)$. First, we will center our discussion on the extreme case with $l^2 \ll k^2$ (i.e., $al \ll 1$ or $Na \ll U$). Note that for bell-shaped mountains, we assume $k \approx 1/a$, instead of $k = 2\pi/a$ for sinusoidal mountains. As discussed earlier, the flow becomes a potential flow in which the buoyancy plays a negligible role. In this case, (5.2.4) can be approximated by

$$w'(x, z) \approx 2\mathrm{Re}\left[U \int_0^\infty ik\, \hat{h}(k) e^{-kz} e^{ikx} dk \right]$$
$$= 2\mathrm{Re}\left[U \int_0^\infty ik \left(\frac{h_m a}{2} \right) e^{-ka} e^{-kz} e^{ikx} dk \right]. \qquad (5.2.7)$$

Since $w = U \partial \eta / \partial x$, the Fourier transform of η can be obtained from that of $\hat{w}$,

$$\hat{\eta}(k, z) = \frac{\hat{w}(k, z)}{ikU}. \qquad (5.2.8)$$

Substituting (5.2.7) into (5.2.8) leads to

$$\eta(x, z) = h_m a\, \mathrm{Re} \int_0^\infty e^{-k(z+a-ix)} dx = \frac{h_m a(z+a)}{x^2 + (z+a)^2}. \qquad (5.2.9)$$

Therefore, similar to the sinusoidal mountain case, the flow pattern is symmetric with respect to the center of the mountain ridge ($x = 0$). However, the amplitude decreases with height linearly, instead of exponentially. The flow pattern is depicted in Fig. 5.3a.

Second, let us consider another extreme case: $l^2 \gg k^2$ (i.e., $al \gg 1$ or $Na \gg U$). As discussed in Section 5.1, the vertical acceleration due to the buoyancy force plays a dominant role. In this case, the solution (5.2.4) can be approximated by

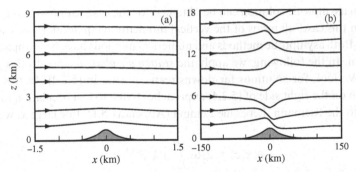

Fig. 5.3 Streamlines of steady-state flow over an isolated, bell-shaped mountain when (a) $l^2 \ll k^2$ (or $Na \ll U$), where a is the half-width of the mountain, or (b) $l^2 \gg k^2$ (or $Na \gg U$). (Adapted after Durran 1990.)

$$w'(x, z) \approx 2\mathrm{Re}\left[U \int_0^\infty i k \hat{h}(k) e^{ilz} e^{ikx} dk \right] = 2\mathrm{Re}\left[U \int_0^\infty ik \left(\frac{h_m a}{2} \right) e^{-ka} e^{ilz} e^{ikx} dk \right].$$

$$(5.2.10)$$

Similarly, the vertical displacement can be obtained,

$$\eta(x, z) = 2\mathrm{Re} \int_0^\infty \frac{h_m a}{2} e^{-ka} e^{i(kx+lz)} dk = \frac{h_m a (a \cos lz - x \sin lz)}{x^2 + a^2}. \qquad (5.2.11)$$

This type of flow is characterized as a hydrostatic mountain wave. The disturbance confines itself over the mountain in horizontal, but repeats itself in vertical with a wavelength of $2\pi U/N$. Without the Boussinesq approximation, the above solution becomes

$$\eta(x, z) = \left(\frac{\rho_s}{\rho(z)} \right)^{1/2} \left[\frac{h_m a (a \cos lz - x \sin lz)}{x^2 + a^2} \right], \qquad (5.2.12)$$

where ρ_s is the air density near surface. Equation (5.2.12) indicates that the wave amplitude will increase with a decreased air density of the basic flow. That is, the wave amplitude will increase at higher altitudes since air density decreases with height in a stably stratified flow. This helps explain the wave amplification in the higher atmosphere, such as large-amplitude gravity waves in the stratosphere. As described in Section 5.1, other fields can be obtained by the governing equations, (5.1.1)–(5.1.4) with $U_z = 0$. The wave drag on the mountain surface in this hydrostatic limit can be obtained by applying the Parseval theorem (Appendix 5.1),

$$\mathcal{D} = \int_{-\infty}^{\infty} p'(x, z = 0) \frac{dh}{dx} dx = \int_{-\infty}^{\infty} p'(x, 0) \frac{dh^*}{dx} dx = \frac{\pi}{4} \rho_0 U N h_m^2 \qquad (5.2.13)$$

where h^* is the complex conjugate of h. The momentum is transferred to a level where the wave breaks down, which is not included in the linear theory.

Third, an asymptotic solution can be obtained for the case with $l^2 \approx k^2$ (i.e., $al \approx 1$ or $Na \approx U$). In this case, all terms of the vertical momentum equation, (5.1.2) are equally important. Both asymptotic methods and numerical methods have been applied to solve the problem. In the following, we apply the *stationary phase method* to this particular problem. We look for solutions far downstream, $x \to \infty$ in (5.2.4). In this limit, the second term on the right side of (5.2.4) approaches 0 due to fast oscillation of $\exp(ikx)$, according to the Riemann–Lebesgue Lemma (Appendix 5.1). For large x, we have

$$\eta(x, z) \approx 2\text{Re} \int_0^l \hat{h}(k) e^{i\phi(k)} dk, \tag{5.2.14}$$

where

$$\phi(k) = \sqrt{l^2 - k^2} z + kx \tag{5.2.15}$$

is a phase function. Based on the stationary phase method, we will look for a particular k^* such that

$$\frac{d\phi}{dk} = 0 \quad \text{at} \quad k = k^*, \tag{5.2.16}$$

where k^* is called the *point of stationary phase*. With large x or z, $\exp(i\phi)$ will oscillate rapidly and, therefore, η will approach 0, according to the Riemann-Lebesgue Lemma. However, near k^*, the contribution to the integration by $\exp(i\phi)$ still remains because ϕ is approximately constant. Substituting the phase function (5.2.15) into (5.2.16) leads to the *influence function*,

$$\frac{z}{x} = \frac{\sqrt{l^2 - k^{*2}}}{k^*}, \tag{5.2.17}$$

in the region near k^*. Taking the Taylor's series expansion of $\phi(k)$ near k^* gives

$$\phi(k) = \phi(k^*) + \left[\frac{\partial \phi}{\partial k}\right]_{k^*} \tilde{k} + \frac{1}{2!} \frac{\partial^2 \phi}{\partial k^2} \tilde{k}^2 + \cdots, \tag{5.2.18}$$

where $\tilde{k} = k - k^*$. The second term on the right side of the above equation disappears due to the definition of k^* in (5.2.16). Thus, (5.2.14) becomes

$$\eta(x, z) = 2\text{Re}\left[\hat{h}(k^*) e^{i\phi(k^*)} \int_0^l e^{i\phi_{kk}\tilde{k}^2/2} dk\right]. \tag{5.2.19}$$

For a bell-shaped mountain,

$$\eta(x, z) = \sqrt{2\pi}\, h_m a e^{-k^* a} \left[\frac{(l^2 - k^{*2})^{3/4}}{lz^{1/2}}\right] \cos\left(\sqrt{l^2 - k^{*2}} z + k^* x - \frac{\pi}{4}\right), \tag{5.2.20}$$

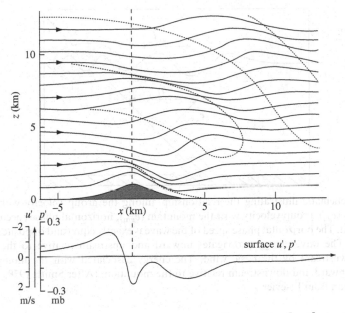

Fig. 5.4 Flow over a two-dimensional ridge of intermediate width ($l^2 \approx k^2$ or $al = Na/U = 1$) where the buoyancy force is important, but not so dominant that the flow is hydrostatic. The waves on the lee aloft are the *dispersive tail* of the nonhydrostatic waves ($k < l$, but not $k \ll l$). (Adapted after Queney 1948.)

where

$$k^* = \frac{l}{\sqrt{(z/x)^2 + 1}}. \qquad (5.2.21)$$

Figure 5.4 shows an example of a flow over a ridge of intermediate width ($l^2 \approx k^2$) where the buoyancy force is important, but not so dominant that the flow becomes hydrostatic. The nearly periodic waves located to the upper right of the mountain are the *dispersive tail* of nonhydrostatic waves with k less than, but not much less than l.

In fact, the influence function, (5.2.17), is related to the energy propagation associated with the mountain waves. The group velocity (c_{gm}) in the frame of reference fixed with the mountain can be obtained from (3.5.11),

$$c_{gm} = \left(U + \frac{\partial \omega}{\partial k} \right) i + \frac{\partial \omega}{\partial m} k, \qquad (5.2.22)$$

where $_m$ stands for mountain and

$$\omega = \frac{-Nk}{\sqrt{k^2 + m^2}}. \qquad (5.2.23)$$

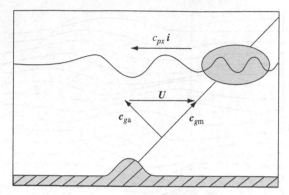

Fig. 5.5 A schematic illustrating the relationship among the group velocity with respect to (w.r.t) the air (c_{ga}), group velocity w.r.t the mountain (c_{gm}), horizontal phase speed ($c_{px}i$), and the basic wind. The horizontal phase speed of the wave is exactly equal and opposite to the basic wind speed. The wave energy propagates upward and upstream relative to the air, but is advected downstream by the basic wind. The energy associated with the mountain waves propagates upward and downstream relative to the mountain. (After Smith 1979, reproduced with permission from Elsevier.)

Substituting (5.2.23) into (5.2.22) leads to

$$c_{gm} = Ui + c_{ga} = \left[U - \frac{Nm^2}{(k^2 + m^2)^{3/2}} \right] i + \left[\frac{Nkm}{(k^2 + m^2)^{3/2}} \right] k, \qquad (5.2.24)$$

where c_{ga} is the group velocity relative to the air. Furthermore, the requirement of stationary waves, $c_{px} + U = 0$, implies

$$U = \frac{N}{\sqrt{k^2 + m^2}}. \qquad (5.2.25)$$

In (5.2.23), the negative sign is chosen in order to obtain positive c_{gz} by assuming positive k and m due to the use of one-sided Fourier transform. The relationship among $c_{px}i$, c_{gm} and c_{ga} is sketched in Fig. 5.5. The upstream phase speed of the mountain wave is exactly equal to and opposite of the basic wind speed. The wave energy propagates upward and upstream relative to the air, but is advected downstream by the basic wind. Thus, relative to the mountain, the energy associated with the mountain waves propagates upward and downstream. The slope of the group velocity can be obtained by substituting U of (5.2.25) into (5.2.24) and then calculating the slope,

$$\frac{c_{gz}}{c_{gx}} = \frac{m}{k} = \sqrt{\frac{N^2}{U^2 k^2} - 1} = \frac{\sqrt{l^2 - k^2}}{k} = \frac{z}{x}. \qquad (5.2.26)$$

In deriving the second equality, we have used (5.2.25), while in deriving the last equality, we have used (5.2.17) near the *point of stationary phase*, i.e. $k = k^*$.

Therefore, the point of stationary phase is the value of k corresponding to a wave with a group velocity beam as shown in Fig. 5.5. Waves are found downstream since the horizontal group velocity is less than the phase speed.

For general cases, such as $l^2 < k^2$ or $l^2 > k^2$, it is not easy to obtain analytical solutions from (5.2.4). With the advancement of numerical techniques, such as the Fast Fourier Transform (FFT), and computers, solutions can be approximately obtained numerically with the implementation of proper boundary conditions.

5.2.2 Basic flow with variable Scorer parameter

In the real atmosphere, the basic wind and stratification normally vary with height. To study the mountain waves produced by this type of basic flow, we assume that the Scorer parameter, (5.1.6), is a slowly varying function of z. In this situation, we expect to find a solution of (5.2.1) in form of,

$$\hat{w}(k, z) = A(k, z) e^{i\phi(k,z)}, \tag{5.2.27}$$

where $A(k, z)$ is a slow-varying amplitude function, and $\phi(k, z)$ is the slow-varying phase function. Substituting (5.2.27) into (5.2.1) yields

$$\left[-A\phi_z^2 + (l^2 - k^2)A\right] + i(A\phi_{zz} + 2A_z\phi_z) + A_{zz} = 0. \tag{5.2.28}$$

The last term makes a minor contribution and can be neglected, since $A(k, z)$ is a slow-varying function of z. Thus, the above equation reduces to

$$\phi_z = \sqrt{l^2 - k^2}, \quad \text{and} \tag{5.2.29}$$

$$\frac{\partial}{\partial z}(A^2\phi_z) = 0. \tag{5.2.30}$$

Combining the above two equations leads to

$$A^2\sqrt{l^2 - k^2} = constant. \tag{5.2.31}$$

For long (hydrostatic) waves ($l^2 \gg k^2$), the above equation reduces to

$$A^2 l = constant. \tag{5.2.32}$$

This implies that the amplitude of the vertical velocity increases (decreases) significantly in regions of weak (strong) stratification or strong (weak) wind. For example, the mountain wave tends to steepen when it propagates to the region below a jet stream or a jet streak since the basic wind speed increases there. Note that in applying (5.2.27) to solve the problem, and in neglecting the last term of (5.2.28), we have implicitly adopted

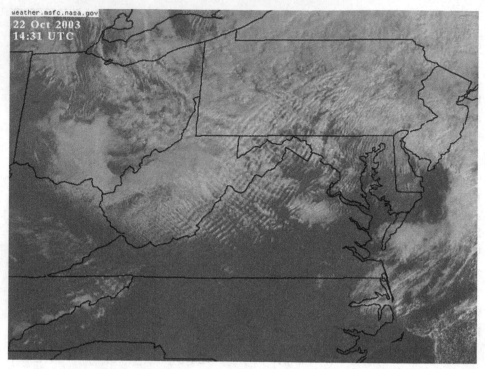

Fig. 5.6 Satellite imagery for lee wave clouds observed at 1431 UTC October 22, 2003 over western Virginia. Clouds originate at the Appalachian Mountains. (Courtesy of NASA.)

a first-order WKBJ (Wentzel–Kramers–Brillouin–Jeffreys) approximation. A second-order WKBJ approximation has been used to calculate wind profile effects on mountain wave drag (e.g., Teixeira and Miranda 2006). It is necessary to extend the WKBJ approximation to second order for these effects to be taken into account.

Based on (5.2.32) and previous discussions, waves may amplify in certain layers due to: (a) weaker stratification, (b) stronger wind, such as a jet stream or jet streak, (c) nonlinear steepening, and (d) abrupt decrease in the mean density, leading to an increase of $\sqrt{\rho_s/\rho(z)}$, in (5.2.12).

5.2.3 Trapped lee waves

One of the most prominent features of mountain waves is the long train of wave clouds over the lee of mountain ridges in the lower atmosphere, such as those shown in Fig. 5.6. This type of wave differs from the dispersive tails in Fig. 5.4 in that it is located in the lower atmosphere and there is no vertical phase tilt. It will be shown below that this type of trapped lee waves, or resonance waves, occurs when the Scorer parameter decreases rapidly with height (Scorer 1949).

The dynamics of trapped lee waves may be understood by considering a two-layer stratified fluid system. The wave equations for the vertical displacement in Fourier space may be written in a form similar to (5.2.1),

$$\hat{\eta}_{zz} + (l_1^2 - k^2)\hat{\eta} = 0 \text{ for } -H \le z < 0 \text{ and} \tag{5.2.33a}$$

$$\hat{\eta}_{zz} - (k^2 - l_2^2)\hat{\eta} = 0 \quad \text{for} \quad 0 \le z. \tag{5.2.33b}$$

In this two-layer fluid system, we have assumed that $l_2^2 < k^2 < l_1^2$. For convenience, the ground and the interface of the lower and upper layers are assumed to be located at $z = -H$ and $z = 0$, respectively. The free wave solutions may be written as

$$\hat{\eta}_1(k, z) = C\left[\cos \mu z - \frac{\lambda}{\mu}\sin \mu z\right] \text{ and} \tag{5.2.34a}$$

$$\hat{\eta}_2(k, z) = C\, e^{-\lambda z}, \tag{5.2.34b}$$

where $\mu = \sqrt{l_1^2 - k^2}$, $\lambda = \sqrt{k^2 - l_2^2}$ and C is a constant coefficient to be determined by the lower boundary condition. The boundedness upper boundary condition has been applied to exclude the $\exp(\lambda z)$ term, and the kinematic and dynamic boundary conditions at the interface, i.e. the continuities of $\hat{w}$ and $\hat{w}_z$ at $z = 0$, have also been applied. Without enforcing a lower boundary condition, (5.2.34) represents free waves associated with this two-layer fluid system. The resonance waves are obtained by seeking the zeros of (5.2.34a) with $z = -H$,

$$\cot \mu H = -\lambda/\mu. \tag{5.2.35}$$

The resonance wave number (k_r^*) may be obtained by solving the above equation either numerically or graphically. The criterion for the existence of one or more resonance waves may be obtained (Scorer 1949):

$$l_1^2 - l_2^2 \ge \frac{\pi^2}{4H^2}. \tag{5.2.36}$$

A more general criterion for resonance waves of the nth mode is

$$\left[\frac{(2n+1)\pi}{2H}\right]^2 \ge (l_1^2 - l_2^2) \ge \left[\frac{(2n-1)\pi}{2H}\right]^2. \tag{5.2.37}$$

The above criterion implies that *in order to have resonance (lee) waves, the Scorer parameter in the lower layer must be much greater than that in the upper layer.* In other words, the lower layer must be more stable or with a much slower basic wind speed than the upper layer.

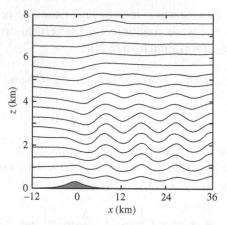

Fig. 5.7 Lee waves simulated by a nonlinear numerical model for a two-layer airflow over a bell-shaped mountain. Displayed are the quasi-steady state streamlines. In the lower layer (below 5 km apporximately) $l^2 = 9x10^{-7}$ m^{-2}, while in the upper layer, $l^2 = 2x10^{-7}$ m^{-2}. (Adapted after Durran 1986b.)

In order to obtain a complete solution of the boundary value problem for a specific obstacle, we may apply the linear lower boundary condition,

$$\hat{\eta}_1(k, -H) = \hat{h}(k). \tag{5.2.38}$$

Substituting the above equation into (5.2.34) and taking the inverse Fourier transform of $\hat{\eta}(k, z)$ leads to the forced wave solution in the lower layer,

$$\eta_1(x, z) \doteq 2\mathrm{Re} \int_0^\infty \frac{\hat{h}(k)(\cos \mu z - (\lambda/\mu) \sin \mu z)e^{ikx}}{(\cos \mu H + (\lambda/\mu) \sin \mu H)} \, dk. \tag{5.2.39}$$

The singularity in the denominator of the above equation corresponds to the resonance mode that will produce lee waves. Equation (5.2.39) can be solved asymptotically or numerically with a given mountain-shape function (Scorer 1949; Smith 1979). Figure 5.7 shows lee waves simulated by a nonlinear numerical model for a two-layer airflow over a bell-shaped mountain. Due to the co-existence of the upward propagating waves and downward propagating waves, there exists no phase tilt in the lee waves.

Once lee waves form, regions of reversed cross-mountain winds near the surface beneath the crests of the lee waves may develop due to the presence of a reversed pressure gradient force. In the presence of surface friction, a sheet of vorticity parallel to the mountain range forms along the lee slopes, originates in the region of high shear within the boundary layer. The vortex sheet separates from the surface, ascends into the crest of the first lee wave, and remains aloft as it is advected downstream by the undulating flow in the lee waves (Doyle and Durran 2004). The vortex with recirculated air is known as *rotor* and the process that forms it is known as *boundary layer separation*, which will be further discussed in Subsection 5.4.2 along with lee vortices. These rotors are often observed to the lee of steep mountain ranges such as over the Owens Valley, California, on the eastern slope of Sierra Nevada (e.g., Grubišić and Lewis 2004). Occasionally, a turbulent, altocumulus cloud forms with the rotor and is referred to as *rotor cloud*.

5.3 Nonlinear flows over two-dimensional mountains

As discussed in Sections 5.1 and 5.2, the response of a stably stratified flow over a two-dimensional mountain ridge has been studied extensively since the 1960s. In particular, the linear dynamics are fundamentally understood, especially due to the development of linear theories in earlier times. Linear theory, however, begins to break down when the perturbation velocity (u') becomes large compared with the basic flow (U) in some regions, so that the flow becomes stagnant. This happens when the mountain becomes very high, the basic flow becomes very slow, or the stratification becomes very strong. In other words, flow becomes more nonlinear when the Froude number, $F = U/Nh$, becomes small. For simplicity, the mountain height is denoted by h. Thus, in order to fully understand the dynamics of nonlinear phenomena, such as upstream blocking, wave breaking, severe downslope winds and lee vortices, we need to take a nonlinear approach. Note that the reciprocal of the Froude number, Nh/U, has also been used as a control parameter and is known as *nondimensional mountain height*. In the text, we will use these two parameters interchangeably.

Nonlinear response of a continuously stratified flow over a mountain is very complicated since the nonlinearity may come from the basic flow characteristics, the mountain height, or the transient behavior of the internal flow, such as wave steepening. In this section, we will begin with the discussion of a nonlinear theory developed by Long (1953), then discuss the two-dimensional flow regimes for a continuously stratified flow over a two-dimensional mountain with the help of nonlinear numerical models, and the generation mechanisms of severe downslope winds and wave breaking.

5.3.1 Nonlinear flow regimes

The governing equation for the finite-amplitude, steady state, two-dimensional, inviscid, continuously and stably stratified flow may be derived (Long 1953)

$$\nabla^2 \delta + \frac{1}{e}\frac{de}{dz}\left[\frac{\partial \delta}{\partial z} - \frac{1}{2}(\nabla \delta)^2\right] + \frac{N^2}{U^2}\delta = 0, \qquad (5.3.1)$$

where $\delta(x, z) = z - z_0$ is the streamline deflection at (x, z) from its far upstream, undisturbed height z_0; U and N are the far upstream basic flow speed and Brunt–Vaisala frequency, respectively, at height z_0, and $e = (1/2)\rho_0 U^2$ is the kinetic energy of the upstream flow. In deriving (5.3.1), it has been assumed that there is no streamline deflection far upstream. In order to solve the above nonlinear equation, (5.3.1), we must specify e. Under the special situation $de/dz = 0$ and when the flow is Boussinesq, which assumes that ρ is approximately constant and $U(z)$ and $N(z)$ are effectively constant, (5.3.1) becomes a linear *Helmholtz equation*,

$$\nabla^2 \delta + l^2 \delta = 0, \tag{5.3.2}$$

where $l = N/U$ is the Scorer parameter of the basic flow far upstream. The nonlinear lower boundary condition for (5.3.2) is given by

$$\delta(x, z) = h(x) \quad \text{at} \quad z = h(x), \tag{5.3.3}$$

where $h(x)$ is the height of the mountain surface. In other words, the nonlinear lower boundary condition is applied on the mountain surface, instead of approximately applied at $z = 0$ as in the linear lower boundary condition, such as (5.1.10). Equation (5.3.2) with the lower boundary condition (5.3.3) forms Long's model, in which the steady-state nonlinear flow is remarkably described by a linear differential equation with constant coefficients. In fact, (5.3.2) is exactly the same differential equation which applies to infinitesimal perturbations adopted in many linear theories and discussed earlier in this chapter. The appropriate upper boundary condition for a semi-infinite fluid, such as the atmosphere, is the radiation or boundedness condition, similar to (5.2.3) in the Fourier space for a uniform basic flow over an infinitesimal mountain.

Following the procedure for treating linear flow over small-amplitude mountains, we make the Fourier transform of (5.3.2),

$$\hat{\delta}_{zz} + (l^2 - k^2)\hat{\delta} = 0. \tag{5.3.4}$$

The general solution for the above equation is

$$\hat{\delta} = \hat{\delta}(k, 0)e^{imz} \quad \text{for } l > k \quad \text{and} \tag{5.3.5a}$$

$$\hat{\delta} = \hat{\delta}(k, 0)e^{-\lambda z} \quad \text{for } l < k, \tag{5.3.5b}$$

where $m = \sqrt{l^2 - k^2}$, and $\lambda = \sqrt{k^2 - l^2}$. Note that the upper radiation and boundedness conditions have been applied to (5.3.5a) and (5.3.5b), respectively, while the linear lower boundary condition has been applied at $z = 0$, instead of at $z = h(x)$. The streamline deflection in the physical space can then be obtained by taking the inverse Fourier transform

$$\delta(x, z) = \text{Re}\left[\int_0^l \hat{\delta}(k, 0)e^{imz}e^{ikx}dk + \int_l^\infty \hat{\delta}(k, 0)e^{-\lambda z}e^{ikx}dk\right], \tag{5.3.6}$$

which may be obtained numerically, as with the Fast Fourier Transform numerical technique. Other dynamical variables may be derived,

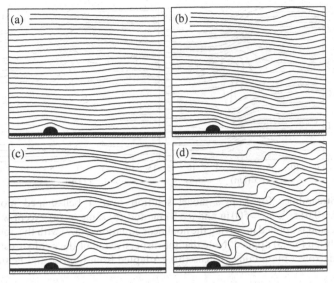

Fig. 5.8 Streamlines of Long's model solutions for uniform flow over a semi-circle obstacle with $Nh/U = $ (a) 0.5, (b) 1.0, (c) 1.27, and (d) 1.5, where Nh/U is the nondimensional mountain height. Note that the streamlines become vertical in (c) and overturn in (d). (Adapted from Miles 1968.)

$$u = \frac{\partial \psi}{\partial z}; \, w = -\frac{\partial \psi}{\partial x}; \, \rho = \rho_0 \left(1 - \frac{N^2}{gU} \psi \right); \, N^2 = -\frac{g}{\rho_0} \frac{\partial \rho}{\partial z}, \quad (5.3.7)$$

where ψ is the streamfunction defined as $U(z - \delta)$. The exact nonlinear lower boundary condition, (5.3.3), can be implemented using an iterative method (e.g., Laprise and Peltier 1989a).

Figure 5.8 shows streamlines of analytical solutions for flow over a semi-circle obstacle for the nondimensional mountain heights $Nh/U = 0.5$, 1.0, 1.27, and 1.5. As mentioned earlier, the nondimensional mountain height is a measure of the nonlinearity of the continuously stratified flow, which equals the reciprocal of the Froude number (U/Nh). When Nh/U is small, such as $Nh/U = 0.5$, the flow is more linear. When Nh/U increases to 1.27, the flow becomes more nonlinear and its streamlines become vertical at the first level of wave steepening. For flow with $Nh/U > 1.27$, the flow becomes statically and shear unstable (Laprise and Peltier 1989b). The vertical streamline marks the approximate limit of applicability of Long's model. For the hydrostatic solution of Long's model with a bell-shaped mountain subject to a nonlinear lower boundary condition, this critical value is $Nh/U = 0.85$ (Miles and Huppert 1969). Thus, for a continuously stratified, hydrostatic flow over a bell-shaped mountain, the flow may be classified as supercritical flow when $U/Nh > 1.18$ ($Nh/U < 0.85$) and as subcritical flow when $U/Nh < 1.18$ ($Nh/U > 0.85$). Note that in the literature it is often misquoted $U/Nh = 1$ as the regime boundary for supercritical and subcritical regimes for continuously stratified flow over mountains.

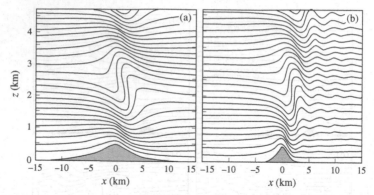

Fig. 5.9 (a) Streamlines for Long's model solution over a bell-shaped mountain with $U = 5\,\text{m s}^{-1}$, $N = 0.01\,\text{s}^{-1}$, $h_\text{m} = 500\,\text{m}$ and $a = 3\,\text{km}$; and (b) same as (a) except with $a = 1\,\text{km}$. An iterative method is applied in solving the nonlinear equation (5.3.2) with the nonlinear lower boundary condition (5.3.3) applied. Note that the dispersive tail of the nonhydrostatic waves is present in the narrower mountain case (case(b)). (Adapted from Laprise and Peltier 1989a.)

As discussed in Section 3.3, there exist five flow regimes in a one-layer shallow-water system, based on the shallow water Froude number, $F = U/\sqrt{gH}$, and the nondimensional mountain height, $M = h_\text{m}/H$ (Fig. 3.3). In a nonrotating, continuously stratified flow over a two-dimensional, bell-shaped mountain, three nondimensional control parameters may be identified: U/Nh, h/a, and Na/U. However, only two of them are independent. The parameter h/a measures the steepness of mountain, and Na/U is the nondimensional mountain width which measures the degree of hydrostatic effect (the larger the more hydrostatic). In the hydrostatic limit ($Na/U \to \infty$), the sole control parameter is the Froude number.

Figure 5.9a shows streamlines for Long's model for flow over a bell-shaped mountain with a half-width (a) of 3 km with the nonlinear lower boundary condition (5.3.3) applied. Internal waves tend to overturn in regions of reversed density gradient (statically unstable), $\partial\rho/\partial z > 0$, which corresponds to $\partial\delta/\partial z > 1$ from (5.3.7). The heights of critical steepening levels differ slightly from those predicted by linear theory for hydrostatic waves, $z_\text{o} = (n + 3/4)(2\pi U/N)$, where n is an integer, just over the crest of the topography (Laprise and Peltier 1989a). In Fig. 5.9a, the first steepening level for nonlinear, hydrostatic waves is about 2.36 km. With a narrower mountain, such as $a = 1\,\text{km}$ (Fig. 5.9b), a dispersive tail, caused by nonhydrostatic dispersion, is produced. The downstream displacement of the steepened region is caused by both the nonhydrostatic effect and the nonlinearity of the interior flow and the lower boundary condition. When $Na/U \gg 1$, the flow approaches the hydrostatic limit. This control parameter can be obtained by comparing the scales of $\partial^2 w'/\partial x^2$ and $N^2 w'/U^2$ of (5.1.5). Direct comparison of $\partial^2 w'/\partial x^2$ and $\partial^2 w'/\partial z^2$ terms by scale analysis leads to the conclusion that h/a is a control parameter of nonhydrostatic effect. The Froude number, U/Nh, can also be derived by comparing the scales of $\partial^2 w'/\partial z^2$ and $N^2 w'/U^2$ of (5.1.5).

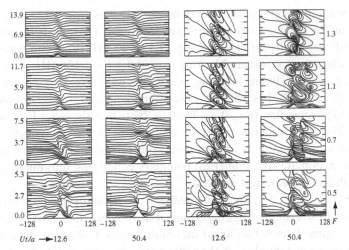

Fig. 5.10 Nonlinear flow regimes for a two-dimensional, hydrostatic, uniform flow over a bell-shaped mountain as simulated by a numerical model, based on the Froude number ($F = U/Nh$). F varies from 0.5 to 1.3, which gives four different flow regimes as discussed in the text. Displayed are the θ fields (left two columns) and the u' fields (right two columns) for two nondimensional times $Ut/a = 12.6$ and 50.4. The dimensional parameters are: $N = 0.01\,\mathrm{m\,s^{-1}}$, $h = 1\,\mathrm{km}$, $a = 10\,\mathrm{km}$, and $U = 5, 7, 11$, and $13\,\mathrm{m\,s^{-1}}$ correspondig to $F = 0.5, 0.7, 1.1$, and 1.3, respectively. A constant nondimensional physical domain height of $1.7\lambda_z$ (where $\lambda_z = 2\pi U/N$) is used. Both the abscissa and ordinate in the small panels are labeled in km. (Adapted after Lin and Wang 1996.)

Long's nonlinear theory advances our understanding of orographically forced flow considerably. However, the constant upstream condition assumed by Long may not be necessarily consistent with the flow established naturally by transients, especially when blocking occurs (Garner, 1995). In the real atmosphere, turbulence will come into play and produce vertical mixing in a subcritical (overturning) flow. To simulate a subcritical flow, one may consider using a laboratory tank experiment or adopting a nonlinear numerical model. As mentioned earlier, flow may become stagnant, where the total horizontal wind speed reduces to zero, in essentially two regions: in the interior of the fluid over the mountaintop or on the lee slope and along the upstream slope of the mountain.

Flow stagnation in a two-dimensional flow is responsible for *flow recirculation*, while stagnation in a three-dimensional flow is responsible for *flow splitting*. Flow stagnanation in the interior of the fluid is due to nonlinear *wave steepening*, which may lead to *wave breaking* and *wave overturning* over the lee slope, while the flow stagnation at the upstream surface of the mountain is called *flow blocking*. Although the two-dimensional, nonrotating, hydrostatic flow may be simply classified as supercritical and subcritical regimes, as discussed above, the transient flow behavior becomes much more complicated. Figure 5.10 shows the time evolution for the θ and u' fields for a hydrostatic flow over a two-dimensional, bell-shaped mountain simulated by a numerical model at nondimensional times $Ut/a = 12.6$ and 50.4 for F ranging from 0.5 to 1.3. Four regimes are identified: (I) flow with neither wave breaking aloft nor upstream blocking (e.g., $1.12 \leq F$), (II) flow with wave breaking aloft in the absence of upstream blocking (e.g., $0.9 \leq F < 1.12$), (III) flow with both wave breaking and upstream blocking, but where wave breaking occurs first (e.g., $0.6 \leq F < 0.9$), and (IV) flow with both wave breaking and upstream blocking, but where blocking occurs first (e.g., $0.3 \leq F < 0.6$). Note that the exact Froude numbers separating these flow regimes might be different in other numerically simulated results because these numbers are sensitive to some numerical factors, such as the grid resolution, domain size, numerical boundary conditions, and numerical scheme adopted in different numerical models.

In regime I (e.g., $F = 1.3$ in Fig. 5.10), neither wave breaking nor upstream blocking occurs, but an upstream propagating *columnar disturbance* does exist. The basic flow structure in regime I resembles linear mountain waves. Columnar disturbances are wave modes with constant phase in the vertical, which permanently alter the upstream temperature and horizontal velocity fields as they pass through the fluid (e.g., Pierrehumbert and Wyman 1985). A columnar disturbance may be generated by a sudden imposition of a disturbance, such as the impulsive introduction of a mountain in a uniform flow. Regime II (e.g., $F = 1.1$ in Fig. 5.10) resembles weakly nonlinear mountain waves. In this flow regime, an internal jump forms at the downstream edge of the wave-breaking region above the mountain, propagates downstream, and then becomes quasi-stationary. The region of wave breaking also extends downward toward the lee slope. After the internal jump travels farther downstream, a stationary mountain wave becomes established in the vicinity of the mountain above the dividing streamline, which is induced by wave breaking. A high-drag state is predicted in this flow regime. In addition, a vertically propagating hydrostatic gravity wave is generated by the propagating jump and travels with it. Along the lee slope, a strong downslope wind develops. Static and shear instabilities may occur locally in the region of wave breaking. The computed critical Froude number for wave breaking is about 1.12, which agrees well with the value 1.18 found by Miles and Huppert (1969).

In regime III (e.g., $F = 0.7$ in Fig. 5.10), the internal jump over the lee slope propagates downstream in the early stage and then becomes quasi-stationary. Note that the propagation of the downstream internal jump is sensitive to the upstream numerical boundary condition, which may cause the internal jump to retrogress upstream. To avoid this artificial effect from the numerical model, the upstream boundary should be placed far enough so as to effectively reduce its impact. Also, the layer depth of blocked fluid upstream is independent of the Froude number. In regime IV (e.g., $F = 0.5$ in Fig. 5.10), a significant portion of the upstream flow is blocked by the mountain. The presence of wave breaking aloft is not a necessary condition for upstream blocking to occur. A vertically propagating gravity wave is generated by the upstream reversed flow and travels with it. The speed of the upstream reversed flow is proportional to h/a. The surface drag increases abruptly from regime I to II, while it decreases gradually from regime II (III) to III (IV).

Flow regimes may also be classified in different ways, depending upon particular characteristics. For example, two-dimensional, uniform flow over an isolated mountain has been classified as either a quasi-linear regime, high-drag state, or blocked state, based on Nh/U and NU/g (Stein 1992). In addition, the flow response of a three-dimensional flow over a long ridge is very different from that of a two-dimensional flow when Nh/U is large. For example, the onset of wave breaking and the transition to the high-drag state in the three-dimensional flow was found to be accompanied by an abrupt increase in deflection of the low-level flow around the ridge (Epifanio and Durran 2001). The increased flow deflection is produced at least in part by upstream-propagating columnar disturbances forced by the transition to the high-drag state.

5.3.2 Generation of severe downslope winds

Severe downslope winds over the lee of a mountain ridge have been observed in various places around the world, such as the *chinook* over the Rocky Mountains, *foehn* over the Alps, *bora* over the Dinaric Alps, *zonda* over the Agentina mountains, *berg wind* in South Africa, *Canterbury-nor'wester* in New Zealand, *halny wiatr* in the mountains of Poland, *Santa Ana winds* in southern California, and *Diabolo winds* in San Francisco Bay Area. One well-known event is the January 11, 1972 windstorm which occurred in Boulder, Colorado, and which reached a peak wind gust as high as $60 \, \mathrm{m \, s^{-1}}$ and produced severe damage in the Boulder area (Fig. 3.4a).

The basic dynamics of the severe downslope wind can be understood from the following two major theories: (a) *resonant amplification theory* (Clark and Peltier 1984), and (b) *hydraulic theory* (Smith 1985), along with later studies on the effects of instabilities, wave ducting, nonlinearity, and upstream flow blocking. These will be reviewed in the following.

a. Resonant amplification theory

Idealized nonlinear numerical experiments indicate that a high-drag (severe-wind) state occurs after an upward propagating mountain wave breaks above a mountain, such as happens in Fig. 5.10 ($F = 0.5, 0.7$, and 1.1), in which severe downslope winds develop in a uniform flow over a bell-shaped mountain. The wave-breaking region is characterized by strong turbulent mixing (where $R_i < 0.25$), with a local wind reversal on top of it. As mentioned in Section 3.8, the *wind reversal level* coincides with the critical level for a stationary mountain wave, and thus is also referred to as the *wave-induced critical level*. The lowest wave-induced critical level starts to develop at the height $z = 3\lambda_z/4 \approx 2.36$, 3.30, and $5.18 \, \mathrm{km}$ for cases of $F = 0.5, 0.7$, and 1.1, respectively (at $Ut/a = 12.6$ in Fig. 5.10), where $\lambda_z \ (= 2\pi U/N)$ is the hydrostatic vertical wavelength. A supercritical

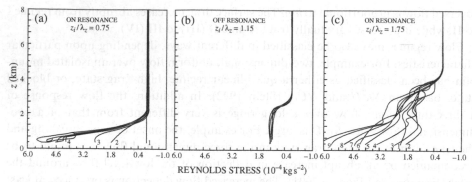

REYNOLDS STRESS ($10^{-4}\,\mathrm{kg\,s^{-2}}$)

Fig. 5.11 Resonant amplification mechanism for severe downslope winds developed for basic flow with a prescribed critical level (z_i) over a bell-shaped mountain. Displayed are the evolution of the Reynolds stress, $\langle \rho_0 u'w' \rangle$, profile for on-resonance flows with: $z_i/\lambda_z =$: (a) 0.75 and (c) 1.75 and for the off-resonance flow with (b) $z_i/\lambda_z = 1.15$, where $\lambda_z = 2\pi U_0/N$. The flow and orographic parameters are: $N = 0.02\,\mathrm{s}^{-1}$, $U(z) = U_0 \tanh[(z-z_i)/b]$ with $U_0 = 8\,\mathrm{m\,s}^{-1}$, $b = 600\,\mathrm{m}$, $R_{i\min} = N^2/(U_0/b)^2 = 2.25$ (minimum R_i), $h = 300\,\mathrm{m}$, and $a = 3\,\mathrm{km}$. The Froude number (U_0/Nh) is 1.33. Height (z) is in km. The profiles in the figure range in time from 0 to $2880\Delta t$, 1440 to $2880\Delta t$, and 2080 to $4240\Delta t$ for panels (a) to (c), respectively, where $\Delta t = 5\,\mathrm{s}$. Some of the profiles are sequentially numbered from earliest to latest (labeled by small numbers). (After Clark and Peltier 1984.)

flow with a severe downslope wind can be found over the lee slope under the wave breaking region, which undergoes a transition from subcritical flow over the upwind slope. The maximum perturbation wind over the lee slope is much higher than those predicted by linear and weakly nonlinear theories. At a later stage, the *well-mixed layer* (*wake*) deepens, the depth of *internal hydraulic jump* (*critically steepened streamlines*) extends to a great depth, the flow above the initial wave-induced critical level is less disturbed compared to that in the lower layer, and severe winds develop over the lee slope and below the well-mixed layer ($Ut/a = 50.4$ in Fig. 5.10).

The above example implies that the wave breaking region aloft acts as an internal boundary which reflects the upward propagating waves back to the ground and produces a high-drag state through partial resonance with the upward propagating mountain waves. This is shown by performing nonlinear numerical simulations for stratified flow over a bell-shaped mountain (Fig. 5.11). In these simulations, the basic flow reverses its direction at a prescribed critical level (z_i). In the absence of shear instability associated with the basic flow, and when the basic-flow critical level is located at a nondimensional height of $z_i/\lambda_z = 3/4 + n$ (n is an integer) above the surface, nonlinear resonant amplification occurs between the upward propagating waves generated by the mountain and the downward propagating waves reflected from the critical level. This leads to an extremely large Reynolds stress or surface drag and severe downslope winds (Figs. 5.11a and 5.11c). In other words, the flow is *on resonance*. On the other hand, when the basic flow critical level is located at a

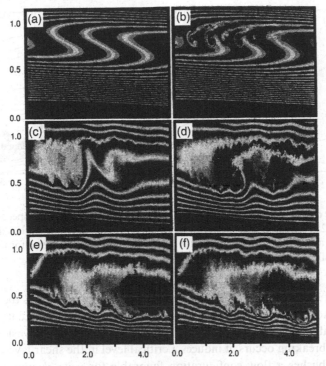

Fig. 5.12 Three distinct stages for the development of severe downslope winds, as revealed by the triply-nested numerical solutions. See text for details. Long's (1953) nonlinear analytical solution is used to initiate the flow. Displayed are the potential temperature fields over the lee slope at model times of (a) 0, (b) 20, (c) 66, (d) 96, (e) 160, and (f) 166 min. The grid resolutions for the outer, middle, and inner domains are 500 m, 50 m, and $16^2/_3$ m respectively. A bell-shaped mountain with $h = 165$ m and $a = 3$ km is used. The upstream flow parameters are $U = 3.3 \, \text{m s}^{-1}$ and $N = 0.02 \, \text{s}^{-1}$. Thus, $F = U/Nh = 1$. (Adapted after Scinocca and Peltier 1993.)

nondimensional height off $z_i/\lambda_z = 3/4 + n$, such as 1.15, there is no wave resonance and no severe downslope winds generated (Fig. 5.11b). Because the severe downslope winds are developed by resonance between upward and downward waves, this mechanism is referred to as the *resonant amplification mechanism*.

Based on numerical simulations with finer-grid resolutions, three distinct stages for the development of severe downslope winds are identified (Scinocca and Peltier 1993). (1) Local static (buoyancy) instability develops when the wave steepens and overturns, thus producing a pool of well-mixed air aloft (Figs. 5.12a–b). (2) A well-defined large-amplitude stationary disturbance is generated over the lee slope. In time, small-scale secondary Kelvin–Helmholtz (K–H) (shear) instability develops in local regions of enhanced shear associated with flow perturbations caused by the large-amplitude disturbance (Figs. 5.12c–d). (3) The region of enhanced wind on the lee slope expands downstream, eliminating the perturbative structure associated with the large-amplitude

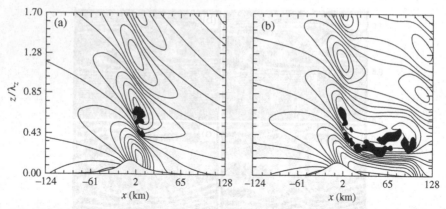

Fig. 5.13 Wave ducting as revealed by the time evolution of horizontal wind speeds and regions of local $R_i < 0.25$ (shaded) for a flow with uniform wind and constant static stability over a mountain ridge at $Ut/a =$ (a) 12.6, and (d) 50.4. The Froude number of the uniform basic wind is 1.0. (Adapted after Wang and Lin 1999.)

stationary disturbance (Figs. 5.12e–f). The K–H instability dominates the flow in this mature windstorm state. Thus, static instability helps explain the initiation of wave-induced critical level and the downstream expansion of the severe downslope winds.

Once wave breaking occurs, it induces a critical level in the shear layer with low R_i and thus establishes a flow configuration favorable for *wave ducting* in the lower uniform flow layer, similar to that in case 3 of Table 4.1 and Fig. 4.12b. Effects of the wave ducting on the development of high-drag states for a flow with uniform wind and constant static stability are illustrated in Fig. 5.13. Shortly after the occurrence of wave breaking, regions with local $R_i < 0.25$ form in the vicinity of the wave breaking (Fig. 5.13a). This turbulent mixing region is expanding downward and downstream due to strong nonlinear effects on the flow with low Richardson number near the critical level (Fig. 5.14a). The turbulent mixing region expands downward by wave reflection, overreflection, and ducting from the wave-induced critical level and accelerates downstream by the nonlinear advevction (Fig. 5.13b).

Effects of wave reflection and/or overreflection are evidenced by the fact that the wave duct with severe downslope wind is located below the region of the turbulent mixing region. Note that the expansion of the turbulent mixing region provides a maintenance mechanism for the existence of the wave duct below it and above the lee slope, because the reflectivity in this region is about 1 according to linear theory. Without this almost perfect reflector, the wave below cannot be maintained and would lose most of its energy due to dispersion. In fact, wave overreflection can occur, according to the wave ducting theory discussed in Chapter 4, through the extraction energy from the well-mixed region and thus contribute to the acceleration of down-slope winds. In the absence of nonlinearity (Fig. 5.14b), the wavebreaking region does not expand downward to reduce the depth of the lower uniform wind layer. This, in

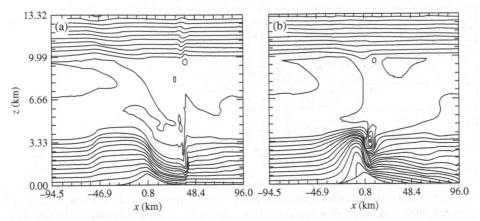

Fig. 5.14 Effects of nonlinearity on the development of severe downslope winds: (a) Potential temperature field from nonlinear numerical simulations for a basic flow with $R_i = 0.1$ and $F = 2$; (b) Same as (a) except from linear numerical simulations. The contour interval is 1 K in both (a) and (b). (Adapted after Wang and Lin 1999.)

turn, prohibits the formation of the severe downslope wind and internal hydraulic jump. These results indicate that the nonlinear wave ducting has contributed to the downward and downstream expansion of the turbulent mixing region.

b. Hydraulic theory

Based on the similarity of flow configurations of severe downslope windstorms and finite-depth, homogeneous flow over a mountain ridge, a hydraulic theory was proposed to explain the development of severe downslope winds (Smith 1985). The hydraulic theory attributes the high-drag (severe-wind) state to the interaction between a smoothly stratified flow and the deep, well-mixed, turbulent "dead" region above the lee slope in the middle troposphere. When a high-drag state develops, a *dividing streamline* encompasses this well-mixed region of uniform density (ρ_c in Fig. 5.15a). Assuming the upstream flow is uniform in U and N and the general flow is smooth, nondissipative, hydrostatic, Boussinesq and steady (Fig. 5.15a), the non-linear, hydrostatic governing equation can be simplified from (5.3.2),

$$\delta_{zz} + l^2\delta = 0, \tag{5.3.8}$$

The horizontal velocity can be derived from (5.3.7),

$$u = U(1 - \delta_z). \tag{5.3.9}$$

The lower boundary condition is given by (5.3.3). By assuming no disturbance above the upper dividing streamline (H_o), the pressure at $z = H_o$ is constant, i.e. $p(x, H_o) = p^*$. If the air in the turbulent region is hydrostatic in the mean and well mixed with a density

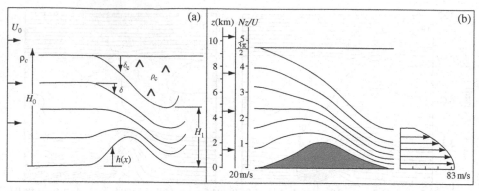

Fig. 5.15 A severe downslope windstorm simulated by a hydraulic theroy. (a) Schematic of an idealized high-drag state flow configuration. A certain critical streamline divides and encompasses a region of uniform density (ρ_c), which is called the dividing streamline. H_o and H_1 denote the heights of upstream dividing streamline and the downstream lower dividing streamline, respectively. (b) An example of transitional flow over a mountain. The dimensional values of the flow and orographic parameters are $U = 20\,\mathrm{m\,s^{-1}}$, $N = 0.01\,\mathrm{s^{-1}}$, $H_o = 9.42\,\mathrm{km}$, and $h = 2\,\mathrm{km}$. This gives $F = U/Nh = 1$. (Adapted after Smith 1985.)

of ρ_c, the pressure along the lower branch of the dividing streamline is $p(x, H_o + \delta_c) = p^* - \rho_c g \delta_c$, where δ_c is the vertical displacement of the lower dividing streamline (H_1). For a steady-state flow, the Bernoulli equation along $z = H_o + \delta_c$ can be written

$$p + (1/2)\rho u^2 + \rho_c g z = constant. \tag{5.3.10}$$

At $z = H_o + \delta_c$, we have

$$\delta_z = 0. \tag{5.3.11}$$

By assuming a wave-like solution in the vertical,

$$\delta(x, z) = A(x) \cos lz + B(x) \sin lz, \tag{5.3.12}$$

the nonlinear solution for high-drag state can be obtained,

$$\tilde{h} = \tilde{\delta}_c[\cos(\tilde{H}_o + \tilde{\delta}_c - \tilde{h})], \tag{5.3.13a}$$

$$\tilde{A} = \tilde{\delta}_c \cos(\tilde{H}_o + \tilde{\delta}_c), \text{ and} \tag{5.3.13b}$$

$$\tilde{B} = \tilde{\delta}_c \sin(\tilde{H}_o + \tilde{\delta}_c), \tag{5.3.13c}$$

where $h(x)$ is the terrain height function and all coefficients and parameters are nondimensionalized by $l (= N/U)$ and denoted by tildes "$\sim$". The above solution can be solved graphically or numerically as long as $\tilde{h}$ and $\tilde{H}_o$ are known.

Figure 5.15 shows an example of a severe downslope windstorm simulated by a hydraulic theory with $F = 1$. The descent of the lower dividing streamline begins over the point where the mountain begins to rise and becomes more rapid over the mountain peak. The final downward displacement of the dividing streamline is a large fraction of the initial layer depth. The flow speed after transition to supercritical flow over the lee slope from subcritical flow over the upslope is greatest near the surface and is several times the upstream value. The flow shown in Fig. 5.15b is qualitatively similar to the 1972 Boulder windstorm observations (Fig. 3.4a). In addition to the above solution, the strength of the transitional flow can be measured by the pressure drag on the mountain per unit length,

$$\mathcal{D} = \frac{\rho_0 N^2}{6} (H_0 - H_1)^3. \qquad (5.3.14)$$

The hydraulic theory of severe downslope winds was confirmed by numerical experiments of stratified fluid flow (e.g., Durran and Klemp 1987; Bacmeister and Pierrehumbert 1988) and laboratory tank experiments (e.g., Rottman and Smith 1989). Note that in order to apply the hydraulic theory to the prediction of the steady-state flow over a mountain, it is necessary to specify the initial height of the dividing streamline line. Thus, the dividing streamline height cannot be determined a priori if the critical level is induced by wave breaking. This, in turn, implies that the hydraulic model is limited to the consistent check of a severe wind state and cannot be used for prediction.

c. Applications of resonant amplification and hydraulic theories

Some discrepancies have been found between the resonant amplification and hydraulic theories of severe downslope windstorms. One discrepancy is the different critical level heights for high-drag (severe wind) states predicted by these two theories. The resonant amplification theory predicts the wave-induced critical (wave breaking) level at a height of $z/\lambda_z = 3/4 + n$ (n an integer), which helps produce severe downslope winds at later times. On the other hand, the hydraulic theory predicts critical level heights falling within the range of $z/\lambda_z = 1/4 + n$ to $3/4 + n$ during a high-drag state. This discrepancy appears to be caused by different stages of the severe downslope wind state being used for prediction. In fact, in earlier stages of a high-drag state, the resonant amplification theory is consistent with weakly nonlinear theories which indicate that the initiation of a high-drag transitional flow begins with linear resonance (Grimshaw and Smyth 1986), and with nonlinear numerical simulations which indicate that the lowest initial wave-induced critical level is near 3/4 (Lin and Wang 1996). It can also be seen clearly from Fig. 5.10 that the wave-induced critical level for a severe-wind state is shifted to a lower level at later time. Therefore, it appears that the resonant amplification theory focuses on the ealier stage of severe downslope wind development, while the hydraulic theory focuses on the later stage.

Part of the discrepancies may be related to the usage of critical level height as the control parameter to determine a high-drag state, as often adopted in many previous

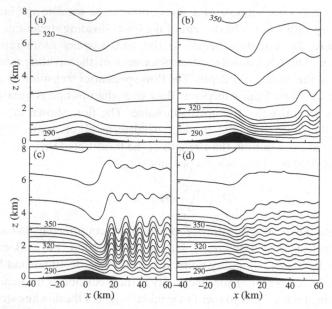

Fig. 5.16 The dependence of high-drag states on the lower-layer depth, as revealed by the isentropes for airflow in a two-layer atmosphere at $Ut/a = 25$, when $N_1 h/U = 0.5$, where N_1 is the Brunt–Vaisala frequency of the lower layer, and the depth of the lowest, most stable layer (U/N_1) is: (a) 1, (b) 2.5, (c) 3.5, and (d) 4. The lower layer resembles: (a) supercritical flow, (b) a propagating hydraulic jump, (c) a stationary jump, and (d) subcritical flow. (After Durran 1986a.)

studies. Based on some numerical experiments, the lower uniform flow layer depth appears to be a more appropriate scale to use (Wang and Lin 1999). Figure 5.16 indicates that the high-drag state is sensitive to the lower stable layer depth (Durran 1986a). Using the lower layer depth as the control parameter, predictions of both high- and low-drag states from several previous numerical studies are shown to be consistent, and the high-drag state does depend on the mountain height, which is consistent with the hydraulic theory. In addition, some discrepancies among previous studies result from the choice of different Richardson numbers and basic flow velocity profiles (e.g., Teixeira *et al.* 2005).

5.4 Flows over three-dimensional mountains

Although the two-dimensional mountain wave theories discussed in previous sections helped explain some important flow phenomena generated by infinitely long ridges, such as upward propagating mountains waves, lee waves, wave overturning and breaking, and severe downslope winds, in reality most of the mountains are of three-dimensional, complex form. The basic dynamics of flow over complex terrain can be understood by considering flow over an idealized, three-dimensional, isolated mountain. In this section, we will discuss a linear theory of a stratified flow past an isolated mountain, as well as the generation of lee vortices in a nonlinear flow over an isolated mountain.

5.4.1 Linear theory

In the following, the two-dimensional, linear mountain wave theory developed in Section 5.2.1 is extended to three-dimensional flow over an isolated mountain. Consider a steady-state, small-amplitude, adiabatic, inviscid, nonrotating, stratified, Boussinesq fluid flow with uniform basic velocity (U) and Brunt–Vaisala frequency (N) over a three-dimensional topography $h(x, y)$. The governing linear equations can be derived from (5.1.1)–(5.1.4),

$$U\frac{\partial u'}{\partial x} + \frac{1}{\rho_0}\frac{\partial p'}{\partial x} = 0, \tag{5.4.1}$$

$$U\frac{\partial v'}{\partial x} + \frac{1}{\rho_0}\frac{\partial p'}{\partial y} = 0, \tag{5.4.2}$$

$$U\frac{\partial w'}{\partial x} - g\frac{\theta'}{\theta_0} + \frac{1}{\rho_0}\frac{\partial p'}{\partial z} = 0, \tag{5.4.3}$$

$$\frac{\partial u'}{\partial x} + \frac{\partial v'}{\partial y} + \frac{\partial w'}{\partial z} = 0, \tag{5.4.4}$$

$$U\frac{\partial \theta'}{\partial x} + \frac{N^2\theta_0}{g}w' = 0. \tag{5.4.5}$$

Using (5.1.19), the above equations can be combined into a single equation of the vertical displacement (η),

$$\nabla^2\eta_{xx} + \frac{N^2}{U^2}\nabla_H^2\eta = 0. \tag{5.4.6}$$

Equation (5.4.6) can be solved by taking the double Fourier transform in x and y to obtain

$$\hat{\eta}_{zz} + m^2\hat{\eta} = 0, \tag{5.4.7}$$

where

$$m^2 = K^2(N^2/k^2U^2 - 1), \tag{5.4.8}$$

and $K = \sqrt{k^2 + l^2}$ is the horizontal wave number. The double Fourier transform pair is defined as

$$\hat{\eta}(k,l,z) = \frac{1}{4\pi^2} \int_{-\infty}^{\infty} \int_{-\infty}^{\infty} \eta(x,y,z) \mathrm{e}^{-\mathrm{i}(kx+ly)} \mathrm{d}x \mathrm{d}y, \text{ and} \qquad (5.4.9a)$$

$$\eta(x,y,z) = \int_{-\infty}^{\infty} \int_{-\infty}^{\infty} \hat{\eta}(k,l,z) \mathrm{e}^{\mathrm{i}(kx+ly)} \mathrm{d}k \mathrm{d}l. \qquad (5.4.9b)$$

The solution to (5.4.7) in the Fourier space can be found

$$\hat{\eta}(k,l,z) = \hat{\eta}(k,l,0) \mathrm{e}^{\mathrm{i}m(k,l)z}. \qquad (5.4.10)$$

Similar to the two-dimensional mountain wave theory, as discussed in Section 5.2, there exist two flow regimes: (I) $N^2/k^2U^2 > 1$, and (II) $N^2/k^2U^2 < 1$. For upward propagating waves (regime I), the sign of m must be the same as the sign of k, in order to satisfy the upper radiation condition. On the other hand, for evanescent waves (regime II), the positive root of (5.4.8) must be chosen, i.e.

$$\hat{\eta}(k,l,z) = \hat{\eta}(k,l,0) \mathrm{e}^{-m_i(k,l)z}, \qquad (5.4.11)$$

where m_i is defined as $K\sqrt{1 - N^2/k^2U^2}$. The linear lower boundary condition is

$$\eta(x,y,z=0) = h(x,y), \qquad (5.4.12)$$

which can be transformed into the Fourier space,

$$\hat{\eta}(k,l,0) = \hat{h}(k,l). \qquad (5.4.13)$$

From the definition of inverse Fourier transform and (5.4.13), we have

$$\eta(x,y,z) = \int_{-\infty}^{\infty} \int_{-\infty}^{\infty} \hat{h}(k,l) \mathrm{e}^{\mathrm{i}m(k,l)z} \mathrm{e}^{\mathrm{i}(kx+ly)} \mathrm{d}k \mathrm{d}l. \qquad (5.4.14)$$

Now let us consider a three-dimensional (circular) bell-shaped mountain

$$h(x,y) = \frac{h}{(r^2/a^2 + 1)^{3/2}}; \quad r = \sqrt{x^2 + y^2}, \qquad (5.4.15)$$

where h and a are the mountain height and horizontal scale, respectively. The Fourier transform of (5.4.15) is,

$$\hat{h}(k,l) = \frac{ha^2}{2\pi} \mathrm{e}^{-aK}. \qquad (5.4.16)$$

The problem may be further simplified by using the hydrostatic approximation, i.e. neglecting the first term of (5.4.3). Note that under the hydrostatic approximation, we require that $Na/U \gg 1$. The solution, (5.4.14), may be reduced to a single integration by converting it into cylindrical coordinates, and asymptotic solutions for the flow aloft and the flow near the ground may thus be obtained (Smith 1980). Substituting (5.4.16) into (5.4.14) and nondimensionalizing it according to

$$(\tilde{x}, \tilde{y}) = (x/a, y/a), (\tilde{z}, \tilde{\eta}) = (Nz/U, N\eta/U), (\tilde{k}, \tilde{l}, \tilde{K}) = (ka, la, Ka), \quad (5.4.17)$$

yields

$$\tilde{\eta}(\tilde{x}, \tilde{y}, \tilde{z}) = \frac{1}{F} \int_{-\infty}^{\infty} \int_{-\infty}^{\infty} e^{-\tilde{K}} e^{i\tilde{m}\tilde{z}} e^{i(\tilde{k}\tilde{x}+\tilde{l}\tilde{y})} d\tilde{k} d\tilde{l}, \quad (5.4.18)$$

where F is the Froude number, as defined earlier. As discussed earlier, the linear theory holds for a large Froude number flow. On the other hand, for a small Froude number flow, nonlinear effects become more important and cannot be ignored. This will be discussed in the next subsection.

Equation (5.4.18) or (5.4.14) can also be solved numerically by applying a two-dimensional numerical FFT algorithm. Figure 5.17 shows an example of a linear hydrostatic flow passing over a bell-shaped mountain with a Froude number of 100. Near the surface, the pattern of vertical displacement resembles the surface topography, (5.4.15), as required by the lower boundary condition. Slightly aloft from the surface at $\tilde{z} = \pi/4$ (Fig. 5.17a), a region of downward displacement forms a U-shaped disturbance over the lee slopes of the mountain and extends some distance downstream. At a level further aloft, such as $\tilde{z} = \pi$ (Fig. 5.17b), the region of downward displacement widens, moves upstream, and is replaced by a U-shaped pattern of upward displacement. The general upstream shift of downward and upward displacement is caused by the upstream phase tilt of upward propagating hydrostatic waves. At greater heights, the zone of disturbance continues to broaden, the disturbance directly in the lee of the mountain disappears, the patterns of upward and downward displacement become more wavelike, due to wave dispersion.

The U-shaped patterns of vertical displacements are explained by a group velocity argument (Smith 1980). The dispersion relation for internal gravity waves in a stagnant Boussinesq fluid may be reduced from (3.6.10)

$$\omega = \frac{\pm NK}{\sqrt{k^2 + l^2 + m^2}}. \quad (5.4.19)$$

With the hydrostatic approximation the above equation becomes

$$\omega = \pm \frac{NK}{m}. \quad (5.4.20)$$

As discussed in Chapter 4, the energy propagation can be described by the group velocity components, which are

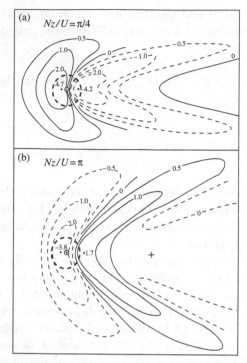

Fig. 5.17 Three-dimensional, linear, hydrostatic stratified flow over a bell-shaped mountain (5.4.15) with $F = U/Nh = 100$. Displayed are the nondimensional vertical displacements at $\tilde{z} = Nz/U =:$ (a) $\pi/4$ and (b) π. U-shaped disturbances are associated with the upward propagating wave energy. Solid and dashed curves represent positive and negative values of vertical displacement. The cross marks the position of the mountain peak. The bold, dashed circle is the topographic contour at $r = a$, where r is the distance (radius) from the center of the mountain. These wave patterns are computed by evaluating (5.4.18) numerically using a two-dimensional FFT. (Adapted after Smith 1980.)

$$c_{gx} = \frac{\partial \omega}{\partial k} = \pm \frac{Nk}{mK}; \; c_{gy} = \frac{\partial \omega}{\partial l} = \pm \frac{Nl}{mK}; \; c_{gz} = \frac{\partial \omega}{\partial m} = \mp \frac{NK}{m^2} \qquad (5.4.21)$$

For steady-state waves on a basic flow, replacing ω by the intrinsic frequency Uk in (5.4.20) leads to

$$m = \frac{\pm NK}{Uk}. \qquad (5.4.22)$$

Adding U to c_{gx}, the components of the group velocity in a frame fixed with the Earth become

$$c_{gmx} = \frac{Ul^2}{K^2}; \; c_{gmy} = \frac{-Ukl}{K^2}; \; c_{gmz} = \frac{U^2 k^2}{NK}. \qquad (5.4.23)$$

In the coordinates fixed with the mountain or Earth, wave energy propagates from the energy source, i.e. the mountain, along straight lines with slopes

$$\frac{x}{z} = \frac{c_{gmx}}{c_{gmz}}; \quad \frac{y}{z} = \frac{c_{gmy}}{c_{gmz}}; \quad \frac{y}{x} = \frac{c_{gmy}}{c_{gmx}}. \tag{5.4.24}$$

The slope on the horizontal plane y/x may be evaluated from (5.4.23) and (5.4.24),

$$\frac{y}{x} = -\frac{k}{l}, \tag{5.4.25}$$

which is the geometric condition that the phase lines passing through the point (x,y) are radial lines from the origin. Using (5.4.23)–(5.4.24) again gives

$$y^2 = \frac{Nxz}{UK}. \tag{5.4.26}$$

Since the mountain is the source of forcing, the horizontal wavenumber may be approximated by the mountain scale, i.e. $K \approx 1/a$, which yields

$$y^2 = \left(\frac{Nax}{U}\right)z. \tag{5.4.27}$$

Thus, the energy concentrates in a parabola or a U-shaped pattern at a certain height, as shown in Fig. 5.17.

In the above theory, the basic flow speed and Brunt–Vaisala frequency are assumed to be constant with height. In the real atmosphere, they normally vary with height. As in the two-dimensional mountain wave problem, a rapid decrease of the Scorer parameter with height leads to the formation of *trapped lee waves*. The formation of three-dimensional trapped lee waves is similar to that of *Kelvin ship waves* over the water surface. Figure 5.18 shows an example of the cloud streets associated with three-dimensional trapped lee waves produced by airflow past a mountainous island. The wave pattern is generally contained within a wedge with the apex at the mountain. The three-dimensional trapped lee waves are composed by *transverse waves* and *diverging waves*, as depicted in Fig. 5.19. The transverse waves lie approximately perpendicular to the flow direction, and are formed by waves attempting to propagate against the basic flow but that have been advected to the lee. The formation mechanism of transverse waves is the same as that of the two-dimensional trapped lee waves. Unlike the transverse waves, the diverging waves attempt to propagate laterally away from the mountain and have been advected to the lee. Also, the diverging waves have crests that meet the incoming flow at a rather shallow angle. Both of the transverse and diverging waves are mathematically associated with a stationary phase point, and the significant disturbance is confined within a wedge angle of about $19°28'$ with the x-axis.

Fig. 5.18 Satellite imagery of three-dimensional trapped lee waves induced by the South Sandwich Islands in the southern Atlantic Ocean on September 18, 2003. The wave pattern is similar to that of the ship waves sketched in Fig. 5.19. (From Visible Earth, NASA.)

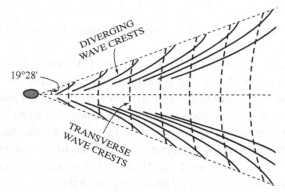

Fig. 5.19 Schematic of transverse (bold-dashed) and diverging (solid) phase lines for a deep water ship wave. (Adapted after Sharman and Wurtele 1983.)

5.4.2 Generation of lee vortices

The above linear theory of three-dimensional, stratified flow over mountains provides an in-depth understanding of the dynamics, but it is valid only for high Froude number flow due to the limitations of the small-amplitude (linear) assumption. When the Froude number decreases, the perturbations generated by the mountain become larger and the flow becomes more nonlinear. Due to mathematical intractability, many observed phenomena associated with nonlinear flow over mountains,

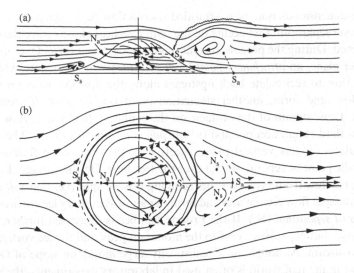

Fig. 5.20 (a) Side view of the mean surface shear stress pattern and streamlines on the center plane of symmetry for a three-dimensional, stratified, viscous flow with $F = U/Nh = 0.4$ past an obstacle with circular contours (e.g., bold solid curve in (b)). In the figure, N and S denote nodes and saddle points, respectively, and subscripts a and s denote attachment and separation, respectively. (b) As (a) but for a plane view of the pattern of surface stress. (Adapted after Hunt and Snyder 1980.)

such as flow recirculation, stagnation points, flow splitting, and lee vortices, have been carried out in tank experiments and by nonlinear numerical simulations.

a. Boundary layer separation

The flow pattern produced by laboratory tank experiments for three-dimensional, stratified flow with relatively large Froude numbers (e.g., $F > 2$) past a bell-shaped mountain is similar to that predicted by linear theory as described in Subsection 5.4.1. Flow patterns are dramatically different for flow with smaller Froude numbers. Figure 5.20 shows a stratified flow with $F = 0.4$ past an isolated mountain in a tank experiment. The most eye-catching phenomenon is a pair of counter-rotating vortices formed in the lee of the obstacle. The formation of this pair of lee vortices is attributed to the *boundary layer separation* mechanism (Batchelor 1967; Hunt and Snyder 1980), as briefly summarized in the following. When the Reynolds number (R_e) is sufficiently high (where $R_e = UL/\nu$, U is the velocity scale, L length scale and ν kinematic viscosity), the boundary-layer flow develops a region of flow reversal near the surface due to an opposing pressure gradient in the direction of flow. The reversed flow meets the incoming flow and forms a stagnation point at which the streamline breaks away from the surface of the obstacle. This process is known as *boundary layer separation*. Mathematically, the streamline of boundary layer separation is a line whose points are singular points of the solutions of the equations of motion in the boundary layer.

For three-dimensional, nonlinear, stratified viscous flow past a symmetric mountain, boundary layer separation first occurs on the center vertical plane before the mountain peak is reached. During the process, several singular points can form. Over the upslope on the center plane, an *attachment point* (*node of attachment*) N_a forms, which forces part of the flow to recirculate back upstream along the upslope, where it meets the incoming flow, and forms another stagnation point (*saddle point of separation*) S_s (Fig. 5.20a). Downstream of the obstacle on the center vertical plane, flow separates and forms a third stagnation point(s) (*saddle points of attachment*) S_a. The separated flow recirculates on this vertical plane, meets with the downslope flow and forms another saddle point of separation (S_s) over the lee slope. On the surface (Fig. 5.20b), the recirculated flow from N_a forces the incoming flow to split (i.e. *flow splitting*) at S_s and part of the split flow recirculates and forms a pair of stationary lee vortices centered at the *nodes of separation* (N_s). If the Froude number is decreased further, this flow pattern persists, but N_a moves closer to the mountain peak and the lee vortices expand further downstream. Although an unrealistically large mountain slope of O(1), compared to that in the real world is often used in laboratory experiments, the simulated flow features are very similar to those observed in the real atmosphere.

b. Generation of lee vortices in an inviscid fluid

Using a nonlinear numerical model with free-slip lower boundary condition, a pair of counter-rotating vortices was found to form on the lee of an isolated mountain when a low-Froude number (e.g., $F = 0.66$, Fig. 5.21a), three-dimensional, stratified, uniform flow passes over the mountain (Smolarkiewicz and Rotunno 1989). The simulated results agree fairly well with laboratory tank experiments as shown in Fig. 5.20. The free-slip lower boundary condition implies no explicit surface friction is included in the model atmosphere. Although linear theory breaks down, at least locally, the vertical displacement field (Fig. 5.21c) still resembles the U-shaped pattern found in the linear theory described in Subsection 5.4.1 (Fig. 5.17). A large-amplitude mountain wave develops over the mountain peak (Fig. 5.21e). The trough of the vertically propagating gravity waves in Fig. 5.21e shifts upstream and becomes narrower, indicating a tendency toward collapse of the isentropic surfaces on the lee slopes of the mountain, which is also in agreement with the linear theory. Since the air parcels are able to flow almost directly across the mountain, this flow regime is characterized as the *flow-over regime*.

When the Froude number is reduced to approximately below 0.5, such as $F = 0.22$ (Fig. 5.21b), a pair of counter-rotating vortices forms on the lee side and a saddle point of separation and a node of attachment are produced on the upstream side of the mountain, strikingly similar to the results obtained in laboratory experiments (Fig. 5.20b). The region of downward displacement is enlarged (Fig. 5.21d). The gravity wave response is drastically reduced, as much of the airflow is diverted around the flanks of the mountain and the disturbance appears to be much more horizontal (Fig. 5.21f). Below the mountain top, there is a recirculating flow associated with the

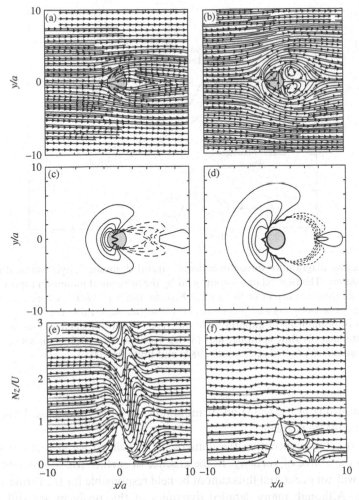

Fig. 5.21 Three-dimensional, stratified, uniform flow with no surface friction over a bell-shaped mountain simulated by a nonlinear numerical model. Surface streamlines, vertical displacements at $Nz/U = \pi/4$, and streamlines in the vertical plane $y/a = 0$ after $Ut/a = 9$ are shown in (a), (c), and (e), respectively, for the case with $F = 0.66$. The same flow fields but for $F = 0.22$ are shown in the right panels ((b), (d) and (f)). The simulated flow fields have reached quasi-steady state. The flow and orographic parameters are: $U = 10\,\mathrm{m\,s^{-1}}$ or $3.3\,\mathrm{m\,s^{-1}}$, $N = 0.01\,\mathrm{s^{-1}}$, $h = 1.5\,\mathrm{km}$, and $a = 10\,\mathrm{km}$, which give $F = 0.66$ or 0.22, respectively. The bell-shaped mountain is prescribed by (5.4.15.). (Adapted after Smolarkiewicz and Rotunno 1989.)

lee vortices. This flow regime is characterized as the *flow-around regime*. Based on the nondimensional mountain height (or inverse Froude number – Nh/U) and horizontal mountain aspect ratio (b/a), four classes of wave and flow phenomena of importance in three-dimensional, stratified, uniform, hydrostatic flow past an isolated mountain

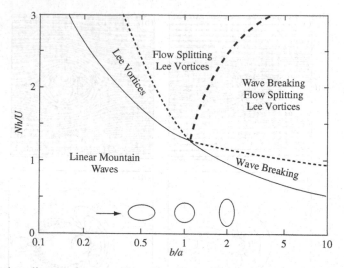

Fig. 5.22 Regime diagram for three-dimensional, stratified, uniform, hydrostatic flow over an isolated mountain. The flow regime is controlled by the horizontal mountain aspect ratio (b/a) and the nondimensional height or the inverse Froude number (Nh/U), where a and b are the mountain scales in along (x) and perpendicular (y) to the basic flow directions, respectively. Four classes of phenomena of importance in this type of flow are: (1) linear mountain waves, (2) wave breaking, (3) flow splitting, and (4) lee vortices. The circles/ellipses represent the mountain contours. (Adapted after Smith 1989a and Epifanio 2003.)

can be identified (Fig. 5.22): (1) linear mountain waves, (2) wave breaking, (3) flow splitting, and (4) lee vortices.

The key question concerning the numerically simulated lee vortices as shown in Fig. 5.21 is the source of vorticity. In the absence of surface friction, boundary layer separation will not occur and thus cannot be held responsible for the formation of the lee vortices. Although many detailed dynamics of this problem are still topics of current research, the basic dynamics for the generation of lee vortices can be understood through the following two major theories: (1) *tilting of baroclinically generated vorticity* (Smolarkiewicz and Rotunno 1989) and (2) *generation of internal potential vorticity* by turbulence dissipation in numerical simulations (Smith 1989b; Schär and Smith 1993a, b).

c. Tilting of baroclinically generated vorticity

The mechanism of baroclinically generated vorticity tilting can be understood by taking cross differentiations of (2.2.1)–(2.2.3) to yield the inviscid vorticity equation

$$\frac{\partial \boldsymbol{\omega}}{\partial t} = -\boldsymbol{V} \cdot \nabla \boldsymbol{\omega} + (\boldsymbol{\omega} \cdot \nabla)\boldsymbol{V} + \frac{\nabla \rho \times \nabla p}{\rho^2}, \qquad (5.4.28)$$

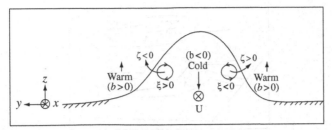

Fig. 5.23 A schematic diagram showing the generation of leeside vorticity by the vertical tilting of baroclinically generated horizontal vorticity (Smolarkiewicz and Rotunno 1989). The downward (upward) arrow below the adiabatically-induced cold (warm) region denotes downward (upward) motion. A negative x-vorticity, $\xi < 0$, is produced over the right-hand side of the upslope baroclinically by the relatively cold air ($b < 0$) along the center line and the relatively warm air to the right (facing downstream), as indicated by (5.4.29). This negative x-vorticity is then swept downstream and produces a positive vertical vorticity, $\zeta > 0$, on the right-hand side of the lee due to the vertical tilting of the x-vorticity, as implied by (5.4.28).

where $\boldsymbol{\omega} = \nabla \times \boldsymbol{V} = (\xi,\ \eta,\ \zeta)$ is the three-dimensional *vorticity vector*. The last term on the right side of the above equation represents the generation of vorticity by baroclinicity. Once local vorticity anomalies are generated, they are advected by the flow field through the first term or tilted and stretched through the second term on the right-hand side of (5.4.28). For mountains with small aspect ratio of the obstacle height and horizontal width, the baroclinicity term reduces to (e.g., Epifanio 2003):

$$\frac{\nabla \rho \times \nabla p}{\rho^2} \approx -\boldsymbol{k} \times \nabla b, \qquad (5.4.29)$$

where b is the buoyancy.

Figure 5.23 shows a schematic diagram depicting the generation of leeside vorticity by the vertical tilting of baroclinically generated horizontal vorticity. A negative x-vorticity, $\xi < 0$, is generated on the right upslope baroclinically by the relatively cold air along the center line and the relatively warm air to the right (facing downstream), as indicated by (5.4.29). This negative x-vorticity is then swept downstream and produces a positive vertical vorticity, $\zeta > 0$, to the lee by the vertical tilting of the x-vorticity, as implied by (5.4.28). Similarly, a positive x-vorticity anomaly generated over the left upslope is tilted into a negative vertical vorticity to the lee. As these vertical vorticity anomalies intensify, recirculating warm-core eddies develop as a result of reconnection. This mechanism dominates during the rapid start-up, early stage, over a nondimensional time $Ut/a = O(1)$, in which the flow is essentially inviscid and adiabatic and the potential vorticity (PV) is conserved (Schär and Durran 1997).

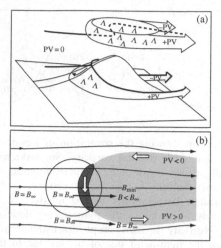

Fig. 5.24 (a) A conceptual model depicting potential vorticity (PV) generation by turbulence dissipation at stagnation points associated with wave breaking aloft and upstream blocking. The symbol "∧∧∧" denotes areas of turbulence generated by wave breaking or blocking. (Adapted after Smith 1989a.) (b) Schematic depiction of the relationship between PV generation and Bernoulli function on an isentropic surface in steady-state, stratified flow over the wave breaking region. Thin lines are streamlines and dark-shaded area over the lee slope denotes a localized region of dissipation due to wave breaking, a hydraulic jump or blocking. The gray shaded area extending downstream denotes a reduced Bernoulli function. Open arrows denote the PV flux J associated with the Bernoulli gradient on the isentropic surface as described by (5.4.35). (From Schär and Durran 1997.)

d. Generation of potential vorticity by turbulence dissipation

At a later stage, the associated thermal anomalies generated by baroclinicity are eroded by dissipative and diffusive processes, whereby the warm surface anomalies are converted into PV. During this stage, the flow is controlled by dissipation and is accompanied by the PV generation over a nondimensional time of $O(10)$–$O(100)$ (Schär and Durran 1997). Note that the *conservation of potential vorticity* is violated in regions of *flow stagnation*, such as in the region of upstream blocking where the isentropic surface intersects the ground, and the region of wave breaking above the lee slope where turbulence occurs (Fig. 5.24a). The dynamics of dissipative generation of PV is directly linked to the reduction in the Bernoulli function within the wake, as demonstrated in steady shallow-water flow past an obstacle (Schär and Smith 1993a). The shallow-water theory can be extended to stratified fluid flow by considering the PV (q) which satisfies a conservative equation of the form (Haynes and McIntyre 1990):

$$\frac{\partial(\rho q)}{\partial t} + \nabla \cdot \boldsymbol{J} = 0, \tag{5.4.30}$$

where q is defined as

$$q = \frac{\nabla\theta \cdot \boldsymbol{\omega}_a}{\rho},$$ (5.4.31)

and the total *PV flux* (*J*) is given by

$$\boldsymbol{J} = \rho q \boldsymbol{V} - (\dot{Q}\boldsymbol{\omega}_a + \boldsymbol{F} \times \nabla\theta).$$ (5.4.32)

In the above equation, $\dot{Q}$ ($\equiv \mathrm{D}\theta/\mathrm{D}t$) is the diabatic heating, $\boldsymbol{\omega}_a$ the three-dimensional absolute vorticity vector, and $\boldsymbol{F}$ the viscous force per unit mass. In this section, we have assumed that the Earth rotation is negligible thus $\boldsymbol{\omega}_a = \boldsymbol{\omega}$.

It can be shown that

$$\boldsymbol{J} = \nabla\theta \times \left(\nabla B + \frac{\partial \boldsymbol{V}}{\partial t}\right) - \boldsymbol{\omega}\frac{\partial\theta}{\partial t},$$ (5.4.33)

where

$$B = \boldsymbol{V} \cdot \boldsymbol{V}/2 + c_p T + gz$$ (5.4.34)

is the *Bernoulli function*. In a steady-state flow, the Bernoulli function is conserved following the flow. In addition, (5.4.33) reduces to

$$\boldsymbol{J} = \nabla\theta \times \nabla B = \frac{\partial\theta}{\partial n}\boldsymbol{n} \times \nabla B,$$ (5.4.35)

where $\boldsymbol{n}$ is a unit vector oriented perpendicular to the isentropic surface and pointing toward warm air. The *generalized Bernoulli's theorem* (Schär 1993), (5.4.35), indicates that non-zero PV fluxes must be present where there is a variation in the Bernoulli function along any isentropic surface. Figure 5.24b shows a schematic of PV generation by turbulence dissipation on an isentropic surface in steady-state stratified flow past an isolated mountain. The narrow dissipative region may be produced by turbulence associated with wave breaking, a hydraulic jump or blocking, and generates Bernoulli function deficit in the wake extending downstream. Based on Fig. 5.24 and (5.4.34), PV is generated in the dissipative region and advected downstream along the edge of the wake. A pair of counter-rotating vortices may form in the wake if the vertical vorticity associated with the generated PV is sufficiently strong.

It appears that the above PV analysis is able to explain the close relationship between dissipative turbulence and PV generation for a low-Froude number, stratified flow over an isolated mountain. The causality, however, is still unclear due to the steady-state assumption. In addition, an assumption of balance is required in order to infer the structure of the flow from the distribution of PV (Hoskins *et al.* 1985). In the near field of the wake, these balance constraints are constantly strongly violated due to the presence of the strong surface temperature gradient over the lee slope, which results from the upstream blocking (Epifanio and Rotunno 2005). Therefore, although the PV generation

Fig. 5.25 A von Kármán vortex street that formed to the lee of the Guadalupe Island, off the coast of Mexico's Baja Peninsula, revealed by MISR images from June 11, 2000 detected by NASA satellite Terra. (From Visible Earth, NASA.)

may have important implications on the downstream evolution of orographic wakes and lee vortices, a fundamental understanding of the wake formation is still needed.

When the wake flow in which the lee vortices are embedded becomes unstable, the vortices tend to shed downstream and form a *von Kármán vortex street*. A von Kármán vortex street is a repeating pattern of alternate and swirling vortices along the center line of the wake flow, and is named after the fluid dynamicist, Theodore von Kármán. This process is also known as *vortex shedding*. Any noise, impulsive disturbance, or asymmetric forcing in the wake flow can trigger an instability, which gives way to a vortex street or vortex shedding. Figure 5.25 shows an example of a von Kármán vortex street formed in the atmosphere to the lee of a mountainous island. The von Kármán vortex street or vortex shedding has also been simulated by many nonlinear numerical models, such as that shown in Fig. 5.28a.

5.5 Flows over larger mesoscale mountains

5.5.1 Rotational effects

In the previous sections, effects of Earth's rotation are neglected. This is approximately valid for flow with Rossby number ($R_o = U/fL$, where L is the horizontal scale of the mountain) much larger than 1. However, for flow over mountains with $R_o = O(1)$ or smaller, the effects of Earth's rotation cannot be ignored. In this situation, the advection time for an air parcel to pass over the mountain is too large to be ignored compared to the period of inertial oscillation due to Earth's rotation ($2\pi/f$). Flow past many mesoscale mountain ranges, such as the European Alps, US Rockies, Canadian Rockies, Andes, the

Scadinavian mountain range, the New Zealand Alps, and the Central Mountain Range of Taiwan, belong to this category. For flow over a very broad large-scale mountain, which gives a very small Rossby number, the flow can be approximated by the quasi-geostrophic theory. In general, however, for flow over mesoscale mountains, the Rossby numbers are on the order of 1, for which the influence of the Coriolis force is too large to be ignored, but too small to be approximated by the quasi-geostrophic theory.

In considering the small-amplitude, two-dimensional ($\partial/\partial y = 0$), steady-state, inviscid, rotating, uniformly stratified, Boussinesq flow over an infinitely long ridge on an f-plane, the governing equations can be derived from (3.6.1)–(3.6.5),

$$\left(U^2 \frac{\partial^2}{\partial x^2} \right) \left(\frac{\partial^2 w'}{\partial x^2} + \frac{\partial^2 w'}{\partial z^2} \right) + f^2 \frac{\partial^2 w'}{\partial z^2} + N^2 \frac{\partial^2 w'}{\partial x^2} = 0. \tag{5.5.1}$$

Similar to the procedure for obtaining solutions described in Subsection 5.2.1, the vertical displacement can be obtained by,

$$\eta(x, z) = 2\text{Re} \left[\int_0^\infty \hat{h}(k) e^{i(kx+mz)} dk \right], \tag{5.5.2}$$

where $\hat{h}(k)$ is the one-sided Fourier transform of the mountain shape $h(x)$ (Appendix 5.1)

$$\hat{h}(k) = \frac{1}{2\pi} \int_{-\infty}^\infty h(x) e^{-ikx} dx, \tag{5.5.3}$$

and m is the vertical wave number

$$m = \sqrt{\frac{k^2(N^2 - k^2 U^2)}{k^2 U^2 - f^2}}. \tag{5.5.4}$$

Note that (5.5.4) is similar to that defined in (3.6.10), in which $\Omega \equiv \omega - kU$ for a time-dependent problem. Several regimes exist for the mountain waves in a rotating stratified atmosphere, similar to those of inertial gravity waves discussed in Chapter 4 and summarized in Table 3.2. We have already discussed two flow regimes in earlier sections, such as the *nonrotating evanescent flow regime* for $k^2 U^2 > N^2 \gg f^2$, and the *hydrostatic gravity wave regime* for $N^2 > k^2 U^2 \gg f^2$.

To demonstrate the rotational effects, we consider an extreme case of a flow past a very broad mountain ridge. Assuming a bell-shaped mountain range, (5.2.5), its Fourier transform can be obtained: $\hat{h}(k) = (h_m/2)ae^{-ka}$ if $k > 0$ [(5.2.6)]. In this situation, we have $N^2 > f^2 \gg k^2 U^2$ and the vertical wave number reduces to $m = ikN/f$. As discussed in Chapter 4, this is equivalent to making the quasi-geostrophic approximation. Substituting $\hat{h}(k)$ and m into (5.5.2) leads to the solution for a quasi-geostrophic flow over a very broad bell-shaped mountain ridge,

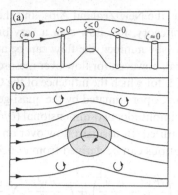

Fig. 5.26. A sketch of a stratified, quasi-geostrophic flow over a circular mountain. (a) The vertical vorticity associated with the deformation of vortex tubes is shown. The solid lines represent trajectories on the vertical cross section through the center of the mountain. (b) The streamline pattern near the surface for the flow associated with (a) as seen from above. (Adapted from Smith 1979 and Buzzi and Tibaldi 1977.)

$$\eta(x, z) = \frac{h_{\mathrm{m}} a^2 (Nz/fa + 1)}{x^2 + a^2 (Nz/fa + 1)^2}. \tag{5.5.5}$$

The above solution can be extended to quasi-geostrophic flow over a three-dimensional, isolated mountain, as shown in Fig. 5.26a. The lifting of θ-surface aloft is less than the mountain height, but the lifting is more widespread. Assuming a non-rigid lid top boundary, the air column upstream of the mountain is stretched slightly, producing weak cyclonic vorticity. Over the mountain, the air column is shortened producing anticyclonic vorticity. This anticyclonic vorticity is associated with a mountain-induced high pressure or anticyclone. On the lee side, the air column is slightly stretched again, producing weak cyclonic vorticity. Figure 5.26b shows the streamline pattern near the surface on a horizontal plane for the flow associated with Fig. 5.26a. When a straight incoming flow approaches a mountain ridge in the Northern Hemisphere, it turns cyclonically (left facing downstream) slightly upstream of the mountain, anticyclonically over the mountain, and cyclonically slightly on the immediate lee side in response to the slight vorticity stretching, major vorticity shrinking, and slight vorticity stretching of the vertical vorticity, in order to conserve the potential vorticity (Fig. 5.26a). This gives the streamline pattern as depicted in Fig. 5.26b. The perturbation velocity and pressure field decay away from the mountain. The upwarping of isentropic surfaces and the cold-core anticyclone near the ground, caused by a cold air mass at the surface, is analogous to a mountain anticyclone. When a preexisting cyclone passes over a mountain, its low pressure or cyclonic vorticity is weakened by the mountain-induced high pressure or even completely suppressed (filled in) on a weather map. After it passes over the mountain, the

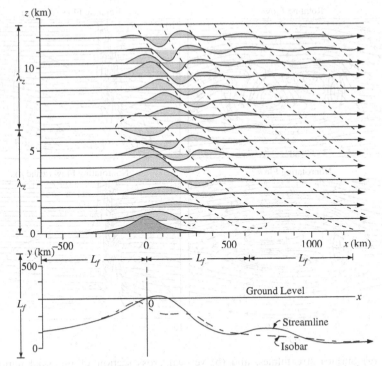

Fig. 5.27 Stratified rotating hydrostatic flow over a two-dimensional ridge with the parameters: $a = 100\,\mathrm{km}$, $f = 10^{-4}\,\mathrm{s}^{-1}$, $U = 10\,\mathrm{m\,s}^{-1}$, and $N = 0.01\,\mathrm{s}^{-1}$. Equation (5.5.2) multiplied by the non-Boussinesq factor $\sqrt{\rho_0/\rho(z)}$ is solved with an asymptotic method. This flow belongs to an intermediate regime between the nonrotating flow regime (e.g., Fig. 5.3b) and the quasi-geostrophic flow regime (e.g., Fig. 5.26a). The vertical wavelength of the wave is $\lambda_z \approx 2\pi U/N \approx 6.28\,\mathrm{km}$. The bottom figure depicts the lateral deflection of the streamlines. In the figure, $L_f \approx 2\pi U/f \approx 600\,\mathrm{km}$ near the surface. (Adapted after Queney 1948.)

cyclone reappears on the lee side. In addition, the cyclone track is often deflected by the mountain-induced anticyclone. This will be addressed in the next subsection.

Due to mathematical intractability, asymptotic methods or numerical techniques are often employed to solve (5.5.2). Figure 5.27 shows an asymptotic solution of (5.5.2) multiplied by the non-Boussinesq factor $\sqrt{\rho_s/\rho(z)}$. This flow belongs to an *intermediate* or *mesoscale mountain wave regime* between the *nonrotating hydrostatic wave regime* (e.g., Fig. 5.3b) and the *quasi-geostrophic flow regime* (e.g., Fig. 5.26a). In this *intermediate regime* with $U/a \approx f$, both the inertia force and Coriolis force play comparable roles in generating and maintaining the mesoscale mountain waves. The influence of the Coriolis force is evident in the lateral deflection of the streamlines and in the dispersive tail of the longer waves trailing behind the mountain at upper levels. The horizontal wavelength is $L_f \approx 2\pi U/f \approx 600\,km$ near the surface and the vertical wavelength is $\lambda_z \approx 2\pi U/N \approx 6.28\,km$. Unlike nonrotating flows, the upstream blocked

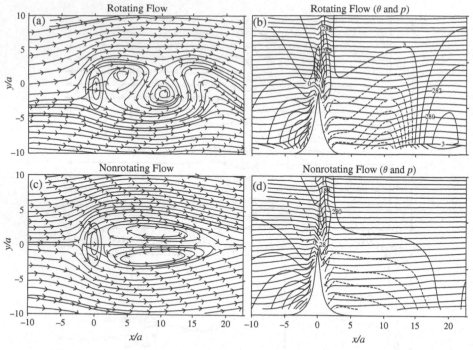

Fig. 5.28 (a) Surface streamlines, and (b) vertical cross section of potential temperature (horizontal curves) and pressure perturbations at the centerline ($y/a = 0$) for a uniform westerly flow over idealized topography after a nondimensional time $Ut/a = 40$. The Froude number and the Rossby number are 0.4 and 2.15, respectively. The dimensional values of the flow and orographic parameters are: $U = 10\,\mathrm{m\,s^{-1}}$, $f = 5.8 \times 10^{-5}\,\mathrm{s^{-1}}$, $a = 40\,\mathrm{km}$, and $h = 2.5\,\mathrm{km}$. In estimating the Rossby number, $2a$ is used. Panels (c) and (d) are the same as (a) and (b) except with no rotation. The x and y scales are nondimensionalized by the mountain half-width in the x direction. (Adapted after Lin *et al.* 1999.)

decelerated zone produced by a mountain ridge on a uniform, stratified flow on an f plane does not propagate to infinity (Pierrehumbert and Wyman 1985). Instead, it attains a maximum extent on the order of the *Rossby radius of deformation* (Nh/f).

Some rotational effects on three-dimensional stratified airflow past an isolated, mesoscale mountain are illustrated in Fig. 5.28. With rotation, a significant portion of the upstream fluid particles move leftward (facing downstream) in response to the excessive pressure gradient force versus the reduced Coriolis force because the weakened easterly flow is out of balance geostrophically (Fig. 5.28a), compared with the case without rotation (Fig. 5.28c). The separation point is shifted to the right-hand side in the rotating flow (Fig. 5.28a), instead of located along the centerline ($y/a = 0$) (Fig. 5.28c). Downstream of the mountain, vortex shedding occurs. Over the mountain peak, the hydrostatic mountain wave is very weak because most of the flow is blocked by the high mountain and is deflected to the right or left, instead of climbing up and passing over the mountain (Fig. 5.28b). On the cross section along the center-line ($y/a = 0$), both rotating and nonrotating flow produce blocking, high pressure and cold air over the upslope and low pressure and warm air over the downslope (Figs. 5.28b and 5.28d). A southerly *barrier jet* forms along the upslope through geostrophic adjustment (see Chapter 1) in response to this reversed (against the basic flow) pressure gradient force in a rotating atmosphere (Fig. 5.28a). In contrast, there is no barrier jet in a nonrotating flow (Fig. 5.28c). Due to the strong shear flow passing over the left tip of the mountain, a jet forms through the *corner effect* caused by the *Bernoulli effect* (Figs. 5.28a and 5.28c). In the absence of rotation (Figs. 5.28c), a pair of cyclonic and anticylonic vortices is generated on the lee, but these are symmetric with respect to the centerline. The vortex shedding in a rotating flow also produces pressure and potential temperature perturbations far downstream (Fig. 5.28b). Three-dimensional flow past an isolated mountain in a rotating frame of reference has also been simulated using many other nonlinear numerical models.

In addition to the above rotational effects, a *topographic Rossby wave* can be produced for a flow over a mesoscale mountain. Consider an air column in a barotropic atmosphere that is pushed up northward against a sloping surface of an east–west oriented mountain range. The air column tends to aquire an anticyclonic vorticity ($\zeta < 0$) in order to conserve the potential vorticity ($\approx (\zeta + f)/H$), where H is the effective depth of the air column, due to decreasing H. An air parcel in the air column then turns anticyclonically, assuming f remains constant and the surface friction is negligible, and moves downslope. This forces the air column to stretch, thus acquiring cyclonic vorticity due to increasing H, and turn upslope. This wavelike motion is called a topographic Rossby wave because the sloping surface plays an analogous role to that of the β effect.

5.5.2 Lee cyclogenesis

The lee sides of mesoscale or large-scale mountains, such as the Alps, Rocky Mountains, the East Asian mountains and the Andes, are favorable regions of cyclogenesis. This type of cyclogenesis is known as *lee cyclogenesis*, and can be defined as the formation of a cyclone with strong positive vertical vorticity or an appreciable fall in pressure with a closed circulation formed in the lee of a mountain that then drifts away. Figure 5.29 shows the cyclogenesis frequency in the Northern Hemisphere during wintertime from 1899 to 1939. Frequency maxima in the figure are located to the lee of major mountain ranges, such as south of the Alps over the Gulf of Genoa and to the east of the Rockies.

Similar to diabatic heating, major mountain ranges serve as sources of stationary wave trains, which produce strong baroclinic zones favorable for *storm tracks* (Blackmon *et al.* 1977). Two well-known storm tracks, the *Pacific trough* and the *Atlantic trough*, are associated with stationary Rossby wave trains emanating from the Himalayas and Rockies, respectively. Based on the above definition of lee cyclogenesis, the weak stationary low (pressure depression) associated with these stationary

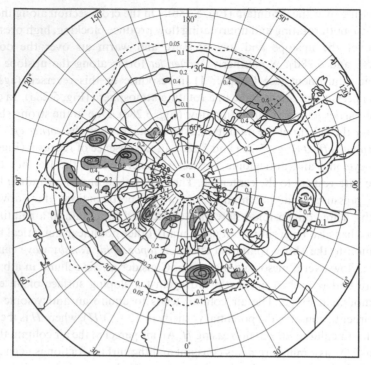

Fig. 5.29 Percentage frequency of cyclogenesis during winter from 1899 to 1939 in the Northern Hemisphere in squares of $100\,000$ km². (Adapted after Petterssen 1956.)

Rossby wave trains in the lee of mountains should be distinguished from the *true* lee cyclogenesis (Chung *et al.* 1976). Observations indicate that lee cyclogenesis may be influenced by different factors, such as the synoptic conditions, as well as the orientation and geometry of the mountain range to the prevailing wind. For example, the Alps are more isolated and parallel to the baroclinic flow, while the Rockies are more elongated and normal to the baroclinic flow. In the following, we will focus the discussion on the theories of lee cyclogenesis over two major mesoscale mountain ranges, the Alps and Rockies. These theories can be applied to lee cyclogenesis over a number of mountain ranges in other parts of the world, depending upon the synoptic conditions and terrain geometry.

a. Alpine lee cyclogenesis

As depicted in Fig. 5.29, a frequency maximum of lee cyclogenesis can be found to the south of the Alps over the Mediterranean Sea. Lee cyclogenesis over the Alps has been studied extensively, compared to that over other mountain ranges, partially due to the Alpine Experiment (ALPEX) field project held in 1982. The general characteristics of the Alpine lee cyclogenesis can be summarized as follows (e.g., Tibaldi *et al.* 1990):

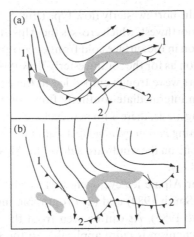

Fig. 5.30 Two types of lee cyclogenesis over the Alps: (a) Southwesterly flow (Vorderseiten) type, and (b) Northwesterly flow (Überströmungs) type, based on upper-level flow. The bold, solid lines indicate the upper-level flow and the surface fronts are plotted for two consecutive times, as denoted by 1 and 2. The shaded areas represent the terrain higher than 1 km. The mountains on the right hand side are the Alps and those on the left are the Pyrenees. (Adapted after Pichler and Steinacker 1987.)

(a) Lee cyclogenesis often occurs in association with a preexisting synoptic-scale trough or cyclone that interacts with the orography.

(b) The development of the lee cyclone starts before the strong thermal contrasts associated with cold frontal penetration take place in the lee.

(c) The scales of mature and deep lee cyclones are on the order of the Rossby radius of deformation (NL_z/f, where L_z is the vertical scale of motion). The influence of orography takes the form of a high-low dipole, which has the same horizontal and vertical scales of the cyclone.

(d) A two-phase deepening process is found. In the first phase, a cyclone deepens very rapidly, but is shallow. In the second phase, the cyclone develops less rapidly, but extends through the whole troposphere with horizontal scales comparable to the Rossby radius of deformation.

Based on upper-level flow, two types of lee cyclogenesis over the Alps can be identified: the *southwesterly flow (Vorderseiten) type* and the *northwesterly flow (Überströmungs) type* (Pichler and Steinacker 1987). Both types are accompanied by blocking and splitting of low-level cold air by the Alps. One part of the low-level cold air flows anticyclonically around the Eastern Alps, often leading to a *bora* (bura) event over the Adriatic Sea, while another part of the cold air flows around the Alps cyclonically and leads to the *mistral wind*, blowing through the Rhone Valley towards the Gulf of Lyons (to the south of France). For the southwesterly flow type (Fig. 5.30a), cyclogenesis starts with the advance of an eastward moving trough, where a wave at the corresponding surface cold front always forms in about the same place, i.e. the Gulf of Genoa (to the south of Italy and north of Corsica), due to the blocking and flow splitting effect of the Alps. A warm front is often generated over northern Italy. For the northwesterly flow type (Fig. 5.30b), the upper-level flow blows generally from the northwest, which crosses the Alps and generates a cyclone on the lee. The wave formation in the surface cold front occurs predominantly in the Gulf of Genoa, the same location as for the southwesterly flow type, due to the shape of the Alpine mountains. A short wave trough embedded in the general northwesterly flow can, however, produce an intermediate southerly flow component. In the northwesterly flow type of lee cyclogenesis, there is no new cyclone formed in the lee of the Alps. Usually, there is a preexisting *parent cyclone* upstream northwest of the Alps, which is weakened by the anticyclone on the windward side of the Alps and intensified in the lee in a favorable baroclinic flow.

Two major theories for Alpine lee cyclogenesis have been proposed in the past: (1) baroclinic lee waves (Smith 1984) and (2) orographic modification of baroclinic instability (Speranza *et al.* 1985). As can be seen from the following description, it appears that the first mechanism is more applicable to the southwesterly flow type, while the second mechanism is more applicable to the northwesterly flow type.

The basic dynamics of the *baroclinic lee wave theory* can be understood by considering a quasi-geostrophic flow with an idealized, linear shear over an isolated mountain (Fig. 5.31). The basic wind represents an incoming cold front or baroclinic zone (with cold air to the west), similar to the southwesterly flow type (Fig. 5.30a). This basic wind profile allows for the growth of a standing baroclinic wave in the lee (south) of the mountain. Although a *baroclinic wave* is often used to imply a growing, baroclinically unstable wave, here we define it more generally as a surface-trapped wave in a baroclinic current whose restoring force is associated with temperature advection at the boundary. Based on the quasi-geostrophic approximation, the geostrophic, Boussinesq potential vorticity equation on an *f*-plane can be written as (Smith 1984):

$$\frac{D_g}{Dt}\left(\nabla^2 P + \frac{f^2}{N^2}P_{zz}\right) = 0, \tag{5.5.6}$$

where P is the pressure. The terms inside the bracket are the potential vorticity, $D_g/Dt = \partial/\partial t + u_g\partial/\partial x + v_g\,\partial/\partial y$, $u_g = -P_y/(\rho_o f)$, $v_g = P_x/(\rho_o f)$) and the thermodynamic equation at the surface is,

$$\frac{D_g\theta}{Dt} + w\frac{\partial\theta}{\partial z} = 0, \quad \text{at} \quad z = 0, \tag{5.5.7}$$

where θ is the potential temperature. Assuming $u_g = U(z) + u'_g$ and $v_g = V(z) + v'_g$ and the basic flow velocity is a linear function of height, i.e. $U(z) = U_s + U_z z$ and $V(z) = V_s + V_z z$, then the potential vorticity in (5.5.6) vanishes everywhere. Here U_s and V_s are the basic wind speeds at the surface. Thus, any perturbation to this basic state must obey,

$$\nabla^2 p' + \frac{f^2}{N^2}p'_{zz} = 0 \tag{5.5.8}$$

and

$$\left(\frac{\partial}{\partial t} + U_s \frac{\partial}{\partial x} + V_s \frac{\partial}{\partial y}\right)\frac{\partial p'}{\partial z} - U_z \frac{\partial p'}{\partial x} - V_z \frac{\partial p'}{\partial y} + \rho_0 N^2 w' = 0 \quad \text{at} \quad z = 0, \quad (5.5.9)$$

where $N^2 = (g/\theta_0)(\partial\bar\theta/\partial z)$. The above two equations form a closed system as long as w' is known at $z = 0$, which can be specified by the linear lower boundary condition, $w'(x, y, z = 0) = U_s \partial h/\partial x + V_s \partial h/\partial y$, where $h(x, y)$ is the mountain height function.

If one considers a two-dimensional, steady-state flow over a bell-shaped mountain ridge, $h(x) = ha^2/(x^2 + a^2)$, an analytical solution can be obtained by applying the Fourier transform to both (5.5.8) and (5.5.9) and an upper boundedness boundary condition (which differs from the rigid lid boundary condition) in a semi-infinite vertical plane. The solution can be written as

$$p'(x, z) = \rho_0 U_s N^2 \left[2\mathrm{Re}\int_0^\infty \frac{\hat{h}(k)e^{-z/H}e^{ikx}}{U_s/H + U_z}dk\right], \quad (5.5.10)$$

where $H(k) = f/Nk$ and $\hat{h}(k) = (ha/2)e^{-kx}$. In deriving (5.5.10), we have applied the one-sided Fourier transform. According to the Riemann–Lebesgue Lemma (Appendix 5.1), the integral will go to 0 as $|x| \to \infty$ due to the rapid oscillation of the term e^{ikx}. In this case, the airflow will be disturbed only near the mountain. However, if the denominator of the integrand vanishes for some value of k, the integral will not vanish. This happens when there is a back shear. For example, when $U_s > 0$, this requires $U_z < 0$ and the basic wind vanishes at height $H^* = f/Nk^*$ (where $k^* = NH^*/f$ and H^* is the wind reversal level). Thus, the vanishing of the basic wind $U(z)$ at some height is the condition for obtaining a disturbance away from the mountain. Physically, this is an orographically forced standing baroclinic wave which has a zero phase speed. Evaluating (5.5.10) gives

$$p'(x, z) = 0, \quad \text{for } x < 0 \quad \text{and} \quad (5.5.11a)$$

$$p'(x, z) = -(4\pi\rho_0 Nf)\hat{h}(k^*)e^{-z/H^*}\sin k^* x \quad \text{for } x > 0. \quad (5.5.11b)$$

Equation (5.5.11b) describes a train of standing baroclinic lee waves on the lee of the mountain. With typical flow and orographic parameters for Alpine lee cyclogenesis, such as $\rho_0 = 1\,\mathrm{kg\,m^{-3}}$, $N = 0.01\,\mathrm{s^{-1}}$, $f = 10^{-4}\,\mathrm{s^{-1}}$, $h = 3 \times 10^3\,\mathrm{m}$, $a = 2.5 \times 10^5\,\mathrm{m}$, $H^* = 5 \times 10^3\,\mathrm{m}$, and $k^* = f/(NH^*) = 2 \times 10^{-6}\,\mathrm{m^{-1}}$, a pressure perturbation $p' = 28\,\mathrm{hPa}$ can be generated. The wavelength of the baroclinic lee wave is $\lambda = 2\pi/k^* \approx 3000\,\mathrm{km}$, which is comparable to observed baroclinic waves over the Alps.

To obtain the transient solution, we take the Fourier transform of (5.5.8), apply the upper boundedness boundary condition, and then substitute the solution to (5.5.9) with the linear lower boundary condition. This leads to

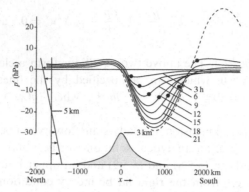

Fig. 5.31 Baroclinic lee wave theory of Alpine lee cyclogenesis: an example of the steady-state solution (dashed curve), (5.5.11), and transient solution (solid curves), (5.5.12) for an initially undisturbed, back-sheared flow over a bell-shaped mountain, $h(x) = ha^2/(x^2 + a^2)$. The flow and orographic parameters are: $\rho_o = 1\,\mathrm{kg\,m^{-3}}$, $N = 0.01\,\mathrm{s^{-1}}$, $f = 10^{-4}\,\mathrm{s^{-1}}$, $h = 3\,\mathrm{km}$, $a = 2.5 \times 10^5\,\mathrm{m}$, $U_s = 20\,\mathrm{m\,s^{-1}}$ and $H^* = 5 \times 10^3\,\mathrm{m}$. (Adapted after Smith 1984.)

$$\hat{p}(t) = A\mathrm{e}^{-Bt} + \frac{Hg\rho_o\bar{\theta}_z}{\theta_o}\frac{ikU_s\hat{h}}{B}, \tag{5.5.12}$$

where

$$B = ik(U_s + HU_z), \tag{5.5.13}$$

and A is a complex coefficient to be determined by the initial condition. The solution in physical space can then be obtained by taking the inverse Fourier transform of (5.5.12). One can apply a FFT algorithm to obtain the numerical solution. Figure 5.31 shows examples of the steady-state solution, (5.5.11), and the transient solution of (5.5.12) with an initially undisturbed flow, $p'(t = 0) = 0$. This leads to rapid lee cyclogenesis with pressure tendencies from 4 to 6 mb/3 h. The fluid is, in effect, trying to form the first trough of a standing baroclinic lee wave. Compared with the steady-state solution, the transient solution produces the simple pattern of a wave growing away from its source. The above system has been extended to obtain the following (Smith 1986): (a) three-dimensional numerical solutions for Alpine cyclogenesis, which show qualitative agreement with observations; (b) an upshear phase tilt with height which indicates that the growing baroclinic wave is able to extract energy from the available potential energy stored in the baroclinic shear flow; (c) with a rigid lid added to the model, there is a rapid two-phase lee cyclone growth supported by baroclinic instability of the Eady type (1949); (d) the baroclinic lee waves belong to the regime of mixed inertia-gravity waves and trapped baroclinic lee waves (Section 3.8), thus making the theory more applicable to flow with moderate R_o and moderate R_i, or small R_o and moderate R_i; and (e) the basic features predicted by the linear theory are consistent with those simulated by nonlinear numerical models (Lin and Perkey 1989; Schär 1990).

In the theory of *orographic modification of baroclinic instability* for Alpine lee cyclogenesis, the formation of lee cyclones is attributed to the baroclinically unstable modes modified by orography. To derive the governing equations, we assume a small-amplitude mountain and a basic flow with constant stratification and vertical shear, and a zonal wind that vanishes at $z = 0$, bounded between $z = 0$ and $z = H$ in the vertical and by lateral walls at $y = \pm L_y/2$. For this type of flow, the linear, nondimensional governing equations and boundary conditions can be written as (Speranza *et al.* 1985):

$$\nabla^2 \psi + \psi_{zz} = 0,$$

$$\psi_{zt} - \psi_x = -J_{yz}(\psi, h/R_o) \qquad \text{at } z = 0,$$

$$\psi_{zt} - \psi_x + \psi_{xz} = 0 \qquad \text{at } z = 1, \text{ and} \qquad (5.5.14)$$

$$\psi = 0 \qquad \text{at } y = \pm L_y/2,$$

where ψ is the streamfunction, J the Jacobian (defined as $J_{yz}(\alpha, \beta) = \alpha_y \beta_z - \beta_y \alpha_z$), h the mountain height, and R_o the Rossby number. Note that (5.5.14) is equivalent to (5.5.8) of the baroclinic lee wave theory except that the streamfunction is used. Assuming $\psi(t, x, y, z) = \phi(x, y, z)e^{-i\omega t}$ and applying the Fourier series expansion to the above system reduces it to an eigenvalue problem, which allows one to find the most unstable baroclinic wave mode from the eigenvalues. Figure 5.32 shows one example of Alpine lee cyclogenesis based on the above theory. The disturbance at the surface (Figs. 5.32a and 5.32b) is more localized near the mountain than that at the mid-troposphere (Figs. 5.32c and 5.32d). The total streamfunction fields (Figs. 5.32b and 5.32d) appear to be able to capture some basic structures of Alpine lee cyclones in the mature stage. Lee cyclogenesis is explained as the intensification of the incident baroclinic wave taking place on the southern (warm) side of the mountain, together with a weakening of the wave amplitude on the northern (cold) side.

When a vertical wall is used to represent the Alps and to block all meridional flow below the crest level, a lee cyclone develops through baroclinic instability (Egger 1972; Pierrehumbert 1986), as in Fig. 5.32, and so the situation illustrated in Fig. 5.32 may thus be considered to belong to the mechanism of orographic modification of baroclinic instability. In fact, the above two theories for Alpine lee cyclogenesis are related because both of them are built upon the baroclinic instability theory of Eady (1949), as can be seen from (5.5.8) and (5.5.14), but use different initial and upper boundary conditions and basic wind profiles. In the baroclinic lee wave theory, an undisturbed flow or a localized baroclinic wave is used to initiate the process, while a fully developed baroclinic wave is used as the initial condition in the theory of orographic modification of baroclinic instability. In addition, the theory of orographic modification of baroclinic instability uses a rigid lid upper boundary condition, as in Eady's model, while the baroclinic lee wave theory does not impose a rigid lid but uses a back-sheared north–south basic wind profile. Based on the above discussions, it appears that the baroclinic lee wave mechanism is more applicable to the southwesterly flow type, while the

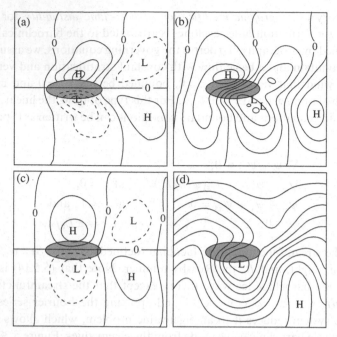

Fig. 5.32 Orographic modification of baroclinic instability for Alpine lee cyclogenesis: an example of cyclogenesis generated by a baroclinic wave in the continuous Eady model, in a long periodic channel with isolated orography. The mountain has a bi-Gaussian form $(e^{-x^2/a^2}e^{-y^2/b^2})$ and the dimensional values of a and b are 1500 km and 500 km, respectively. The shaded oval denotes the orographic contour of e^{-1} of the maximum height. (a) Streamfunction of orographic perturbation at $z = 0$ (nondimensional); (b) total streamfunction of the modified baroclinic wave at $z = 0$; (c) and (d) are as (a) and (b), respectively, but at the middle level, $z = 0.5$. The basic zonal wind is added in (d). (Adapted after Speranza *et al.* 1985.)

mechanism of orographic modification of baroclinic instability is more applicable to the northwesterly flow type which normally contains a preexisiting cyclone.

Some other physical processes may also play important roles in the Alpine lee cyclogenesis, such as nonlinear processes (Tafferner and Egger 1990), geostrophic adjustment to upper-level potential vorticity advection (Bleck and Mattock 1984), and orographic modification of the baroclinic parent cyclone (Orlanski and Gross 1994), under certain synoptic situations. The two-phase deepening process has been attributed to the rapid formation of a low-level orographic vortex, followed by its baroclinic and diabatic interaction with an approaching upper-level trough (Aebischer and Schär 1998). In the first phase, orographically generated elongated bands of positive and negative PV, i.e. *PV banners or streamers*, may wrap up and subsequently contribute to the low-level PV anomaly within the developing cyclone. The second phase is interpreted as mere orographic modification of baroclinic instability.

b. Rockies lee cyclogenesis

The major difference that sets apart the Rockies lee cyclogenesis from the Alpine lee cyclogenesis is that the Rocky Mountains are elongated in a direction normal to the basic wind, while the Alps are elongated in a direction parallel to the thermal wind (basic wind shear). In addition, the Rocky Mountains have a much larger horizontal scale than the Alps in the direction of the low-level flow. Three stages can be identified for the lee cyclogenesis over the Rocky Mountain cordillera:

(a) Stage I: A parent cyclone approaches the Rocky Mountain cordillera from the northwest. As the cyclone moves closer to the mountains, it decelerates and curves to the north.
(b) Stage II: During the passage of the parent cyclone over the cordillera, the low-level portion of the cyclone weakens or fills in (disappears) over the mountain, while the upper-level trough continues to move over the mountain, but with some track deflection;
(c) Stage III: A new surface cyclone forms in the lee of the mountain range and to the south of the original track. In the meantime, the upper-level trough, which is associated with the parent cyclone, has progressed over the mountain and helps strengthen the surface cyclone through baroclinic development.

Figure 5.33 shows an example of cyclogenesis which occurred in the lee of the US Rocky Mountain cordillera on March 18–20, 1994. At 1200 UTC March 18 (3/18/12Z), a parent cyclone was located over the southern Oregon and was approaching the Rockies. It subsequently moved eastward over southern Idaho by 3/19/00Z (Stage I). Stage II occurs between 3/19/00Z and 3/19/12Z. At 3/19/12Z, a new cyclone forms over eastern Colorado, which is to the southeast of the incidence of the parent cyclone. As the upper-level trough begins moving east of the mountains, the phase shift between the upper trough and the lee cyclone begins to decrease and classical baroclinic development takes place.

Similar to Alpine lee cyclogenesis, the Rockies lee cyclogenesis has also been studied extensively in the past and several mechanisms have been proposed. Due to the fact that a parent cyclone is almost a prerequisite for the Rockies lee cyclogenesis and that the flow is highly baroclinic, the formation of lee cyclones can be reasonably explained by the theory of orographic modification of baroclinic instability, described by (5.5.14). The major difference is that the low-level tracks of the Rocky Mountains cyclones are strongly influenced by the mountain geometry, as will be discussed in Subsection 5.5.3, and by the upper-level trough's effect on the intensity of the lee cyclone.

The theory of orographic modification of baroclinic instability, as described in (5.5.14), can be extended to avoid the limitations of gentle sloping and small amplitude mountains by allowing for the interaction of growing normal modes in a baroclinic channel with isolated topography of finite height (Buzzi *et al.* 1987). In addition, the isolated mountain may be elongated in the north–south direction to mimic the Rockies topography. Figure 5.34 shows the time evolution of the fastest growing baroclinic mode in the lower layer of the two-layer channel model. The three stages

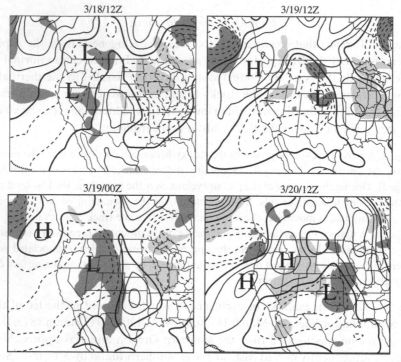

Fig. 5.33 An example of Rockies lee cyclogenesis. Three stages, as discussed in the text, are clearly shown in the perturbation geopotential fields on 860 mb at 12-h intervals from 3/18/12Z (1200 UTC 18 March) to 3/20/12Z 1994. Solid (Dashed) contours denote positive (negative) values, while bold solid contours denote zero geopotential perturbations. Darker (Lighter) shading denotes warm (cold) areas at 900-mb $\theta' \geq 4$ K ($\theta < 4$ K). (Adapted after Davis 1997.)

summarized earlier are reasonably reproduced by this linear stability model. The track deflection will be explained in Subsection 5.5.3.

Note that nonlinearity should be included in the above lee cyclogenesis theory when applied to the real atmosphere (Tafferner and Egger 1990). Various physical processes have been emphasized in other theories of the orographic modification of baroclinic instability mechanism, such as: (1) superposition of a growing baroclinic wave and a steady mountain wave (e.g., Hayes *et al.* 1993), and (2) Eady edge waves, i.e., an orographically modified surface potential temperature distribution leading to the upstream northward deviation of the baroclinic waves (Davis 1997).

c. Mesoscale lee cyclogenesis

A mesoscale vortex formed on the lee of a mesoscale mountain, as discussed in Subsection 5.4.2, can further develop into a lee cyclone by aquiring low pressure at the center of the vortex through various processes, such as latent heating associated with moist

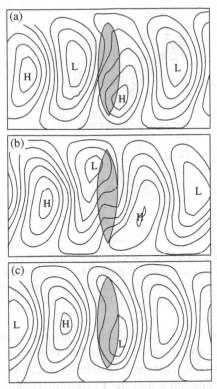

Fig. 5.34 Application of a linear theory of orographic modification of baroclinic instability to the Rockies lee cyclogenesis. The streamfunction of the fastest growing mode in the lower layer of a two-layer channel model on a β plane, at three times of evolution of the baroclinic mode are shown, which also reflects the three stages of Rockies lee cyclogenesis (see text). The channel length and width are 10 000 and 6000 km, respectively. The basic state mean velocity is 17 m s^{-1} in the upper layer and 0 m s^{-1} in the lower layer. The area of the mountain with height above he^{-1} is shaded, where h is the mountain height (2.5 km). (Adapted after Buzzi *et al.* 1987.)

convection, surface heating over a warm sea surface, merging with a nearby mesolow, and interaction with an approaching upper-level trough. This type of mesoscale cyclogenesis has been observed in the lee of some mesoscale mountains and simulated by numerical models. Examples are the Denver cyclone (Crook *et al.* 1990), the mesocyclones to the east of Taiwan's CMR (Sun *et al.* 1991; Lin *et al.* 1992) , and the lee cyclones generated by the northern Korean mountain complex over the East (Japan) Sea (Lee *et al.* 1998).

5.5.3 *Orographic influence on cyclone track*

Similar to what is shown in Fig. 5.33, track deflection of cyclones by mountains also occurs in other mesoscale mountain ranges, such as the Appalachians and Greenland,

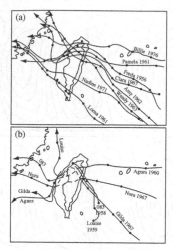

Fig. 5.35 Tropical cyclones traversing the Central Mountain Range (CMR) of Taiwan with (a) continuous tracks, and (b) discontinuous tracks. A cyclone track is defined as discontinuous when the original cyclone (i.e. a low pressure and closed cyclonic circulation) and a lee cyclone simultaneously co-exist. (Adapted after Wang 1980 and Chang 1982.)

as long as the mountain range has appreciable extension in the direction normal to the prevailing wind in which the cyclone is embedded. Orographic influence on cyclone tracks has also been observed for tropical cyclones passing over mesoscale mountains, such as the northern Philippines and the Caribbean Islands, the Sierra Madre of Mexico, and Taiwan. Figure 5.35 shows examples of track deflection of tropical cyclones past the CMR of Taiwan. Depending upon the intensity, steering wind speed, size of the cyclone, landfalling location and impinging angle, the tracks over CMR can be classified as continuous or discontinuous.

Figure 5.36 illustrates the orographic influence on cyclone track through the turning of the basic flow and the outer circulation of the cyclone. The orography is assumed to be shallow, so that the blocking is relatively weak and most of the low-level air is able to pass over the mountain. The track deflection of a cyclone embedded in a straight incoming flow on a mesoscale mountain ridge, as shown in Fig. 5.36a, can be explained by either the force balance or the conservation of potential vorticity, as discussed earlier along with Fig. 5.26a.

Cyclone tracks can also be affected by the orographic influence on the outer circulation of the cyclone. This process is illustrated in Fig. 5.36b. Air ascends in the southwest and northeast quadrants and descends in the northwest and southeast quadrants with respect to the intersection of the original (west–east) undisturbed track and the mountain crest. This produces a positive local vorticity tendency ($\partial \zeta / \partial t > 0$) in the northwest and southeast quadrants due to vorticity stretching, $(\zeta + f)\partial w/\partial z$ (e.g., part of the second term on the right side of (5.4.28) with the Coriolis term added). Similarly, $\partial \zeta / \partial t < 0$ is produced in the southwest and northeast corners due to vorticity shrinking. This helps migrate the cyclone

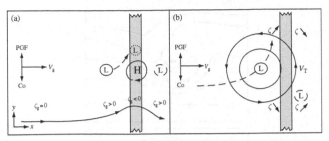

Fig. 5.36 Effects of shallow orography on cyclone track deflection through the turning of (a) the basic flow and (b) outer circulation of the cyclone. The solid, dot, and dash-circled L denote the parent and new cyclones, respectively. The incoming flow is geostrophically balanced. H denotes the mountain-induced high pressure. The lower curve in (a) denotes the trajectory of an upstream air parcel. The cyclone is steered to the northeast by the basic flow velocity V_g. ζ_g is the vertical vorticity associated with the basic flow. In (b), the dashed curve associated with L denotes the cyclone track. The new cyclone forms on the lee due to vorticity stretching associated with the cyclonic circulation that is, in turn, associated with the parent cyclone. V_T is the maximum or a characteristic tangential wind speed of the parent cyclone. The arrow adjacent to ζ denotes the local vorticity tendency ($\partial\zeta/\partial t$). (Panel (b) is adapted after Carlson 1998.)

toward the northwest corner and generate a secondary (new) cyclone in the southeast quadrant. The lee cyclone develops quickly once it is coupled with the upper portion of the parent cyclone or trough. Note that this process will produce a lee cyclone that is at most of the same strength as the parent cyclone. In order to generate a stronger lee cyclone, other processes are required, such as barolinic instability or upstream flow blocking.

The continuity of cyclone tracks and degree of track deflection is directly related to the degree of orographic blocking. Figure 5.37 shows orographic influence on cyclone track deflection for a drifting cyclone in a uniform flow past an isolated, finite-length mountain on an f-plane. With weak blocking (panel a), the cyclone is deflected to the right (facing downstream) and a weak secondary cyclone forms on the lee. The track is continuous in this case. With strong blocking (panel b), the cyclone is deflected to the left and a relatively strong secondary cyclone forms on the lee. In this case, the track is discontinuous. It is found that the track deflection is associated with the local vorticity tendency ($\partial\zeta/\partial t$) which is dominated by the vorticity advection and vorticity stretching while the cyclone is crossing over a mesoscale mountain range (Lin *et al.* 1999; 2005). For example, for the case of Fig. 5.37a (weak blocking), the track deflection is dominated by vorticity advection. For the case of Fig. 5.37b, strong blocking tends to generate a strong northerly jet along the upwind slope of the mountain range, which, in turn, produces a local vorticity tendency to the left (north) of the undisturbed cyclone track by vorticity advection. The strong secondary vortex on the lee is generated by the vorticity stretching. The degree of orographic blocking can be measured by nondimensional control parameters, such as the basic-flow Froude number (U/Nh), the vortex Froude number (V_T/Nh), and the relative scale of the cyclone and the orography in the direction perpendicular to the steering flow (R/L_y), where U is the basic flow speed, V_T the characteristic tangential wind speed of the

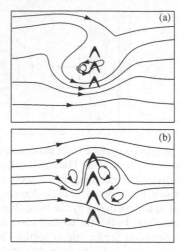

Fig. 5.37 Orographic influence on cyclone track deflection for a drifting cyclone in a uniform flow past an isolated, finite-length mountain (denoted by the mountain symbol) on an f-plane: (a) weak blocking, and (b) strong blocking. (Adapted after Lin *et al.* 2005.)

parent cyclone (such as V_{max}), N the Brunt–Vaisala frequency, h the mountain height, and R the cyclone scale represented by its radius of maximum tangential wind.

5.6 Other orographic effects

In addition to the phenomena discussed in previous sections, orography also plays a significant role in influencing the general airflow and weather systems, such as frontal passage, the formation of coastally trapped disturbance, cold-air damming, gap flow, orographic precipitation and foehn wind. The dynamics of orographic precipitation and foehn wind require backgrounds in thermally forced flow, mesoscale instabilities, and moist convection. Thus, these will be discussed in Chapter 11. The orographic effects on other phenomena are briefly described in the following.

5.6.1 Effects on frontal passage

When a front passes over a mesoscale mountain range, it is often distorted by the mountain. As soon as the front impinges on the mountain, an anticyclonic circulation, associated with the mountain-induced high pressure, is produced in the cold post-frontal air and the front advances faster on the left (facing the frontal movement). Figure 5.38 shows a conceptual model of a two-dimensional front passing over a mesoscale mountain ridge. The mountain blocks and decelerates the approaching front at the surface while the upper-level potential vorticity anomaly associated with the front is affected to a much lesser degree as it moves across the mountain. When the blocking is strong, the surface front remains trapped on the windward slope and the

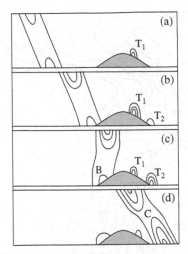

Fig. 5.38 A conceptual model of a two-dimensional front propagating over a mesoscale mountain, which indicates the low-level blocking (denoted by B) of the front along the windward slope, the development of the lee trough (T) and secondary trough (T_2) along the lee slope, the separation of the upper-level and lower-level frontal waves, and the coupling (C) of the upper-level frontal wave with the secondary trough in the lee of the mountain. (Adapted after Dickinson and Knight 1999.)

frontal propagation is discontinuous across the ridge. During frontal passage over a mountain, several interesting phenomena can occur, such as the development of a lee trough or of a secondary trough along the lee slope, the separation of the upper-level and lower-level frontal waves, and the coupling of the upper-level frontal wave with the secondary trough in the lee of the mountain.

The passage of three-dimensional fronts over mesoscale mountains is more complicated than that of a two-dimensional front. Figure 5.39a shows a schematic of a cold front passing over the Alps, in which the northeastern portion of the front accelerated eastward, while the southwestern portion of the front was retarded at least initially. In other words, the front passed over the mountain anticyclonically. This type of deformation of the surface fronts has also been observed for Mei-Yu fronts passing over the CMR of Taiwan (Fig. 5.39b). The Mei-Yu (Baiu or Changma) front is a nearly stationary, east–west-oriented weak baroclinic zone in the lower troposphere occurring from mid- to late spring through early to midsummer, and typically stretching from eastern China to Taiwan, Japan or Korea. When an east–west-oriented Mei-Yu front passes over the CMR (which is oriented from north-northeast to south-southwest) from the north, the Mei-Yu front at the surface splits into eastern and western flanks, with an eastern flank that accelerates and a western flank that is retarded. In addition to the mountain-induced anticyclonic circulation, the retardation of the left flank of the Mei-Yu front is enhanced by the southwesterly monsoon. Recent nonlinear numerical experiments are able

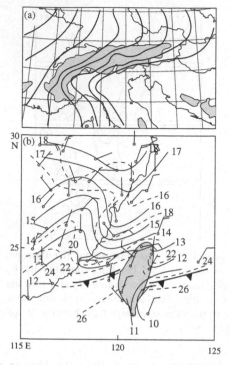

Fig. 5.39 Examples of three-dimensional frontal passage over mesoscale mountains: (a) Schematic of 3-h surface isochrones of a cold front passing over the Alps (shaded). Note that as the front passes over the mountain range it is turned anticyclonically. (Adapted after Steinacker 1981.) (b) Surface analysis of a Mei-Yu front passing over the CMR of Taiwan (shaded) at 1800 UTC May 13, 1987. Solid and dashed curves denote the sea level pressures (mb; variation from 1000 mb) and isotherms (°C), respectively. (Adapted after Chen and Hui 1992.)

to: (a) reproduce the orographically induced high pressure, front splitting, and anticyclonic circulation during the passage of Mei-Yu front over the CMR; (b) show the important contribution of the nonlinear advection, Coriolis force, pressure gradient force, and friction to the local rate of change of horizontal momentum; and (c) indicate that the front bears no dynamical resemblance to either an orographically trapped density current or a Kelvin wave (Sun and Chern 2006).

The dynamics of orographic distortion of fronts can be understood by taking a passive scalar approach, in which an initially straight front comes under the influence of the anticyclonic circulation induced by the mountain, and experiences an anticyclonic turning. This is analogous to the passage of a cyclone over a high mountain, as discussed in the previous section (Section 5.5). As mentioned earlier, as soon as the front impinges on the mountain, an anticyclonic circulation is produced in the cold postfrontal air and the front advances faster on the left (facing the frontal movement) side of the mountain and is retarded on the right side. Consequently, weakening of the

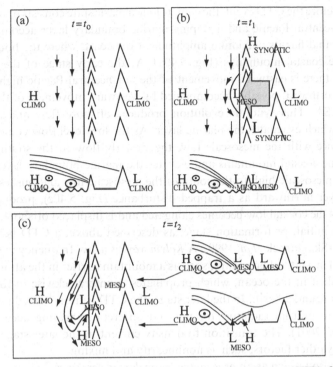

Fig. 5.40 Schematic of an idealized coastally trapped disturbance (CTD) formation and evolution. See text for detailed description. In the figure, "CLIMO" stands for climatological basic state, while "SYNOPTIC" and "MESO" stand for synoptic and mesoscale states, respectively. (Adapted after Skamarock *et al.* 1999.)

front (*frontolysis*) is dominant on the left side and lee side, while strengthening of the front (*frontogenesis*) occurs on the right side. It is found that the degree of blocking on a front along the upwind slope is determined by the Froude number ($F = U/Nh$) and Rossby number ($R_o = U/fa$, where a is the half-width or horizontal scale of the mountain) (e.g., Davies 1984; Gross 1994).

5.6.2 Coastally trapped disturbances

A *coastally trapped disturbance* (*CTD*) is a mesoscale disturbance associated with a high pressure surge along a coastal mountain range, confined laterally against the mountain barrier and vertically by stratification, and propagating along the mountain such that the mountain is on the right in the Northern Hemisphere. On the US west coast, a CTD forms in spring and summer and propagates northward, and appears as a transition from the prevailing northerly flow. The CTD has a length scale of about 1000 km along the shore, 100 km across the shore, and 0.5 km in the vertical. It normally lasts for two to three days. Figure 5.40 illustrates the formation and evolution of CTD along the US west coast. Before the CTD event, the climatological basic state consists of high pressure in the north-central Pacific and a sloping marine boundary layer accompanied by a northerly jet that has its maximum amplitude at the coast, where the boundary layer intersects the coastal mountains (Fig. 5.40a). At the early stage of the CTD event (Fig. 5.40b), there is eastward movement of the northcentral Pacific high onshore to the Pacific Northwest, possibly accompanied by westward movement of the low in the southwest USA. This synoptic evolution produces offshore flow and a mesoscale coastal low which evacuates the marine layer. As the low-level flow comes into geostrophic balance with the mesoscale low, the westerly flow to the south of the low encounters the coastal mountains and elevates the marine boundary layer. This produces a new mesoscale high to the south of the mesoscale low and pushes the marine boundary layer northward as a trapped disturbance (Fig. 5.40c), producing coastal southerlies as the coastal low becomes elongated and is displaced offshore.

During its initial, or formation stage, as described above, a CTD behaves like a Kelvin wave (Skamarock *et al.* 1999). A *Kelvin wave* is a low-frequency inertia-gravity wave trapped in a lateral boundary, such as a mountain barrier in the atmosphere or a continental shelf in the ocean, which propagates counterclockwise in the Northern Hemisphere around a basin. In the later stage, the CTD is transformed into a density current due to diurnal radiative forcing and differential heating across the coast (Reason *et al.* 2001). The transition to density current in the later stage might also be affected by other factors, such as nonlinearity and mixing.

A CTD is also known as an *orographic jet* or *ducted coastal ridging* in meteorology and is called *coastally trapped density current* or *coastal jet* in oceanography. CTDs have also been observed in southeastern Australia, southern Africa, and the southwest coast of South America (Holland and Leslie 1986). Ahead of the coastal ridging in Australia, there often exists a surge of cold air with squally winds. This phenomenon is called the *southerly buster*.

5.6.3 Cold-air damming

When a cold anticyclone is located to the north of an approximately north–south oriented mountain range in winter, a pool of cold air may become entrenched along the eastern slope and form a cold dome capped by a sloping inversion underneath warm easterly or southeasterly flow. This phenomenon has been observed over the Appalachian Mountains and the Front Range of the Rocky Mountains.

Figure 5.41a shows the surface geopotential temperature field at 1200 UTC March 22, 1985 at the mature phase of a cold-air damming event that occurred to the east of the US Appalachian Mountains. In the initiation phase, a surface low pressure system moved from the Great Lakes northeastward and the trailing anticyclone moved southeastward to central New York. The cold, dry air was then being advected by the northeasterly flow associated with the anticyclone. The air parcels ascending the mountain slopes experienced adiabatic cooling and formed a *cold dome*, while those over the

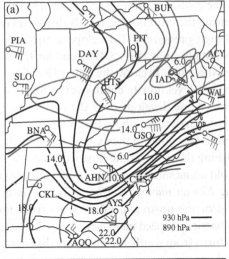

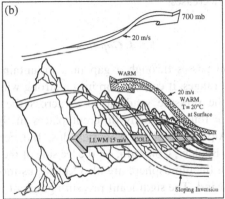

Fig. 5.41 (a) The 930 hPa height (in m, dark solid) and 890 hPa potential temperature (°C, grey solid) fields valid at 1200 UTC March 22, 1985. Winds are in m s^{-1} with one full barb and one pennant representing 5 m s^{-1} and 25 m s^{-1}, respectively. (b) Conceptual model of a mature cold-air damming event. LLWM stands for low-level wind maximum. (Adapted after Bell and Bosart 1988.)

ocean were subjected to differential heating when they crossed the Gulf Stream toward the land. At the mature phase, the surface anticyclone remained relatively stationary in New York and the cold air was advected southward along the mountain slopes within the cold dome. Figure 5.41b shows a conceptual model of the cold-air damming occurred to the east of the Applachian Mountains at the mature phase. The low-level wind maximum (LLWM in the figure) moved southward at a speed of about 15 m s^{-1} within the cold dome. The cold dome was capped by an inversion and an easterly or southeasterly flow associated with strong warm advection into the warm air existed above the dome. Moving further aloft to 700 mb, the wind flows from south or southwest associated with the advancing short-wave trough west of the Appalachians.

As described above, the cold-air damming process can be divided into the initiation and mature phases. During the initiation phase, the low-level easterly flow adjusts to the mountain-induced high pressure and develops a northerly barrier jet along the eastern slope of the mountain range (Fig. 5.41b), similar to that shown in Fig. 5.28a except for the basic flow direction. Because the upper-level flow is from the southwest, it provides a cap for the low-level flow to climb over to the west side of the mountain. In the mean time, a cold dome develops over the eastern slope due to the cold air supplied by the northerly barrier jet, adiabatic cooling associated with the upslope flow, and/or the evaporative cooling. In the mature stage, the frictional force plays an essential role in establishing the steady-state flow with the cold dome (Xu *et al.* 1996). It was found that the cold dome shrinks as the Froude number (U/Nh) increases or, to a minor degree, as the *Ekman number* ($\nu/(fh^2)$, where ν is the coefficient of eddy viscosity) decreases and/or the upstream inflow veers from northeasterly to southeasterly. The northerly barrier jet speed increases as the Ekman number decreases and/ or the upstream inflow turns from southeasterly to northeasterly or, to a lesser degree, as the Froude number decreases.

5.6.4 Gap flow

When a low-level wind passes through a gap in a mountain barrier or a channel between two mountain ranges, it can develop into a strong wind due to the acceleration associated with the pressure gradient force across the barrier or along the channel. Gap flows are found in many different places in the world, such as the Rhine Valley of the Alps, Senj of the Dinaric Alps, Independence, California, in the Sierra Nevada, and Boulder, Colorado, on the lee side of the Rockies (Mayr 2005). Gap flows occurring in the atmosphere are also known as mountain-gap wind, jet-effect wind or canyon wind. The significant pressure gradient is often established by (a) the geostrophically balanced pressure gradient associated with the synoptic-scale flow and/or (b) the low-level temperature differences in the air masses on each side of the mountains.

Based on Froude number three gap-flow regimes can be identified: (1) linear regime (large *F*): with insignificant enhancement of the gap flow; (2) mountain wave regime (mid-range *F*): with large increases in the mass flux and wind speed within the exit region due to downward transport of mountain wave momentum above the lee slopes, and where the highest wind occurs near the exit region of the gap; and (3) upstream-blocking regime (small *F*): where the largest increase in the along-gap mass flux occurs in the entrance region due to lateral convergence (Fig. 5.42; Gaberšek and Durran 2004). Gap flows are also influenced by frictional effects which imply that: (1) the flow is much slower, (2) the flow accelerates through the gap and upper part of the mountain slope, (3) the gap jet extends far downstream, (4) the slope flow separates, but not the gap flow, and (5) the highest winds occur along the gap (Zängl 2002).

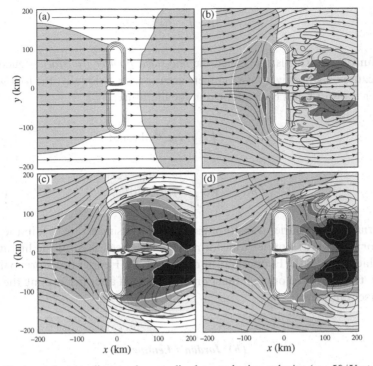

Fig. 5.42 Horizontal streamlines and normalized perturbation velocity $(u - U)/U$ at $z = 300$ m and $Ut/a = 40$, for flow over a ridge with a gap when the Froude number (U/Nh) equals (a) 4.0, (b) 0.72, (c) 0.36, and (d) 0.2. The contour interval is 0.5; dark (light) shading corresponds to negative (positive) values. Terrain contours are every 300 m. (From Gaberšek and Durran 2004.)

Appendix 5.1: Some mathematical techniques and relations

(a) Fourier Transform

The Fourier transform of $f(x)$ is defined as

$$\hat{f}(k) = \frac{1}{2\pi} \int_{-\infty}^{\infty} f(x) e^{-ikx} dx \quad \text{and} \tag{A5.1.1a}$$

$$f(x) = \int_{-\infty}^{\infty} \hat{f}(k) e^{ikx} dk. \tag{A5.1.1b}$$

One of the advantages of the Fourier transform is that it is able to distinguish the up- and down-going waves. An alternative Fourier transform pair may be defined as

$$\hat{f}(k) = \frac{1}{2\pi} \int_{-\infty}^{\infty} f(x) e^{-ikx} dx \quad \text{and} \tag{A5.1.2a}$$

$$f(x) = 2\mathrm{Re} \int_0^\infty \hat{f}(k)e^{ikx}\mathrm{d}k, \tag{A5.1.2b}$$

for real function f. This is also called the *one-sided Fourier transform* (Queney *et al.* 1960) because only positive wave numbers are used. Occasionally, the following pair of Fourier transform has been adopted in the literature,

$$\hat{f}(k) = \frac{1}{\pi} \int_{-\infty}^\infty f(x)e^{-ikx}\mathrm{d}x \text{ and} \tag{A5.1.3a}$$

$$f(x) = \mathrm{Re} \int_0^\infty \hat{f}(k)e^{ikx}\mathrm{d}k. \tag{A5.1.3b}$$

Other variations of Fourier transform pairs, such as using e^{ikx} (e^{-ikx}) in the forward (inverse or backward) Fourier transform, have also been used in the literature. No matter which form of the Fourier transform is used, the Fourier transform pair should be able to transform the original function back to itself after performing the forward and inverse transforms.

(b) Jordan's Lemma

If

$$\lim_{R\to\infty} |f(z)| = 0,$$

then

$$\lim_{R\to\infty} \int_{\Gamma_R} f(z)e^{ikz}\mathrm{d}z = 0, \tag{A5.1.4}$$

for $k > 0$ and where Γ_R is an counterclockwise contour of a semicircle above the $\mathrm{Re}\,z$ axis with radius R. Jordan's Lemma is very useful for converting a line integral to a contour integral.

(c) Riemann–Lebesgue Lemma

$$\lim_{x\to\infty} \int_{-\infty}^\infty \hat{f}(k)e^{ikx}\mathrm{d}k = 0 \quad \text{if} \quad \hat{f}(k) \text{ is smooth.} \tag{A5.1.5}$$

A smooth function here means that the function is ordinary and absolutely integrable. The above conclusion is reached by the reasoning of cancellation.

(d) Parseval Theorem

$$\frac{1}{2\pi} \int_{-\infty}^{\infty} f(x)g^*(x)\mathrm{d}x = \int_{-\infty}^{\infty} \hat{f}(k)\hat{g}^*(k)\mathrm{d}k, \qquad (A5.1.6)$$

where " $\wedge$ " indicates the Fourier transformed functions. The Parseval theorem is useful for computing wave energy and momentum flux.

References

Aebischer, U. and C. Schär, 1998. Low-level potential vorticity and cyclogenesis to the lee of the Alps. *J. Atmos. Sci.*, **55**, 186–207.

Bacmeister, J. T. and R. T. Pierrehumbert, 1988. On high-drag states of nonlinear stratified flow over an obstacle. *J. Atmos. Sci.*, **45**, 63–80.

Batchelor, G. K., 1967. *An Introduction to Fluid Dynamics*. Cambridge University Press.

Bell, G. D. and L. F. Bosart, 1988. Appalachian cold-air damming. *Mon. Wea. Rev.*, **116**, 137–61.

Blackmon, M., J. M. Wallace, N. C. Lau, and S. Mullen, 1977. An observational study of the Northern Hemisphere wintertime circulation. *J. Atmos. Sci.*, **34**, 1040–53.

Bleck, R. and C. Mattocks, 1984. A preliminary analysis of the role of potential vorticity in Alpine lee cyclogenesis. *Beitr. Phys. Atmos.*, **57**, 357–68.

Buzzi, A. and S. Tibaldi, 1977. Inertial and frictional effects on rotating stratified flow over topography. *Quart. J. Royal Meteor. Soc.*, **103**, 135–50.

Buzzi, A., A. Trevisan, S. Tibaldi, and E. Tosi, 1987. A unified theory of orographic influences upon cyclogenesis. *Meteor. Atmos. Phys.*, **36**, 91–107.

Carlson, T. N., 1998. *Mid-Latitude Weather Systems*. Amer. Meteor. Soc.

Chang, S. W.-J., 1982. The orographic effects induced by an island mountain range on propagating tropical cyclones. *Mon. Wea. Rev.*, **110**, 1255–70.

Chen, Y.-L. and N. B.-F. Hui, 1992. Analysis of a relatively dry front during the Taiwan Area Mesoscale Experiment. *Mon. Wea. Rev.*, **120**, 2442–68.

Chung, C. S., K. Hage, and E. Reinelt, 1976. On lee cyclogenesis and airflow in the Canadian Rocky Mountains and the East Asian Mountains. *Mon. Wea. Rev.*, **104**, 878–91.

Clark, T. L. and W. R. Peltier, 1984. Critical level reflection and the resonant growth of nonlinear mountain waves. *J. Atmos. Sci.*, **41**, 3122–34.

Crook, N. A., T. L. Clark, and M. W. Moncrieff, 1990. The Denver cyclone. Part I: Generation in low Froude number flow. *J. Atmos. Sci.*, **47**, 2725–42.

Davis, C. A., 1997. The modification of baroclinic waves by the Rocky Mountains. *J. Atmos. Sci.*, **54**, 848–68.

Davies, H. C., 1984. On the orographic retardation of a cold front. *Beitr. Phys. Atmos.*, **57**, 409–18.

Dickinson, M. J. and D. J. Knight, 1999. Frontal interaction with mesoscale topography. *J. Atmos. Sci.*, **56**, 3544–59.

Doyle, J. D. and D. R. Durran, 2004. Recent developments in the theory of atmospheric rotors. *Bull. Amer. Meteor. Soc.*, **85**, 337–42.

Durran, D. R., 1986a. Another look at downslope windstorms. Part I: The development of analogs to supercritical flow in an infinitely deep, continuously stratified fluid. *J. Atmos. Sci.*, **43**, 2527–43.

Durran, D. R., 1986b. Mountain waves. In *Mesoscale Meteorology and Forecasting*, P. S. Ray (ed.), Amer. Meteor. Soc., 472–92.

Durran, D. R., 1990. Mountain waves and downslope winds. *Atmospheric Processes over Complex Terrain*, Meteor. Monogr., No. 45, Amer. Meteor. Soc., 59–81.

Durran, D. R. and J. B. Klemp, 1987. Another look at downslope winds. Part II: Nonlinear amplification beneath wave-overturning layers. *J. Atmos. Sci.*, **44**, 3402–12.

Eady, E. T., 1949. Long waves and cyclone waves. *Tellus*, **1**, 33–52.

Egger, 1972. Numerical experiments on lee cyclogenesis in the Gulf of Genoa. *Beitr. Phys. Atmos.*, **45**, 320–46.

Epifanio, C. C., 2003. Lee vortices. In *Encyclopedia of the Atmospheric Sciences*, Cambridge University Press, 1150–60.

Epifanio, C. C. and D. R. Durran, 2001. Three-dimensional effects in high-drag-state flows over long ridges. *J. Atmos. Sci.*, **58**, 1051–65.

Epifanio, C. C. and R. Rotunno, 2005. The dynamics of orographic wake formation in flows with upstream blocking. *J. Atmos. Sci.*, **62**, 3127–50.

Garner, S. T., 1995. Permanent and transient upstream effects in nonlinear stratified flow over a ridge. *J. Atmos. Sci.*, **52**, 227–46.

Gaberšek, S. and D. R. Durran 2004. Gap flows through idealized topography. Part I: Forcing by large-scale winds in the nonrotating limit. *J. Atmos. Sci.*, **61**, 2846–62.

Grimshaw, R. H. and N. Smyth, 1986. Resonant flow of a stratified fluid over topography. *J. Fluid Mech.*, **169**, 429–64.

Gross, B. D., 1994. Frontal interaction with isolated topography. *J. Atmos. Sci.*, **51**, 1480–96.

Grubišić, V. and J. M. Lewis, 2004. Sierra Wave Project revisited. *Bull. Amer. Meteor. Soc.*, **85**, 1127–42.

Hayes, J. L., R. T. Williams, and M. A. Rennick, 1993. Lee cyclogenesis. Part II: Numerical studies. *J. Atmos. Sci.*, **50**, 2354–68.

Haynes, P. H. and M. E. McIntyre, 1990. On the conservation and impermeability theorems for potential vorticity. *J. Atmos. Sci.*, **47**, 2021–31.

Holland, G. J. and L. M. Leslie, 1986. Ducted coastal ridging over S. E. Australia. *Quart. J. Roy. Meteor. Soc.*, **112**, 731–48.

Hoskins, B. J., M. E. McIntyre, and A. W. Robertson, 1985. On the use and significance of isentropic potential vorticity maps. *Quart. J. Roy. Meteor. Soc.*, **111**, 877–946.

Hunt, J. C. R. and W. H. Snyder, 1980. Experiments on stably and neutrally stratified flow over a model three-dimensional hill. *J. Fluid Mech.*, **96**, 671–704.

Laprise, R. and W. R. Peltier, 1989a. On the structural characteristics of steady finite-amplitude mountain waves over bell-shaped topography. *J. Atmos. Sci.*, **46**, 586–95.

Laprise, R. and W. R. Peltier, 1989b. The linear stability of severe downslope windstorms. *J. Atmos. Sci.*, **46**, 545–64.

Lee, T.-Y., Y.-Y. Park, and Y.-L. Lin, 1998. A numerical modeling study of mesoscale cyclogenesis to the east of the Korean Peninsula. *Mon. Wea. Rev.*, **126**, 2305–29.

Lin, Y.-L. and D. J. Perkey, 1989. Numerical modeling studies of a process of lee cyclogenesis. *J. Atmos. Sci.*, **46**, 3685–97.

Lin, Y.-L. and T.-A. Wang, 1996. Flow regimes and transient dynamics of two-dimensional stratified flow over an isolated mountain ridge. *J. Atmos. Sci.*, **53**, 139–58.

Lin, Y.-L., N.-H. Lin, and R. P. Weglarz, 1992. Numerical modeling studies of lee mesolows, mesovortices, and mesocyclones with application to the formation of Taiwan mesolows. *Meteor. Atmos. Phys.*, **49**, 43–67.

Lin, Y.-L., J. Han, D. W. Hamilton, and C.-Y. Huang, 1999. Orographic influence on a drifting cyclone. *J. Atmos. Sci.*, **56**, 534–62.

Lin, Y.-L., S.-Y. Chen, C. M. Hill, and C.-Y. Huang, 2005. Control parameters for track continuity and deflection associated with tropical cyclones over a mesoscale mountain. *J. Atmos. Sci.*, **62**, 1849–66.

Long, R. R., 1953. Some aspects of the flow of stratified fluids. Part I. A theoretical investigation. *Tellus*, **5**, 42–58.

Mayr, G. J., 2005. Gap flows – our state of knowledge at the end of MAP. *Croatian Meteor. J.*, **40**, 6–10.

Miles, J. W., 1968. Lee waves in a stratified flow. Part 2: Semi-circular obstacle. *J. Fluid Mech.*, **33**, 803–14.

Miles, J. W. and H. E. Huppert, 1969. Lee waves in a stratified flow. Part 4: Perturbation approximation. *J. Fluid Mech.*, **35**, 497–525.

Orlanski, I. and B. D. Gross, 1994. Orographic modification of cyclone development. *J. Atmos. Sci.*, **51**, 589–611.

Petterssen, S., 1956. *Weather Analysis and Forecasting*. Vol. I, McGraw Hill.

Pichler, H. and R. Steinacker, 1987. On the synoptics and dynamics of orographically induced cyclones in the Mediterranean. *Meteor. Atmos. Phys.*, **36**, 108–17.

Pierrehumbert, R. T., 1986. Lee cyclogenesis. In *Mesoscale Meteorology and Forecasting*, P. S. Ray (ed.), Amer. Meteor. Soc., 493–515.

Pierrehumbert, R. T. and B. Wyman, 1985. Upstream effects of mesoscale mountains. *J. Atmos. Sci.*, **42**, 977–1003.

Queney, P., 1948. The problem of air flow over mountain: A summary of theoretical studies. *Bull. Amer. Meteor. Soc.*, **29**, 16–26.

Queney, P., G. Corby, N. Gerbier, H. Koschmieder, and J. Zierep, 1960. *The Airflow Over Mountains*. WMO Tech. Note, 34.

Reason, C. J. C., K. J. Tory, and P. L. Jackson, 2001. A model investigation of the dynamics of a coastally trapped disturbance. *J. Atmos. Sci.*, **58**, 1892–906.

Rottman, J. W. and R. B. Smith, 1989. A laboratory model of severe downslope winds. *Tellus*, **41A**, 401–15.

Schär, C., 1990. Quasi-geostrophic lee cyclogenesis. *J. Atmos. Sci.*, **47**, 3044–66.

Schär, C., 1993. A generation of Bernoulli's theorem. *J. Atmos. Sci.*, **50**, 1437–43.

Schär, C. and D. R. Durran, 1997. Vortex formation and vortex shedding in continuously stratified flows past isolated topography. *J. Atmos. Sci.*, **54**, 534–54.

Schär, C. and R. B. Smith, 1993a. Shallow-water flow past isolated topography. Part I: Vorticity production and wake formation. *J. Atmos. Sci.*, **50**, 1373–1400.

Schär, C. and R. B. Smith, 1993b. Shallow-water flow past isolated topography. Part II: Transition to vortex shedding. *J. Atmos. Sci.*, **50**, 1401–12.

Scinocca, J. F. and W. R. Peltier, 1993. The instability of Long's stationary solution and the evolution toward severe downslope windstorm flow. Part I: Nested grid numerical simulations. *J. Atmos. Sci.*, **50**, 2245–63.

Scorer, R. S., 1949. Theory of waves in the lee of mountains. *Quart. J. Roy. Meteor. Soc.*, **75**, 41–56.

Scorer, R. S., 1954. Theory of airflow over mountains: III – airstream characteristics. *Quart. J. Roy. Meteor. Soc.*, **80**, 417–28.

Sharman, R. D. and M. G. Wurtele, 1983. Ship waves and lee waves. *J. Atmos. Sci.*, **40**, 396–427.

Skamarock, W. C., R. Rotunno, and J. B. Klemp, 1999. Models of coastally trapped disturbances. *J. Atmos. Sci.*, **56**, 3349–65.

Smith, R. B., 1979. The influence of mountains on the atmosphere. *Adv. Geophys.*, **21**, 87–230.

Smith, R. B., 1980. Linear theory of stratified hydrostatic flow past an isolated mountain. *Tellus*, **32**, 348–64.

Smith, R. B., 1984. A theory of lee cyclogenesis. *J. Atmos. Sci.*, **41**, 1159–68.

Smith, R. B., 1985. On severe downslope winds. *J. Atmos. Sci.*, **42**, 2597–2603.

Smith, R. B., 1986. Further development of a theory of lee cyclogenesis. *J. Atmos. Sci.*, **43**, 1582–1602.

Smith, R. B., 1989a. Hydrostatic airflow over mountains *Adv. Geophys.*, **31**, 1–41.

Smith, R. B., 1989b. Comments on "Low Froude number flow past three-dimensional obstacles. Part I: Baroclinically generated lee vortices." *J. Atmos. Sci.*, **46**, 3611–13.

Smolarkiewicz, P. K. and R. Rotunno, 1989. Low Froude number flow past three-dimensional obstacles. Part I: Baroclinically generated lee vortices. *J. Atmos. Sci.*, **46**, 1154–64.

Speranza, A., A. Buzzi, A. Trevisan, and P. Malguzzi, 1985. A theory of deep cyclogenesis in the lee of the Alps: Part I, Modification of baroclinic instability by localized topography. *J. Atmos. Sci.*, **42**, 1521–35.

Stein, J., 1992. Investigation of the regime diagram of hydrostatic flow over a mountain with a primitive equation model. Part I: Two-dimensional flows. *Mon. Wea. Rev.*, **120**, 2962–76.

Steinacker, R., 1981. Analysis of the temperature and wind fields in the Alpine region. *Geophys. Astrophys. Fluid Dyn.*, **17**, 51–62.

Sun, W.-Y. and J. D. Chern, 2006. Numerical study of the influence of Central Mountain Ranges in Taiwan on a cold front. *J. Meteor. Soc. Japan*, **84**, 27–46.

Sun, W.-Y., J. D. Chern, C. C. Wu, and W.-R. Hsu, 1991. Numerical simulation of mesoscale circulation in Taiwan and surrounding area. *Mon. Wea. Rev.*, **119**, 2558–73.

Tafferner, A. and J. Egger, 1990. Test of lee cyclogenesis: ALPEX cases. *J. Atmos. Sci.*, **47**, 2417–28.

Teixeira, M. A. C. and P. M. A. Miranda, 2006. A linear model of gravity wave drag for hydrostatic sheared flow over elliptical mountains. *Quart. J. Roy. Met. Soc.*, **132**, 2439–58.

Teixeira, M. A. C., P. M. A. Miranda, J. L. Argain, and M. A. Valente, 2005. Resonant gravity wave drag enhancement in linear stratified flow over mountains, *Quart. J. Roy. Met. Soc.*, **131**, 1795–1814.

Tibaldi, S., A. Buzzi, and A. Speranza, 1990. Orographic cyclogenesis. In *Extratropical Cyclones*, C. Newton and E. O. Holopainen (eds.), Amer. Meteor. Soc., 107–28.

Wang, S.-T., 1980. *Prediction of the movement and strength of typhoons in Taiwan and its vicinity*. Res. Rep., 108, National Sci. Council Tech. Rep., Taipei, Taiwan.

Wang, T.-A. and Y.-L. Lin, 1999. Wave ducting in a stratified shear flow over a two-dimensional mountain. Part II: Implication for the development of high-drag states for severe downslope windstorms. *J. Atmos. Sci.*, **56**, 437–52.

Xu, Q., S. Gao, and B. H. Fiedler, 1996. A theoretical study of cold air damming with upstream cold air inflow. *J. Atmos. Sci.*, **53**, 312–26.

Zängl, G., 2002. Stratified flow over a mountain with a gap: Linear theory and numerical simulations. *Quart. J. Roy. Meteor. Soc.*, **128**, 927–49.

Problems

5.1 Derive the solutions, (5.1.21)–(5.1.23). Verify the maxima and minima of different variables denoted in Fig. 5.1a.

5.2 Show that the vertical energy flux, $\overline{p'w'}$, for terms C in (5.1.28) is positive.

5.3 Prove (5.1.35).

5.4 (a) Based on (5.2.12) and using the Boussinesq approximation, find the solutions of other variables, w', u', p', and θ', for a hydrostatic mountain wave. (b) Determine the maxima and minima of u', p', and θ' near the mountain surface.

5.5 Find the wavelength of the resonance waves of the Scorer's (1949) model with $H = 2.7$ km, $l_1^2 = 2.12 \times 10^{-6}$ m^{-2}, and $l_2^2 = 0.33 \times 10^{-6}$ m^{-2}, by solving (5.2.35) graphically for k^* using a graphics calculator or spreadsheet program. [Hint: When plotting the functions with either a graphics calculator or graphics software, you must recall from the text that $l_2^2 < k^2 < l_1^2$ and restrict your plotting range to between l_1 and l_2.]

5.6 Solve (5.2.34) with the interface conditions to obtain the solution (5.2.39).

5.7 (a) What is the corresponding stratified-flow Froude number (F) and nondimensional mountain height (M) for the flow in Figure 5.9a? To find M, you must estimate H_0 from Fig. 5.9a, noting that $H = H_0$. With the values of F and M you calculate, determine the flow regime from Fig 3.3. (b) Using Fig. 5.15 and based on Smith's theory (Eq. (5.3.14)), calculate the pressure drag (D) on the mountain per unit length, assuming $\rho_0 = 1$ kg m^{-3} and a depth of lower downstream dividing streamline, $H_1 = 3.14$ km.

5.8 Show that with vertical wind shear, (5.4.6) becomes $(\nabla^2 + l^2)w'_{xx} + (N^2/U^2)w'_{yy} = 0$, where l is the Scorer parameter [(5.1.6)].

5.9 Prove (5.5.5). Based on (5.5.5) with $h_m = 500$ m, $a = 2000$ km, $N = 0.01$ s^{-1}, $f = 10^{-4}$ s^{-1}, make a plot and compare with the sketch of Fig. 5.26.

5.10 (a) Derive (5.5.9) from (5.5.7) and the geostrophic wind and thermal wind relations. Let $u_g = U + u'_g$, $v_g = V + v'_g$, $U = U_s + U_z z$, and $V = V_s + V_z z$. (b) Derive (5.5.10) by assuming the flow is steady and two-dimensional and taking the Fourier transform to the system (5.5.8) and (5.5.9).

6

Thermally forced flows

Some of the basic dynamics of a number of thermally forced flows can be understood by using a known function to represent diabatic heating or cooling. For example, in theoretical studies of sea breeze circulations, the differential heating associated with the land–sea temperature contrast is prescribed as a periodic function at the ground level. This approach makes the mathematical problem more tractable and applicable to other problems related to mesoscale circulations. These problems include heat island circulations, sea and land breezes, mountain-plain solenoidal circulations, density current generation and propagation, formation of thunderstorm cloud tops, as well as circulations and gravity waves that are generated by diabatic heating associated with coastal frontogenesis, moist convection, and orographic precipitation systems.

In this chapter, we will discuss the problems listed above and will place a greater emphasis on the basic dynamics involved. Section 6.1 details the responses of a uniform, steady, continuously stratified flow to a mesoscale heat source or sink. This will make the fundamental dynamics easier to understand. This section will also discuss sinusoidal and isolated heat sources, transient flow response to a mesoscale heat source, pulse heating and steady heating. In addition, this section will include applications of the thermally forced circulation theory to various types of mesoscale circulations. In Section 6.2, we will discuss three-dimensional flow and shear flow over heat sources or sinks. Sections 6.1 and 6.2 will aid in understanding the dynamics of sea and land breezes. Mountain–plain solenoidal circulations are discussed in Sections 6.3 and 6.4.

6.1 Two-dimensional flows

6.1.1 Steady flows over a sinusoidal heat source

For uniform flows that are nonrotating, inviscid, Boussinesq, with a small-amplitude and at steady state, with constant buoyancy frequency over a two-dimensional mesoscale heat source, (2.2.14)–(2.2.18) reduce to

$$U\frac{\partial u'}{\partial x} + \frac{1}{\rho_o}\frac{\partial p'}{\partial x} = 0, \qquad (6.1.1)$$

$$U\frac{\partial w'}{\partial x} - g\frac{\theta'}{\theta_o} + \frac{1}{\rho_o}\frac{\partial p'}{\partial z} = 0,$$ (6.1.2)

$$\frac{\partial u'}{\partial x} + \frac{\partial w'}{\partial z} = 0, \text{ and}$$ (6.1.3)

$$U\frac{\partial \theta'}{\partial x} + \frac{N^2\theta_o}{g}w' = \frac{\theta_o}{c_p T_o}q'.$$ (6.1.4)

The fluid system expressed in the equations above is similar to that discussed for orographically forced flow, except that they are characterized by thermal forcing instead of orographic forcing. Equations (6.1.1)–(6.1.4) can be combined into a single equation for the vertical velocity:

$$\frac{\partial^2 w'}{\partial x^2} + \frac{\partial^2 w'}{\partial z^2} + l^2 w' = \frac{g}{c_p T_o U^2}q',$$ (6.1.5)

where $l^2 = N^2/U^2$ is the square of the *Scorer parameter*, as defined in (5.1.6), for a uniform basic flow ($U_z = 0$). As a first approximation, we may assume a separable heating function,

$$q'(x,z) = Q_o f(x)g(z),$$ (6.1.6)

where $g(z)$ is normalized according to

$$\int_0^\infty g(z)\mathrm{d}z = 1,$$ (6.1.7)

so that

$$\rho_o \int_0^\infty q'(x,z)\mathrm{d}z = \rho_o Q_o f(x),$$ (6.1.8)

which represents the total thermal energy added to a vertical column of the atmosphere per unit time. Note that the net heating involved in diabatic processes in a steady state fluid flow tends to produce a vertical displacement that continues to increase downstream. To avoid the *net heating problem*, we impose the constraint

$$\int_{-\infty}^\infty f(x)\mathrm{d}x = 0,$$ (6.1.9)

at every level.

We then apply the *Green's function method* by assuming the heating is concentrated at height $z = 0$ in an unbounded atmosphere,

$$q'(x, z) = Q_o f(x) \delta(z = 0).$$ (6.1.10)

At the interface, $z = 0$, the *kinematic boundary condition* (3.7.11) requires that the vertical velocity be continuous, i.e.

$$w'(z = 0^+) = w'(z = 0^-),$$ (6.1.11)

where $z = 0^+$ and $z = 0^-$ denote the heights just above and below $z = 0$, respectively. Substituting (6.1.10) into (6.1.5) and integrating it from $z = 0^-$ to $z = 0^+$ gives the second interface condition, i.e. the *dynamic boundary condition* (3.7.13),

$$w'_z(0^+) - w'_z(0^-) = \frac{g Q_o f(x)}{c_p T_o U^2}.$$ (6.1.12)

The above condition is equivalent to the continuity of perturbation pressure across the interface (Chapter 4). Away from $z = 0$, (6.1.5) reduces to the *Scorer's equation*, i.e. (5.1.5),

$$w'_{xx} + w'_{zz} + l^2 w' = 0.$$ (6.1.13)

The mathematical problem associated with appropriate upper and lower boundary conditions in (6.1.11)–(6.1.13) is similar to problems encountered in mountain wave theory (Sections 5.1 and 5.2).

For simplicity, we consider a spatially sinusoidal heating function

$$q'(x, z) = Q_o \cos kx \, \delta(z = 0),$$ (6.1.14)

and look for solutions in the form of

$$w'(x, z) = w_1(z) \cos kx + w_2 \sin kx.$$ (6.1.15)

Thus, Scorer's equation, which governs solutions for w_i, $i = 1, 2$ becomes

$$w_{izz} + (l^2 - k^2) w_i = 0, \quad i = 1, 2.$$ (6.1.16)

As in mountain wave theory, two regimes are associated with (6.1.16), namely, (1) $k^2 > l^2$: the evanescent flow regime, and (2) $k^2 < l^2$: the vertically propagating wave regime. For $k^2 > l^2$, the solutions can be written as

$$w_i(x, z) = A_i(x) e^{-\lambda z} + B_i(x) e^{\lambda z}, \quad \text{for } z \geq 0,$$ (6.1.17a)

$$w_i(x, z) = C_i(x) e^{-\lambda z} + D_i(x) e^{\lambda z}, \quad \text{for } z < 0,$$ (6.1.17b)

where $\lambda = \sqrt{k^2 - l^2}$. Terms B_i and C_i represent disturbances that increase in the vertical away from the heating level, and should be eliminated, i.e, $B_i = C_i = 0$, in order to satisfy the boundedness condition, at infinity far from the energy source located at $z = 0$. Applying the interface conditions (6.1.11) and (6.1.12) to (6.1.17) leads to

$$w'(x, z) = \left(\frac{-gQ_0}{2c_p T_0 U^2}\right) \frac{\cos kx}{\sqrt{k^2 - l^2}} e^{-\sqrt{k^2 - l^2}|z|}, \quad \text{for } k^2 > l^2. \tag{6.1.18}$$

As demonstrated in Chapters 3 and 5, the condition $k^2 > l^2$ corresponds to $2\pi/N > L/U$, i.e. a relatively stronger wind with weaker stability over a narrower heat source. When $k^2 \gg l^2$, the buoyancy force becomes extremely weak and can be ignored. In this limit, the disturbance will approach a *potential (irrotational) flow*,

$$w'(x, z) = \left(\frac{-gQ_0}{2c_p T_0 U^2 k}\right) \cos kx \, e^{k|z|}. \tag{6.1.19}$$

The flow field of the evanescent waves is simply a negative cosine function at the heating level ($z = 0$), that exponentially decays with height away from this level. The negative phase of the vertical motion and heating will be explained later.

When $k^2 < l^2$, the solution for (6.1.16) can be written as

$$w_i(z) = A_i \sin mz + B_i \cos mz, \quad i = 1, 2, \tag{6.1.20}$$

where $m^2 = l^2 - k^2$. Combining this with (6.1.15), the above solution can be rewritten as

$$w'(x, z) = C \cos(kx + mz) + D \cos(kx - mz)$$
$$\tag{6.1.21a}$$
$$+ E \sin(kx + mz) + F \sin(kx - mz), \quad \text{for } z \geq 0,$$

$$w'(x, z) = C' \cos(kx + mz) + D' \cos(kx - mz)$$
$$\tag{6.1.21b}$$
$$+ E' \sin(kx + mz) + F' \sin(kx - mz), \quad \text{for } z < 0,$$

As in mountain wave theory, terms with argument $kx + mz$ have an upstream phase tilt with height and represent upward energy propagation for $z > 0$, while terms with argument $kx - mz$ have a downstream phase tilt and represent downward energy propagation for $z < 0$. Since the energy source is located at $z = 0$, the upper and lower radiation conditions require $D = F = C' = E' = 0$. Applying the interface conditions (6.1.11) and (6.1.12) to (6.1.21), we obtain

$$w'(x, z) = \left(\frac{gQ_0}{2c_p T_0 U^2}\right) \left(\frac{\sin\left(kx + \sqrt{l^2 - k^2}|z|\right)}{\sqrt{l^2 - k^2}}\right), \quad \text{for } k^2 < l^2 \tag{6.1.22}$$

The above solution represents *vertically propagating waves*, which satisfy the radiation conditions at $z = \pm \infty$. Again, the condition $k^2 < l^2$ corresponds to $2\pi/N < L/U$, i.e. a relatively weaker wind with stronger stability over a broader heat source. The flow response predicted by (6.1.22) becomes hydrostatic for $k^2 \ll l^2$. In this limit, the above equation reduces to

$$w'(x,z) = \left(\frac{gQ_o}{2c_p T_o U^2 l}\right) \sin(kx + l|z|), \qquad (6.1.23)$$

As in mountain wave theory, the above solution at $x = 0$ repeats itself with a vertical wavelength of $2\pi U/N$, which is referred to as the *hydrostatic vertical wavelength*. With a typical atmospheric situation of $U = 10$ m s^{-1} and $N = 0.01$ s^{-1}, the vertical wavelength of the forced wave is about 6.28 km. As is the case in the mountain wave theory, when the vertical displacement (η) is in a steady-state and the flow is linear, η is related to w' through

$$w' = U\frac{\partial \eta}{\partial x}. \qquad (6.1.24)$$

Figure 6.1 shows an example for an unbounded, hydrostatic, stratified airflow over a periodic heating and cooling concentrated at the $z = 0$ level. Vertically propagating waves are evident above and below the heating level with upstream phase tilting.

Note that *the vertical displacement at the heating level is exactly out of phase with the heating and cooling* in Fig. 6.1. That is, the air parcel is displaced downward (upward) in the heating (cooling) region. A similar phenomenon has been observed over heat

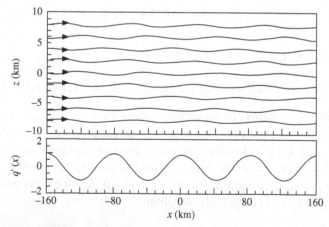

Fig. 6.1 The vertical displacement of an unbounded, hydrostatic stratified airflow for periodic heating and cooling concentrated at the $z = 0$ level, as implied by (6.1.23). The parameters used are: $U = 10$ m s^{-1}, $N = 0.01$ s^{-1}, $L = 2\pi/k = 80$ km, $Q_o = 1200$ J kg^{-1} m s^{-1}, $T_o = 287$ K. The normalized heating rate function, $q'(x, z) = Q_o \cos kx\, \delta\, (z = 0)$, at $z = 0$ is shown in the lower panel. The unit of q' is J kg^{-1}s^{-1}. Vertically propagating waves are evident from the upstream phase tilting above and below the heating level. (From Smith and Lin 1982.)

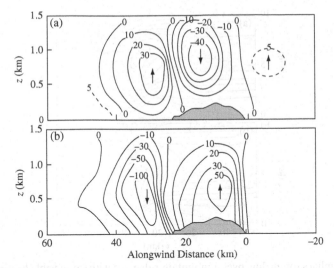

Fig. 6.2 Vertical velocity fields observed over Barbados during the summers of 1968 and 1969 by: (a) the day and (b) the night (DeSouza 1972). The basic wind is from right. The unit of w' is $0.01\ \mathrm{m\ s^{-1}}$. (Adapted after Garstang *et al.* 1975.)

islands. For example, Fig. 6.2 shows that, during the daytime, there is a downward motion over Barbados followed by an upward motion over the ocean on the downwind side. This is also consistent with other theoretical studies on stratified flow over a diabatic heat source or sink. It will be discussed in Section 6.2 that the responses to diabatic forcing are closely related to the flow speed or the Froude number associated with thermal forcing. Another example is that mountain waves may be strengthened by sensible cooling (Fig. 6.3b) and weakened by sensible heating. Responses to diabatic forcing are closely related to the flow speed or the Froude number associated with thermal forcing which will be discussed in Section 6.2.

Similar to (5.1.25), the vertical flux of horizontal momentum can be calculated by

$$D = \rho_0 \int_0^L u'w'\mathrm{d}x, \qquad (6.1.25)$$

where $L\ (=2\pi/k)$ is the horizontal wavelength of the heating. From (6.1.23) and (6.1.25), we have

$$\frac{D}{L} = \frac{-\rho_0}{2kl}\left(\frac{gQ_0}{2c_pT_0U^2}\right)^2\frac{z}{|z|}. \qquad (6.1.26)$$

The transport of mechanical energy away from the layer of thermal forcing is accompanied by a flux of horizontal momentum towards the layer.

To examine the effect of vertical momentum flux, we consider the time-dependent nonlinear horizontal momentum equation

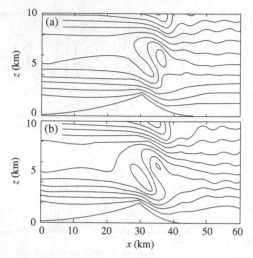

Fig. 6.3 Streamlines for airflow over a mountain ridge: (a) with no sensible heating or cooling (adiabatic), and (b) with boundary layer sensible cooling. Note that the adiabatic mountain waves are strengthened by the sensible cooling. A similar result has also been found in Olfe and Lee (1971). (Adapted after Raymond 1972.)

$$\frac{\partial u'}{\partial t} + U \frac{\partial u'}{\partial x} + \frac{1}{\rho_0} \frac{\partial p'}{\partial x} = -u' \frac{\partial u'}{\partial x} - w' \frac{\partial u'}{\partial z}. \qquad (6.1.27)$$

Taking the horizontal integration over one wavelength yields

$$\frac{\partial \overline{u}}{\partial t} = -\frac{\partial}{\partial z} \overline{uw}, \qquad (6.1.28)$$

where

$$\overline{(\ )} = \frac{1}{L} \int_0^L (\ ) \mathrm{d}x. \qquad (6.1.29)$$

Thus, the convergence of the vertical momentum flux tends to accelerate the flow. Note that this acceleration is not explicitly accounted for in linear theory since nonlinear terms are neglected. This acceleration may be relevant to the problems of moist convection, heat islands, and orographic precipitation.

6.1.2 Steady flows over an isolated heat source

A useful localized heating function may be chosen

$$q'(x, z) = Q_0 \left(\frac{b_1^2}{x^2 + b_1^2} - \frac{b_1 b_2}{x^2 + b_2^2} \right) \delta(z - z_{\mathrm{H}}), \qquad (6.1.30)$$

where z_H is the level of thermal forcing, b_1 is the half-width of the heating function and b_2 is the horizontal scale of the compensated cooling. In addition, unlike in the previous problem, here we consider a semi-infinite atmosphere with a flat surface located at $z = 0$. Again, we can apply the Green's function method to obtain the solution since the heating is concentrated at a particular level. Substituting the above equation into (6.1.5) leads to

$$w'_{xx} + w'_{zz} + l^2 w' = \left(\frac{gQ_o}{c_p T_o U^2}\right)\left(\frac{b_1^2}{x^2 + b_1^2} - \frac{b_1 b_2}{x^2 + b_2^2}\right)\delta(z - z_H). \tag{6.1.31}$$

Applying the one-sided Fourier transform of $w'(x,z)$ in x to (6.1.31) yields (Appendix 5.1),

$$\tilde{w}_{zz} + (l^2 - k^2)\tilde{w} = \left(\frac{gQ_o b_1}{2c_p T_o U^2}\right)\left(e^{-kb_1} - e^{-kb_2}\right)\delta(z - z_H). \tag{6.1.32}$$

For $z \neq z_H$, (6.1.32) reduces to the Scorer's equation, (6.1.13), in the Fourier space,

$$\tilde{w}_{zz} + (l^2 - k^2)\tilde{w} = 0. \tag{6.1.33}$$

For a hydrostatic wave ($k^2 \ll l^2$), the solution may be written as

$$\tilde{w}(k, z) = A e^{ilz} + B e^{-ilz} \quad \text{for } 0 \leq z < z_H, \tag{6.1.34a}$$

$$\tilde{w}(k, z) = C e^{ilz} + D e^{-ilz} \quad \text{for } z_H \leq z. \tag{6.1.34b}$$

The lower boundary condition requires $w' = 0$ over a flat surface at $z = 0$. As in the previous discussion and in the mountain wave theory, the upper radiation boundary condition requires $D = 0$. The interface boundary conditions may be determined in a similar manner to that used for the periodic heat source problem (Subsection 5.1.1) using (6.1.11) and (6.1.12). Once the interface boundary conditions have been determined, applying them to (6.1.34), using (6.1.24) and taking the inverse Fourier transform allows us to obtain the vertical displacement for a heating concentrated at $z = z_H$,

$$\eta(x, z) = -A_1 \sin lz\{T_1(x)\cos lz_H + L_1(x)\sin lz_H\}, \quad \text{for } 0 \leq z < z_H, \tag{6.1.35a}$$

$$\eta(x, z) = -A_1 \sin lz_H\{T_1(x)\cos lz + L_1(x)\sin lz\}, \quad \text{for } z_H \leq z, \tag{6.1.35b}$$

where

$$A_1 = \frac{gQ_o b_1}{c_p T_o U^3 l}, \quad T_1(x) = \tan^{-1}\left(\frac{x}{b_1}\right) - \tan^{-1}\left(\frac{x}{b_2}\right), \quad L_1(x) = \frac{1}{2}\ln\left(\frac{x^2 + b_2^2}{x^2 + b_1^2}\right).$$

The vertical displacement for heating distributed in a layer can be determined by superposition of the above solution through integration.

The surface pressure perturbation can be calculated from (6.1.35) using *Bernoulli's equation*, which is obtained by substituting (6.1.3) and (6.1.24) into (6.1.1),

$$p'(x,0) = -(\rho_0 U^2 l)A_1\{T_1(x) \cos lz_H + L_1(x) \sin lz_H\}. \qquad (6.1.36)$$

The vertical momentum flux associated with (6.1.35) is

$$\mathcal{D} = 0, \qquad\qquad \text{for } 0 \leq z < z_H, \qquad (6.1.37a)$$

$$\mathcal{D} = -\pi\rho_0 U A_1 \ln\left\{(b_1 + b_2)^2/4b_1 b_2\right\}, \qquad \text{for } z_H \leq z, \qquad (6.1.37b)$$

The vertical displacement and surface pressure perturbation produced by the thermally induced gravity waves satisfy the rigid lower boundary condition $w' = 0$ at $z = 0$, thus causing complete reflection of the downward propagating wave produced by the elevated heating. Flux cancellation of the upward and downward propagating waves results in a vertical momentum flux of zero between the heating level and the surface. This gives the disturbance no vertical phase tilt. The flow response is sensitive to the heating level since the upward propagating and downward propagating waves may cancel each other out. If the heating is added very near the surface, $lz_H < 1$, the resulting disturbance is extremely small and may be neglected. From (6.1.35), cancellation of the direct upward propagating wave and the reflected upward propagating wave above z_H can also occur at $lz_H = 0, \pi, 2\pi, \ldots, n\pi$. This effect is less evident if the heating is spread over a layer of finite depth.

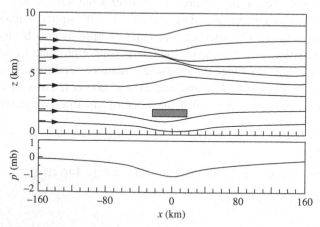

Fig. 6.4 Hydrostatic response of a stratified airstream to an isolated heating-widespread cooling function (6.1.30) with $Q_0 = 900$ W m kg^{-1}, $b_1 = 20$ km, $b_2 = 100$ km, $U = 10$ m s^{-1} and $N = 0.01$ s^{-1}: (a) vertical displacement and (b) perturbation pressure. The lower surface is flat. A low pressure is generated directly below the region of concentrated heating (shaded). (From Smith and Lin 1982.)

Figure 6.4 shows an example of the hydrostatic response to isolated heating and widespread cooling [(6.1.30)] added at $lz_H = \pi/2$. A downward displacement, similar to that of the sinusoidal heating function, is produced near the region where the heating occurs. This relationship will be explained in Section 6.2. The upstream phase tilt of the thermally forced gravity waves is evident above the heating level. The vertical displacement at the heating level is repeated every 6.28 km ($2\pi/l$). The surface perturbation pressure is shown in the lower panel of Fig. 6.4. The hydrostatic equation indicates that the surface pressure is an integral measure of the temperature or density anomaly aloft. The thermodynamic equation implies that the heating directly causes the temperature anomaly, while thermally induced vertical motion causes it indirectly. The relationship between the system's thermal response and the basic flow will be discussed in the following section as it is related to a transient flow over a heat source.

6.2 Transient flows

6.2.1 Flow responses to pulse heating

The basic dynamics of flow responses to pulse heating can be studied by considering a two-dimensional, inviscid, nonrotating, hydrostatic, Boussinesq uniform flow over a heat source. The governing equation can be deduced from (3.2.1),

$$\left(\frac{\partial}{\partial t} + U\frac{\partial}{\partial x}\right)^2 \frac{\partial^2 w'}{\partial z^2} + N^2 \frac{\partial^2 w'}{\partial x^2} = \left(\frac{g}{c_p T_o}\right)\frac{\partial^2 q'}{\partial x^2}. \tag{6.2.1}$$

The above equation is similar to (4.2.1). To solve (6.2.1), we again apply *Green's function method* in the vertical direction. Taking the Fourier transform in x and the Laplace transform in t of the above equation (which transforms $w'(t,x,z)$ into $\hat{w}(s,k,z)$; see Appendix 6.1) yields

$$\frac{\partial^2 \hat{w}}{\partial z^2} + \lambda^2 \hat{w} = \frac{g\lambda^2}{c_p T_o N^2}\hat{q}, \tag{6.2.2}$$

where $\lambda \equiv iNk/(s+iUk)$, and $\mathrm{Re}(s) > 0$ is assumed. Assuming the heating is released in a short pulse at a single level, $z = 0$, in an unbounded fluid,

$$q'(t, x, z) = \left(\frac{Q_o b^2}{x^2 + b^2}\right)\delta(z)\delta(t), \tag{6.2.3}$$

taking the Fourier and Laplace transforms of the above heating function and substituting it into (6.2.2) gives

$$\frac{\partial^2 \hat{w}}{\partial z^2} + \lambda^2 \hat{w} = \frac{gQ_o b\lambda^2 e^{-bk}\delta(z)}{c_p T_o N^2}. \tag{6.2.4}$$

Similar to the steady-state problem, the solution for the vertical displacement can be obtained by applying the appropriate upper, lower, and interface conditions, yielding

$$\hat{\eta}(s, k, z) = \left(\frac{gQ_\mathrm{o}b}{2c_pT_\mathrm{o}N}\right)\frac{ke^{-bk}e^{-Nk|z|/(s+iUk)}}{(s+iUk)^2}. \tag{6.2.5}$$

In the above equation, the vertical displacement is related to the vertical velocity through $w' = D\eta/Dt = \partial\eta/\partial t + U\partial\eta/\partial x$. Taking the inverse Laplace transform in s (Appendix 6.1) and the inverse Fourier transform in k (Appendix 5.1) of the above equation, leads to (Lin and Smith 1986)

$$\eta(t, x, z) = \left(\frac{gQ_\mathrm{o}bte^{-bK}}{2c_pT_\mathrm{o}N}\right)\{(b^2 - X^2)\cos(KX) + 2bX\sin(KX)\}, \tag{6.2.6}$$

where $K = Nt|z|/(X^2 + b^2)$ and $X = x - Ut$ is the horizontal coordinate in the reference frame moving with the basic wind. The flow response in a moving frame is analogous to a pulse heating in a quiescent fluid, as discussed in Section 6.1.

As in the steady-state problem, the solution for heating that is distributed in a layer can be obtained through the superposition of the heating at a particular level within the heating layer, while the solution for a half-plane with a rigid lower surface can be obtained from that in an unbounded fluid, (6.2.6), by the *method of images* (e.g. Hildebrand 1976). In the method of images, the solution is obtained by superimposing a solution to the original forcing and that of its mirror image at the same distance below the rigid boundary ($z = 0$ in this case). The solution for a heating layer can be obtained by integrating the solution of a single level across the heating layer. The vertical displacement at the center of the heating layer ($z = z - d$ to $z + d$), where d is the half-depth of the heating layer, in the unbounded fluid can be derived to be

$$\tilde{\eta} = \left(\frac{1}{\tilde{x}^2 + 1}\right)\left\{1 - e^{-\tilde{t}/(\tilde{x}^2+1)}\left[\tilde{x}\sin\left(\frac{\tilde{x}\tilde{t}}{\tilde{x}^2 + 1}\right) + \cos\left(\frac{\tilde{x}\tilde{t}}{\tilde{x}^2 + 1}\right)\right]\right\}, \tag{6.2.7}$$

where $\tilde{x} = (x - Ut)/b$, $\tilde{t} = Ndt/b$, and $\tilde{\eta} = (c_pT_\mathrm{o}N^2/gQ_\mathrm{o})\eta(t, x, z = 0)$. We are interested in the flow response to pulse heating in two regions: (a) the region of drifting heated air and (b) the region of the initial heating. Figure 6.5a shows the evolution of the vertical displacement around the center of drifting disturbance. This displacement occurs in a reference frame that is moving with the basic wind. The fluid's early response to the heating is an upward displacement at the drifting center and downward displacements at the upstream and downstream sides of the growing disturbance. The weak downward displacements are necessary because they compensate for the upward motion at the center. Mass continuity requires this downward displacement.

The vertical displacement at the drifting center ($\tilde{x} = 0$) grows according to the function $\exp(1 - e^{-\tilde{t}})$. The final vertical displacement $\tilde{\eta}(t = \infty, x, z)$ is proportional

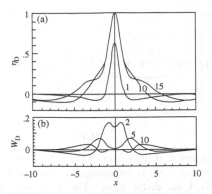

Fig. 6.5 (a) The vertical displacements at the heating level of the pulse heat source in a frame moving with the basic wind. The solution is given by (6.2.7). The three curves represent $\tilde{\eta}$ at $\tilde{t} = 1$, 10, and 15. Note that there is a strong updraft near the drifting center and two regions of compensated downdrafts to the sides of the growing updraft. (b) The normalized vertical velocities at $\tilde{t} = 2$, 5 and 10. The horizontal coordinate is nondimensionalized as in (6.2.7). (Adapted after Lin and Smith 1986.)

at all points to the total amount of heat received by that air parcel. This displacement can be found by letting $\tilde{t} \to \infty$ in (6.2.7)

$$\tilde{\eta}_D(t = \infty, x, z) = \frac{1}{\tilde{x}^2 + 1}, \quad \text{for } z_0 - d \le z \le z_0 + d, \tag{6.2.8}$$

which is proportional at all points to the total amount of heat received by the air parcel. Figure 6.5b shows the nondimensional vertical velocity at the center of the pulse heating at three times. Once the updraft at the drifting center weakens, the fluid in the adjacent regions can rise. Subsequently, two updrafts develop and propagate outward. This action is analogous to the left and right moving waves in a two-dimensional shallow water system. These updrafts will overcome the downward displacement produced earlier and generate upward displacement at a later time, as can be seen from Fig. 6.5a. When this time is reached, the original disturbance will have split in two.

To find the flow response at the heating center, we set $x = 0$ in $\tilde{x}$ in (6.2.7). The nondimensional vertical displacement at the origin of the initial pulse heating can thus be obtained

$$\tilde{\eta}_0 = \frac{1 - \exp[-\tilde{t}/F_r(\tilde{t}^2 + 1)]}{\tilde{t}^2 + 1}\left\{\tilde{t}\sin\left(\frac{\tilde{t}^2}{F_r(\tilde{t}^2 + 1)}\right) + \cos\left(\frac{\tilde{t}^2}{F_r(\tilde{t}^2 + 1)}\right)\right\}, \tag{6.2.9}$$

where $F_r \equiv U/Nd$ is the *thermal Froude number*. The above equation reduces to $\tilde{\eta}_0 \approx (1/\tilde{t})\sin(1/F_r)$ when $\tilde{t} \to \infty$. Therefore, the response of the flow at the heating center ($x = 0$) is strongly dependent on F_r, and changes sign at $F_r = 1/n\pi$. The vertical displacement at the heating center decreases as $1/\tilde{t}$ when $\tilde{t} \to \infty$, as shown in Fig. 6.6. The response of the flow at the origin of the pulse heating is an

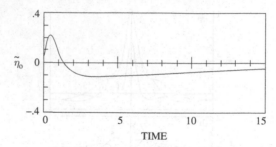

Fig. 6.6 The vertical displacement at the origin of the pulse heating. The solution is given by (6.2.9). Note that the response of $\tilde{\eta}_0$ is an upward displacement followed by a downward displacement, as the heated air drifts away. The time and vertical displacement are nondimensionalized. (From Lin and Smith 1986.)

upward displacement, followed by a downward displacement as the heated air drifts away. For a stronger basic flow (larger F_r), the downward displacement produced later in the process is associated with the compensating downdraft as the growing updraft drifts downstream due to the advection effect. On the other hand, for a weaker basic flow (smaller F_r), the advection effect is weaker and the growing updraft dominates the flow response near the heating center and produces an upward displacement. This *advection mechanism* helps explain the negative phase relationship between the vertical displacement and heating shown in Figs. 6.1 and 6.4.

6.2.2 Flow responses to steady heating

As shown in Figs. 6.1 and 6.4, steady heating in a moving airstream produces a curiously negative relationship between heating and vertical displacement. That is, heating (cooling) produces a downward (upward) displacement in the vicinity of the heat source (sink). This effect is directly related to the steadiness of the heating and can be explained as follows:

Taking $b \rightarrow 0$ and keeping $Q_0 b$ constant in (6.2.6) gives the vertical displacement of a moving stratified airstream to a point heat source that is released initially as a pulse

$$\eta(t, x, z) = \frac{-Q_0 bt}{2\pi N \tilde{x}^2} \cos\left(\frac{Nzt}{\tilde{x}}\right), \qquad (6.2.10)$$

where $\tilde{x} = (x - Ut)/b$. Since we are interested in the response near the origin of the heating, $(x, z) = (0, 0)$, the above equation reduces to

$$\eta(t, 0, 0) = \frac{-Q_0 b^3}{2\pi N U^2 t}. \qquad (6.2.11)$$

Thus, the air is displaced downward proportional to t^{-1} at the origin of a pulse heating, consistent with (6.2.9) as $\tilde{t} \rightarrow \infty$. The steady-state heat source may be regarded as a

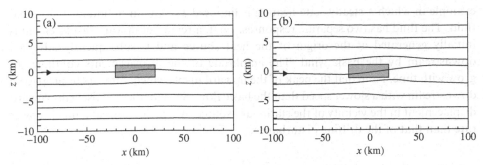

Fig. 6.7 Vertical displacement of an unbounded, hydrostatic atmosphere to a steady state heating (shaded) imposed at $t=0$. Two times are shown: (a) 3000 s, and (b) 7000 s. The solution can be obtained by integrating (6.2.6) for a layer from $z=z_o-d$ to z_o+d and in time. The flow parameters used are: $U=10 \text{ m s}^{-1}$, $N=0.01 \text{ s}^{-1}$, $T_o=273 \text{ K}$, $Q_o=1 \text{ J kg}^{-1} \text{ s}^{-1}$, $b=20 \text{ km}$, $d=1 \text{ km}$, and $z_o=0 \text{ km}$. (From Lin and Smith 1986.)

succession of very short heat pulses, which leads to an accumulated downward displacement by individual pulse heating events at the origin of heat source. This advection mechanism is consistent with the group velocity argument (Bretherton 1988). In the group velocity argument, the $1/t$ decay of the displacement is shown to be a geometrical consequence of dispersion in two dimensions. The growing response to a steady heating is understood as the result of energy being pumped into the gravity wave modes, whose group velocity is near zero, faster than it can spread in physical space due to dispersion.

In addition to the advection mechanism and group velocity argument, an alternative explanation of the negative relationship between vertical displacement and thermal forcing can be made using the energy budget. A linearized steady state energy equation may be derived through a method similar to that used in deriving (4.4.5). Thus, by excluding the basic shear terms and including the diabatic heating term, we obtain

$$\frac{\partial}{\partial x}(EU + p'u') + \frac{\partial}{\partial z}(p'w') = \left(\frac{\rho_0 g^2}{c_p T_0 N^2 \theta_0}\right)\theta' q', \qquad (6.2.12)$$

where $E = (\rho_0/2)[u'^2 + (g\theta'/N\theta_0)^2]$ is the perturbation wave energy in a hydrostatic atmosphere. According to the above equation, the addition of thermal energy to the system requires that steady heating be added where the air is warm or the air density is low. This condition implies that the perturbation flow field must adjust itself so that the regions of negative density anomaly (negative displacement) may receive the heat.

The gravity waves produced by a pulse heating in an unsheared flow are symmetric about the heating center and impart no net momentum flux to the flow. Thus, no vertically propagating gravity waves are produced. However, a steady heating or cooling can generate vertically propagating gravity waves. The vertical displacement for the steady state heating can be obtained by integrating (6.2.6). Figure 6.7 shows an example in which a steady state heating is imposed $t = 0$ in an unbounded stratified fluid. The fluid has two separate responses. First, a region of upward displacement is initially generated at the origin of the heat source and is subsequently advected downstream by the basic wind. The amplitude of the displacements continues to grow with time. Note that the peak of the upward displacement appears to propagate downstream with a slower speed than the basic flow. Second, there is also a downward displacement in the vicinity of the stationary source, which develops at a much slower rate than that of the drifting disturbance.

6.3 Applications to mesoscale circulations

6.3.1 Density current formation and propagation

Even though more realistic representations of evaporative cooling have been adopted to simulate the cold-air outflows and density currents in nonlinear numerical models, prescribed cooling has shown to be a useful way to help understand the formation and propagation of density currents in both theoretical and numerical models. In studying the response of a predominantly unstratified environment to evaporative cooling in a fully nonlinear numerical model, it is found that the flow can be characterized by the ratio of the basic flow speed to the spreading speed of the density current in quiescent fluid (Thorpe *et al.* 1980). Since this ratio is directly proportional to the F_r defined in (6.2.9), we may use F_r to represent it. The upstream propagation speed of the density current is reduced when F_r increases. When $F_r \approx 1$, the forcing associated with the density current is comparable to that associated with the basic flow. At this point, the density current reaches a critical condition and becomes stationary.

Figure 6.8 shows the response of a two-dimensional, hydrostatic, stratified flow to a prescribed cooling, which represents a quasi-steady cooling due to rain evaporation, in the subsaturated layer beneath a thunderstorm, as simulated by a nonlinear numerical model. The upward vertical displacement in the vicinity of the heat sink resembles the flow structure near the front of a squall line, and may provide a possible mechanism for its maintenance. The negative phase relationship between the cooling and the vertical displacement (Fig. 6.8a) can be explained by the advection mechanism and group velocity argument discussed earlier. The vertical velocity field indicates that flow convergence near the surface causes an upward motion upstream of the heat sink due to mass continuity, a motion that is responsible for the upward vertical displacement. Note that a hydrostatic gravity wave is generated above the heat sink. The density field shows that a pool of cold air exists near the stationary heat sink (Fig. 6.8b). The strong density gradient in front of the heat sink ($x = -35$ km) may be regarded as an upstream *gust front* produced by a density current. The high-density region is related to the *mesohigh*, often observed under the strong downdraft region. On the downstream side, the density gradient is weaker than on the upstream side due to the advection of the cold air.

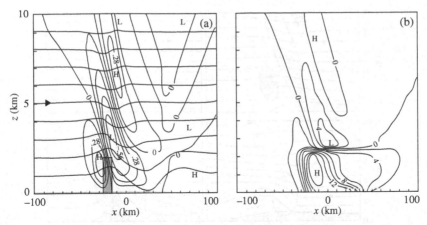

Fig. 6.8 Response of a two-dimensional, hydrostatic, stratified flow to a prescribed cooling, which represents a quasi-steady evaporative cooling of rainfall under a thunderstorm, as simulated by a nonlinear numerical model at 4000 s: (a) vertical displacement (horizontal curves) and vertical velocity (contours) and (b) perturbation density. The flow parameters used are: $U = 15$ m s^{-1} and $N = 0.01$ s^{-1}. The prescribed cooling is concentrated in the shaded region in (a). (Adapted after Lin and Smith 1986.)

In fact, for a uniform, stably stratified flow with prescribed cooling, flow responses are controlled by two nondimensional parameters, $U/(Q_o Ld)^{1/3}$ and $\pi U/Nd$, where U is the basic flow speed, $Q_o = (g/c_p T_o)q'$ is the cooling rate, and d and L are the depth and width of the cooling region, respectively (Raymond and Rotunno 1989). The second nondimensional number $(\pi U/Nd)$ is directly proportional to the thermal Froude number $(F_r = U/Nd)$ defined in (6.2.9). Based on these two nondimensional numbers, four flow regimes can be identified (Raymond and Rotunno 1989): (a) supercritical relative to both gravity waves and cold air outflow, (b) subcritical relative to gravity waves and supercritical relative to cold air outflow, (c) subcritical to both gravity waves and cold air outflow, and (d) supercritical relative to gravity waves and subcritical relative to cold air outflow. If the flow is subcritical relative to the gravity waves, then the gravity waves can propagate upstream against the basic flow. Similarly, if the flow is subcritical relative to cold air outflow, then a density current can form and propagate upstream against the basic flow. These four flow regimes are illustrated in Fig. 6.9. This study is extended (Lin *et al.* 1993) to investigate the flow response and the interaction between gravity waves and cold air outflows. It is found that a density current is able to form or is suppressed depending on the interaction between the traveling gravity wave and the cold air outflow.

6.3.2 *Heat island circulations*

A *heat island* is defined as a local area which is significantly warmer than its surroundings. When this happens in a metropolitan area, it is referred to as an *urban heat island*. The

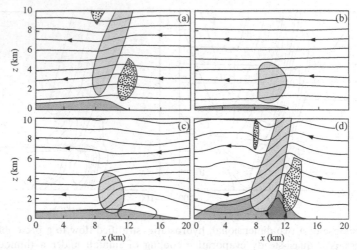

Fig. 6.9 Based on $F = U/(Q_o Ld)^{1/3}$ and $G = \pi U/Nd$, four flow regimes are identified: (a) supercritical relative to both gravity waves and cold air outflow (both F and G are large), (b) subcritical relative to gravity waves and supercritical relative to cold air outflow (F small, G large), (c) subcritical to both waves and cold air outflow (both F and G are small), and (d) supercritical relative to gravity waves and subcritical relative to cold air outflow (F large, G small). Streamlines are denoted by solid lines, while regions of significant vertical motion are shaded. Dotted (light shaded) area indicates upward (downward) motion. Cold pools are denoted by dark shading. (Adapted after Raymond and Rotunno 1989.)

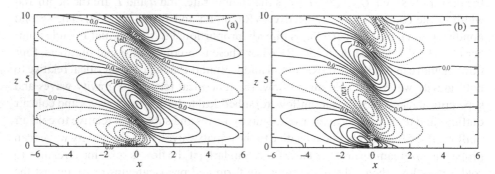

Fig. 6.10 Response of a uniform flow passing over a heat source located at $x = 0$: (a) zeroth order and (b) first order of perturbation vertical velocity obtained from a weakly nonlinear theory. All variables are nondimensionalized (e.g., x by the half-width of the heating and z by U/N) and the heating depth is 1. Upward (downward) motion is denoted by solid (dashed) lines. (From Chun and Baik 1994.)

linear theory discussed in the previous section has been extended to the case of weakly nonlinear flow (Chun and Baik 1994; Baik and Chun 1997). The nonlinear component of the solution indicates that the downward or upward motion downstream depends on the heating depth, or the thermal Froude number. When F_r is small (Fig. 6.10) but still within a valid range for the perturbation expansion to be valid, the linear and weakly nonlinear effects work together constructively to produce enhanced upward motion on the downstream side of the heat island. This motion occurs not far from the heating center. These findings might explain to a greater extent the precipitation enhancement observed downstream of a heat island than linear effects alone (Hjelmfelt 1982). The effect of rainfall enhancement is often explained by the addition of condensation nuclei introduced by the urban heat island into the airstream. However, from the above theories, the urban heat island effect can create a stationary heating source that produces ascent under a certain range of the thermal Froude numbers. Thus, a combined study of the addition of condensation nuclei and of thermal effects may be required to fully understand this phenomenon.

6.3.3 Moist convection

Application of the linear theory involving prescribed cooling is useful for studying the dynamics of snow melting-induced mesoscale circulations. Although the latent heat of melting is eight times smaller than the latent heat of evaporation, melting is concentrated in a shallow layer of near 0°C temperature. Thus, the cooling rate induced by melting can be comparable to that induced by evaporation (Robichaud and Lin 1989). Most rainfall in the midlatitudes is initiated through ice formation processes followed by melting. The transition from the solid to liquid phase normally takes place in an atmospheric layer that appears as a layer of enhanced reflectivity called a *bright band*. Theoretical studies of flow responses of a stratified flow to a prescribed temperature perturbation were found to be useful for understanding the dynamics of snow melting-induced mesoscale circulations (e.g., Szeto *et al.* 1988; Lin and Stewart 1991). It is found that thermally induced circulations have a length scale similar to that of the temperature perturbations that produce them. The updraft branch of the thermally induced circulation may enhance precipitation in a saturated environment.

The transient flow responses to prescribed heating and cooling obtained from linear theories can also help us understand the mean flow and shear effects in the adjustment to latent heating in cumulus cloud fields and, consequently, to improve the schemes of cumulus parameterization (Bretherton 1993).

6.3.4 Gravity wave generation and propagation

Section 4.2 introduced examples of theoretical studies of stratified flow over prescribed heating or cooling to gravity wave generation. These types of studies can also be applied to wave propagation. For example, based on analytical solutions of the transient linear response of a quiescent, two-dimensional, nonrotating atmosphere to prescribed low-level steady heating, it is found that two modes exist in the flow when the atmosphere is bounded above by a rigid top: (1) a deep fast-moving mode,

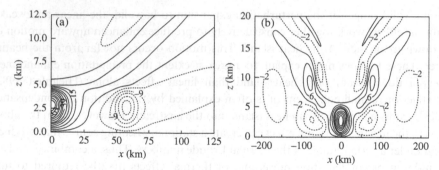

Fig. 6.11 (a) Linear hydrostatic solution for a quiescent, semi-infinite atmosphere with constant stratification to a prescribed low-level steady heating; and (b) nonlinear, nonhydrostatic numerical simulation of a semi-infinite atmosphere with variable buoyancy frequency (stable in the lower troposphere, less stable in the upper troposphere, and more stable in the stratosphere) to a prescribed low-level steady heating. The vertical velocity field (0.01 m s^{-1}) at 3600 s is shown in both panels. (Adapted after Nicholls *et al.* 1991.)

responsible for subsidence warming throughout the depth of troposphere, and (2) a slower moving mode, corresponding to midlevel inflow and lower- and upper-level outflows (Fig. 6.11a; Nicholls *et al.* 1991). The linear hydrostatic solution shows a region of upward motion extending in a jetlike flow from the top of the heat source (2.5 km). Upward propagating gravity waves are clearly shown by the vertically tilted phase lines. Regions of compensating weak downward motion are thus produced and propagated outward from the heating center. Note that the positive phase relationship between the vertical motion and diabatic heating is anticipated for a low thermal Froude number flow, which is 0 in this extreme case because $U = 0$. The nonlinear, numerically simulated flow response for an atmosphere with constant stratification is similar to the response obtained through linear theory (Fig. 6.11b). The numerical simulation is more realistic: the atmosphere is stable in the lower troposphere, less stable in the upper troposphere and more stable in the stratosphere. Significant wave reflection occurs under these conditions.

For a rigid-lid solution, the outward-propagating downward motion occurs at $x = \pm 160$ km corresponding to the $n = 1$ mode, while vertically oriented gravity waves with shorter vertical wavelengths occur at $x = \pm 60$ km below 5 km corresponding to the $n = 2$ mode. These propagating waves are similar in structure to the gravity waves produced in two-dimensional numerical simulations of convection occurring over the Florida peninsula. Another example of gravity wave generation and propagation studied by prescribing diabatic cooling is shown in Fig. 4.3, in which a non-hydrostatic model is used to simulate a density current in an environment characterized by a complex stratification and vertical wind. It is found that the density current generated gravity waves, which then propagated along the current itself. Figure 4.4 provides an additional example of gravity waves generated in the stratosphere by moist convection.

6.4 Effects of shear, three dimensionality, and rotation

6.4.1 Two-dimensional shear flows

The effects of vertical shear on thermally forced flow, such as squall lines and convective cloud bands, are well documented. One example is the moist convection associated with midlatitude squall lines, such as shown in Fig. 6.12. In particular, this squall line exhibits a critical level near 6 km. Note that the critical level coincides with the wind reversal level in a steady-state flow because the phase speed is 0 (Section 3.8; Section 5.3). In this section, we will investigate the effects of vertical shear in a thermally forced flow in a simple environment. This will help us understand its effects in more complicated mesoscale circulations.

The effects of shear flows with thermal forcing can be discerned by considering a two-dimensional, unbounded, steady-state, small-amplitude, nonrotating, inviscid, Boussinesq, shear flow with constant buoyancy frequency over a heat source,

$$U^2 \frac{\partial^2}{\partial x^2}\left(\frac{\partial^2 w'}{\partial x^2} + \frac{\partial^2 w'}{\partial z^2}\right) - UU_{zz}\frac{\partial^2 w'}{\partial x^2} + N^2 \frac{\partial^2 w'}{\partial x^2} = \left(\frac{g}{c_p T_o}\right)\frac{\partial^2 q'}{\partial x^2}. \qquad (6.4.1)$$

The above equation is a special form of the *Taylor–Goldstein equation*, (3.7.19) except that $\partial/\partial t = 0$ and with the addition of the diabatic heating term. To simplify the problem, we further assume the basic wind varies linearly with height,

$$U(z) = \alpha z, \qquad (6.4.2)$$

where $\alpha = \partial U/\partial z$ and $z = 0$ is the critical (wind reversal) level. The mathematics can be further simplified by assuming that the heating function is separable in x and z directions,

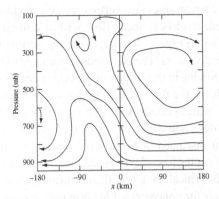

Fig. 6.12 Streamlines relative to a midlatitude squall line observed in Oklahoma on May 22, 1976 in the vertical plane with x representing distance ahead of the leading edge of the squall line. Three important features are shown: (a) upshear tilt of the updraft, (b) downdraft fed by the front-to-rear flow, and (c) flow overturning in the middle layer. (From Ogura and Liou 1980.)

$$q'(x, z) = Q_o f(x) g(z). \tag{6.4.3}$$

and the flow is hydrostatic ($\partial^2/\partial x^2 \ll \partial^2/\partial z^2$). Substituting (6.4.2) and (6.4.3) into (6.4.1) and taking the Fourier transform in x yields

$$\frac{\partial^2 \hat{w}}{\partial z^2} + \left(\frac{N}{\alpha z}\right)^2 \hat{w} = \left(\frac{gQ_o}{c_p T_o \alpha^2 z^2}\right) \hat{f}(k) g(z). \tag{6.4.4}$$

Once the heating function is known, the above problem can be solved in a way similar to that used to solve the uniform flow problem discussed in Sections 3.8 and 6.1. For example, if we take a prescribed isolated heating contained in the layer from the surface to a level below the critical level H_1, $q'(x) = Q_o f(x)$ for $|z| \leq H_1$, the general solution of (6.4.4) is

$$\hat{w}(k, z) = Az^{1/2+i\mu}, \quad z < -H_1 \tag{6.4.5a}$$

$$\hat{w}(k, z) = Bz^{1/2+i\mu} + Cz^{1/2-i\mu} + \left(\frac{gQ_o}{c_p T_o N^2}\right) \hat{f}(k), \quad |z| \leq H_1 \tag{6.4.5b}$$

$$\hat{w}(k, z) = Dz^{1/2-i\mu}, \quad H_1 < z, \tag{6.4.5c}$$

where $\mu^2 = R_i - 1/4$ and $R_i (= N^2/\alpha^2)$ is the *Richardson number*. Similar to the boundary conditions discussed in Section 3.8, the upper radiation condition has been imposed in the lower and upper layer. Following a similar procedure as in Section 6.1, the interface conditions at $z = -H_1$ can be derived. They require both $\hat{w}$ and $\hat{w}_z$ to be continuous across the interface. Special attention should be paid to the solution at the critical level, which is similar to that discussed in Section 3.8.

Figure 6.13 shows the streamlines and vertical velocity field for a two-dimensional, steady-state, unbounded, stratified, shear flow with a critical level over an isolated heat source, similar to that in Fig. 3.14b except that the atmosphere is unbounded and there is heating at the critical level. The Richardson number of the basic flow is $R_i = 1$. The presence of thermal forcing in the vicinity of the critical level significantly modifies the flow. Below the critical level, the fluid particle undergoes a downward motion upstream (on the left side) of the heating center, followed by an ascending motion downstream (Fig. 6.13a). This broad descending motion can be explained by the advection effect and group velocity argument discussed in Section 6.2. In the vicinity of the critical level, however, the fluid particle in the lower layer experiences a strong upward motion near the heating center, crosses the critical level, and then returns to the left of the domain in the upper layer. The flow near the concentrated heating region is strongly dominated by an upward motion (Fig. 6.13b). The in-phase relationship between the vertical motion and the heating, which is to be explained in the next paragraph, is important in order to maintain the convection. Away from the concentrated heating region, the upstream phase

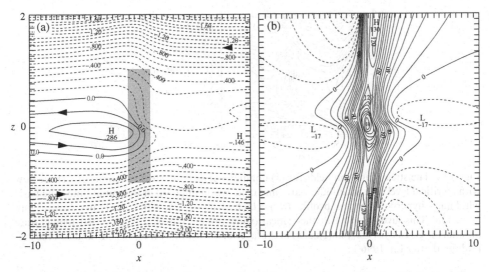

Fig. 6.13 (a) Streamlines for two-dimensional, steady-state, unbounded, stratified, shear flow with a critical level over an isolated heat source that is concentrated within the shaded area; (b) same as (a) except for the vertical velocity. All the quantities displayed are nondimensionalized. The Richardson number of the basic flow is 1. Solid (dashed) contours denote positive (negative) values. (Adapted after Lin 1987.)

tilt induces upward (downward) propagation of wave energy in the upper (lower) layer. When applied to the dynamics of moist convection, the condensational heating in the vicinity of the critical level appears to facilitate the flow interaction above and below the critical level. In fact, the flow circulation in the vicinity of the heating area is somewhat similar to that associated with the squall line observed on May 22, 1976 (Fig. 6.12).

The in-phase relationship between the vertical motion and the heating in the heating region at the critical level can be explained by considering the steady-state thermodynamic equation,

$$U\frac{\partial \theta'}{\partial x} + \frac{N^2 \theta_0}{g} w' = \frac{\theta_0}{c_p T_0} q'. \tag{6.4.6}$$

At the critical level ($z = 0$), w' is proportional to q' because the basic wind disappears. This also means that no temperature anomalies are produced directly by the heating at the critical level. In fact, at the critical level, the nondimensional form of the relationship between vertical velocity and diabatic heating may be expressed as $\tilde{w} = \tilde{q}/R_i$. Thus, the vertical velocity at the critical level increases as R_i decreases for a constant heating rate. As discussed in Section 3.8, for $0.25 < R_i < 2.0$, a significant amount of the wave energy is reflected and some of it is also transmitted through the critical level. This also helps explain the strong interaction between the flow above and

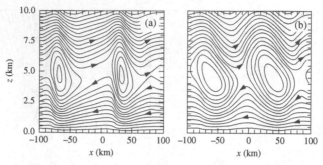

Fig. 6.14 The streamfunction fields for (a) nonlinear and (b) linear simulations of a shear flow with a critical level over a periodic warm and cold region imposed at the initial time. The basic wind has a hyperbolic tangent profile, $U(z) = U_o \tanh{[(z-z_i/h]}$, with $U_o = 15$ m s^{-1}, $z_i = 5$ km, $h = 2$ km. This gives $R_i = 0.1$ at the critical (wind reversal) level, z_i. The initial temperature anomaly is given by $\theta_i = \theta_{io} \sin(2\pi x/b)$ in the layer of 4 to 6 km with $\theta_{io} = 0.3$ K and $b = 100$ km. (Adapted after Lin 1996.)

below the critical level for cases where $R_i = 1$ (Figs. 3.14b and 6.13). For $R_i < 0.25$, *overreflection* can occur.

Note that a small perturbation in the horizontal velocity field may easily exceed the basic horizontal flow velocity in the vicinity of the critical level, thereby resulting in a highly nonlinear flow. In order to study the nonlinear effects, a nonlinear numerical simulation is conducted with a setting similar to that in Fig. 6.13. Figure 6.14a shows the nonlinear response in a thermally forced flow. The basic wind has a hyperbolic tangent profile with $R_i = 0.1$ at the wind reversal level ($z_i = 5$ km). The nonlinear transient response (Fig. 6.14a) is similar to the steady-state linear solution for a linear shear flow over isolated heating (Fig. 6.13) and the transient linear flow response (Fig. 6.14b). The closed circulation centered at the critical level resembles the squall line observation shown in Fig. 6.12 and simulated by the wave-CISK model (Fig. 4.19b). The major difference of the nonlinear response relative to the linear response is that the former is stronger and more compact. The following energy argument can be used to explain the *upshear phase tilt* of the upward motion in the shear layer, as shown in Fig. 6.14. Since there is instability shear, the local rate of change in total perturbation energy is positive, which, in turn, requires the vertical momentum flux term (integral of $-\rho_o U_z \overline{uw}$) to be negative in the whole domain for $U_z > 0$. This condition requires an upshear tilt of the vertical motion. This aspect will be further discussed in Section 7.1.

This type of approach for treating thermally forced flow in simple environments has been taken to help understand various mesoscale circulation problems in the atmosphere. Related developments include: (a) squall line initialization in numerical models (Crook and Moncrieff 1988), (b) the response of a nonlinear shear flow with a critical level to a steady cooling (Lin and Chun 1991), (c) amplification mechanisms for melting-induced circulations (Robichaud and Lin 1989), (d) the effects of evaporative cooling in a three-layer flow with a critical level (e.g. Chun and Lin 1995), (e) the role of internal gravity waves in modifying the behavior and structure of a simulated squall line (Schmidt and Cotton 1990), and the effects of variable wind shear on the mesoscale circulations (Reuter and Jacobsen 1993), (f) the transient responses in a shear flow with critical level (Baik *et al.* 1999), and (g) the convective gravity wave drag parameterization (Song and Chun 2005).

6.4.2 Three-dimensional nonrotating flows

The small-amplitude equation governing the vertical velocity for a steady state, three-dimensional, stratified, incompressible, Boussinesq, nonrotating flow can be reduced from (2.2.14)–(2.2.18) to the following,

$$\left(U\frac{\partial}{\partial x} + V\frac{\partial}{\partial y} + \nu \right)^2 \nabla^2 w' - \left(U\frac{\partial}{\partial x} + V\frac{\partial}{\partial y} + \nu \right)\left(U_{zz}\frac{\partial}{\partial x} + V_{zz}\frac{\partial}{\partial y} \right)w' \frac{\partial^2 w'}{\partial x^2}$$

$$+ N^2 \nabla_{\mathrm{H}}^2 w' = \left(\frac{g}{c_p T_o} \right) \nabla_{\mathrm{H}}^2 q'. \tag{6.4.7}$$

The procedure involved in obtaining the analytical solution is also similar to that in the mountain wave theory described in Section 5.4.1 except that the *Green's function method* is employed here to solve the problem with interface boundary conditions.

By considering a special case with uniform wind $U(z) = U$ and $V = 0$, and a bell-shaped heat source with circular contours,

$$q'(x, y, z) - \frac{Q_o \delta(z)}{(x^2/b^2 + y^2/b^2 + 1)^{3/2}}, \tag{6.4.8}$$

in an unbounded atmosphere, the problem can be solved analytically in the Fourier space and then transformed back to the physical space by applying a fast Fourier transform (FFT) algorithm.

Figure 6.15 shows the nondimensional vertical displacement of a three-dimensional, hydrostatic, continuously stratified, uniform flow over an isolated shallow heat source added at $\tilde{z} = 0$. The surface, assumed to be flat, is located at $\tilde{z} = -\pi$. The basic flow is directed from left to right. The fluid response to the heating at the heating level ($\tilde{z} = 0$) is a downward displacement upstream of the prescribed heat source followed by an upward displacement downstream (Fig. 6.15b). This response is similar to that given by the thermally induced two-dimensional flow with high Froude number discussed earlier in this chapter. The region of disturbance generally widens as one moves above and below the heating level. At $\tilde{z} = \pi/2$, a V-shaped pattern, which is similar to the U-shaped disturbance as shown in Fig. 5.17, in the region of upward displacement forms above the heating center (Fig. 6.15c). This region of upward displacement shifts upstream as one moves higher as required by the upper radiation

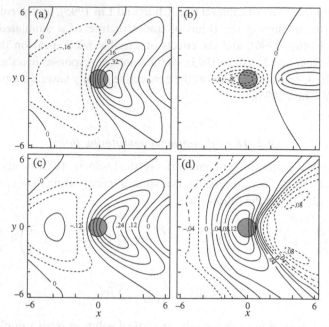

Fig. 6.15 Nondimensional vertical displacement ($\tilde{\eta} = N\eta/U$) associated with the response of a three-dimensional, hydrostatic, continuously stratified, uniform flow over an isolated shallow heat source (shaded), which is added at $\tilde{z} = Nz/U = 0$, with a flat surface at $\tilde{z} = -\pi$. Four levels are shown at $\tilde{z} =:$ (a) $-\pi/2$, (b) 0, (c) $\pi/2$, and (d) π. The basic flow is directed from left to right and the flow speed and the buoyancy frequency are $U = 10$ m s^{-1} and $N = 0.01$ s^{-1}, respectively. Both x and y axes are nondimensional. (Adapted after Lin 1986.)

condition. This upstream movement allows the heating-generated energy to propagate upward to infinity. At $Nz/U = \pi$ (Fig. 6.15d), the V-shaped region of upward displacement shifts further upstream and widens in the cross-wind direction. In addition, a region of downward displacement also forms downstream of the V-shaped region of upward displacement. The response is almost periodic in the vertical and the flow fields at $Nz/U = -\pi/2$ and $\pi/2$ (Figs. 6.15a and 6.15c) are quite similar.

The flow behavior is similar to the three-dimensional flow over an isolated mountain and the formation of a V-shaped pattern of upward displacement can be explained using a group velocity argument for the U-shaped disturbance for mountain waves (Fig. 5.17; Smith 1980). The pattern is produced by basic wind advection of the heating-generated upward propagating gravity waves, while the wave energy is concentrated in the V-shaped region trailing downstream. The hydrostatic group velocities (5.4.23) can be extended for a nonhydrostatic flow,

$$c_{gx} = U\left(\frac{l^2 + k^4 U^2/N^2}{k^2 + l^2}\right),$$

(6.4.9a)

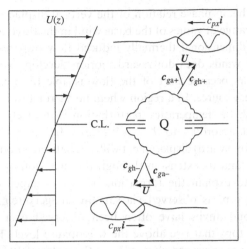

Fig. 6.16 The propagation of wave energy associated with steady waves forced by a heat source in an unbounded, nonhydrostatic shear flow. Symbols c_{px}, c_{ga+} (c_{ga-}), c_{gh+} (c_{gh-}) represent the horizontal phase speed with respect to (w.r.t.) the heat source, upward (downward) group velocity w.r.t. the air, and upward (downward) group velocity w.r.t. the heat source, respectively. (From Lin and Li 1988.)

$$c_{gy} = -U\left(\frac{kl(1 - k^2 U^2/N^2)}{k^2 + l^2}\right), \tag{6.4.9b}$$

$$c_{gz} = \frac{U^2 k^2 (1 - k^2 U^2/N^2)^{1/2}}{N(k^2 + l^2)^{1/2}}. \tag{6.4.9c}$$

The propagation of the wave energy associated with quasi-steady waves forced by a prescribed heating in an unbounded, nonhydrostatic, shear flow is sketched in Fig. 6.16. Above the heating, the energy propagates upward and upstream relative to the air (c_{ga+}), but is then advected downstream by the basic flow. Thus, the wave energy is along the direction of c_{gh+} or c_{gh-} relative to the heat source. The nonhydrostatic effects are the main cause of the formation of the disturbance's repeating, damped oscillations. The wave energy is concentrated in the region enclosed by the parabola:

$$y^2 = \left(\frac{Nz}{U}\right)\left\{\frac{[1 - (kU/N)^2]^{3/2}}{(k^2 + l^2)^{1/2}(1 + k^4 U^2/N^2 l^2)}\right\}x. \tag{6.4.10}$$

The response to a heating layer from z_1 to z_2 can be obtained by taking a continuous superposition (i.e. integration) of $w'(x,y,z_0)$ with respect to z_0, where $z_1 \le z_0 \le z_2$, and $w'(x,y,z_0)$ is the Green's function solution for heating applied at a single level z_0. In the absence of a basic wind ($U = 0$), the vertical velocity distribution corresponds to the axisymmetric shape of the heat source, thus resulting in upward motion over the region of heating. This in-phase relation of the vertical displacement and heating is present over a wide range of values of the basic wind in the three-dimensional flow. In contrast to the two-dimensional thermally induced flow responses discussed earlier, only in strong basic winds does downward motion develop over the heating layer. Physically, this occurs because part of the flow is able to circumvent the isolated three-dimensional heat source. In a region where no thermal forcing exists, such as in Figs. 6.15a, c, and d, the temperature perturbation is inversely proportional to the vertical displacement according to $\theta' = -(N^2\theta_o/g)\eta$. In other words, the V-shaped regions of upward (downward) displacement will adiabatically produce cold (warm) air.

The above theory and its extension through the inclusion of vertical wind shear have been applied to explain the formation of the V-shaped cloud anvil over a thunderstorm cloud top, as observed via satellite imagery (e.g., Adler and Mack 1986). V-shaped cloud anvils have often been observed over severe midwestern thunderstorm cloud tops that rise above the tropopause level. These thunderstorm cloud tops normally are measured on a scale on the order of 10 km and are typically embedded in more diffuse V-shaped anvils with lateral widths of over 100 km and lengths of several hundred kilometers. Figure 6.17a shows the V-shaped cloud top features based on the analysis of a satellite image. Prominent features of this thunderstorm top are: (a) a region of lowest cloud top temperatures associated with the updraft air overshooting the tropopause, (b) a V-shaped region of lower equivalent blackbody temperature with the point of the V either at or above the cloud top, and

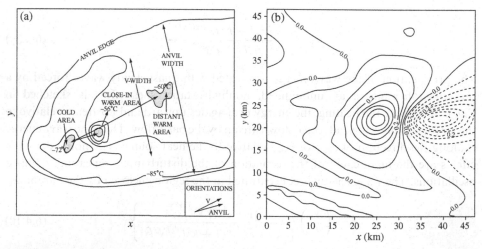

Fig. 6.17 (a) V-shaped anvil, cold area, and close-in warm area over a severe thunderstorm cloud top analyzed from the 2334 UTC May 2, 1979 GOES IR imagery. The unit of temperature is °C. (Adapted from Heymsfield and Blackmer 1988.) (b) Vertical velocity field (m s^{-1}) at $z = 14$ km solved by a linear theory of hydrostatic flow in response to a prescribed heating distributed from $z = 1.5$ km to 12 km. A clockwise directional vertical shear of the basic wind is assumed (From Lin and Li 1988.)

(c) a region of higher temperatures 20–40 km downwind of the cloud top (i.e. the closed-in warm area), forming a cold-warm thermal couplet with (a). Figure 6.17b shows the vertical velocity field at $z = 14$ km predicted by a linear theory of hydrostatic flow over a prescribed heating distributed from $z = 1.5$ km to 12 km and a linear unidirectional vertical shear of the basic wind. The V-shaped region of upward (downward) displacement corresponds to the cold (warm) region resulting from adiabatic cooling (warming), which mimics the observed V-shaped cloud anvils over severe thunderstorms (e.g. Fig. 6.17a). The asymmetric, skewed V-shaped cloud anvils result from the advection effects of directional shear flow. In addition to the thermal forcing, other mechanisms, such as the mechanical forcing by the cloud top (e.g., Heymsfield and Blackmer 1988), the mixing of stratospheric and cloud air (e.g., Adler and Mack 1986), and injection of the plume of cloud water vapor into the stratosphere (e.g., Wang 2003), have been proposed.

6.4.3 Three-dimensional rotating flows

Equations (5.1.1)–(5.1.4) can be extended to investigate the three-dimensional, steady-state, small-amplitude response of an inviscid, stratified, hydrostatic, Boussinesq flow on an f plane to a thermal forcing,

$$R_o u_{\tilde{x}} - v + \pi_{\tilde{x}} = 0, \tag{6.4.11}$$

$$R_o v_{\tilde{x}} + u + \pi_{\tilde{y}} = 0, \tag{6.4.12}$$

$$\pi_{\tilde{z}} - b = 0, \tag{6.4.13}$$

$$u_{\tilde{x}} + v_{\tilde{y}} + R_o w_{\tilde{z}} = 0, \tag{6.4.14}$$

$$b_{\tilde{x}} + w = \dot{q}, \tag{6.4.15}$$

where $R_o = U/fa$ is the *Rossby number*, π' $(= p'/\rho_o)$ the *perturbation kinematic pressure*, b' $(= g\theta'/\theta_o)$ the *perturbation buoyancy* and a the half-width (horizontal scale) of the heat source or sink. The nondimensional variables are defined as

$$(\tilde{x}, \tilde{y}) = (x/a, y/a); \qquad \tilde{z} = z/H;$$

$$(u, v) = (u'/U, v'/U); \qquad w = w'b/R_o UH;$$

$$\pi = \pi'/(fUa) = p'/(\rho_o fUa); \quad b = b'H/(fUa) = g\theta'H/(\theta_o fUa);$$

$$\dot{q} = q' gH/(c_p T_o U^2 f),$$

$$\tag{6.4.16}$$

where $H = fa/N$ is the *deformation depth*.

Equations (6.4.11)–(6.4.15) can be combined into a single equation of w (with "$\sim$" dropped)

$$R_o^2 \frac{\partial^4 w}{\partial z^2 \partial x^2} + \frac{\partial^2 w}{\partial z^2} + \nabla_H^2 w = \nabla_H^2 \dot{q}. \tag{6.4.17}$$

Again, the mathematical problem can be solved using the Green's function method. For simplicity, we prescribe the surface heating as

$$\dot{q}(x, y, z) = h(x, y) e^{-z/\gamma}, \tag{6.4.18}$$

where γ is the nondimensional heating depth. Taking the double Fourier transform (Appendix 5.1) of the above equation in both x and y directions leads to

$$\frac{\partial^2 \hat{w}}{\partial z^2} + \left(\frac{K^2}{R_o^2 k^2 - 1} \right) \hat{w} = \frac{K^2 \hat{h} e^{-z/\gamma}}{R_o^2 k^2 - 1}. \tag{6.4.19}$$

The general solution of the above equation is

$$\hat{w} = A e^{iKz/(R_o^2 k^2 - 1)^{1/2}} + B e^{-iKz/(R_o^2 k^2 - 1)^{1/2}} + \left(\frac{\gamma^2 K^2}{(R_o^2 k^2 - 1) + \gamma^2 K^2} \right) \hat{h} e^{-z/\gamma}. \tag{6.4.20}$$

There exist two flow regimes for (6.4.20): (a) $R_o^2 k^2 > 1$ and (b) $R_o^2 k^2 < 1$.

For $R_o^2 k^2 > 1$, the upper radiation boundary condition requires $B = 0$ in order to allow the energy to radiate upward. Over a flat surface, the lower boundary condition requires $w = 0$ at $z = 0$. Applying the lower boundary condition, the solution may be written as

$$\hat{w} = \left(\frac{-\gamma^2 K^2 \hat{h}}{(R_o^2 k^2 - 1) + \gamma^2 K^2} \right) \left[e^{iKz/(R_o^2 k^2 - 1)^{1/2}} - e^{-z/\gamma} \right] \text{ for } R_o^2 k^2 > 1. \tag{6.4.21}$$

For $R_o^2 k^2 < 1$, the upper boundary condition requires the solution to decrease with height. Thus it requires $A = 0$ and the solution becomes

$$\hat{w} = \left(\frac{\gamma^2 K^2 \hat{h}}{(1 - R_o^2 k^2) - \gamma^2 K^2} \right) \left[e^{-Kz/(1 - R_o^2 k^2)^{1/2}} - e^{-z/\gamma} \right] \text{ for } R_o^2 k^2 < 1. \tag{6.4.22}$$

As long as the heating function, $h(x,y) \exp(-z/\gamma)$, is known, the solutions in the physical space can be obtained by performing inverse double Fourier transforms on (6.4.21) and (6.4.22). Because the inverse Fourier transform tends to be analytically intractable, a numerical method, such as a FFT subroutine, is often employed.

Consider a bell-shaped warm region as

$$\Theta'(x, y) = \frac{\Theta_o}{(x^2/a^2 + y^2/a^2 + 1)^{3/2}}, \tag{6.4.23}$$

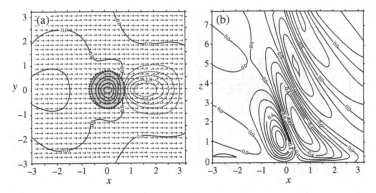

Fig. 6.18 The response of an inviscid, hydrostatic, westerly flow to a prescribed, isolated warm region (shaded): (a) vertical velocity (solid positive; dashed negative) and horizontal wind vector fields at $\tilde{z} = 0.25$ ($z = 250\,\text{m}$), and (b) the vertical velocity on the cross section $\tilde{y} = 0$. The Rossby number is 1 and the flow belongs to the regime of $R_o^2 k^2 > 1$. All the quantities displayed are nondimensionalized (with "$\sim$" dropped). (From Lin 1989.)

where Θ_o is the maximum temperature anomaly of the warm region.

To a first approximation, the diabatic heating rate associated with this specified warm region in a basic flow can be prescribed by $q'(x,y,0) = h'(x,y) \approx (c_p T_o/\theta_o)\, U\partial\Theta'/\partial x$ if the diabatic heating is mainly created and maintained by the horizontal temperature advection. Thus, a prescribed warm region, as described by (6.4.23), implies that a coupled diabatic heating and cooling are present in the basic flow. A more realistic representation of the sensible heating in the boundary layer could be used in a parameterization (see Chapter 14), but this would require a numerical model to solve the problem.

Figure 6.18a shows the response of an inviscid, hydrostatic, westerly flow to an isolated warm region near the surface. A region of upward motion concentrated in the warm region followed by a region of downward motion downstream is produced. Note that the descent that occurs upstream and over the heating region, as found in the two-dimensional case for a large R_o flow, is absent for most values of R_o in a three-dimensional flow due to lateral deflection. In addition, a cyclonic flow exists on the lee side of the warm region. The relatively strong advection effect moves the thermally forced cyclonic flow out of phase with the warm region. Note that the positive vorticity or cyclonic flow is not in phase with the low pressure for such a relatively high Rossby number flow. This occurs because the vertical motion still plays a significant role in the vorticity equation, $\hat{\zeta} = -K^2\hat{\pi} - iR_o^2 k\hat{w}_z$. The vertical velocity field on the vertical plane across the heating center has a strong upstream phase tilt, indicating that the wave energy produced by the diabatic heating and cooling associated with the prescribed warm region is able to propagate upward (Fig. 6.18b).

Figure 6.19 shows a case similar to that of Fig. 6.18 but with $R_o = 0.2$, which falls mostly in the regime $R_o^2 k^2 < 1$. A region of upward motion is produced over the heat

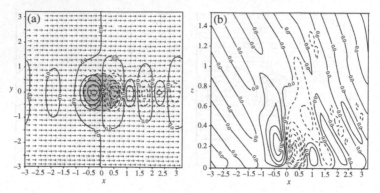

Fig. 6.19 Same as Fig. 6.18 except $R_o = 0.2$. The flow falls in the regime $R_o^2 k^2 < 1$. (From Lin 1989.)

source and followed by a downward motion to the downstream side of the heat source. These regions are a direct consequence of the heating and cooling that the prescribed heating function produces (Fig. 6.19a). The positive vorticity or cyclonic flow is more in phase with the warm region ($\theta' > 0$), and the low pressure, than in Fig. 6.18a. The flow response is similar to that predicted by the quasi-geostrophic theory except that the advection effect makes the disturbance stronger and allows a small amount of energy to propagate upward, as shown in Fig. 6.19b. The oscillating pattern on the lee side of the heat source (Fig. 6.19a) is caused by the dispersion associated with the evanescent inertia-gravity waves near the surface (Fig. 6.19b). For $R_o k \ll 1$, the first term inside the square bracket on the right-hand side of (6.4.22) exponentially decays with height, therefore reducing the above fluid flow system to the *quasi-geostrophic flow* (Table 3.2). This is consistent with the results shown in Figs. 6.18 and 6.19.

The above three-dimensional theory has been extended to investigate various mesoscale problems with surface sensible heating, such as (a) lake effects on snowstorms in the vicinity of Lake Michigan (e.g. Hsu 1987; Sousounis and Shirer 1992), (b) coastal cyclogeneis in a baroclinic flow using a semi-geostrophic model (Lin 1990), and (c) coastal frontogenesis using a linear theory (Riordan and Lin 1992) and a nonlinear primitive equation model (Xie and Lin 1996). For example, Fig. 6.20 shows a conceptual model under various basic wind conditions for the coastal frontogenesis associated with confluence zones and low-level jets. The coastal front forms at the confluence zone which is located at different points relative to the Gulf Stream front (i.e., the major axis of the heat source). Advection and rotational effects play a significant role in this thermally forced flow. This approach helps us understand the basic dynamics of coastal frontogenesis as revealed in observational analyses (e.g. Riordan 1990), and sophisticated numerical simulations (e.g. Doyle and Warner 1993).

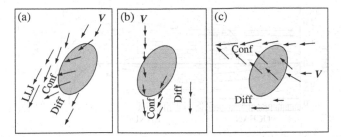

Fig. 6.20 A conceptual model for the coastal frontogenesis associated with confluence/diffluence (Conf/Diff) zones and low-level jets (LLJ) under various basic wind conditions, based on nonlinear simulations of baroclinic flow with a prescribed heat source (shaded) representing the Gulf Stream to the east of the Carolina coast. The upstream basic wind velocity is denoted by V, and is from (a) northeast, (b) north, and (c) east. (Adapted after Xie and Lin 1996.)

6.5 Dynamics of sea and land breezes

Sea and land breezes are the atmosphere's response to the differential surface heating across coastlines or shores of large lakes. They have been recognized among fishermen for several centuries and have been studied extensively by meteorologists for several decades to the present day. Figure 6.21 shows an example of a lake-(sea-) breeze circulation observed near Chicago. Aside from the difference in forcing and circulation scales, the basic dynamics of sea-breeze circulations and lake-breeze circulations are identical. During the day, a smaller heat capacity causes the land to heat up more rapidly than the adjacent water surface. As a result, the air above the land surface expands and rises. At a height of about 1 km or the top of the convective boundary layer, the rising air spreads outward, creating an area of low pressure near the surface of the land. Less heating takes place over the adjacent water, thus causing the air pressure to be greater over water than over land. A sea breeze then develops as cooler air over the sea or lake is pushed toward the land by the pressure gradient force. As a sea breeze advances toward land, a distinct boundary forms between cooler maritime air and the continental warmer air it displaces. This boundary is called the *sea breeze front*, and is characterized by often producing an abrupt drop in temperature by as much as 5 to 10°C as it passes overhead. The cooling effect of the sea breezes may reach a maximum distance of 100 km inland in the tropics and 50 km in midlatitudes. Across lakeshores, the scale of sea breezes is smaller. At night, the situation reverses: the land cools more rapidly than the sea and a *land breeze* develops.

The intensity and reach of sea and land breezes depends on location and time of year. For example, sea breezes are more frequent and intense in the tropics due to intense solar heating throughout the year. In the midlatitudes, sea breezes are more frequent during the warmer season, but the land breezes are often missing because the land does not always cool below the ocean temperature. In higher latitudes, the atmospheric circulations are often dominated by high- and low-pressure

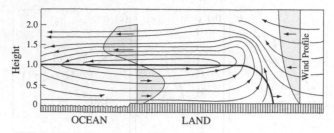

Fig. 6.21 An example of a lake (sea) breeze circulation observed near Chicago. Shown are the streamlines calculated from pilot balloons. The ordinate is in km. (Adapted after Lyons and Olsson 1972.)

systems, making sea and land breezes less noticeable. Sea-breeze circulations can be described in terms of the depth between the lower current and the upper "return" current, and their horizontal extent. Sea-breeze depth ranges from just over 100 m to 1 km or higher. Dynamically, sea and land breezes are influenced by the diurnal variation of differential heating across the coastline, diffusion of heat, stability, the Coriolis parameter, and friction. Ideally, these effects can be understood and predicted by theoretical and numerical models. In the following, we will make a brief description of their fundamental dynamics.

6.5.1 Linear theories

The basic dynamics of sea and land breezes can be understood by considering the following set of equations governing the two-dimensional (across the coastline), small-amplitude, Boussinesq fluid flow,

$$\frac{\partial u'}{\partial t} + U\frac{\partial u'}{\partial x} - fv' = -\frac{1}{\rho_0}\frac{\partial p'}{\partial x} + F_{r1}, \tag{6.5.1}$$

$$\frac{\partial v'}{\partial t} + U\frac{\partial v'}{\partial x} + fu' = F_{r2}, \tag{6.5.2}$$

$$\frac{\partial w'}{\partial t} + U\frac{\partial w'}{\partial x} = -\frac{1}{\rho_0}\frac{\partial p'}{\partial z} + g\frac{\theta'}{\theta_0} + F_{r3}, \tag{6.5.3}$$

$$\frac{\partial u'}{\partial x} + \frac{\partial w'}{\partial z} = 0, \tag{6.5.4}$$

$$\frac{\partial \theta'}{\partial t} + U\frac{\partial \theta'}{\partial x} + \frac{N^2\theta_0}{g}w' = \frac{\theta_0}{c_p T_0}q', \tag{6.5.5}$$

where the F_{r1}, F_{r2}, and F_{r3} terms represent the viscous forces in the x, y, and z, directions, respectively. The above equation set is similar to (6.1.1)–(6.1.4) except that they are time-dependent, and contain viscosity and Coriolis force terms, as well as the y-momentum equation. A simple and common approach used to represent the viscous force in the planetary boundary layer is to assume the *Fickian diffusion*, $(F_{r1}, F_{r2}, F_{r3}) = \nu\nabla^2(u', v', w')$, or $(F_{r1}, F_{r2}, F_{r3}) = \nu(\partial^2/\partial z^2)(u', v', w')$, where $\nabla^2 = \partial^2/\partial x^2 + \partial^2/\partial z^2$ and ν is the *eddy viscosity*.

Combining (6.5.1)–(6.5.5) with the frictional terms neglected and the thermal forcing term regarded as a known function leads to a single governing equation for w'

$$\left(\left(\frac{\partial}{\partial t} + U\frac{\partial}{\partial x} \right)^2 + N^2 \right) \frac{\partial^2 w'}{\partial x^2} + \left(\left(\frac{\partial}{\partial t} + U\frac{\partial}{\partial x} \right)^2 + f^2 \right) \frac{\partial^2 w'}{\partial z^2} = \frac{\partial^2 \dot{Q}}{\partial x^2}, \qquad (6.5.6)$$

where $\dot{Q} = (g/c_p T_0)\dot{q}'$. Equation (6.5.6) reduces to (6.2.1) for a nonrotating and hydrostatic flow. Assuming

$$(w', \dot{Q}) = (\hat{w}, \hat{Q})e^{i(kx-\omega t)}, \qquad (6.5.7)$$

and substituting it into (6.5.6) yields

$$\frac{\partial^2 \hat{w}}{\partial z^2} + \left(\frac{N^2 - \Omega^2}{\Omega^2 - f^2} \right) k^2 \hat{w} = \left(\frac{k^2}{\Omega^2 - f^2} \right) \hat{Q}, \qquad (6.5.8)$$

where $\Omega = \omega - kU$ is the *Doppler-shifted frequency*. The above equation is similar to (3.6.8) except with the assumption of two-dimensionality and the addition of diabatic heating. Equation (6.5.8) contains a thermally forced mode and a free mode. Thus, as discussed in Chapter 3, the free mode includes the following three major flow regimes: (1) $\Omega > N > f$: *high-frequency evanescent flow regime*; (2) $N > \Omega > f$: *vertically propagating inertia-gravity wave regime*; and (3) $N > f > \Omega$: *low-frequency evanescent flow regime*. The disturbance decays exponentially with height and is confined to its neighborhood for evanescent flow regimes. It is able to propagate vertically as inertia-gravity waves for the vertically propagating inertia-gravity wave regime.

In applying (6.5.8) to sea-breeze circulations with no basic wind ($U = 0$), the thermal forcing is controlled by the diurnal cycle of sensible heating, which has an intrinsic frequency (ω) of 7.272×10^{-5} s^{-1} ($= 2\pi/24$ h). Since ω is generally much smaller than N, (6.5.8) approximately reduces to

$$\frac{\partial^2 \hat{w}}{\partial z^2} + \left(\frac{N^2}{\omega^2 - f^2} \right) k^2 \hat{w} = \left(\frac{k^2}{\omega^2 - f^2} \right) \hat{Q}, \qquad (6.5.9)$$

and only two flow regimes exist in the system, i.e. the vertically propagating inertia-gravity wave regime ($\omega > f$), and the low-frequency evanescent flow regime ($\omega < f$).

In addition, the regime where $\omega = f$ should be considered (Rotunno 1983). In this particular flow regime, friction needs to be considered; otherwise (6.5.9) is singular.

This inviscid theory gives different flow regimes for latitudes higher and lower than 30°. In reality, however, the effects of friction and thermal diffusion influence the critical latitude that separates the vertically propagating inertia-gravity wave regime from the low-frequency evanescent flow regime. By prescribing the heating function as an arc tangent function and introducing a streamfunction ψ ($u' = \partial\psi/\partial z$; $w' = -\partial\psi/\partial x$), a mathematical problem similar to that governed by (6.5.9) has been solved analytically (Rotunno 1983). Figure 6.22 shows the nondimensionalized horizontal velocity and vertical velocity fields for $f > \omega$. The horizontal scale of the sea breeze is confined within a distance of order $Nd/\sqrt{f^2 - \omega^2}$, where d is the vertical scale of heating. When $f < \omega$, the response associated with the sea breeze circulations is in the form of inertia-gravity waves.

The above theory was extended to include Rayleigh friction $[(F_{r1}, F_{r2}, F_{r3}) = -\alpha(u',v',w')]$ and Newtonian cooling ($\dot{Q} = -\alpha\theta'$). It is found (Dalu and Pielke 1989) that: (a) when friction is small, periodicity in the forcing enhances the intensity and the horizontal scale of the breeze; (b) when the friction e-folding time is of the order of one day, the opposite is true; (c) when the dissipation is small ($\alpha^2 < \omega^2 - f^2$), waves might occur after a few days of sea breeze below the latitude $\sin^{-1}\left(\sqrt{\omega^2 - \alpha^2}/2\omega\right)$, which is lower than the 30° derived from inviscid theory; and (d) wave patterns below a latitude of 30° predicted by inviscid linear theory are likely to be rare.

Although friction controls the diffusion of momentum, it is not necessarily important for producing the sea-breeze circulations (Niino 1987). Friction is important in satisfying the no-slip lower boundary condition at the ground level and producing a realistic wind profile near the ground. Thus, the vertical scale of the heating is a function of viscosity and cannot generally be prescribed as in the above linear theory. With the effects of friction included, it is found (Niino 1987) that: (a) the singularity at 30° latitude vanishes, (b) the horizontal extent of the sea breeze is controlled by $N\kappa^{1/2}\,\omega_*^{-3/2}\,g(f)$, where N is the buoyancy frequency, κ the

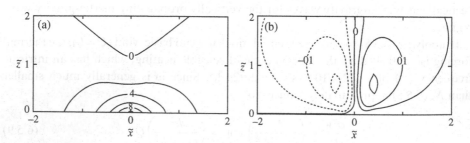

Fig. 6.22 (a) Horizontal velocity and (b) vertical velocity fields of the sea breeze circulation obtained with (6.5.9) for a case with $f > \omega$. All independent and dependent variables are nondimensionalized (solid positive; dashed negative). (From Rotunno 1983.)

eddy thermal diffusivity, ω, the frequency of sea breeze toward the ground, and $g(f)$ is a universal function which remains constant (about 2.1) for latitudes below 30° and decreases rapidly to 0.9 at the North or the South Pole; and (c) nonhydrostatic effects are significant in the immediate neighborhood of the coastline. Sun and Orlanski (1981a, b) solved both linearized and nonlinear equations as initial value problems and confirmed that the two-day-waves can be easily excited by the diurnal oscillation of the land–sea contrast at lower latitudes (<15 degrees). On the other hand, a combination of 1-day and 2-day waves may coexist up to 30 degrees. These waves may correspond to the mesoscale cloud bands observed along coastlines with a space interval of a few tens to few hundred kilometers.

6.5.2 Nonlinear numerical studies

Figure 6.23 shows an example of the structure of a sea-breeze front, as simulated by a three-dimensional nonlinear numerical model. At 1000 local time (Fig. 6.23a), the sea breeze is about 400 m deep with a maximum horizontal velocity of about 6 m s^{-1} and a vertical velocity of about 1.5 m s^{-1} above the sea breeze front. The front has advanced a distance of about 6 km inland in spite of resistance from the offshore basic wind of 10 ms^{-1}. Since the sea breeze front has a scale of only 200 m, there is a need for treating it in greater detail. This is approached by utilizing higher horizontal resolution for more accurate and detailed simulation of the front. At 1200 local time (Fig. 6.23b), the sea breeze front advances to about 60 km inland and has developed to a depth

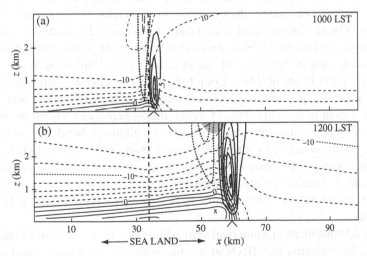

Fig. 6.23 An example of two-dimensional numerical simulation of a sea breeze front. Horizontal wind velocities (2 m s^{-1} contours; thin solid and dashed) perpendicular to the coastline and vertical velocities (0.5 m s^{-1} contours; thick solid and dotted) at (a) 1000 and (b) 1200 local time are shown on the vertical cross sections. Shaded regions denote the cloudy areas. The horizontal grid interval of the model is 1 km. (After Dailey and Fovell 1999.)

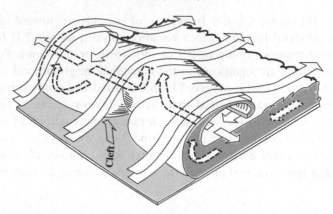

Fig. 6.24 Schematic diagram of the three-dimensional flow at a sea-breeze front. The sea-breeze front behaves like a density current and its head is divided into a series of lobes and clefts. Some of the warmer air is overrun and ingested in the cleft in the center of the lobe. The clefts may help the development of horizontal convective rolls which align in the direction of the low-level shear vector (denoted by the long arrows from left to right). (From Simpson 1994.)

of about 800 m. In three-dimensional simulations, *horizontal convective rolls* (HCR) tend to develop over land in response to strong daytime surface heating with a parallel alignment to the vertical wind shear vector (Dailey and Fovell 1999). The sea-breeze front, along with the horizontal convective rolls parallel to the front, are thus able to initiate deep convection (Fovell 2005).

At the nose of the sea-breeze front, the denser sea-breeze air overruns the less dense land airmass, an occurrence that extends to a height of approximately 100 m (Fig. 6.24). The sea-breeze front thus begins to behave like a density current. Its head is divided into a series of lobes and clefts. Some of the warmer air is overrun and ingested in the cleft in the center of the lobe, as depicted in Fig. 6.24. The spacing between the clefts is about 1 km. Longitudinal bands aligned with environmental shear vectors are the preferred mode of convection for small-amplitude perturbations over a flat terrain in both dry and saturated atmospheres, as revealed in theoretical studies (Asai 1972), although sometimes, the longitudinal band may coexist with transverse bands associated with gravity waves (Sun 1978). In addition to the sea-breeze front, *Kelvin–Helmholtz billows* have been observed; they are caused by the development of shear instability, as depicted in the schematic diagram of Fig. 6.25. These features of the sea-breeze front have also been reproduced in laboratory experiments.

Further advancement in numerical simulations of the sea and land breezes have been made by exploring the effects of diurnal variation, land breeze, isolated lakes and islands, basic wind shear, differences between land breeze and sea breeze, combined effects of sea breeze and mountain solenoidal circulations, initiation of and interaction with deep convection, air pollutant transport by sea breezes, and mountain effects.

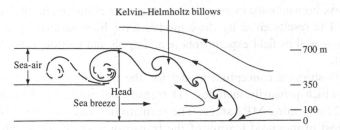

Fig. 6.25 Conceptual model of a sea-breeze density current. In the moving frame, the warm air from the right is forced to flow over the head of the sea-breeze front. Kelvin–Helmholtz billows form along the interface between the warm air above and the cold air of the sea-breeze density current, grow and eventually become unstable and break down. (From Simpson 1994.)

6.6 Dynamics of mountain–plains solenoidal circulations

The dynamics of *mountain–plains solenoidal (MPS) circulations* is a little explored area of orographically influenced flow and weather phenomena. This is mainly due to the complicated interactions between orographic and thermal forcings. Taking into consideration sensible heating or cooling over elevated terrain results in a considerably more complex flow than has been considered until now. The classical view of orographically and thermally forced winds in mountains includes the slope and mountain-valley winds. During the day, the mountain serves as an elevated heat source due to the sensible heat released by the mountain surface. In a quiescent atmosphere, this can induce mountain *upslope flow or upslope wind*, which in turn may initiate cumuli or thunderstorms over the mountain peak and produce orographic precipitation. At night, the opposite occurs: surface cooling produces downslope *drainage flow*.

Based on observations, four stages in the development of a thermally forced circulation generated by solar heating in a mountain valley have been identified (e.g., Banta 1990):

(a) Before sunrise, the nocturnal inversion layer contains drainage flow, which generally blows in a different direction from the winds above the inversion. Just prior to sunrise, this very stable layer remains adjacent to the surface.
(b) After sunrise, surface sensible heating erodes the inversion layer and produces a shallow *convective boundary layer (CBL)* below the inversion layer and the upslope flow.
(c) The shallow CBL or upslope layer deepens as the surface heating continues.
(d) After the nocturnal inversion layer disappears during the afternoon, a deep, well-mixed CBL is created.

Linear theories described in Sections 6.1 and 6.2 have been applied to study the combined effects of orographic and thermal forcing for mesoscale mountain flow (e.g., Raymond 1972; Smith and Lin 1982). Numerical modeling studies of the combined orographical and thermal forcing have been explored as early as the 1960s (e.g., Orville 1964, 1968). More sophisticated numerical models with a variety of initial conditions have been adopted in the more recent studies of mountain–plains solenoidal circulations. The results given by these models have been verified by conventional observations as well as field experiments (e.g. Tripoli and Cotton 1989; Wolyn and McKee 1994).

Figure 6.26 shows a conceptual model for the daytime evolution of the MPS circulation, which primarily includes: (1) transitional stage, (2) developing MPS stage, and (3) migrating MPS stage. The transitional stage occurs when the sun rises. The most pronounced feature of the transition stage is the katabatic jetlike flow down the east side of the mountain (Fig. 6.26a). The slowing of the nocturnal jet on the eastern plains produces a convergence that lifts the cold air, thus creating a *stable core* that is shallower farther east of the barrier. This nocturnal katabatic flow weakens as it is affected by the surface heating, and is replaced by a mesoscale solenoidal circulation 3–4 h after sunrise (Fig. 6.26b). A shallow CBL is produced below the inversion layer and an upslope flow is produced by the horizontal pressure gradient force toward the slope in response to the buoyancy associated with the surface sensible heating. The main upward motion of the solenoidal circulation occurs in a narrow zone over the eastern slope of the mountain, and is called the *leeside convergence zone* (LCZ). The LCZ lifts the air into the ambient air above, creating the cold core (denoted by "C" in Fig. 6.26b). A strong sinking motion occurs to the east of the cold core, creating a pressure trough in which the center of the solenoid is located. The horizontal pressure gradient associated with the cold core and the trough to the east produces a horizontal wind speed maximum. A broad region of sinking motion is located to the east of the solenoid center. At the later time of this stage, the sinking and horizontal warm-air advection immediately east of the solenoid center is able to warm the air enough to create a negative pressure gradient in the stable core above the CBL. The final stage of the mountain–plains solenoidal circulation is characterized by the eastward migration (Fig. 6.26c). Convergence (divergence) near the height of the wind maximum region and divergence (convergence) near the surface tend to produce sinking motion ahead (behind) the horizontal wind maximum located beneath the leading edge of the cold core,. The solenoid center is located in a pressure trough beneath the eastward-moving leading edge of the cold core, while the LCZ remains anchored over the lee slopes. Only the migrating MPS may be defined as a disturbance, and as thus can significantly affect the atmosphere on the plains located east of the system during the daytime circulations. The CBL grows explosively and the depth of the upslope flow increases when the solenoid passes a location.

The MPS has been shown to be responsible for producing a strong updraft, which in turn generated the dominant wave of the second episode of gravity waves observed on 11–12 July 1981 during the CCOPE (Koch *et al.* 2001). A gravity wave was generated as the updraft impinged upon a stratified shear layer above the deep, well-mixed boundary layer developed by strong sensible heating over the Absaroka Mountains. Explosive convection developed directly over the remnant gravity wave

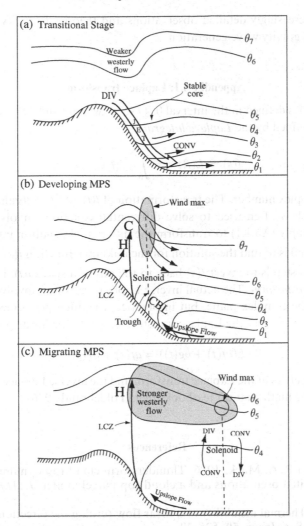

Fig. 6.26 Conceptual model of the daytime evolution of the mountain–plains solenoidal (MPS) circulation east of a mesoscale mountain under conditions of clear skies, steady-state synoptic-scale situation, and light basic westerly wind (e.g., 5 m s^{-1}). Three primary stages may be identified: (a) transition stage, (b) developing MPS stage, and (c) migrating MPS stage. Symbols DIV, CONV, JET, C, H, CBL, and LCZ denote divergence, convergence, katabatic jetlike flow, cold core, higher pressure, convective boundary layer, and leeside convergence zone, respectively. Regions of wind maximum are shaded. Solid lines are the isentropes. (Adpated after Wolyn and McKee 1994.)

as an eastward-propagating density current, produced by a rainband generated within the MPS leeside convergence zone, merged with a westward-propagating density current in eastern Montana. The complicated interactions of differing sensible heat contributions from complex terrain, gravity waves, and convection indicate the need for increasingly detailed observations and theories to verify existing MPS hypotheses and gravity wave generation.

Appendix 6.1: Laplace transform

If a function $f(t)$ is defined in the interval $0 \leq t < \infty$, where t and $f(t)$ are real, then the function $\hat{f}(s)$, defined by the *Laplace integral*

$$\hat{f}(s) = L(f(t)) = \int_0^\infty f(t)e^{-st}dt, \tag{A6.1.1}$$

where s is a complex number. The transformation of $f(t)$ into $\hat{f}(s)$ is called the *Laplace transform*, which is often used to solve differential equations involving time. The first step is to apply (A6.1.1) to transform the differential equation into the Laplace space. The second is to find the solution for the unknown function $\hat{f}(s)$ in the Laplace space. The third step is to invert $\hat{f}(s)$ back to the physical space $f(t)$, i.e., to take the *inverse Laplace transform*. The actual inverse Laplace transform involves the contour integration in the complex plane, but in practice it is often performed by applying some known properties of the Laplace transform, such as the linear property,

$$L(af(t) + bg(t)) = a\hat{f}(s) + b\hat{g}(s). \tag{A6.1.2}$$

Some basic properties of the Laplace transform and the inverse Laplace transform can be found in some mathematical textbooks (e.g., Hildebrand 1976).

References

Adler, R. F. and R. A. Mack, 1986. Thunderstorm cloud top dynamics inferred from satellite observations and a cloud top parcel model. *J. Atmos. Sci.*, **43**, 1945–60.

Asai, T., 1972. Thermal instability of a shear flow turning the direction with height. *J. Meteor. Soc. Japan*, **50**, 525–32.

Baik, J.-J. and H.-Y. Chun, 1997. A dynamical model for urban heat islands. *Bound.-Layer Meteor.*, **83**, 463–77.

Baik, J.-J., H.-S. Hwang, and H.-Y. Chun, 1999. Transient, linear dynamics of a stably stratified shear flow with thermall forcing and a critical level. *J. Atmos. Sci.*, **56**, 483–99.

Banta, R. M., 1990. The role of mountain flows in making clouds. *Atmospheric Processes over Complex Terrain, Meteor. Monogr.*, 45, Amer. Meteor. Soc., 229–83.

Bretherton, C., 1988. Group velocity and the linear response of stratified fluids to internal heat or mass sources. *J. Atmos. Sci.*, **45**, 81–93.

Bretherton, C., 1993. The nature of adjustment in cumulus cloud fields. In *The Representation of Cumulus Convection in Numerical Models, Meteor. Monogr.*, 46, Amer. Meteor. Soc., 63–74.

Chun, H.-Y., and J.-J. Baik, 1994. Weakly nonlinear response of a stably stratified atmosphere to diabatic focing in a uniform flow. *J. Atmos. Sci.*, **51**, 3109–21.

Chun, H.-Y., and Y.-L. Lin, 1995. Enhanced response of an atmospheric flow to a line type heat sink in the presence of a critical level. *Meteor. Atmos. Phys.*, **55**, 33–45.

Crook, N. A., and M. W. Moncrieff, 1988. The effect of large-scale convergence on the generation and maintenance of deep moist convection. *J. Atmos. Sci.*, **45**, 3606–24.

Dailey, P. S. and R. G. Fovell, 1999. Numerical simulation of the interaction between the sea-breeze front and horizontal convective rolls. Part I: Offshore ambient flow. *Mon. Wea. Rev.*, **127**, 858–78.

Dalu, G. A. and R. A. Pielke, 1989. An analytical study of the sea breeze. *J. Atmos. Sci.*, **46**, 1815–25.

DeSouza, R. L., 1972. A study of atmospheric flow over a tropical island. M. S. thesis, Dept. of Meteor., Florida State University.

Doyle, J. D. and T. T. Warner, 1993. Nonhydrostatic simulations of coastal mesoscale vortices and frontogenesis. *Mon. Wea. Rev.*, **121**, 3371–92.

Fovell, R. G. 2005. Convective initiation ahead of the sea-breeze front. *Mon. Wea. Rev.*, **133**, 264–78.

Garstang, M., P. D. Tyson, and G. D. Emmitt, 1975. The structure of heat islands. *Rev. Geophys. Space Phys.*, **13**, 139–65.

Heymsfield, G. M. and R. H. Blackmer, Jr., 1988. Satellite-observed characteristics of Midwest severe thunderstorm anvils. *Mon. Wea. Rev.*, **116**, 2200–24.

Hildebrand, F. B., 1976. *Advanced Calculus for Applications*. 2nd edn., Prentice-Hall Inc., USA.

Hjelmfelt, M. R., 1982. Numerical simulations of the effects of St. Louis on mesoscale boundary layer airflow and vertical air motion: Simulations of urban vs non-urban effects. *J. Appl. Meteor.*, **31**, 1239–57.

Hsu, H.-M., 1987. Mesoscale lake-effect snowstorms in the vicinity of Lake Michigan: Linear theory and numerical simulations. *J. Atmos. Sci.*, **44**, 1019–40.

Kimura, R., and T. Eguchi, 1978. On dynamical processes of sea- and land-breeze circulation. *J. Meteor. Soc. Japan*, **56**, 67–85.

Koch, S. E., F. Zhang, M. L. Kaplan, Y.-L. Lin, R. P. Weglarz, and C. M. Trexler, 2001: Numerical simulations of a gravity wave event over CCOPE. Part III: The role of a mountain-plains solenoid in the generation of the second wave episode. *Mon. Wea. Rev.*, **129**, 909–33.

Lin, C. A. and R. E. Stewart, 1991. Diabatically forced mesoscale circulations in the atmosphere. *Adv. Geophys.*, **33**, 267–305.

Lin, Y.-L., 1986. Calculation of airflow over an isolated heat source with application to the dynamics of V-shaped clouds. *J. Atmos. Sci.*, **43**, 2736–51.

Lin, Y.-L., 1987. Two-dimensional response of a stably stratified flow to diabatic heating. *J. Atmos. Sci.*, **44**, 1375–93.

Lin, Y.-L., 1989. Inertial and frictional effects on stratified hydrostatic airflow past an isolated heat source. *J. Atmos. Sci.*, **46**, 921–36.

Lin, Y.-L., 1990. A theory of cyclogenesis forced by diabatic heating. Part II: A semi-geostrophic approach. *J. Atmos. Sci.*, **47**, 1755–77.

Lin, Y.-L., 1996. Structure of dynamically unstable shear flow and their implications for shallow internal gravity waves. Part II: Nonlinear response. *Meteor. Atmos. Phys.*, **59**, 153–72.

Lin, Y.-L. and H.-Y. Chun, 1991: Effects of diabatic cooling in a shear flow with a critical level. *J. Atmos. Sci.*, **48**, 2476–91.

Lin, Y.-L. and S. Li, 1988. Three-dimensional response of a shear flow to elevated heating. *J. Atmos. Sci.*, **45**, 2987–3002.

Lin, Y.-L. and R. B. Smith, 1986. Transient dynamics of airflow near a local heat source. *J. Atmos. Sci.*, **43**, 40–9.

Lin, Y.-L., T.-A. Wang, and R. P. Weglarz, 1993. Interaction between gravity waves and cold air outflows in a stably stratified uniform flow. *J. Atmos. Sci.*, **50**, 3790–816.

Lyons, W. A. and L. E. Olsson, 1972. The climatology and prediction of the Chicago lake breeze. *J. Appl. Meteor.*, **11**, 1254–72.

Nicholls, M. E., R. A. Pielke, and W. R. Cotton, 1991. Thermally forced gravity waves in an atmosphere at rest. *J. Atmos. Sci.*, **48**, 1869–84.

Niino, H., 1987. The linear theory of land and sea breeze circulation. *J. Meteor. Soc. Japan*, **65**, 901–21.

Ogura, Y. and M.-T. Liou, 1980. The structure of a midlatitude squall line: A case study. *J. Atmos. Sci.*, **37**, 553–67.

Olfe, D. B. and R. L. Lee, 1971. Linearized calculation of urban heat island convection effects. *J. Atmos. Sci.*, **28**, 1374–88.

Orville, H. D., 1964. On mountain upslope winds. *J. Atmos. Sci.*, **21**, 622–33.

Orville, H. D., 1968. Ambient wind effects on the initiation and development of cumulus clouds over mountains. *J. Atmos. Sci.*, **25**, 385–403.

Raymond, D. J., 1972. Calculation of airflow over an arbitrary ridge including diabatic heating and cooling. *J. Atmos. Sci.*, **29**, 837–43.

Raymond, D. J. and R. Rotunno, 1989. Response of a stably stratified flow to cooling. *J. Atmos. Sci.*, **46**, 2830–37.

Reuter, G. W. and O. Jacobsen, 1993. Effects of variable wind shear on the mesoscale circulation forced by slab-symmetric diabatic heating. *Atmosphere-Ocean*, **31**, 451–69.

Riordan, A. J., 1990. Examination of the mesoscale features of the GALE coastal front of 24–25 January 1986. *Mon. Wea. Rev.*, **118**, 258–82.

Riordan, A. J. and Y.-L. Lin, 1992. Mesoscale wind signatures along the Carolina coast. *Mon. Wea. Rev.*, **120**, 2786–97.

Robichaud, A. and C. A. Lin, 1989. Simple models of diabatically forced mesoscale circulations and a mechanism for amplification. *J. Geophys. Res.*, **94**, D3, 3413–26.

Rotunno, R., 1983. On the linear theory of the land and sea breeze. *J. Atmos. Sci.*, **40**, 1999–2009.

Schmidt, J. M. and W. R. Cotton, 1990. Interactions between upper and lower atmospheric gravity waves on squall line structure and maintenance. *J. Atmos. Sci.*, **47**, 1205–22.

Simpson, J. E., 1994. *Sea Breeze and Local Winds*. Cambridge University Press.

Smith, R. B., 1980. Linear theory of stratified hydrostatic flow past an isolated mountain. *Tellus*, **32**, 348–64.

Smith, R. B. and Y.-L. Lin, 1982. The addition of heat to a stratified airstream with application to the dynamics of orographic rain. *Quart. J. Roy. Meteor. Soc.*, **108**, 353–78.

Song, I.-S. and H.-Y. Chun, 2005. Momentum flux spectrum of convectively forced internal gravity waves and its application to gravity wave drag parameterization. Part I: Theory. *J. Atmos. Sci.*, **62**, 107–24.

Sousounis, P. J. and H. N. Shirer, 1992. Lake aggregate mesoscale disturbances. Part I: Linear analysis. *J. Atmos. Sci.*, **49**, 80–96.

Sun, W.-Y., 1978. Stability analysis of cloud streets. *J. Atmos. Sci.*, **35**, 466–83.

Sun, W.-Y. and I. Orlanski, 1981a. Large mesoscale convection and sea breeze circulation. Part I: Linear stability analysis. *J. Atmos. Sci.*, **38**, 1675–93.

Sun, W.-Y. and I. Orlanski, 1981b. Large mesoscale convection and sea breeze circulation. Part 2: Non-Linear numerical model. *J. Atmos. Sci.*, **38**, 1694–706.

Szeto, K. K., C. A. Lin, and R. E. Steward, 1988. Mesoscale circulations forced by melting snow. II: Application to meteorological features. *J. Atmos. Sci.*, **45**, 1642–50.

Thorpe, A. J., M. J. Miller, and M. W. Moncrieff, 1980. Dynamical models of two-dimensional downdraughts. *Quart. J. Roy. Meteor. Soc.*, **106**, 463–84.

Tripoli, G. J. and W. R. Cotton, 1989. Numerical study of an observed mesoscale convective system. Part I: Simulated genesis and comparison with observations. *Mon. Wea. Rev.*, **117**, 273–304.

Wang, P. K., 2003. Moisture plumes above thunderstorm anvils and their contributions to cross-tropopause transport of water vapor in midlatitudes. *J. Geophys. Res.*, **108**, D6, 4194–208.

Wolyn, P. G. and T. B. McKee, 1994. The mountain-plains circulation east of a 2-km-high north-south barrier. *Mon. Wea. Rev.*, **122**, 1490–508.

Xie, L. and Y.-L. Lin, 1996. Responses of low-level flow to an elongated surface heat source with application to coastal frontogenesis. *Mon. Wea. Rev.*, **124**, 2807–27.

Problems

6.1 Prove that the first term $w'(x,z) = C \cos(kx + mz)$ in (6.1.21a) represents upward energy propagation. [Hint: Use the continuity equation and the momentum equation and integrate $p'w'$ with respect to x from $x = 0$ to $2\pi/k$]

6.2 Calculate the perturbation pressure at the surface, (6.1.36), from (6.1.35) using *Bernoulli's equation* [Hint: Substitute (6.1.3) and (6.1.24) into (6.1.1)]

6.3 (a) Consider a two-dimensional, inviscid, nonrotating, hydrostatic, Boussinesq uniform flow over a heat source, and derive the governing equation for w' to be written as shown in (6.2.1). (b) Show that for a nonhydrostatic flow, the term $\partial^2 w'/\partial z^2$ should be replaced by $\partial^2 w'/\partial x^2 + \partial^2 w'/\partial z^2$.

6.4 (a) Derive (6.2.2) by taking the Fourier transform in x and the Laplace transform in t of (6.2.1) (see Appendix 6.1 for the Laplace transform).

(b) Based on the flow and thermal forcing parameters estimate the Scorer parameter and dominant wave number for the case of Fig. 6.4, and explain the wave propagation character. [Hint: $k = 2\pi/b_1 = 3.14 \times 10^{-4} \text{ m}^{-1} < l = N/U = 0.001 \text{ m}^{-1}$]

(c) In order to generate a flow response in the evanescent flow regime, how weak does the Brunt–Vaisala frequency have to be for the case of Fig. 6.4, if other flow and thermal forcing parameters remain unchanged? Sketch the flow configuration for this regime.

6.5 Derive (6.2.12). [Hint: Use (6.1.1)–(6.1.4)]

6.6 Based on (2.2.14)–(2.2.18), apply appropriate approximations to derive the governing equation (6.4.1).

6.7 (a) Extend (6.1.1)–(6.1.4) to three-dimensional, rotating flow and nondimensionalize them to obtain (6.4.11)–(6.4.15).

   (b) Combine (6.4.11)–(6.4.15) into a single equation for w, (6.4.17)

   (c) Taking the double Fourier transform (Appendix 5.1) of (6.4.17) and (6.4.18), obtain (6.4.19).

6.8 (a) Based on (6.4.11)–(6.4.15), derive the $\hat{b}$ and $\hat{\pi}$ fields in terms of $\hat{w}$ in the Fourier space. (b) Derive the vorticity equation in Fourier space, $\hat{\zeta} = -K^2\hat{\pi} - iR_o^2 k\hat{w}_z$.

7

Mesoscale instabilities

It is well documented that instabilities play an important role in triggering mesoscale circulations, such as squall lines, mesoscale rainbands, mesoscale convective complexes, frontogenesis, mesoscale cyclogenesis, clear air turbulence, billow clouds and orographic precipitating systems. For example, Fig. 7.1 shows *mesoscale cellular convection* observed over the Atlantic Ocean off the southeast coast of the United States on February 19, 2002. This type of cloud is an atmospheric manifestation of *Rayleigh–Bénard convection*, which includes both unstable updrafts and downdrafts and turbulent eddies triggered by the thermal instability between two horizontal plates heated from below, in the atmosphere. Mesoscale cellular convection normally forms as cold air passes over warmer ocean waters, as seen in the wake of a cold front over a warm ocean current during the winter. Hexagonal *open cells* are produced in the cloud-topped boundary layer and convective clouds exist at the vertices of the hexagonal cells. In these cells, air goes up on the rim and descends in the center where the skies are clear. On the other hand, the circulation associated with a *closed cell* is opposite to that of an open cell. Normally, moist convection in the atmosphere is associated with localized heat source, and thus behaves quite differently from the mesoscale cellular convection shown in Fig. 7.1. Localized moist convection is characterized by: (1) strong, compact, turbulent, unstable, upward motion, (2) weak, compensated, laminar, stable, downward motion over a wide area of the surroundings, and (3) inertia-gravity waves generated outside and which propagate away from the convective region (Emanuel and Raymond 1984). Localized dry convection, such as that triggered by forest fire, also behaves like localized moist convection.

As discussed in Chapter 1, atmospheric instabilities may serve as energy conversion mechanisms for mesoscale circulations and weather systems. Although instabilities associated with the mean wind velocity or thermal structure of the atmosphere are a rich energy source for atmospheric disturbances, the maximum growth rates of most idealized atmospheric instabilities are either on the large scale, such as baroclinic and barotropic instabilities, or on the microscale, such as Kelvin–Helmholtz and static (gravitational) instabilities. Conditional, potential (convective), inertial, and symmetric instabilities may occur on the mesoscale range. In this chapter, we will discuss these mesoscale instabilities as well as instabilities with both large-scale and microscale maximum growth rates that have important impacts on mesoscale circulations and weather systems.

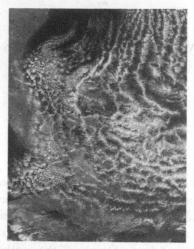

Fig. 7.1 Open cell cloud formation observed over the Atlantic Ocean off the southeast coast of the United States on February 19, 2002, as viewed by satellite Terra. Notice the hexagonal open cells produced in the cloud-topped boundary layer and the convective clouds at the vertices of the hexagonal cells. (From Visible Earth, NASA.)

7.1 Wave energy transfer through instabilities

The linear energy equation can be used to study the energy transfer among different forms, such as potential energy and kinetic energy, and through different hydrodynamic instabilities. The governing equations for a small-amplitude, inviscid, Boussinesq atmosphere on a planetary f-plane can be combined into a single equation for the vertical velocity, such as (3.2.1),

$$
\begin{aligned}
\frac{D}{Dt} \Bigg\{ & \frac{D^2}{Dt^2}\nabla^2 w' + f^2 w'_{zz} - \left(U_{zz}\frac{D}{Dt} + fV_{zz} \right) w'_x - \left(V_{zz}\frac{D}{Dt} - fU_{zz} \right) w'_y \\
& + N^2\nabla_H^2 w' + 2f\left(U_z w'_{yz} - V_z w'_{xz} \right) \Bigg\} + 2fU_z V_z \left(w'_{xx} - w'_{yy} \right) + 2f\left(V_z^2 - U_z^2 \right) w'_{xy} \\
& - 2f^2\left(U_z w'_{xz} + V_z w'_{yz} \right) = \frac{g}{c_p T_o}\frac{D}{Dt}\nabla_H^2 q'.
\end{aligned}
$$

$$(7.1.1)$$

As also mentioned in earlier chapters, this equation represents a mesoscale atmospheric system which may contain the generation mechanisms of: (a) pure and inertia-gravity waves, (b) static instability, (c) conditional instability, (d) potential (convective) instability, (e) shear (Kelvin–Helmholtz) instability, (f) symmetric instability, (g) inertial instability, (h) baroclinic instability, and (i) wave-CISK. The theory of wave-CISK has been discussed in Subsection 4.3.3 and various places in the text, and will not be repeated here.

The energy transfer equation for the above system, that has no meridional basic state wind component ($V = 0$), but includes a meridionally sheared zonal flow ($U_y \neq 0$) can be derived

$$\left(\frac{\partial}{\partial t} + U\frac{\partial}{\partial x}\right)E + \rho_o u'w'U_z + \rho_o u'v'U_y - \left(\frac{\rho_o g f}{N^2 \theta_o}\right)v'\theta'U_z + \nabla \cdot (p'v')$$

$$= \left(\frac{\rho_o g^2}{c_p T_o N^2 \theta_o}\right)\theta' q' \qquad (7.1.2)$$

where

$$E = \frac{\rho_o}{2}\left[(u'^2 + v'^2 + w'^2) + \left(\frac{g}{N\theta_o}\right)^2 \theta'^2\right], \qquad (7.1.3)$$

is the total perturbation energy. The total perturbation energy consists of the perturbation kinetic energy and the perturbation potential energy, which are respectively represented by the first and second terms inside the square bracket of (7.1.3). Integrating horizontally (7.1.2) over a single wavelength for a periodic disturbance or from $-\infty$ to $+\infty$ for a localized disturbance in both x and y directions gives (with primes dropped):

$$\frac{\partial \overline{E}}{\partial t} = -\rho_o(\overline{uw})U_z - \rho_o(\overline{uv})U_y + \left(\frac{\rho_o g f}{N^2 \theta_o}\right)(\overline{v\theta})U_z - \frac{\partial}{\partial z}(\overline{pw}) + \left(\frac{\rho_o g^2}{c_p T_o N^2 \theta_o}\right)(\overline{\theta q})$$

$$\qquad (7.1.4)$$

Now, integrating the above equation from the surface, $z = z_s$, to the top of either the physical domain or that of a finite-area numerical model, $z = z_T$, with which the physical phenomena under consideration are investigated or simulated, yields

$$\frac{\partial E_T}{\partial t} = -\rho_o \int_{z_s}^{z_T} \overline{uw}U_z dz - \rho_o \int_{z_s}^{z_T} \overline{uv}U_y dz + \left(\frac{\rho_o g f}{N^2 \theta_o}\right)\int_{z_s}^{z_T} \overline{v\theta}U_z dz$$

$$\quad (1) \qquad\qquad (2) \qquad\qquad (3) \qquad\qquad (4)$$

$$\qquad (7.1.5)$$

$$-\overline{pw}(z_T) + \overline{pw}(z_s) + \left(\frac{\rho_o g^2}{c_p T_o N^2 \theta_o}\right)\int_{z_s}^{z_T} \overline{\theta q}\, dz.$$

$$\quad (5) \qquad\quad (6) \qquad\qquad (7)$$

where E_T is the domain-integrated total perturbation energy.

Term 1 of (7.1.5) represents the local rate of change in the total perturbation energy of the system. Term 2 represents the vertical momentum flux transfer between the kinetic energy of the basic current and the perturbation energy. When shear instability occurs, the energy is transferred from the basic state shear flow to the perturbation, which results in a net loss of kinetic energy of the basic state. In order to show the phase relationship of the basic shear and the perturbation velocities, we define a *perturbation streamfunction* ψ in a two-dimensional (x, z) flow as

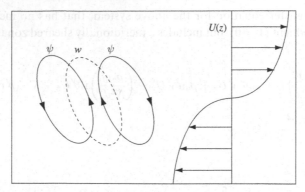

Fig. 7.2 A sketch of the basic wind profile and the upshear tilt of the perturbation streamfunctions (solid) and updraft (dashed) associated with an unstable growing gravity wave in a stably stratified flow. The perturbation wave energy is converted from the basic flow shear. (Adapted after Lin and Chun, 1993.)

$$u = \frac{\partial \psi}{\partial z}; \quad w = -\frac{\partial \psi}{\partial x}. \tag{7.1.6}$$

Note that the above relationships satisfy the incompressible continuity equation. With this definition, uwU_z can be written as

$$uwU_z = -\left[\frac{\partial \psi}{\partial x} \Big/ \frac{\partial \psi}{\partial z}\right]\left(\frac{\partial \psi}{\partial z}\right)^2 U_z = \left[\frac{\partial z}{\partial x}\right]_\psi \left(\frac{\partial \psi}{\partial z}\right)^2 U_z, \tag{7.1.7}$$

because $(\partial \psi/\partial x)dx + (\partial \psi/\partial z)dz = d\psi = 0$ on a constant streamline ψ. This in turn implies that if shear instability exists (i.e. term $1 > 0$) and $U_z > 0$, then we have on average on the vertical plane

$$\left[\frac{\partial z}{\partial x}\right]_\psi < 0. \tag{7.1.8}$$

Therefore, the growing wave in a stably stratified flow must have a phase tilt in an opposite direction of the shear vector, i.e., an upshear phase tilt. This also implies that the updraft has an upshear tilt since it is out of phase with the streamfunction by a factor of $\pi/2$. This phase relationship is sketched in Fig. 7.2.

Term 3 of (7.1.5) represents the horizontal momentum flux transfer between the kinetic energy of the basic state horizontal shear and the perturbation wave energy. The energy transfer to horizontal momentum occurs when *inertial instability* is present. The argument for horizontal phase tilt is similar to the one presented above for shear instability. Term 4 represents the energy exchange between the basic state vertical shear and the perturbation heat flux. The basic state vertical shear is supported by the baroclinicity, i.e. the horizontal temperature gradient through thermal wind balance. In this situation, the available potential energy stored in the system is transferred to perturbation kinetic energy when *baroclinic instability* occurs. Term 5 represents the forcing from the upper boundary condition, while Term 6 represents the forcing from the lower boundary. Term 7 represents the contribution from diabatic sources or sinks to the total perturbation energy.

7.2 Integral theorems of stratified flow

7.2.1 Governing equations

Consider the governing equations of (2.2.14)–(2.2.18) for a nonrotating, Boussinesq fluid system with all the nonlinear and inhomogeneous terms added and lumped into source terms on the right-hand side of the governing equations,

$$\frac{\partial u'}{\partial t} + U\frac{\partial u'}{\partial x} + U_z w' + \frac{1}{\rho_0}\frac{\partial p'}{\partial x} = F_1, \tag{7.2.1}$$

$$\frac{\partial v'}{\partial t} + U\frac{\partial v'}{\partial x} + \frac{1}{\rho_0}\frac{\partial p'}{\partial y} = F_2, \tag{7.2.2}$$

$$\frac{\partial w'}{\partial t} + U\frac{\partial w'}{\partial x} - g\frac{\theta'}{\theta_0} + \frac{1}{\rho_0}\frac{\partial p'}{\partial z} = F_3, \tag{7.2.3}$$

$$\frac{\partial u'}{\partial x} + \frac{\partial v'}{\partial y} + \frac{\partial w'}{\partial z} = 0, \tag{7.2.4}$$

$$\frac{\partial \theta'}{\partial t} + U\frac{\partial \theta'}{\partial x} + \frac{N^2\theta_0}{g}w' = H, \tag{7.2.5}$$

where

$$F_1 = -\frac{\partial(u'u')}{\partial x} - \frac{\partial(u'v')}{\partial y} - \frac{\partial(u'w')}{\partial z}, \tag{7.2.6}$$

$$F_2 = -\frac{\partial(v'u')}{\partial x} - \frac{\partial(v'v')}{\partial y} - \frac{\partial(v'w')}{\partial z}, \tag{7.2.7}$$

$$F_3 = -\frac{\partial(w'u')}{\partial x} - \frac{\partial(w'v')}{\partial y} - \frac{\partial(w'w')}{\partial z}, \tag{7.2.8}$$

$$H = \left(\frac{\theta_0}{c_p T_0}\right)q' - \left[\frac{\partial(u'\theta')}{\partial x} + \frac{\partial(v'\theta')}{\partial y} + \frac{\partial(w'\theta')}{\partial z}\right]. \tag{7.2.9}$$

In the above equations, $F = (F_1, F_2, F_3)$ is the perturbation *Reynolds stress*, and H is the *effective heating*, which is composed of the diabatic heating and *turbulent heat fluxes*. Note that H and F are significant only in small regions, such as turbulent or convective areas, while the homogeneous part of the equations apply to broader regions. However, the horizontal nonlinear advection terms may become significant even outside regions of convection or turbulence.

Considering a two-dimensional flow on x–z plane and taking the normal mode approach by assuming

$$\phi = \hat{\phi} e^{ik(x-ct)}, \tag{7.2.10}$$

where ϕ is a generic variable that can be either u, v, w, θ or p, and substituting it into (7.2.1)–(7.2.5) yields a single equation for the amplitude of w:

$$\frac{\partial^2 \hat{w}}{\partial z^2} + m^2 \hat{w} = \frac{g}{\theta_o} \frac{\hat{H}}{(c-U)^2} + \left(\frac{1}{c-U}\right)(\hat{F}_{1z} - ik\hat{F}_3), \tag{7.2.11}$$

where

$$m^2 = \frac{1}{c-U} \frac{d^2 U}{dz^2} + \frac{N^2}{(c-U)^2} - k^2. \tag{7.2.12}$$

In order to simplify (7.2.11), we may define a new variable h as

$$h \equiv \frac{\hat{w}}{c-U}. \tag{7.2.13}$$

Note that, physically, $-h/ik$ is the vertical displacement. By rearranging (7.2.11) and (7.2.12), the term $[1/(c-U)]d^2U/dz^2$ can be eliminated. A new governing equation in terms of h can then be obtained

$$\frac{\partial}{\partial z}\left[(c-U)^2\frac{\partial h}{\partial z}\right] + \left[N^2 - k^2(c-U)^2\right]h = \frac{g}{\theta_o}\frac{\hat{H}}{c-U} + \hat{F}_{1z} - ik\hat{F}_3. \tag{7.2.14}$$

Taking Im $\int_0^\infty h^* \cdot (7.2.14) dz$ yields

$$\text{Im} \int_0^\infty h^* \frac{\partial}{\partial z}\left[(c-U)^2\frac{\partial h}{\partial z}\right]dz - \text{Im} \int_0^\infty k^2(c-U)^2|h|^2 dz$$
$$= \frac{g}{\theta_o}\text{Im}\int_0^\infty \frac{\hat{w}^*\hat{H}}{|c-U|^2}dz + \text{Im}\int_0^\infty h^*(\hat{F}_{1z} - ik\hat{F}_3)dz, \tag{7.2.15}$$

where h^* is the complex conjugate of h, which is assumed to be 0 at $z = 0, \infty$. Note that the term involving N^2 does not appear in (7.2.15) because it has no imaginary part. Substituting $c = c_r + ic_i$ into (7.2.15) and taking the imaginary part yields

$$-2 \int_0^\infty c_i(c_r - U) \left[\left| \frac{\partial h}{\partial z} \right|^2 + k^2 |h|^2 \right] dz = \frac{g}{\theta_o} \text{Im} \int_0^\infty \frac{\hat{w}^* \hat{H}}{|c - U|^2} dz$$

$$+ \text{Im} \int_0^\infty h^* (\hat{F}_{1z} - ik\hat{F}_3) dz. \tag{7.2.16}$$

Based on the above equation, two cases can be considered: (i) with no forcing ($H = 0$, $F = 0$), and (ii) with only thermal forcing ($H \neq 0$, $F = 0$).

For case (i), because there is no forcing ($H = 0$, $F = 0$), (7.2.16) reduces to

$$-2 \int_0^\infty c_i(c_r - U) \left[\left| \frac{\partial h}{\partial z} \right|^2 + k^2 |h|^2 \right] dz = 0. \tag{7.2.17}$$

Substituting $c = c_r + ic_i$ into (7.2.10) gives $\phi = \hat{\phi} \exp(ik(x - c_r t)) \exp(kc_i t)$, which indicates that in order for instability to occur, c_i must be greater than 0. Based on (7.2.17), term $(c_r - U)$ must change sign at some level between $z = 0$ and ∞ because the term in the square bracket is always positive. Therefore, *in order for instability to occur in a two-dimensional, nonrotating flow, it requires the existence of a critical level at which $U = c_r$.*

For case (ii), i.e., with thermal forcing only, (7.2.16) reduces to

$$-2 \int_0^\infty c_i(c_r - U) \left[\left| \frac{\partial h}{\partial z} \right|^2 + k^2 |h|^2 \right] dz = \frac{g}{\theta_o} \text{Im} \int_0^\infty \frac{\hat{w}^* \hat{H}}{|c - U|^2} dz. \tag{7.2.18}$$

In general, we can assume that the phase of the forcing H differs from that of the vertical velocity (w') by an amount α, or say,

$$H = Qe^{i\alpha} w', \tag{7.2.19}$$

where Q and α are real and may vary with height. Assuming a wave-like disturbance for w' and substituting (7.2.19) into the right-hand side of (7.2.18) gives

$$\text{Im} \int_0^\infty \frac{\hat{w}^* \hat{H}}{|c - U|^2} dz = \int_0^\infty \frac{Q|\hat{w}|^2 \sin \alpha}{|c - U|^2} dz. \tag{7.2.20}$$

If the heating is in phase with vertical velocity everywhere, i.e. $\alpha = 0$, then the right hand side of (7.2.18) becomes zero. Then, (7.2.18) implies that either $c_i = 0$, where there is no amplification or instability, or $c_r = U$ at a certain level, there exists a critical (steering) level. This condition leads to the conclusion that *if a diabatic forcing is to generate an amplifying, non-steering level perturbation, the forcing must be somewhere out of phase with the vertical velocity* (Bolton 1980). This condition was also noted by Moncrieff (1978).

7.2.2 Miles' theorem

By considering a two-dimensional, nonrotating, stratified shear flow with no forcing, (7.2.14) reduces to

$$\frac{\partial}{\partial z}\left[(U-c)^2\frac{\partial h}{\partial z}\right] + \left(N^2 - k^2(U-c)^2\right)h = 0, \tag{7.2.21}$$

where $h(z_1) = h(z_2) = 0$. If the flow is unstable ($c_i > 0$), then c will be complex and $U - c \neq 0$ for any height z. We then can choose one branch of $\sqrt{U-c}$ for the interval $[z_1, z_2]$, which will be as differentiable as U. Assuming $h = G/\sqrt{U-c}$ and substituting it into the above equation gives

$$\frac{\partial}{\partial z}\left((U-c)\frac{\partial G}{\partial z}\right) - \left(\frac{1}{2}U_{zz} + k^2(U-c) + \frac{U_z^2/4 - N^2}{U-c}\right)G = 0, \quad z_1 \leq z \leq z_2, \tag{7.2.22}$$

with $G(z_1) = G(z_2) = 0$. Multiplying the above equation by the complex conjugate of G (i.e. G^*) and integrating from z_1 to z_2 yields

$$\int_{z_1}^{z_2} (U - c_r - ic_i)\left[\left|\frac{\partial G}{\partial z}\right|^2 + k^2|G|^2\right]dz + \frac{1}{2}\int_{z_1}^{z_2} U_{zz}|G|^2 dz + \int_{z_1}^{z_2} (U - c_r - ic_i)\left|\frac{G}{U-c}\right|^2\left(\frac{1}{4} - \frac{N^2}{U_z^2}\right)U_z^2\,dz = 0. \tag{7.2.23}$$

The imaginary part of (7.2.23) gives

$$c_i\left(\int_{z_1}^{z_2}\left[\left|\frac{\partial G}{\partial z}\right|^2 + k^2|G|^2\right]dz + \int_{z_1}^{z_2}\left|\frac{G}{U-c}\right|^2 U_z^2\left(R_i - \frac{1}{4}\right)dz\right) = 0, \tag{7.2.24}$$

where $R_i \equiv N^2/U_z^2$ is the *Richardson number*, also known as the *gradient Richardson number* as mentioned in Chapter 3. Since the flow is assumed to be unstable, it requires that $c_i > 0$. The above equation therefore implies that it is necessary for $R_i < 1/4$ at some level between z_1 and z_2 for instability to occur. This also leads to the *Miles' Theorem* (Miles 1961): *If a stratified fluid system has $R_i \geq 1/4$ everywhere, then it is stable.*

7.2.3 Howard's semicircle theorem

Consider a stratified fluid system governed by (7.2.11) with no forcing,

$$\frac{\partial^2 \hat{w}}{\partial z^2} + m^2\hat{w} = 0. \tag{7.2.25}$$

Define the vertical displacement η as

$$w' = \frac{D\eta}{Dt} = \frac{\partial \eta}{\partial t} + U\frac{\partial \eta}{\partial x}, \tag{7.2.26}$$

which gives

$$\hat{\eta} = \frac{\hat{w}}{ik(U-c)}, \tag{7.2.27}$$

if we assume a wave-like disturbance. Substituting (7.2.27) into (7.2.25) yields

$$\frac{\partial}{\partial z}\left[(U-c)^2\frac{\partial \hat{\eta}}{\partial z}\right] + [N^2 - k^2(U-c)^2]\hat{\eta} = 0. \tag{7.2.28}$$

The boundary conditions are $\hat{\eta} = 0$ at $z = z_1$ and z_2. Multiplying (7.2.28) by the complex conjugate of $\hat{\eta}$ (i.e. $\hat{\eta}^*$) and then integrating from z_1 to z_2 leads to

$$\int_{z_1}^{z_2} (U-c)^2\left(\left|\frac{\partial \hat{\eta}}{\partial z}\right|^2 + k^2|\hat{\eta}|^2\right)dz - \int_{z_1}^{z_2} N^2|\hat{\eta}|^2 dz = 0. \tag{7.2.29}$$

Substituting $c = c_r + ic_i$ into the above equation and then separating the real and imaginary parts yields

$$\text{Real Part} = \int_{z_1}^{z_2}\left[(U-c_r)^2 - c_i^2\right]R\,dz - \int_{z_1}^{z_2} N^2|\hat{\eta}|^2 dz = 0, \tag{7.2.30}$$

$$\text{Imaginary Part} = 2c_i\int_{z_1}^{z_2}(U-c_r)R\,dz = 0. \tag{7.2.31}$$

where

$$R = \left|\frac{\partial \hat{\eta}}{\partial z}\right|^2 + k^2|\hat{\eta}|^2. \tag{7.2.32}$$

Again, for instability to occur, we require $c_i > 0$. In turn, (7.2.31) leads to

$$\int_a^b UR\,dz = c_r\int_a^b R\,dz. \tag{7.2.33}$$

Equations (7.2.30) and (7.2.33) imply that

$$\int_a^b U^2 R\,dz = (c_r^2 + c_i^2)\int_a^b R\,dz + \int_a^b N^2|\hat{\eta}|^2 dz. \tag{7.2.34}$$

Now supposing $a \le U(z) \le b$ (where $a = U_{\min}$ and $b = U_{\max}$), we then have

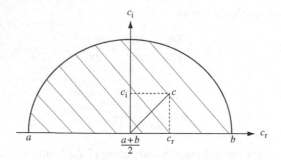

Fig. 7.3 A sketch of Howard's semicircle theorem, based on (7.2.36). In the figure, the basic wind profile $U(z)$ satisfies $a \le U(z) \le b$ (where $a = U_{\min}$ and $b = U_{\max}$), and $c = c_r + ic_i$ is the phase speed.

$$\left([c_r - (a+b)/2]^2 + c_i^2 - [(a-b)/2]^2\right) \int_a^b R\,dz + \int_a^b N^2 |\hat{\eta}|^2 dz \le 0. \qquad (7.2.35)$$

Since the integrations in the above equation are both positive, we require that

$$[c_r - (a+b)/2]^2 + c_i^2 \le [(a-b)/2]^2. \qquad (7.2.36)$$

This leads us to the *Howard's semicircle theorem* (Howard 1961): *The complex phase speed, c, of an unstable normal mode must lie within the semicircle enclosed by $U_{\min}$ and $U_{\max}$* (Fig. 7.3).

7.3 Static, conditional, and potential instabilities

7.3.1 Static instability

Static instability, which is also known as *buoyant* or *gravitational instability*, describes an atmospheric state in which an air parcel accelerates away from its initial level due to the density difference between the air parcel and its environment. The concept of static instability can be understood by applying parcel theory to the vertical momentum equation,

$$\frac{Dw}{Dt} = -\frac{1}{\rho}\frac{\partial p}{\partial z} - g, \qquad (7.3.1)$$

where ρ and p are the density and pressure of the air parcel, respectively. In *parcel theory*, we assume that:

(a) the pressure of the air parcel adjusts immediately to the pressure of its environment ($\bar{p}$), i.e. $p = \bar{p}$, when it moves away from its initial level;
(b) the environment of the air parcel is in hydrostatic balance;
(c) no compensating motions exist in the parcel's environment; and
(d) the air parcel does not mix with its environment and so retains its original identity.

Applying condition (a) to the air parcel leads to

$$\frac{Dw}{Dt} = g\left(\frac{\bar{p} - \rho}{\rho}\right) \equiv b, \tag{7.3.2}$$

where b is the *buoyancy*, more exactly the buoyancy force per unit mass. The above equation indicates that the vertical acceleration of the air parcel is controlled by the *buoyancy force*, $g(\bar{p} - \rho)/\rho$. Note that

$$\left(\frac{\bar{p} - \rho}{\rho}\right) \approx \frac{\theta - \bar{\theta}}{\bar{\theta}} \approx \frac{T - \bar{T}}{\bar{T}}. \tag{7.3.3}$$

For an infinitesimal vertical displacement, η, (7.3.2) becomes

$$\frac{D^2\eta}{Dt^2} + N^2\eta = 0, \tag{7.3.4}$$

where

$$N^2 = -\frac{\partial b}{\partial z} = \frac{g}{\bar{\theta}}\frac{\partial\bar{\theta}}{\partial z} = \frac{g}{T}(\Gamma_d - \gamma). \tag{7.3.5}$$

Here, $\gamma \equiv -\partial T/\partial z$ is the observed *environmental (actual) lapse rate*, and $\Gamma_d = g/c_p$ is the parcel or *dry adiabatic lapse rate*. The overbar represents values of the environmental air. The dry adiabatic lapse rate is the decrease of temperature experienced by a dry air parcel when it ascends adiabatically, which is approximately $9.76 \times 10^{-3}\,\mathrm{K\,m^{-1}}$. In deriving (7.3.5), we have used *Poisson's equation*, (2.2.7).

Equation (7.3.4) implies that the criteria for static stability, static neutrality, and static instability are $N^2 > 0$, $N^2 = 0$, and $N^2 < 0$, respectively. The mathematical solutions of (7.3.4) can be obtained by taking the real part of $\eta(t) = A\exp(iNt) + B\exp(-iNt)$ for an atmosphere with static stability; $\eta(t) = C + Dt$ for an atmosphere with static neutrality; and $\eta(t) = E\exp(N_i t) + F\exp(-N_i t)$, where $N_i^2 = -N^2$, for an atmosphere with static instability. The constant coefficients are determined by two initial conditions: η and $D\eta/Dt = w$ at $t = 0$. The static instability acts on a horizontal scale of tens to thousands of meters. Its criteria for dry air may also be determined by the vertical gradient of the environmental potential temperature:

$$N^2 > 0, \quad \gamma < \Gamma_d, \quad \text{or} \quad \partial\bar{\theta}/\partial z > 0: \textit{absolutely stable},$$
$$N^2 = 0, \quad \gamma = \Gamma_d, \quad \text{or} \quad \partial\bar{\theta}/\partial z = 0: \textit{dry neutral}, \tag{7.3.6}$$
$$N^2 < 0, \quad \gamma > \Gamma_d, \quad \text{or} \quad \partial\bar{\theta}/\partial z < 0: \textit{dry absolutely unstable}.$$

These criteria have been listed in Table 7.1, along with the criteria for other instabilities. Parcel theory assumes that the air parcel derives its buoyancy from the density or temperature difference between itself and the air that surrounds it. Thus, buoyancy is regarded as a relative quantity that depends on the choice of a basic-state environmental

Table 7.1 *Criteria for different types of instabilities. (Adapted after Schultz and Schumacher 1999.)*

	Static (Gravitational)	Inertial	Symmetric
Dry	Absolute Instability $\partial\bar\theta/\partial z<0$; $\gamma>\Gamma_d$	Inertial Instability $\partial\bar{M}/\partial x<0$; $\zeta_{ga}+f<0$	Symmetric Instability $(\partial\bar\theta/\partial z)_{\bar{M}}<0$; $(\partial\bar{M}/\partial x)_{\bar\theta}<0$ $\Gamma_d\|_{\bar{M}}<\gamma\|_{\bar{M}}$ $|\partial z/\partial x|_{\bar{M}}<|\partial z/\partial x|_{\bar\theta}$ $PV_g<0$
Moist	Moist Absolute Instability (MAI) $\Gamma_s<\gamma_s$	N/A	N/A
Conditional [#]	Conditional Instability (CI) $\partial\bar\theta_e^*/\partial z<0$ $\Gamma_s<\gamma<\Gamma_d$ (parcel lifted above LFC)	N/A	Conditional Symmetric Instability (CSI) $\partial\bar\theta_e^*/\partial z|_{\bar{M}}<0$; $\partial\bar{M}/\partial z|_{\bar\theta_e^*}<0$ $\Gamma_s\|_{\bar{M}}<\gamma\|_{\bar{M}}<\Gamma_d\|_{\bar{M}}$ $|\partial z/\partial x|_{\bar{M}}<|\partial z/\partial x|_{\bar\theta_e^*}$ $MPV_g^*<0$ (parcel lifted above LFC)
Potential [##]	Potential Instability (PI) $\partial\bar\theta_e/\partial z<0$	N/A	Potential Symmetric Instability (PSI) $\partial\bar\theta_e/\partial z|_{\bar{M}}<0$; $\partial\bar{M}/\partial z|_{\bar\theta_e}<0$ $|\partial z/\partial x|_{\bar{M}}<|\partial z/\partial x|_{\bar\theta_e}$ $MPV_g<0$

[#]at saturation, $\bar\theta_e=\bar\theta_e^*$; [##] $\bar\theta_w$ can be used equivalently to $\bar\theta_e$.

Meanings of symbols: (1) γ: observed environmental lapse rate; (2) Γ_d: dry lapse rate; (3) Γ_s: moist lapse rate; (4) γ_s: observed environmental saturated lapse rate; (5) $\bar\theta$: environmental potential temperature; (6) $\bar\theta_e$: environmental equivalent potential temperature; (7) $\bar\theta_e^*$: environmental saturation equivalent potential temperature; (8) $\bar{M}$: environmental geostrophic absolute momentum; (9) PV_g: geostrophic potential vorticity (PV); (10) MPV_g: moist geostrophic PV; and (11) MPV_g^*: saturated geostrophic PV (7.6.13e).

profile. With this method of calculation, the buoyancy refers to the *Archimedean buoyancy force* acting on an object, which is the weight of the fluid displaced minus the weight of the object. Under this approach, the vertical perturbation pressure gradient force is not taken into account. In order to take this effect into account, a non-Achimedean approach is required (Das 1979; Davies-Jones 2003). When the vertical perturbation pressure gradient is included, the vertical momentum equation (7.3.2) becomes (Doswell and Markowski 2004),

$$\frac{Dw}{Dt} = -\frac{1}{\rho}\frac{\partial p'}{\partial z} - g\frac{\rho'}{\rho},$$ (7.3.7)

where $\rho' = \rho - \bar{\rho}$. The density ($\rho$) in the denominators of the above equation is replaced by $\bar{\rho}$ in the derivations of parcel theory in some textbooks, which implicitly involves making either an anelastic [$\bar{\rho} = \bar{\rho}(z)$] or Boussinesq ($\bar{\rho} = \rho_0 = $ constant) approximation. The first term on the right-hand side, also called the *thermal buoyancy* (Kessler 1985), is associated with the vertical gradient of the perturbation pressure and is often ignored. The second term is commonly recognized as *buoyancy*. Most cloud models adopt a form of the vertical momentum equation similar to (7.3.7) and take into account the effects of the vertical perturbation pressure gradient force.

The vertical acceleration is contributed by the *dynamic pressure* (p'_d) and the *buoyancy pressure* (p'_b). The dynamic pressure arises from the flow field differences created by the fluid motion, while the buoyancy pressure is generated by the vertical buoyancy gradient. Taking these effects into consideration, the vertical momentum equation can be written as

$$\frac{Dw}{Dt} = -\frac{1}{\bar{\rho}}\frac{\partial p'_d}{\partial z} + \left(b - \frac{1}{\bar{\rho}}\frac{\partial p'_b}{\partial z}\right),$$ (7.3.8)

where b is the buoyancy and p'_d and p'_b are derived from the Bossinesq form of the momentum equation (Rotunno and Klemp 1985; Emanuel 1994),

$$-\frac{1}{\bar{\rho}}\nabla^2 p'_d = |D|^2 - |\boldsymbol{\omega}|^2,$$ (7.3.9)

$$-\frac{1}{\bar{\rho}}\nabla^2 p'_b = -\frac{\partial b}{\partial z}.$$ (7.3.10)

and where $|D|$ is the magnitude of the total deformation (see (8.4.9) for the mathematical definition) and $\boldsymbol{\omega}$ is the three-dimensional vorticity vector. The first term on the right-hand side of (7.3.8) occurs due to the dynamic perturbation pressure, which is independent of the thermodynamic base state. The buoyancy should include both terms inside the bracket on the right-hand side of (7.3.8), instead of just b (Doswell and Markowski 2004). The partition of perturbation pressure into dynamic pressure and buoyancy pressure has helped in the understanding of storm dynamics, which will be elaborated on in Chapter 8.

The buoyancy of saturated air is affected by vertical displacements, because the displacement causes the water to undergo a phase change. For a reversible vertical displacement of cloudy air, the *saturated moist Brunt–Vaisala frequency* (N_m) can be derived as (Lalas and Einaudi 1974)

$$N_m^2 = \frac{g}{\overline{T}}\left(\frac{\partial \overline{T}}{\partial z} + \Gamma_s\right)\left(1 + \frac{Lq_{vs}}{R_d \overline{T}}\right) - \left(\frac{g}{1 + q_w}\right)\frac{\partial q_w}{\partial z}, \tag{7.3.11}$$

where Γ_s is the *moist adiabatic lapse rate*, q_{vs} is the *saturation mixing ratio of water vapor*, R_d is the *ideal gas constant for dry air*, L is the *latent heat of condensation* or *evaporation*, and q_w is the *total water mixing ratio*. An equivalent expression of N_m^2 in terms of moist conservative variables can also be derived as

$$N_m^2 = \left(\frac{1}{1 + q_w}\right)\left\{\left[\frac{\Gamma_s(c_p + c_l q_w)}{\overline{\theta}_e}\right]\frac{\partial \overline{\theta}_e}{\partial z} - \left[g - \Gamma_s c_l \ln\left(\frac{\overline{\theta}_e}{\overline{T}}\right)\right]\frac{\partial q_w}{\partial z}\right\}, \tag{7.3.12}$$

where c_l is the specific heat capacity at constant pressure of liquid water and $\overline{\theta}_e$ is the *equivalent potential temperature* of the saturated environmental air. The above expression of N_m^2 is identical to those derived in the literature (e.g., Emanuel (1994); Kirshbaum and Durran 2004), but has the advantage of being able to separate the vertical gradients $\overline{\theta}_e$ and q_w completely. The Brunt–Vaisala frequency (N) in (7.3.4) and (7.3.5) should be replaced by N_m for a saturated cloudy air parcel. Thus, the cloudy air is statically stable if $N_m^2 > 0$ (Durran and Klemp 1982). On the other hand, if $N_m^2 < 0$, then the saturated cloudy air will be statically unstable, more exactly *moist absolutely unstable* to an infinitesimal vertical displacement. *The criterion for moist absolute instability* (MAI) is $\Gamma_s < \gamma_s$, where Γ_s and γ_s are the moist lapse rate and environmental saturated lapse rate, respectively. According to (7.3.12), this criterion is equivalent to $\partial \overline{\theta}_e/\partial z < 0$ only if the total water mixing ratio (q_w) is uniform with height. The *equivalent potential temperature* (θ_e) can be approximately calculated using the following equation:

$$\theta_e = \theta e^{Lq_v/c_p T_{LCL}}, \tag{7.3.13}$$

where T_{LCL} is the parcel temperature at the lifting condensation level (LCL). Physically, the equivalent potential temperature is defined as the potential temperature that an air parcel would have if all of its moisture were condensed and the resulting latent heat released warmed the parcel, or the parcel followed a *pseudoadiabatic process*. Note that for a saturated air parcel, $T_{LCL} = T$ and $q_v = q_{vs}$ in the above definition. Therefore, θ_e is a conserved quantity for both a dry adiabatic process and a pseudoadiabatic process. The equivalent potential temperature can also be obtained graphically using a thermodynamic diagram. In a moist atmosphere, parcel theory is modified so that the parcel follows a dry adiabatic process until saturated, after which it continues through a pseudoadiabatic process.

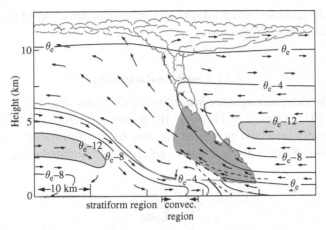

Fig. 7.4 A schematic of the formation of a moist absolutely unstable layer (MAUL) for moist absolute instability (MAI) through slab convective overturning. Wind vectors relative to the outflow boundary are denoted by arrows, cloud boundaries are denoted by scalloped lines, θ_e contours are denoted by solid lines, the outflow boundary or frontal zone is denoted by heavy solid line, the midlevel layer of low θ_e is highlighted by light shading, and the MAUL is denoted by dark shading. (Adapted after Bryan and Fritsch 2000.)

Note that $\partial\bar{\theta}_e/\partial z < 0$ is a necessary condition for potential (convective) instability to occur, as will be discussed in Subsection 7.3.3. When $\Gamma_s c_l \ln(\bar{\theta}_e/\overline{T})$ is less than g, (7.3.12) indicates that the increase of q_w with height in a saturated layer means that the layer may become moist absolutely unstable even though it is convectively stable (i.e., $\partial\bar{\theta}_e/\partial z > 0$). While the stability measure given in (7.3.12) provides a useful assessment of stability within clouds, however, it is not useful for evaluating the stability of unsaturated airstreams which lead to the development of clouds themselves. It has been observed in the atmosphere that moist absolute instability can be created and maintained as mesoscale convective systems develop (Bryan and Fritsch 2000). A strong, mesoscale, ascent not driven by buoyancy, such as mechanical lifting along surface-based outflow layers, frontal zones, or elevated terrain, may bring a conditionally unstable environmental layer to saturation relatively fast. Figure 7.4 illustrates how, in an idealized mesoscale convective system, a *moist absolutely unstable layer* (MAUL) forms ahead of a density current created through slab convective overturning. Ahead of the mesoscale convective system, the atmosphere is *conditionally unstable*, which means $\Gamma_s < \gamma < \Gamma_d$. The mesoscale convective system contains a well-mixed boundary layer. The air parcels rising from the well-mixed layer are unable to overcome the *convective inhibition* (CIN) and cannot reach their *level of free convection* (LFC). These issues related to conditional instability will be discussed in Subsection 7.3.2. However, as the outflow boundary of the mesoscale convective system approaches, the environmental air is lifted strongly to the saturation level, thus creating a MAUL. Unlike other instabilities, such as conditional and potential instabilities, which require upward displacement, MAI occurs for both upward and downward displacement.

7.3.2 Conditional instability

The method for determining static instability, as discussed above, assumes either no saturation or 100% saturation. However, the situation is significantly different if the air parcel becomes saturated as is lifted upward. In an unsaturated atmosphere, the *unsaturated moist Brunt–Vaisala frequency* (N_w) can be estimated using the following formula:

$$N_w^2 = \frac{g}{\bar{\theta}_v} \frac{\partial \bar{\theta}_v}{\partial z},$$ (7.3.14)

where θ_v is the *virtual potential temperature* of the environmental air and is related to the *virtual temperature* (T_v) by

$$\theta_v = T_v \left(\frac{p_s}{p}\right)^{R_d/c_p},$$ (7.3.15)

where

$$T_v = \left(\frac{1 + q/\varepsilon}{1 + q}\right) T \approx \left(\frac{1 + 1.609q}{1 + q}\right) T \approx (1 + 0.61q)T.$$ (7.3.16)

In the above equation, q is the water vapor mixing ratio and ε is the ratio of the molecular weight of water vapor (m_v) to that of dry air (m_d), which has a value of 0.622.

When an unsaturated air parcel is lifted upward, its temperature follows a dry adiabat path until it reaches the LCL (Fig. 7.5). Further lifting will result in condensation and cause the temperature of the air parcel to follow a moist adiabat. The latent heat released from the condensation will warm the air parcel and slow down the moist adiabatic lapse rate (Γ_s), making it slower than the *dry lapse rate* (Γ_d). Suppose the value of the observed environmental lapse rate (γ) is between Γ_s and Γ_d (i.e. $\Gamma_s < \gamma < \Gamma_d$), and that the forcing is strong enough to lift the air parcel past its LCL. A continued lifting will then force the air parcel to cool at a rate of Γ_s, eventually reaching a level where the temperature of the air parcel and the environment are equal. Further lifting will cause the air parcel's temperature to surpass its environment temperature, causing it to accelerate upward due to the buoyancy force. Since the buoyancy force acts on the parcel during this step, no additional forcing is needed. In other words, the air parcel is now experiencing free convection. This level is known as the LFC. Since this type of air parcel instability is subjected to a finite-amplitude displacement from its initial level to the LFC, it is referred to as *conditional instability*. Thus, *the necessary conditions for conditional instability to occur are: (a)* $\Gamma_s < \gamma < \Gamma_d$ *and (b) a lifting of the air parcel past its LFC.*

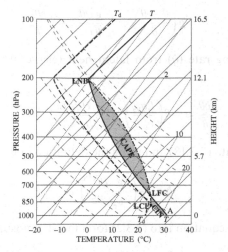

Fig. 7.5 Example of an idealized sounding with conditional instability displayed on a *skew-T log-p* thermodynamic diagram. The isotherms, saturation mixing ratio, and dry sadiabat are denoted by slanted solid, dotted, and dashed curves. The environmental temperature (T) and dew point profile (T_d) are denoted by thick solid and dashed curves, respectively. The lifting condensation level (LCL), level of free convection (LFC), and level of neutral buoyancy (LNB) for the air parcel originated at A are denoted in the figure. The convective available potential energy (CAPE) is the area enclosed by the temperature profile and the moist adiabat (thick dot-dashed curve) in between LFC and LNB, while the convective inhibition (CIN) is the area enclosed by the temperature profile and the dry adiabat (below LCL) and the temperature profile and the moist adiabat (in between LCL and LFC).

The criterion for conditional instability can also be determined via the vertical gradient of the *saturation equivalent potential temperature* (θ_e^*) which is defined as the equivalent potential temperature of a hypothetically saturated atmosphere at the initial level. This hypothetical atmosphere has been set to mimic the thermal structure of the actual atmosphere. In other words, θ_e^* can be defined as the equivalent potential temperature that the air parcel would have if it were saturated initially at the same pressure and temperature, and can be calculated by

$$\theta_e^* = \theta e^{Lq_{vs}/c_p T}. \tag{7.3.17}$$

In order to derive the criterion for conditional instability, we consider an air parcel lifted from $z_o - \delta z$ to z_o. At $z_o - \delta z$, the air parcel is assumed to have the same potential temperature as that of the environment, $\bar{\theta} - (\partial\bar{\theta}/\partial z)\eta$, where $\bar{\theta}$ is the potential temperature of the environmental air at z_o, and $\eta = \delta z$ is the vertical displacement. The potential temperature of the air parcel experiences a change of $\delta\theta$ when it is lifted from $z_o - \eta$ to z_o, i.e. $[\bar{\theta} - (\partial\bar{\theta}/\partial z)\eta] + \delta\theta$. Thus, the buoyancy of the air parcel at z_o is

$$b = g\left(\frac{\theta - \bar{\theta}}{\bar{\theta}}\right) = -\frac{g}{\bar{\theta}}\frac{\partial\bar{\theta}}{\partial z}\eta + g\frac{\delta\theta}{\bar{\theta}}. \tag{7.3.18}$$

Substituting the heating rate $(\dot{q})$ from latent heat release, $\dot{q} = -L(Dq_{vs}/Dt)$, into (2.2.5) gives

$$\frac{\delta\theta}{\bar{\theta}} \approx -\delta\left(\frac{Lq_{vs}}{c_p T}\right) \approx -\frac{\partial}{\partial z}\left(\frac{Lq_{vs}}{c_p T}\right)\eta. \tag{7.3.19}$$

Substituting (7.3.19) into (7.3.18) and using the definition of θ_e^* leads to

$$b \approx -\left(\frac{g}{\bar{\theta}_e^*}\frac{\partial\bar{\theta}_e^*}{\partial z}\right)\eta. \tag{7.3.20}$$

Substituting the above equation into (7.3.2) for a moist atmosphere, gives us

$$\frac{D^2\eta}{Dt^2} + \left(\frac{g}{\bar{\theta}_e^*}\frac{\partial\bar{\theta}_e^*}{\partial z}\right)\eta = 0. \tag{7.3.21}$$

Therefore, *the conditional stability criterion for a saturated layer of air becomes*

$$\frac{\partial\bar{\theta}_e^*}{\partial z}\begin{cases} >0 & \text{conditionally stable} \\ =0 & \text{conditionally neutral} \\ <0 & \text{conditionally unstable} \end{cases} \tag{7.3.22}$$

Note that in addition to $\partial\bar{\theta}_e^*/\partial z < 0$, the release of conditional instability requires the air parcel to be lifted above its LFC. This requirement is not included in the above derivation (e.g. Sherwood 2000; Schultz *et al.* 2000). Parcel theory also neglects the effects of mass continuity and pressure perturbation (Xu 1986), as also known from dry static instability.

Figure 7.5 illustrates the concept of conditional instability, where an idealized sounding is plotted on a *skew-T log-p* thermodynamic diagram. The LCL, LFC, and LNB for the air parcel originating at A are denoted in the figure. The amount of energy available for free convection is called the *convective available potential energy* (CAPE), which is defined as the work done by the buoyancy force in lifting an air parcel from its LFC to LNB,

$$\text{CAPE} = \int_{z_{LFC}}^{z_{LNB}} b\, dz = \int_{z_{LFC}}^{z_{LNB}} g\left(\frac{\bar{\rho} - \rho}{\rho}\right) dz$$
$$= \int_{z_{LFC}}^{z_{LNB}} g\left(\frac{T - \bar{T}}{\bar{T}}\right) dz = \int_{z_{LFC}}^{z_{LNB}} g\left(\frac{\theta - \bar{\theta}}{\bar{\theta}}\right) dz. \tag{7.3.23}$$

In a thermodynamic diagram, CAPE is proportional to the area enclosed by the environmental temperature curve and the moist adiabat of the air parcel in between the LFC and LNB (Fig. 7.5). The moist adiabat follows a *saturated adiabatic process*, which assumes all of the condensates remain in the air parcel. The *moist adiabatic lapse rate* can be derived to be

$$\Gamma_s \equiv -\frac{dT}{dz} = \frac{\Gamma_d}{1 + (L/c_p)(dq_{vs}/dT)}. \tag{7.3.24}$$

The saturation water vapor mixing ratio is defined as the ratio of the mass of water vapor to the mass of dry air containing the vapor at saturation. *A saturated adiabatic process is almost identical to a pseudoadiabatic process.* This is because the heat carried by the condensates is negligible compared to that carried by the air parcel. The moist adiabatic lapse rate can be approximated by

$$\Gamma_s \equiv \frac{\Gamma_d[1 + (Lq_{vs}/R_d T)]}{1 + \varepsilon L^2 q_{vs}/(c_p R_d T^2)}, \tag{7.3.25}$$

where $\varepsilon = m_v/m_d \approx 0.622$ as defined earlier. Observed values of Γ_s show that it is about $4\,\mathrm{K\,km^{-1}}$ near the ground in humid conditions, increases to 6 to $7\,\mathrm{K\,km^{-1}}$ in the middle troposphere and is nearly equal to the dry lapse rate of $9.8\,\mathrm{K\,km^{-1}}$ at high altitudes where the air is colder and holds less water vapor.

The total amount of potential energy for an air parcel lifted upward from a certain level z_i to its LNB can be calculated as follows:

$$\mathrm{CAPE}_i = \int_{z_i}^{z_{LNB}} g\left(\frac{T - \overline{T}}{\overline{T}}\right) dz. \tag{7.3.26}$$

The CAPE is also referred to as *static potential energy* or *available buoyant energy* and is represented as the *positive area* (PA) on a thermodynamic diagram if z_i is assumed to be at LFC. The positive area is defined as

$$\mathrm{PA} = \int_{p_{LNB}}^{p_{LFC}} R_d(T - \overline{T})d(\ln p) = \int_{z_{LFC}}^{z_{LNB}} g\left(\frac{T - \overline{T}}{\overline{T}}\right) dz. \tag{7.3.27}$$

On the other hand, the *negative area* (NA) on a thermodynamic diagram is the area confined by the dry adiabat (below LCL) or the moist adiabat (above LCL) to the left, and the sounding to the right, from the initial level to the LFC (Fig. 7.5). The negative area represents the energy needed to lift an air parcel vertically and dry adiabatically or pseudoadiabatically to its LFC and is also known as the *convective inhibition* (CIN). Mathematically, CIN is defined as

$$\mathrm{CIN} = \int_{z_i}^{z_{LFC}} g\left(\frac{\overline{T} - T}{\overline{T}}\right) dz. \tag{7.3.28}$$

In practice, the surface height (z_s) is used for z_i. It can be shown that $\mathrm{CAPE}_i = \mathrm{PA} - \mathrm{NA}_i$, where CAPE_i and NA_i are the CAPE and NA at z_i. Thus, *a positive* $CAPE_i$ *is a necessary condition for conditional instability to occur*, so that the air parcel has potential energy for convection. In the absence of horizontal advection, the maximum vertical velocity that can be realized by an air parcel occurs when all the potential energy is converted into kinetic energy, i.e. $w_{max} = \sqrt{2CAPE}$, because $Dw/Dt \approx w\partial w/\partial z$.

When rain evaporates in sub-saturated air or when a solid precipitate (snow or hail) melts at the melting level, a downdraft will be generated by the cooled air. The maximum downdraft can be estimated as $-w_{max} = \sqrt{2DCAPE_i}$, where $DCAPE_i$ is the *downdraft convective available potential energy* and is defined as

$$DCAPE_i = \int_{z_s}^{z_i} g\left(\frac{\bar{T} - T}{\bar{T}}\right) dz, \qquad (7.3.29)$$

where z_s is normally the surface or the level at which an air parcel descends from the initial level z_i, allowing a neutral buoyancy to be achieved.

Therefore, in addition to the commonly adopted lapse-rate definition of conditional instability, i.e. the environmental lapse rate lies between the dry and moist adiabatic lapse rates, an *available-energy* definition has also been proposed, i.e. an air parcel must possess positive buoyant energy (i.e., $CAPE_i > 0$). More precisely, for an unsaturated air parcel, the stability can be classified as: (a) No CAPE: stability for all vertical displacements, (b) CAPE > 0: instability for some finite vertical displacements, which contains two subcategories: (i) CAPE > CIN and (ii) CIN > CAPE (Schultz *et al.* 2000). The available-energy definition is more consistent with the concept of *subcritical instability* (Sherwood 2000), which is defined as an instability that requires a finite amplitude perturbation exceeding a critical amplitude (Drazin and Reid 1981). Thus, it has been suggested that the term "conditional instability" should be reserved only for the lapse-rate concept, and the term "*latent instability*" for the energy-based concept (Schultz *et al.* 2000).

In summary (also see Table 7.1), there exist six static stability states for dry and moist air:

$$
\begin{array}{lll}
(1) \text{ absolutely stable} & \gamma < \Gamma_s, & \\
(2) \text{ saturated neutral} & \gamma = \Gamma_s, & \\
(3) \text{ conditionally unstable} & \Gamma_s < \gamma < \Gamma_d, & \\
(4) \text{ dry neutral} & \gamma = \Gamma_d, & (7.3.30) \\
(5) \text{ dry absolute unstable} & \gamma > \Gamma_d, & \\
(6) \text{ moist absolutely unstable} & \gamma_s > \Gamma_s, &
\end{array}
$$

where γ_s is the saturated lapse rate of the environmental air. Note that moist absolute instability is not equivalent to conditional instability. In a typical conditionally unstable situation, an initially unsaturated air parcel is lifted to saturation in an unsaturated environment. The air parcel will then follow a moist adiabat, and will become unstable after further lifting. However, under certain circumstances, an initially conditionally unstable atmosphere may become moist absolute unstable after lifting (Bryan and Fritsch 2000).

Parcel theory has been used to assess the instabilities discussed above. We must note that this theory is generally developed in terms of buoyancy, namely the density or temperature difference between an ascending air parcel and its environmental air. This makes parcel theory incapable of dealing with instabilities that depend on horizontal pressure gradients, such as baroclinic and barotropic instabilities (Emanuel 1994). In addition, parcel theory does not take into account the following: (1) the entrainment of the environmental air, (2) the compensating vertical motions of the environmental air in response to the intrusion of clouds, and (3) the effect of condensate on buoyancy. The first effect tends to reduce the buoyancy. It may be assumed that as saturated cloud air ascends a distance δz, a mass δm of environmental air is entrained. This will require heat from the air parcel (cloudy air) to warm this entrained air. In addition, the condensed water in the cloud will evaporate in response to the entrainment of drier environmental air. Taking these effects into account, (7.3.19) becomes

$$\frac{\delta\theta}{\theta} \approx -\frac{\partial}{\partial z}\left(\frac{Lq_{vs}}{c_p T}\right)\delta z - \left(b + \frac{L}{c_p T}(q_{vs} - \overline{q})\right)\frac{\delta m}{m}, \qquad (7.3.31)$$

where $\delta z = \eta$ and $\overline{q}$ is the mixing ratio of the entrained environmental air. The effect of compensating vertical motions of the environmental air can be improved by the *slice method* (Bjerknes 1938), which takes into account of the area of compensated downward motion through mass continuity in estimating the stability. Finally, the melting of snow or evaporative cooling of rainfall may also reduce the buoyancy. Representing these effects more completely and accurately would normally require the use of a numerical model to solve the mathematical problem involving (7.3.31).

7.3.3 Potential instability

Potential instability, also known as *convective instability*, describes an atmospheric state in which an atmospheric layer becomes unstable statically after lifting. This means that dense air lies on top of less dense air within the layer. An entire atmospheric layer may be lifted by a density current, a broad mountain range, a frontal surface, or a cyclone. Under these situations, *layer theory* is more appropriate for assessing the instability of the lifted layer, as opposed to parcel theory for conditional instability.

In order to understand potential instability, we first consider a dry layer of the atmosphere extending from pressure level p_1 up to p_2, and assume that: (a) the mass is conserved within the layer and (b) the atmosphere is in hydrostatic balance. Since the pressure difference is directly proportional to the mass per unit area contained in the air column, it can be approximated by $p_1 - p_2 = \rho_{av}g\delta z$, where ρ_{av} is the averaged air density of the layer and $\delta z = z_2 - z_1$. Therefore, layer lifting (sinking) tends to make the column stretch (shrink), because ρ_{av} decreases (increases) with height. It follows that $\partial\overline{\theta}/\partial z$ decreases (increases) for the lifting (sinking) of a dry or unsaturated layer with

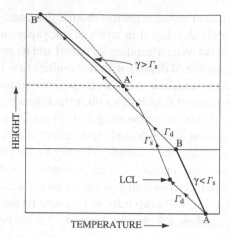

Fig. 7.6 Illustration of potential (convective) instability by lifting an initially absolutely stable layer AB with $\partial \bar{\theta}_e / \partial z < 0$. The top of the layer (B) follows a dry adiabat to saturation at B', while the bottom of the layer becomes saturated earlier (at LCL) and then follows a moist adiabat to A'. The lapse rate of the final saturated layer (A'B') is greater than the moist adiabat, thus is unstable. (Adapted after Darkow 1986.)

no condensation since $\delta \bar{\theta} \approx (\partial \bar{\theta} / \partial z) \delta z$ and $\delta \bar{\theta}$ is conserved. In other words, lifting tends to make a layer less stable in a dry or unsaturated atmosphere, while sinking tends to make the layer more stable. One exception to this occurs when the layer of air is neutrally stable, in which case $\partial \bar{\theta} / \partial z = 0$ before and after the vertical displacement.

The situation is completely different if saturation occurs during the layer lifting, since lifting may make an initially stable layer absolutely unstable. The process leading to instability ($\gamma > \Gamma_s$) can be illustrated by using a thermodynamic diagram, such as Fig. 7.6. We lift an initially absolutely stable layer ($\gamma < \Gamma_s$) with $\partial \bar{\theta}_e / \partial z < 0$. In this situation, the top of the layer follows a dry adiabat. After lifting, the layer (A'B') becomes saturated and unstable because $\gamma > \Gamma_s$. Similar graphical operations indicate that an initially absolutely stable layer can become saturated and stable (neutral) when $\partial \bar{\theta}_e / \partial z > 0$ ($\partial \bar{\theta}_e / \partial z = 0$). Thus, the criteria for potential (convective) instability can be expressed in terms of the equivalent potential temperature, since this is a conserved quanity for the moist layer,

$$\frac{\partial \bar{\theta}_e}{\partial z} \begin{cases} > 0 & \text{potentially stable} \\ = 0 & \text{potentially neutral} \\ < 0 & \text{potentially unstable} \end{cases} \qquad (7.3.32)$$

Since the *wet-bulb potential temperature* (θ_w) is a conserved variable under dry and pseudoadiabatic processes, it has also been used, replacing θ_e in the above criteria (7.3.32), to determine potential stability, neutrality, or instability. Note that the wet-bulb potential temperature is the temperature that the air parcel would have if it descends following a moist adiabat from the LCL to 1000 hPa. If the vertical gradient of the cloud water mixing ratio ($\partial q_w / \partial z$) is taken into account, then the

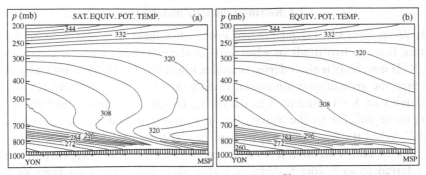

Fig. 7.7 (a) Saturation equivalent potential temperature ($\bar{\theta}_e^*$) and (b) equivalent potential temperature ($\bar{\theta}_e$) fields on a cross section from Dauphin, Manitoba (YDN) to Minneapolis-St. Paul, Minnesota (MSP), as predicted by the Rapid Update Cycle (a numerical model) for a 9-h forecast valid for 0000 UTC January 3, 1998. Note that above the surface cold front, the air is conditionally unstable ($\partial\bar{\theta}_e^*/\partial z < 0$), but is potentially stable ($\partial\bar{\theta}_e/\partial z > 0$). The ordinate is pressure (mb). (Adapted after Schultz and Schumacher 1999.)

condition for potential instability to occur becomes more restrictive than (7.3.32), according to (7.3.12).

Note that a layer that is conditionally unstable need not be potentially unstable, nor is a potentially unstable layer necessarily conditionally unstable (Rogers and Yao 1989). This is illustrated by Fig. 7.7, which shows the $\bar{\theta}_e^*$ and $\bar{\theta}_e$ fields on a cross section across an equatorward-moving cold front over the northern US from Dauphin, Manitoba (YDN) to Minneapolis-St. Paul, Minnesota (MSP), forecast by a numerical model for 0000 UTC January 3, 1998. There is a significant difference between these two fields. Although the air above the frontal surface is conditionally unstable since $\partial\bar{\theta}_e^*/\partial z < 0$ (Fig. 7.7a), it is potentially stable at the same location since $\partial\bar{\theta}_e/\partial z > 0$ (Fig. 7.7b). Unlike conditional instability, potential instability does not imply a reservoir of potential energy for convection (Emanuel 1994). Thus, a continuous external supply of energy, such as continuous forcing by a density current, cold front, mountain, or a cyclone, is required to convert the potential instability into actual instability. When the air is saturated, the vertical gradients of $\bar{\theta}_e$ and $\bar{\theta}_e^*$ are equivalent, and therefore conditional instability and potential instability are equivalent (Schultz and Schumacher 1999). In addition, unlike the parcel-lifting process, the layer-lifting process is not typically associated with the development of isolated upright deep moist convection. If it were, layer lifting would initially produce stable stratiform clouds, which would then develop into deep moist convection (Schultz *et al.* 2000). As will be discussed in Subsection 9.1.2 and Fig. 9.5 (Chapter 9), the embedded convective cells in Fig. 9.5a appear to be the result of layer lifting (potential instability), while the single convective cell in Fig. 9.5b appears to be the result of parcel lifting (conditional instability). The exact relationship between these two instabilities deserves a further investigation.

7.4 Kelvin–Helmholtz instability

Kelvin–Helmholtz (K–H) instability may occur when there is vertical shear across the interface between two fluids or when vertical shear is present within a continuous fluid. K–H instability is also referred to as shear instability in the literature. Figure 7.8 illustrates the growth of a sinusoidal disturbance in a homogeneous shear flow associated with K–H instability. In the figure, two layers of fluid, one beneath the other, move parallel to the *x*-axis, but in opposite directions and at the same speed *U*. When the fluid system is undisturbed, the interface of these two layers of fluid is horizontal. This interface is regarded as a thin layer of strong vorticity and is also referred to as a vortex sheet. A *vortex sheet* is defined as a layer of fluid consisting of small, discrete vortices rotating in the same direction. In the figure, the curved arrows indicate the direction of the self-induced movement of the vorticity in the sheet, as well as (a) the general rotation about points like C and (b) the accumulation of vorticity at points like C. The positive vorticity accumulating at points like C will induce clockwise velocity around these points, thereby amplifying the sinusoidal disturbance of the vortex sheet. This process of vorticity accumulation at points like C will continue, leading to exponential growth of the disturbance. The spatial form of the disturbance will not change as long as the disturbance is small enough not to significantly change the basic state. The basic mechanism of K–H instability is the conversion of the available kinetic energy embedded in the basic shear flow into the kinetic energy of the disturbance, whereby the fluid aquires the potential energy needed to lift or lower fluid parcels when $N^2 > 0$ ($d\bar{\rho}/dz < 0$) everywhere. Therefore, vertical shear tends to destabilize the flow while buoyancy tends to stabilize it.

The K–H instability can be understood by using the energy argument presented below (as characterized by Chandrasekhar 1961). To quantify the effects of buoyancy and vertical shear, we consider two neighboring fluid parcels of equal volumes at

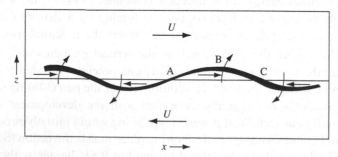

Fig. 7.8 A sketch illustrating the growth of a sinusoidal disturbance associated with shear or Kelvin–Helmholtz instability. The initially uniform vortex sheet (thick line) has positive vorticity in the *y* direction (into the paper). The local strength of the vortex sheet is represented by the thickness of the sheet and the curved arrows indicate the direction of the movement induced by the vorticity in the sheet. (Adapted after Batchelor 1967.)

heights z and $z + \delta z$ and interchange them. The work δW that must be done to incur this interchange against gravity is given by

$$\Delta W = -g \, \delta \bar{\rho} \, \delta z, \qquad (7.4.1)$$

where $\delta \bar{\rho} = (d\bar{\rho}/dz)\delta z$ is the difference in the basic density at the two heights. The kinetic energy per unit volume available to do this work is given by

$$\Delta K = \underbrace{\left\{ \frac{1}{2}\rho U^2 + \frac{1}{2}\rho(U + \delta U)^2 \right\}}_{1 \qquad\qquad 2} - \underbrace{\frac{\rho}{2}\frac{[U + (U + \delta U)]^2}{2}}_{3} = \frac{1}{4}\bar{\rho}(\delta U)^2. \qquad (7.4.2)$$

In (7.4.2), terms 1 and 2 respectively represent the kinetic energies of air parcels 1 and 2 before the interchange, while term 3 represents the kinetic energy of the air parcels after the interchange. A necessary condition for this interchange, and thus for instability, is that $\Delta W \le \Delta K$ somewhere in the flow:

$$R_i \equiv \frac{N^2}{U_z^2} = \frac{-(g/\bar{\rho})d\bar{\rho}/dz}{(dU/dz)^2} < \frac{1}{4}. \qquad (7.4.3)$$

This leads to the same criterion as stated in Miles' theorem, as discussed in Subsection 7.2.2. Thus, K–H instability tends to occur when the airflow experiences marked vertical shear and weak thermal stratification. Note that the example given in Fig. 7.8 is in a homogeneous fluid with $R_i = 0$, which indicates the K–H instability is purely produced by the shear effect and may thus be defined as *shear instability*. When $R_i < 0$, either buoyancy (N) or the shear (U_z) effect may lead to K–H instability (because $R_i = N^2/U_z^2$). In this situation, the K–H instability appears to be a mixed instability, but may be more dominated by the shear effect (shear instability) when R_i is less negative (e.g., $-0.25 < R_i < 0$) and by the buoyancy effect (static or buoyant instability) when R_i is more negative (e.g., $R_i < -0.25$).

K–H instability may produce large-amplitude gravity waves and clear air turbulence (CAT). CAT is a major cause of atmospheric turbulence and an extraordinarily challenging subject, long studied by atmospheric scientists, computational fluid dynamicists, and aerospace engineers. CAT is of crucial interest to aviators because of the significant impact of this phenomenon on aviation safety. Occasionally, K–H instability becomes visible when there is enough moisture in the atmosphere and air parcels are lifted above their LFCs by the K–H waves. Figure 7.9 shows an example of such K–H *billow clouds*.

7.5 Inertial instability

As discussed earlier, for a statically stable atmosphere ($N^2 > 0$), the buoyancy force acts as a restoring force which makes a vertically displaced air parcel return to its original position and oscillate vertically. Analogous to static stability, for inertial

Fig. 7.9 An example of breaking Kelvin–Helmholtz waves in clouds (billow clouds) formed over Laramie, Wyoming, USA. (Photo by Brooks Martner, NOAA Environmental Technology Laboratory.)

stability, the Coriolis force acts as a restoring force which makes a horizontally displaced air parcel in a horizontally sheared flow return to its original position and oscillate horizontally. However, under certain conditions, the horizontally displaced parcel may accelerate away from its original position. In this case, the atmosphere is called inertially unstable.

The criterion of inertial instability can be derived by considering a horizontally displaced air parcel embedded in a geostrophically balanced, inviscid basic flow on an f-plane. Assume the basic flow $(u, v) = (0, v_g(x))$, then the approximate equations of horizontal motion can be written as

$$\frac{Du}{Dt} = f(v - v_g) = f(M - \overline{M}),\qquad(7.5.1)$$

$$\frac{Dv}{Dt} = -fu - \frac{1}{\rho}\frac{\partial p}{\partial y} = -fu = -f\frac{Dx}{Dt},\qquad(7.5.2)$$

where $M = v + fx$ is the *absolute momentum* and $\overline{M} = v_g + fx$ is the *geostrophic absolute momentum* of the basic state of the fluid. In (7.5.2), we have used the geostrophic wind relation $-(1/\rho)\partial p/\partial y = fu_g = 0$. Now consider a parcel whose initial location is at $x = x_o$ and moves with the basic flow. If the air parcel is displaced a small distance (δx) to $x = x_o + \delta x$, the new meridional velocity can be obtained by integrating (7.5.2),

$$v(x_o + \delta x) = v_g(x_o) - f\delta x.\qquad(7.5.3)$$

The geostrophic wind at the new location can be approximated by

$$v_g(x_o + \delta x) = v_g(x_o) + \frac{\partial v_g}{\partial x}\delta x.\qquad(7.5.4)$$

Substituting (7.5.3) and (7.5.4) into (7.5.1) at $x_o + \delta x$ leads to,

$$\frac{D^2 \delta x}{Dt^2} + f\left(\frac{\partial v_g}{\partial x} + f\right)\delta x = 0, \qquad (7.5.5)$$

Three flow regimes are contained in the solutions of (7.5.5), depending upon the sign of $f(\partial v_g/\partial x + f)$. If $f(\partial v_g/\partial x + f) > 0$, the solution is $\delta x = A\exp[i\sqrt{f(\partial v_g/\partial x + f)}t] + B\exp[-i\sqrt{f(\partial v_g/\partial x + f)}t]$, and gives $Du/Dt = D^2(\delta x)/Dt^2 < 0$ for an eastward displacement ($\delta x > 0$). The parcel is therefore stable because it will accelerate back to its initial position. In fact, the air parcel will oscillate about the initial position, making the system *inertially stable*. If $f(\partial v_g/\partial x + f) = 0$, the solution becomes $\delta x = C + Dt$. In this situation, the air parcel will not accelerate and the system will be *inertially neutral*. If $f(\partial v_g/\partial x + f) < 0$, the solution becomes $\delta x = E\exp[-\sqrt{-f(\partial v_g/\partial x + f)}t] + F\exp[\sqrt{-f(\partial v_g/\partial x + f)}t]$, giving $Du/Dt = D^2(\delta x)/Dt^2 > 0$ for an eastward displacement. In this case, the air parcel will accelerate away from its initial position and the system becomes *inertially unstable*. Thus, the criteria for inertial stability, neutrality, and instability are:

$$f\frac{\partial \overline{M}}{\partial x} = f\left(\frac{\partial v_g}{\partial x} + f\right) \quad \begin{cases} > 0 \ \text{inertially stable} \\ = 0 \ \text{inertially neutral} \\ < 0 \ \text{inertially unstable} \end{cases} \qquad (7.5.6)$$

In the Northern Hemisphere ($f > 0$), the flow is inertially stable if the absolute vorticity of the basic flow, $\partial \overline{M}/\partial x = \partial v_g/\partial x + f$, is positive. When an air parcel moves eastward across the meridional geostrophic flow, a force imbalance is created between the Coriolis force and the pressure gradient force. The conservation of absolute momentum (M) gives the parcel less meridional velocity and a lower Coriolis force than that required to balance the local pressure gradient force at the new position. This imbalance forces the air parcel to accelerate back toward its original position, thus creating an inertially stable system.

The inertial instability criterion can also be generalized to two dimensions, i.e. $\partial v_g/\partial x - \partial u_g/\partial y + f = \zeta_{ga} < 0$ in the Northern Hemisphere, where ζ_{ga} is the absolute geostrophic vorticity. The argument is independent of horizontal coordinates. Thus, the absolute momentum has also been defined in Cartesian coordinates as $M \equiv u - fy$ or $M \equiv fy - u$. The absolute momentum corresponds to angular momentum in a circular motion. With these definitions, the inertial instability criterion is $-f\partial \overline{M}/\partial y = f(f - \partial u_g/\partial y) < 0$ or $f\partial \overline{M}/\partial y = f(f - \partial u_g/\partial y) < 0$, respectively. One special case of an inertially stable flow is the *inertial oscillation*, in which a parcel follows an anticyclonic circular trajectory in a quiescent atmosphere. In the atmosphere, the release of inertial instability is characterized by strong divergent anticyclonic flow. Anticyclonic relative vorticity can result from either strong anticyclonic shear or strong anticyclonic curvature, which sometimes exists equatorward of jet maxima or in subsynoptic-scale ridges in the midlatitudes, respectively. In the atmosphere, inertial instability occurs on scales of tens to hundreds of kilometers. Observations show that the flow is nearly always inertially stable for extratropical synoptic-scale systems. Low potential vorticity and a low Richardson number, which are typical of anticyclonically shearing jet streams embedded within strong frontal systems, are also directly related to inertial instability. Inertial instability thus provides a possible mechanism for turbulence development in the vicinity of jet entrance regions, fronts, regions of low Richardson number, regions with strong deformation, and vorticity (Stone 1966), and could cause clear air turbulence by promoting gravity wave genesis and breaking (Knox 1997). Observations also indicate that inertial instability exists in regions of low upper-tropospheric absolute vorticity above tornado outbreaks (MacDonald 1977). Based on a climatological study, it is found that: (a) a climatology of inertial instability at the 250-hPa level showed that inertially unstable conditions is most common in the subtropical upper troposphere; and (b) upper-level inertial instability is not present often enough to consider using it as a tool in forecasting severe weather (Schumacher and Schultz 2001). Observational and modeling studies also show that: (1) Rossby wave breaking in the tropics acts as a trigger for the onset of equatorial inertial instability; (2) inertial instability can enhance the outflow from mesoscale convective systems such as thunderstorms, tropical plumes, and hurricanes; and (3) the divergence–convergence couplets of inertial instability appear to determine the location of near-equatorial convection and the mean latitude of the ITCZ (Knox 2003).

7.6 Symmetric instability

In a statically (gravitational) stable atmosphere ($N^2 > 0$), an air parcel displaced vertically will return to its original position. Similarly, in an inertially stable atmosphere ($f \partial \overline{M} / \partial x = f(\partial v_g / \partial x + f) > 0$), an air parcel displaced horizontally will return to its original position. However, under certain conditions, it is possible for an air parcel to accelerate away from its original position if it is displaced along a slantwise path even if the atmosphere is both statically and inertially stable. The classical work on this area was called "the stability of a baroclinic circular vortex," which has been studied in the 1960s and earlier (Fjortooft 1950, Eliassen 1957, Ooyama 1966). This type of instability is also called *symmetric instability* (Kuo 1954, Stone 1966), and is named so because the argument is independent of the horizontal coordinates. Due to this reason, either $\partial / \partial x = 0$ or $\partial / \partial y = 0$ has been used to describe this state in the literature. In Section 7.5, we assumed $\partial / \partial y = 0$ and thus defined $M = v + fx$. Symmetric instability has also been referred to as *isentropic inertial instability* in the literature.

In a dynamic sense, static, potential, inertial, and symmetric instabilities are very closely related (e.g. Ooyama 1966; Emanuel 1983; Xu and Clark 1985). Each of the instabilities can be thought of as resulting from an unstable distribution of body forces acting on a fluid element. The responsible body forces are: the gravitational force for static and potential instabilities, the Coriolis force for inertial instability, and a combination of gravitational forces in the vertical and Coriolis forces in the horizontal for symmetric instability. Due to the combination of these two forces, symmetric instability occurs when the motion is slantwise.

7.6.1 Dry symmetric instability

Symmetric instability in a dry atmosphere can be analyzed by applying parcel theory. For convenience, consider a Boussinesq fluid in which the basic flow and disturbances are independent of the y direction, i.e. $\partial/\partial y = 0$, as was assumed in the discussion of inertial instability. The air parcel can now be viewed as a tube of air extending to infinity in the y direction. To demonstrate that symmetric instability may occur when the basic flow is both statically and inertially stable, we consider a basic state of the atmosphere as illustrated in Fig. 7.10. The basic flow is also assumed to be in hydrostatic and geostrophic balance, i.e. in thermal wind balance, as given by

$$f\frac{\partial v_g}{\partial z} = \frac{g}{\theta_o}\frac{\partial \bar{\theta}}{\partial x}, \tag{7.6.1}$$

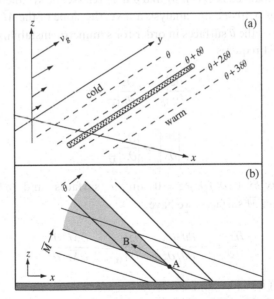

Fig. 7.10 Schematic of a mean state with symmetric instability. (a) A meridional, steady baroclinic flow. The unidirectional geostrophic wind v_g is in thermal wind balance. One may imagine an idealized broad front (dotted tube) that is aligned with the y-axis and located in between the cold and warm region. In this situation, v_g is the along-front wind. (Adapted after Emanuel 1994.) (b) The $\bar{M}$ and $\bar{\theta}$ surfaces on the x–z plane are tilted such that both of them increase upward and eastward. Displacement of an air parcel upward anywhere within the shaded area from point A is symmetrically unstable. (Adapted after Houze 1993, reproduced with permission from Elsevier.)

where $\bar{\theta}(x)$ and θ_o are the mean potential temperature and a constant reference potential temperature, respectively. Imagine that an idealized front is aligned with the y-axis and located between the cold and warm regions, as sketched in Fig. 7.10a. In this situation, v_g becomes the along-front wind. The $\overline{M}$ and $\bar{\theta}$ surfaces on the x–z plane are tilted such that both $\bar{\theta}$ and $\overline{M}$ increase upward and eastward (Fig. 7.10b). Since $\partial\bar{\theta}/\partial z > 0$, an air parcel displaced vertically from A will accelerate back to its origin. Analogously, since $\partial\overline{M}/\partial x > 0$, an air parcel displaced horizontally in the negative x direction from point A acquires a positive value of $(M - \overline{M})$ and is accelerated back toward A, according to (7.5.1).

If an air parcel is displaced slantwise upward anywhere within the shaded wedge in Fig. 7.10b, however, it will be subjected to an upward and leftward acceleration in the direction of its initial displacement since $\bar{\theta}$ decreases and $\overline{M}$ increases along the displacement AB. Thus, the air parcel will accelerate away from A. This is called *symmetric instability* (SI) or more precisely *dry symmetric instability*, and can be easily explained by viewing it as either dry static instability on an $\overline{M}$ surface (i.e. $\partial\bar{\theta}/\partial z|_{\overline{M}} < 0$) or inertial instability on an isentropic surface (i.e. $\partial\overline{M}/\partial x|_{\bar{\theta}} < 0$). Thus, any slantwise displacement within the wedge expanded between $\overline{M}$ and $\bar{\theta}$ may release the symmetric instability.

According to Fig. 7.10 and our analysis above, the slope of the $\overline{M}$ surfaces must be less than the slope of the $\bar{\theta}$ surfaces in order for symmetric instability to occur. Thus, the occurrence of SI requires

$$-\left.\frac{\partial z}{\partial x}\right|_{\bar{\theta}} > -\left.\frac{\partial z}{\partial x}\right|_{\overline{M}}, \text{ or} \qquad (7.6.2a)$$

$$\left|\frac{\partial z}{\partial x}\right|_{\bar{\theta}} > \left|\frac{\partial z}{\partial x}\right|_{\overline{M}}. \qquad (7.6.2b)$$

Since $\delta\bar{\theta} = (\partial\bar{\theta}/\partial x)\delta x + (\partial\bar{\theta}/\partial z)\delta z = 0$ on $\bar{\theta}$ surfaces and $\delta\overline{M} = (\partial\overline{M}/\partial x)\delta x + (\partial\overline{M}/\partial z)\delta z = 0$ on $\overline{M}$ surfaces, we have

$$\left.\frac{\partial z}{\partial x}\right|_{\bar{\theta}} = -\frac{\partial\bar{\theta}/\partial x}{\partial\bar{\theta}/\partial z}; \quad \left.\frac{\partial z}{\partial x}\right|_{\overline{M}} = -\frac{\partial\overline{M}/\partial x}{\partial\overline{M}/\partial z}. \qquad (7.6.3)$$

Substituting (7.6.3) into (7.6.2a) leads to

$$\frac{(\partial\bar{\theta}/\partial x)(\partial\overline{M}/\partial z)}{(\partial\bar{\theta}/\partial z)(\partial\overline{M}/\partial x)} > 1. \qquad (7.6.4)$$

Based on the thermal wind relation, (7.6.1), the definition of the buoyancy frequency in a Boussinesq fluid, $N^2 = (g/\theta_o)(\partial\bar{\theta}/\partial z)$, and on the definition of $\overline{M}$, the necessary condition for the dry symmetric instability can be obtained as

$$\frac{R_i}{f}\frac{\partial \overline{M}}{\partial x} < 1, \quad \text{or} \quad R_i < \frac{f}{\zeta_g + f} \tag{7.6.5}$$

where $R_i = N^2/(\partial v_g/\partial z)^2$ is the Richardson number. Thus, symmetric instability is favored by low static stability, strong basic state vertical wind shear, and anticyclonic relative vorticity. If no horizontal wind shear exists, the above criterion is reduced to $R_i < 1$. An alternative method to derive the criterion for dry symmetric instability is to consider a small displacement δx along a constant θ surface [i.e., $(\delta x)_\theta$] and to then use (7.5.1) and (7.5.2) with the assumption of thermal wind balance (7.6.1), which leads to

$$\frac{D^2(\delta x)_\theta}{Dt^2} + f^2\left[\frac{\zeta_g + f}{f} - \frac{1}{R_i}\right](\delta x)_\theta = 0. \tag{7.6.6}$$

The derivation of the above equation is similar to that of (7.5.5) for inertial instability. The necessary condition for symmetric instability to occur requires the term in square brackets of (7.6.6) to become negative, which leads to the criterion in (7.6.5). It can be shown that the critical (maximum) Richardson number for instability becomes $f/(\zeta_g + f)$.

The *geostrophic potential vorticity*, as defined by

$$q_g \equiv \frac{g}{\theta_o}(\boldsymbol{\omega}_g \cdot \nabla\theta) = (\zeta_g + f)N^2\left(1 - \frac{f}{\zeta_g + f}\frac{1}{R_i}\right), \tag{7.6.7}$$

becomes 0 when the Richchardson number becomes critical $(R_i = f/(\zeta_g + f))$. Hence symmetric instability occurs when $q_g < 0$. In (7.6.7), $\boldsymbol{\omega}$ is the total vorticity vector and the subscript g denotes the geostrophic value. The second equality is derived by expanding the first equality, dropping the y derivatives, and using the thermal wind balance (e.g., Emanuel and Raymond 1984).

Strictly speaking, symmetric instability should be determined by hydrostatically and geostrophically balanced basic states (Xu 1992). In observational studies, the geostrophically balanced basic state wind is often approximated by the total wind, and the hydrostatically balanced basic state $\bar{\theta}$ is often approximated by the observed θ. In assessing the condition for dry symmetric instability, replacing $\bar{\theta}$ by θ is generally a valid approximation, while replacing $\overline{M}$ by M is a poor approximation (Schultz and Schumacher 1999). Another complicated issue is the *partitioning problem* that occurs when the observed flow fields are partitioned into basic (geostrophic) and perturbation states.

In practice, the geostrophic wind is directly estimated from the observed total pressure or height field. This value is then subtracted from the total wind to obtain the perturbation wind. This implies that the perturbation pressure caused by the unbalanced part of the flow, such as static (gravitational) convection, symmetric instability, or gravity waves, is negligible (Schultz and Schumacher 1999). Purely symmetric instability theories consider only two-dimensional perturbations with their banded structures along the basic shear. The y-component of the Earth rotation ($2\Omega\cos\phi$; ϕ is the latitude) can significantly impact the growth rate of symmetric instability in the lower latitudes, because the x-component wind (u) can enhance or decrease the z-component acceleration through the $(2\Omega\cos\phi)u$ term in the momentum equation (Sun 1995). Other three-dimensional perturbations may also occur in symmetrically unstable flows (Jones and Thorpe 1992; Gu *et al.* 1998).

7.6.2 Moist symmetric instability

The addition of moisture and its associated effects, such as latent heat release, considerably complicates the theory of dry symmetric instability. Applying the parcel and layer theories to symmetric instability in a moist atmosphere (i.e. *moist symmetric instability*) leads to the necessary conditions for *conditional symmetric instability* (CSI) and *potential symmetric instability* (PSI) to occur, $\partial\bar{\theta}_e^*/\partial z|_{\overline{M}}<0$ and $\partial\bar{\theta}_e/\partial z|_{\overline{M}}<0$, respectively (Ooyama 1966; Bennetts and Hoskins 1979). The subscript $\overline{M}$ means along the $\overline{M}$ surface. This is analogous to the necessary conditions $\partial\bar{\theta}_e^*/\partial z<0$ and $\partial\bar{\theta}_e/\partial z<0$ for CI and PI to occur, respectively. Analogous to (7.6.2b), the necessary condition for CSI to occur can be derived as $|\partial z/\partial x|_{\overline{M}}<|\partial z/\partial x|_{\bar{\theta}_e^*}$. However, the most unstable mode has a larger wavelength when condensation occurs in the ascending branch only (Sun 1984).

Figure 7.11 illustrates the motion of an air tube in a simple, conditional-symmetrically unstable atmosphere with idealized $\bar{\theta}_e^*$ and $\overline{M}$ surfaces. The air tube extends from negative to positive infinities along the y direction and the basic atmospheric state is assumed to be in thermal wind balance. Above $z = H$, there is no vertical shear, so $\overline{M}$

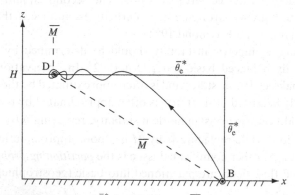

Fig. 7.11: Idealized configuration of $\bar{\theta}_e^*$ (solid line) and $\overline{M}$ (dashed line) surfaces on a x–z plane for a tube of air in a simple conditional-symmetrically unstable atmosphere. The basic state of the atmosphere is in thermal wind balance. Above $z = H$, the basic state flow is barotropic (no shear), thus the $\overline{M}$ surfaces are oriented vertically. The $\bar{\theta}_e^*$ surfaces are oriented vertically (horizontally) below (above) $z = H$. The tube of air extends from negative to positive infinity along the y direction (into the paper). The trajectory of the air tube is denoted by the solid curve with arrow. (Adapted after Emanuel 1983.)

surfaces are oriented vertically. The $\bar{\theta}_e^*$ surfaces are oriented vertically (horizontally) below (above) $z = H$. When the air tube is pushed upward and to the left, it accelerates in the same direction until it reaches the $\bar{\theta}_e^*$ surface at $z = H$. After reaching $z = H$, the air tube loses upward acceleration due to the lack of buoyancy. However, it continues to accelerate to the left, but oscillates vertically about the $\bar{\theta}_e^*$ surface until it reaches the $\bar{M}$ surface again. Once there, it will oscillate about the intersection between the $\bar{\theta}_e^*$ surface and the $\bar{M}$ surface. The trajectory of the air tube is denoted in the figure by the solid curve with the arrow.

Analogous to CAPE for conditional instability, the *slantwise convective available potential energy* (SCAPE) can be defined to calculate the total amount of kinetic energy converted from potential energy that is available from CSI. SCAPE is defined as the work done in displacing an air tube from its LFC to LNB along an $\bar{M}$ surface,

$$\text{SCAPE} = \int_{z_{\text{LFC}}}^{z_{\text{LNB}}} g\left(\frac{\theta_e^* - \bar{\theta}_e^*}{\bar{\theta}_e^*}\right)(\text{d}z)_{\bar{M}}, \tag{7.6.8}$$

Since the work done in displacing the air tube from its initial point to the final point is path independent, however, we can also express (7.6.8) along any $\bar{s}$ surface (e.g., the solid curve with arrow in Fig. 7.11),

$$\text{SCAPE} = \int_{x_{\text{LFC}}}^{x_{\text{LNB}}} f(M - \bar{M})(\text{d}x)_{\bar{s}}, \tag{7.6.9}$$

or in general as

$$\text{SCAPE} = \int_{x_{\text{LFC}}}^{x_{\text{LNB}}} f(M - \bar{M})\text{d}x + \int_{z_{\text{LFC}}}^{z_{\text{LNB}}} g\left(\frac{\theta_e^* - \bar{\theta}_e^*}{\bar{\theta}_e^*}\right)\text{d}z. \tag{7.6.10}$$

If the atmosphere is barotropic, then SCAPE is identical to CAPE since the $\bar{M}$ surfaces are vertical. Similarly, the *slantwise convective inhibition* (SCIN), the work needed to lift an air tube adiabatically and pseudoadiabatically from its original level to LFC for CSI to occur, can be defined as

$$\text{SCIN} = \int_{z_i}^{z_{\text{LFC}}} g\left(\frac{\bar{\theta}_e^* - \theta_e^*}{\bar{\theta}_e^*}\right)(\text{d}z)_{\bar{M}} \tag{7.6.11}$$

The necessary condition for dry symmetric instability in terms of potential vorticity, $q_g < 0$ can be extended to

$$q_g^* = \text{MPV}_g^* = \frac{g}{\theta_o}\left(\boldsymbol{\omega}_{\text{ga}} \cdot \nabla\theta_e^*\right) < 0, \tag{7.6.12}$$

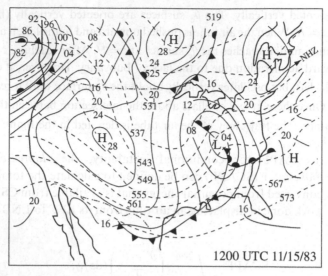

Fig. 7.12 Surface pressure (in hPa) and 1000–500 hPa thickness (dm) at 1200 UTC November 15, 1983, when a cyclone was developing over the Ohio Valley. (After Emanuel 1988.)

where $q_g^*(= \mathrm{MPV}_g^*)$ is the *saturated geostrophic potential vorticity*. Therefore, using the three-dimensional form of MPV_g^* to assess CSI does not require strict adherence to the two-dimensionality assumption as using the $\overline{M} - \overline{\theta}_e^*$ relationship for cross sections.

In summary, based on parcel theory, the necessary conditions for conditional symmetric instability to occur is

$$\partial \overline{\theta}_e^* / \partial z|_{\overline{M}} < 0, \tag{7.6.13a}$$

$$\partial \overline{M} / \partial z|_{\overline{\theta}_e^*} < 0, \tag{7.6.13b}$$

$$\Gamma_s|_{\overline{M}} < \gamma|_{\overline{M}} < \Gamma_d|_{\overline{M}}, \tag{7.6.13c}$$

$$|\partial z / \partial x|_{\overline{M}} < |\partial z / \partial x|_{\overline{\theta}_e^*}, \text{ or} \tag{7.6.13d}$$

$$q_g^* \equiv \mathrm{MPV}_g^* = g(\boldsymbol{\omega}_{ga} \cdot \nabla \theta_e^*) < 0, \tag{7.6.13e}$$

provided that the air tube has been lifted above its LFC and the environment is conditionally and inertially stable.

Figure 7.12 shows the surface pressure and 1000–500 hPa thickness associated with a cyclone developing over the Ohio Valley at 1200 UTC November 15, 1983. The cyclone was weaker and in Tennessee. It then moved to New England while

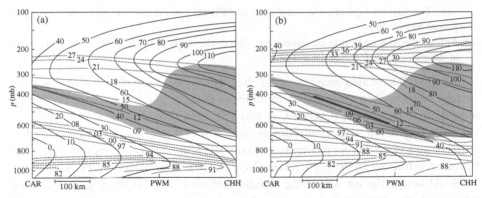

Fig. 7.13: (a) $\overline{M}$ (solid; m s^{-1}) and $\overline{\theta}_e^*$ (dotted; K, with first digit omitted) surfaces on a cross section from Caribou, Maine (CAR) through Portland, Maine (PWM) to Chatham, Massachusetts (CHH) (shown in Fig. 7.12), constructed from rawinsonde observations at 0000 UTC November 16, 1983. The region with $\Gamma_s|_{\overline{M}} \leq \gamma|_{\overline{M}}$ is shaded, within which a large portion is nearly moist adiabatic along $\overline{M}$ surfaces. (b) Same as (a) except for $\overline{\theta}_e$ instead of $\overline{\theta}_e^*$. The heavy line descending through 500 hPa between CAR and PWM shows the partial path of the $\overline{M}$-surface aircraft sounding. (After Emanuel 1988.)

intensifying. An aircraft-observed sounding along $\overline{M}$ surfaces was very close to moist-neutral, down to about 650 hPa. However, the vertical sounding indicates the atmosphere is statically (gravitationally) stable. Thus, it appears that the atmosphere tends to prefer slantwise neutrality. The $\overline{M}$ and $\overline{\theta}_e^*$ surfaces constructed from 0000 UTC November 16 soundings are plotted on the cross section from Caribou, Maine (CAR) through Portland, Maine (PWM) to Chatham, Massachusetts (CHH) as shown in Fig. 7.13. Regions with a lapse rate along $\overline{M}$ surfaces that is greater than or equal to moist adiabatic, i.e. $\Gamma_s|_{\overline{M}} \leq \gamma|_{\overline{M}}$, are shaded in the figure. In fact, a large portion of the shaded region follows moist adabatic lapse rates along $\overline{M}$ surfaces. Note that slantwise convective neutrality includes both conditional symmetric neutrality (Fig. 7.13a) and potential symmetric neutrality (Fig. 7.13b). Similarly, potential symmetric instability (PSI) can be assessed by replacing θ_e^* with θ_e in (7.6.13).

A closed form analytical solution was obtained for CSI modes in Xu (1986), which revealed and quantified several important features of CSI that are absent or modified in dry SI and purely moist SI theories. The major features are briefly summarized below:

(a) The moist updraft tends to be very narrow (but the dry subsidence must be sufficiently wide) to obtain the maximum growth rate in the inviscid limit, because the CSI circulation gains (loses) energy from (to) the basic state in the moist updraft (dry subsidence) region.
(b) The slope of the moist updraft is along the inertial acceleration surface which is related to the growth rate. Since the growth rate is related to the width of the moist updraft, the slope and width of the moist updraft are functions of each other and also functions of the growth

rate. The relationships of these functions are different from those derived from the dry (or moist) SI theories.

(c) As the inviscid growth rate increases, the width of the moist updraft decreases and the slope of the moist updraft increases.

CSI and PSI have also been misused in the literature occasionally, so readers should exercise caution when using these terminologies (Schultz and Schumacher 1999). It is important to distinguish between CSI and PSI, even though both tend to produce slantwise convection when conditions are met. For example, Fig. 7.7 shows θ_e^* and θ_e surfaces on a vertical cross section from Dauphin, Manitoba to Minneapolis-St. Paul based on a forecast by Rapid Update Cycle (RUC), valid at 0000 UTC January 3, 1998. Although the air above the frontal surface is conditionally unstable since $\partial \theta_e^*/\partial z < 0$ (Fig. 7.7a), it is potentially stable since $\partial \bar{\theta}_e/\partial z > 0$ (Fig. 7.7b). This also happens along the $\overline{M}$ surfaces, i.e. $\partial \bar{\theta}_e^*/\partial z |_{\overline{M}} < 0$ and $\partial \bar{\theta}_e/\partial z |_{\overline{M}} > 0$, which indicates

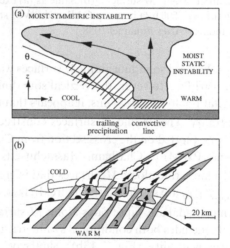

Fig. 7.14 (a) A conceptual model of upscale convective-symmetric instability in a midlatitude mesoscale convective system. The bold curve encloses the cloud (shaded). The arrows represent the direction of circulation. The labeled solid line represents the orientation of typical potential-temperature contours in the cool air. The hatched area is proportional to precipitation intensity. The upright updraft caused by the release of static (gravitational) instability (CI or PI) is followed by slantwise convection, caused by the release of symmetric instability (CSI or PSI), which then produces downdrafts that descend following sloping (dry and/or moist) isentropes. (Adapted after Seman 1992 and Schultz and Schumacher 1999.) (b) Conceptual elevator–escalator model of warm-frontal ascent. The warm southerly airstream (warm conveyor – flat shaded arrows) rises over the cold easterly polar airstream (cold conveyor – tubular dashed arrow). Mesoconvective strong sloping ascent (the elevator, solid arrows; labeled "1") and convective clouds (shaded with white anvils) are shown at regular intervals between regions of gentler slantwise ascent (the escalator, labeled "2"). The elevator is triggered by PI, while the escalator is triggered by PSI. (Adapted after Neiman *et al.* 1993.)

that CSI may occur, but not PSI. At saturation, the vertical gradient of θ_e, θ_e^* and θ_w are equivalent in this case, therefore, CI and PI, as well as CSI and PSI are equivalent.

In the real atmosphere, CI/PI and CSI/PSI may coexist, creating the *convective-symmetric instability*, which is responsible for producing some mesoscale convective systems (e.g. Jascourt *et al.* 1988). Two mechanisms, namely *upscale development* and *downscale development*, have been proposed to explain the convective systems triggered by convective-symmetric instability. For the former mechanism, small-scale moist static (gravitational) convection develops first, followed by mesoscale banded organization of clouds due to the release of symmetric instability as the environment stabilizes gravitationally (Xu 1986). Figure 7.14a shows a conceptual model of the upscale convective-symmetric instability in a midlatitude mesoscale convective system. The upright updraft caused by static (gravitational) instability (CI or PI) is followed by slantwise convection associated with the release of symmetric instability (CSI or PSI), which then produces downdrafts that follow sloping (dry and/or moist) isentropes on descent. On the other hand, a downscale convective-symmetric instability may also occur, such as the elevator–escalator conceptual model of a warm-frontal ascent shown in Fig. 7.14b. The mesoscale circulation is composed of regions of (1) strong sloping ascent tilted about $45°$ to the horizontal with a width of about 10 km (i.e. elevator triggered by PI), and (2) weaker regions of gentler slantwise ascent tilted about $10°$ from the horizontal with a width of about 15 km (i.e. escalator triggered by PSI).

The criteria for static (gravitational) instability, conditional instability, potential instability, inertial instability, symmetric instability, conditional symmetric instability, and potential symmetric instability are summarized in Table 7.1.

7.7 Baroclinic instabiltity

A *baroclinic atmosphere* is one for which the density is a function of both the temperature and the pressure and for which the vertical shear of geostrophic wind is supported by the horizontal temperature gradient via the thermal wind balance. In contrast to this, in a *barotropic atmosphere* the density is a function of pressure only or say, there is no horizontal temperature variation on isobaric surfaces. In a baroclinic atmosphere, available potential energy is stored in the mean flow associated with the horizontal temperature gradient on isobaric surfaces, which may be converted to kinetic energy. The available potential energy stored in a baroclinic atmosphere is analogous to the potential energy stored in a water system where a divider separates two sides of the water with different temperatures. When the divider is lifted, the colder (denser) fluid will flow toward the warmer (less dense) fluid near the surface. Similarly, under certain conditions, the available potential energy stored in a baroclinic atmosphere will be released and converted into perturbation available potential energy and then into perturbation kinetic energy. The release of perturbation kinetic energy may cause spontaneous growth of small disturbances. This particular atmospheric state is then referred to as having *baroclinic instability*. Baroclinic instability is

considered as the major mechanism responsible for midlatitude cyclogenesis and frontogenesis. For example, the lee cyclogenesis mechanisms discussed in Chapter 5 are
based on either baroclinic instability or baroclinic lee wave generation. Fronts usually
evolve from the westerly basic flow via some manifestation of baroclinic instability,
which leads to their association with developing baroclinic wave disturbances.

Conversion of available potential energy into kinetic energy requires that
$\overline{(v\theta)}\,U_z > 0$ on average, based on Term 4 of (7.1.5). That is, the meridional heat flux
$\overline{(v\theta)}$ must be positive (northward) for a forward westerly shear flow ($U_z > 0$). Since the
perturbation meridional velocity and perturbation potential temperature are related
to the perturbation pressure as

$$v' = \left(\frac{1}{f\rho_0}\right)\frac{\partial p'}{\partial x}; \quad \text{and} \quad \theta' = \left(\frac{\theta_0}{g\rho_0}\right)\frac{\partial p'}{\partial z}, \tag{7.7.1}$$

it can be derived that

$$\frac{\partial E_T}{\partial t} \propto \frac{-1}{\rho_0 N^2}\int_0^{z_T}\overline{\left[\frac{\partial z}{\partial x}\right]_p\left(\frac{\partial p'}{\partial z}\right)^2}U_z\,dz, \tag{7.7.2}$$

based on (7.1.5). Thus, converting the available potential energy to perturbation
kinetic energy requires an upshear tilt of the constant phase of perturbation pressure,
i.e. $[\partial z/\partial x]_p < 0$, on average.

The dynamics of baroclinic instability can be elucidated using the Eady model
(1949) which assumes: (1) quasi-geostrophic approximation, (2) Boussinesq approximation, (3) f-plane approximation, (4) constant vertical shear of the basic wind (i.e.,
U_z is constant), and (5) rigid upper and lower boundaries at $z = 0$ and H, the perturbation potential vorticity equation and thermodynamic energy equation can be derived
from (5.5.6) and (5.5.9),

$$\left(\frac{\partial}{\partial t}+U\frac{\partial}{\partial x}\right)\left[\nabla^2 p'+\frac{f^2}{N^2}\frac{\partial^2 p'}{\partial z^2}\right]=0, \quad \text{and} \tag{7.7.3}$$

$$\left(\frac{\partial}{\partial t}+U\frac{\partial}{\partial x}\right)\frac{\partial p'}{\partial z}-U_z\frac{\partial p'}{\partial x}+\rho_0 N^2 w'=0 \quad \text{at} \quad z=0,H, \tag{7.7.4}$$

where p' is the perturbation pressure and ρ_0 is a constant reference density under the
Boussinesq approximation. As mentioned in Chapter 5, the term inside the square
bracket of (7.7.3) is the perturbation potential vorticity. To solve the problem, we
assume a wave-like solution as

$$p'(x,y,z,t) = P(z)\cos(ly)\exp[ik(x-ct)], \tag{7.7.5}$$

where $P(z)$ and c are the wave amplitude and phase speed, respectively, and are
complex numbers. Substituting (7.7.5) into (7.7.3) and (7.7.4) yields

$$\frac{\partial^2 P}{\partial z^2} - m^2 P = 0, \tag{7.7.6}$$

$$\frac{\partial P}{\partial z} - \left(\frac{1}{z - c/U_z}\right) P = 0 \quad \text{at} \quad z = 0, H, \tag{7.7.7}$$

where $m^2 = (k^2 + l^2)N^2/f^2$ is the square of the vertical wave number. The general solution of (7.7.6) can be written as

$$P(z) = A \sinh mz + B \cosh mz, \tag{7.7.8}$$

where A and B are constant coefficients. Applying the upper and lower boundary conditions (7.7.7) to (7.7.8) leads to

$$mcA + U_z B = 0, \tag{7.7.9a}$$

$$\begin{aligned}[m(H - c/U_z)\cosh mH - \sinh mH]A + \\ [m(H - c/U_z)\sinh mH - \cosh mH]B = 0.\end{aligned} \tag{7.7.9b}$$

A nontrivial solution requires the determinant of the coefficients of A and B to vanish, which gives

$$c = \frac{U_z H}{2}\left[1 \pm \left(1 - \frac{4\cosh mH}{mH \sinh mH} + \frac{4}{m^2 H^2}\right)^{1/2}\right]. \tag{7.7.10}$$

In general, the phase speed c is a complex number, which can be expressed as $c = c_r + ic_i$, and this allows us to rewrite (7.7.5) as

$$p'(x, y, z, t) = P(z)\cos(ly)\exp[ik(x - c_r t)]\exp(kc_i t). \tag{7.7.11}$$

When $c_i > 0$, disturbances grow exponentially with time and the flow becomes baroclinically unstable. On the other hand, when $c_i < 0$, disturbances decay exponentially and the flow is baroclinically stable. Thus, from (7.7.10) the necessary condition for baroclinic instability to occur is

$$1 - \frac{4\cosh mH}{mH \sinh mH} + \frac{4}{m^2 H^2} < 0. \tag{7.7.12}$$

The critical condition for baroclinic instability to occur may be found by setting instead the left-hand side of (7.7.12) to 0, which yields (Eady 1949)

$$\left[\frac{m_c H}{2} - \tanh\left(\frac{m_c H}{2}\right)\right]\left[\frac{m_c H}{2} - \coth\left(\frac{m_c H}{2}\right)\right] = 0, \tag{7.7.13}$$

where m_c represents the critical vertical wave number. This leads to

$$(k^2 + l^2) < (mf/N)^2 \approx 5.76/L_R^2, \tag{7.7.14}$$

Mesoscale instabilities

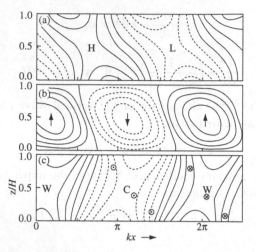

Fig. 7.15 A sketch of the vertical structure of the most unstable Eady wave. (a) Perturbation pressure; H and L denote high and low pressures, respectively (positive solid and negative dotted). (b) Vertical velocity; up and down arrows denote maximum upward and downward motion, respectively. (c) Perturbation potential temperature; W and C denote warmest and coldest regions, respectively. Northward and southward meridional velocities are denoted by circles with cross and dot, respectively. (Adapted after Holton 2004, reproduced with permission from Elsevier.)

where $L_R \equiv NH/f$ is the Rossby radius of deformation for a continuously stratified fluid (but the fluid is bounded between $z = 0$ and $z = H$), which has a value of about 1000 km for a typical atmosphere with $N = 0.01\ \mathrm{s}^{-1}$, $H = 10\ \mathrm{km}$, and $f = 10^{-4}\ \mathrm{s}^{-1}$. For waves with $k = l$, the wavelength of maximum growth rate (for which kc_i is a maximum) is found to be $L_{max} = 2\sqrt{2}\pi L_R/(Hm_{max}) \cong 5500$ km. However, in the real atmosphere with non-linearity and latent heating present, the wavelength of the maximum growth rate may be shorter than that predicted by the Eady model. Substituting this m_{max} into (7.7.8) and eliminating B with A by using (7.7.9), one may obtain the vertical structure of the most unstable wave, as sketched in Fig. 7.15. Both the constant phases of perturbation pressure and vertical motion tilt upshear (westward) with height (Figs. 7.15a–b), a condition that is required in the conversion of available potential energy into perturbation kinetic energy, (7.7.2). The warmest and coldest regions, however, tilt downshear (eastward) with height. At all levels, air moving poleward is generally warmer than air moving equatorward (Fig. 7.15c), which results in a net poleward (positive) heat flux $(\overline{v\theta})$.

References

Batchelor, G. K., 1967. *An Introduction to Fluid Dynamics*. Cambridge University Press.
Bennetts, D. A. and B. J. Hoskins, 1979. Conditional symmetric instability – a possible explanation for frontal rainbands. *Quart. J. Roy. Meteor. Soc.*, **105**, 945–62.

Bjerknes, J., 1938. Saturated-adiabatic ascent of air through dry-adiabatically descending environment. *Quart. J. Roy. Meteor. Soc.*, **64**, 325–30.

Bolton, D., 1980. Application of the Miles theorem to forced linear perturbations. *J. Atmos. Sci.*, **37**, 1639–42.

Bryan, G. H. and J. M. Fritsch, 2000. Moist absolute instability: The sixth static stability state. *Bull. Amer. Meteor. Soc.*, **81**, 1207–30.

Chandrasekhar, S., 1961. *Hydrodynamic and Hydromagnetic Stability*. Oxford University Press (reprinted by Doves Publications 1981).

Darkow, G. L., 1986. Basic thunderstorm energetics and thermodynamics. In *Thunderstorm Morphology and Dynamics*, E. Kessler (ed.), University of Oklahoma Press, 59–72.

Das, P., 1979. A non-Achimedean approach to the equation of convection dynamics. *J. Atmos. Sci.*, **36**, 2183–90.

Davies-Jones, R., 2003. An expression for effective buoyancy in surroundings with horizontal density gradients. *J. Atmos. Sci.*, **60**, 2922–5.

Doswell, C. A. and P. M. Markowski, 2004. Is buoyancy a relative quantity? *Mon. Wea. Rev.*, **132**, 853–63.

Drazin, P. G. and W. H. Reid, 1981. *Hydrodynamic Stability*. Cambridge University Press.

Durran, D. R. and J. B. Klemp, 1982. On the effects of moisture on the Brunt-Vaisala frequency. *J. Atmos. Sci.*, **39**, 2152–8.

Eady, E. T., 1949. Long waves and cyclone waves. *Tellus*, **1**, 33–52.

Emanuel, K. A., 1983. On assessing local conditional symmetric instability from atmospheric soundings. *Mon. Wea. Rev.*, **111**, 2016–33.

Emanuel, K. A., 1988. Observational evidence of slantwise convective adjustment. *Mon. Wea. Rev.*, **116**, 1805–16.

Emanuel, K. A., 1994. *Atmospheric Convection*. Oxford University Press.

Emanuel, K. A. and D. J. Raymond, 1984. Dynamics of Mesoscale Weather Systems, J. B. Klemp (ed.), NCAR.

Fjortooft, R., 1950. Application of integral theorems in deriving criteria of stability for laminar flows and for the baroclinic circular vortex. *Geofys. Publikasjoner*, **17** (6).

Gu, W., Q. Xu, and R. Wu, 1998. Three-dimensional instability of nonlinear viscous symmetric circulations. *J. Atmos. Sci.*, **55**, 3148–58.

Holton, J. R., 2004. *Introduction to Dynamic Meteorology*. 4th edn., Elsevier Academic Press, Inc.

Houze, R. A. Jr., 1993. *Cloud Dynamics*. Academic Press.

Howard, L. N., 1961. Note on a paper of John Miles. *J. Fluid Mech.*, **10**, 509–12.

Jascourt, S. D., S. S. Lindstrom, C. J. Seman, and D. D. Houghton, 1988. An observation of banded convective development in the presence of weak symmetric stability. *Mon. Wea. Rev.*, **116**, 175–91.

Jones, S. C. and A. J. Thorpe, 1992. The three-dimensional nature of "symmetric" instability. *Quart. J. Roy. Meteor. Soc.*, **118**, 227–58.

Kessler, E., 1985. Severe weather. In *Handbook of Applied Meteorology*, D. Houghton (ed.), John Wiley and Sons, 133–204.

Kirshbaum, D. J. and D. R. Durran, 2004. Factors governing cellular convection in orographic precipitation. *J. Atmos. Sci.*, **61**, 682–98.

Knox, J. A.: 1997. Possible mechanisms of clear air turbulence in strongly anticyclonic flows. *Mon. Wea. Rev.*, **125**, 1251–9.

Knox, J. A., 2003: Inertial Instability. *Encyclopedia of Atmospheric Sciences*, J. R. Holton *et al.* (eds.), 1004–13, Academic Press.

Kuo, H. L., 1954. Symmetric disturbances in a thin layer of fluid subject to a horizontal temperature gradient and rotation. *J. Meteor.*, **11**, 399–411.

Lalas, D. P. and F. Einaudi, 1974. On the correct use of the wet adiabatic lapse rate in the stability criteria of a saturated atmosphere. *J. Appl. Meteor.*, **13**, 318–24.

Lin, Y.-L. and H.-Y. Chun, 1993. Structures of dynamically unstable shear flows and their implications for shallow internal gravity waves. *Meteor. Atmos. Phys.*, **52**, 59–68.

MacDonald, A. E., 1977. On a type of strongly divergent steady state. *Mon. Wea. Rev.*, **105**, 771–85.

Miles, J. W., 1961. On the stability of heterogeneous shear flow. *J. Fluid Mech.*, **10**, 496–508.

Moncrieff, M. W., 1978. The dynamical structure of two-dimensional steady convection in constant vertical shear. *Quart. J. Roy. Meteor. Soc.*, **104**, 543–567.

Neiman, P. J., M. A. Shapiro, and L. S. Fedor, 1993. The life cycle of an extratropical marine cyclone. Part II: Mesoscale structure and diagnostics. *Mon. Wea. Rev.*, **121**, 2177–99.

Ooyama, K., 1966. On the instability of baroclinic circular vortex: a sufficient criterion for instability. *J. Atmos. Sci.*, **23**, 43–5.

Rogers, R. R. and M. K. Yao, 1989. *A Short Course in Cloud Physics*. 3rd edn., Pergamon Press.

Rotunno, R. and J. B. Klemp, 1985. On the rotation and propagation of simulated supercell thunderstorms. *J. Atmos. Sci.*, **42**, 271–92.

Schultz, D. M. and P. N. Schumacher, 1999. The use and misuse of conditional symmetric instability. *Mon. Wea. Rev.*, **127**, 2709–32; Corrigendum, **128**, 1573.

Schultz, D. M., P. N. Schumacher, and C. A. Doswell III, 2000. The intricacies of instabilities. *Mon. Wea. Rev.*, **128**, 4143–8.

Schumacher, R. S. and D. M. Schultz, 2001. Inertial instability: climatology and possible relationship to severe weather predictability. 9th Conf. on Mesoscale Processes, Amer. Meteor. Soc., Fort Lauderdale, FL, 30 July–2 August.

Seman, C. J., 1992. On the role of nonlinear convective-symmetric instability in the evolution of a numerically simulated mesoscale convective system. Preprints, Fifth Conf. on Mesoscale Processes, Atlanta, GA, Amer. Meteor. Soc., 282–7.

Sherwood, S. C., 2000. On moist instability. *Mon. Wea. Rev.*, **128**, 4139–42.

Smith, R. B., 1984. A theory of lee cyclogenesis. *J. Atmos. Sci.*, **41**, 1159–68.

Stone, P. H., 1966. On non-geostrophic baroclinic stability. *J. Atmos. Sci.*, **23**, 390–400.

Sun, W.-Y., 1984. Rainbands and symmetric instability. *J. Atmos. Sci.*, **41**, 3412–26.

Sun, W.-Y., 1995. Unsymmetrical symmetric instability. *Quart. J. Roy. Meteor. Soc.* **121**, 419–31.

Xu, Q., 1986. Conditional symmetric instability and mesoscale rainbands. *Quart. J. Roy. Meteor. Soc.*, **112**, 315–34.

Xu, Q., 1992. Formation and evolution of frontal rainbands and geostrophic potential vorticity anomalies. *J. Atmos. Sci.*, **49**, 629–48.

Xu, Q. and J. H. E. Clark, 1985. The nature of symmetric instability and its similarity to convective and inertial instability. *J. Atmos. Sci.*, **42**, 2880–3.

Problems

7.1 Derive (7.2.11).

7.2 Derive (7.2.35).

7.3 Show that the buoyancy force, $g[(\bar{\rho} - \rho)/\rho]$ may be approximated by $g[(\theta - \bar{\theta})/\bar{\theta}]$.

7.4 Considering a fluid system controlled by (7.3.4), do the following.

    (a) Assuming $N = 0.01\ \text{s}^{-1}$, find the vertical displacement of an air parcel (η) after 10 minutes if the initial vertical displacement is 0 m and the initial vertical velocity is $1\ \text{m s}^{-1}$.

    (b) Find the vertical displacement of an air parcel after 2 minutes for an atmosphere with $N^2 = -10^{-6}\ \text{s}^{-2}$, assuming $\eta = 5\ \text{m}$ and $w = 2\ \text{m s}^{-1}$ at $t = 0\ s$.

    (c) Sketch the vertical displacement of the air parcels versus time and discuss the properties of the parcel motion in (a) and (b).

7.5 Prove that the positive area (PA) defined in (7.3.27) (the first equality) is equivalent to the CAPE defined in (7.3.26) with $z_i = z_{\text{LFC}}$.

7.6 Prove the necessary condition for symmetric instability from (7.6.4),

7.7 Following the derivation of (7.5.5) for inertial instability, derive (7.6.6) for symmetric instability. (Hint: Approximate v_g by $v_{g0} + (\partial v_g / \partial z)_o \delta z + (\partial v_g / \partial x)_o \delta x$ and use $d\bar{\theta} = 0 = (\partial \bar{\theta} / \partial x) \delta x + (\partial \bar{\theta} / \partial z) \delta z$ for displacement along a constant $\bar{\theta}$ surface.

7.8 Derive (7.7.2).

8

Isolated convective storms

Isolated convective storms are generally considered to be cumulonimbus clouds that produce thunder and lightning, heavy rainfall, wind gusts, and occasionally large hail and tornadoes. Understanding the dynamics of isolated convective storms is important because these storms form the building blocks of much more complicated mesoscale convective systems, such as squall lines and mesoscale convective complexes. Aside from their hazards to society, cumulonimbus clouds also play an important role in providing needed rainfall to many regions of the Earth, participating in the general circulation by transporting moisture and sensible and latent heat to the upper troposphere and lower stratosphere, and composing a significant part in the radiative budgets of the atmosphere.

Various methods have been proposed to classify different storm types according to their internal structure, rainfall intensity, potential severity, longevity, and propagation properties. In this text, we adopt standard nomenclature and categorize them as single-cell, multicell and supercell storms. The ordinary, *single-cell storm* consists of only one convective cell, which is short-lived and often associated with weak vertical shear. When a cluster of single cells at various stages exists simultaneously within a storm, it is called a *multicell storm*. Multicell storms normally exist in an environment with moderate shear, and have longer lifetimes than single-cell storms since new cells continue to develop along the gust front as the older cells are dissipating. A *supercell storm* consists of a single rotating updraft, which often exists in an environment with strong shear and has a lifetime much longer than a single-cell storm, often lasting for several hours. Supercell thunderstorms produce the majority of severe weather.

Since the characteristics of these three types of isolated convective storms have been discussed in detail in several textbooks and review articles on cloud dynamics, this chapter will focus on the basic dynamics of convective storms and important recent findings in the field.

8.1 Dynamics of single-cell storms and downbursts

Compared to multicell and supercell storms, the dynamics of single-cell storms are better understood partially due to their less complicated characteristics and the wealth

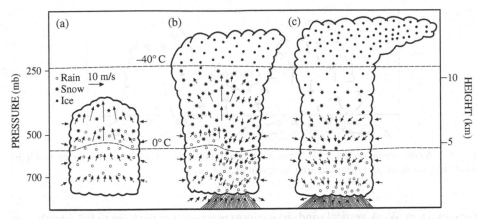

Fig. 8.1 Three stages for the life cycle of a single-cell storm: (a) developing stage, (b) mature stage, and (c) dissipation stage. (Adapted after Byers and Braham 1949.)

of information from earlier field observations. The basic characteristics of a single-cell storm may be summarized as the following:

(1) The storm is composed of one single updraft or updraft–downdraft pair.
(2) The updraft may penetrate the tropopause and have a horizontal scale on the order of several kilometers.
(3) The environmental vertical wind shear is weak, often smaller than $10 \, \mathrm{m \, s^{-1}}$ in the lowest 4 km.
(4) The storm normally moves with the mean wind in the lower to middle troposphere.
(5) The storm has a lifetime on the order of 30 minutes.
(6) The storm is usually not strong enough to produce severe weather, but is occasionally accompanied by thunder and lightning.

The structure and life cycle of a single-cell storm are sketched in Fig. 8.1. A single-cell storm is often accompanied by thunder and lightning and is referred to as an *air mass thunderstorm*. This type of thunderstorm was the target of the Thunderstorm Project, which was held over Florida and Ohio during the late 1940s. Three distinctive stages may be identified for a single-cell storm: (a) *developing stage*, (b) *mature stage*, and (c) *dissipating stage*. The *developing stage* is characterized by a towering cumulus cloud, which consists entirely of a warm, strong updraft (Fig. 8.1a). This updraft develops in a weak shear environment and its velocity may exceed $10 \, \mathrm{m \, s^{-1}}$. Entrainment occurs at the lateral cloud boundaries. At this stage, precipitation starts to form as raindrops or ice particles in the upper portion of the cumulus cloud, however no significant rainfall occurs in the subcloud layer. During the *mature stage*, the cloud continues to grow and precipitation particles begin to fall below the cloud base (Fig. 8.1b). Evaporation from falling precipitation particles cools the unsaturated air below the cloud base, thus forming a cold pool which, hydrostatically, leads to a mesoscale region of high pressure (*mesohigh*) near the surface. The maximum vertical velocity of the storm is located at the middle of the cloud. Detrainment occurs above that level while divergence occurs near

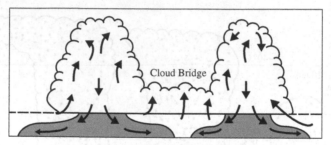

Fig. 8.2 A conceptual model of cloud merging through downdraft interaction in the case of light wind and weak shear. (Adapted after Simpson *et al.* 1980.)

the cloud top. Weak vertical wind shear causes precipitation particles to fall directly into the updraft, which results in a downdraft that effectively shuts off the updraft. This triggers the *dissipation stage*.

Single-cell storms are usually not strong and do not last long enough to produce severe weather. The Thunderstorm Project observations also found that only 20% of the condensed water in the updraft actually reaches the ground. This means that on average the *precipitation efficiency*, i.e. the ratio of total precipitation to total available moisture of a cloud system, for air mass thunderstorms is low. On the other hand, the precipitation efficiency of a squall line may in average reach as high as 50% due to less entrainment and a longer lifetime (Newton 1966). Larger precipitation efficiency is often associated with: (a) a weaker wind shear, (b) a longer lifetime of a cloud system, (c) a lower cloud base, (d) a larger cloud base area, (e) a larger cloud base mixing ratio, and (f) a smaller CIN (e.g., Market *et al.* 2003).

The most intense and persistent convective clouds, as has been well documented, are often associated with the merging of two or more adjacent convective cells. *Cloud merging* may increase cloud size significantly, resulting in more rainfall than in unmerged cells. Cumulus downdrafts and their associated cold outflow have often been proposed to play a critical role in the cloud merging process (Fig. 8.2). The downdrafts produce cold outflows, which approach each other and collide to form a cloud bridge. New towers then surge upward from the bridge to fill the gap. The process of cloud bridge formation often precedes the radar echo merger, and therefore cannot be detected through radar echo analysis. In addition to the cloud merging process, cold outflow also plays an essential role in the generation of new cells in multicell storms and the generation, maintenance, and propagation of squall lines, as will be discussed in the rest of this chapter and in Chapter 9.

Almost all intense convective storms are able to produce strong horizontal winds at the ground level. A *downburst* refers to a large area of damaging winds induced by a strong downdraft on or near the ground. The severe wind associated with downbursts and gust fronts is often referred to as a *straight-line wind* in order to distinguish it from the twisting nature of the damaging winds associated with tornadoes. Normally, a

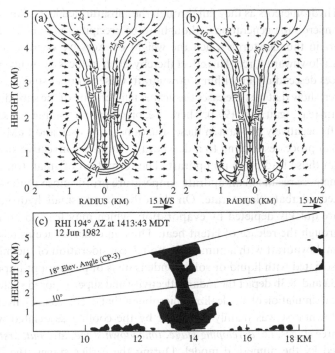

Fig. 8.3 The numerically simulated radar reflectivity and superimposed vector wind field for an idealized microburst that occurred in Denver on July 14, 1982 at (a) 22 min and (b) 23 min (Proctor 1989). The contour values are in units of dBZ. The funnel, shaft and arrow-head shapes of the simulated radar echo mimic those of an observed microburst as shown in (c), which was observed during the field experiment JAWS. (Adapted after Fujita and Wakimoto 1983.)

downburst covers a horizontal area with a diameter from less than 1 km to 10 km. Depending on the intensity and characteristics of the downburst, it may be further classified into: (a) *macroburst*: a downburst that occurs over a horizontal area greater than 4 km and lasts for 5–30 min; or (b) *microburst*: a downburst that occurs over a horizontal area less than 4 km and lasts for 2–5 min. The formation of a microburst may be classified into three stages: (1) developing stage; (2) touchdown stage; and (3) outburst stage. Figures 8.3a and 8.3b show an observed and numerically simulated microburst which occurred in Denver on July 14, 1982.

The microburst may be further classified into (Wakimoto 1985): (a) dry microburst: a microburst accompanied by little or no precipitation during the period of outflow and usually associated with virga from mid-level altocumuli or high-based cumulonimbi; and (b) wet microburst: a microburst often accompanied by heavy precipitation during the period of outflow and is usually associated with strong precipitation shafts from thunderstorms. The intense downdrafts produced within thunderstorms are mainly caused by the negative buoyancy of the air parcels. The negative buoyancy is produced by the cooling associated with the evaporation of raindrops and/or melting and sublimation of frozen precipitation hydrometeors as they fall through the subsaturated layer. Another primary cause of wet microburst generation is thought to be precipitation loading. *Precipitation loading* occurs in thunderstorms when the weight of excessive condensate within the cloud creates a downward force. This effect thereby either induces a downward current of air or enhances descending air within a downdraft. Observational studies of downdraft cores, however, show that in some cases the virtual temperature of the subsiding air can be relatively warm (e.g., Igau *et al.* 1999). Therefore, precipitation loading can serve to make the downdrafts negatively buoyant in these situations (Jorgensen and LeMone 1989).

The physical processes leading to a downdraft within a convective storm are very different from those leading to an updraft. The formation of small (on the order of mm) hydrometeors results in an updraft temperature profile close to the saturated or supersaturated adiabatic lapse rate. On the other hand, small hydrometeors in a downdraft are quickly depleted by evaporation, which may lead to an unsaturated lapse rate through the release of latent heat. Thus, in order to accurately simulate a thunderstorm downdraft with a numerical model, consideration of the microphysical processes associated with liquid or solid condensate is of paramount importance.

Figures 8.3a and 8.3b depict the radar reflectivity and superimposed vector wind field for a numerical simulation of an idealized microburst that occurred in Denver on July 14, 1982. This downburst was mainly produced by the cooling associated with melting snow. The three stages, i.e. *developing stage*, *touchdown stage*, and *outburst stage*, were well simulated by the numerical model. During the earlier stages, the falling snow particles produced a radar echo which has been typical in other snow-produced downburst events. The downburst base is formed by small rain droplets that are swept into the microburst's ring-vortex circulation (Figs. 8.3a–b). Later, the base disappears as the ring vortex stretches outward behind the expanding outflow (Fig. 8.3c). In a melting-induced downburst, the maximum radar reflectivity is often located at the melting level. In this particular case, throughout the life cycle of the microburst, the maximum radar reflectivity due to melting snow is located about 2.4 km above the ground.

A majority of the outburst winds from outflow associated with microbursts are straight. Therefore, the streamlines of outburst winds spread out radially and do not rotate around the center of the outflow. However, some microbursts are associated with small-scale cyclonic circulations aloft and the rotation of the microburst may strengthen or weaken it, depending upon the height of the minimum pressure perturbation. When the minimum pressure perturbation is located at or near the surface, it tends to strengthen the microburst; when it is located substantially above the surface, it tends to weaken the microburst.

8.2 Dynamics of multicell storms

A multicell storm is composed of several convective cells throughout various stages of their life cycles. The existence of stronger shear in the environment prevents circulations associated with individual convective cells from interfering with each other, thus

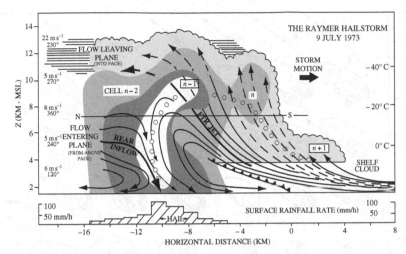

Fig. 8.4 A schematic diagram for a multicell storm in the vertical plane along the direction of the storm's movement. This storm has been referred to as the Raymer hailstorm. A series of convective cells, denoted as $n-2$, $n-1$, n, and $n+1$, were generated at the gust front and moved to the left as they developed. The solid lines represent storm-relative streamlines on the vertical plane; the broken lines on the left and right sides of the figure represent flow into and out of the plane and flow remaining within a plane a few kilometers closer to the reader, respectively. FTR JET and RTF JET stand for front-to-rear jet and rear-to-front jet (rear inflow), respectively, and are denoted by thick streamlines. The open circles represent the trajectory of a hailstone during its growth from a small droplet at cloud base. (Adapted after Browning *et al.* 1976.)

allowing the overall multicell storm system to last for several hours. The individual embedded cells are normally short-lived single cells that are generated by the quasi-steady updraft cell over the leading edge of the cold pool of outflow, i.e. the *gust front*. Normally, a new cell then forms in or near the gust front updraft, developing vertically and moving rearward within the *front-to-rear jet*. The front-to-rear jet is an ascending airstream from the low-level environmental inflow that resides above the subcloud cold pool and flows toward the rear portion of the storm and over the *rear-to-front jet*. The rear-to-front jet or rear inflow is the jet with low-θ_e air, flowing from the rear part of the storm into the rear part of the cold pool. Both the front-to-rear and rear-to-front jets are denoted by the bold streamlines in Fig. 8.4. The new cells are generated, develop, and are cut off from the gust front updraft in a discrete manner and in a periodic fashion every 10–15 minutes.

Figure 8.4 shows a schematic diagram for a multicell storm along the direction of the storm's movement. The storm produced hail and is often referred to as the Raymer hailstorm, which occurred on July 9, 1973 at Raymer, Colorado. A series of convective cells, denoted as $n-2$, $n-1$, n, and $n+1$, were generated as the gust front propagated rightward. These four individual convective cells are at four different stages of their life cycles. Cell $n-2$ is in the decaying stage, which is characterized at most levels by weak downdrafts and a residual weak updraft in places aloft. Part of the mature cell

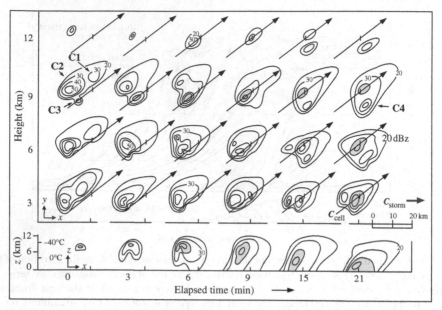

Fig. 8.5 Schematic diagram for the propagation and evolution of an ordinary multicell storm (Raymer hailstorm, Fig. 8.4). Both the horizontal and vertical radar reflectivity contours (at 10 dBZ intervals) are sketched. Horizontal cross sections are illustrated for four altitudes at six different times. Individual cell motions are steered by a midlevel wind toward the northeast (denoted by c_{cell}), while the entire storm moves towards the east (denoted by c_{storm}). New convective cells, such as cells 3 and 4 (denoted as C3 and C4), are generated to the south of the storm against the low-level wind. (After Chisholm and Renick 1972.)

$n-2$ has already been converted into a vigorous downdraft. Cell $n-1$ is in its mature stage, and has almost reached its maximum intensity. Cell n is just starting to grow out of the shelf cloud. A *shelf cloud* is a low cloud that protrudes horizontally from a thunderstorm cloud in association with a gust front. A distinct new Cell $n+1$ forms as a shelf cloud about 15 minutes later. Figure 8.5 shows a schematic diagram of individual cell movement within an ordinary multicell storm, which is often observed in the central USA. New convective cells, such as Cells 3 and 4, are generated to the south of the multicell storm and against the low-level wind where the gust front is located. The clockwise veering of the environmental wind increases with height causing the mid-tropospheric wind to steer individual cells to the northeast once they form. The entire multicell system, however, moves to the east. This type of multicell storm is often referred to as the *right-moving multicell storm*. However, individual cell regeneration may also cause multicell storm movement to deviate substantially from the mean wind direction, resulting in different configurations of cell and storm movements and mean wind direction. As described above, a new cell forms in or near the gust front updraft of a multicell storm, develops vertically, and moves rearward within the front-to-rear jet. The new cells are generated in a discrete manner and in a

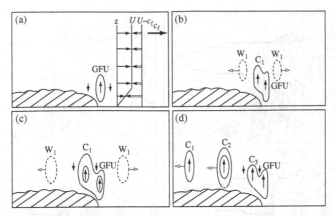

Fig. 8.6 A conceptual model of the advection mechanism for cell regeneration, development and propagation within a multicell storm in the reference frame moving with the storm (Lin *et al.* 1998). Three stages are found for cell regeneration: (a) Stage I: the gust front updraft (GFU) formed by low-level shear and cold outflow. Two compensating downdrafts flank the GFU. (b) Stage II: rearward advection of the upper portion of the growing GFU. (c) Stage III: Cutting off of the growing cell (C_1) from the GFU by the upstream compensating downdraft. Panel (d) shows cell regeneration of cell C_3 and rearward movement of cells C_1 and C_2. C_2 is at the end of the growing stage (maximum vertical velocity) and C_1 is moving more quickly within the propagating stage. W_1's are the gravity waves generated by the GFU and propagate away from it. The moving speed of the gust front is denoted by c_f. (Adapted after Lin *et al.* 1998.)

remarkably periodic fashion with a period of about 10–15 minutes. The cell's life cycle, which is composed of the developing stage, mature stage, and dissipating stage, as described above, occurs typically in less than an hour or so.

Based on idealized, systematic numerical experiments using nonlinear, nonhydrostatic cloud models, two major mechanisms have been proposed to explain the cell regeneration within a multicell storm: (a) advection mechanism (Lin *et al.* 1998) and (b) buoyancy circulation mechanism (Fovell and Tan 1998). The advection mechanism includes three stages, which are depicted in the schematic diagram of Figs. 8.6a–c. In Stage I (Fig. 8.6a), the gust front updraft (GFU) is forced by the cold outflow on the low-level inflow, analogous to the upward motion over the upslope of a mountain if the cold outflow is viewed as an obstacle in a moving frame of reference. This has been shown by replacing the cold pool by a plateau in a numerical simulation of multicell storms (Lin and Joyce 2001). The pressure gradient force established by the high pressure at the gust front near the ground and low pressure above the cold outflow induces the rearward and upward acceleration of the inflow, which may also contribute to the GFU formation (Parker and Johnson 2004). The GFU may generate gravity waves that propagate away from the density current, although they may be too weak to be detected in the real atmosphere. Two compensating downdrafts flank the GFU. An interaction between the low-level shear and the cold outflow is important and is explained by the buoyant pressure field, which will be discussed later in Section 8.3.

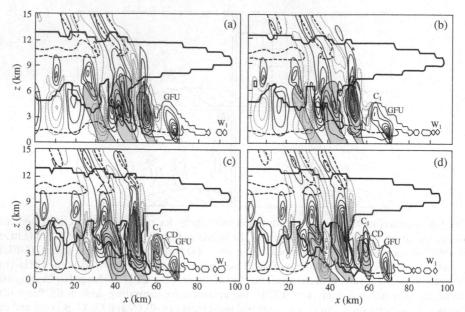

Fig. 8.7 Cell regeneration within a two-dimensional multicell storm, as simulated by a cloud model – ARPS (Xue *et al.* 2001). Four times the vertical velocity field in the reference frame moving with the gust front for the case with basic wind speed of $U = 10 \, \text{m s}^{-1}$ above 2.5 km and linearly sheared below it are shown: (a) 252, (b) 254, (c) 256, and (d) 258 min. Positive (negative) values of vertical velocity are denoted by solid (thin dotted) contours with intervals of $1 \, \text{m s}^{-1}$. The cloud boundary is bold contoured ($>0 \, \text{g kg}^{-1}$). The density current is roughly represented by the $-1 \, \text{K}$ potential temperature perturbation contour (bold dashed) near the surface. The rainwater is shaded ($>5 \times 10^{-4} \, \text{g kg}^{-1}$). The symbols GFU, W_1, C_1, and CD represent the gust front updraft, gravity wave 1, convective cell 1, and compensating downdraft, respectively. (Adapted after Lin *et al.* 1998.)

In Stage II (Fig. 8.6b), as the upper portion of the GFU grows, it is advected rearward relative to the gust front, which propagates faster than the mean wind. Relatively stronger gravity waves (W_1) are produced by this growing cell (C_1), which is still attached to the GFU. In Stage III (Fig. 8.6c), the compensating downdrafts grow stronger as the convective cell develops further and the upstream (right) downdraft cuts off the growing cell (C_1) from the GFU. During this stage, maximum perturbation potential temperature maxima are collocated with the updrafts in the middle and lower layers. Stages I–III repeat and individual cells continue to be generated as the GFU keeps pushing against the mean wind.

Figure 8.7 demonstrates the advection mechanism simulated by a cloud model for cell regeneration within a two-dimensional multicell storm. At $t = 252$ min (Fig. 8.7a), the GFU begins to develop vertically, signaling the generation of a new convective cell. The GFU is flanked by two downdrafts, which correspond to the stage represented in Fig. 8.6a. As the upper portion of the GFU or the new cell embryo moves rearward,

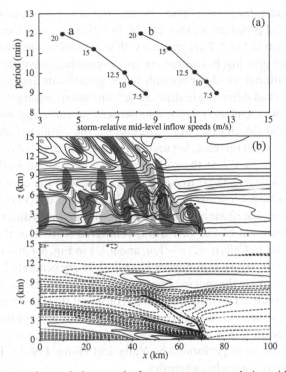

Fig. 8.8 (a) Cell regeneration period versus the far upstream storm relative mid-level inflow speeds (curve a) and the cell generation period versus the 2.5–5.5 km layer averaged storm relative mid-level inflow speeds (curve b). The numbers to the left of the curves denote the uniform basic wind speeds above 2.5 km. (b) Perturbation potential temperature fields versus vertical velocity field at $t = 262$ min in the moving frame with the gust front for the case of Fig. 8.7. Regions with $w < 1 \, \mathrm{m \, s^{-1}}$ are dark shaded, while regions of $w < -1 \, \mathrm{m \, s^{-1}}$ are light shaded. Regions of $\theta' > 0 \, \mathrm{K}$ ($\theta' < 0 \, \mathrm{K}$) are denoted by solid (thin solid) contours. The density current is approximately denoted by the bold contour near the surface. (c) Time-averaged, storm-relative, total u-wind component for the case of (b). Positive (negative) values are solid (dashed) contoured with an interval of $2 \, \mathrm{m \, s^{-1}}$. Critical levels are denoted by bold curves. (Adapted after Lin *et al.* 1998.)

compensating downdrafts begin to form on either side (Fig. 8.7b), corresponding to the stage represented in Fig. 8.6b. The strengthening downdraft on the upstream (right) side aids in the separation of the new cell from the GFU at later times (Figs. 8.7c–d). After separation, the cell strengthens and begins to precipitate as it moves into the modified air at the rear of the system, corresponding to the stage depicted in Fig. 8.6c. Thus, the life cycle depicted in Fig. 8.4 can be explained by the advection mechanism.

This numerical simulation reproduces cell regeneration and the inverse relationship between the period of cell regeneration and the mid-level, storm-relative wind speed (Fig. 8.8a). After formation, the convective cells propagate rearward with respect to the storm movement. The cell's propagation speed increases in the trailing stratiform region.

Two distinctive modes, the *growing mode* and the *propagating mode*, can be identified for cell development and propagation, respectively. For example, cells located in the region of 40 km $< x <$ 70 km in Fig. 8.7 are growing with time, while those located in the region of $x <$ 40 km are not growing, but instead are just propagating downstream. The growing mode and propagation mode differ not only in cell growth rate and propagation speed, but also have a marked difference in their phase relationship among flow variables.

Figure 8.8b shows the phase relationship between perturbation potential temperature (θ') and vertical velocity (w') at $t = 262$ min for the simulated idealized multicell storm. The basic wind, in this case, is a uniform speed of 10 m s^{-1} above $z = 2.5$ km and linearly decreases to 0 m s^{-1} at the surface. From Fig. 8.8b, in both the lower and middle tropospheric layers, the θ' maxima are in phase with the updrafts downstream of the gust front (40 km $< x <$ 70 km) and are behind the updrafts farther downstream ($x <$ 40 km). In short, the phase relationship between θ' and w' is different in the region immediately downstream of the gust front, compared to that in the region farther downstream. Similar to Cells $n - 2$, $n - 1$, n, and $n + 1$ in Fig. 8.4 and Cells C_1, C_2, and C_3 in Fig. 8.6, the convective cells depicted in Fig. 8.8c are generated at the gust front and move downstream (in the moving frame with GFU; to the left in the figure) as they develop. These individual convective cells are at different stages of their life cycles. The phase relationships between the perturbation potential temperature and vertical velocity as revealed in Fig. 8.8b are consistent with the numerical simulation of a squall-type multicell storm performed by Yang and Houze (1995). The propagating mode exhibits gravity wave characteristics.

The different phase relations among flow variables between the growing mode and propagating mode can be explained by the following simple argument. In a quiescent, Boussinesq fluid in the absence of diabatic heating, the thermodynamic equation, (2.2.18) reduces to

$$\frac{\partial \theta'}{\partial t} = -\frac{N^2 \theta_o}{g} w'.   \tag{8.2.1}$$

Multiplying by θ' on both sides and taking an average over an interval of time leads to

$$\frac{1}{2}\frac{\partial \overline{\theta'^2}}{\partial t} = -\frac{N^2 \theta_o}{g}\overline{\theta' w'}.   \tag{8.2.2}$$

For the growing mode of a convective cell, $\partial \overline{\theta'^2}/\partial t > 0$, we have $\overline{\theta' w'} > 0$ since $N^2 < 0$ for an unstable flow (where N is the Brunt–Vaisala frequency). This implies that θ' is in phase with w'. For the propagating mode of a convective cell, θ' is out of phase with w' since $\partial \overline{\theta'^2}/\partial t = 0$. These phase relationships are depicted in Fig. 8.8b. The upstream phase tilt (with respect to the basic wind and cell propagation) and phase relationships among flow variables of the disturbance (Fig. 8.8b) in the stratosphere ($z > 10$ km) represent an upward propagating gravity wave in a stable layer.

The *buoyant circulation mechanism* (summarized in Fig. 10 of Fovell and Tan 1998), however, emphasizes the importance of a local circulation induced by the heating on the strength of the forced updraft above the gust front. In the absence of such a local buoyancy-induced circulation, a persistent vertical updraft at the gust front would exist when a balance existed between a baroclinically generated cold pool circulation (with negative vorticity) and a circulation associated with the low-level vertical wind shear (with positive vorticity), as proposed in the RKW theory (Rotunno *et al.* 1988; see Subsection 8.3.1 for details). Where one circulation dominates, the forced updraft tilts either upshear or downshear and is less intense. The local circulation induced by the buoyant heating in the cell developing above the gust front modulates the cold pool/shear balance, and this modulation can be divided into two phases. In the first phase, the horizontal part of the buoyant circulation opposes the upper part of the cold pool circulation, diminishing the effect of the cold pool to push the cell rearward. The result of this is stronger forced lifting above the gust front and so the cell grows. In the second phase, the cell moves rearward, as the cold pool circulation still dominates. It is as the cell moves rearward that the horizontal branch of the buoyant circulation reinforces the upper part of the cold pool circulation, driving air parcels rearward and leading to weaker forced lifting. As the cell moves further rearward, the effect of the buoyant circulation on the cold pool/shear balance at the gust front diminishes, the forced updraft intensifies again, and the process repeats. A primary difference between the advection mechanism and this buoyant circulation mechanism is the importance of the compensating subsidence in the former and the modulation of the cold pool/shear balance in the latter.

8.3 Effects of shear and buoyancy

8.3.1 Effects of shear on cold outflow

Observations indicate that ambient wind and precipitation play important roles in the dynamics of severe convective storms. The ambient wind, and in particular the vertical wind shear, may affect the organization, development, and propagation of a severe convective storm. The precipitation associated with a severe convective storm may produce a density current via evaporative cooling, which in turn may affect the strength and propagation of the gust front, eventually influencing a new storm's development and propagation. Another factor which affects precipitation directly is the buoyancy of the environmental air. In a conditionally unstable atmosphere, the buoyancy can be measured by the CAPE, which is required for storms to develop. Observations of convective storms indicate that vertical wind shear increases for each ascending level of storm type: single-cell, multicell, and supercell, as shown in Fig. 8.9. Effects of environmental shear on severe convective storms and mesoscale convective systems have long been recognized by observations from upper-air soundings. The most severe and long-lasting storms typically

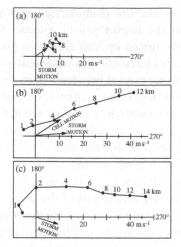

Fig. 8.9 Typical wind hodographs observed during the Alberta Hail Studies project for (a) single cell, (b) multicell, and (c) supercell storms. (From Chisholm and Renick 1972.)

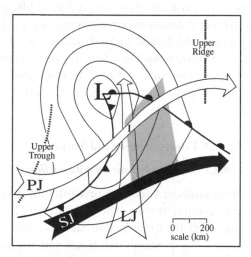

Fig. 8.10 A schematic of the synoptic environment for a severe convective storm outbreak. Symbols are denoted as follows: LJ, low-level jet; PJ, polar jet; SJ, subtropical jet; I, intersection of PJ and LJ; and light-shaded area: area of severe convection. Surface features are denoted by conventional symbols. (After Barnes and Newton 1986.)

form in strongly sheared environments. Severe convective storms often occur in the vicinity of upper-level polar jets and/or subtropical jets, and low-level jets (e.g. Fig. 8.10), which may entail the vertical wind shear in the synoptic environment. In addition, it has been proposed that vertical wind shear plays an important role in maintaining squall lines.

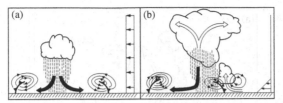

Fig. 8.11 Schematic diagram for convective cell development in an environment without shear (a) and with low-level shear (b). In (b), the presence of low-level shear allows the shear-induced circulation to balance the cold-outflow-induced circulation and produce deeper lifting on the downshear side. Note that downshear means the downstream side of the shear vector or the right-hand side in the case of panel b. This is known as the RKW (Rotunno, Klemp, and Weisman 1988) theory.

For a better understanding of the basic dynamics of squall lines, we will first discuss the shear effects on convective lines in a two-dimensional framework. The longevity of a two-dimensional squall line may be discerned through the schematic depicted in Fig. 8.11. A two-dimensional convective cell in an environment with no vertical shear would, via evaporative cooling, produce a pair of surface outflows that move rapidly away from the convective cell. The convective cell would dissipate quickly in a way similar to the ordinary single cell shown in Fig. 8.1. If the low-level shear exists in the ambient wind, it may prevent the cold pool from developing into a density current that moves away from the convective cell, and support the original cell to develop into deep convection (Thorpe *et al.* 1982). Based on a vorticity argument (Rotunno *et al.* 1988), in the absence of low-level shear (Fig. 8.11a), the circulation of a spreading cold pool would inhibit deep lifting, which would prevent the triggering of a convective cell. On the other hand, the presence of low-level shear (Fig. 8.11b) would allow the shear-induced circulation to counteract the cold-outflow-induced circulation. When these circulations are in balance, i.e. the optimal state, a deeper lifting is produced on the downshear side of the cold pool. The long-lived state is characterized not by a long-lived cell, but by a long-lived system of convective cells, such as the multicells discussed earlier, that are constantly generated at the gust front on the downshear side. This mechanism is referred to as the *RKW theory* for squall lines.

The RKW theory addresses the question of how the shear influences the transformation of ordinary thunderstorms into a long-lasting convective system where cells are continuously generated along a line. This theory identifies three stages in the evolution of a convective system (Fig. 8.12). The vorticity dynamics of each stage can be deduced from the y component of the vorticity (η) in a two-dimensional Boussinesq fluid flow

$$\bar{\rho}\frac{D}{Dt}\left(\frac{\eta}{\bar{\rho}}\right) = -\frac{\partial b}{\partial x},\tag{8.3.1}$$

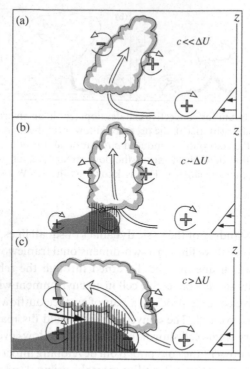

Fig. 8.12 Three stages in the evolution of a convective system, based on the RKW theory (Rotunno *et al.* 1988). (a) When $c \ll \Delta U$, the updraft tilts downshear (to the right) with height. (b) When $c \approx \Delta U$, the updraft becomes upright due to the balance between the ambient shear and the circulation induced by the cold pool. (c) When $c > \Delta U$, the updraft tilts upshear. Symbols c and ΔU represent the speed of the density current generated by the cold pool and the ambient low-level vertical wind shear, respectively. The updraft current is denoted by the thick, double-lined flow vector, and the rear-inflow in (c) is denoted by the thick solid wind vector. The surface cold pool is shaded and areas of rainfalls are depicted by vertical lines. Regions of significant horizontal vorticity are denoted by thin, circular arrows. Clouds are outlined by thick grey curves. (Adapted after Weisman 1992.)

where

$$\eta = \frac{\partial u}{\partial z} - \frac{\partial w}{\partial x}, \qquad (8.3.2)$$

and $\bar{\rho}(z)$ is the basic-state density, b the buoyancy, u the cross-line velocity, and w the vertical velocity. In the early stage of convective development, no significant cold pool is yet produced and the stage is characterized by $c \ll \Delta U$, where c and ΔU represent the density current speed associated with the cold pool and the ambient wind shear, respectively. Air parcels ascend from the boundary layer and have an initial positive vorticity, $\eta_i = \mathrm{d}U(z)/\mathrm{d}z$, where U is the basic-state velocity. Because there is little or no outflow, there is no low-level baroclinic modification of η. Therefore, at higher levels, a net positive bias prevails, causing the axis of the updraft to tilt in the downshear direction (Weisman and Rotunno 2004).

During the second stage, a cold pool is produced from evaporative cooling associated with the rainfall (Fig. 8.12b). The system then evolves to a state where $c \approx \Delta U$. The cold pool generates an additional negative η of comparable strength to the positive η at the downshear (right) side of the cold pool. The generation of this additional negative vorticity helps make the main updraft of the convective system upright. In the final stage, the cold pool continues to intensify. The system evolves to the state for $c > \Delta U$ (Fig. 8.12c). New cells are generated continuously at the gust front of the downshear (right) propagating outflow boundary. On the upshear side, the surface cold pool and heating aloft baroclinically generate a rear-inflow jet. The system tilts upshear at this stage. Most squall lines tend to evolve through all three stages. Rotunno *et al.* (1988) claim that the second stage, where $c/\Delta U \approx 1$, represents an *optimal state* for convective system development. At this optimal state, the system that maintains an upright configuration and the deepest lifting is produced at the leading edge of the cold pool. This helps explain the observed longevity of some squall lines, although it does not exclude the fact that squall lines can be long lived even when they are not in an optimal state.

It can be derived that for a steady balanced state, the speed for a two-dimensional density current generated by a square cold pool with a finite volume is (Rotunno *et al.* 1988)

$$c = \sqrt{-2bH} = \Delta U, \qquad (8.3.3)$$

where H is the height of the density current and b is the buoyancy

$$b = \frac{g\Delta\theta}{\theta_0}. \qquad (8.3.4)$$

In (8.3.4), $\Delta\theta$ is the potential temperature deficit of the cold pool from its environment. The density current speed estimated by (8.3.3) is identical to that estimated by Benjamin (1968). The above vorticity argument for shear and cold pool interaction can be illustrated through a series of idealized numerical experiments for a dry, unstratified flow with low-level shear passing over a cold pool of air in the two-dimensional volume of $|x| \leq 1$ and $0 \leq z \leq 1$. A two-dimensional vorticity–streamfunction model may be constructed to solve the following governing equations in the x–z plane (Weisman and Rotunno 2004):

$$\frac{D\eta}{Dt} = -\frac{\partial b}{\partial x} + \nu\nabla^2\eta, \qquad (8.3.5)$$

$$\frac{Db}{Dt} = \nu\nabla^2 b, \qquad (8.3.6)$$

$$\nabla^2\psi = \eta, \qquad (8.3.7)$$

where $D/Dt \equiv \partial/\partial t + u\partial/\partial x + w\partial/\partial z$, b is the buoyancy, ψ the streamfunction, and ν the kinematic viscosity. The velocities (u, w) are related to the streamfunction by $\partial\psi/\partial z$

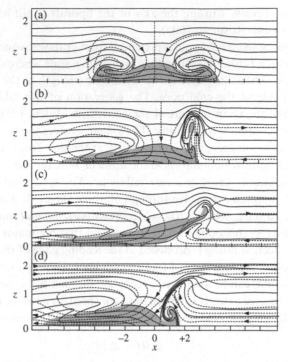

Fig. 8.13 Effects of the interaction between the vertical wind shear and cold pool on the flow circulation and density current, as illustrated through a series of idealized numerical experiments for a dry, unstratified flow with low-level shear passing over a prescribed cold pool of air in the box of $|x| \leq 1$ and $0 \leq z \leq 1$ after a nondimensional time $t = 4$. The basic flow is assumed to be 0 at the surface, increasing to: (a) $\Delta U =$: (a) 0, (b) 0.85, and (c) 1.5 at $z = 0.5$, and remains constant above $z = 0.5$. In (d), a deep shear is used. The basic flow speed increases linearly from 0 at the surface to $\Delta U = 2.0$ at $z = 2$. Areas with buoyancy $b < -0.5$ are shaded; the streamfunction ψ is denoted with dashed lines; and the tracer fields are denoted by heavy solid lines. (Adapted after Weisman and Rotunno 2004.)

and $-\partial \psi / \partial x$, respectively. All the variables associated with this model and in relevant results are nondimensionalized.

Figures 8.13a–c illustrate the impacts of the interaction between low-level shear and the cold pool on the flow after a nondimensional simulation time $t = 4$. The nondimensional time may be defined as $t = \tilde{U}\tilde{t}/\tilde{a}$, where $\tilde{U}$ is the dimensional average wind speed in the shear layer ($= \Delta \tilde{U}/2$), $\tilde{t}$ the dimensional time, and $\tilde{a}$ the dimensional half-width of the initial cool pool. For $\tilde{U} = 10 \text{ m s}^{-1}$ and $\tilde{a} = 10 \text{ km}$, a nondimensional time of $t = 4$ corresponds to 4000 s. At this time, the basic flow is assumed to be 0 at the surface, increasing to $\Delta U = 0$, 0.85, and 1.50 at $z = 0.5$, and remaining constant above $z = 0.5$. In the case of no vertical shear (Fig. 8.13a), the cold pool initially spreads in both $+x$ and $-x$ directions, but eventually spreads to the whole domain in the absence of ambient shear. In the case of $\Delta U = 0.85$, the cold pool tilts downshear initially due to the presence of $+y$-vorticity generated vertical shear. However, by $t = 4$ (Fig. 8.13b), this tendency is balanced by the cold-pool generated $-y$-vorticity on the downshear (right side in the figure) edge, which results in deep lifting. When the shear increases to $\Delta U = 1.5$ (Fig. 8.13c), the vorticity generated by the vertical shear becomes too strong, upsetting the balance with the cold-pool generated vorticity. This imbalance causes the circulation to tilt downshear, resulting in a more shallow lifting than in the case of $\Delta U = 0.85$. The case in which the vorticity generated by the ambient shear is balanced by the vorticity generated by the density current $\Delta U = 0.85$ in Fig. 8.13b corresponds to the *optimal state* in RKW theory. When the depth of the shear layer increases to the entire domain, $z = 2$ (Fig. 8.13d), the lifting is weaker but otherwise similar to that in the optimal state (Fig. 8.13b). Application of the RKW theory to squall-line dynamics will be discussed in Chapter 9.

8.3.2 Effects of buoyancy

In addition to the vertical wind shear, storm development may also be influenced by buoyancy. When the available buoyancy increases, the intensity of induced convection tends to strengthen. This, in turn, will produce heavier rainfall and increase evaporative cooling, thereby strengthening the surface outflow. Strengthening the surface outflow or density current tends to increase the propagation speed of the gust front. As defined in (7.3.2), the buoyancy will contribute to vertical acceleration or deceleration. Because CAPE is an integrated quantity of parcel buoyancy, it may serve as a measure of the effect of buoyancy. It can be approximated that the theoretical maximum speed of an updraft is $w_{max} = (2CAPE)^{1/2}$ based on the assumption that all the CAPE is converted into the air parcel's kinetic energy.

Three examples, based on idealized numerical simulations, are given in Fig. 8.14, in which the initial $q_{vo} = 11$, 14 and 16 g kg^{-1} corresponds to CAPE = 1000, 2200, and 3500 m^2 s^{-2}, respectively. These examples demonstrate that the environmental vertical shear and buoyancy play essential roles in determining the storm types, such as a short-lived ordinary single cell storm, a multicell storm or a supercell storm. As discussed in Section 8.2, the advection effect may cause the maximum vertical velocity of the initial updraft to decrease as the low-level vertical shear increases. Note that the increase of low-level shear may also produce stronger mixing which helps decrease the updraft. When shear strength is kept constant and CAPE increases, the maximum vertical velocity of the initial storm increases (Fig. 8.14a). The threshold CAPE required to sustain moist convection for the given initial impulse is about 1000 m^2 s^{-2}. As shown in Fig. 8.14b, the development of the secondary storm occurs when the environment has high CAPE and low-to-moderate shear. A second storm will not develop if no shear exists. This effect may be explained by the RKW theory: the balance between the storm relative inflow, which is controlled by the low-level shear (ΔU), and storm-induced surface outflow, which is controlled by the CAPE, prevents the development

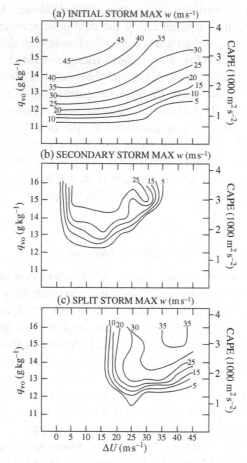

Fig. 8.14 Effects of the vertical shear (ΔU) and CAPE on storm development in idealized numerical experiments. The basic wind speed ($U(z)$) follows (8.4.1) with $z_o = 3$ km. The maximum vertical velocity (m s^{-1}) in the domain for (a) initial storms (in the first 30 min), (b) secondary storms and (c) split storms is plotted on the control parameters ΔU and CAPE. The w_{max} for (a) is obtained in the initial updraft which occurs approximately 30 min into the simulation. In these simulations, CAPE has been altered by varying the surface water vapor mixing ratio. (After Weisman and Klemp 1982.)

of a secondary storm. The supercell regime, which induces storm splitting, exists at only moderate to high shear flows (Fig. 8.14c). The optimal development of a supercell storm, i.e. the maximum intensities of a split storm, is reached when both CAPE and low-level shear increase.

Based on the above discussions, the dependence of the simulated storm types on environmental buoyancy and low-level shear may be consolidated and generalized in terms of a nondimensional control parameter, namely the *bulk Richardson number*,

$$R_B \equiv \frac{\text{CAPE}}{(\Delta U)^2/2},$$ (8.3.8)

where ΔU is the difference between the midlevel (e.g., 6-km) density-weighted mean wind speed and the mean near surface layer (e.g., 500-m) wind speed. For a directional shear, the denominator is replaced with $[(\Delta U)^2 + (\Delta V)^2]/2$. The numerator is a measure of potential updraft strength but is also indirectly a measure of potential downdraft and surface outflow strength. The denominator may be interpreted as a measure of the inflow kinetic energy made available to the storm by the vertical wind shear. Operationally, R_B is less used; instead, the CAPE and shear are used separately to help making prediction of the occurance of severe storms.

Based on idealized numerical experiments, the nondimensional storm strength may be defined as

$$S = \frac{w_{\max}}{\sqrt{2\text{CAPE}}},$$ (8.3.9)

and can be displayed as a function of the bulk Richardson number (Fig. 8.15). Note that $w_{\max}$ is the model simulated maximum vertical velocity within the domain, and the denominator is the theoretical maximum vertical velocity for a certain CAPE, as discussed in Chapter 7. However, the air parcel may not normally attain this value in the real atmosphere due to some assumptions made in the parcel theory, such as neglecting mixing and entrainment, and the immediate adjustment of the air parcel pressure to its environmental pressure. The initial storm strength S increases as R_B increases (Fig. 8.15a). Updrafts cease to develop into storms for small R_B, such as $R_B < 10$, under the same initial forcing. This may be explained by the RKW theory by assuming that the strength of the cold pool is proportional to the CAPE. Thus, for a fixed CAPE, small R_B implies that shear is too strong for $C \ll \Delta U$ to be balanced by the cold pool (C). Figure 8.15b shows the multicell storm strength S as a function of R_B for the secondary storm. No secondary storm development occurs for $R_B < 35$.

The values of R_B for supercell storm development are concentrated in the range of $15 < R_B < 35$, based on the numerical simulations shown in Fig. 8.15c. For $R_B < 10$, the shear is too strong to allow for storm growth, while for $R_B > 50$, strong buoyancy produces a density current that is too strong for the shear to balance. These results suggest that unsteady, multicellular convection occurs for $R_B > 35$ and that supercell convection occurs when $10 < R_B < 50$. Note that the critical values of R_B for different storm regimes are model-dependent, where a model's output may be affected by numerical schemes used, such as microphysics parameterization schemes and numerical diffusion. For example, the model used for generating results shown in Fig. 8.15 is very diffusive. The numerical diffusivity may smooth out disturbances. Thus, the critical R_B values, such as 10, 35, and 50, discussed above should be used for reference only.

Several limitations are encountered in representing storm characteristics with one or two nondimensional control parameters, such as the bulk Richardson number (R_B)

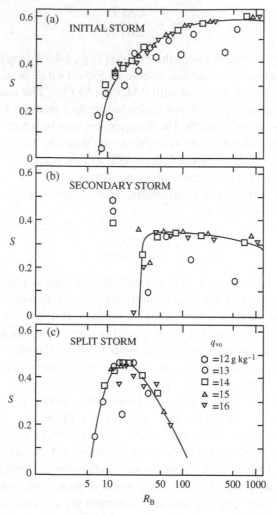

Fig. 8.15 Simulated storm strength S, defined in (8.3.9), versus the bulk Richardson number R_B for (a) initial storms, (b) secondary storms, and (c) split storms. (After Weisman and Klemp 1982.)

and nondimensional storm strength (S) in the above experiments. R_B represents the ratio of potential energy to kinetic energy, or indirectly represents the force balance created between the surface cold outflow and vertical wind shear. The CAPE in S indirectly represents the strength of the surface cold outflow or density current. However, the strength of the surface cold outflow may also be influenced by other factors, such as vertical distribution of moisture content and detailed microphysical processes. In fact, the strength of the surface cold outflow is arguably more directly related to a parameter called the *downdraft convective available potential energy* (DCAPE), than the CAPE. Similar to (7.3.29), DCAPE may also be defined as

$$\text{DCAPE} = g \int_{z_s}^{z_i} \frac{\bar{\theta}(z) - \theta(z)}{\bar{\theta}(z)} \, dz. \tag{8.3.10}$$

DCAPE is physically equivalent to the kinetic energy gained by a parcel descent from a certain height (z_i) to the surface. The parcel temperature is obtained by cooling it to saturation via the wet-bulb process and then lowering the parcel saturated- or pseudo-adiabatically with just enough evaporation to keep it saturated. When rain evaporates in subsaturated air or solid precipitation (snow or hail) melts at the melting level or sublimates, the cooled air generates a downdraft. Thus, the maximum downdraft may be estimated by $-w_{max} = \sqrt{2\text{DCAPE}}$. Idealized numerical simulations indicate that storm development is also controlled by DCAPE and the mid-tropospheric moisture content (dryness) (e.g., Gilmore and Wicker 1998). Although DCAPE may more accurately measure the strength of the surface cold outflow, it does not address how much precipitation will be produced. Finding a small set of control parameters for representing storm characteristics continues to be a challenging task.

8.4 Dynamics of supercell storms

8.4.1 General characteristics

A *supercell* storm is defined as a convective storm that possesses a persistent, deep, rotating updraft. These rotating updrafts are often found in conjunction with mesocyclones. A *mesocyclone* is a cyclonically rotating vortex, around 2 to 10 km in diameter, that has a vorticity on the order of $10^{-2}\,\text{s}^{-1}$ or greater. Most supercell storms are characterized by the following features:

(a) They are situated in an environment with strong vertical wind shear.
(b) A mesocyclone, with a diameter of several kilometers, associated with the rotating updraft is often embedded in a supercell storm.
(c) They often propagate in a direction dictated by the mean environmental wind.
(d) They tend to last for several hours due to strong vertical shear in their environment.
(e) Although some supercell storms have a size comparable to that of multicell storms, their cloud structure, flow circulation, and formation processes of precipitation are organized by a single massive updraft–downdraft pair.
(f) Sometimes a pair of supercells are produced through a splitting process with a cyclonic supercell moving toward the right (with respect to the mean wind), and an anticyclonic supercell moving toward the left.
(g) The rightward propagation of a supercell storm is related to the veering of the environmental wind shear vector.
(h) Most precipitation falls downshear from the main storm updraft, which lies above the intersection of the forward and rear flank gust fronts.
(i) Supercell storms tend to produce the most intense, long-lasting tornadoes and damaging hail through complicated processes.

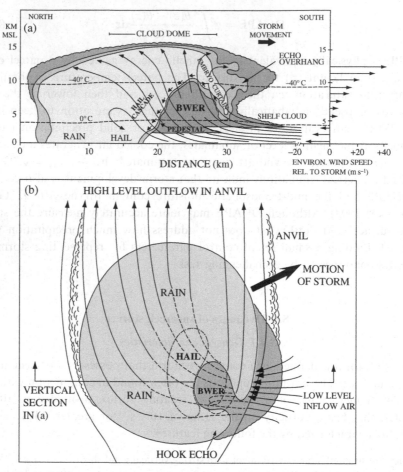

Fig. 8.16 Vertical and horizontal structure of a mature supercell storm. (a) The vertical cross section is taken along the direction of storm movement through the center of the main updraft as depicted in (b). The shaded regions represent two levels of radar reflectivity. Areas of weak echo region or bounded weak echo region (BWER) are shaded. Arrows to the right of the figure indicate the environmental wind relative to the storm movement. Arrows within the figure denote projections of streamlines of airflow relative to the storm movement. (b) The horizontal view at the height of 5.2 km of (a) seen from above. The major region of radar reflectivity in (b) is light shaded. (Adapted after Browning and Foote 1976.)

(j) Tornadoes may develop in regions where the environmental inflow and storm outflow meet beneath the mesocyclone or along the nose of the gust front.

(k) A supercell usually has a very organized internal structure that enables it to propagate continuously.

Figure 8.16 shows the schematics of vertical and horizontal structures of a mature supercell storm observed by radar in northeast Colorado. This cross section is taken

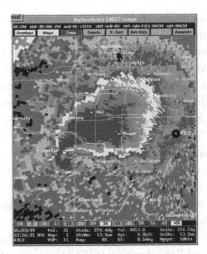

Fig. 8.17 A hook echo shown in a radar image taken at 2356 UTC May 3, 1999 from KTLX, the Oklahoma City WSR-88D. The storm produced a 1/4 to 1/2 mile wide F5 tornado in the Oklahoma community of Bridge Creek shortly after the image was taken. (Courtesy of NWS Norman, Oklahoma. Image was provided by C. J. Ringley.)

along the direction of storm movement. The supercell's massive updraft (Fig. 8.16a) is very intense ($\sim$10–40 m s^{-1}), and significantly larger than than that of a single-cell storm or multicell storm (Fig. 8.4). The cloud droplets within the updraft are swept upward to a level of about $-40\ ^{\circ}$C and may penetrate to tropopause. Since the cloud droplets within this intense updraft have insufficient time to grow, they cannot produce strong radar echoes. Instead, they form a *hook echo* (as seen in radar images) that wraps around a so-called *weak echo region* (WER) or *bounded weak echo region* (BWER), often located along the right flank, facing the direction of the storm movement (Fig. 8.16b). The largest amount of hail falls to the ground in a narrow band behind the BWER, while rain falls on the ground in a wider region behind the region of hail. An *echo overhang* can be found on the forward flank of the supercell storm.

A hook echo wraps around the BWER at the midtropospheric horizontal plane. The hook's cyclonic shape results from the mesocyclone's cyclonic winds. The hook itself is actually a result of the mesocyclone, a region of rotation with cyclonic vortex and rising air. Many of the violent tornadoes associated with supercells exhibit a distinct hook echo in radar images. To a weather forecaster, a hook echo is one clue that a supercell has the potential to produce a tornado. Figure 8.17 shows a hook echo in a radar image taken at 2356 UTC May 3, 1999 from the Oklahoma City's Doppler radar. The storm was responsible for a 1/4 to 1/2 mile wide tornado that was rated F5 on the Fujita Scale of tornado intensity in Bridge Creek, Moore, and Oklahoma City, Oklahoma after the image was taken. Note that many supercells have hook echos, but only a small fraction produce tornadoes. While there are many factors that signal the formation of a tornado, the hook echo is the only one that can be captured by conventional radar. However, it has also been found that Doppler velocities serve as a significantly better indicator of tornadoes than the hook echo.

A typical supercell storm evolves through three stages: (a) initial stage, (b) developing stage, and (c) mature stage. At the initial stage, the supercell storm is no different from that of a single convective cell in an ordinary single-cell storm or embedded in a multicell storm. At this stage, the storm moves in the mean wind direction and no weak echo regions or overhang echoes are present. A WER develops in the region of strengthening updraft. Approximately 90 minutes into the storm lifetime, the supercell reaches its quasi-steady mature stage. At this stage, a BWER develops from the WER, which coincides with the strong updraft. A BWER is usually associated with a strong updraft.

The basic structure of a supercell includes several important features such as a shelf cloud, wall cloud, tornado, tail cloud, hail and rain locations, virga, anvils, mammatus cloud, and penetrating cloud top. In addition, during tornadogenesis, particular flow features such as the forward flank downdraft (FFD) and rear flank downdraft (RFD) may form within a supercell storm. Major features associated with this type of supercell storm have been reasonably well simulated by nonlinear, nonhydrostatic cloud models. The dynamics of mesocyclones and tornadogenesis, and their associated flow features will be discussed later in this chapter.

Based on precipitation patterns and storm characteristics, supercell storms may be classified as: (a) *classic supercell*, (b) *high precipitation* (HP) *supercell*, and (c) *low-precipitation* (LP) *supercell*. Among these three types of supercell storms, classic supercells are the most extensively studied. In a classic supercell, most of the precipitation falls downwind from the main storm updraft. The updraft lies above the interaction of the forward flank and rear flank gust fronts. Tornadoes usually develop in regions where the environmental inflow and storm outflow meet beneath the mesocyclone or along the nose of the gust front. When there is a balance between low-level outflow and inflow, long-lived or multiple tornadoes tend to occur because the mesocyclone does not occlude rapidly (Wakimoto *et al.* 1996).

HP supercells occur less frequently in the High Plains of the United States, but are predominant elsewhere, such as Ohio valley, Gulf coast and Southeastern USA. The HP supercell exhibits the following characteristics: (a) it tends to be larger than the classic supercell; (b) the mesocyclone and tornadoes created by the storm are largely embedded in precipitation; (c) it exhibits a distinctively kidney or S-shaped echo patterns, inflow notches, and persistent weak-echo regions; (d) it may contain multiple reflectivity cores; (e) hook echoes, if present, are often very broad; (f) many HP supercell storms undergo a life cycle in which they evolve from one form to another, transitioning from classic to high-precipitation supercell storm, or HP to bow echo; (g) the classic supercell may appear to evolve into a high precipitation supercell as it decays, but the HP supercell distinguishes itself from the dissipating classic supercell by sustaining its rotation; and (h) HP supercells tend to be outflow dominated, which undercuts the mesoscyclone and limits the potential for long-lived tornadoes.

The LP supercell storms normally form in the High Plains of the USA. They possess the following common features: (a) very few radar signatures compared to those observed for classic supercell storms, (b) little (if any) precipitation falling under the cloud base; no evidence of strong downdraft at the surface, (c) intense updrafts at the storm's rear flank and weak-to-moderate-intensity downdrafts, (d) large hail falling outside the main cumuliform tower, (e) few tornadoes produced because they have little to no RFD, and (f) high LCL height may retard deep stretching of vorticity associated with mesocyclone. Major features associated with HP and LP storms have been reasonably well simulated by nonlinear, nonhydrostatic cloud models.

A typical synoptic scale environment conducive to an outbreak of severe storms in the midlatitudes is depicted in Fig. 8.10. Severe convective storms are considered most likely to occur in the warm sector of a surface cyclone (represented as the light-shaded area in the figure). The region along and just north of the warm front is favorable for tornadogenesis. Severe storms tend to form near the intersection of the polar jet and low-level jet (denoted as "I" in the figure). There is an upper-level synoptic wave oriented from southwest to northeast with the inflection point located right over the surface cyclone. Supercell forecasting has advanced to be more quantitative, which is far beyond the conceptual model as shown in Fig. 8.10. Since this is a rapidly evolving subject of current research, it is only appropriate to provide a very brief summary. Some important findings related to supercell storms are (e.g., Doswell 2001):

(a) Multiple control parameters are needed for making more accurate forecasting of supercell storms.
(b) Some control parameters include the storm relative helicity (SRH), vertical shear (e.g., $\Delta U = U_{6\,km} - U_s$, i.e. surface–6 km shear), CAPE, bulk Richardson number (R_B), buoyant RFD, low LCL (enhancing the ingest of high θ_e air into low-level mesocyclone), super composite parameter (combination of mid-level CAPE, 0–6 km shear and 0–1 km SRH).
(c) RFD has to remain buoyant (Markowski *et al.* 2002).
(d) Supercell environments are characterized by:
  (i) $\Delta U = U_{6\,km} - U_s > 15\text{–}20\,\mathrm{m\,s^{-1}}$,
  (ii) $10 < R_B < 50$,
  (iii) $1000\,\mathrm{m^2\,s^{-2}} < \mathrm{CAPE} < 3500\,\mathrm{m^2\,s^{-2}}$.

8.4.2 *Effects of unidirectional shear*

If convection is initialized by an isolated thermal instead of a line thermal, as used in the Section 8.3, the flow responses are more complicated. To elucidate this, consider a three-dimensional conditionally unstable flow with unidirectional shear. The vertical temperature and dew point profiles from these experiments are shown in Fig. 8.18a, where the surface water vapor mixing ratio is $q_{vo} = 14\,\mathrm{g\,kg^{-1}}$. The sounding has a CAPE of $2200\,\mathrm{J\,kg^{-1}}$, which represents an environment of moderate conditional instability. The vertical wind profiles are assumed to be

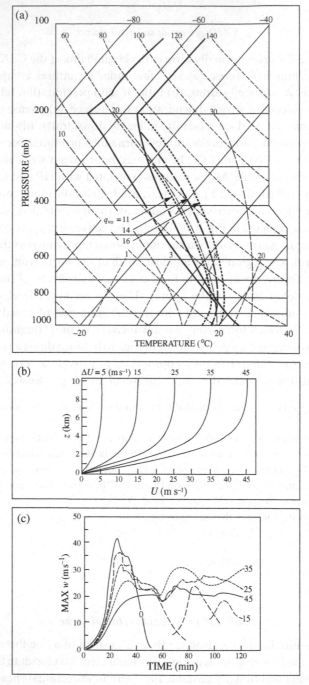

Fig. 8.18 Storm development in a three-dimensional, conditionally unstable, unidirectional shear flow. (a) Temperature and dew point profiles (heavy solid lines) on a skew-T log-p thermodynamic diagram. The heavy dashed line denotes the moist adiabat following the air parcel ascent from the surface with a mixing ratio of $q_{vo} = 14$ g kg^{-1}. Heavy dotted lines denote similar parcel ascents for $q_{vo} = 11$ g kg^{-1} and $q_{vo} = 16$ g kg^{-1}. Tilted solid lines are isotherms, short dashed lines are dry adiabats, and long dashed lines are moist adiabats. (b) Wind profiles defined by (8.4.1). (c) Time series of maximum vertical velocities for wind shear experiments, in which the vertical wind profiles are depicted in (b) and $q_{vo} = 14$ g kg^{-1}. (After Weisman and Klemp 1982.)

$$U(z) = \Delta U \tanh (z/z_0), \tag{8.4.1}$$

where $z_0 = 3$ km, a constant throughout all the simulations, and the basic wind shear is concentrated in the lower layer (Fig. 8.18b). Six wind profiles with ΔU varying from 0 to 45 m s^{-1} are used to investigate the sensitivity of the flow responses to differing shear strength. The convection is initiated by an isolated, spherical thermal with a radius of 10 km in the horizontal, a radius of 1.4 km in the vertical, and a $+2$ K temperature perturbation. The sensitivity to shear is illustrated by the time evolution of the maximum vertical velocities simulated for the above six cases (Fig. 8.18c). For zero vertical shear, the maximum w increases rapidly in the first 25 minutes and then decreases to almost 0 m s^{-1} in the next 25 minutes. For moderate shear, such as $\Delta U = 15$ m s^{-1}, distinct new convective cells are regenerated at the gust front of the density current, revealing a multicellular storm structure. The regeneration of convective storms can be explained by the RKW theory. The initial storm weakens rapidly as a result of precipitation loading associated with the rainfall, which cuts off the supply of potentially warm air needed for maintaining the updraft. Surface convergence is initially strongest on the right and left flanks of the storm; however, as the surface outflow spreads, deep lifting occurs along the gust front directly downshear (east) of the storm. This leads to the formation of the second storm, which propagates downshear and reaches its maximum strength. Following the same redevelopment process, a third storm forms consecutively and reaches its maximum strength, but is weaker than the first and second storms. The redevelopment process is controlled by the competing forcings from the low-level shear and the density current, as discussed in the RKW theory in Section 8.3.

For a strong shear, such as $\Delta U \geq 25$ m s^{-1}, the storm redevelopment is replaced by a regime of *storm splitting*. The storm splitting dynamics will be discussed in subsection 8.4.4. In this regime, the initial updraft is split into right- (facing downshear) and left-moving storms. A pair of equal, self-maintaining storms are produced in a relatively continuous fashion on both the right and left (facing downshear) flanks of the original storm's outflow boundary. The updraft of the right (left) moving storm is flanked by a cyclonic (anticyclonic) vorticity. The right-moving storm (Fig. 8.19) behaves like a supercell storm. The maximum vertical velocity of the initial updraft decreases as shear increases, which may be attributed to the advection effect by the basic wind (Fig. 8.18c). The maximum w of a split storm oscillates at about the same value at later times compared to initial times, in response to the redevelopment process of the storm. The storm splitting process can be clearly seen for $\Delta U = 35$ m s^{-1} (Fig. 8.19). Unlike the weak shear case, no westerly surface outflow has developed behind the gust front (Figs. 8.19b–c). This lack of outflow weakens the ability for the secondary storm to redevelop. On the other hand, the increased shear has enhanced the process of storm splitting, as evidenced by the stronger maximum vertical velocity (Fig. 8.18). With a unidirectional shear, the left-moving storm is simply a mirror image of the right-moving storm as shown in the figure. In order for a steady split storm to develop, the storm-relative inflow shear must be strong enough to keep the density current from propagating away from

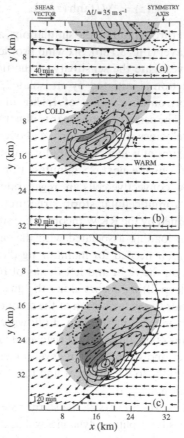

Fig. 8.19 Storm development in a three-dimensional, conditionally unstable, strong shear flow where $\Delta U = 35\,\mathrm{m\,s^{-1}}$ and $q_{vo} = 14\,\mathrm{g\,kg^{-1}}$ (see Fig. 8.16a) at (a) 40 min, (b) 80 min, and (c) 120 min. Displayed in the figures are: the horizontal wind vectors at low-level ($z = 178\,\mathrm{m}$), surface rain indicated by light shading, the dark shading indicates rain areas $> 4\,\mathrm{g\,kg^{-1}}$, mid-level ($z = 4.6\,\mathrm{km}$) w contoured every $5\,\mathrm{m\,s^{-1}}$ ($2\,\mathrm{m\,s^{-1}}$) for upward (downward) motion, and the surface gust front (represented by $-0.5\,^{\circ}\mathrm{C}$ temperature perturbation and denoted by the solid barbed line). Positive signs denote the location of local low-level w_{max}. Only the southern half of the domain on a moving frame of reference is shown. Note that the storm splits into right- (southward) and left-moving storms. (After Weisman and Klemp 1982.)

the updraft. The right-moving storm becomes dominant when the vertical shear is directional and clockwise and is also referred to as the *severe right* (*SR*) *storm* or *right mover*.

8.4.3 Storm splitting

As discussed earlier, storm redevelopment falls into the regime of *storm splitting*, as the unidirectional vertical shear increases (Figs. 8.14c and 8.19). Near the end of the initial

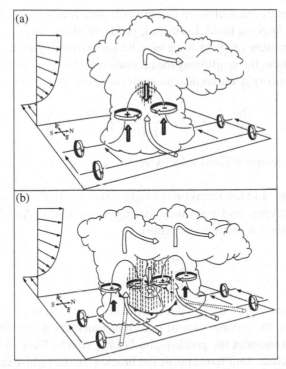

Fig. 8.20 A schematic depicting rotation development and the storm splitting. (a) Rotation development: in the early stage, a pair of vortices forms through tilting of horizontal vorticity associated with the (westerly) environmental shear. (b) Storm splitting: in the later stage, the updraft is split into two convective cells by the upward pressure gradient forces. See text for details. Cylindrical arrows denote the direction of the storm-relative airflow, and heavy solid lines represent vortex lines with the sense of rotation denoted by circular arrows. Shaded arrows represent the forcing promoting new updraft and downdraft acceleration. Vertical dashed lines denote regions of precipitation. Frontal symbols at the surface mark the boundary of cold air outflow. (After Klemp 1987. Reprinted, with permission, from the *Annual Review of Fluid Mechanics*, Vol. 19 @ 1987 by Annual Reviews.)

updraft, a pair of equal, self-maintaining storms is produced in a relatively continuous fashion on both the right and left (facing downshear) flanks of the original storm's updraft. Figure 8.20 contains a three-dimensional schematic depicting storm splitting and rotation development. During the early stage of storm development, tilting of horizontal vorticity associated with the environmental shear (westerly shear in this case) causes a pair of vortices to form. During the later stage of storm development, precipitation accumulates within the updraft, thus resulting in increasingly negative buoyancy that produces a downdraft within the cloud. If the environmental shear is strong, then the upward pressure gradient forces reinforce new updraft growth on the southern and northern flanks of the original updraft center. The upward pressure gradient forces then gradually split the updraft into two convective cells that move rightward (southward) and leftward (northward) with respect to the shear vector. At this stage, vortex lines are tilted downward, producing two vortex pairs.

The upward pressure gradient force may be understood by analyzing the vertical momentum equation. By partitioning the pressure into the buoyancy pressure (p'_b) and the dynamic pressure (p'_d), the vertical momentum equation (7.3.8) may be rewritten as

$$\frac{\partial w}{\partial t} = -\boldsymbol{v} \cdot \nabla w - \frac{1}{\bar{\rho}}\frac{\partial p'_d}{\partial z} - \left(\frac{1}{\bar{\rho}}\frac{\partial p'_b}{\partial z} - b\right). \tag{8.4.2}$$

where $\boldsymbol{v}$ is the three-dimensional velocity vector, $\bar{\rho}$ is the basic state density and b is the buoyancy.

The first, second, and last terms on the right-hand side of (8.4.2) are the advection term, dynamic forcing, and buoyancy forcing, respectively. As often adopted in storm dynamics, the full *buoyancy* is defined as

$$b \equiv g\left[\frac{\theta'}{\bar{\theta}} + 0.61 q'_v - q_T\right], \tag{8.4.3}$$

where θ' is the perturbation potential temperature, q'_v is the perturbation water vapor mixing ratio, q_T is the mixing ratio of total hydrometeors in the air. The last term inside the bracket includes the *precipitation loading* effects. Note that the buoyancy forcing should include both terms inside the bracket on the right-hand side of (8.4.2) because $b > 0$ includes a compensating downward buoyant pressure gradient force. Taking the divergence of the nonrotating equation of motion

$$\frac{\partial \boldsymbol{v}}{\partial t} = -\boldsymbol{v} \cdot \nabla \boldsymbol{v} - \frac{1}{\bar{\rho}}\nabla p' + b\boldsymbol{k}, \tag{8.4.4}$$

and using the anelastic continuity equation,

$$\nabla \cdot (\bar{\rho}\boldsymbol{v}) = 0, \tag{8.4.5}$$

leads to

$$\nabla^2 p' = -\nabla \cdot (\bar{\rho}\boldsymbol{v} \cdot \nabla \boldsymbol{v}) + \frac{\partial(\bar{\rho}b)}{\partial z}. \tag{8.4.6}$$

With the partition of $p' = p'_d + p'_b$, we have

$$\nabla^2 p'_d = -\nabla \cdot (\bar{\rho}\boldsymbol{v} \cdot \nabla \boldsymbol{v}), \tag{8.4.7}$$

$$\nabla^2 p'_b = \frac{\partial(\bar{\rho}b)}{\partial z}, \tag{8.4.8}$$

subjected to the Neumann boundary conditions $\partial p'_d/\partial z = 0$ and $\partial p'_b/\partial z = b$ at $z = 0$, z_T, where z_T is the top of the domain considered. Applying the anelastic continuity equation (8.4.5) to (8.4.7) gives

$$\nabla^2 p'_d = \bar{\rho}(|\boldsymbol{\omega}|^2 - |D|^2),$$
$$|\boldsymbol{\omega}|^2 = (w_y - v_z)^2 + (u_z - w_x)^2 + (v_x - u_y)^2,$$
$$|D|^2 = (u_x^2 + u_y^2 + u_z^2) + (v_x^2 + v_y^2 + v_z^2) + (w_x^2 + w_y^2 + w_z^2)$$
$$-(d\ln \bar{\rho}/dz)\boldsymbol{v} \cdot \nabla w - (d^2 \ln \bar{\rho}/dz^2)w^2.$$

$$(8.4.9)$$

In (8.4.9), $\boldsymbol{\omega} \equiv \nabla \times \boldsymbol{v}$ is the three-dimensional vorticity vector and $|D|$ is the magnitude of the *total deformation* which is composed by *shear deformation* (e.g., u_y and u_z) and *stretching deformation* (e.g., u_x). In a thunderstorm with strong vertical vorticity, such as a supercell storm, the right-hand side is dominated by $v_x u_y$. For a horizontal wind field with pure rotation, in which $v_x = -u_y$, then $v_x u_y \approx -1/4\,\zeta^2$, where ζ is the vertical vorticity. Assuming sinusoidal variations in a flow's interior, the Laplacian of a variable is roughly proportional to the negative of the variable itself, i.e. $\nabla^2 p'_d \propto -p'_d$, if the vertical variations of density and virtual potential temperature are neglected on the left side of (8.4.9). This leads to

$$p'_d \propto -\zeta^2. \qquad (8.4.10)$$

Thus, a dynamically perturbed low pressure is associated with a vortex, regardless of whether the vortex is cyclonic or anticyclonic. In other words, to a first-order approximation, the flow tends to adjust to *cyclostrophic flow* balance, where, in absence of the Earth's rotation, the pressure gradient force is balanced by the centrifugal force. Note that an upward motion is associated with the convergence near a flat lower surface through vertical stretching as required by the mass continuity in an anelastic or incompressible fluid. This upward motion does not require a vertical acceleration (Dw/Dt) from the imbalance between the vertical pressure gradient force and the buoyancy force in the vertical momentum equation. On the other hand, the relationship (i.e., mass continuity) between the vertical motion and near surface convergence is diagnostic, thus the causality is unclear and may deserve a further study. During the early stage of supercell storm development, a pair of midlevel vortices forms through tilting of horizontal vorticity associated with the strong environmental shear (Fig. 8.20a). In accordance with (8.4.10), strong rotation, a characteristic of the individual midlevel vortex, lowers pressure at the vortex center, thereby inducing acceleration in the updraft on each flank of the storm, according to (8.4.2). The strengthening updrafts then split the initial updraft into two, creating the storm split (Fig. 8.20b). When the storm matures, the precipitation-induced strengthening downdraft tilts the vortex line downward and helps produce an additional pair of vortices. This facilitates propagation of the split storm away from the original storm center.

Based on observations and the schematic in Fig. 8.20, one may suspect that precipitation-induced downdraft causes storm splitting and propagation. However, numerical simulations have demonstrated that even in the absence of precipitation (which prevents formation of the central downdraft), updraft splitting still occurs.

8.4.4 Storm rotation and propagation

To illustrate rotational development in a supercell storm, consider the momentum equation, (8.4.4), and the anelastic continuity equation, (8.4.5). If we take the curl of (8.4.4) and note that the curl of the gradient vanishes, we then find the three-dimensional vorticity equation

$$\frac{\partial \boldsymbol{\omega}}{\partial t} = \nabla \times (\boldsymbol{v} \times \boldsymbol{\omega}) + \nabla \times (b\boldsymbol{k}). \tag{8.4.12}$$

The local rate of change of the vertical vorticity, $\zeta = \boldsymbol{k} \cdot \boldsymbol{\omega}$, can then be obtained by taking $\boldsymbol{k} \cdot (8.4.12)$

$$\frac{\partial \zeta}{\partial t} = -\boldsymbol{v} \cdot \nabla \zeta + \left(\frac{\partial u}{\partial z} \frac{\partial w}{\partial y} - \frac{\partial v}{\partial z} \frac{\partial w}{\partial x} \right) + \zeta \frac{\partial w}{\partial z}. \tag{8.4.13}$$

The first, second, and third terms on the right-hand side of (8.4.13) represent the vorticity advection, tilting of the horizontal vorticity, and vertical vorticity stretching, respectively. Both the advection and stretching terms require pre-existence of the vertical vorticity. The only initial source of vertical vorticity in a nonrotating system is therefore the tilting of the horizontal vorticity. The rotation of the Earth does not play an important role in storm dynamics and is therefore omitted from the basic equations. The horizontal vorticity is contained in the environmental shear and is characterized by the environmental wind shear vector, $\boldsymbol{S} = \mathrm{d}(\boldsymbol{V} - \boldsymbol{c})/\mathrm{d}z$, where $\boldsymbol{V} - \boldsymbol{c}$ is the storm-relative environmental wind velocity. The shear vector consists of two components, namely the *speed shear* and the *directional shear*. Speed shear is a change in wind speed with height while directional shear is a change in wind direction with height. Mathematically, the storm-relative wind vector can be expressed as $\boldsymbol{V} - \boldsymbol{c} = |\boldsymbol{V} - \boldsymbol{c}|\boldsymbol{p}$, where $\boldsymbol{p}$ is the unit vector in the direction of $\boldsymbol{V} - \boldsymbol{c}$, and the shear vector becomes

$$\boldsymbol{S} = \frac{\mathrm{d}(\boldsymbol{V} - \boldsymbol{c})}{\mathrm{d}z} = \frac{\mathrm{d}|\boldsymbol{V} - \boldsymbol{c}|}{\mathrm{d}z} \boldsymbol{p} + |\boldsymbol{V} - \boldsymbol{c}| \frac{\mathrm{d}\boldsymbol{p}}{\mathrm{d}z}. \tag{8.4.14}$$

The speed shear and directional shear are represented by the first and second terms on the right-hand side of (8.4.14). The directions of the shear vector in a composite hodograph are illustrated in Fig. 8.21. Figure 8.22a shows an example of a shear vector that contains only the speed shear. In this figure, $\boldsymbol{S}$ is oriented parallel to the line LMH. Directional shear, such as that shown in Fig. 8.22b, plays an essential role in affecting the storm rotation and propagation, which will be discussed later.

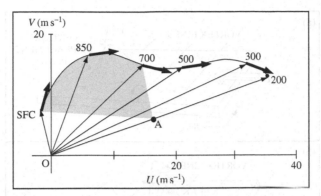

Fig. 8.21 A composite sounding for wind hodograph from 62 tornado outbreak cases. The winds are computed at each level relative to the estimated storm motion. Heavy arrows denote the direction of the shear vector at each level (in hPa). The estimated mean storm motion is denoted by the vector OA. The shaded area is proportional to the 0–3 km storm-relative environmental helicity (Davies-Jones 1984, adapted after Maddox 1976.)

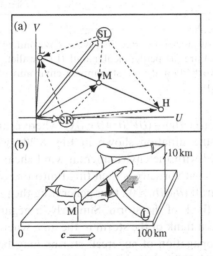

Fig. 8.22 Conceptual model for a right-moving supercell storm (SR) proposed by Browning (1964). (a) Horizontal wind plot showing low-(L), middle-(M), and high-(H) level winds relative to the ground (solid arrows) and relative to the storm (dashed arrows). The SR moves in the direction denoted by the open arrow. The possible left-moving supercell storm (SL) and winds relative to SL are also shown. (b) Three-dimensional storm-relative airflow within the SR storm viewed from the south-southeast. Area of precipitation at the ground is shaded and the position of gust front is denoted by a barbed line and, when present, the tornado symbol. (Adapted after Browning 1964.)

Storm dynamics are highly nonlinear and complex, which prohibits a closed form of analytical theory even in its linear form. However, a few basic dynamics of storms can still be understood by studying the linear dynamics. For example, vertical tilting of the horizontal vorticity can be understood by considering the linearized form of the tilting

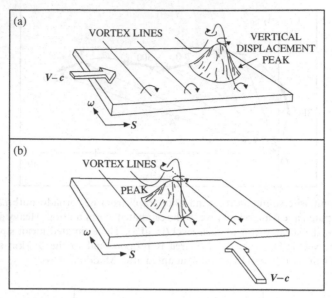

Fig. 8.23 Effects of a localized updraft, represented by the vertical displacement "peak" (i.e. "hump" in isentropic surface), on vortex lines when environmental vorticity ω and storm-relative mean flow $(V - c)$ are (a) perpendicular and (b) parallel, which will produce pure crosswise environmental vorticity and pure streamwise environmental vorticity, respectively. (From Davies-Jones 1984.)

term in (8.4.13), $(\partial U/\partial z)(\partial w/\partial y) - (\partial V/\partial z)(\partial w/\partial x)$, where U and V are the basic wind speeds. Now consider the situation shown in Fig. 8.20a where positive horizontal vorticity is associated with the environmental wind shear $(\partial U/\partial z > 0; V = 0$ and $\partial V/\partial z = 0$). This horizontal vorticity is then tilted into the vertical when the vortex tube encounters the updraft $(\partial w/\partial y > 0)$, thus generating the positive vertical vorticity shown at the southern flank of the storm. Similarly, a negative vertical vorticity is generated on the northern flank of the storm in the area where $\partial w/\partial y < 0$.

The rotation and propagation of supercell storms may be understood from the streamwise vorticity and helicity point of view. When an isentropic surface bulges up by an updraft, it may be represented by the "peak" of a θ_e surface, as sketched in Fig. 8.23. For the situation in Fig. 8.23a, the storm-relative mean flow $(V - c)$ is perpendicular to the environmental vorticity vector (ω), which is called *crosswise vorticity*. Note that c is the storm translation velocity or motion vector. The vortex lines are forced to go up over the peak, giving rise to a cyclonic vorticity on the right-hand side (facing downshear) of the peak and an anticyclonic vorticity on the left-hand side of the peak. The updraft (downdraft) is located upstream (downstream) of the peak in the storm-relative moving frame. In this case, there is no correlation between vertical velocity and vertical vorticity. On the other hand, in the situation of Fig. 8.23b, the storm-relative mean flow $(V - c)$ is parallel to the environmental vorticity vector (ω), which is called *streamwise vorticity*. The upslope (downslope) side of the peak is also the cyclonic (anticyclonic) side. The vertical velocity and vertical vorticity are correlated. Thus, the supercell storm motion may lead to the generation of helical updraft rotation via vertical tilting of the *environmental streamwise vorticity*, which is defined as

$$\zeta_s = \frac{(V - c) \cdot \omega}{|V - c|}. \tag{8.4.15}$$

In fact, the tendency of an environment to produce a rotating storm can be measured by a quantity called the *storm-relative helicity* (Lilly 1986) which is defined as the integration of the streamwise vorticity times the storm-relative flow speed through the depth of the primary inflow layer of the storm h:

$$H(c) = \int_0^h (V - c) \cdot \omega \, dz. \tag{8.4.16}$$

Helicity is conserved for an inviscid and homogeneous fluid. A rotating storm requires a high correlation between the storm-relative velocity and the streamwise vorticity, which therefore requires a high helicity for the environmental wind. Since $\omega \approx k \times dV/dz$, the *storm-relative environmental helicity (SREH)*, (8.4.16), may also be defined as (Davies-Jones 1984):

$$H(c) = -\int_0^h k \cdot (V - c) \times \frac{dV}{dz} \, dz. \tag{8.4.17}$$

The 0–3 km SREH can be calculated as twice the area swept out by the storm-relative wind vector between the surface and 3 km in a hodograph (shaded area in Fig. 8.21). Approximate ranges of SR helicity favorable for weak, strong and violent right movers are found to be 150–299 $m^2 s^{-2}$, 300–449 $m^2 s^{-2}$ and greater than 450 $m^2 s^{-2}$, respectively. In addition, the concurrent presence of high SR helicity values and the existence of long-lived updrafts imply strong right movers, which may favor tornadogenesis. Thus, SREH may serve as an index for supercell forecasting and potential tornado warning.

8.4.5 Effects of directional shear

As briefly mentioned in Section 8.4.1, cyclonically rotating, right moving storms, often referred to as the severe right (SR) storms or right movers, are dominant when the vertical shear is directional and clockwise. In fact, anticyclonically rotating, left-moving storms are rarely observed. Observations indicate that the majority of super-cell storms in the United States rotate cyclonically. This bias is not directly due to the

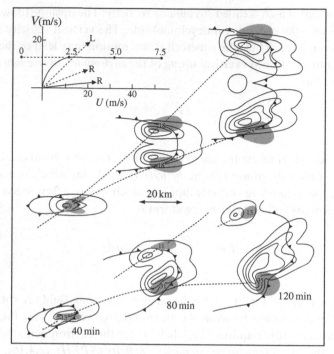

Fig. 8.24 Evolution of numerically simulated storm splitting in a unidirectional shear (upper panel) and a directional shear (lower panel). The hodographs for these two cases are shown in the upper-left corner. The unidirectional shear, i.e. the straight-line hodograph, produces two identical, mirror-image right- and left-moving storms, while the directional shear, i.e. the curved-line hodograph, enhances the cyclonically rotating, right-moving storm. The rainwater content is contoured at $2 \, g \, kg^{-1}$ intervals and the midlevel (6 km) updrafts (in $m \, s^{-1}$) are shaded. The barbed lines denote the surface cold-air outflow boundary. The dotted lines denote the tracks of the updraft center. Arrows in the hodograph denote the supercell propagation velocities for the unidirectional (dashed) and directional (solid) shear profiles. (From Klemp 1987. Reprinted, with permission, from the *Annual Review of Fluid Mechanics*, Vol. 19 @ 1987 by Annual Reviews.)

Coriolis force, however. Instead, it results from the fact that the low-level shear vector turns anticyclonically with height throughout most of the midlatitudes. This orientation of the wind profiles, in turn, is influenced by the sign of the Coriolis parameter. This directional shear is clearly shown in the composite sounding of tornadic thunderstorms in the central United States (Fig. 8.21).

Figure 8.24 shows, for two completely different wind profiles, how a numerically simulated storm evolves in the absence of the earth's rotation and surface drag. One profile has a unidirectional shear, i.e. a straight-line hodograph, and the other exhibits a directional shear, i.e., a curved-line hodograph, which is representative of a clockwise turning of the wind-shear vector with height. In the case of unidirectional shear, the initial storm splits into two storms after 40 min. The storms move apart as they propagate towards the northeast. As can be seen from the figure, they evolve into two almost identical, mirror-image right- and left-moving storms. When there is clockwise turning of the wind-shear vector with height, however, the cyclonically rotating, right-moving storm evolves into an intense storm while the anticyclonically rotating, left-moving storm cannot develop any further and is short-lived. The key factor in storm longevity is the turning of the wind-shear vector, as opposed to the wind vector. The wind vector turns in a clockwise direction in both cases, and does not directly influence the length of the storm's lifetime.

The means by which directional shear enhances the right-moving storm can be explained by examining the linear dynamic pressure equation of (8.4.9),

$$\nabla^2 p'_d = -2\bar{\rho} \mathbf{S} \cdot \nabla_H w, \tag{8.4.18}$$

where $\mathbf{S}$ is the shear vector as defined in (8.4.14). Since the Laplacian of a variable is roughly proportional to the negative of the variable itself in the flow interior, (8.4.18) implies

$$p'_d \propto \mathbf{S} \cdot \nabla_H w. \tag{8.4.19}$$

Therefore, within the context of linear theory, the interaction between the updraft and the shear flow produces high (low) pressure on the upshear (downshear) side of the maximum region of the updraft, as seen in the directional shear figure (Fig. 8.25b). In other words, a high-to-low pressure gradient develops across the updraft in the direction of local shear vector. Since the vertical velocity and the gradient of w bounding it are strongest at midtroposphere, this pressure perturbation is also strongest at that level. The cumulative strength of all three produces an upward pressure-gradient acceleration in the lower troposphere on the downshear side of the storm. In the unidirectional shear case (Fig. 8.25a), an eastward (downshear) high-to-low pressure gradient force is produced because the shear vector points eastward at all levels. As the updraft intensifies, these pressure perturbations reinforce inflow on the downshear (east) side. However, they do not contribute to any preferential growth on either side of the storm.

8.5 Tornado dynamics

8.5.1 Supercell tornadogenesis

Based on measured local circulations and wind speeds, the tornado is the most intense type of circulation observed in the atmosphere. A *tornado vortex* has a typical diameter of a few hundred meters and a maximum tangential wind speed of up to $140 \, \text{m s}^{-1}$. Observations show that most intense tornadoes are produced by supercell storms, although not all supercell storms produce tornadoes. Out of the 5322 individual mesocyclones detected by Doppler radars, only 26% were associated with tornadoes (e.g., Trapp *et al.* 2005). A tornado occurring in a rotating supercell

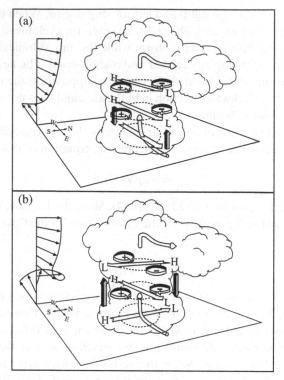

Fig. 8.25 Pressure and vertical vorticity perturbed by the interaction of an updraft in a supercell storm with the environmental wind shear for (a) unidirectional shear and (b) directional shear. The high (H) to low (L) horizontal pressure gradients parallel to the shear vectors (flat arrows) are labeled along their preferred locations of cyclonic (+) and anticyclonic (−) vorticity. The vertical high-to-low pressure gradients are depicted by shaded arrows. (Adapted from Rotunno and Klemp 1982 and Klemp 1987. Reprinted, with permission, from the *Annual Review of Fluid Mechanics*, Vol. 19 @ 1987 by Annual Reviews.)

storm is often referred to as a *supercell tornado*. Supercell tornadoes are preceded by a deep persistent mesocyclone of a mean diameter 3–9 km and a vertical vorticity that is typically greater than 0.01 s^{-1}. Tornadoes that occurr in the absence of mesocyclones are referred to as *nonsupercell tornadoes*.

When a supercell storm transitions into its tornadic phase, the storm's circulation and quasi-steady structure experience a significant alteration. Based on observations, the following processes have been found to be associated with the evolution of the mesocyclone and updraft: (1) a rapid increase in low-level rotation, (2) a decrease in updraft intensity, (3) a small-scale downdraft forming behind the updraft, and (4) a low-level flow in which outflow and inflow air spiral around the circulation center. A low-level mesocyclone acquires its rotation by vertically tilting the baroclinically generated horizontal vorticity via the updraft and a combination of both the updraft

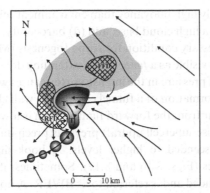

Fig. 8.26 Schematic of a low-level flow during supercell tornadogenesis. Radar echoes with higher intensity are shaded; the boundary between the warm inflow and cold outflow and the "occluded" gust front are denoted by frontal symbols. Region of low-level updraft is gradient-shaded; forward flank downdraft (FFD) and rear flank downdraft (RFD) region are cross-hatched; storm-relative surface streamlines are denoted by arrows and the likely locations of tornadoes are shown by T's. (Adapted after Lemon and Doswell 1979.)

and downdraft. The baroclinicity and updraft come from the forward flank outflow boundary.

During tornadogenesis, the low-level flow behaves in a similar manner to that shown in Fig. 8.26. Tornadogenesis within a supercell storm is often preceded by the development of a *rear flank downdraft* (RFD) on the upshear side of the updraft. On the forward (downshear) flank of the updraft, another area of downdraft, i.e. *forward flank downdraft* (FFD), is generated by evaporative cooling. Along the edge of the FFD, horizontal vorticity is baroclinically produced along the boundary between warm and cold air (on the south and north side, respectively). The vorticity is tilted upward into the updraft (denoted by encircled "T" in Fig. 8.26). The vertical vorticity associated with the low-level mesocyclone amplifies as the air column is stretched by the updraft. The boundary between RFD and the warm inflow from the southeast, and the FFD and the warm inflow from the east, behaves like a storm-scale frontal system. As the RFD intensifies the downdraft outflow progresses cyclonically around the center of rotation which circles around the updraft. The updraft and rotation associated with the mesocyclone may develop to tornado intensity and form a tornado. Recent *in situ* surface observations indicate that: (a) tornadogenesis is more likely and tornado intensity and longevity increase as the surface buoyancy, potential buoyancy (measured by CAPE), and equivalent potential temperature in the rear RFD increase, and as the convective inhibition associated with RFD parcels at the surface decreases; (b) evaporative cooling and entrainment of midlevel potentially cold air may play smaller roles in the development of RFDs associated with tornadic supercells compared to nontornadic supercells; (c) environments characterized by very moist boundary layer air and a low cloud base may be more conducive to RFDs associated with relatively high buoyancy than environments characterized by very dry boundary layer air and a high cloud base; and (d) baroclinity at the surface within the hook echo is not a necessary condition for tornadogenesis (Markowski *et al.* 2002).

The *tornado vortex* is visible as a *funnel cloud* extending down from the rotating *wall cloud* base. Lowering of pressure in the intense vortex causes water vapor to condense, which then causes the formation of a funnel cloud. The wall cloud forms at the lifting condensation level of air from the forward flank, which is lower than that of environmental inflow. An intense supercell updraft prevents precipitation from forming until the air parcels have ascended to higher levels. A hook-shaped *echo* (*hook echo*) appears in radar images (Figs. 8.16 and 8.17). Sometimes, if the updraft is intense, the WER becomes bounded and is referred to as BWER. A hook echo (Fig. 8.17) may form as the mesocyclone advects precipitation away from the precipitation core and around the updraft. The formation of a hook echo serves as a precursor for predicting the location of a tornado. Since the winds of a mesocyclone are cyclonic, the reflectivity signature of a hook echo will have a cyclonically shaped hook. Inside the hook, the only area free from reflectivity is the updraft and inflow notch region of the supercell. Many of the violent tornadoes associated with classic supercells will show a distinct hook echo. Before tornado touchdown at the surface, Doppler radar may detect a local horizontal shear region, which is referred to as the *tornadic vortex signature* (TVS). Not all tornadoes have detectable TVS, especially for those that build upward (Trapp *et al.* 1999). In this type of tornado, the mesocyclone occludes as the RFD cuts off the supply of warm moist air, causing the original updraft to weaken. Although the supercell storm can persist in a quasi-steady configuration for up to several hours, the tornadogenesis process usually occurs in less than 10 minutes.

Many observed features of tornadic storms have been reproduced by cloud model simulations. Figure 8.27 shows the flow fields near the surface as simulated by a cloud model. The model uses a horizontal resolution of 120 m and a stretched grid in the vertical with a resolution of 120 m near the surface. At this time of simulation, the tornado is at its most intense stage, which is located at $(x, y) = (28.8, 36.0 \, km)$ (Fig. 8.27a). There are two maximum areas of vertical velocity, one associated with the tornado and the other associated with the gust front. The rear-flank downdraft, initially located to the northwest of the mesocyclone, merges with the occlusion downdraft. An *occlusion downdraft* is a small-scale downdraft that occurs during the collapse phase of a supercell as the gust front occludes with the stationary boundary, where warm inflow air is meeting outflow from the downdraft at the front flank of the supercell (Klemp *et al.* 1981). It is hypothesized that the RFD brings rotation to the surface initially, and then the gust front occludes. The occlusion process causes a tightening of low-level cyclonic vorticity, intensifying the preexisting downdraft (Markowski *et al.* 2002). A column of vertical vorticity greater than $0.125 \, s^{-1}$ extends from the surface to a height of about 4 km (Fig. 8.27b). A surface mesolow of $-17.3 \, hPa$ is produced (Fig. 8.27c) and collocated with the peak vorticity where the low-level mesocyclone is located. The maximum storm-relative wind at this level

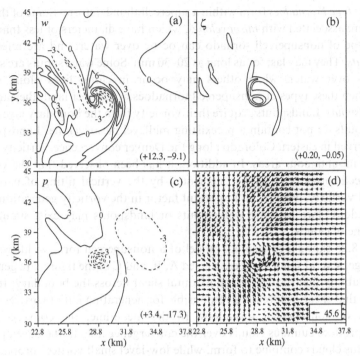

Fig. 8.27 Numerically simulated tornadogenesis. The low-level horizontal flow fields at 87 min of simulation are shown: (a) vertical velocity with a contour interval of $3\,\mathrm{m\,s^{-1}}$; (b) vertical vorticity with contour intervals of 0.01, 0.05, 0.1, 0.2, and $0.3\,\mathrm{s^{-1}}$; (c) the pressure field with contour interval of 3 hPa; and (d) the horizontal wind vector field. All fields, except for the vertical velocity field, are at 100 m height. The maximum and minimum values for every field are shown at the lower right corner. (From Wicker and Wilhelmson 1995.)

(100 m) exceeds $45\,\mathrm{m\,s^{-1}}$, while the maximum ground-relative wind speed exceeds $60\,\mathrm{m\,s^{-1}}$ (Fig. 8.27d). Prior to tornadogenesis, updrafts in the northern half of the mesocyclone intensify rapidly, causing the mesocyclone circulation to shrink in diameter. The tornado's decay begins when the surrounding updrafts weaken and the occlusion downdraft moves in on the tornado.

8.5.2 Nonsupercell tornadogenesis

Although the most intense long-lasting tornadoes are found to form within a supercell thunderstorm, tornadoes may also form within a nonsupercell thunderstorm. In other words, a mid-level mesocyclone is not a necessary condition for tornadogenesis. Nonsupercell tornadoes often occur along a stationary or slowly moving front, or in a horizontal windshift line. They are also referred to as gust-front tornadoes, or *type II tornadoes*; *type I tornadoes* form within a supercell thunderstorm. Some of their parent vortices are associated with *misocyclones*, which have diameters of less than 4 km.

This type of nonsupercell tornado can occur over water, and are referred to as *waterspouts*. They may last for as long as 20–30 min. Some waterspouts are spun up by supercells over water, while others may occur in cloud lines along sea-breeze fronts.When these types of nonsupercell tornadoes occur over land, they are referred to as *landspouts*. Landspouts acquire their vorticity from the boundary layer and their parent clouds do not contain a preexisting midlevel mesocyclone. Landspouts have been observed in eastern Colorado along the Denver convergence-vorticity zone, and also over mountainous areas. In addition, it has been observed that the vorticity of nonsupercell tornadoes may be augmented by the vertical tilting of baroclinically generated vorticity, which is an important factor in the vorticity generation of supercell tornadoes. Two or more waterspouts or landspouts may exist simultaneously along a line of clouds.

Figure 8.28 shows a conceptual model of a nonsupercell tornado lifecycle. In the initial stage, several small vortices, labeled A, B, and C in the figure, are generated by shear instability that results from horizontal shear across the boundary. It has been proposed that shear instability is responsible for generating initial vortices for different types of nonsupercell tornadoes. In the meantime, the convergence-forced updrafts produce cumulus clouds over the convergence line. In the developing stage, the cumulus clouds continue to form, while low-level small vortices propagate along the convergence line. These vortices then interact and merge with each other to create a misocyclone within which landspouts or waterspouts ultimately develop. These vortices derive their vertical vorticity mainly from the stretching of the preexisting vertical vorticity by $\partial w/\partial z$ term in the vertical vorticity equation; although the vertical tilting of baroclinically induced horizontal vorticity may also help enhance the low-level vorticity of initial vortices. In the mature stage, one of the vortices, such as vortex C, is collocated with the adjacent updraft of a towering cumulus by vorticity stretching due to convection.

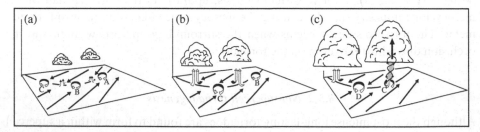

Fig. 8.28 A conceptual model of the life cycle of nonsupercell tornadoes. The radar detectable convergence line is denoted by a bold solid line; low-level vortices are labeled by letters. Three stages are shown: (a) initial stage, (b) developing stage, and (c) mature stage. (From Wakimoto and Wilson 1989.)

8.5.3 Tornado vortex dynamics

In the developing stage of a tornado vortex, a spiral updraft strengthens from the near-surface inflow to the base of the funnel cloud. The vortex circulation is invisible below the funnel cloud unless there is dust or debris. At the mature stage, the circulation surrounding the tornado vortex is rather complicated. In the following, we will qualitatitvely describe the basic tornado vortex dynamics. Conceptually, the flow structures of the tornado vortex circulation may be classified into four major regions (Fig. 8.29). Region Ib is the *tornado core*, which surrounds the rotating axis of the tornado vortex and extends out to the radius of maximum tangential wind. The core is approximately in *solid-body rotation*. Region Ia is the *outer flow*, which consists of converging air. Region II is the *boundary layer*, where the frictional force destroys the cyclostrophic flow balance (i.e. the balance between the pressure gradient force and centrifugal force). The net inward force of these three forces drives strong radial inflow into Region III, which is the *corner flow* region, where a strong updraft is produced and flows into the tornado core. Region IV is the *buoyant updraft*, which caps the vortex within the parent cloud. Tornado vortex dynamics have been studied extensively since the 1970s via observations, theories, tornado vortex chamber experiments and numerical modeling simulations.

Flow regime of the vortex circulation changes as the *swirl ratio* ($S = v_c/w_c$ at the radius of updraft or core radius r_c, where v_c is the tangential velocity of the inflow at r_c, and w_c is average vertical velocity of the updraft at the domain top) increases. If S is very small, then a one-cell vortex is formed (Figs. 8.30a and 8.30b). Note that *boundary layer separation*, as discussed in subsection 5.4.2, occurs in the corner flow for a viscous fluid with a very small swirl ratio (Fig. 8.30a). When the swirl ratio increases to a moderate value, the dynamic low pressure associated with the vortex near the rotation center, as indicated by (8.4.10), may induce a downward motion due

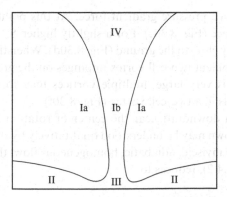

Fig. 8.29 Different flow regions of a tornado vortex: (a) Region Ib: the tornado core, (b) Region Ia: the outer flow, (c) Region II: the boundary layer, (d) Region III: the corner flow region, and (e) Region IV: the buoyant updraft. (From Davies-Jones *et al.* 2001 and Lewellen 1976.)

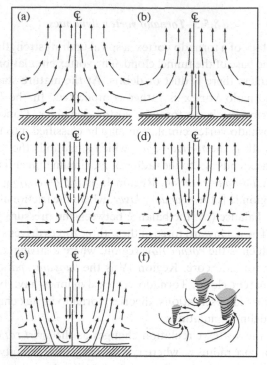

Fig. 8.30 Effects of the swirl ratio (S) on tornadic vortex circulation: (a) very small S: boundary layer separation, (b) low S: one-cell vortex, (c) moderate S: vortex breakdown above the surface, (d) large S: two-cell vortex with downdraft touches down on the surface, (e) larger S: than (d): turbulent two-cell vortex, and (f) very large S: multiple vortices rotating about the annulus separating the two cells in (c). (From Davies-Jones *et al.* 2001.)

to the strong downward pressure gradient force. At this point, a *vortex breakdown* occurs above the surface (Fig. 8.30c). For a slightly higher S, the vortex breakdown stagnation point is very close to the ground (Fig. 8.30d). When the swirl ratio increases to a large value, a turbulent two-cell vortex impinges on the ground (Fig. 8.30e), and when the swirl ratio is very large, multiple vortices may form about the annulus separating the two cells in a two-cell vortex (Fig. 8.30f).

The formation of a downdraft near the center of rotation and a two-cell vortex during vortex breakdown may be understood qualitatively by the following argument (Rotunno 1984). In an inviscid, adiabatic, homogeneous flow, the vertical momentum equation, (7.3.8) or (8.4.2), reduces to

$$\frac{Dw}{Dt} = -\frac{1}{\rho_o}\frac{\partial p}{\partial z}. \tag{8.5.1}$$

Taking $\partial/\partial r$ of the above equation and using the cyclostrophic flow relation

$$\frac{v^2}{r} = \frac{1}{\rho_0}\frac{\partial p}{\partial r}, \qquad (8.5.2)$$

leads to

$$\frac{\partial}{\partial r}\left(\frac{Dw}{Dt}\right) = -\frac{\partial}{\partial z}\left(\frac{v^2}{r}\right). \qquad (8.5.3)$$

Thus, the vertical acceleration decreases toward the center of rotation when the tangential velocity decreases with height. In reality, the boundary layer dynamics cannot be ignored, since its influence helps to produce the flow in the boundary layer and corner flow regions (Figs. 8.29). The four different flow regions of a tornado vortex, as mentioned above, may be understood by considering a few special solutions, such as the steady state, for the above set of equations.

In the boundary layer away from the axis of rotation (Region II of Fig. 8.29), an inbalance between the inward pressure gradient force and the reduced centrifugal force causes the surface friction to retard the rotating flow and induce a radial inflow within the boundary layer. A tornado's boundary layer has a depth of about 100 m.

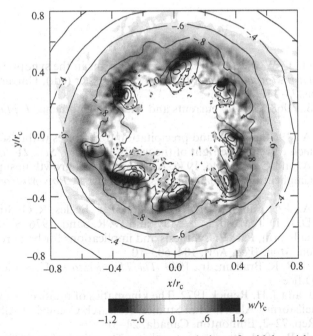

Fig. 8.31 Effects of turbulence in a large eddy simulation of a high swirl tornado vortex. Instantaneous fields of perturbation pressure (solid contours) and normalized vertical velocity (grayscale) are shown at a height of $0.2r_c$, where r_c is the core radius (e.g., 200 m). The tangential velocity at r_c is v_c. Negative vertical velocity areas are enclosed by dashed lines. (Adapted after Lewellen *et al.* 2000; courtesy of D. C. Lewellen.)

Since the buoyant updraft of a tornado vortex (Region IV of Fig. 8.29) is embedded in the parent storm, the dynamics of the buoyant updraft are therefore strongly influenced by those of the parent storm.

In the boundary layer near the axis of rotation, the radial inflow associated with the secondary circulation produces strong convergence and then a vertical jet due to mass continuity. This region is called the *corner flow* (Region III in Fig. 8.29). With a laminar flow, the details of the corner flow are rather sensitive to the constant eddy viscosity coefficients used. In other words, the corner flow associated with the tornado vortex is quite sensitive to turbulence. Figure 8.31 shows one example of the effect of turbulence in a large eddy simulation of a tornado vortex with a high swirl ratio. The instantaneous pressure and vertical velocity near the surface at a height of $r_c = 0.2$, where r_c is the core radius (e.g. 200 m), are shown. A large amount of turbulent kinetic energy is contained in the region. Several (seven in this case) secondary vortices rotating about the main vortex are clearly depicted in the figure. The strong updraft and downdraft couplet associated with each secondary vortex is largely due to its tilt. These secondary vortices provide a net angular momentum transport directed inward into the center because the vortex flow is nearly a solid body rotation so the angular momentum increases outward.

References

Barnes, S. L. and C. W. Newton, 1986. Thunderstorms in the synoptic setting. *Thunderstorm Morphology and Dynamics*, E. Kessler (ed.), University of Oklahoma, 75–112.

Benjamin, T. B., 1968. Gravity currents and related phenomena. *J. Fluid Mech.*, **31**, 209–48.

Browning, K. A., 1964. Airflow and precipitation trajectories within severe local storms which travel to the right of the winds. *J. Atmos. Sci.*, **21**, 634–9.

Browning, K. A. and G. B. Foote, 1976. Airflow and hail growth in supercell storms and some implications for hail suppression. *Quart. J. Roy. Meteor. Soc.*, **102**, 499–534.

Browning, K. A., J. C. Fankhauser, J.-P. Chalon, P. J. Eccles, R. G. Straugh, F. H. Merrem, D. J., D. J. Musil, E. L. May, and W. R. Sand, 1976. Structure of an evolving hailstorm, Part V: Synthesis and implications for hail growth and hail suppression. *Mon. Wea. Rev.*, **104**, 603–10.

Byers, H. R. and R. R. Braham, Jr., 1949. *The Thunderstorm*. U. S. Government Printing Office.

Chisholm, A. J. and J. H. Renick 1972. The kinematics of multicell and supercell Alberta hailstorms. Alberta Hail Studies, Research Council of Alberta Hail Studies, Rep. 72–2, Edmonton, Canada, 24–31.

Davies-Jones, R. P., 1984. Streamwise vorticity: The origin of updraft rotation in supercell storms. *J. Atmos. Sci.*, **41**, 2991–3006.

Davies-Jones, R. J. Trapp, and H. B. Bluestein, 2001. Tornadoes and tornadic storms. *Severe Convective Storms*, C. A. Doswell (ed.), Meteor. Monogr., 28, Amer. Meteor. Soc., 167–221.

Doswell, C. A. III, 2001: *Severe Convective Storms.* Meteor. Monogr., 28, Amer. Meteor. Soc.

Fovell, R. G. and P.-H. Tan, 1998. The temporal behavior of numerically simulated multicell-type storms. Part II: The convective cell life cycle and cell regeneration. *Mon. Wea. Rev.*, **126**, 551–77.

Fujita, T. T. and R. M. Wakimoto, 1983. Microbursts in JAWS depicted by Doppler radars, PAM, and aerial photographs. Preprints, 21st Conf. on Radar Meteorology, Edmonton, Canada, Amer. Meteor. Soc., 638–45.

Gilmore, M. S. and L. J. Wicker, 1998. The influence of midtropospheric dryness on supercell morphology and evolution. *Mon. Wea. Rev.*, **126**, 943–58.

Igau, R. C., M. A. LeMone, and D. Wei. 1999. Updraft and downdraft cores in TOGA COARE: Why so many buoyant downdraft cores? *J. Atmos. Sci.*, **56**, 2232–45.

Jorgensen, D. P. and M. A. LeMone. 1989. Vertically velocity characteristics of oceanic convection. *J. Atmos. Sci.*, **46**, 621–40.

Klemp, J. B., 1987. Dynamics of tornadic thunderstorms. *Annu. Rev. Fluid Mech.*, **19**, 369–402.

Klemp, J. B., R. B. Wilhelmson, and P. S. Ray, 1981. Observed and numerically simulated structure of a mature supercell thunderstorm. *J. Atmos. Sci.*, **38**, 1558–80.

Lemon, L. R. and C. A. Doswell III, 1979. Severe thunderstorm evolution and mesocyclone structure as related to tornadogenesis. *Mon. Wea. Rev.*, **107**, 1184–97.

Lewellen, W. S., 1976. Theoretical models of the tornado vortex. Proc. Symp. On Tornadoes: Assessment of Knowledge and Implications for Man, Lubbock, TX, Texas Tech University, 107–43.

Lewellen, W. S., 1993. Tornado vortex theory. In *The Tornado: Its Structure, Dynamics, Prediction, and Hazards*, C. Church *et al.* (eds.), Amer. Geophys. Union, Geophys. Monogr., 79, 19–40.

Lewellen, D. C., W. S. Lewellen, and J. Xia, 2000. The influence of a local swirl ratio on tornado intensification near the surface. *J. Atmos. Sci.*, **57**, 527–44.

Lilly, D. K., 1986. The structure, energetics and propagation of rotating convective storms. Part II: Helicity and storm stabilization. *J. Atmos. Sci.*, **43**, 126–40.

Lin, Y.-L. and L. E. Joyce, 2001. A further study of the mechanisms of cell regeneration, propagation, and development within two-dimensional multicell storms. *J. Atmos. Sci.*, **58**, 2957–88.

Lin, Y.-L., R. L. Deal, and M. S. Kulie, 1998. Mechanisms of cell regeneration, development, and propagation within a two-dimensional multicell storm. *J. Atmos. Sci.*, **55**, 1867–86.

Maddox, R. A., 1976. An evaluation of tornado proximity wind and stability data. *Mon. Wea. Rev.*, **104**, 133–42.

Market, P., A. Allen, R. Scofield, R. Kuligowski, and A. Gruber, 2003. Precipiation efficiency of warm-season Midwestern mesoscale convective systems. *Wea. Forec.*, **18**, 1273–85.

Markowski, P. M., J. M. Straka, and E. N. Rasmussen, 2002. Direct surface thermodynamic observations within the rear-flank downdrafts of nontornadic and tornadic supercells. *Mon. Wea. Rev.*, **130**, 1692–721.

Newton, C. W., 1966. Circulations in large sheared cumulonimbus. *Tellus*, **18**, 699–712.

Parker, M. D. and R. H. Johnson, 2004. Structures and dynamics of quasi-2D mesoscale convective systems. *J. Atmos. Sci.*, **61**, 545–67.

Proctor, F. H., 1989. Numerical simulation of an isolated microburst. Part II: Sensitivity experiments. *J. Atmos. Sci.*, **46**, 2143–65.

Rotunno, R., 1984. An investigation of a three-dimensional asymmetric vortex. *J. Atmos. Sci.*, **41**, 283–98.

Rotunno, R. and J. B. Klemp, 1982. The influence of the shear-induced pressure gradient on thunderstorm motion. *Mon. Wea. Rev.*, **110**, 136–51.

Rotunno, R., J. B. Klemp, and M. L. Weisman, 1988 (*RKW Theory*). A theory for strong, long-lived squall lines. *J. Atmos. Sci.*, **45**, 463–85.

Simpson, J., N. E. Westcott, R. J. Clerman, and R. A. Pielke, 1980. On cumulus mergers. *Arch. Meteor. Geophys. Bioklim.*, Ser. **A29**, 1–40.

Thorpe, A. J., M. J. Miller, and M. W. Moncrieff, 1982. Two-dimensional convection in nonconstant shear: a model of mid-latitude squall lines. *Quart. J. Roy. Meteor. Soc.*, **108**, 739–62.

Trapp, R. J., E. D. Mitchell, G. A. Tipton, D. W. Effertz, A. I. Watson, D. L. Andra Jr., and M. A. Magsig, 1999. Descending and nondescending tornadic vortex signatures detected by WRS-88Ds. *Wea. Forec.*, **14**, 625–39.

Trapp, R. J., G. J. Stumpf, and K. L. Manross, 2005. A reassessment of the percentage of tornadic mesocyclones. *Wea. Forec.*, **20**, 680–7.

Wakimoto, R. M., 1985. Forecasting dry microburst activity over the High Plains. *Mon. Wea. Rev.*, **113**, 1131–43.

Wakimoto, R. M. and J. W. Wilson, 1989. Non-supercell tornadoes. *Mon. Wea. Rev.*, **117**, 1113–40.

Wakimoto, R. M., W.-C. Lee, H. B. Bluestein, C.-H. Liu, and P. H. Hildebrand, 1996. Eldora observations during VORTEX 95. *Bull. Amer. Meteor. Soc.*, **77**, 1465–81.

Weisman, M. L., 1992. The role of convectively generated rear-inflow jets in the evolution of long-lived mesoconvective systems. *J. Atmos. Sci.*, **49**, 1826–47.

Weisman, M. L. and J. B. Klemp, 1982. The dependence of numerically simulated convective storms on vertical wind shear and buoyancy. *Mon. Wea. Rev.*, **110**, 504–20.

Weisman, M. L. and R. Rotunno, 2004. "A theory for strong long-lived squall lines" revisited. *J. Atmos. Sci.*, **61**, 361–82.

Wicker, L. J. and R. B. Wilhelmson, 1995. Simulation and analysis of tornado development and decay within a three-dimensional supercell thunderstorm. *J. Atmos. Sci.*, **52**, 2675–703.

Xue, M., K. K. Droegemeier, V. Wong, A. Shapiro, K. Brewster, F. Carr, D. Weber, Y. Liu, and D.-H. Wang, 2001. The Advanced Regional Prediction System (ARPS) – A multi-scale nonhydrostatic atmospheric simulation and prediction tool. Part II: Model physics and applications. *Meteor. Atmos. Phys.*, **76**, 143–65.

Yang, M.-J. and R. A. Houze, Jr., 1995. Multicell squall line structure as a manifestation of vertically trapped gravity waves. *Mon. Wea. Rev.*, **123**, 641–61.

Problems

8.1 Derive (8.4.13).

8.2 The pressure deepening rate of a Rankine vortex is given by:

$$\left(\frac{\partial \bar{P}}{\partial t}\right)_{max} = -\frac{\rho_0 v_{max}^2 c}{r_c},$$

where c is the ground speed of the tornado and other variables are described in the text. Estimate the pressure deepening rate (in $hPa\,s^{-1}$) at a certain location for a tornado with $v_{max} = 150\,m\,s^{-1}$, $r_c = 100\,m$, which moves to this location at a speed of $10\,m\,s^{-1}$. [A Rankine vortex is defined as $v = (C/2\pi)(r/r_c^2)$ for $r \le r_c$, and $v = (C/2\pi r)$ for $r > r_c$, where v_{max} and C are the maximum tangential velocity and circulation, respectively, at the core (critical) radius r_c].

9

Mesoscale convective systems

A *mesoscale convective system* (MCS) is an organized cluster of thunderstorms, which persists at least for several hours and produces a continguous precipitation area. An MCS may be linear or circular in shape, and is often used to refer to a cluster of thunderstorms that does not satisfy the definition of a mesoscale convective complex (MCC) (Section 9.2). MCSs include squall lines and MCCs in the midlatitudes, and tropical storms (cyclones) and cloud clusters in the tropics. The term *mesoscale convective line* has also been used to represent a linear form of mesoscale convective systems.

MCSs have a horizontal scale greater than that of an individual thunderstorm, but smaller than the Rossby radius of deformation. The Rossby radius of deformation is defined in (4.2.12) as $L_R = NL_z/f$, where L_z is the vertical scale of the motion. In the midlatitudes, L_R is approximately 1000 km, assuming $N = 0.01$ s^{-1}, $L_z = 10$ km, and $f = 10^{-4}$ s^{-1}. These conditions require the horizontal scale of a MCS to be approximately 100 km. Observations have indicated that MCSs have a typical lifetime of 3 hours or more, and their accompaning stratiform clouds may be sustained, or at least remain, for several days. Dynamically, this implies that the Coriolis acceleration is at least comparable to the other terms of the momentum equations, such as the inertial acceleration terms. This gives a time scale of f^{-1} (e.g., see Table 1.1), i.e. at least 3 h for midlatitude MCSs, which is consistent with observations. In addition to influencing mesoscale weather, MCSs also affect the general circulation. This effect occurs due to an energy transfer associated with heating that is comparable in scale to L_R, as discussed in Chapter 1 and revealed in Fig. 1.1.

MCSs are often composed of ordinary, multicell and/or supercell thunderstorms (the building blocks of the MCS), which exhibit deep, moist convective overturning adjacent to or embedded within a mesoscale vertical circulation that is at least partially driven by convective processes. Furthermore, stratiform clouds are often generated and maintained by these MCSs, which may incur heavy stratiform rain, in addition to the heavy rain showers from the convective region (Houze *et al.* 1989). The dynamics of MCSs are rather complicated because in addition to the complicated dynamics of individual, isolated thunderstorms, the interactions among these thunderstorms and the generation and maintenance of the associated stratiform clouds and precipitation have to be considered.

9.1 Squall lines and rainbands

9.1.1 Squall line classifications

A squall line is an organized line of thunderstorms that produces a contiguous precipitation area. The organized convection associated with a squall line is created by deep convective cells on a scale of O(1–10 km), embedded in the mesoscale circulations. Based on synoptic environments and flow structures, different conceptual models of squall lines have been proposed: (a) deep-shear model, (b) shallow-shear model, and (c) jet-like shear model (Fig. 9.1). These models show that a low-level flow carrying warm, moist, high θ_e air overruns density current-associated low θ_e air. The lifting of this high θ_e low-level flow by the density current helps raise the low-level air parcels or layer to the realization of conditional or potential instability.

The deep-shear model (Fig. 9.1a) was the earliest conceptual model of midlatitude squall lines. This model assumes the existence of deep shear layer, which extends to the middle or upper troposphere. Strong updrafts tend to tilt against the basic wind shear (upshear tilt), which allows the updraft to unload its precipitation upshear and permits the updraft and downdraft circulations to continue free of interference. Note that this occurs only when the cold pool is strong. This type of quasi-steady two-dimensional convective cell was thought to be the basic building block for a squall line with an upshear tilt that contributes to its longevity by sustaining cool air downdrafts rearward of it. As briefly mentioned in the multicell storm section (Section 8.2), the gust front updraft (GFU) is forced by the cold outflow on the low-level inflow, analogous to the upward motion over the upslope of a mountain because the cold outflow acts more or less like an obstacle to the incoming airflow in a moving

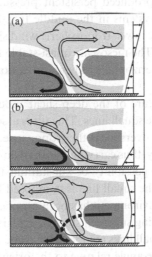

Fig. 9.1 Major conceptual models of squall lines derived from case studies: (a) deep-shear model (Ludlam 1963; Newton 1963); (b) shallow-shear model (Carbone 1982); and (c) jet-like shear model (Zipser 1977). Areas of high (low) θ_e air are lightly (darkly) shaded.

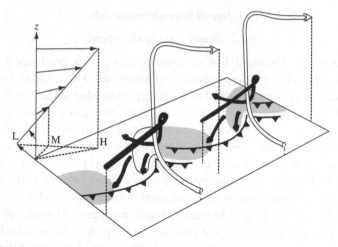

Fig. 9.2 Maintenance of a squall line by three-dimensional effect. See text for details. The environmental wind profile is shown on the left. The relative winds at low, middle, and high levels are denoted by L, M, and H, respectively. The shaded regions indicate the radar image hook echo formed by the rain areas and the barbed lines denote the gust fronts. (After Rotunno *et al*. 1988, based on Lilly 1979.)

frame of reference. The pressure gradient force established by the high pressure at the gust front near the ground and the low pressure above the cold outflow induces the rearward and upward acceleration of the inflow, which may also contribute to the GFU formation. Some important dynamics of this type of squall lines may be summarized as follows: (1) A front-to-rear mid-tropospheric inflow jet develops, mainly due to a buoyancy-produced persistent pressure minimum that occurs on the line-leading side of the systems in the middle troposphere. (2) This midtropospheric inflow jet decreases the lower-tropospheric wind shear and increases the upper-tropospheric wind shear. (3) The mesoscale quasi-stable flow field feeds back into the transient accelerations via horizontal gradients in a nonlinear part of the dynamic pressure perturbation.

The most noteworthy feature of the deep-shear model is the upshear tilt of the updraft found in most mature squall lines. This tilt is associated with the low-level air that rises over the cold pool, overshoots the tropopause, penetrates the lower stratosphere, and turns horizontally, thus carrying the anvil downshear for several thousand kilometers. Considering the possibility of substantial three-dimensional motion, a conceptual model in which the squall line is composed by a line of supercell storms in a deep, strongly sheared environment has been proposed. The deep, strong shear is required to support the rotating supercell storms, explaining the longevity characteristics of the squall line. In squall lines composed by supercells in deep shear (Fig. 9.2), the shear vector forms a large angle relative to the orientation of the convective line, thus helping to avoid the interference of circulations between two adjacent supercells. Idealized numerical simulations indicate that: (a) shear oblique to (say, $\sim 45°$ from) the convective line is most apt for supporting neighboring cyclonic supercells within a squall line, while at the same time supporting an anticyclonic supercell at the down-shear end of the line; (b) shear normal to the line of forcing is favorable for the maintenance of a squall line with isolated supercells at either end; and (c) shear parallel to the line of forcing is favorable for isolated supercells only on the downshear side of the line (Bluestein and Weisman 2000).

A squall line may also be composed by multicell storms. The inflow into the squall line extends from the surface to the 400 hPa level, thus occupying a deep layer. The flow structure is similar to that of the deep-shear model (Fig. 9.1a), except that the air coming from the front of the squall line also feeds the downdraft in the rear part of the storm. In fact, many observed squall lines have the updraft and downdraft both feeding from the front. Careful examination of the flow fields also indicates that the equivalent potential temperature is generally not constant along streamlines, implying the presence of significant cross-stream mass exchange and a strongly time-dependent flow.

Unlike the deep-shear squall lines, many squall lines form in an environment with shallow shear. In this type of squall line, the shear is focused in the layer comparable to that of the cold pool. For example, Fig. 9.1b shows a conceptual model of these squall lines, based on an analysis of a California squall line. Similar to the deep-shear model, the updraft over the density current is tilted upshear (rearward) in this model. The major difference is that nearly the entire anvil portion of the clouds is advected rearward. Two-dimensional numerical simulations have success-fully reproduced the upshear-tilted updraft of this type of squall lines. The upshear tilt of the updraft can be explained by the interaction between the shear-induced and the cold-outflow induced circulations when, $c > \Delta U$ as discussed in Section 8.3 (Fig. 8.11c).

A third type of squall line posseses jet-like shear, but with no steering (critical) levels, as shown in Fig. 9.1c. During the mature stage of this type of squall lines, there is front-to-rear flow at all levels. In other words, the squall line moves so rapidly (~ 10–$15\,\mathrm{m\,s^{-1}}$) that there exists no critical level in the moving frame of reference. Both tropical and midlatitude squall lines of this type share similar flow structures, such as that: (1) both are formed by a line of multicells or supercells and (2) both have pronounced front-to-rear flow that rises at the squall line gust front.

Mesoscale convective lines, which include linear MCSs and squall lines, may also be classified by their precipitation characteristics and associated flow structures. Based on the analysis of eighty-eight linear MCSs which occurred in the central United States during May 1996 and May 1997, the mesoscale convective lines are classified into three types (Parker and Johnson 2000): (a) convective line with *trailing stratiform precipitation* (TS), (b) convective line with *leading stratiform precipitation* (LS), and (c) convective line with *parallel stratiform precipitation* (PS). Idealized patterns of radar reflectivity at initiation, development, and maturity stages for these mesoscale con-vective lines are depicted in Fig. 9.3. Similar mesoscale convective line types have also

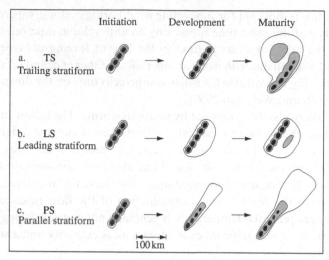

Fig. 9.3 Three types of mesoscale convective lines and their life cycles are classified: (a) convective line with trailing stratiform precipitation (TS); (b) convective line with leading stratiform precipitation (LS); and (c) convective line with parallel stratiform precipitation (PS). Approximate time interval between 3 life cycles are 3–4 h for TS, 2–3 h for LS and PS. Levels of shading roughly correspond to 20, 40, and 50 dBZ. (After Parker and Johnson 2000.)

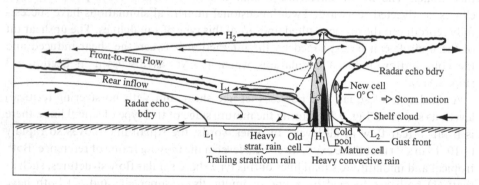

Fig. 9.4 Conceptual model of a mesoscale convective line with trailing stratiform precipitation (TS). Intermediate and strong radar reflectivity is indicated by medium and dark shading. The asterisk denotes ice particles falling from the front-to-rear flow. Mesolows and mesohighs are denoted by L and H, respectively. The bold horizontal arrows denote the environmental flow of the steering type of Moncrieff's (1978) squall line model (Adapted after Houze et al. 1989.)

been observed in other parts of the world. Observations indicate that about 60% of observed mesoscale convective lines belong to TS convective line type, while about 20% of the mesoscale convective lines belong to the LS convective line type, with the remaining 20% belonging to the PS convective line type. The schematic in Fig. 9.4

represents the mature stage of a mesoscale convective line. Major features of the TS mesoscale convective lines are as follows:

(1) During the formation and intensification stages, the trailing stratiform rain is not present, but the convective cells are more intense.
(2) The presence of the leading cloud overhang and precipitation shown on the right-hand side of Fig. 9.4 depends on the strength of the environmental wind shear. For example, when there is strong relative rearward flow at upper levels, there is no upper-level cloud overhang and precipitation.
(3) The front-to-rear flow is located above the rear inflow and advects rearward ice particles falling from the convective cells (denoted by asterisk).
(4) The ice particles collect smaller ice particles and aggregate each other to form large snow aggregates that melt at $0°C$ level to produce bright band radar echoes and a region of heavier stratiform rain directly below.

The LS system mentioned above could be further classified into *front-fed LS* (FFLS) and *rear-fed LS* (RFLS), which are sustained by inflow of high-θ_e air from ahead and behind the system, respectively. Three most common flow structures for observed mesoscale convective lines are: front-fed TS, front-fed LS and rear-fed LS (Parker and Johnson 2004a). They are dynamically unique, and the front-fed TS and the rear-fed LS convective lines have some similar characteristics. Even though there are some differences between the front-fed LS and rear-fed LS, they are kinematically similar and the flow circulations are more or less mirror images of each other. The basic structure of LS convective lines is similar to that of the deep-shear model (Fig. 9.1a) except that it contains a front-to-rear flow or jet. The flow circulation of the TS type is similar to that of the shallow-shear model (Fig. 9.1b). In fact, the flow circulation associated with LS squall lines represents very well the convective line observed in central Oklahoma on May 22, 1976 as shown in Fig. 6.12. PS convective line circulation looks like a mixture of the LS and TS types, although with much weaker anvil advection in the cross-line direction due to weaker wind in that direction when compared to that in the along-line direction. Based on idealized numerical simulations of FFLS convective lines, it is found that (Parker and Johnson 2004a,b):

(a) The middle- and upper-tropospheric wind shear are important to the updraft tilt and overall structure of the simulated systems.
(b) In time, simulated FFLS convective lines tend to evolve toward a convective line with TS precipitation structure because they tend to decrease the line-perpendicular vertical shear nearby.
(c) This, along with gradual increases in the system's cold pool strength, contributes to more rearward-sloping updrafts.

In summary, observations and numerical simulations indicate that the flow circulations and precipitation characteristics of a convective line are controlled by the cold pool strength and the environmental shear. Aside from the flow structure and characteristics of precipitation discussed above, other major surface features of midlatitude squall lines have been observed. These include: (a) presquall mesolow ahead of the convective line (L_2 in Fig. 9.4), (b) squall mesohigh under the convective line (H_1 in Fig. 9.4), (c) wake low associated with the trailing stratiform rain area (L_1 in Fig. 9.4), and (d) mesoscale convective (bookend) vortices at the ends of a squall line.

MCSs may produce *derechos*, which are characterized as severe, straight-line windstorms. The derecho is defined to include any family of downburst clusters associated with an extratropical mesoscale convective system, and has been classified into: (a) *serial derecho*, which is produced by several separate convective elements and is favored under strong large-scale forcing, and (b) *progressive derecho*, which is produced by a single, curved squall line or mesoscale convective system oriented perpendicular to, and with a bulge in the general direction of the mean midtropospheric flow. Both significant upper-tropospheric shear above 5 km and low-level shear play important roles in maintaining the strength of long-lived squall lines. Derecho-type systems may produce severe damage swaths of 100 km or more in width and 500–1000 km in length. A bulging convective line, called a *bow echo* in radar image, has the ability to produce strong, diverging outflow and is typically associated with swaths of damaging surface winds near the apex of the bow.

9.1.2 Formation mechanisms

Mesoscale convective systems are initiated in a variety of large-scale environments. Some form in moist environments with strong wind shear in the form of squall lines or warm-sector rainbands, while others, such as MCCs and a few warm-front rainbands, occur in moist environments with no strong wind shear. Several common ingredients found for the formation of squall lines are: (1) an environment with sufficient moisture for the existence of one or more instabilities, such as conditional and potential instabilities; (2) a mechanical and/or thermal forcing which provides lifting; and (3) an environmental vertical shear that will help organize the convective cells or thunderstorms.

The importance of larger-scale convergence for the formation of cloud convection and mesoscale deep moist convection is well documented. In addition to the required upward motion for moist convection, other important effects of large-scale convergence are its ability to bring a wide region of the atmosphere to saturation before convection begins and to continuously destabilize the environment after convection has been initiated. For example, Fig. 9.5a shows a numerical model simulation of moist convection initiated by momentum forcing associated with large-scale convergence. As noted in Subsection 8.4.3, the upward motion associated with the large-scale convergence is able to produce upward motion near the rigid flat surface through the mass continuity. In the presence of large-scale convergence, the convective system is not as critically dependent on the formation of a low-level cold pool as those initiated by

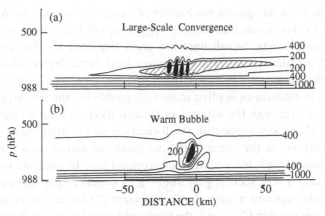

Fig. 9.5 Vertical displacement in (m) needed for air to be lifted to its lifting condensation level for moist convection initiated by (a) large-scale convergence and (b) warm bubble. Cloud areas are shaded and regions where the air has to be lifted less than 100 m are hatched. (Adapted after Crook and Moncrieff 1988)

a warm bubble (Fig. 9.5b). The embedded convective cells in Fig. 9.5a appear to be the result of layer lifting, while the single convective cell in Fig. 9.5b appears to be the result of parcel lifting. The potential instability may play a more dominant role in the layer-lifting situation, while the conditional instability may play a more dominant role in the parcel-lifting situation.

Previous studies have demonstrated that environmental wind shear and cold pools play important roles in the development of squall lines and convective cloud bands by producing circulations that maintain individual thunderstorms and squall lines. The interactions between the low-level shear and cold pool on isolated storms have been discussed in Section 8.3, and the theories are applicable to the situation of squall lines. Occasionally, a critical level exists in a shear flow over a heat source or sink. One example is the squall line shown in Fig. 6.12, which exhibits a critical (steering) level near 6 km. Note that the critical level coincides with the wind reversal level in a steady-state flow because the phase speed is 0 there (Chapter 5). This implies that in a moving frame of reference of a quasi-steady squall line, the wind reversal level corresponds to the critical level. Thus, as implied from (6.4.6), the addition of latent heat at the critical level contributes to the upward motion directly, without losing its energy to the advection, because the mean wind speed in the moving frame at that location is 0. This will help produce optimal growth of the disturbance, equivalent to the addition of heating in a quiescent atmosphere.

Observational, theoretical and numerical studies indicate that squall lines may form by: (1) frontal forcing, (2) orographic forcing, (3) thermal forcing, (4) upper-level forcing, (5) gravity waves, and (6) mesoscale instabilities. Frontal forcing may be associated with either synoptic fronts or mesoscale fronts. Synoptic front forcing is probably the most recognized mechanism of squall line formation. Although both cold and warm fronts can theoretically trigger squall lines, more squall lines are triggered by cold fronts. Squall lines are often triggered by the ascent of warm, moist, high θ_e air (the so-called *warm conveyor belt*) forced by an advancing surface cold front in the warm sector of a midlatitude cyclone. Due to the deep layer of vertical wind shear, this mechanism is often more responsible for the development of deep-shear squall lines through the adjustment toward thermal-wind balance (Fig. 9.1a). This shear is also a key element in the formation of LS squall lines (9.3b), where the inflow overturns in the upper levels and produces and advects the anvil clouds downshear, generating leading stratiform precipitation. In addition to the synoptic fronts, mesoscale fronts, such as gust fronts, may produce near-surface convergence and its associated upward motion that help trigger the formation of squall lines. In particular, as discussed in Chatper 8, the lifting associated with a density current plays an essential role in regenerating convective cells within a multicell storm squall line. The circulations generated by the gust front or cold pool may interact with those generated by the vertical shear and affect the generation of squall lines (e.g., Figs. 8.11 and 8.12).

Orographic and/or sensible heating may produce lines of upward motion for convection and subsequently squall lines. It is well documented that combined orographic and thermal forcing over mesoscale mountain ridges, boundary-layer forcing, and differential heating associated with a sea-breeze front, can produce lines of low level convergence and moist convection through the release of instability if there is abundant moisture and the environmental air is unstable. Figure 9.6 shows an example of an orographically and thermally forced squall line observed on June 21, 1993 over the Rocky Mountain Front Range. The initial convection in this squall line began at 1730 UTC (1030 MST). Convection between 1000 and 1300 MST along the Front Range is common during the summer months and is typically driven by upslope flow. As discussed in Section 6.6, the upslope flow is produced by the buoyancy associated with air heated by the mountain surface. This buoyancy generates a horizontal pressure gradient force toward the mountain surface and creates an upslope flow. This upslope flow provides the upward motion needed for convection and a subsequent squall line. The low-level upward motion may be enhanced by the passage of an upper-level disturbance, such as a short-wave trough, that may provide favorable upper-level divergence which acts to enhance and organize the convection. For example, the short-wave trough that existed at 1900 UTC June 21, 1993 upstream from the genesis region over the Colorado Front Range appears to strengthen and organize the squall line depicted in Fig. 9.6.

Deep convection is also observed to develop in the absence of large-scale forcing. Numerical simulations indicate that convection can be initiated by free tropospheric gravity waves being excited by boundary layer eddies penetrating the stable layer (Redelsperger and Clark 1989). Based on idealized numerical simulations, a density current is able to form through the interaction of a traveling gravity wave

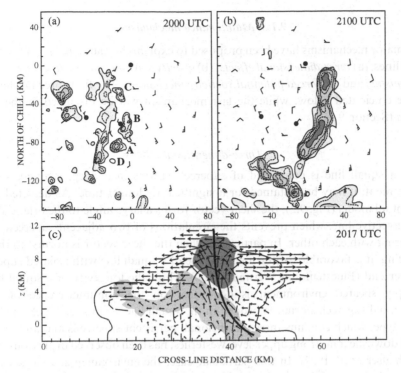

Fig. 9.6 An orographically and thermally induced squall line observed on June 21, 1993. Constant altitude maps of reflectivity fields at 3.0 km above the ground observed by a Doppler radar at (a) 2000 (1300) and (b) 2100 (1400) UTC (MST) June 21, 1993. Contours are in 7.5-dBZ increments beginning at 15 dBZ (dotted). Shading intervals are 15 and 45 dBZ. Surface winds (full barb = 5 m s^{-1}) from the mesonet are also plotted. Four new cells are labeled A, B, C, and D in panel (a). Three radar sites are denoted by dots. The Rocky Mountain Front Range is oriented roughly in north–south direction with the terrain height at the leftmost radar site is about 1.6 km rising westward to 2.8 km in a horizontal distance of about 60 km. The wind vectors on the east-southeast to west-northwest cross section through cell A at 2017 UTC (1317 MST) is plotted in (c). Dark (light) shading indicates divergence (convergence). The dark curve denotes the axis of maximum updraft. Dashed contours denote the radar reflectivity in 10-dBZ increments beginning with 1-dBZ. (Adapted after Grady and Verlinde 1997.)

and cold pool, depending upon the convergence produced by the wave and whether or not the cold pool and the traveling wave are in or out of phase (Lin *et al.* 1993). This indicates that convection may be initiated by gravity waves and their interactions with the cold pool in a moist atmosphere. Release of moist instability is the most effective way to initiate moist convection associated with squall lines. Different types of mesoscale instabilities are able to serve this purpose, such as conditional instability (e.g., Takeda 1971), potential instability (e.g., Crook and Moncrieff 1988), and moist absolute instability (e.g., Bryan and Fritsch 2000).

9.1.3 Maintenance mechanisms

Four major mechanisms have been proposed to explain the maintenance of long-lived squall lines: (a) *three-dimensional effects*, (b) *gravity wave mechanisms*, (c) *local balance mechanisms*, and (d) *forced potential vorticity mechanism*. The first three mechanisms will be elucidated below, while the last mechanism will be discussed in the MCC section (Section 9.2).

a. Three-dimensional effects

When a squall line is composed of supercell storms in a deep, strongly sheared environment, it may be sustained for a significantly longer time. As depicted in the conceptual model (Fig. 9.2), the shear vector forms a large angle to the orientation of the convective line, which prevents the circulations of two adjacent supercells from interfering with each other. In particular, when the shear vector is normal to the line of forcing, it is favorable for the maintenance of a squall line with isolated supercells at either end (Bluestein and Weisman 2000). Thus, the longevity of a squall line in a deeply sheared environment is dependent on the non-interference of three-dimensional supercell storms embedded in the squall line, as in Fig. 9.2. A squall line of this type, which contains three-dimensional, persistent supercells and an anvil that spread downshear in strong upper-level westerlies, has been observed in previous studies (Fankhauser *et al.* 1992). In this particular case, the environmental wind shear was linear at about $4\,m\,s^{-1}\,km^{-1}$ through the troposphere. Although this three-dimensional maintenance mechanism of squall lines theoretically provides a reasonable interpretation of squall line longevity, most squall lines are not composed of supercell thunderstorms (Bluestein and Jain 1985). In this case, the squall lines may be maintained by other mechanisms as described below.

b. Gravity wave mechanism

Convective heating can produce a series of forced internal gravity waves. These waves can, in turn, generate regions of low-level convergence and divergence. The resulting concentration of convection into convergent regions provides a coherent driving mechanism for the wave itself. In this manner, the convection provides the energy and the wave provides the scale selection and organization of convection, thus creating a self-sustaining convective system, also known as a wave-CISK system. This interaction may help explain the longevity of mesoscale convective systems. As depicted in Fig. 4.20, the wave packet consists of one convergent and one divergent region with a dominant wavelength comparable to the scale of the squall line or mesoscale convective system. Convective plumes develop in the convergent region, passing into the divergent region as they decay and produce precipitation. When the squall line or the mesoscale convective system moves at the same speed as the region of surface convergence or the forced gravity waves, then the latent heating can be supported by the upward vertical motion produced by the surface convergence in a coherent manner. The weakness of this mechanism is that many MCSs do not move at the speed of the internal gravity waves.

c. Local balance mechanism

As discussed in Section 8.3, the RKW theory may be used to explain the longevity of a squall line. Based on the RKW theory, the vorticity produced by the density current may counter that produced by the low-level shear on the downshear side of a convective line. Ideally, the long-lived (optimal) state is characterized not by a long-lived cell, but by a long-lived system of convective cells that are constantly generated at the gust front on the downshear side.

In order to elucidate the effects of shear and cold pool interaction on the structure of squall line, three-dimensional simulations in variable surface-based unidirectional shear flow are shown in Figs. 9.7 and 9.8. The basic flow is assumed to have zero velocity at the surface, which increases linearly from 0 to 10, 20, and 30 m s^{-1}, but remains uniform above 5 km. A line thermal in a homogeneous state initiates the convection. In the case of no vertical shear, i.e. $\Delta U = 0$ m s^{-1}/5 km (Figs. 9.7a and 9.8a), the convective cells are scattered, highly disorganized, and short-lived behind the gust front. When the shear increases to $\Delta U = 10$ m s^{-1}/5 km, the simulated squall line is more organized and tilted in the lower layer upshear (leftward in the figure; Fig. 9.7b). In the upper layer, a branch is extended from the upshear-tilted updraft over the density current, while another branch of the updraft is tilted downshear.

The existence of a critical level ($z \approx 6$ km) of the relative flow in the moving frame of reference with the gust front provides a dynamic explanation of the above circulation, as discussed in Subsection 9.1.2. The overturning or downshear-tilted branch of the flow, often referred to as the leading anvil or *anvil outflow*, is formed from forward advection above the critical level. A larger number of convective cells are scattered behind the gust front (Fig. 9.8b). These cells, however, are still more organized than the cells in the case of $\Delta U = 0$ m s^{-1}/5 km. This case corresponds to the mesoscale convective line with trailing stratiform precipitation (TS, Figs. 9.3a and 9.4). When the shear increases to $\Delta U = 20$ m s^{-1}/5 km, the system updraft becomes stronger and more upright compared to the $\Delta U = 10$ m s^{-1}/5 km case (Fig. 9.7c). A stronger relative basic wind in the upper layer results in a stronger forward anvil outflow. The updraft and rain cells in the lower layer form a more continuous line both along and directly behind the gust front (Fig. 9.8c). This case represents an *optimal state*, which corresponds to that represented in Figs. 8.12b and 8.13b, which may last significantly longer than other cases due to the balance between the shear and cold pool generated circulations. When the shear increases to $\Delta U = 30$ m s^{-1}/5 km, the system-scale updraft remains upright to downshear-tilted throughout most of the simulation (Fig. 9.7d). This case corresponds to the mesoscale convective line with leading stratiform precipitation (LS, Fig. 9.3b). Strong and highly three-dimensional

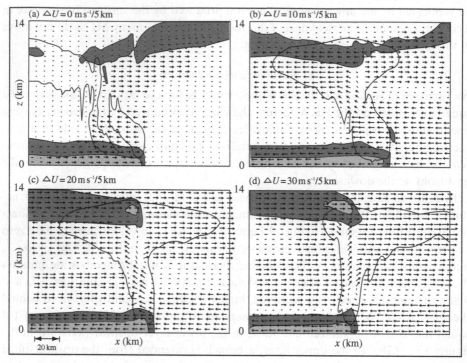

Fig. 9.7 Effects of shear and cold pool interaction on the structure of a convective line after 4 h as illustrated through simulations for a series of three-dimensional idealized numerical experiments with variable strength of surface-based shear flow. The convection is initialized by a line heat source in a homogeneous state characterized by a single profile of temperature and moisture, as used by Weisman *et al.* (1988). Line-averaged system-relative flow vector wind field, negative buoyancy field (shaded), and outline of the cloud field (thick line) on a vertical cross section through the convective line are shown for different shear strengths: (a) $\Delta U = 0\,\mathrm{m\,s^{-1}}$, (b) $\Delta U = 10\,\mathrm{m\,s^{-1}}$, (c) $\Delta U = 20\,\mathrm{m\,s^{-1}}$, and (d) $\Delta U = 30\,\mathrm{m\,s^{-1}}$ in the lowest 5 km. Buoyancy values between -0.01 and $-0.1\,\mathrm{m\,s^{-2}}$ are darkly shaded and values less than $-0.1\,\mathrm{m\,s^{-2}}$ are lightly shaded. The horizontal domain plotted is 240 km. (After Weisman and Rotunno 2004.)

convective cells with supercellular characteristics form along the gust front in a narrow band (Fig. 9.8d).

The interaction between the thunderstorm outflow, in the form of a density current, with the environmental shear, has also been studied using numerical simulations (Xu 1992; Xue 2000). In these models, quasi-steady-state solutions are found for the density current depth and propogation speed relative to the environmental flow that may have non-constant shear. The general conclusion is that the environmental shear, either at the low or upper levels, promotes deeper density currents and steeper slopes of the updraft circulation, although low-level shear tends to be more effective in doing so.

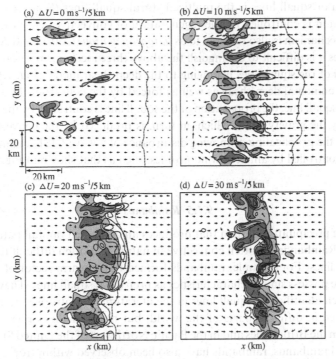

Fig. 9.8 Same as Fig. 9.7 except that this figure shows the system-relative flow vector wind, rainwater mixing ratio, and positive vertical velocity fields on the horizontal cross section at $z = 3$ km. The vertical velocity is contoured at an interval of $3 \, \text{m s}^{-1}$ (solid lines) and the rainwater is lightly shaded for 0.001 to $0.004 \, \text{g kg}^{-1}$ and darkly shaded for rainwater greater than $0.004 \, \text{g kg}^{-1}$. The gust front or cold pool boundary is depicted by $\theta' = -1$ K (dashed line). The domain plotted is 80 km x 80 km. (After Weisman and Rotunno 2004.)

9.1.4 Squall line movement

Based on observations, numerical simulations, and theoretical studies, squall line movement is mainly controlled by the following three mechanisms: (1) advection, (2) externally forced movement, and (3) internally forced movement. In fact, the movement of a squall line is closely related to its maintenance mechanism. In the advection mechanism, the squall line or mesoscale convective system is advected or steered by the environmental mean wind. Examples of squall lines belonging to this category are depicted by the deep-shear model (Fig. 9.1a), and the steering type of squall line model (Moncrieff 1978, also see Fig. 9.4). The externally forced propagation mechanism refers to the sustained squall line or mesoscale convective system by some external forcing usually larger in scale than the convective system itself. Observed external forcing include convergence associated with cold fronts, rainbands, sea-breeze fronts, and gravity waves. One example of an internally forced propagation of squall lines is the multicell storm squall lines (Section 8.2), in which its propagation is controlled by the regeneration of convective cells at its gust front. The third mechanism is also referred to as *autopropagation* (Cotton and Anthes 1989).

Sometimes it is difficult to distinguish the above three mechanisms. For example, the propagation of the squall lines belonging to the deep-shear model (Fig. 9.1a) is steered by the environmental deep shear as well as forced by the cold front. In a wave-CISK maintained squall line, the propagation of the squall line is controlled by the convergence generated by the gravity wave, which, in turn, is produced by the convective heating. In addition, the propagating or moving disturbance of a wave-CISK maintained squall line is also controlled by ambient wind profiles, such as that shown in Fig. 4.19b.

9.1.5 Rainbands

A cloud and precipitation system that produces an elongated rainfall pattern is called a *rainband*. Rainbands are often convective in nature and often difficult to distinguish from squall lines. For example, rainbands formed in the warm sector of an extratropical cyclone possess similar characteristics as prefrontal squall lines. They have therefore also been identified as squall lines in the literature. Extratropical rainbands have been observed in different parts of the world, but much of the earlier research was carried out in the British Isles and in the Pacific Northwest of the United States. Besides extratropical rainbands, rainbands have also been observed within tropical cyclones.

Six types of rainbands have been identified to form within extratropical cyclones (Houze 1981):

(1) *warm-frontal rainbands*, located either ahead of or straddling the warm front and have a width of about 50 km;
(2) *warm-sector rainbands*, typically up to 50 km wide, located in the warm sector and oriented parallel to the surface cold front;
(3) *wide cold-frontal rainbands*, oriented parallel to the cold front and either straddle or are behind the surface cold front, and have a width of about 50 km;
(4) *narrow cold-frontal rainbands*, coincide with the position of the surface cold front and are about 5 km wide;
(5) *prefrontal cold-surge rainbands*, associated with surges of cold air preceding the cold fronts; and
(6) *postfrontal rainbands*, lines of convective clouds that form well behind and generally parallel to the cold front.

The general locations of these rainbands are depicted in Fig. 9.9.

The wide variety of precipitation patterns has led to the proposal of different mechanisms to explain the formation of different types of rainbands. The precipitation associated with the warm-frontal (type 1), wide cold-frontal (type 3) and prefrontal cold-surge (type 5) rainbands are the enhancement of stratiform precipitation. These rainbands appear to be triggered by the release of potential instability or moist

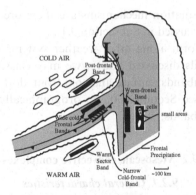

Fig. 9.9 Rainband types associated with a mature extratropical cyclone: (1) warm-frontal rainband; (2) warm-sector rainband; (3) wide cold-frontal rainband; (4) narrow cold-frontal rainband; (5) prefrontal cold-surge rainband; and (6) postfrontal rainband. (Adapted after Houze 1981; modified by permission of American Geophysical Union.)

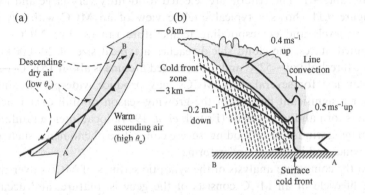

Fig. 9.10 Schematic of airflow and *narrow cold-frontal rainband* in a cold front: (a) horizontal projection; and (b) vertical cross section along AB as denoted in (a). Outlined trajectories denote flow of warm moist airstream into and through the cloudy zone. Dashed lines indicate precipitation. Dashed trajectories denote cold, dry airstream. (Adapted after Browning 1986.)

(potential or conditional) symmetric instability either directly from the forced ascent of moist air moving over the stable layer, or from differential advection in the middle troposphere. The moist (potential or conditional) symmetric instability has been discussed in Chapter 7 and the differential advection mechanism will be discussed in Subsection 11.3.2. The narrow cold-frontal rainbands (type 4) are forced convection by the density current associated with the low-level leading edge of the cold front, as shown in Fig. 9.10. The rising air is often associated with the poleward-moving warm air current and with the high equivalent potential temperature which ascends at the surface cold front and forms the line convection. The warm-sector (type 2) and postfrontal (type 6) rainbands often consist of lines of convective cells or thunderstorms. Their formation mechanisms are therefore similar to those of pre-frontal squall lines, as discussed in Subsection 9.1.2.

Rainbands may also form along other weather systems or near forcing regions. Along with the rainbands discussed above, we may classify the rainbands as the following: (a) frontal rainbands (surface and cold front aloft – CFA, see Chapter 10), (b) cyclone rainbands (see Section 9.3), (c) orographically forced rainbands, and (d) thermally forced rainbands.

9.2 Mesoscale convective complexes

9.2.1 General characteristics

A *mesoscale convective complex* (MCC) is defined as a mesoscale convective system (MCS) that, in satellite observations, exhibits a large, circular, long-lived, and cold cloud shield (Maddox 1980). Based on the infrared satellite imagery and Maddox's definition, the cold cloud shield of an MCC must exhibit the physical characteristics specified in Table 9.1. The criteria are selected to identify very large and long-lived MCCs. Figure 9.11 shows a typical satellite view of an MCC with key infrared enhancement levels and corresponding temperature ranges. For MCCs over the Central United States, the cloud shield reaches a typical size of $200\,000\,\mathrm{km}^2$ for a temperature threshold of $-52°C$ and the averaged lifetime is about 15 h. Occasionally an MCC can last for several days. MCCs may produce widespread rainfall that accounts for a significant portion of the growing-season rainfall over much of the United States corn and wheat belts (Fritsch *et al.* 1994), local intense rainfalls, wide-spread damaging winds, flash flooding, severe convective phenomena, such as torna-does, hail, wind, and intense thunderstorms.

Based on the composite analysis of the synoptic settings of MCCs over the Great Plains, the lifecycle of an MCC consists of the genesis, mature, and decay stages (Maddox 1983). During the *genesis stage*, a typical MCC forms in the vicinity of a weak surface front with a pronounced southerly low-level jet transporting warm,

Table 9.1 *Characteristics of the cold cloud shield of an MCC (Maddox 1980).*

Size	A – Cloud shield with continuously low infrared (IR) temperature $\leq -32°C$ must have an area $\geq 100\,000\,\mathrm{km}^2$.
	B – Interior cold cloud region with temperature $\leq -52°C$ must have an area $\geq 50\,000\,\mathrm{km}^2$.
Initiate	Size definitions A and B are first satisfied.
Duration	Size definitions A and B must be met for a period $\geq 6\,\mathrm{h}$.
Maximum extent	Contiguous cold cloud shield (IR temperature $\leq -32°C$) reaches maximum size.
Shape	Eccentricity (minor axis/major axis) ≥ 0.7 at time of maximum extent.
Terminate	Size definitions A and B no longer satisfied.

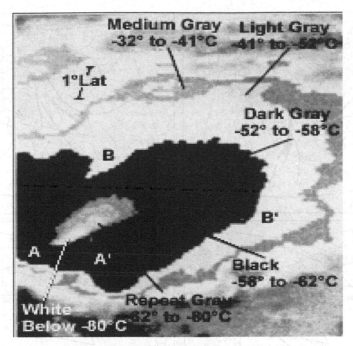

Fig. 9.11 Example of satellite view of an MCC with key infrared enhancement levels and corresponding temperature ranges. (After Maddox 1980.)

moist air into the region. A typical MCC develops and organizes within a precursor mesoscale region of upward motion (e.g., 0.5–1 Pa s^{-1} at 500 hPa) ahead of a midlevel short-wave trough. While MCC formation is linked to a weak, eastward propagating short-wave trough at midlevels, the primary forcing is due to low-level warm advection. A conditionally unstable thermodynamic structure over a large region is located ahead and to the right of the advancing MCC. Layer lifting of a potentially unstable air over a synoptic front may also provide the low-level forcing.

Based on the composite analysis of the synoptic settings of MCCs over the Great Plains of the USA, the synoptic environment conducive to the MCC genesis may be summarized as:

(1) a broad trough exists over the western states, a ridge over the southeast, and a weak high over the Great Lakes;
(2) a weak surface front, with a pronounced, southerly low-level jet transports warm, moist (high θ_e) air into the region (Fig. 9.12a);
(3) strong warm air advection is present at all of the lower levels of the atmosphere;
(4) conditionally unstable air over a large region is located ahead and to the right of the advancing MCC. Layer lifting of a potentially unstable air over synoptic front may also provide the low-level forcing;
(5) at 700 mb, winds typically veer (turn clockwise) and weaken with height (Fig. 9.12b);

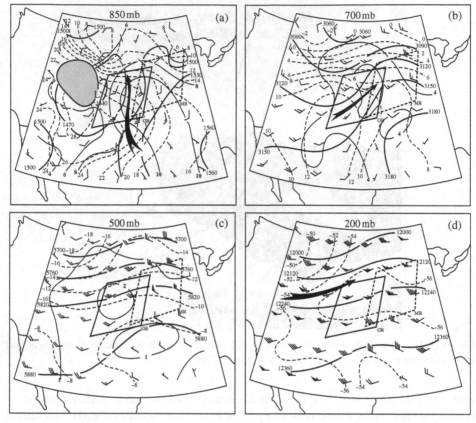

Fig. 9.12 Synoptic environment conducive to MCC genesis, based on the composite analysis of the synoptic settings of MCCs over the Great Plains of the US. Height (m, heavy solid contours), isotherms (°C, dashed), and winds (full barb = 5 m s^{-1}, flag = 25 m s^{-1}) on (a) 850 mb. Dark arrow shows axis of maximum winds; (b) 700 mb; (c) 500 mb; and (d) 200 mb. The shaded region in (a) indicates terrain elevations above the 850 mb level. (Adapted after Maddox 1983.)

(6) normally a downwind trough signal with weak upwind shortwave moving into the initiation area;

(7) the 500 mb flow characteristically out of the southwest, with an increase in wind gradient from south to north (Fig. 9.12c);

(8) a weak jet streak to the north of the genesis region on 200 mb flow (Fig. 9.12d); and

(9) characteristically weak diffluent 200 mb-flow over the genesis region.

During the *mature stage*, as mentioned above, an MCC is characterized by a warm-core updraft with an embedded mesoscale convective vortex, a cold-core anticyclone near the tropopause, and a cold, divergent mesohigh due to outflow near the surface. The warm advection pattern shifts eastward across the Great Plains as low-level winds increase and veer during the night, while low to mid-level convergence is associated with a strong upward mass flux. Note that the veering and increase of the mean wind is associated with warm advection through the thermal wind relstionship, $V_T = (R/f) \ln(p_1/p_2)(k \times \nabla \overline{T})$, where V_T is the thermal wind velocity, p_1 and p_2 are the lower- and upper-level pressures of the layer, and $\overline{T}$ is the mean temperature of the layer. The mesoscale environment reflects the intense convection. A deep, warm-core updraft develops, which is, in turn, overlain by an intense anticyclonic outflow. The *decay stage* of the MCC results from moving into a more stable and convectively less favorable environment. During its lifecycle, the MCC produces a chilled surface mesohigh, a deep, moist tropospheric layer, an amplified midlevel short-wave, and a cold anticyclonic outflow. The modified environment may persist for several days and affect the evolution of meteorological features over much of the Eastern United States, and this lifecycle definition is similar to that for some Oklahoma squall lines (Ogura and Chen 1977).

MCCs have been documented not only in the Central Plains of the United States, but also in other parts of the world (Laing and Fritsch 1997), such as Central and South America, the western Pacific region and Australia, the Indian subcontinent, Europe, and Africa. Understanding the formation and propagation of MCCs may help meteorologists forecast heavy rainfall, often a precursor to flash flooding, especially over the western Pacific region, and Europe. A significant number of tropical cyclones that have occurred over the eastern Atlantic Ocean, particularly near the Cape Verde Islands, can be traced back to the African continent as MCCs. Thus, understanding the formation and development of MCCs over northern Africa around $10°$ N has important implications in elucidating the dynamics of tropical cyclogenesis over the tropical eastern Atlantic Ocean.

9.2.2 Formation and development mechanisms

Observations show that many MCCs in the United States originate from thunderstorm activity over the Rocky Mountains or their eastern slopes. Many of these storms develop over the Rocky Mountains and eventually grow into large nocturnal storms over the Great Plains. Many MCCs also result from mergers and interactions between groups of thunderstorms that develop in different locations. Some MCCs are initially smaller squall lines that gradually acquire MCC characteristics as they persist and grow in size. MCCs forming over the Rockies tend to do so early in the afternoon, moving eastward into the central Great Plains by the middle of the night and being advected by the mean wind of the mid-tropospheric (e.g., 700–500 hPa) layer. The low-level nocturnal southerly jet provides moisture and unstable airstream as well as lifting through warm air advection, which may help explain the prevalent occurrence of MCCs at night. Figure 9.13 shows the tracks of MCCs documented during 1978–82 from May 16 to May 31. Some of the MCCs form over the Rocky Mountains, while the remaining form over the Great Plains. Only a few MCCs in the United States form over the Gulf of Mexico. A similar situation occurs in other parts of the world. For example, approximately 30% of the MCCs in South America formed over the eastern slopes of the Andes, whereas the rest developed over the plains of the Plata River basin.

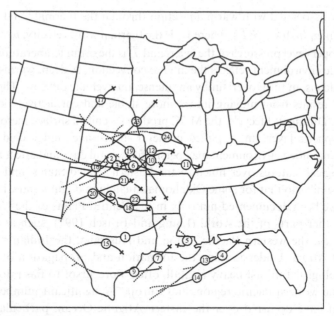

Fig. 9.13 Tracks of MCCs documented during 1978–82 for the period of May 16–31. MCCs initiated at locations at the beginning of the solid lines and terminated at the end of the solid lines denoted by x. The MCC "initiation" and "termination" are defined in Table 9.1. The "maximum extent" of the MCCs is circle-numbered. Dotted line denotes tracks of a developing thunderstorm prior to MCC initiation. (After Maddox *et al.* 1986.)

The tendency for MCCs to be concentrated only in preferred regions, such as in the United States, South America, Africa, India, China, and Australia, suggests that the large-scale environment in these regions must be structured in a fashion that favors the formation of these convective weather systems. In fact, the MCC genesis environments for regions outside North America are very similar and exhibit many of the same dynamic and thermodynamic structures (Laing and Fritch 2000). In the western Pacific region, areas favorable to MCC genesis are: (i) northern Australia, (ii) New Guinea, (iii) northeast India/Bangladesh, (iv) mainland China, and (v) South China Sea (Miller and Fritsch 1991). In summary, MCC systems in most regions tend to form: (1) in the lee of mountain ranges (Fig. 9.14), (2) over land in areas frequented by low-level jets which transport high-θ_e air, and (3) where convective available potential energy is large relative to surrounding areas. MCCs in the tropical easterlies tend to form in the Sahelian region west of the North African mountains, while MCC genesis in the midlatitude westerlies occur east of the South African escarpment (Laing and Fritsch 1993). These mountain ridges are still able to initiate convection, although they are lower than mountains in the Americas and Tibet. Many African MCCs tend to develop over mountains as MCSs, merge with lee vortices, and embed within African easterly waves and could even be traced to the Ethiopian Highlands (e.g., Fig. 9.22).

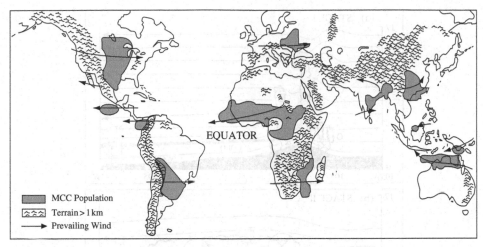

Fig. 9.14 Relationship among MCC population centers, elevated terrain, and prevailing mid-level flow. (After Laing and Fritsch 1997.)

Based on the dynamical mechanism responsible for producing the convection and the large cold cloud shields, an MCC may be categorized into two types. Type 1 results when a mesoscale ribbon of low-level potentially unstable air is forced to ascend in a frontal or baroclinic zone. Type 2 occurs in more barotropic environments and depends on the moist-downdraft production of a surface-based cold pool. The interaction between the cold pool and the ambient vertical wind shear produces mesoscale ascent, as discussed in Section 8.3, and the associated large stratiform cloud shield characteristic of MCCs. In either case, the genesis region must have CAPE to release the conditional instability, layer lifting to release the potential instability, mechanical and/or thermal forcing which helps trigger individual thunderstorms, and a synoptic or mesoscale disturbance that helps organize the thunderstorms into a larger convective system.

Based on observations and numerical simulations, a three-stage conceptual model of orogenic MCC formation may be proposed (Fig. 9.15):

(1) Stage I: A dry mountain-plains solenoidal (MPS) circulation develops over the lee slope. A group of deep convective clouds forms over the mountain peak (Fig. 9.15a).
(2) Stage II: These deep convective clouds organize into a meso-β-scale convective system, which moves eastward. The meso-β-scale convective system expands into a meso-α-scale convective system while it is moving eastward (Fig. 9.15b);
(3) Stage III: The meso-α-scale convective system continues to grow. As the scale of the unsteady MCS increases to be larger than the Rossby radius of deformation, it becomes more geostrophically balanced and a meso-α-scale upward motion is induced because of geostrophic adjustment (Fig. 9.15c).

Dynamically, an MCC has a Rossby number on the order of 1 and exhibits a horizontal scale comparable to the *Rossby radius of deformation*. Similar to (4.2.12),

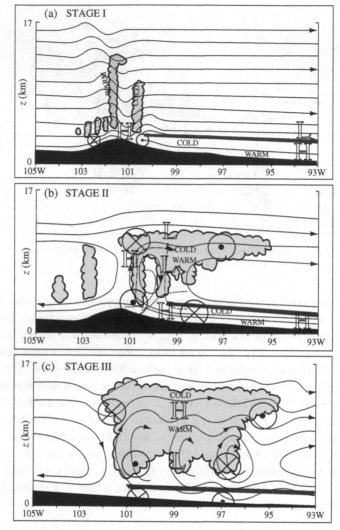

Fig. 9.15 A three-stage conceptual model of orogenic MCC formation. (a) Stage I, (b) Stage II, and (c) Stage III. See text for details. Topography and the plain inversion are dark-shaded. Regions of clouds are light-shaded. Circles denote flow perpendicular to the plane (X: into page, •: out of page). The high and low pressure perturbations are denoted by H and L, respectively, while the cold and warm perturbations are also denoted. (Adapted after Tripoli and Cotton 1989.)

the Rossby radius of deformation for a circular motion in a stratified fluid may be defined as (Schubert *et al.* 1980):

$$L_R = \frac{NL_z}{(\zeta+f)^{1/2}(2V/R+f)^{1/2}}, \qquad (9.2.1)$$

where ζ is the vertical relative vorticity, and V the tangential wind speed at the *radius of local curvature R* of the circulation. Observational analysis indicates a dynamic radius of about 300 km. The L_R estimated from numerical simulations suggests that MCC is an inertially stable balanced MCS (Olsson and Cotton 1997). The unbalanced flow is found to be composed largely of divergent circulations of gravity waves. MCSs typically develop a strong positive vorticity maximum and associated potential vorticity anomaly in the lower and middle troposphere. As an MCC develops this positive vorticity maximum, it achieves greater inertial stability, and hence a smaller L_R. The inertial stability is associated with the horizontal restoring force that resists lateral displacement and typically takes about 3 to 6 hours for the Coriolis effects to become significant at midlatitudes. Based on the above discussion of MCC genesis, Stage III is characterized by a significant transformation from an unsteady MCS with a scale smaller than L_R into a more balanced MCC with a scale larger than the Rossby radius of deformation, although the convection is still characterized by multicells, i.e. the convection is still unsteady. The CAPE released in the deep convection in an MCS excites gravity waves. However, for an MCS with a scale larger than L_R, most of the gravity wave energy is confined locally within the storms and is projected into the rotational flow.

Figure 9.16 depicts the structure and development mechanism of the mesoscale convective vortex (MCV) in a mature MCC, which includes a shallow, cold, long-lived anticyclone in the lower troposphere (below 850 hPa), a strong updraft with a warm core, an MCV with low pressure in the middle troposphere, and a large, cold, shallow, short-lived anticyclone with high pressure in the upper troposphere. The conceptual model for mature MCCs was constructed, based on the forced potential vorticity theory of Raymond and Jiang (1990), on the assumption of an environment with a low-level jet and a deep layer of weak flow and weak shear in the middle and upper troposphere (Fritsch *et al.* 1994). Below the level of maximum low-level jet (e.g., 850 mb), the strengthening of shear-induced vorticity and cold pool-induced vorticity on the upshear (left) side of the cold pool will produce an overtaking flow tilted more downshear (rightward) over the cold outflow surface. Thus, the overtaking air over the cold ouflow would be unlikely to reach its level of free convection (LFC) in the rear (upshear) of the cold pool until it flows isentropically into the interior of the convective system. Once the air parcel reaches its LFC inside the convective cloud, it would flow more vertically due to the buoyancy. Thus, the overtaking air can initiate new convection beneath the PV maximum and the convective system can feed from the rear (left-hand side in Fig. 9.16) instead of the front.

The structure of PV and temperature anomalies depicted in Fig. 9.16 can be interpreted by considering the Ertel's PV definition

$$q = \frac{1}{\rho} [(\boldsymbol{\omega} + f\boldsymbol{k}) \cdot \nabla\theta], \tag{9.2.2}$$

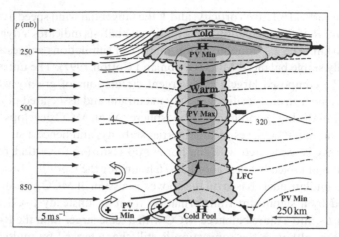

Fig. 9.16 A conceptual model of the structure and development of the mesoscale warm core vortex of a mature MCC. Thin arrows along the ordinate denote the environmental wind. Open arrows with either a plus or a minus sign denote the sense of the horizontal vorticity generated by the environmental wind shear and cold pool. Thick solid arrows indicate the wind inside and around the convective system. Frontal symbols depict outflow boundaries. Dashed lines are potential temperature (5 K intervals) and solid lines are potential vorticity (PV) (interval of 2×10^{-7} $m^2 s^{-1} K kg^{-1}$). The convective system is propagating from left to right at about 5 to $8 m s^{-1}$ and is being overtaken by high-θ_e air in the low-level jet. Air overtaking the vortex ascends through isentropic surfaces, reaches its level of free convection (LFC), thereby initiating deep convection (shaded). (Adapted after Fritsch *et al.* 1994, Thiao *et al.* 1993, and Jiang and Raymond 1995.)

and the PV tendency equation,

$$\frac{Dq}{Dt} = \frac{1}{\rho} \left[(\boldsymbol{\omega} + f\boldsymbol{k}) \cdot \nabla\dot{\theta} + \nabla \times \boldsymbol{F} \cdot \nabla\theta \right], \tag{9.2.3}$$

where $\boldsymbol{F}$ is the frictional force vector and $\dot{\theta}$ is the diabatic heating rate. Neglecting the frictional force, the PV tendency describing cumulus convection may be approximated by

$$\frac{Dq}{Dt} \approx \left(\frac{\zeta + f}{\rho} \right) \frac{\partial\dot{\theta}}{\partial z}. \tag{9.2.4}$$

Assuming the diabatic heating associated with cumulus convection has a maximum at the midtroposphere, a pair of positive and negative PV anomalies will result below and above the maximum heating level, respectively. This, in turn, will generate a pair of cyclonic and anticyclonic circulations in the lower and upper parts of the MCC, respectively.

MCVs have been observed in the mature-to-decaying MCCs as well as in tropical mesoscale convective systems. An MCV embedded within a tropical MCC has a warm core similar to that of a tropical cyclone and may provide an enhanced inertial stability that makes the diabatic heating more efficient in producing a balanced, rotational flow (Olsson and Cotton 1997). The cyclonic circulation associated with the mesoscale vortex embedded within the MCC at the midtroposhere is related to the development of a stratiform precipitation region that accompanies the MCC (Johnson and Bartels 1992). A midtropospheric short-wave propagating in the ambient flow is able to further enhance the diabatic processes by organizing the thunderstorms and increasing the background relative vorticity. Note that the above MCV generation mechanism differs from that responsible for the vortices (which are also called *bookend vortices*) depicted in Fig. 9.8c, in which the MCVs are formed by the tilting of either ambient or baroclinically(cold-pool)-generated horizontal vorticity at the ends of finite-length convective lines. Once an MCC moves into a region with unfavorable conditions for sustaining its convection, the disintegration of the MCC is inevitable. As the system dissipates, the precipitation intensity and rate continue to decrease, but the strong upward motion and cyclonic circulation in the midtroposphere and anticyclonic circulation in the upper troposphere may still persist during the early part of the MCC dissipation stage.

An accurate quantitative precipitation forecast accompanying an MCC is heavily dependent on a successful forecast of the MCC's movement. In this regard, the movement of the radar-observed meso-β scale convective elements (MBEs) embedded within the MCC, rather than the movement of the cold cloud shield centroid, is more responsible for the heaviest rainfall. The movement of an MCS is composed of the sum of an advection component and a propagation component. The advection component is controlled by the mean motion of the convective cells or the mean flow in the deep tropospheric layer (roughly between 850 to 300 mb). The propagation component depends on the rate and direction at which the new convective cells are growing, which is, in turn, related to the low-level jet shear and the cold pool.

9.3 Tropical cyclones

9.3.1 General characteristics

Tropical cyclones (TCs) are non-frontal synoptic-scale "warm-core" low-pressure systems that originate over the tropical or subtropical oceans and contain organized deep convection and a well-defined cyclonic surface wind circulation. Since the radius of gale-force winds of a mature tropical cyclone is about several hundred km and the radius of intense convection and strong winds is about 100 km, tropical cyclones may also be viewed as mesoscale convective systems. Although the cloud and precipitation characteristics are similar to those of mesoscale convective systems, as discussed in previous sections, moist convection associated with a tropical cyclone is strongly linked to the tropical cyclone vortex, which arises from convergence of planetary vorticity. The centrifugal force plays a significant role in the force balance in such a system since it has very high wind speeds and relatively small horizontal scales. To a first approximation, the radial pressure gradient force, Coriolis force, and centrifugal

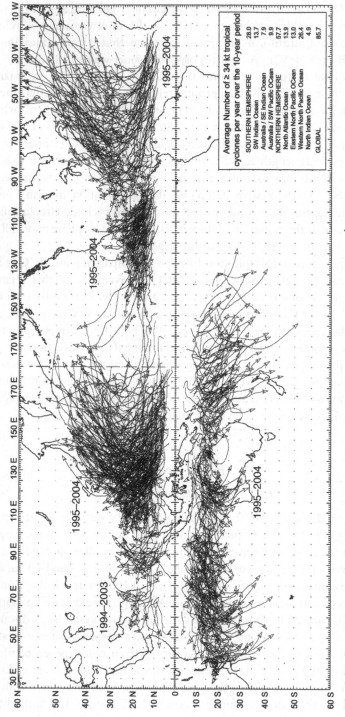

Fig. 9.17 Genesis locations and tracks of tropical cyclones with wind speeds of at least $17\,\mathrm{m\,s^{-1}}$ for the period of 1995–2004. (Adapted after Neumann 1993; Courtesy of Dr. Charles Neumann.)

Average Number of ≥ 34 kt tropical cyclones per year over the 10-year period	
SOUTHERN HEMISPHERE	28.0
SW Indian Ocean	13.7
Australia / SE Indian Ocean	7.9
Australia / SW Pacific OCean	9.9
NORTHERN HEMISPHERE	57.7
North Atlantic Ocean	13.9
Eastern North Pacific OCean	13.0
Western North Pacific Ocean	26.4
North Indian Ocean	4.9
GLOBAL	85.7

force in a quasi-steady-state tropical cyclone vortex are in *gradient wind* balance. In addition to strong horizontal winds, excessive rain produced by deep convection, supported by the presence of a deep layer of very moist air (e.g., $q_v > 20\,\mathrm{g\,kg^{-1}}$, where q_v is the water vapour mixing ratio) and strong upward motion with vertical velocity exceeding $1\,\mathrm{m\,s^{-1}}$ produced by strong convergence, may cause flooding and landslides. The average rainfall rate in the radial ring of about 200 km is about 100 mm $\mathrm{d^{-1}}$, but the maximum rainfall rate may exceed 500 mm $\mathrm{d^{-1}}$. Despite the damage that the high winds and extremely heavy rainfall tropical cyclones may bring, the rainfall supplied by weak and moderate hurricanes or typhoons are essential to agriculture and fresh water supply to some areas.

Tropical cyclones tend to affect the eastern and equatorward portions of the continents. Figure 9.17 shows genesis locations and tracks of tropical cyclones with wind speeds of at least $17\,\mathrm{m\,s^{-1}}$ for the period of 1995–2004. Based on intensity, tropical cyclones in the Atlantic, eastern and western Pacific are classified as a *tropical depression* for a weaker system with $V_{max} \leq 17\,\mathrm{m\,s^{-1}}$ ($\sim 62\,\mathrm{km\,h^{-1}} \sim 38\,\mathrm{mph}$), a *tropical storm* for a moderate system with $18\,\mathrm{m\,s^{-1}} \leq V_{max} \leq 32\,\mathrm{m\,s^{-1}}$, and a *hurricane* or *typhoon* for a strong system with $V_{max} \geq 33\,\mathrm{m\,s^{-1}}$. Hurricanes are called *major hurricanes* when $V_{max} \geq 50\,\mathrm{m\,s^{-1}}$ ($\sim 178\,\mathrm{km\,h^{-1}} \sim 111\,\mathrm{mph}$) and typhoons are classified as *super typhoons* when $V_{max} \geq 67\,\mathrm{m\,s^{-1}}$ ($\sim 241\,\mathrm{km\,h^{-1}} \sim 150\,\mathrm{mph}$). The maximum wind speed of a strong tropical cyclone may exceed $100\,\mathrm{m\,s^{-1}}$, which may produce storm surge by driving an ocean rise of several meters along the coast.

The characteristics of tropical cyclones differ from those of extratropical cyclones in many ways. Tropical cyclones derive their energy from sensible and latent heat fluxes from the ocean; while the extratropical cyclones derive their energy from available potential energy associated with baroclinicity. Because tropical cyclones are barotropic, they are more symmetric and circular than extratropical cyclones which are often associated with cold fronts, warm fronts and occluded fronts. During the development stage, tropical cyclones are oriented vertically, while extratropical cyclones are normally tilted into the prevailing shear. Structurally, tropical cyclones have their strongest winds near the Earth's surface, while extratropical cyclones have their strongest winds near the tropopause. This is caused by the thermal wind balance responding to the warm core of tropical cyclones and the cold core of extratropical cyclones. A tropical cyclone may transform into an extratropical cyclone as it interacts with midlatitude baroclinicity (e.g., Atallah and Bosart 2003). Occasionally, an extratropical cyclone may lose its frontal features, develop convection near the center of the storm and transform into a full-fledged tropical cyclone (e.g., Davis and Bosart 2003). Such a process is most common in the North Atlantic and Northwest Pacific basins.

9.3.2 Tropical cyclogenesis

Tropical cyclogenesis is often preceded by some disturbances or precursors, such as *easterly* (*tropical*) *waves*, *African easterly waves* (*West African disturbance lines* – WADLs), *tropical upper tropospheric troughs* (TUTTs), and old frontal boundaries. Once a disturbance forms and sustained convection develops, it can become more organized and eventually evolve into a tropical depression given an abundant supply of heat and moisture from the ocean, background cyclonic vorticity associated with the Earth's rotation, preexisting disturbance with sufficient vorticity, and weak wind shear.

Several necessary conditions for tropical cyclogenesis have been proposed (e.g., Riehl 1948; Palmén 1948; Gray 1968, 1998):

(1) A warm and large body of ocean water of at least 26.5°C, throughout a sufficient depth at least 50 m to provide the thermal energy for the tropical cyclone.
(2) Sufficiently large planetary vorticity, i.e. a distance of at least 500 km from the equator, so that the near gradient wind balance can occur.
(3) A near-surface preexisting disturbance with sufficient vorticity and convergence, such as easterly waves, to generate a convergent low-level wind, triggering convection, and helping to organize and sustain the mesoscale convective systems and cloud clusters.
(4) Divergence associated with an upper-level trough overlaying the low-level cyclonic disturbance to enhance and sustain the deep upward motion.
(5) A potentially unstable atmosphere to allow the heat stored in the ocean to be released for the tropical cyclone development.
(6) Relatively moist middle troposphere for allowing the continuous development of widespread thunderstorm activity.
(7) Weak vertical wind shear, with less than $10\,\mathrm{m\,s}^{-1}$ between the ocean surface and the tropopause, to avoid the disruption of the vertical development of deep convection around the cyclone center.

The formation of tropical cyclones differs significantly from that of extratropical cyclones. A tropical cyclone forms through complicated scale interactions between the small-scale cumulus clouds, mesoscale organization of cumulus clouds, and large-scale environment. Tropical cyclone formation is one of the most challenging problems in dynamic meteorology and has been studied since 1940s. Since the early 1960s, several major mechanisms have been proposed to explain tropical cyclogenesis: (a) cooperative intensification mechanism, (b) linear CISK mechanism, (c) wind-induced surface-heat exchange (WISHE) mechanism, (d) vortex interaction mechanism, and (e) hot-tower mechanism. These mechanisms are discussed below.

a. Cooperative intensification mechanism

A cooperative intensification mechanism between cumulus convection and moisture convergence in the planetary boundary layer of a quasi-balanced cyclonic vortex has been proposed to explain tropical cyclogenesis (Ooyama 1964, 1969). This cooperative intensification mechanism assumes the release of latent heat by moist convection given off by the conditionally unstable moist air at the bottom of the vertical column. The cooperative intensification mechanism includes the following processes: (1) The heating effects of deep cumulus clouds are represented in terms of a mass flux from the boundary layer to upper layer wherever there is resolved-scale boundary-layer line.

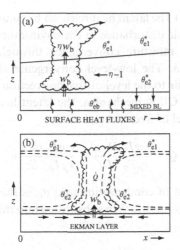

Fig. 9.18 Schematics of flow configurations of: (a) cooperative intensification mechanism (Ooyama 1964, 1969) and (b) linear CISK mechanism (Charney and Eliassen 1964). Dashed lines denote constant saturation potential temperatures (θ_e^*). In (a), the latent heating is proportional to $\eta w_b'$, where η is the *entrainment parameter* which is defined as $\eta = 1 + (\theta_{eb} - \theta_{e1}^*)/(\theta_{e1}^* - \theta_{e2})$. In (b), the latent heating ($\dot{Q}$) is proportional to w_b' at the top of Ekman layer and w', in turn, is proportional to relative vorticity at the same level. Surface heat and moisture fluxes are represented in the formulation in mechanism (a), but not in mechanism (b). (Adapted after Smith 1997.)

(2) The deep cumuli will entrain ambient air from the middle layer as they rise through it and detrain in the upper layer. (3) The entrainment rate is determined as a function of time to satisfy energy conservation, assuming that the air detrained into the upper layer follows the saturation equivalent potential temperature of ambient air in that layer. (4) Although a relatively large amount of CAPE is used in the initial sounding in a nonlinear model incorporating this mechanism (Ooyama 1969), this is not required for the mechanism to work (Dengler and Reeder 1997). This mechanism is illustrated in Fig. 9.18a.

The nonlinear model simulations, based on cooperative intensification mechanism, demonstrated that (a) the latent and sensible heat fluxes from the warm ocean were crucial to vortex intensification and that (b) the progressive reduction of the local Rossby radius of deformation occurs as the inertial stability of the TC vortex increases in the inner-core region. The latter effect is nonlinear, and helps reduce the scale separation between the deep cumuli and the balanced tangential circulation of the vortex, so that the individual cumuli become more and more under the control of the balanced dynamics.

b. Linear CISK mechanism

A linear *Conditional Instability of the Second Kind* (*CISK*) mechanism has been proposed to explain tropical cyclogenesis (Charney and Eliassen 1964), which includes the following processes: (1) The latent heat released by cumulus convection produces a lower-tropospheric cyclonic disturbance in the environment. (2) The disturbance, in turn, produces low-level moisture convergence through boundary layer friction (i.e., Ekman pumping). (3) The low-level convergence then lifts the conditionally unstable boundary layer air to the level of free convection and leads to the release of latent heat. In this linear CISK mechanisn, the latent heating is represented by the vertical velocity at the level of free convection z_b,

$$\dot{Q} = \frac{-gL_v}{c_p T_0} \frac{Dq_{vs}}{Dt} \approx \frac{-gL_v}{c_p T_0} \frac{\partial q_{vs}}{\partial z} w', \tag{9.3.1}$$

where L_v is the latent heat of condensation and q_{vs} is the saturated mixing ratio of water vapor. The above equation is then substituted into the thermodynamic equation to yield

$$\frac{\partial b'}{\partial t} + N_m^2 w' = 0, \tag{9.3.2}$$

where b' is the perturbation buoyancy ($g\theta'/\theta_0$), and

$$N_m^2 = \frac{g}{\bar{\theta}_e^*} \frac{d\bar{\theta}_e^*}{dz} \approx \frac{g}{\theta_0} \frac{d\bar{\theta}_e}{dz} + \frac{gL_v}{c_p T_0} \frac{\partial q_{vs}}{\partial z}. \tag{9.3.3}$$

In the above equation, N_m is the *moist buoyancy frequency* defined as $N_m^2 = (g/\bar{\theta}_e^*)(d\bar{\theta}_e^*/dz)$, where $\bar{\theta}_e^*$ is the saturation equivalent potential temperature of the environmental air. This represents a self-exciting process as illustrated in Fig. 9.18b.

The Charney–Eliassen linear CISK theory received wide support for explaining tropical cyclogenesis for more than two decades after it was proposed in the early 1960s, but has suffered heavy criticism since the early 1980s (e.g., Ooyama 1982; Emanuel 1994; see review in Smith 1997). The major deficiencies of the Charney–Eliassen linear CISK mechanism are as follows. (1) The heating function (9.3.1) is not supported by the fact that the θ_e of rising air does not increase with increasing vertical motion. (2) The growth rates of CISK unstable modes are rather uniform over a broad range of horizontal scales. (3) The theory fails to take into account the nonlinear processes that are needed to explain the dynamics of a mature tropical cyclone. (4) The representation of cumulus heating is unable to capture the feedback of convection on the large-scale flow.

c. WISHE mechanism

In light of the drawbacks of linear CISK theory, it has been proposed that tropical cyclone intensification results from a finite-amplitude instability involving the feedback between the cyclone and WISHE (Emanuel 1986). In other words, a tropical cyclone can intensify and maintain its circulation against dissipation entirely by self-induced anomalous fluxes of *moist enthalpy* from the sea surface with virtually no

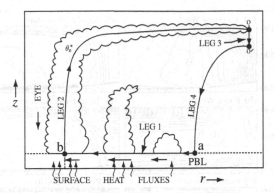

Fig. 9.19 Idealized Carnot cycle for a steady-state, mature tropical cyclone, based on the WISHE mechanism (Emanuel 1986, 1989). Solid curves represent hypothetical air trajectory in a Carnot cycle. (Adapted after Emanuel 1997.)

contribution from preexisting CAPE. The *specific enthalpy* is defined as $h = c_p T + L_v q_v$ and the change in *enthalpy* becomes $\delta h = c_p \delta T + L_v \delta q_v$.

The energy cycle of a mature tropical cyclone is analogous to a *Carnot heat engine*, as illustrated in Fig. 9.19. The energy cycle of a mature tropical cyclone consists of the following four processes: isothermal expansion (with the addition of enthalpy), adiabatic expansion, isothermal compression, and adiabatic compression. Imagine that an air parcel moves from the outer region near sea surface toward the eyewall along Leg 1 (a to b in the figure), it experiences a drop in pressure and an increase in sensible and latent heat fluxes. This will induce an increase in *moist entropy*. The change in moist entropy (δs) can be calculated from the following relationship:

$$T\delta s = c_p \delta T - \alpha \delta p + L_v \delta q_v. \tag{9.3.4}$$

Equation (9.3.4) indiates that lowering surface pressure while maintaining air temperature near the sea surface ($\delta T = 0$) will lead to an increase of moist entropy, assuming that $\delta q_v = 0$. In the meantime, the frictional force exerted by the sea surface decreases the angular momentum, as the air parcel spirals in along Leg 1. Leg 1 is regarded as an isothermal expansion process as in the Carnot cycle because the temperatures are nearly constant ($\delta T = 0$). At the eyewall (b in Fig. 9.19), the air parcel turns upward and closely follows surfaces of constant entropy and angular momentum as it ascends upward and outward near the storm top along Leg 2 (b to o). Leg 2 can be approximately regarded as an adiabatic expansion process in a Carnot cycle. In reality, an air parcel would exchange entropy and momentum with its environment due to mixing. Thus, the tropical cyclone is actually not a closed cycle. Along leg 3 (o to o'), the air parcels lose entropy by radiation acquired along Leg 1 and gain angular momentum by mixing with the environment when individual parcels are descending and mixing with the environmental air. Leg 3 is entirely isothermal, and can be regarded as an isothermal compression process as in the Carnot cycle. The

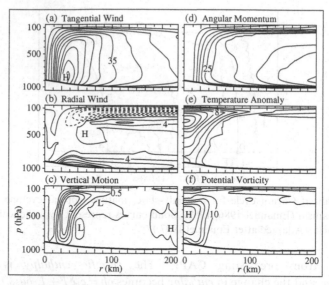

Fig. 9.20 Numerically simulated mesoscale structure of an idealized tropical cyclone at the mature stage after 90 h of model integration. Shown are the height-radius cross sections of the azimuthal mean of (a) tangential wind (contour interval of $7\,\mathrm{m\,s^{-1}}$), (b) radial winds (contour interval of $4\,\mathrm{m\,s^{-1}}$, inflow positive), (c) vertical motion (contour interval of $0.5\,\mathrm{m\,s^{-1}}$), (d) angular momentum (contour interval of $6.25\,\mathrm{m^2\,s^{-1}}$), (e) temperature anomaly from the undisturbed environment (contour interval of $4\,\mathrm{K}$), and (f) potential vorticity (contour interval of $15\,\mathrm{PVU}$; $1\,\mathrm{PVU} = 10^{-6}\,\mathrm{K\,m^2\,kg^{-1}\,s^{-1}}$). The finest horizontal grid interval is $5\,\mathrm{km}$. (After Wang 2002.)

Carnot cycle is closed by leg 4 (o' to a in Fig. 9.19). Along leg 4, the angular momentum is conserved, although entropy is lost and gained. The irreversible entropy source is produced in its entirety by mixing of moist and dry air. Leg 4 can approximately be viewed as an adiabatic compression process.

The WISHE mechanism is consistent with the cooperative intensification mechanism discussed above. Numerical simulations (Rotunno and Emanuel 1987) demonstrated that a hurricane-like vortex indeed amplifies as a result of air–sea interaction instability in an atmosphere that is neutral to cumulus convection (i.e. no CAPE). Figure 9.20 shows a numerically simulated idealized tropical cyclone at the mature stage, which depicts the detailed core structure reasonably well. The subsidence in the eye with its maximum around the inner edge of the eyewall originates from the returning flow at the top of the eyewall (Figs. 9.20a–c). The motion in the radial-vertical plane of the mean cyclone is mainly along the constant absolute angular momentum surface (Fig. 9.20d), consistent with the WISHE mechanism. The cyclone has a warm core with a maximum temperature anomaly of about $18\,°\mathrm{C}$ at about $300\,\mathrm{hPa}$ (Fig. 9.20e). The maximum PV (Fig. 9.20f) is located just inside the radius of maximum wind (RMW), below about $800\,\mathrm{hPa}$, which is due to concentration of PV associated with convective heating in the eyewall, and these features appear to be realistic.

The difference between the cooperative intensification mechanism and WISHE mechanism appears to be the degree of emphasis placed by the respective authors on the relative importance of convection and surface fluxes in the intensification of hurricanes (Smith 1997). The physics implied in the cooperative intensification and WISHE mechanism should be represented in contemporary models if the heat transfer between convection and air–sea interaction heat is to be properly treated.

d. Vortex interaction mechanism

The vortex interaction mechanism, as originally proposed by Fujiwhara (1921), was revived in the early 1990s and has been evolving since then. Since a two-dimensional turbulent flow has a tendency to form finite-amplitude, coherent vortex structures in the absence of external forcing, a random distribution of vorticity will concentrate into isolated vortices due to an upscale merger process in which smaller vorticity perturbations in the flow merge into a larger one. This provides a potential mechanism for the development and growth of mesoscale vortices and initial development of tropical cyclones in the atmosphere in the absence of thermal forcing. Thus, tropical cyclogenesis can arise from a constructive interaction between the larger-scale environmental and mesoscale vortex dynamics (Ritchie and Holland 1993; Hendricks *et al.* 2004). Figure 9.21 shows an example of this type of vortex interaction mechanism,

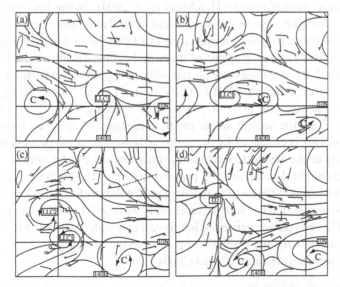

Fig. 9.21 The vortex interaction mechanism for tropical cyclone intensification is demonstrated by the interaction of two low-level vortices, denoted as "LLC1" and "LLC2", which merged to form a tropical depression (TD) and finally developed into Typhoon Irving (1992). Shown are manual surface streamline analyses at 0000 UTC for: (a) 25 (b) 27, (c) 29, and (d) 31 July 1992. Latitude and longitude line intervals are 10 degrees. A full wind barb is $5\,\mathrm{m\,s^{-1}}$. (Adapted after Ritchie and Holland 1997.)

in which two low-level vortices (LLC1 and LLC2) merged to form a tropical depression (TD), which eventually developed into Typhoon Irving (1992). Note that this mechanism may produce a precursor vortex, but convection is still required for amplification.

e. Hot-tower mechanism

A preexisting surface vortex of finite amplitude or sufficient strength is needed for tropical cyclogenesis to occur via the WISHE mechanism. Observations indicate that cumulus convection is an essential ingredient for tropical cyclogenesis because tropical cyclones form out of mesoscale convective complexes that are embedded in tropical or easterly waves. Thus, the WISHE theory implicitly assumes that a transformation of a tropical disturbance to a surface vortex of sufficient strength has already taken place (Montgomery and Farrell 1993). The problem arises from how a weak-amplitude tropical disturbance is transformed into a finite-amplitude surface vortex, assuming that WISHE mechanism works. Since this is still a current topic of research, a brief summary is only appropriate.

Based on numerical simulations, it has been proposed that *vortical hot towers*, i.e. tall cumulonimbi with strong vertical vorticity, are important ingredients for forming tropical cyclones via the following two-stage evolutionary processes. (1) Multiple small-scale cyclonic PV towers are generated by latent heat release in the lower troposphere and (2) multiple mergers of convectively generated PV or vorticity anomalies from the hot towers (Hendricks *et al.* 2004). In the first stage, a pulsing and locally enhancing cyclonic PV is produced by the latent heat release within the hot tower in the lower troposphere, as indicated by (9.2.4). Note that the PV generation is dominated by the vertical gradient of diabatic heating. In other words, the strong updrafts in the hot towers converge and stretch existing low-level vorticity into intense small-scale vortex tubes. A 64 PVU ($1\,\text{PV unit} = 1 \times 10^{-6}\ \text{m}^2\,\text{s}^{-1}\text{K}\ \text{kg}^{-1}$) may be produced in the layer of $z = 0$ to 5 km, $\partial\dot\theta/\partial z = 45\ \text{K}\ \text{h}^{-1}\text{km}^{-1}$, $\zeta = 10^{-3}\ \text{s}^{-1}$ and $\rho = 0.7\ \text{kg m}^{-3}$. In the second stage, multiple mergers of these cyclonic anomalies occur. The merger is enhanced by the *vortex axisymmetrization* process which means that the decay of the asymmetries favor the creation of a master vortex.

Hot towers might develop in a TC precursor, such as the MCS shown in Fig. 9.22. The MCS developed in the lee of the Ethiopian Highlands (EH) (see Subsection 9.2.2). This MCS was embedded in an African easterly wave and traveled across the African continent. The convective vortices or hot towers associated with the MCS might have served as precursor for the tropical depression over the eastern Atlantic Ocean that eventually became Hurricane Alberto (2000).

f. Preexisting disturbances and a unified mechanism for tropical cyclogenesis

There is a growing body of evidence suggesting that tropical cyclogenesis in the eastern Pacific Ocean occurs in association with easterly waves that have been produced over the African continent and propagated across the Atlantic and Caribbean and into the eastern Pacific (Molinari and Vollaro 2000). We must note that this area is on the lee

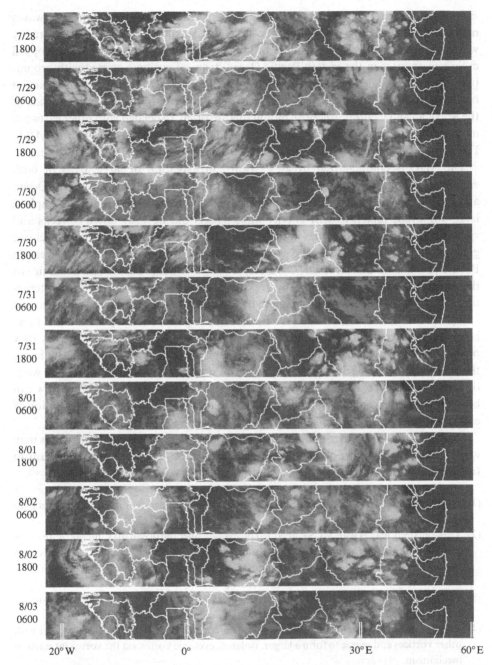

Fig. 9.22 Time-longitude section of METEOSAT-7 infrared satellite imagery for 1800 UTC July 28, 2000 through 1800 UTC August 3, 2000 every 12 h. This MCS developed in the lee of the Ethiopian Highlands, was embedded in an African easterly wave and served as a precursor for the tropical depression that developed into Hurricane Alberto. (From Lin *et al.* 2005; Imagery from EUMETSAT © 2003.)

side of the mountains in Mexico and Central America. The westward moving waves may include different types of synoptic-scale waves, such as mixed-Rossby gravity waves, equatorial Rossby waves, and off-equatorial Rossby waves. In some tropical cyclogenesis cases in the eastern Pacific Ocean, tropical cyclones started with orographically generated lee cyclones. In addition to the westward propagating waves, it has been observed that tropical cyclogenesis over the eastern Pacific on the lee side of the Mexican and Central American mountains are often associated with other eastward moving or stationary systems, such as the Madden–Julian (1994) oscillation (MJO), ITCZ, monsoon trough, or any long-period, slowly moving diabatic heating disturbance.

The MJO is characterized by an eastward propagation of large areas of both enhanced and suppressed tropical rainfall, observed mainly over the Indian Ocean and Pacific Ocean. The oscillation period is approximately 30–60 days. For example, a large-amplitude mixed Rossby-gravity wave packet during 1987 lasted 5 weeks and spawned three tropical cyclones in the western Pacific Ocean, one with each cyclonic gyre within the wave packet as the gyres turned away from the equator in the western Pacific (Dickinson and Molinari 2002; Aiyyer and Molinari 2003). The dynamics of the interactions of the mixed Rossby-gravity waves and the MJO, and their interactions with sea surface temperature anomaly (SSTA) and topography are complicated (e.g., Hsu and Lee 2005), and deserve further investigation. The significance of these larger-scale waves and their interactions with topography lies in the fact that they provide cyclonic vorticity in a broader region for cyclogenesis to occur. This indicates that the tropical cyclogenesis over the western Pacific basin may be predicted a week or earlier in advance, if the geneses of these synoptic waves and background flow can be identified. Thus, emphasis should be placed in the investigation of the interaction of easterly waves, eastward propagating disturbances, SSTA, and orography.

Based on the above description, it appears that some of the above mechanisms may work cooperatively and sequentially to form a tropical cyclone. A unified mechanism for tropical cyclogenesis may work as follows:

(1) A TC precursor begins as an MCS that forms and is embedded within easterly waves over the land or ocean.
(2) These easterly waves with MCS embedded propagate over the ocean may interact with other eastward moving or stationary systems, such as MJO, monsoon trough, ITCZ or Kelvin wave. The TC genesis region must also possess other key ingredients, including high SST, weak vertical wind shear, and away from the equator.
(3) The cumuli embedded in the MCS develop further into vortical hot towers via the hot-tower mechanism.
(4) These hot towers, or convective vortices associated with the MCS, may also interact with other vortices and merge to form a larger, isolated, cyclonic vortex via the vortex interaction mechanism.
(5) Once the surface wind speeds exceed a certain threshold, and if the disturbance is of sufficient size, then the vortex will develop further into a tropical cyclone via the WISHE/cooperative intensification mechanism if other conditions are met.

g. Control parameters of tropical cyclogenesis

The necessary conditions and formation mechanisms discussed above suggest the following control parameters (Gray 1998):

(1) f (Coriolis parameter)
(2) ζ (low-level relative vertical vorticity)
(3) S (tropospheric vertical wind shear)
(4) E (ocean thermal energy, manifested as ocean temperatures greater than 26°C to a depth of 60 m)
(5) $\Delta\theta_e$ (difference in θ_e between the surface and 500 mb), and
(6) RH (relative humidity in the middle troposphere),

which can be combined into a *dynamic potential* ($f\zeta/S$) and a *thermal potential* [$E(\Delta\theta_e)$RH]. Multiplying both dynamic and thermal potentials gives a seasonal TC genesis parameter. Calculations of this parameter from climatological data appear to be well correlated with the regions of long-term frequency of occurrence of global tropical cyclogenesis (McBride 1995). A genesis parameter for evaluating the potential for eastern Atlantic tropical cyclogenesis, which includes the effects of vertical wind shear, barotropic-baroclinic instabilities, and mid-level moisture variable controlling entrainment of environmental air, has also been proposed (DeMaria *et al.* 2005). This genesis parameter appears to be useful for the prediction of tropical cyclogenesis over the eastern Atlantic Ocean. Some common ingredients discussed above were not included in this genesis parameter and could be very useful if a universal set of necessary conditions or nondimensional control parameters for tropical cyclogenesis were to be proposed.

h. Extratropical hurricanes

Storms that are similar to tropical cyclones in structure sometimes occur over extratropical oceans. These storms possess special characteristics usually found in tropical cyclones, such as their occurrence within deep moist adiabatic atmospheres, the small, intense vortices with clear eyes surrounded by deep cumulonimbi, and the presence of warm cores with strong surface winds. These extratropical cyclones, which form over relatively warm water during the winter and possess tropical cyclone characteristics, are known as *extratropical hurricanes* (Bergeron 1954). In particular, some polar lows and Mediterranean cyclones are named, respectively, *arctic hurricanes* (Emanuel and Rotunno 1989) and *Mediterranean hurricanes* (Emanuel 2005).

Polar lows are small synoptic-scale or subsynoptic-scale cyclones forming during the cold season within polar or arctic air masses over open sea, such as the Bering Sea and the Baltic Sea. Typically, they are several hundred kilometers in diameter, and often possess strong winds. Unlike tropical cyclones, some polar lows tend to form in highly baroclinic environments, such as to the north of major frontal zones, and are preceded by strong upper-tropospheric troughs or lows. Typically, satellite imagery of polar lows shows spiral or comma cloud patterns, and sometimes reveals an inner eye and warm core similar to that of tropical cyclones. Although some polar lows have maximum surface winds of $30 \, \mathrm{m \, s^{-1}}$, they normally do not possess hurricane or typhoon strength winds ($V_{max} \geq 33 \, \mathrm{m \, s^{-1}}$).

Numerical studies indicate that both the baroclinic processes and surface fluxes play roles in the polar low formation. Since the environments conducive to polar low formation vary to a certain extent, different mechanisms have been proposed to explain their formation, which include (1) CISK mechanism (e.g., Rasmussen 1979), (2) baroclinic instability (e.g., Reed 1979), (3) WISHE (Emanuel and Rotunno 1989), and (4) coupled forcing mechanism (Montgomery and Farrell 1992; to be explained below). The first and third mechanisms are related to those proposed for tropical cyclogenesis, as discussed earlier in this subsection. The CISK mechanism does not fully describe the polar low formation as well as it does for tropical cyclogenesis.

Baroclinic instability is a well-known mechanism used to explain extratropical cyclogenesis. However, many e-folding timescales, i.e. several days, are usually required in order to have the most unstable normal mode of baroclinic instability develop to intensities of observed polar lows. Observations indicate that the time for rapidly developing polar lows to attain their mature stage can be as short as one day. The WISHE mechanism for tropical cyclogenesis may also explain polar lows, thus proposing that they are indeed arctic hurricanes (Emanuel and Rotunno 1989). The smaller radii of polar lows are due to larger values of the Coriolis parameter, i.e. smaller λ_R. Similar to tropical cyclones, surface flux-driven polar lows require sufficiently well developed disturbances to initiate the air–sea interaction intensification process. The coupled forcing mechanism for polar low development consists of two stages (Montgomery and Farrell 1992). In the first stage (induced self-development), a mobile upper-level trough initiates a rapid low-level circulation in a conditionally neutral baroclinic atmosphere. In the second stage (diabatic destabilization), the development is associated with the production of low-level potential vorticity through latent heat release. The WISHE mechanism helps explain the relatively short (about 1 day) spinup timescale for polar lows, and it appears that polar lows may require more than one mechanism to spin up and that the dominant mechanism(s) is(are) analogous to those involved in tropical cyclogenesis.

9.3.3 Intensity and mesoscale structure

Once a tropical cyclone reaches its *mature stage*, it is maintained by the extraction of sensible and latent heat fluxes from the warm ocean and the transport of the heat released by cloud formation upward to the upper troposphere, as discussed in the previous subsection. At this stage, it consists of mesoscale features such as the *eye*, *eyewall(s)*, *spiral rainbands*, secondary vortices, and upper-level outflow and cloud shields (Fig. 9.23a). The surface winds spiral inward cyclonically and upward toward the center, forming an *outer vortex* with spiral cloud and rainbands. The most intense

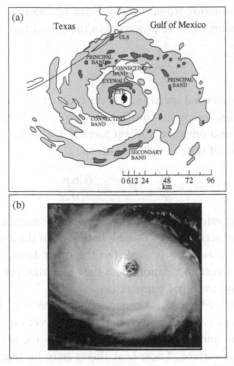

Fig. 9.23 (a) Radar echo pattern seen in Hurricane Alicia (1983). Principal and secondary rainbands, eyewall, and eye are clearly depicted. Contours are for 25 and 40 dBZ. (Adapted after Marks and Houze 1987.) (b) Defense Meteorological Satellite Program (DMSP) imagery of Hurricane Isabel at 1315 UTC September 12, 2003. Within the eye of the hurricane, there are six mesovortices, one at the center and five surrounding it, as found by Kossin and Schubert (2004).

rain and winds occur in the eyewall, which surrounds a nearly symmetric circular eye with downward motion, clear sky, and light winds. Occasionally, multiple *concentric eyewalls* occur, in which the outer eyewall may contract and replace the inner eyewall. Most, but not all, tropical cyclones with maximum winds in excess of $40 \, \mathrm{m \, s^{-1}}$ have eyes visible on satellite imagery. Eye diameters vary from 10 to more than 100 km and average about 30 km. A nearly circular and symmetric spiral current composes the overall wind field associated with a tropical cyclone, in a moving frame of reference.

a. Intensity of a mature tropical cyclone

Prediction of tropical cyclone intensity is extremely important due to the associated high wind and torrential rain. In spite of advanced observational techniques and numerical weather prediction models developed in the past several decades, improvement of intensity prediction is still limited, especially compared with the numerical prediction of tropical cyclone tracks, and remains challenging.

The winds within a mature tropical cyclone can be assumed to be in *gradient wind balance*. This force balance can be expressed in the cylindrical coordinates as

$$\frac{V^2}{r} + fV - \frac{1}{\rho}\frac{\partial p}{\partial r} = 0,$$ (9.3.5)

where V is the tangential wind speed and r is the radial distance from the rotating axis of the cyclone. The first, second, and third terms of (9.3.5) represent the centrifugal force, Coriolis force, and pressure gradient force in the radial direction, respectively. Applying the equation of state to (9.3.5) leads to

$$\frac{V^2}{r} + fV = RT\frac{\partial \ln p}{\partial r}.$$ (9.3.6)

Thus, if the tangential velocity is known at a certain location away from the cyclone center, the pressure reduction can be roughly estimated if the averaged temperature in the area encircled by a radius r is used. On the other hand, if the radial pressure gradient is known at a certain location, then the tangential velocity can be estimated by assuming a constant averaged temperature.

The surface pressure reduction associated with a tropical cyclone cannot be attributed solely to the latent heat released by cumulus convection (Riehl 1948). It is well recognized that the major additional energy source is the sensible and latent heat fluxes from the ocean surface. These heat fluxes may be represented by the moist entropy change, (9.3.4), which indicates that lowering surface pressure while maintaining air temperature near the sea surface will lead to an increase of moist entropy (assuming that $\delta q_v = 0$) for an air parcel that moves from the outer region near sea surface toward the eyewall along Leg 1 of Fig. 9.19. The moist entropy is also related to the equivalent potential temperature (θ_e) by

$$T\delta s \approx c_p T\delta \ln \theta_e.$$ (9.3.7)

Thus, increasing moist entropy leads also to an increase of θ_e. The high θ_e air ascends, warms the column, and lowers the surface pressure. Based on this concept, the *maximum potential intensity* (MPI) of a tropical cyclone can be estimated by δp by combining (9.3.4) and (9.3.7) (Holland 1997),

$$\delta p = \rho[c_p\delta T + L_v\delta q_v - c_p T\delta \ln \theta_e].$$ (9.3.8)

Another approach for estimating the MPI is based on the Carnot cycle concept (Fig. 9.19). The maximum wind speed is approximated by (Emanuel 1997):

$$|V_{max}|^2 \approx \frac{C_h}{C_D}\left(\frac{T_s - T_o}{T_o}\right)(h_o^* - h),$$ (9.3.9)

where C_h and C_D are the exchange coefficients for enthalpy and momentum, respectively, T_s the sea surface temperature, T_o the mean temperature of Leg 3 of the TC

Carnot cycle, h the moist enthalpy, and h_o^* the saturated (with respect to T_s) moist enthalpy of air near the sea surface. The middle term on the right-hand side of (9.3.9) has a form similar to the thermodynamic efficiency, $\epsilon = (T_s - T_o)/T_s$, except that it has the output temperature in the denominator, which reflects the added contribution from the *dissipative heating*. The dissipative heating represents the heat gained by the viscous dissipation of turbulence kinetic energy, which can be approximated by

$$D_h = C_D \rho |V|^3. \tag{9.3.10}$$

Equation (9.3.9) provides an estimate of MPI in terms of maximum tangential wind velocity. The minimum pressure of a tropical cyclone can be estimated by assuming an air parcel that ascends moist adiabatically from the level of free convection to the undisturbed top of a mean tropical atmosphere. The pressure difference between the storm center and ambient atmosphere of a tropical cyclone can be estimated by

$$\delta p_c \approx - \left(\frac{c_p \bar{p}}{R}\right)\left(\frac{T_b - T_o}{T_b}\right)\frac{\delta\theta_{ec}^*}{\bar{\theta}_e}, \tag{9.3.11}$$

where overbars denote the ambient values, T_b is the temperature at the cloud base or top of the boundary layer, T_o is the outflow temperature, and $\delta\theta_{ec}^*$ is the difference of saturation equivalent potential temperature at the storm center.

Prediction of tropical cyclone intensity remains challenging. For example, in the formulation of (9.3.8), a large ambient CAPE may be produced by raising the surface temperature while keeping the remainder of the sounding the same. This may lead to the estimate of an unrealistic low surface pressure (Camp and Montgomery 2001). In the mean time, the formulation of (9.3.11) may lead to an estimate of MPI that is not large enough (Persing and Montgomery 2003), and thus, further research is required to resolve the MPI issue. In addition, in simulations of Hurricane Diana (1984), it was found that almost any intensity can be produced by various choices of "model physics," depending on the selection of grid spacing, initialization time, and model physical parameterizations (Davis and Bosart 2002). In addition to the above physical and numerical factors, tropical cyclone intensity is also affected by factors in its environment, such as vertical wind shear and sea surface temperature. Tropical-cyclone intensity can be affected by shear via *ventilation*, where low-level heat and moisture are advected rapidly away in the upper levels. In order to improve the prediction of MPI, further research is required to improve our understanding of tropical cyclone dynamics.

b. Mesoscale structure of a tropical cyclone

The structure of a tropical cyclone is relatively well understood due to the availability of advanced observational instruments and techniques. For example, Fig. 9.23a shows the radar echo pattern seen in Hurricane Alicia, which clearly depicts the principal and secondary rainbands, eyewall, and eye. Figure 9.23b shows six mesovortices, with one

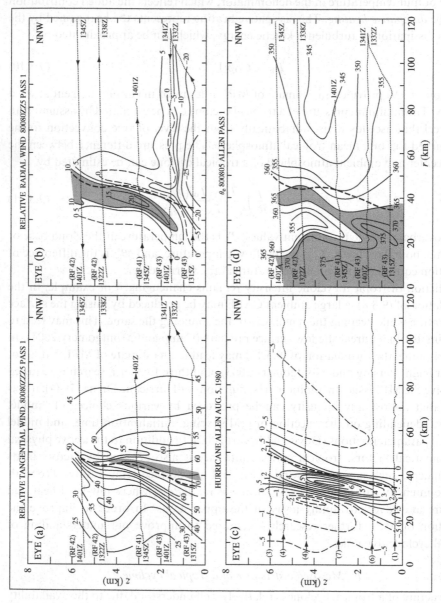

Fig. 9.24 Observed radius–height ($r-z$) cross sections of (a) tangential wind (m s^{-1}) and radar reflectivity greater than 37 dBZ (shaded), radius of maximum wind (RMW) is denoted by the dashed line; (b) relative radial wind (m s^{-1}) and outflow greater than 10 m s^{-1} (shaded); (c) composite mesoscale vertical velocity; and (d) equivalent potential temperature. Regions where $\theta_e > 365$ K are shaded. Data were obtained from the flights into Hurricane Allen on August 5, 1980. (Adapted after Jorgensen 1984.)

located in the center and the other five surrounding it, embedded in the eye of Hurricane Isabel at 1315 UTC September 12, 2003, as revealed in satellite imagery. Theoretical studies indicate that these vortices underwent multiple merger events (Kossin and Schubert 2004). Examples of observed tangential wind, radial wind, composite mesoscale vertical velocity, and equivalent potential temperature fields for Hurricane Allen (1980) can be found in Fig. 9.24. The descending and ascending motions associated with eye and eyewall, respectively, are clearly depicted. The eyewall expands outward with height. The maximum tangential wind is located near the surface, while the maximum vertical velocity is located at about 3.5 km.

The eye and eyewalls are the most prominent features from satellite and radar imagery of tropical cyclones. Within the eye of a tropical cyclone, the atmosphere is characterized by relative warm, light wind, clear or broken clouds, and low surface pressure. The eye is formed by the adiabatic warming associated with the subsidence of air and is surrounded by the eyewall. The eyewall is composed of a ring of cumulonimbus convection, which contains the strongest winds and most intense rainfall of a tropical cyclone. When the core of a tropical cyclone becomes saturated (moist neutral), intensification is controlled by the feedback process between the surface *enthalpy flux* and wind speeds, according to

$$F_h = -C_h \rho \, |V| \, (h_o^* - h), \tag{9.3.12}$$

where F_h is the enthalpy flux from the ocean into the atmosphere. The above equation indicates that the surface heat flux increases with increasing surface wind speeds. In the meantime, however, dissipation increases as the cube of wind speed, according to (9.3.10), which eventually balances energy production and leads the tropical cyclone to a quasi-steady state. The spinup time for tropical cyclone vortex may be estimated by dividing the depth scale of the cyclone by the vertical velocity scale, i.e. the time for air to move vertically through the cyclone (Emanuel 1989). The depth of the cyclone scale can be represented by the atmospheric scale height (H). The vertical velocity scale can be estimated by $C_D V_{max}$, where V_{max} is the maximum wind speed given by (9.3.9). This scale is the same as that which governs surface fluxes (Betts 1983). Combining these two scales together leads to the spinup time scale for a tropical cyclone

$$\tau \approx \frac{H}{C_D V_{max}}. \tag{9.3.13}$$

Without the subsidence in the warm eye, a tropical cyclone cannot reach a central surface pressure as low as that observed by simply lifting low-level air in the clouds surrounding the eye (Riehl 1954). The subsidence may be explained by the fact that the air detrains from the top of the eyewall clouds and sinks inside the eye due to compensating downward motion. The air in the eye descends to the lower troposphere where it is entrained back to the eyewall. This adiabatic warming contributes to the low pressure inside the eye. The amount of warming can be estimated by following a hypothetical air parcel, which mixes with air from the eyewall, descending through the eye. The downward motion inside the eye of a tropical cyclone may be explained by the following vortex dynamics. Subsidence is driven by an adverse, axial gradient of perturbation pressure that is associated principally with the decay and/or radial spread of the tangential wind field with height. Assuming the TC vortex is in gradient wind balance, the perturbation pressure gradient force at the center of the eye is related to the tangential wind by (Smith 1980):

$$-\frac{1}{\rho}\left(\frac{\partial p'}{\partial z}\right)_{r=0} = \frac{1}{\rho}\frac{\partial}{\partial z}\int_0^R \rho\left(\frac{V^2}{r}+fV\right)dr, \qquad (9.3.14)$$

where r is the radius from the vortex center and R is a large radius where perturbation pressure (p') vanishes. The above equation implies that in the layer where tangential wind (V) decreases with height, there exists a downward pressure gradient force. This occurs because the right side of (9.3.14) is negative in average sense, thus producing subsidence inside the eye of the TC vortex. The buoyancy force due to adiabatic warming almost exactly opposes this pressure gradient force. Observations indicate that stronger downward motion in hurricane eyes occur during their intensifying stage, as implied by the mechanism described above. The eye also derives its warm temperature through inward advection and detrainment mixing from the eyewall (Holland 1997). Observations also indicate that the air supplying the subsidence originates both at the tops of eyewall cumuli (hot towers) as well as from air detrained out of the adjacent updraft at the midlevels. This deduction is based on Doppler radar observations of subsiding air, with vertical velocities of several meters per second, that descends at least 9 km and extends horizontally nearly 25 km into the eye interior of Hurricane Bonnie (1998) (Heymsfield *et al.* 2001).

Numerical models with fine horizontal resolution are able to simulate the core structure of tropical cyclones reasonably well. Figure 9.25 shows an example of numerically simulated vertical structure and flow circulation around the eye and eyewall of Hurricane Andrew (1992). The basic structures and circulations in and surrounding the eye are consistent with observations.

Observations indicate that the sea level pressure at the center of an intense tropical cyclone is 50–100 hPa lower than outside the vortex, but only 10–30 hPa of the total pressure fall occurs inside the eye between the eyewall and the vortex center (Willoughby 1998). The pressure difference between the eye and eyewall is due to the adiabatic warming associated with dry subsidence inside the eyewall. Figure 9.26 shows a sounding taken by an aircraft inside the eye of Hurricane Olivia (1994), which indicates warm and very dry air aloft, separated by an inversion from cloudy air below. The dewpoint depressions at the inversion level inside eyewall of tropical cyclones, typically 850–500 hPa, are 10–30 K, much less than about the 100 K that would occur if the air descended from tropopause level without dilution by the environmental clouds. However, the observed temperature and dewpoint distribution

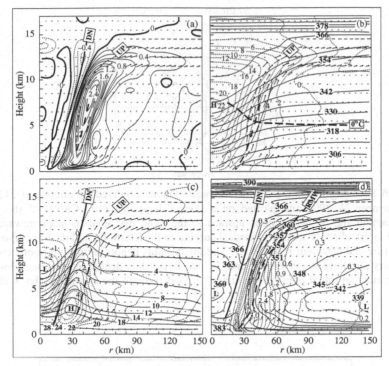

Fig. 9.25 Vertical structure of the eye, eyewall, eyewall updraft, inner-edge dry downdraft, and the radius of maximum wind (RMW). Shown are the numerically simulated radius–height cross sections of the hourly and azimuthally averaged fields of Hurricane Andrew (1992): (a) vertical velocity (m s^{-1}); (b) θ (K) and θ' (K); (c) specific humidity (solid, g kg^{-1}) and its deviations (dashed, g kg^{-1}); and (d) equivalent potential temperature (solid, K) and total cloud hydrometeors (dashed, g kg^{-1}). Wind vectors are superimposed in every panel. These fields are taken from the 56–57-h numerical simulation valid between 2000 and 2100 UTC August 23, 1992. The horizontal resolution of the numerical model is 5 km. Symbols UP, DN, RMW, H, and L denote respectively the eyewall updraft, the inner-edge dry downdraft, the radius of maximum wind, local maximum and local minimum. (From Zhang *et al.* 2002.)

above the inversion can be derived by sinking an undiluted air dry adiabatically 100 hPa from an initial sounding that is somewhat more stable than a moist adiabat. This indicates that the subsidence of the air trapped inside the eyewall and above the inversion is drawn downward toward the inversion level as the air below it flows outward into the eyewall, as depicted in Fig. 9.27.

When low-level air parcels flow inward and spiral upward along the eyewall, their PV increases rapidly. This may form an annular tower of high-PV air against the low-PV air inside the eye, and produces an area of air that is highly unstable barotropically (Schubert *et al.* 1999). When the unstable region is perturbed, the vorticity of the eyewall region pools into discrete areas, creating polygonal eyewalls (Fig. 9.23b) and is proposed to be produced by vortex Rossby waves (e.g., Kuo *et al.* 1999). The vortex

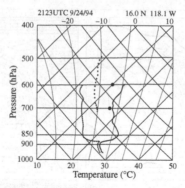

Fig. 9.26 Sounding inside the eye of Hurricane Olivia, observed by aircraft at 2123 UTC on September 24, 1994 and plotted on a skew-T log-p diagram. Curves of this skew-T log-p diagram are: isotherms slope upward to the right; dry adiabats slope upward to the left; and moist adiabats are nearly vertical curving to the left. Solid and dashed curves denote temperature and dewpoint respectively. The smaller dots denote saturation points computed for the dry air above the inversion, and the two larger dots denote temperature observed at the innermost saturated point as the aircraft passed through the eyewall. (After Willoughby 1998.)

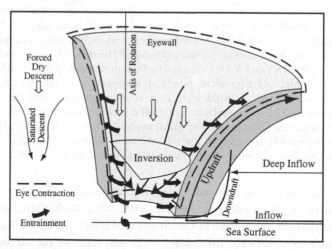

Fig. 9.27 A schematic of the secondary flow around the eye of a tropical cyclone. The frictional inflow feeds the buoyancy-driven primary updraft and outflow in the eyewall cloud. Saturated convective downdrafts in the eyewall and the evaporatively driven descent along the inner edge of the eyewall feed moist air into the volume below the inversion surface. The air column inside the eye is warmed adiabatically by the dynamically driven descent of dry air. The descent is forced as entrainment into the eyewall draws mass from the bottom of the eye into the eyewall. Balance between moist-air production and loss by entrainment determines the rate of rise or fall of the inversion. (Adapted after Willoughby 1998.)

Rossby waves are Rossby waves that propagate azimuthally on the radial gradient of relative potential vorticity in the cyclonic vortex circulation. The radial PV gradient associated with the vortex motion serves as the restoring force for the vortex Rossby waves, analogous to the PV gradient associated with the β effect for the planetary Rossby waves. Occasionally, mesovortices form within the eye, as shown in Fig. 9.23b. This type of vortex Rossby wave may play an important role in determining TC structure and intensity changes. Numerical simulations indicate that the maximum growth of the instabilities of the initial flow occurs at wavenumber 13, and 13 mesovortices are "rolled up" from the vorticity in the eyewall (Kossin and Schubert 2004). As the mesovortices migrate into and around the eye while rotating cyclonically, rapid merger occurs and results in the pentagonal patterns.

A typical hurricane or typhoon consists of an azimuthally averaged symmetric flow and banded asymmetric features. The most prominent banded features associated with a hurricane or typhoon are the *spiral rainbands*, which are often evident in satellite and radar imagery, such as that shown in Fig. 9.23a. There are a couple of ways to classify the rainbands of tropical cyclones. Based on their motion, they can be classified as either *stationary* or *moving rainbands* (Willoughby *et al.* 1984). Some stationary rainbands may be interrelated, and are often referred to as *stationary band complex*. The stationary band complex is composed of the *principal band*, the *connecting band*, and the *secondary band*. Figure 9.23a shows an example of radar echo patterns of *principal* and *secondary rainbands*, which are distinct from the eye wall, formed within a hurricane. Another type of rainband associated with a tropical cyclone can be purely stratiform. The stratiform rainband is formed when a plume of ice particles from somewhere along the eyewall falls through the 0°C level after spiraling outward at upper levels. Based on location and dimension, tropical cyclone rainbands can be classified also as inner and outer rainbands. Two major mechanisms have been proposed to explain the generation of tropical cyclone spiral rainbands: inertia-gravity waves and vortex Rossby waves. In the first mechanism, the TC spiral rainbands are considered to be a manifestation of inertia-gravity waves (e.g. Willoughby *et al.* 1984; Chow *et al.* 2002). In the second mechanism, the inner bands are considered as breaking vortex Rossby waves, and the outer bands as the result of the nonlinear breakdown of the intertropical convergence zone through barotropic instability (e.g., Guinn and Schubert 1993). Therefore, different mechanisms may be responsible for generating different types of TC spiral rainbands.

The cyclonic circulation of a tropical cyclone in the lower troposphere weakens with height, eventually turning into an anticyclonic circulation at the top of the storm, in response to the high pressure built up in the upper troposphere. The TC outflow is quite asymmetric and is often composed of one or two anticyclonically curving jets (e.g., Merrill 1988). Observations show different patterns of the TC outflow, depending upon the interaction of the TC outflow with environmental synoptic flow. The TC outflow provides a channel that carries high potential temperature air from the convective core of the tropical cyclone outward to its far environment, which is then cooled by radiation, as an important process in completing one branch of the tropical cyclone Carnot cycle (Fig. 9.19). The evolution of the tropical cyclone outflow plays important roles in affecting both the TC motion, structural change, and intensity (Holland and Merrill 1984).

9.3.4 Tropical cyclone movement

A tropical cyclone, once it forms, may move as far as several thousand kilometers away from its genesis region. After its formation, a tropical cyclone in the Northern Hemisphere normally moves to the northwest and then recurves to the northeast after moving into the midlatitude westerlies. Prediction of tropical cyclone movement remains one of the most important and challenging tasks in tropical cyclone forecasting. A tropical cyclone's movement is influenced by both large-scale dynamics and its internal dynamics, which are mesoscale in nature. The TC motion is mainly influenced by the following factors: (1) interaction of a TC vortex with an environmental flow or nearby TCs, (2) β-effect, (3) vertical shear, (4) convective heating, and (5) topography.

The fundamental characteristics of TC motion influenced by the interaction between the TC vortex and environmental flow can be understood by considering the nondivergent barotropic vorticity equation on a β plane,

$$\frac{\partial \zeta}{\partial t} = -V \cdot \nabla \zeta - \beta v. \tag{9.3.15}$$

The last term on the right-hand side is the planetary vorticity (f) advection by the meridional flow (v). A local vorticity maximum tends to be advected downstream by the environmental flow, thus the vorticity center of the TC vortex is displaced or steered downstream. Examples of environmental flows in affecting TC motion include the prevailing easterlies in the tropics, prevailing westerlies in the midlatitudes, and winds associated with synoptic systems, such as high pressure systems over Pacific and Atlantic, synoptic fronts, synoptic-scale waves, and even the rotational flow associated with an adjacent tropical cyclone (i.e. the *Fujiwhara* (1921) *effect*, to be explained later). Here, the environmental flows also include mid- and upper-tropospheric flow. These influences may also be dependent on the TC's intensity since the TC may also interact with its environmental flow.

The β-effect on TC movement is associated with the differential advection of planetary vorticity by the meridional wind. This mechanism may be interpreted by considering a vortex in a quiescent fluid, which is depicted in Fig. 9.28. On the west side of the vortex center in the northern hemisphere, northerly wind associated with the rotational flow of the vortex advects larger planetary vorticity from higher latitudes. On the other hand, on the east side of the vortex center, a southerly wind component advects smaller planetary vorticity from lower latitudes. Thus, a local positive (negative) relative vorticity tendency ($\partial \zeta / \partial t$) is produced to the west (east) of the vortex center. This tends to induce a pair of secondary circulations with opposite

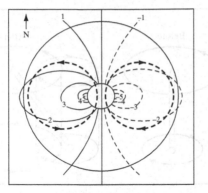

Fig. 9.28 A sketch of relative vorticity tendency ($\partial\zeta/\partial t$) of a symmetric nondivergent vortex on a Northern Hemispheric β plane with no background flow. The vortex is indicated by the inner circle. The solid (dashed) contours in arbitrary units represent positive (negative) relative vorticity produced by northerly (southerly) rotational flow associated with the vortex (i.e. the β effect). The bold-dashed contours denote the secondary circulations induced by the β effect, which tends to advect the vortex northward. (After Holland 1983.)

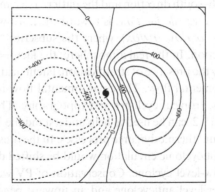

Fig. 9.29 The asymmetric streamfunction (ψ) after integrating (9.3.15) for 24 h. The plot domain is 2400×2400 km. Positive values (solid) indicate anticyclonic streamfunctions while negative values (dashed) indicate cylonic streamfunctions. The center of the vortex is denoted by the tropical cyclone symbol. Contour interval is 10^5 m^2 s^{-1}. (From Fiorino and Elsberry 1989.)

sense to advect the vortex northward (Fig. 9.28). However, the actual linear numerical solution of (9.3.15) for a vortex in a quiescent fluid only gives a westward dispersion with very little displacement of the vortex, similar to the linear Rossby waves, while the nonlinear solution gives a northwestward movement (Chan and Williams 1987). This phenomenon can be explained by the nonlinear interaction between the vortex and the horizontal gradient of the planetary vorticity. An isolated vortex sitting on the Earth's surface tends to produce an asymmetric flow with a dipole structure, which is referred to as a β-gyre, as shown in Fig. 9.29. Although the dipole structure of the β-gyre resembles that depicted in Fig. 9.28, the gyres are oriented counterclockwise

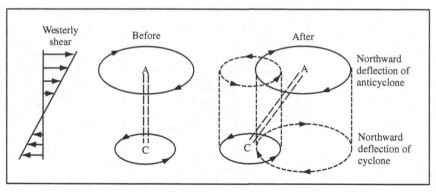

Fig. 9.30 Schematic of PV upward penetration and downward penetration in a shear flow (Wu and Emanuel 1993). The low-level cyclone and upper-level anticyclone are initially stacked in vertical, but are then displaced by a westerly shear flow. This results in a downward (upward) penetration of the anticyclonic (cyclonic) flow, which produces an additional northward "steering" flow. The entire vortex moves to the left (northward to northeastward) of the shear vector (eastward). (After Elsberry 1995, based on Wu and Emanuel 1993.)

about 45° from the north–south direction and the pattern is asymmetric. This asymmetric flow in the center, i.e. the *ventilation flow*, is toward the northwest, as can be found by calculating $(u_{as}, v_{as}) = (-\partial\psi/\partial y, \partial\psi/\partial x)$, where the subscript "as" denotes asymmetric and ψ is the streamfunction. In the real atmosphere, the factors affecting the TC movement are complicated by the presence of stratification, latent heat release, friction, vertical wind shear and environmental flow. The environmental flow is strongly influenced by nearby weather systems, tropical cyclones and topography. The potential vorticity, as defined earlier in Section 9.2, can be used to help understand the TC movement because it includes both dynamic and thermodynamic processes. In the presence of the vertical wind shear, the upper-level anticyclonic circulation would be shifted downstream of the shear vector relative to the low-level cyclone. Conservation of PV would cause a downward penetration of the upper-level anticyclone and an upward penetration of the low-level cyclone that creates an additional steering flow, as sketched in Fig. 9.30. Both upward penetration and downward penetration result in northward deflection of the vortex. Latent heating released by the cloud and precipitation formation associated with a tropical cyclone will produce a negative (positive) PV anomaly at upper (lower) levels, which in turn can induce a downward (upward) penetration of the anticyclonic (cyclonic) circulation and affect the TC motion (Wu and Kurihara 1996). In addition to the effects discussed above, a tropical cyclone's movement may also be influenced by a nearby tropical cyclone, known as the Fujiwhara (1921) effect, inducing a cyclonically rotation around each other due to the interaction of their outer circulations which may even cause an eventual merger. Once a tropical cyclone makes landfall on a continent, the surface friction tends to weaken and change the direction of tropical cyclone motion. If mountains exist on the continent or island, the movement, intensity, and the primary circulation of a tropical cyclone can be strongly influenced (e.g., see Subsection 5.5.3).

References

Aiyyer, A. R. and J. Molinari, 2003. Evolution of mixed Rossby–gravity waves in idealized MJO environments. *J. Atmos. Sci.*, **60**, 2837–55.

Atallah, E. H. and L. F. Bosart, 2003. The extratropical transition and precipitation distribution of Hurricane Floyd (1999). *Mon. Wea. Rev.*, **131**, 1063–81.

Bergeron, T., 1954. The problem of tropical hurricanes. *Quart. J. Roy. Meteor. Soc.*, **80**, 131–64.

Betts, A. K., 1983. Thermodynamics of mixed stratocumulus layers saturation point budgets. *J. Atmos. Sci.*, **40**, 2655–70.

Bluestein, H. B. and M. H. Jain, 1985. Formation of mesoscale lines of precipitation: Severe squall lines in Oklahoma during the spring. *J. Atmos. Sci.*, **42**, 1711–32.

Bluestein, H. B. and M. L. Weisman, 2000. The interaction of numerically simulated supercells initiated along lines. *Mon. Wea. Rev.*, **128**, 3128–49.

Browning, K. A., 1986. Conceptual models of precipitating systems. Wea. Forecasting, **1**, 23–41.

Bryan, G. H. and J. M. Fritsch, 2000. Moist absolute instability: The sixth static stability state. *Bull. Amer. Meteor. Soc.*, **81**, 1207–30.

Camp, J. P. and M. T. Montgomery, 2001. Hurricane maximum intensity: Past and present. *Mon. Wea. Rev.*, **129**, 1704–17.

Carbone, R. E., 1982. A severe frontal rainband. Part I: Stormwide hydrodynamic structure. *J. Atmos. Sci.*, **39**, 258–79.

Chan, J. C. L. and R. T. Williams, 1987: Analytical and numerical studies of the beta-effect in tropical cyclone motion. Part I: zero mean flow. *J. Atmos. Sci.*, **44**, 1257–65.

Charney, J. G. and A. Eliassen, 1964. On the growth of the hurricane depression. *J. Atmos. Sci.*, **21**, 68–75.

Chow, K. C., K. L. Chan, and A. K. H. Lau. 2002. Generation of moving spiral bands in tropical cyclones. *J. Atmos. Sci.*, **59**, 2930–50.

Cotton, W. R. and R. A. Anthes, 1989. *Storm and Cloud Dynamics*. Academic Press.

Crook, N. A. and M. W. Moncrieff, 1988. The effect of large-scale convergence on the generation and maintenance of deep moist convection. *J. Atmos. Sci.*, **45**, 3606–24.

Davis, C. A. and L. F. Bosart, 2002. Numerical simulations of the genesis of Hurricane Diana (1984). Part II: Sensitivity of track and intensity prediction. *Mon. Wea. Rev.*, **130**, 1100–24.

Davis, C. A. and L. F. Bosart, 2003. Baroclinically induced tropical cyclogenesis. *Mon. Wea. Rev.*, **131**, 2730–47.

Dengler, K. and M. J. Reeder, 1997. The effects of convection and baroclinicity on the motion of tropical-cyclone-like vortices. *Quart. J. Roy. Meteor. Soc.*, **123**, 699–725.

DeMaria, M., M. Mainelli, L. K. Shay, J. A. Knaff and J. Kaplan, 2005. Further improvements to the statistical hurricane intensity prediction scheme (SHIPS). *Wea. Forecasting*, **20**, 531–43.

Dickinson, M. and J. Molinari, 2002. Mixed Rossby-gravity waves and western Pacific tropical cyclogenesis. Part I: Synoptic evolution. *J. Atmos. Sci.*, **59**, 2183–96.

Elsberry, R. L. (ed.), 1995. *Global Perspective of Tropical Cyclones*. WMO-/TD No. 693, World Meteor. Org.

Emanuel, K. A., 1986. An air-sea interaction theory for tropical cyclones. Part I: Steady state maintenance. *J. Atmos. Sci.*, **43**, 585–604.

Emanuel, K. A., 1989. The finite-amplitude nature of tropical cyclogenesis. *J. Atmos. Sci.*, **46**, 3431–56.

Emanuel, K. A., 1994. *Atmospheric Convection*. Oxford University Press.

Emanuel, K. A., 1997. Some aspects of hurricane inner-core dynamics and energetics. *J. Atmos. Sci.*, **54**, 1014–26.

Emanuel, K. A., 2005. Genesis and maintenance of "Mediterranean hurricanes". *Adv. Geosci.*, **2**, 217–20.

Emanuel, K. A. and R. Rotunno, 1989. Polar lows as arctic hurricanes. *Tellus*, **41A**, 1–17.

Fankhauser, J. C., G. M. Barnes, and M. A. LeMone, 1992. Structure of a midlatitude squall line formed in strong unidirectional shear. *Mon. Wea. Rev.*, **120**, 237–60.

Fiorino, M. and R. L. Elsberry, 1989. Some aspects of vortex structure related to tropical cyclone motion. *J. Atmos. Sci.*, **46**, 975–90.

Fritsch, J. M., J. D. Murphy, and J. S. Kain, 1994. Warm core vortex amplification over land. *J. Atmos. Sci.*, **51**, 1780–807.

Fujiwhara, S., 1921. The mutual tendency towards symmetry of motion and its application as a principle in meteorology. *Quart. J. Roy. Meteor. Soc.*, **47**, 287–93.

Grady, R. L. and J. Verlinde, 1997. Triple-Doppler analysis of a discretely propagating, long-lived, High Plains squall line. *J. Atmos. Sci.*, **54**, 2729–48.

Gray, W. M., 1968. Global view of the origin of tropical disturbances and storms. *Mon. Wea. Rev.*, **96**, 669–700.

Gray, W. M., 1998. The formation of tropical cyclones. *Meteor. Atmos. Phys.*, **67**, 37–69.

Guinn, T. and W. H. Schubert, 1993. Hurricane spiral bands. *J. Atmos. Sci.*, **50**, 3380–404.

Hendricks, E. A., M. T. Montgomery, and C. A. Davis, 2004. The role of "vortical" hot towers in the formation of tropical cyclone Diana (1984). *J. Atmos. Sci.*, **61**, 1209–32.

Heymsfield, G. M., J. B. Halverson, J. Simpson, L. Tian, and T. P. Bui, 2001. ER-2 Dopper radar investigations of the eyewall of Hurricane Bonnie during the convection and moisture experiment-3. *J. Appl. Meteor.*, **40**, 1310–30.

Hill, C. M. and Y.-L. Lin, 2003. Initiation of a mesoscale convective complex over the Ethiopian Highlands preceding the genesis of Hurricane Alberto (2000). *Geophys. Res. Lett.*, **30**, 1232, doi:10.1029/2002GL016655.

Holland, G. J., 1983. Tropical cyclone motion: environmental interaction plus a beta-effect. *J. Atmos. Sci.*, **40**, 328–42.

Holland, G. J., 1997. The maximum potential intensity of tropical cyclones. *J. Atmos. Sci.*, **54**, 2519–41.

Holland, G. J. and R. T. Merrill, 1984. On the dynamics of tropical cyclone structural changes. *Quart. J. Roy. Meteor. Soc.*, **110**, 723–45.

Houze, R. A., Jr., 1981. Structures of atmospheric precipitation systems – A global survey. *Radio Sci.*, **16**, 671–89.

Houze, R. A., Jr., S. A. Rutledge, M. I. Biggerstaff, and B. F. Smull, 1989. Interpretation of Doppler weather-radar displays in midlatitude mesoscale convective systems. *Bull. Amer. Meteor. Soc.*, **70**, 608–19.

Hsu, H.-H. and M.-Y. Lee, 2005. Topographic effects on the eastward propagation and initiation of the Madden-Julian oscillation. *J. Climate*, **18**, 795–809.

Jiang, H. and D. J. Raymond, 1995. Simulation of a mature mesoscale convective system using a nonlinear balance model. *J. Atmos. Sci.*, **52**, 161–75.

Johnson, R. H. and D. L. Bartels, 1992. Circulations associated with a mature-to-decaying midlatitude mesoscale convective system. Part II: Upper-level features. *Mon. Wea. Rev.*, **120**, 1301–20.

Jorgensen, D. P., 1984. Mesoscale and convective scale characteristics of mature hurricanes. Part I: Inner core structure of Hurricane Allen (1980). *J. Atmos. Sci.*, **41**, 1287–311.

Kossin, J. P. and W. H. Schubert, 2004. Mesovortices in Hurricane Isabel. *Bull. Amer. Meteor. Soc.*, **85**, 151–3.

Kuo, H.-C., R.-T. Williams, and J.-H. Chen, 1999. A possible mechanism for the eye rotation of typhoon Herb. *J. Atmos. Sci.*, **56**, 1659–673.

Laing, A. G. and J. M. Fritsch, 1993. Mesoscale convective complexes in Africa. *Mon. Wea. Rev.*, **121**, 2254–63.

Laing, A. G. and J. M. Fritsch, 1997. The global population of mesoscale convective complexes. *Quart. J. Meteor. Soc.*, **123**, 389–405.

Laing, A. G. and J. M. Fritsch, 2000. The large-scale environments of the global populations of mesoscale convective complexes. *Mon. Wea. Rev.*, **128**, 2756–76.

Lilly, D. K., 1979. The dynamical structure and evolution of thunderstorms and squall lines. *Ann. Rev. Earth Planet. Sci.*, **7**, 117–71.

Lin, Y.-L., K. E. Robertson, and C. M. Hill, 2005. Origin and propagation of a disturbance associated with an African easterly wave as a precursor of Hurricane Alberto (2000). *Mon. Wea. Rev.*, **133**, 3276–98.

Lin, Y.-L., T.-A. Wang, and R. P. Weglarz, 1993. Interactions between gravity waves and cold air outflows in a stably stratified uniform flow. *J. Atmos. Sci.*, **50**, 3790–816.

Ludlam, F. H., 1963. Severe local storms: a review. *Meteor. Monogr.*, **5**, 1–30.

Madden, R. A. and P. R. Julian 1994. Observations of the 40–50 day tropical oscillation – A review. *Mon. Wea. Rev.*, **122**, 814–37.

Maddox, R. A., 1980. Mesoscale convective complexes. *Bull. Amer. Meteor. Soc.*, **61**, 1374–87.

Maddox, R. A., 1983. Large-scale meteorological conditions associated with midlatitude mesoscale convective complexes. *Mon. Wea. Rev.*, **111**, 1475–93.

Maddox, R. A., K. W. Howard, D. L. Bartels, and D. M. Rodgers, 1986. Mesoscale convective complexes in the middle latitudes. In *Mesoscale Meteorology and Forecasting*, P. S. Ray (ed.), Amer. Meteor. Soc., 390–413.

Marks, F. D., Jr., and R. A. Houze, Jr., 1987. Inner-core structure of Hurricane Alicia from airborne Doppler-radar observations. *J. Atmos. Sci.*, **44**, 1296–317.

McBride, J. L., 1995. Tropical cyclone formation. In *Global Perspectives on Tropical Cyclones*. R. Elsberry (ed.), World. Meteor. Org., 63–105.

Merrill, R. T., 1988. Characteristics of the upper-tropospheric environmental flow around hurricanes. *J. Atmos. Sci.*, **45**, 1665–77.

Miller, D. and J. M. Fritsch, 1991. Mesoscale convective complexes in the western Pacific region. *Mon. Wea. Rev.*, **119**, 2978–92.

Molinari, J. and D. Vollaro, 2000. Planetary- and synoptic-scale influence on eastern Pacific tropical cyclogenesis. *Mon. Wea. Rev.*, **128**, 3296–307.

Moncrieff, M. W., 1978. The dynamical structure of two-dimensional steady convection in constant vertical shear. *Quart. J. Roy. Meteor. Soc.*, **104**, 543–67.

Montgomery, M. T. and B. F. Farrell, 1992. Polar low dynamics. *J. Atmos. Sci.*, **49**, 2484–505.

Montgomery, M. T. and B. F. Farrell, 1993. Tropical cyclone formation. *J. Atmos. Sci.*, **50**, 285–310.

Neumann, C., 1993. Global overview. In *Global Guide to Tropical Cyclone Forecasting. WMO/TD-560*, G. J. Holland (ed.), World Meteor. Org., 3.1–3.46.

Newton, C. W., 1963. *Dynamics of severe convective storms. Meteor. Monogr.*, **5**, Amer. Meteor. Soc., 33–58.

Ogura, Y., and Y.-L. Chen, 1977. A life history of an intense mesoscale convective storm in Oklahoma. *J. Atmos. Sci.*, **34**, 1458–76.

Olsson, P. O. and W. R. Cotton, 1997. Balanced and unbalanced circulations in a primitive equation simulation of a midlatitude MCC. Part II: Analysis of balance. *J. Atmos. Sci.*, **54**, 479–97.

Ooyama, K., 1964. A dynamical model for the study of tropical cyclone development. *Geophys. Int.*, **4**, 187–98.

Ooyama, K., 1969. Numerical simulation of the life cycle of tropical cyclones. *J. Atmos. Sci.*, **26**, 3–40.

Ooyama, K., 1982. Conceptual evolution of the theory and modeling of the tropical cyclone. *J. Meteor. Soc. Japan*, **60**, 369–80.

Palmén, E., 1948. On the formation and structure of tropical cyclones. *Geophysics*, **3**, 26–38.

Parker, M. D. and R. H. Johnson, 2000. Organization modes of midlatitude mesoscale convective systems. *Mon. Wea. Rev.*, **128**, 3413–36.

Parker, M. D. and R. H. Johnson, 2004a. Structures and dynamics of quasi-2D mesoscale convective systems. *J. Atmos. Sci.*, **61**, 545–67.

Parker, M. D. and R. H. Johnson, 2004b. Simulated convective lines with leading precipitation. Part I: Governing dynamics. *J. Atmos. Sci.*, **61**, 1637–55.

Persing, J. and M. T. Montgomery, 2003. Hurricane superintensity. *J. Atmos. Sci.*, **60**, 2349–71.

Rasmussen, E. R., 1979. The polar low as an extratropical CISK disturbance. *Quart. J. Roy. Meteor. Soc.*, **105**, 531–49.

Raymond, D. J. and H. Jiang, 1990. A theory for long-lived mesoscale convective systems. *J. Atmos. Sci.*, **47**, 3067–77.

Redelsperger, J.-L. and T. L. Clark, 1989. The initiation and horizontal scale selection of convection over gently sloping terrain. *J. Atmos. Sci.*, **47**, 516–41.

Reed, R. J., 1979. Cyclogenesis in polar air streams. *Mon. Wea. Rev.*, **107**, 38–52.

Riehl, H., 1948. On the formation of typhoons. *J. Meteor.*, **5**, 247–64.

Riehl, H., 1954. *Tropical Meteorology*. McGraw-Hill Co.

Ritchie, E. A. and G. J. Holland 1993. On the interaction of tropical-cyclone scale vortices. II: interacting vortex patches. *Quart. J. Roy. Meteor. Soc.*, **119**, 1363–97.

Ritchie, E. A. and G. J. Holland, 1997. Scale interactions during the formation of Typhoon Irving. *Mon. Wea. Rev.*, **125**, 1377–96.

Rotunno, R. and K. A. Emanuel, 1987. An air-sea interaction theory for tropical cyclones. Part II: Evolutionary study using a nonhydrostatic axiymmetric numerical model. *J. Atmos. Sci.*, **44**, 140–53.

Rotunno, R., J. B. Klemp, and M. L. Weisman, 1988. A theory for strong, long-lived squall lines. *J. Atmos. Sci.*, **45**, 463–85.

Schubert, W. H., J. J. Hack, P. L. Silva Dias, and S. R. Fulton, 1980. Geostrophic adjustment in an axisymmetric vortex. *J. Atmos. Sci.*, **37**, 1464–84.

Schubert, W. H., M. T. Montgomery, R. K. Taft, T. A. Guinn, S. R. Fulton, J. P. Kossin, and J. P. Edwards, 1999. Polygonal eyewalls, asymmetric eye contraction, and potential vorticity mixing in hurricanes. *J. Atmos. Sci.*, **56**, 1197–223.

Smith, R. K., 1980. Tropical cyclone eye dynamics. *J. Atmos. Sci.*, **37**, 1227–32.

Smith, R. K., 1997. On the theory of CISK. *Quart. J. Roy. Meteor. Soc.*, **123**, 407–18.

Takeda, T., 1971. Numerical simulation of a precipitating system convective cloud: The formation of a "long-lasting" cloud. *J. Atmos. Sci.*, **28**, 350–76.

Thiao, W., R. A. Scofield, and J. Robinson, 1993. *The Relationship Between Water Vapor Plumes and Extreme Rainfall Events During the Summer Season*, NOAA Technical Report, NESDIS, 67, 69 pp.

Tripoli, G. J. and W. R. Cotton, 1989. Numerical study of an observed orogenic mesoscale convective system. Part I: Simulated genesis and comparison with observations. *Mon. Wea. Rev.*, **117**, 273–304.

Wang, Y., 2002. Vortex Rossby waves in a numerically simulated tropical cyclone. Part I: Overall structure, potential vorticity, and kinetic energy budgets. *J. Atmos. Sci.*, **59**, 1213–37.

Weisman, M. L., and R. Rotunno, 2004. "A theory for strong long-lived squall lines" revisited. *J. Atmos. Sci.*, **61**, 361–82.

Weisman, M. L., J. B. Klemp, and R. Rotunno, 1988. Structure and evolution of numerically simulated squall lines. *J. Atmos. Sci.*, **45**, 1990–2013.

Willoughby, H. E., 1998. Tropical cyclone eye thermodynamics. *Mon. Wea. Rev.*, **126**, 3053–67.

Willoughby, H. E., F. D. Marks, and R. J. Feinberg, 1984. Stationary and moving convective bands in hurricanes. *J. Atmos. Sci.*, **41**, 3189–211.

Wu, C.-C. and K. A. Emanuel, 1993. Interaction of a baroclinic vortex with background shear: Application to hurricane movement. *J. Atmos. Sci.*, **50**, 62–76.

Wu, C.-C. and Kurihara, 1996. A numerical study of the feedback mechanisms of hurricane-environment interaction on hurricane movement from the potential vorticity perspective. *J. Atmos. Sci.*, **53**, 2264–82.

Xu, Q., 1992. Density currents in shear flows – A two-fluid model. *J. Atmos. Sci.*, **49**, 511–24.

Xue, M., 2000. Density currents in two-layer shear flows. *Quart. J. Roy. Met. Soc.*, **126**, 1301–20.

Zhang, D.-L., Y. Liu, and M. K. Yau, 2002. A multiscale numerical study of Hurricane Andrew (1992). Part V: Inner-core thermodynamics. *Mon. Wea. Rev.*, **130**, 2745–63.

Zipser, E. J., 1977. Mesoscale and convective-scale downdrafts as distinct components of squall-line structure. *Mon. Wea. Rev.*, **105**, 1568–89.

Problems

9.1 Using the equation of state to show that (9.3.5) can be written as

$$\frac{V^2}{r} + fV = RT\frac{\partial \ln p}{\partial r}.$$

At a certain location that is 50 km away from the storm center near the sea surface, the tangential velocity is measured at $50\,\mathrm{m\,s^{-1}}$ and the pressure is 960 mb. Assuming the storm is centered at 15° N and the averaged near-surface air temperature is 28°C, estimate the pressure gradient at this location. If the pressure decreases linearly toward the storm center, what is the center pressure of this tropical cyclone? Is it realistic?

9.2 Use (9.3.4) and the definition of $\theta_e = \theta \exp (L_v q_v / c_p T)$ to prove (9.3.7). You may assume $\delta (L_v q_v / c_p T) \approx (L_v / c_p T) \delta q_v$.

9.3 The average vertical gradient of diabatic heating due to latent heat release in a hot tower over a tropical ocean located at 20° N is 45 K h^{-1} km^{-1} from $z = 0$ to 5 km for 1 h and the mean air density in the lower troposphere is 0.7 kg m^{-3}. The initial relative vertical vorticity is 9.5×10^{-4} s^{-1}. The vertical gradient of potential temperature from the surface to 5 km increases from 12 K to 24 K. Estimate approximate average PV tendency and the final relative vorticity after 1 h.

9.4 Figure 9.26 shows the sounding taken from Hurricane Olivia (1994). Use a skew-T log-p diagram to do the following:

  (a) Estimate the final temperature of an air parcel (denoted by a large dot at 600 mb) that has been lifted moist adiabatically (moist neutral) along the eyewall cumulonimbus up to the top of the hurricane (say 150 mb).

  (b) Now assuming this air parcel detrain from the cloud top and descend adiabatically down to the inversion level ($\sim$ 875 mb), what is the temperature of the air parcel? How does it compare with the observed temperature in the eye (bold solid curve)?

  (c) Do you agree with the conventional point of view described above? If not, then what could be the problem with this hypothesis?

10

Dynamics of fronts and jet streaks

Understanding the dynamics of fronts and jet streaks is important for improving the accuracy of weather forecasts since many severe weather events, such as thunderstorms, squall lines, rainbands, and small-scale disturbance (e.g., gravity waves, Kelvin–Helmholtz instability and clear air turbulence), are associated with fronts and jet streaks. An atmospheric *front* can be defined as an interface or transition zone between two air masses of different characteristics, such as air temperature, moisture, or wind speed and direction. Fronts may be classified as *synoptic-scale fronts* and *mesoscale fronts*. In the literature, "*fronts*" often refer to synoptic-scale fronts near the Earth's surface, as shown on surface weather maps. Synoptic-scale fronts are characterized by large gradients in temperature ($\geq 10\,\mathrm{K}/1000$ km), static stability, vertical wind shear (e.g., ≥ 5–$10\,\mathrm{m\,s^{-1}\,km^{-1}}$), positive vorticity associated with horizontal wind shear, and/or moisture. Typically, synoptic-scale fronts on the Earth's surface have a length of 1000–2000 km, a width of 100–200 km, a depth of 1–2 km but may extend to 5–6 km, and last for several days. Along the front, the Rossby number is usually small and the rotational effects are significant so that the flow is approximately in geostrophic balance across the front.

Major synoptic-scale fronts include *cold fronts, warm fronts, polar fronts, arctic fronts*, and *occluded fronts*. While less dense air rises over denser air for both cold and warm fronts, cold fronts have a more well-defined pressure trough than warm fronts because cold fronts have steeper slopes and stronger temperature gradients than warm fronts. Theoretically, synoptic fronts can develop at any level in the atmosphere, although they often form near the solid or internal boundaries, such as the Earth's surface or tropopause. They are generated by internal dynamical processes, such as horizontal deformation and differential heating, and are often embedded in midlatitude baroclinic waves, which have a horizontal scale of 2000–10 000 km.

Occasionally, the frontal boundaries separating air masses with different characteristics have length and width scales about one order smaller than those of synoptic-scale fronts. This type of front is called the *mesoscale front*. Unlike synoptic scale fronts, mesoscale fronts are shallower and only last a few hours. Mesoscale fronts formed near Earth's surface are often contained within the planetary boundary layer. Examples of mesoscale fronts are *sea-breeze fronts, coastal fronts, gust fronts* and *drylines.* To facilitate the discussion, the *upper-level fronts* will be discussed in the same section as of drylines since both of them possess different characteristics from conventional surface synoptic-scale fronts.

The formation or intensification of a front is called *frontogenesis*. Frontogenesis occurs when an air mass moves toward another air mass that has a different density, thus causing the horizontal temperature gradient to intensify significantly. A prerequisite for synoptic-scale frontogenesis is *baroclinicity* in which there exist horizontal temperature gradients on isobaric surfaces. The weakening of a front is called *frontolysis*. The region of local wind maximum embedded within the jet stream is called the *jet streak*. Jet streaks often form near the tropopause above surface synoptic-scale fronts due to the thermal wind effect. Predicting the location and structure of upper-level fronts and jet streaks is essential to aviation safety since clear-air turbulence often forms in the strong shear environment associated with upper-level fronts and jet streaks. In addition, upper-level fronts may help transport ozone and other material from stratosphere to troposphere.

10.1 Kinematics of frontogenesis

Some basic frontal structures and kinematic properties can be understood through the *wedge model of fronts*, in which the front is viewed as a sloping boundary between two different air masses with different densities, as proposed in the *polar front theory* (Bjerknes 1919). Bjerknes' polar front theory proposes that cyclones form on a preexisting polar front. As the wave grows on the temperature discontinuity, the leading portion of the front is referred to as a warm front and the trailing portion as a cold front.

Figure 10.1 shows an example of a cold front analyzed for 0200 UTC April 26, 1979. The structure of the front at the meso-α scale is well depicted by the analysis. The front is well defined by the strong temperature gradient and relative vorticity (Fig. 10.1a). The potential temperature and along-front wind component (Fig. 10.1b) indicate the low-level frontal zone behind a low-level potential temperature maximum, consistent with the presence of the surface warm tongue (Fig. 10.1a). Low-level northerly and southerly jets are located respectively behind and ahead of the surface front (Fig. 10.1b). The vertical wind shear is distributed through the troposphere and results in a jet core at the tropopause. The kinematically derived vertical motion field indicates that air ascends in the warm region ahead of the front and descends in the cold region behind the front (Fig. 10.1c). From the first approximation, this polar front may be represented by a wedge model shown in Fig. 10.2. The frontal slope for a dry, hydrostatic, inviscid atmosphere can be derived as

$$\frac{dz}{dy} = \lim_{\delta y \to 0} \frac{\delta z}{\delta y} = \frac{(\partial p/\partial y)_c - (\partial p/\partial y)_w}{g(\rho_c - \rho_w)}, \tag{10.1.1}$$

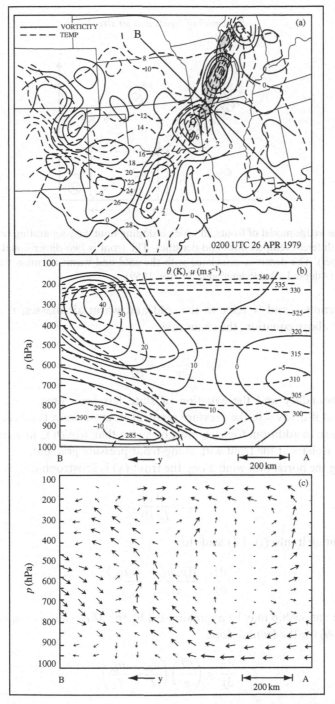

Fig. 10.1 Frontal structure as depicted by a front analyzed for 0200 UTC April 26, 1979: (a) Surface temperature (°C, dashed), relative vorticity (contour interval $2 \times 10^{-5}\,\mathrm{s}^{-1}$, solid). Radar echoes at 0235 UTC April 26, 1979 are shaded. (b) θ (K, dashed) and velocity component normal to the cross section AB depicted in (a) ($\mathrm{m\,s}^{-1}$, solid; southerly positive). (c) Transverse circulation relative to the movement of the front. (Adapted after Ogura and Poris 1982.)

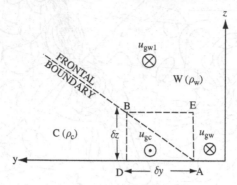

Fig. 10.2 The wedge model of fronts, in which a sloping boundary separating two different air masses with different temperatures and densities. The front is two-dimensional in x direction (into the paper). The densities associated with the cold and warm air masses are ρ_c and ρ_w, respectively. (Adapted after Palmén and Newton 1969)

where subscripts c and w refer to the cold and warm air masses, respectively. In deriving the above equation, the chain rule

$$dp = \frac{\partial p}{\partial y} dy + \frac{\partial p}{\partial z} dz, \qquad (10.1.2)$$

along the boundary and the equation of state has been used. Equation (10.1.1) indicates that if $dz/dy \neq 0$, the pressure gradient in cross-front (y) direction must be discontinuous. In addition, $(\partial p/\partial y)_c$ must be larger than $(\partial p/\partial y)_w$ in value, which will give kinked isobars at the front with along-front pressure gradient.

Assuming the horizontal wind along the front (x) is geostrophic,

$$u_g = \frac{-1}{f\rho} \frac{\partial p}{\partial y}, \qquad (10.1.3)$$

and substituting it into (10.1.1) leads to

$$\frac{dz}{dy} = \frac{f(\rho_w u_{gw} - \rho_c u_{gc})}{g(\rho_c - \rho_w)}. \qquad (10.1.4)$$

The above relationship may be approximated by the *Margules' formula* (Palmén and Newton 1969) for the frontal slope

$$\frac{dz}{dy} \approx \left(\frac{f\overline{T}}{g}\right)\left(\frac{u_{gw} - u_{gc}}{T_w - T_c}\right), \qquad (10.1.5)$$

where $\overline{T}$ is the mean temperature across the front. Equation (10.1.5) indicates that the frontal slope is determined by the relative magnitudes of the geostrophic wind difference and temperature contrast across the front. Note that (10.1.5) does not imply that stronger fronts (larger $T_w - T_c$) are less steep, because they are also likely to have larger vorticity (larger $u_{gw} - u_{gc}$). For a normal frontal situation, $dz/dy > 0$, (10.1.5) leads to

$$u_{gw} - u_{gc} > 0. \tag{10.1.6}$$

This gives a cyclonic geostrophic vorticity across the front, as also depicted in Fig. 10.2. Note that the basic wind is normally set up by the synoptic environment, thus different configurations of the flow field, i.e. positive or negative u_g, across the front may occur. Equation (10.1.5) may be rearranged to give

$$\frac{u_{gw1} - u_{gc}}{\delta z} \approx \left(\frac{g}{f\bar{T}}\right)\left(\frac{T_w - T_c}{\delta y}\right), \tag{10.1.7}$$

which indicates that the vertical shear of the geostrophic wind is controlled by the horizontal temperature gradient or baroclinicity. That is, the flow is in *thermal wind* balance.

Assuming the frontal surface behaves as an internal boundary where the flow moves along it, then the vertical velocity along the frontal surface can be obtained by equating the slope of the motion to that of the front,

$$w = (v - c)\frac{\partial z}{\partial y}, \tag{10.1.8}$$

where v is the cross-frontal velocity component and c is the speed of the front movement. We have assumed that the front behaves more or less like an *effective mountain*, i.e., the frontal surface does not vary in the along-front (x) direction and that the frontal surface does not change shape with time. Applying (10.1.8) to both the cold and warm sides of the front yields

$$w_w - w_c = (v_w - v_c)\frac{\partial z}{\partial y}. \tag{10.1.9}$$

This gives two types of fronts: (a) *anafronts* when $w_w > w_c$ ($v_w > v_c$) and (b) *katafronts* when $w_w < w_c$ ($v_w < v_c$) (Bergeron 1937). Anafronts and katafronts are characterized by post- and pre-frontal cloud bands, respectively. In reality, the temperature or density across the front is continuous, which is not well represented by the wedge model, such as that revealed in Fig. 10.2. Thus, a front should be regarded as a *frontal zone*, which also allows the definition of the strength of a front to be the magnitude of the horizontal temperature gradient. This was why an atmospheric front has also been defined as a sloping zone of pronounced gradient in the thermal and wind fields (Keyser 1986).

Some fundamental kinematics and thermodynamics of the frontogenesis can be quantified by the *frontogenesis (frontogenetical) function*, which not only helps the analysis of frontogenesis from observed data, but in addition helps in the understanding of frontogenesis mechanisms. A three-dimensional frontogenesis function may be defined as (Miller 1948):

$$F = \frac{\mathrm{D}}{\mathrm{D}t}|\nabla\theta|,\tag{10.1.10}$$

which leads to

$$F = \frac{1}{|\nabla\theta|}\left\{\frac{\partial\theta}{\partial x}\underbrace{\left(\frac{\partial\dot{Q}}{\partial x}}_{1} - \underbrace{\frac{\partial u}{\partial x}\frac{\partial\theta}{\partial x}}_{2} - \underbrace{\frac{\partial v}{\partial x}\frac{\partial\theta}{\partial y}}_{3} - \underbrace{\frac{\partial w}{\partial x}\frac{\partial\theta}{\partial z}}_{4}\right)\right.$$

$$+\frac{\partial\theta}{\partial y}\underbrace{\left(\frac{\partial\dot{Q}}{\partial y}}_{5} - \underbrace{\frac{\partial u}{\partial y}\frac{\partial\theta}{\partial x}}_{6} - \underbrace{\frac{\partial v}{\partial y}\frac{\partial\theta}{\partial y}}_{7} - \underbrace{\frac{\partial w}{\partial y}\frac{\partial\theta}{\partial z}}_{8}\right)$$

$$\left.+\frac{\partial\theta}{\partial z}\underbrace{\left(\frac{\partial\dot{Q}}{\partial z}}_{9} - \underbrace{\frac{\partial u}{\partial z}\frac{\partial\theta}{\partial x}}_{10} - \underbrace{\frac{\partial v}{\partial z}\frac{\partial\theta}{\partial y}}_{11} - \underbrace{\frac{\partial w}{\partial z}\frac{\partial\theta}{\partial z}}_{12}\right)\right\}.\tag{10.1.11}$$

In deriving the above equation, the thermodynamics energy equation, (2.2.5), was used to eliminate $D\theta/Dt$ and, unlike in Petterssen's (1956) definition, the height coordinate is adopted. Note that $\dot{Q} \equiv (\theta/c_p T)\dot{q}$, where $\dot{q}$ is the diabatic heating rate per unit mass (in $\mathrm{J\,kg^{-1}\,s^{-1}}$). Terms 1, 5, and 9 are the *differential heating terms*. As mentioned earlier in Chapter 2, diabatic heating includes radiative and latent heating. Sea-breeze fronts and coastal fronts are two examples of frontogenesis induced by differential heating. Term 9 affects only the static stability, but does not induce horizontal frontogenesis. Terms 2 and 7 are the *stretching deformation (confluence)* terms; Terms 3 and 6 are the *shearing deformation terms*. Figures 10.3a and 10.3b show

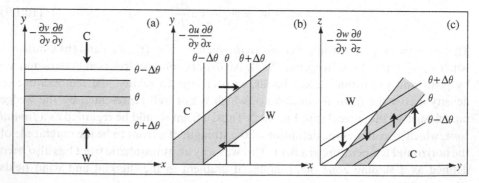

Fig. 10.3 Schematics for (a) stretching deformation $[-(\partial v/\partial y)(\partial\theta/\partial y)]$, (b) shearing deformation $[-(\partial u/\partial x)(\partial\theta/\partial y)]$, and (c) tilting $[-(\partial w/\partial y)(\partial\theta/\partial z)]$ of the frontogenesis function, (10.1.11). The shaded area denotes the frontal zone at a later time after frontogenesis starts. C and W denote initial cold and warm regions.

examples of frontogenesis associated with stretching deformation (confluence) and shearing deformation. Terms 4 and 8 are the *tilting terms* (Fig. 10.3c). Terms 10 and 11 are the *vertical deformation terms* and term 12 is the *vertical divergence term*. The last three terms affect only the static stability, but do not induce horizontal frontogenesis.

Equation (10.1.11) reduces to the two-dimensional *Petterssen frontogenesis function* (Petterssen 1956) if the terms involving the vertical velocity and vertical shear of horizontal winds are neglected and the isobaric coordinates are adopted. If the diabatic terms are also neglected, then (10.1.11) reduces to

$$F = \frac{-1}{|\nabla \theta|} \left[\left(\frac{\partial \theta}{\partial x}\right)^2 \frac{\partial u}{\partial x} + \frac{\partial \theta}{\partial x}\frac{\partial \theta}{\partial y}\frac{\partial v}{\partial x} + \frac{\partial \theta}{\partial x}\frac{\partial \theta}{\partial y}\frac{\partial u}{\partial y} + \left(\frac{\partial \theta}{\partial y}\right)^2 \frac{\partial v}{\partial y} \right]. \qquad (10.1.12)$$

The above equation can be written in terms of vorticity (ζ), divergence (δ) and stretching deformation (D_1) and shearing deformation (D_2),

$$F = \frac{-1}{|\nabla \theta|} \left[\frac{1}{2}(\delta + D_1)\left(\frac{\partial \theta}{\partial x}\right)^2 + \frac{1}{2}(\zeta + D_2)\frac{\partial \theta}{\partial x}\frac{\partial \theta}{\partial y} - \frac{1}{2}(\zeta - D_2)\frac{\partial \theta}{\partial x}\frac{\partial \theta}{\partial y} \right.$$
$$\left. + \frac{1}{2}(\delta - D_1)\left(\frac{\partial \theta}{\partial y}\right)^2 \right], \qquad (10.1.13)$$

where ζ, δ, D_1 and D_2 are defined as

$$\zeta = \frac{\partial v}{\partial x} - \frac{\partial u}{\partial y}; \; \delta = \frac{\partial u}{\partial x} + \frac{\partial v}{\partial y}; \; D_1 = \frac{\partial u}{\partial x} - \frac{\partial v}{\partial y}; \; D_2 = \frac{\partial v}{\partial x} + \frac{\partial u}{\partial y}. \qquad (10.1.14)$$

Thus, frontogenesis is induced by divergence, stretching and shearing deformation. Note that the vorticity terms in (10.1.13) cancel, which implies that the frontogenesis is independent of vorticity. Even though vorticity is produced in the frontal zone, it plays no direct role in frontogenesis. However, vorticity can still play an indirect role by rotating isotherms into alignment with the axis of dilatation. By rotating the isotherms to be along the axis of dilatation, the simplified frontogenesis function (10.1.13) becomes,

$$F = \frac{1}{2}|\nabla \theta|(D\cos 2b - \delta), \qquad (10.1.15)$$

where $D = \sqrt{D_1^2 + D_2^2}$ is the resultant deformation and b is the angle measured from the axis of dilatation to the potential temperature isotherms passing through the point under consideration (Fig. 10.4a). When $0° < b < 45°$, such as that in Fig. 10.4a, the flow field is frontogenetic. When $45° < b < 90°$, the flow field is frontolytic. From the above equation, it becomes clear that convergence tends to induce frontogenesis and divergence tends to induce frontolysis. Figure 10.4b shows an example of effects of deformation on frontogenesis. At location A, a warm front

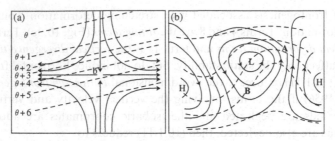

Fig. 10.4 (a) Effect of stretching deformation on frontogenesis. When $0° < b < 45°$, where b is the angle measured from the axis of dilatation (x-axis) to the potential temperature isotherms, there is frontogenesis; when $45° < b < 90°$, there is frontolysis. (Adapted after Petterssen 1956.) (b) An example of warm frontogenesis induced by stretching deformation at A, and cold frontogenesis induced by shear deformation B. (After Hoskins and Bretherton 1972.)

will form later mainly due to stretching deformation and at location B, a cold front will form later mainly due to shearing deformation. Another example is that a *confluent flow*, in which zonal flow is superimposed by horizontal deformation, tends to be frontogenetic. Confluent flow occurs in the entrance region of a jet streak at the tropopause and research has found that North America and East Asia are two favorable regions of frontogenesis due to the existence of confluent flow (Palmén and Newton 1969).

The Petterssen frontogenesis function can be generalized to diagnose *airstream boundaries*, such as drylines, lee troughs, and baroclinic troughs. Unlike fronts, these airstream boundaries are often characterized by abrupt changes in wind speed and direction, and are not necessarily associated with temperature gradients. One approach is to express the time dependence of the separation in horizontal terms between two fluid parcels, i.e. the *separation vector*, in terms of growth rate and rotation rate. The resultant growth rate of the separation vector corresponds to the scalar frontogenesis function and the rotation rate of the separation vector corresponds to the rotational component of the *vector frontogenesis function* (Keyser et al. 1988).

Taking the kinematic approach for calculating the frontogenesis function is useful in diagnostically obtaining distributions of instantaneous rates of change of frontal quantities along parcel trajectories. We can better understand the frontogenesis process through the distribution of these quantities. However, rapid frontogenesis cannot be developed based on the stretching and shearing deformation mechanisms described by these kinematic models. Also, these kinematic models do not reveal the contributions from local rates of change and advection terms to tendencies of frontogenesis. Since the late 1940s, fronts have been viewed as being formed in growing linear baroclinic waves (Charney 1947; Eady 1949). In these baroclinic instability theories, fronts are treated as a smooth rather than discontinuous thermal contrast, and develop in response to wave amplification rather than preceding it.

10.2 Dynamics of two-dimensional frontogenesis

10.2.1 Geostrophic momentum approximation

Significant advances in the understanding of frontogenesis have been made since the 1970s, mainly due to the advent of semi-geostrophic frontogenesis theory, detailed mesoscale field programs, availability of more sophisticated numerical models and more powerful computers. The foundation of the *semi-geostrophic frontogenesis* theory is the *geostrophic momentum approximation* that includes some essential ageostrophic effects. Conceptually, the geostrophic momentum approximation can be analyzed by performing scale analysis on the inertial acceleration and the Coriolis force terms of the momentum equations. In the following discussion, the front is assumed to be stationary and oriented in the x direction. Let L_x and L_y (U and V) respectively represent the along-frontal and cross-frontal length (velocity) scales. We may assume $L_y \ll L_x$ and $V \ll U$, which is consistent with the length and velocity scales for a typical midlatitude synoptic-scale front, i.e., $L_x \sim 1000\,\mathrm{km}$, $L_y \sim 100\,\mathrm{km}$, $U \sim 10\,\mathrm{m\,s^{-1}}$, $V \sim 1\,\mathrm{m\,s^{-1}}$, and $f \sim 10^{-4}\,\mathrm{s^{-1}}$. The scales and magnitudes of the ratios of the inertial acceleration and Coriolis force in the x and y directions are

$$\frac{\mathrm{D}u/\mathrm{D}t}{fv} \sim \frac{U^2/L_x}{fV} \left(\text{or} \frac{UV/L_y}{fV} \right) \sim 1, \text{ and} \tag{10.2.1a}$$

$$\frac{\mathrm{D}v/\mathrm{D}t}{fu} \sim \frac{V^2/L_y}{fU} \left(\text{or} \frac{UV/L_x}{fU} \right) \sim 10^{-2}. \tag{10.2.1b}$$

Thus, the along-front inertial acceleration ($\mathrm{D}u/\mathrm{D}t$) is of first-order importance. On the other hand, the cross-frontal inertial acceleration ($\mathrm{D}v/\mathrm{D}t$) should be much smaller than the Coriolis force, so that the Coriolis force must be balanced by the pressure gradient force,

$$\frac{\mathrm{D}v_\mathrm{g}}{\mathrm{D}t} = -fu - \frac{\partial \phi}{\partial y} = -fu_\mathrm{a} = 0, \tag{10.2.2}$$

where ϕ is the *perturbation kinematic pressure* or a *geopotential function*. Equation (10.2.2) leads to $u = u_\mathrm{g}$ and $v = v_\mathrm{g} + v_\mathrm{a}$, where v_g and v_a are the same order of magnitude. With this approximation, the x-momentum equation, thermodynamic equation and continuity equation in an inviscid, adiabatic, Boussinesq fluid flow can be written as:

$$\frac{\mathrm{D}u_\mathrm{g}}{\mathrm{D}t} - fv_\mathrm{a} = 0, \tag{10.2.3}$$

$$\frac{\mathrm{D}\theta}{\mathrm{D}t} = \dot{Q}, \tag{10.2.4}$$

$$\frac{\partial v_{\mathrm{a}}}{\partial y} + \frac{\partial w}{\partial z} = 0,$$ (10.2.5)

where $\dot{Q} = (\theta_0/c_p T_0)\dot{q}$, $\dot{q}$ is the diabatic heating rate per unit mass $(\mathrm{J\,kg^{-1}\,s^{-1}})$, θ_0 and T_0 are constant reference potential temperature and temperature, respectively, and $\mathrm{D}/\mathrm{D}t$ is a full three-dimensional total (material time) derivative defined as

$$\frac{\mathrm{D}}{\mathrm{D}t} \equiv \frac{\partial}{\partial t} + u\frac{\partial}{\partial x} + v\frac{\partial}{\partial y} + w\frac{\partial}{\partial z} = \frac{\partial}{\partial t} + u_{\mathrm{g}}\frac{\partial}{\partial x} + (v_{\mathrm{g}} + v_{\mathrm{a}})\frac{\partial}{\partial y} + w\frac{\partial}{\partial z}.$$ (10.2.6)

Since the momentum in the momentum equations is replaced by the geostrophic momentum, the approximation is referred to as the *geostrophic momentum approximation* (Sawyer 1956; Eliassen 1962). Note that u_{g} is replaced by $(u_{\mathrm{g}} + u_{\mathrm{a}})$ in a three-dimensional flow.

Mathematically, the geostrophic momentum approximation can be obtained by substituting the geostrophic wind relation (Hoskins 1975)

$$f\mathbf{k} \times \mathbf{V}_{\mathrm{g}} = -\nabla_{\mathrm{H}}\phi,$$ (10.2.7)

into the horizontal momentum equation

$$\frac{\mathrm{D}\mathbf{V}}{\mathrm{D}t} + f\mathbf{k} \times \mathbf{V} + \nabla_{\mathrm{H}}\phi = 0,$$ (10.2.8)

which leads to

$$\frac{\mathrm{D}\mathbf{V}}{\mathrm{D}t} + f\mathbf{k} \times \mathbf{V}_{\mathrm{a}} = 0.$$ (10.2.9)

Taking the cross product of (10.2.9) by $\mathbf{k}$ yields

$$\mathbf{V} = \mathbf{V}_{\mathrm{g}} + \frac{1}{f}\mathbf{k} \times \frac{\mathrm{D}\mathbf{V}}{\mathrm{D}t}.$$ (10.2.10)

Substituting (10.2.10) into itself gives

$$\mathbf{V} = \mathbf{V}_{\mathrm{g}} + \frac{1}{f}\mathbf{k} \times \frac{\mathrm{D}\mathbf{V}_{\mathrm{g}}}{\mathrm{D}t} - \frac{1}{f^2}\frac{\mathrm{D}^2\mathbf{V}_{\mathrm{g}}}{\mathrm{D}t^2} - \cdots$$ (10.2.11)

The above equation can then be nondimensionalized by considering the *Lagrangian Rossby number*, $R_{o\mathrm{L}} = 1/f\tau$, where τ is the *Lagrangian time scale* for parcel acceleration, and assuming V and V_{g} are on the same order of magnitude (i.e., $O(V) = O(V_{\mathrm{g}})$)

$$\tilde{\mathbf{V}} = \tilde{\mathbf{V}}_{\mathrm{g}} + R_{o\mathrm{L}}\mathbf{k} \times \frac{\mathrm{D}\tilde{\mathbf{V}}_{\mathrm{g}}}{\mathrm{D}\tilde{t}} - R_{o\mathrm{L}}^2 \frac{\mathrm{D}^2\tilde{\mathbf{V}}_{\mathrm{g}}}{\mathrm{D}\tilde{t}^2} + \cdots,$$ (10.2.12)

Table 10.1 *Various approximate forms of the inviscid horizontal equations of motion in height coordinate on an f-plane.*

Geostrophic	$0 = -f\mathbf{k} \times V_g - \nabla\phi$
Quasi-Geostrophic	$\dfrac{D_g V_g}{Dt} = \dfrac{\partial V_g}{\partial t} + (V_g \cdot \nabla)V_g = -f\mathbf{k} \times V - \nabla\phi = -f\mathbf{k} \times V_a$
Geostrophic-Momentum	$\dfrac{D V_g}{Dt} = \dfrac{\partial V_g}{\partial t} + (V \cdot \nabla)V_g + w\dfrac{\partial V_g}{\partial z} = -f\mathbf{k} \times V - \nabla\phi = -f\mathbf{k} \times V_a$
Full Primitive Equations	$\dfrac{D V}{Dt} = \dfrac{\partial V}{\partial t} + (V \cdot \nabla)V + w\dfrac{\partial V}{\partial z} = -f\mathbf{k} \times V - \nabla\phi = -f\mathbf{k} \times V_a$

where $\breve{V} = V/U$, $\breve{V}_g = V_g/U$, U is a horizontal velocity scale, and $\tilde{t} = t/\tau$. Neglecting terms higher than $O(R_{oL}^2)$ and restoring the dimensions gives

$$\frac{D V_g}{Dt} \approx -f\mathbf{k} \times V_a = \frac{DV}{Dt}. \tag{10.2.13}$$

Equation (10.2.13) leads to the x-momentum equation under the geostrophic momentum approximation, (10.2.3). The essential difference between (10.2.3) and (10.2.4) and with the corresponding quasi-geostrophic approximation is that the *ageostrophic advection of geostrophic momentum* is retained in the geostrophic momentum approximation. Various approximate forms of inviscid horizontal momentum equations on an f-plane can be found in Table 10.1. The validity of the geostrophic momentum approximation is equivalent to the requirement of $R_{oL} \ll 1$ or $1/f \ll \tau$. Thus, this assumption imposes no direct constraint on the admissible length scales of a given solution. The aforementioned assumption, together with the ageostrophic advection of geostrophic momentum, makes the geostrophic momentum approximation ideal for modeling frontogenesis.

10.2.2 Frontogenesis and cross-frontal circulations

Equations (10.2.3)–(10.2.5) and the following hydrostatic equation and geostrophic wind relations,

$$\frac{\partial\phi}{\partial z} = \frac{g\theta}{\theta_0}, \tag{10.2.14}$$

$$fu_g = -\frac{\partial\phi}{\partial y}; \quad fv_g = \frac{\partial\phi}{\partial x}, \tag{10.2.15}$$

form a closed set of equations, which is referred to as the *geostrophic momentum equations* (Eliassen 1962) in the literature. The thermal wind relations can be obtained from the above two equations,

$$f\frac{\partial u_g}{\partial z} = -\frac{\partial b}{\partial y}; \quad f\frac{\partial v_g}{\partial z} = \frac{\partial b}{\partial x}; \tag{10.2.16}$$

where $b = g\theta/\theta_o$ is the *buoyancy*. Taking $f\, \partial/\partial z$ of (10.2.3) gives

$$\frac{D}{Dt}\left(f\frac{\partial u_g}{\partial z}\right) = Q_2 + f\left(f - \frac{\partial u_g}{\partial y}\right)\frac{\partial v_a}{\partial z} - \frac{\partial v_a}{\partial y}\frac{\partial b}{\partial y}, \qquad (10.2.17)$$

and taking $(-g/\theta_o)\partial/\partial y$ of (10.2.4) gives

$$\frac{D}{Dt}\left(-\frac{\partial b}{\partial y}\right) = -Q_2 + \frac{\partial w}{\partial y}\frac{\partial b}{\partial z} + \frac{\partial v_a}{\partial y}\frac{\partial b}{\partial y} - \frac{g}{\theta_o}\frac{\partial \dot{Q}}{\partial y}. \qquad (10.2.18)$$

Here Q_2 is defined as

$$Q_2 = -\frac{\partial u_g}{\partial y}\frac{\partial b}{\partial x} - \frac{\partial v_g}{\partial y}\frac{\partial b}{\partial y}. \qquad (10.2.19)$$

Note that in the quasi-geostrophic theory, the ageostrophic terms $(\partial u_g/\partial y)(\partial v_a/\partial z)$, $(\partial v_a/\partial y)(\partial b/\partial y)$, and $(\partial w/\partial y)(\partial b/\partial z)$ on the right-hand sides of (10.2.17) and (10.2.18) are neglected.

When $Q_2 > 0$, based on (10.2.17), $\partial u_g/\partial z$ will increase and in turn, $-\partial b/\partial y$ will increase according to (10.2.16). In other words, the above inequality creates a *fronto-lytic* situation. On the other hand, (10.2.18) indicates that $Q_2 > 0$ will decrease $-\partial b/\partial y$ and induce frontogenesis. Thus, the geostrophic wind alone will destroy the thermal wind balance. In the present geostrophic momentum theory or semi-geostrophic theory (to be discussed later), the ageostrophic terms are able to help maintain the thermal wind balance, as explained in the preceding argument.

Equating (10.2.17) and (10.2.18) and using the thermal wind relations, (10.2.16), yields

$$\frac{\partial w}{\partial y}\frac{\partial b}{\partial z} + 2\frac{\partial v_a}{\partial y}\frac{\partial b}{\partial y} - f\left(f - \frac{\partial u_g}{\partial y}\right)\frac{\partial v_a}{\partial z} = 2Q_2 + \frac{g}{\theta_o}\frac{\partial \dot{Q}}{\partial y}. \qquad (10.2.20)$$

Since the ageostrophic circulation lies entirely in the y–z plane, we can define a streamfunction (ψ) such that

$$v_a = -\frac{\partial \psi}{\partial z}; \quad w = \frac{\partial \psi}{\partial y}. \qquad (10.2.21)$$

Substituting the above relations into (10.2.20) leads to the *Sawyer–Eliassen equation* (Sawyer 1956; Eliassen 1962):

$$N^2\frac{\partial^2 \psi}{\partial y^2} + 2S^2\frac{\partial^2 \psi}{\partial y\partial z} + F^2\frac{\partial^2 \psi}{\partial z^2} = 2Q_2 + \frac{g}{\theta_o}\frac{\partial \dot{Q}}{\partial y}, \qquad (10.2.22)$$

where

$$N^2 = \frac{\partial b}{\partial z}, \quad S^2 = -\frac{\partial b}{\partial y} = f\frac{\partial u_g}{\partial z}, \quad F^2 = f\left(f - \frac{\partial u_g}{\partial y}\right) = f\frac{\partial M_g}{\partial y}, \qquad (10.2.23)$$

where M_g is the *geostrophic absolute momentum*, $M_g = fy - u_g$. Note that the geostrophic absolute momentum was defined as $\overline{M} = v_g + fx$ in Chapter 7, which is equivalent to the above relation physically, but with different coordinate systems. The terms of (10.2.23) respectively represent the *static stability, baroclinic stability,* and *inertial stability*. Note that using the thermal wind relations, (10.2.16), Q_2 can be rewritten as

$$Q_2 = f J_{yz}(v_g, u_g), \tag{10.2.24}$$

where J is the *Jacobian differential operator* defined as

$$J_{yz}(\alpha, \beta) = \frac{\partial \alpha}{\partial y}\frac{\partial \beta}{\partial z} - \frac{\partial \beta}{\partial y}\frac{\partial \alpha}{\partial z}, \tag{10.2.25}$$

for any two dependent variables α and β. Substituting (10.2.24) into (10.2.20) indicates that in an inviscid flow, the transverse secondary circulation is solely determined by the intensity and spatial distribution of geostrophic winds and diabatic forcing. Frictional effects can also be represented as a forcing term on the right-hand side of (10.2.22).

If the coefficients and the forcing of (10.2.22) are known, then the response (ψ) can be determined uniquely provided that boundary conditions are specified and an elliptic condition

$$q = N^2 F^2 - S^4 > 0, \tag{10.2.26}$$

is satisfied. In fact, this condition is equivalent to the criterion for symmetric stability, which is related to (7.6.5). In fact, q is the *potential vorticity*. From (10.2.3)–(10.2.5) and (10.2.14), it can be shown that q is conserved following the fluid motion. The Sawyer–Eliassen equation controls the development of transverse secondary circulation. The right-hand side of (10.2.22) represents the effects of geostrophic stretching deformation $[-(\partial v_g/\partial y)(\partial b/\partial y)]$, shearing deformation $[-(\partial u_g/\partial y)(\partial b/\partial x)]$, and differential diabatic heating $(\partial \dot{Q}/\partial y)$.

Effects of ageostrophic advection of geostrophic momentum on frontogenesis can be better analyzed by studying the Sawyer–Eliassen equation in the *geostrophic coordinates*, $(X, Y, Z, \tau) = (x + v_g/f, y - u_g/f, z, t)$. The Sawyer–Eliassen equation in the geostrophic coordinates can be derived by transforming (10.2.3), (10.2.4), and the thermal wind relations into the geostrophic coordinates to obtain

$$\frac{1}{f}\frac{\partial}{\partial Y}\left(q_g \frac{\partial \psi}{\partial Y}\right) + f^2 \frac{\partial^2 \psi}{\partial Z^2} = 2\tilde{Q}_2 + \frac{g}{\theta_o}\frac{\partial \dot{Q}}{\partial Y}, \tag{10.2.27}$$

where

$$\tilde{Q}_2 = -\frac{\partial v_g}{\partial Y}\frac{\partial b}{\partial Y} - \frac{\partial u_g}{\partial Y}\frac{\partial b}{\partial X}, \tag{10.2.28}$$

$$q_g = \zeta_g \frac{\partial b}{\partial Z} = \zeta_g \left(\frac{g}{\theta_0} \frac{\partial \theta}{\partial Z} \right), \tag{10.2.29}$$

and $\zeta_g = f - \partial u_g / \partial y$ is the absolute vertical vorticity. $\tilde{Q}_2$ is the same as Q_2 except in the geostrophic coordinates. Similar to (10.2.24), $\tilde{Q}_2$ may also be written in terms of geostrophic winds,

$$\tilde{Q}_2 = f \left(\frac{\partial v_g}{\partial Y} \frac{\partial u_g}{\partial Z} - \frac{\partial u_g}{\partial Y} \frac{\partial v_g}{\partial Z} \right) = f J_{YZ}(v_g, u_g). \tag{10.2.30}$$

The generation of transverse, ageostrophic, *secondary circulation* associated with the frontogenesis forced by $\tilde{Q}_2$ and differential diabatic heating (i.e. the right-hand side of (10.2.27)) can be understood qualitatively by considering some simple examples. First, let us examine the frontogenesis forced by the stretching deformation, $-(\partial v_g / \partial Y)(\partial b / \partial Y)$ or $-(\partial v_g / \partial Y)(\partial \theta / \partial Y)$, i.e. the first term of $\tilde{Q}_2$. This forcing is also referred to as *Bergeron forcing* in the literature. A simple synoptic situation is sketched in Fig. 10.3a if the large-scale convergent flow is considered to be geostrophic (v_g) and on the Y–Z plane (Fig. 10.5a). Since $\partial v_g / \partial Y < 0$ and $\partial \theta / \partial Y > 0$, we have $\tilde{Q}_2 > 0$. Also, since (10.2.27) is of second order and q_g is assumed to be positive, positive values of the forcing correspond to relative minima (maxima) in the stream-function (ψ) and a circulation with upward (downward) motion to the positive Y-direction of the circulation center since $w = \partial \psi / \partial Y$, as shown in Fig. 10.5a. In other words, the frontogenesis produces a direct transverse, ageostrophic, secondary circulation in the Y–Z plane by a convergent deformation basic flow.

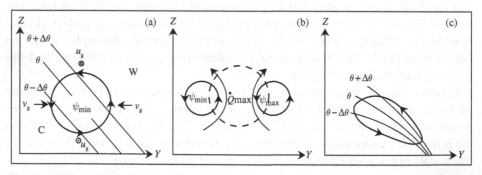

Fig. 10.5 Effects of geostrophic motion and diabatic heating on the generation of transverse, ageostrophic, secondary circulation forced by: (a) stretching deformation, $- (\partial v_g / \partial Y)(\partial b / \partial Y)$(first term of Q_2 in (10.2.30a)), in a synoptic situation of Fig. 10.3a with geostrophic convergent wind (v_g). Based on (10.2.29), a minimum streamfunction (ψ) and a counterclockwise, thermally direct circulation is induced; (b) diabatic heating with maximum heating denoted by $\dot{Q}_{max}$. A pair of counterclockwise and clockwise circulations is generated to the left and right sides of the heating center and an upward motion is induced through the heating center; and (c) transforming the circulation of (a) back to the physical space (y, z).

Based on an equivalent equation of (10.2.18) in geostrophic coordinates, the descending cold air to the left and ascending warm air to the right of the ageostrophic circulation center ($\partial w/\partial Y > 0$) tends to decrease the cross-front thermal contrast through adiabatic heating, which also compensates part of the $\partial b/\partial Y$ increase produced by the large-scale convergent flow. In other words, this ageostrophic secondary circulation tends to bring the atmosphere back toward thermal wind balance by adjusting the buoyancy back toward its original environment before the buoyancy gradient increases. At the same time, easterly and westerly winds are created respectively at upper and lower levels through (10.2.3), which contributes to increasing the vertical shear in the along-front wind component (u_g). This helps maintain the thermal wind balance by adjusting the vertical wind shear to the increased buoyancy (temperature) gradient. Therefore, the ageostrophic motion, (v_a, w), supports a *thermally direct circulation*, with warm air rising and cold air sinking, and acts to restore the atmosphere to hydrostatic and geostrophic balance.

Furthermore, according to (10.2.18), the geostrophic momentum theory also includes the generation of larger buoyancy gradients on the warm side near the surface of the circulation due to the convergence of ageostrophic wind. For example, $\partial v_a/\partial Y < 0$ near the lower right quadrant of the ageostrophic circulation is depicted in Fig. 10.5a. Note that in the quasi-geostrophic theory, terms $(\partial u_g/\partial Y)(\partial v_a/\partial Z)$, $(\partial v_a/\partial Y)(\partial b/\partial Y)$, and $(\partial w/\partial Y)(\partial b/\partial Y)$ are not included in the right-hand sides of (10.2.17) and (10.2.18) of the geostrophic momentum theory. The effect of diabatic heating on the ageostrophic secondary circulation may be illustrated by Fig. 10.5b, in which the maximum diabatic heating is located at $\dot{Q}_{max}$. Based on (10.2.27), a minimum ψ and a maximum ψ are generated respectively to the left- and right-hand side of $\dot{Q}_{max}$, which generates a pair of ageostrophic secondary circulations with an induced upward motion through the maximum heating.

Note that the quasi-geostrophic counterpart of the Sawyer–Eliassen equation (10.2.22),

$$N^2 \frac{\partial^2 \psi}{\partial y^2} + f^2 \frac{\partial^2 \psi}{\partial z^2} = 2Q_2 + \frac{g}{\theta_0} \frac{\partial \dot{Q}}{\partial y}, \qquad (10.2.31)$$

has a similar form to that in the geostrophic coordinates, (10.2.27). Since (10.2.31) does not contain a cross differentiation term on the left side, $\partial^2 \psi/\partial y \partial z$, the circulation does not tilt with height. The terms responsible for the vertical tilt of the secondary circulation in the geostrophic momentum approximation come from terms $(\partial w/\partial y)(\partial b/\partial z)$ and $(\partial v_a/\partial y)(\partial b/\partial y)$ of (10.2.20) which are related to the ageostrophic advection of geostrophic momentum. In addition, the $\partial^2 \psi/\partial z^2$ term does not include the effects of relative vorticity ($\partial v_g/\partial x$) as it does in (10.2.22). Note that one important effect of the ageostrophic motion is to tilt the frontal zone. Transforming back the ageostrophic secondary circulation of Fig. 10.5a to the physical space (y, z) tilts the ellipses along lines of constant Y and produces more intense flow in regions where the relative vorticity $(-\partial u_g/\partial y)$ is large. The tilted ellipses give an intensification of the cross-front flow near the surface and stronger upward motion on the warm side of the front where the relative vorticity is large, as sketched in Fig. 10.5c. In addition, the frontal zone slopes toward the cold air with height. With the ageostrophic winds included, the frontal gradients increase (frontogenetical) near the surface on the warm side and in the upper level on the cold side in Fig. 10.5c, which speeds up the frontogenesis dramatically. In the presence of a boundary, i.e. the surface or the tropopause, the positive feedback from the ageostrophic circulations does not diminish due to adiabatic warming and cooling and is strong enough to produce a discontinuity at the lower right and upper left corners of Fig. 10.5c in a finite time.

10.3 Frontogenesis and baroclinic waves

The geostrophic momentum equations for frontogenesis, (10.2.3)–(10.2.5), (10.2.14) and (10.2.15), in an adiabatic, Boussinesq, hydrostatic fluid flow under the geostrophic momentum approximation can be extended to a three-dimensional flow by adding the ageostrophic wind advection $u_a\partial/\partial x$ to the total derivative (the right-hand side of (10.2.6)) and modifying the y-momentum equation, (10.2.2), to be

$$\frac{\mathrm{D}v_g}{\mathrm{D}t} + fu_a = 0. \tag{10.3.1}$$

Note that the momentum equations under the geostrophic momentum approximation form a balanced system since they do not include gravity waves. They can, however, describe motions in regions of large shear vorticity, such as fronts or jet streaks, in an arbitrary direction, provided that the curvature vorticity is small compared with the planetary vorticity (Coriolis parameter). However, the set of geostrophic momentum equations, (10.2.3)–(10.2.5), (10.2.14), (10.2.15), and (10.3.1), are not very useful for weather prediction because only the geostrophic velocities are predicted. This can be overcome by transforming these equations into the *geostrophic coordinates* (Hoskins 1975):

$$(X, Y, Z, \tau) = (x + v_g/f, y - u_g/f, z, t), \tag{10.3.2}$$

which are so called because the horizontal motion in the geostrophic coordinates is geostrophic in nature,

$$\frac{\mathrm{D}X}{\mathrm{D}t} = u_g \quad \text{and} \quad \frac{\mathrm{D}Y}{\mathrm{D}t} = v_g. \tag{10.3.3}$$

The transformation is valid only if the Jacobian of the transformation,

$$J = J_{xy}(X, Y) = \frac{\zeta_g}{f} + \frac{1}{f^2}\left(\frac{\partial u_g}{\partial x}\frac{\partial v_g}{\partial y} - \frac{\partial v_g}{\partial x}\frac{\partial u_g}{\partial y}\right), \tag{10.3.4}$$

is neither 0 nor infinity. In the above equation, the *absolute geostrophic vorticity* is defined as

$$\zeta_g = f + \left(\frac{\partial v_g}{\partial x} - \frac{\partial u_g}{\partial y}\right). \tag{10.3.5}$$

When the convergence of geostrophic winds are small compared with f and $|\partial v_g/\partial x| \ll f$ and $|\partial u_g/\partial y| \ll f$, the Jacobian terms of (10.3.4) can be ignored. Neglecting the Jacobian terms in (10.3.4) is also consistent with the geostrophic momentum approximation, because $J \approx \zeta_g/f$, we choose to use stretched coordinates for regions of large vorticity.

In the geostrophic space, the *potential vorticity* (q_g) (i.e., geostrophic PV) may be defined as (Hoskins and Draghici 1977):

$$q_g \equiv \boldsymbol{\omega}_g \cdot \nabla b = \left(\frac{g}{\theta_o}\right) \boldsymbol{\omega}_g \cdot \nabla \theta, \tag{10.3.6}$$

where $\boldsymbol{\omega}_g$ is the three-dimensional *geostrophic vorticity vector*,

$$\boldsymbol{\omega}_g = \left(-\frac{\partial v_g}{\partial z}, \frac{\partial u_g}{\partial z}, f + \frac{\partial v_g}{\partial x} - \frac{\partial u_g}{\partial y}\right). \tag{10.3.7}$$

In (10.3.7), some Jacobian terms are neglected, assuming the magnitude of the geostrophic deformation is much smaller than the Coriolis parameter. It can be shown that the *static stability* $[(g/\theta_o)\partial\theta/\partial Z]$ is proportional to the *geostrophic potential vorticity* (q_g),

$$q_g = fJ\frac{\partial b}{\partial Z} \approx \zeta_g \frac{\partial b}{\partial Z}. \tag{10.3.8}$$

In the geostrophic space, if the potential vorticity is constant initially, the effective static stability will be constant because the potential vorticity is conserved following the motion. A *geopotential function* (Φ) can be defined as

$$\Phi = \phi + \frac{1}{2}\left(u_g^2 + v_g^2\right), \tag{10.3.9}$$

such that

$$(fv_g, fu_g, b) = \left(\frac{\partial\Phi}{\partial X}, -\frac{\partial\Phi}{\partial Y}, \frac{\partial\Phi}{\partial Z}\right). \tag{10.3.10}$$

From (10.3.6)–(10.3.10) and the following coordinate transformation relations,

$$\begin{aligned}
\partial/\partial x &= \left(1 + v_{gx}/f\right)\partial/\partial X - \left(u_{gx}/f\right)\partial/\partial Y, \\
\partial/\partial y &= \left(v_{gy}/f\right)\partial/\partial X + \left(1 - u_{gy}/f\right)\partial/\partial Y, \\
\partial/\partial z &= \left(v_{gz}/f\right)\partial/\partial X - \left(u_{gz}/f\right)\partial/\partial Y + \partial/\partial Z, \\
\partial/\partial t &= \left(v_{gt}/f\right)\partial/\partial X - \left(u_{gt}/f\right)\partial/\partial Y + \partial/\partial\tau.
\end{aligned} \tag{10.3.11}$$

It can be derived:

$$q_g = \frac{f\Phi_{ZZ}}{1 - (\Phi_{XX} + \Phi_{YY})/f^2 + (\Phi_{XX}\Phi_{YY} - \Phi_{XY}^2)/f^4}. \tag{10.3.12}$$

Based on the conservation of potential vorticity

$$\left(\frac{\partial}{\partial \tau} + u_g\frac{\partial}{\partial X} + v_g\frac{\partial}{\partial Y} + w\frac{\partial}{\partial Z}\right)q_g = 0, \tag{10.3.13}$$

and (10.3.12), the problem reduces to the advection of q_g in the interior of a fluid system. The geopotential function Φ in (10.3.12) can be inverted by applying appropriate boundary conditions to the thermodynamic equation with $w = 0$ (assuming flat surfaces at lower and upper boundaries),

$$\left(\frac{\partial}{\partial \tau} + u_g\frac{\partial}{\partial X} + v_g\frac{\partial}{\partial Y}\right)\frac{\partial \Phi}{\partial Z} = 0. \tag{10.3.14}$$

For flow over topography, w may be approximated by free-slip boundary condition in an inviscid fluid, as discussed in Chapter 5, which may introduce some complication to the mathematics of the problem.

To study the ageostrophic motion (Hoskins and Draghici 1977), the following new ageostrophic wind components may be defined:

$$u_a^* = u_a + \frac{w}{f}\frac{\partial v_g}{\partial Z}, \quad v_a^* = v_a - \frac{w}{f}\frac{\partial u_g}{\partial Z}, \quad w^* = \frac{w}{J}. \tag{10.3.15}$$

Substituting (10.3.15) into the above semi-geostrophic equations yields

$$\left(\frac{\partial}{\partial \tau} + u_g\frac{\partial}{\partial X} + v_g\frac{\partial}{\partial Y}\right)u_g - fv_a^* = 0, \tag{10.3.16}$$

$$\left(\frac{\partial}{\partial \tau} + u_g\frac{\partial}{\partial X} + v_g\frac{\partial}{\partial Y}\right)v_g + fu_a^* = 0, \tag{10.3.17}$$

$$\left(\frac{\partial}{\partial \tau} + u_g\frac{\partial}{\partial X} + v_g\frac{\partial}{\partial Y}\right)b + \left(\frac{q_g}{f}\right)w^* = 0. \tag{10.3.18}$$

In the geostrophic space, the continuity equation becomes

$$\frac{\partial u_a^*}{\partial X} + \frac{\partial v_a^*}{\partial Y} + \frac{\partial w^*}{\partial Z} = 0. \tag{10.3.19}$$

Equations (10.3.15)–(10.3.19) form a closed set of the *semi-geostrophic equations* which can be solved numerically.

The *semi-geostrophic vorticity equation* can be obtained by taking the cross differentiation of (10.3.16) and (10.3.17),

$$\left(\frac{\partial}{\partial \tau} + u_g \frac{\partial}{\partial X} + v_g \frac{\partial}{\partial Y}\right)\left(\frac{\partial v_g}{\partial X} - \frac{\partial u_g}{\partial Y}\right) = f \frac{\partial w^*}{\partial Z}. \tag{10.3.20}$$

The *semi-geostrophic omega equation* can be derived by manipulating (10.3.16)–(10.3.18) with thermal wind relations and (10.3.19),

$$\frac{1}{f}\nabla_H^2(q_g w^*) + f^2 \frac{\partial^2 w^*}{\partial Z^2} = 2\nabla_H \cdot \tilde{\mathbf{Q}}, \tag{10.3.21}$$

where

$$\tilde{\mathbf{Q}} = (Q_1, Q_2) = \left(-\frac{\partial \mathbf{V}_g}{\partial X} \cdot \nabla_H b, -\frac{\partial \mathbf{V}_g}{\partial Y} \cdot \nabla_H b\right). \tag{10.3.22}$$

In the above equation, ∇_H is the horizontal gradient operator in the semi-geostrophic space.

The semi-geostrophic omega equation, (10.3.21), has a form identical to the *quasi-geostrophic omega equation*,

$$N^2 \nabla_H^2 w + f^2 \frac{\partial^2 w}{\partial z^2} = 2\nabla_H \cdot \mathbf{Q}, \tag{10.3.23}$$

where the quasi-geostrophic $\mathbf{Q}$ *vector* is identical to (10.3.22) except defined in the physical space. The relative importance of the terms on the left-hand side of the quasi-geostrophic omega equation is controlled by the nondimensional parameter, NL_z/fL. For a wave-like disturbance, (10.3.23) implies an upward (downward) motion exists in the region of convergence (divergence) of the $\mathbf{Q}$ vectors. Qualitatively, the $\mathbf{Q}$ vectors can be more easily found by aligning the x-axis with the local isotherm leaving the cold air to the left ($\partial b/\partial y < 0$). In this newly oriented coordinate system, $\partial b/\partial x = 0$, we have

$$\mathbf{Q} = (Q_1, Q_2) = -\left(\frac{\partial v_g}{\partial x}\frac{\partial b}{\partial y}, \frac{\partial v_g}{\partial y}\frac{\partial b}{\partial y}\right). \tag{10.3.24}$$

Thus, Q_1 and Q_2 become the shearing deformation and stretching deformation of the three-dimensional frontogenesis function, (10.1.11), respectively. In the new coordinate system, the $\mathbf{Q}$ vector can be written as

$$\mathbf{Q} = -\left|\frac{\partial b}{\partial y}\right| \mathbf{k} \times \frac{\partial \mathbf{V}_g}{\partial x}. \tag{10.3.25}$$

With (10.3.23) and (10.3.25), the $\mathbf{Q}$ vector can be obtained by: (1) evaluating the vectorial change of $\mathbf{V}_g$ along the isotherm with cold air to the left, (2) rotating the resulting change vector clockwise by $90°$, and (3) multiplying the resulting vector by $|\partial b/\partial y|$.

As mentioned earlier, the polar front theory views cyclogenesis as a process triggered by the instability associated with frontal surface discontinuity (Bjerknes 1919). On the other hand, in the baroclinic instability theories (Charney 1947; Eady 1949), fronts are viewed as being formed in growing baroclinic waves. In these baroclinic instability theories, the fronts are treated as thermal contrast (which is smooth in QG theories but can become discontinuous in SG theories), and develop in response to baroclinic wave amplification rather than precede it. In the following, we will use two examples to demonstrate the formation of baroclinic waves and fronts by solutions of the semi-geostrophic system of equations. First, let us consider the classical *Eady problem* (Eady 1949) using the semi-geostrophic equations. The basic state zonal wind is assumed to be

$$U_g(Z) = U_z(Z - H/2), \qquad (10.3.26)$$

where U_z is the constant vertical shear of the zonal wind and H is the depth of the fluid between the upper and lower rigid boundaries. Following Eady (1949), and unlike in Sections 10.1 and 10.2, the along-front direction is now chosen to be in the y direction. In this special case, the semi-geostrophic potential vorticity equation, (10.3.12), reduces to a much simpler form on an X–Z plane

$$\frac{\partial^2 \Phi}{\partial X^2} + \frac{f^2}{N^2}\frac{\partial^2 \Phi}{\partial Z^2} = f^2, \qquad (10.3.27)$$

where the buoyancy frequency (N) in the geostrophic space is approximately proportional to q_g,

$$N^2 = \frac{q_g}{f}. \qquad (10.3.28)$$

Equation (10.3.27) is a second-order, linear, partial differential equation with constant coefficients if N and f are constants. Similar to (10.3.14), the upper and lower boundary conditions for flow over rigid flat surfaces become

$$\left[\frac{\partial}{\partial \tau} + U_z(Z - H/2)\frac{\partial}{\partial X}\right]\frac{\partial \Phi}{\partial Z} = U_z\frac{\partial \Phi}{\partial X}, \quad \text{on } Z = 0, H. \qquad (10.3.29)$$

The horizontal advection of the basic potential temperature gradient is retained and the thermal wind relation has been used in deriving (10.3.29). The inverse transformation back to the physical space is given by:

$$(x, z, t) = (X - f^{-2}\partial \Phi/\partial X, Z, \tau). \qquad (10.3.30)$$

The semi-geostrophic version of the Eady problem, (10.3.27) and (10.3.29), is equivalent to the quasi-geostrophic version, (7.7.3) and (7.7.4), except that it is in semi-geostrophic space. They can be solved analytically (Hoskins and Bretherton 1972). If no deformation

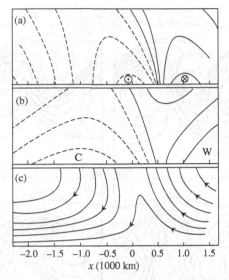

Fig. 10.6 The baroclinic wave and frontogenesis derived from the semigeostrophic Eady problem, given by (10.3.27) and (10.3.29), in a vertical cross section view in physical space for the following: (a) along-front velocity, (b) and (c) cross-frontal streamfunction. C and W denote the cold and warm regions, respectively, and the arrow tail and head denote +v and −v, respectively. (From Hoskins and Bretherton 1972.)

exists in the initial flow, then the development of a baroclinic wave and the subsequent frontogenesis within the baroclinic wave can be examined. Figure 10.6 shows the baroclinic wave and frontogenesis revealed by the analytic solution of the above Eady problem in the physical space. A strong gradient of along-front velocity and a region of concentrated vertical vorticity forms at about $x = 500$ km (Fig. 10.6a) which implies the formation of a cold front, evidenced by the formation of strong potential temperature gradient (Fig. 10.6b). Compared to the quasi-geostrophic solution of the Eady problem, Fig. 7.15, the temperature gradient at the front is much stronger. The streamfunction field (Fig. 10.6c) indicates a thermally direct circulation across the front, i.e. air rises on the warm side of the cold front and descends on the cold side behind the front. All of the phases of vorticity, temperature, and streamfunction tilt upshear (westward) with height and this is required for the conversion of available potential energy into perturbation kinetic energy, as discussed in Section 7.7 (e.g., (7.7.2) and Fig. 7.15).

In the second example, unlike the Eady problem, a basic zonal flow at the upper boundary ($z = H$) is portrayed with the following jet structure in the y direction,

$$u(y, z) = U \left\{ \frac{z}{H} - \frac{1}{2} \left(\frac{z}{H} + \frac{\sinh(lz/H)}{\sinh l} \cos(ly/L_y) \right) \right\}, \tag{10.3.31}$$

where $U = 29.4$ m s^{-1}, $H = 9$ km, $l = 1.1358$, $L_y = 5623$ km. The basic zonal flow is zero at the lower boundary and has a jet structure in the y direction. The jet has a

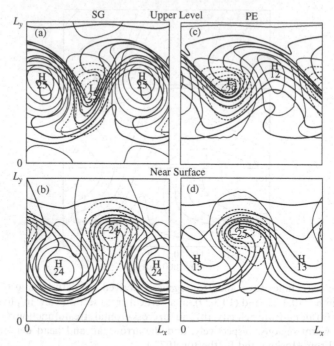

Fig. 10.7 (a) The geopotential perturbation (contour interval $500\,\text{m}^2\,\text{s}^{-2}$) and potential temperature (bold-solid horizontal curves; interval $5\,\text{K}$) at $z = 8.775\,\text{km}$ for day 6.3 simulated by a semigeostrophic model. (b) Same as (a) except at $z = 0.225\,\text{km}$. Panels (c) and (d) same as (a) and (b), respectively, except for results simulated by a primitive-equation model. The horizontal domain is: $(L_x, L_y) = (4090\,\text{km}, 5623\,\text{km})$. (After Snyder *et al.* 1991.)

maximum wind speed of $29.4\,\text{m}\,\text{s}^{-1}$ at $y = 0$ on the upper boundary. The potential vorticity associated with this basic jet flow is zero. The vertical shear is supported by a baroclinic field with cold air to the north. At the initial time, a very small-amplitude wave disturbance of the most unstable normal mode is added to the basic flow. The wavelength of this normal mode is $4090\,\text{km}$ and the double time of this wave mode is 1.34 days. The initial amplitude of v_g is $1.7\,\text{m}\,\text{s}^{-1}$.

Figures 10.7a and 10.7b show the potential temperature and geopotential fields near the upper rigid boundary ($z = 8.775\,\text{km}$) and near the surface ($z = 0.225\,\text{km}$) after a 6.3 day simulation of a semi-geostrophic model. Before day 6.3, a baroclinic wave has developed near the surface. At this time, the simulated baroclinic wave is similar to a linear baroclinic wave, with equal magnitudes for the maximum and minimum perturbation geopotentials as well as for the maximum and minimum geopotential gradients. The lows and highs are tilted westward with height, showing a typical unstable baroclinic wave behavior, as shown in the Eady problem (Fig. 10.6) and discussed in the baroclinic instability section (Section 7.7). At day 6.3, the baroclinic wave is well developed near the surface (Fig. 10.7b). A cold front and a warm front form to the southwest and southeast, respectively, of the low. The low (high) is shifted to the north (south) of its original latitude compared with that at earlier time. Strong vorticity is concentrated in the vicinity of the low and along the cold front. Cold and warm fronts also form in the upper layer near the lid (Fig. 10.7a), although they are oriented in different directions from the surface fronts. Again, the upper-level low is located upshear (to the west) of the surface low. Although lacking the dynamic simplicity of *balanced models*, such as the quasi-geostrophic (QG) and semi-geostrophic (SG) models, the primitive-equation (PE) models retain more terms from the governing equations and thus produce more accurate results than balanced models. Figures 10.7c and 10.7d shows a PE model simulated potential temperature and geopotential fields. Compared with the SG model simulated fields, one may find that the lows simulated by the PE model are deeper and the highs are weaker, and the nonlinear PE wave produces a cyclonic wrapping of the temperature contours on both lower and upper levels and has an associated "bent-back" frontal structure at the surface. The differences are attributed to the lack of the ageostrophic vorticity in the SG model. The accuracy of baroclinic wave and frontogenesis simulated by the SG theory can be improved by extending the QG theory to higher powers in Rossby number. Note that although these realistic reproductions of surface fronts and upper-level fronts (to be discussed later in this chapter) are given in the context of baroclinic instability theory, they are generated as a consequence of low-level cyclogenesis (upper-level wave amplification), rather than appearing in advance of these processes (Keyser and Shapiro 1986).

10.4 Moist and frictional effects on frontogenesis

Although theoretical and numerical studies indicate that frontogenesis may occur without the presence of moisture and friction, it may also be enhanced by the latent heat release associated with cloud formation on the rising branch of the frontal circulation, evaporative cooling associated with rainfall on the cold side of the front, and the convergence associated with boundary layer process along the frontal zones. Basically, the impacts of the addition of latent heat on frontogenesis is twofold: one is to reduce the static stability, thus enhancing the front development, and the other is to decrease the most unstable wavelength for growing baroclinic waves, thus widening the spectrum of unstable wave numbers. In addition to the elevated heating associated with latent heat release, the surface heating or cooling can also influence the frontogenesis.

The dry Eady problem, as discussed in Section 10.3, may be extended to include diabatic heating. Assuming the velocity perturbations are independent of Y, the basic equation for q_g may be written as (Thorpe and Emanuel 1985):

$$\left(\frac{\partial}{\partial \tau} + u_g \frac{\partial}{\partial X} + w \frac{\partial}{\partial Z}\right) q_g = \left(\frac{g}{\theta_o}\right) \zeta_g \frac{\partial \dot{Q}}{\partial Z} \qquad (10.4.1)$$

where $\dot{Q} = (\theta_o/c_p T_o)\dot{q}$, as defined in (10.2.4). The diabatic heating due to latent heating may be represented by (Emanuel *et al.* 1987):

$$\dot{Q} = w\left(\frac{\partial \theta}{\partial Z} - \frac{\Gamma_m}{\Gamma_d}\frac{\partial \theta_e}{\partial Z}\right), \tag{10.4.2}$$

where θ_e is the equivalent potential temperature. Substituting (10.4.2) into (10.4.1) gives

$$\left(\frac{\partial}{\partial \tau} + u_g\frac{\partial}{\partial X} + w\frac{\partial}{\partial Z}\right)q_g = \frac{\zeta_g}{f}\frac{\partial}{\partial Z}\left[w\left(q_g - \frac{\Gamma_m}{\Gamma_d}q_{ge}\right)\right], \tag{10.4.3}$$

in saturated regions, where q_{ge} is the equivalent potential vorticity. In unsaturated regions, q_g is conserved. In deriving the above equation,

$$q_g = \left(\frac{g}{\theta_o}\frac{\partial \theta}{\partial Z}\right)\zeta_g \approx f\left(\frac{g}{\theta_o}\frac{\partial \theta}{\partial Z}\right) \tag{10.4.4}$$

have been used. Equation (10.4.3) can be solved numerically to determine the time evolution of the potential vorticity q_g. The current distribution of q_g is determined by inverting (10.3.27),

$$\frac{1}{f^2}\frac{\partial^2 \Phi}{\partial X^2} + \frac{g}{q}\frac{\partial^2 \Phi}{\partial Z^2} = 1, \tag{10.4.5}$$

where the potential vorticity q is defined as

$$q = \begin{cases} q_g = \left(\frac{g}{\theta_o}\right)\zeta_g \frac{\partial \theta}{\partial Z}, \text{ unsaturated} \\ \frac{\Gamma_m}{\Gamma_d}q_{ge} = \left(\frac{g}{\theta_o}\right)\frac{\Gamma_m}{\Gamma_d}\zeta_g\frac{\partial \theta_e}{\partial Z}, \text{ saturated.} \end{cases} \tag{10.4.6}$$

Under certain conditions, the q_{ge} conservation equation can be inverted to determine the flow and the temperature distribution, which is known as the *invertibility principle* (Hoskins *et al.*, 1985). Equation (10.4.5) can be integrated numerically with the boundary conditions, (10.3.29). The adiabatic ageostrophic circulation can be solved numerically from the Sawyer–Eliassen equation, similar to (10.2.27) except on the X–Z plane,

$$f^2\frac{\partial^2 \psi}{\partial Z^2} + \frac{\partial}{\partial X}\left(q\frac{\partial \psi}{\partial X}\right) = -2\frac{g}{\theta_o}\left(\frac{\partial u_g}{\partial X}\frac{\partial \theta}{\partial X} + \frac{\partial v_g}{\partial X}\frac{\partial \theta}{\partial Y}\right). \tag{10.4.7}$$

When applied to the current Eady problem, the first term on the right-hand side of (10.4.7) vanishes due to the assumption of no Y-variation of velocity perturbations, and the second term can be replaced by $-2U_z\partial v_g/\partial X$.

Figures 10.8a and 10.8b show the meridional flow (v_g), the potential temperature fields, and the streamfunction (ψ) for dry frontogenesis after 3 days in a two-dimensional,

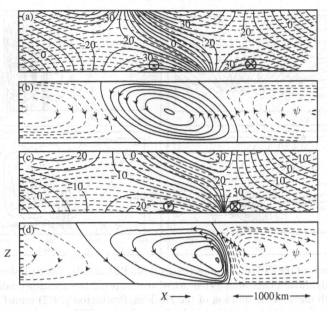

Fig. 10.8 Effects of moisture on frontogenesis: the dry Eady wave with wavelength 4554 km after 3 day simulation is shown in (a) and (b), while the corresponding moist Eady wave after 2 day simulation is shown in (c) and (d). The numerical model solves the semigeostrophic equation set involving (10.4.3). The dry Eady wave corresponds to a dry normal mode of small amplitude. (a) The v_g (solid lines with contour interval of $5\,\mathrm{m\,s^{-1}}$) and θ (dashed lines with contour interval of 4 K) fields for dry flow. (b) The streamfunction (ψ) (contour interval of $4000\,\mathrm{m^2\,s^{-1}}$) for the ageostrophic dry flow. (c) and (d) are the same as (a) and (b), respectively, except for the corresponding moist flow. (After Emanuel *et al.* 1987.)

constant-shear flow with initial wavelength of 4554 km. The wavelength corresponds to the fastest growing Eady mode. The corresponding flow fields for moist frontogenesis after 2 days are shown in Figs. 10.8c and 10.8d. With latent heating included, the updraft becomes stronger and narrower, while the downdraft becomes weaker and wider. In particular, the updraft is intense and very narrow near the lower boundary. The geostrophic vorticity strengthens in a narrower region when moisture is included in the system. The potential vorticity, which acts as the *effective static stability*, is reduced in saturated air due to diabatic effects, such as the latent heating in the present case. The resulting regions of small symmetric stability provide a stronger response on a smaller scale to a given frontogenetic forcing, allowing formation of intense mesoscale rainbands. The relatively larger scale forcing associated with the frontogenesis may help the release of moist symmetric instability, as discussed in Subsection 7.6.2, by lifting an air parcel to its *level of free slantwise convection*. The lifting of air parcels to the level of free slantwise convection may produce banded precipitation accounting for the generation of frontal rainbands (e.g., Xu 1992; Schultz and Schumacher 1999).

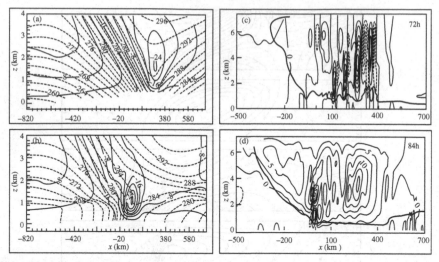

Fig. 10.9 Effects of the PBL on frontogenesis. (a) No PBL effects: The θ (dashed) and ω (solid; 4 mb h^{-1}) fields are simulated after 84 h by a hydrostatic primitive-equation model. The model is initialized with the analytic solution of the Hoskins–Bretherton (1972) model (Fig. 10.6). (b) Effects of PBL: Same as (a) except that the bulk-aerodynamic PBL parameterization is activated at 60 h. (c) Effects of PBL and moisture: The 72-h w field (contour of 0.5 cm s^{-1}) is simulated by a nonhydrostatic model. The initial condition is similar to that of (a). A TKE PBL and a grid-explicit microphysical parameterization scheme are used. The horizontal grid interval is 5 km. A region of convective instability ($\partial\theta_e/\partial z < 0$) is depicted by the bold solid lines. (d) Same as (c) except for 84 h. ((a) and (b) after Keyser and Anthes 1982; (c) and (d) after Bénard *et al.* 1992).

Most theoretical studies of frontogenesis within baroclinic waves have ignored the effects of surface friction. Surface drag tends to suppress the development of secluded warm sectors and weakens the warm front while the cold front remains strong. In addition to the surface friction, another effect of the planetary boundary layer (PBL) that acts on the atmosphere is the surface sensible heating. Numerical simulations by primitive-equation models indicate that vertical fluxes of sensible heating at levels above the surface tend to drive the lapse rates on both sides of the front toward neutral, but that differential thermal advection dominates the forcing. Figures 10.9a and 10.9b demonstrate the effects of the PBL, via a bulk-aerodynamic parameterization in a hydrostatic, nonlinear primitive-equation numerical model, on the surface frontogenesis. The basic flow is based on the Hoskins and Bretherton (1972) model (Fig. 10.6). The numerical model is initialized with analytic solutions computed at 60 h. The basic features in the Hoskins–Bretherton model are reasonably well simulated by the primitive-equation model (Fig. 10.9a). With the inclusion of the PBL, several realistic features are simulated (Fig. 10.9b): (a) an intense, narrow updraft at the top of the PBL above the sea-level pressure trough at the warm edge of the frontal zone; (b) a stable layer capping the PBL to the rear of the frontal zone; and (c) slightly unstable or neutral lapse rates in the PBL behind the front and stable lapse rates in the PBL ahead of the front. The intense, narrow jet of rising air is caused by the PBL convergence associated with the u_a field in the region of near-zero static stability.

Using the same initial condition, but with a nonhydrostatic model, the combined PBL and moisture effects are shown in Fig. 10.9c. In this model, a turbulence kinetic energy parameterization scheme is adopted to represent the PBL processes and a grid-explicit microphysical parameterization is adopted to represent the moist processes. The horizontal grid interval is 5 km, which is much finer than that used in Figs. 10.9a and 10.9b. During the earlier (convective) stage (Fig. 10.9c) the initial large-scale ascent turns into small-scale convective cells. At a later (nonconvective) stage (Fig. 10.9d), the upright convection vanishes, and instead four types of rainbands are produced by the combined effects of PBL and moisture: (1) narrow cold-frontal rainband, (2) narrow free-atmosphere rainbands, (3) wide cold-frontal rainbands, and (4) wide warm-sector rainbands. A zone of convective instability ($\partial \theta_e / \partial z < 0$) is continuously maintained in the frontal PBL. Therefore, moisture and the PBL (friction and surface heating) play important roles in frontogenesis.

10.5 Other types of fronts

Mesoscale fronts have length and width scales about one order smaller than those of synoptic-scale fronts. In addition, mesoscale fronts are shallower and only last for a few hours. As mentioned in the beginning of this chapter, dynamically, a mesoscale front behaves more like a density current since the Earth's rotation plays a less significant role than in synoptic-scale fronts. For mesoscale fronts, semi-geostrophic theories fail at earlier stages in simulating the frontogenesis. The background deformation field is often nongeostrophic, and analytical solutions are not appropriate to describe important physical processes. *Sea-breeze fronts*, *coastal fronts*, *gust fronts*, and *drylines* are some examples of mesoscale fronts. Since sea-breeze fronts and coastal fronts have been discussed in Chapter 6 and gust fronts have been discussed along with the convective systems in Chapters 8 and 9, we will only discuss the *upper-level fronts* and *drylines* in this section.

10.5.1 Upper-level frontogenesis

The *upper-level front*, also known as an *upper-tropospheric front* or a *middle-upper tropospheric front*, is a zone often associated with the polar-front jet stream characterized by a large horizontal temperature gradient and high static stability in the upper and middle troposphere. Upper-level fronts were first observed as early as in the 1930s, and more information about detailed frontal structures were found in later times using more sophisticated observational instruments and a finer upper-air sounding network. Based on observational analyses, characteristics of upper-level fronts may be summarized as follows (Reed 1955; Keyser and Shapiro 1986):

(a) Upper-level fronts are often associated with jet streams. They are narrow (~100 km) across the front and extend downward from the tropopause, with a thin slice of stratospheric air extruded into the midtroposphere along the front, as evidenced by high values of potential vorticity and ozone mixing ratio.

(b) The upper-level front often forms downstream from an upper-level ridge and upstream from an amplifying upper-level trough, in a synoptic-scale environment characterized by sinking motion.

(c) The divergence patterns associated with upper-level frontal systems and their accompanying jet streaks play an active role in midlatitude cyclogenesis by contributing to low-level geopotential height (mass) changes.

(d) The vertical circulations associated with upper-level and surface frontal zones often contribute to the development and organization of midlatitude, mesoscale cloud and precipitation systems.

(e) Upper-level fronts are preferred regions of small-scale mixing by a variety of mesoscale phenomena including gravity waves and clear-air turbulence that often produce hazardous aviation conditions for commercial and military aircrafts.

(f) Upper-level fronts are regions of significant mass exchange (including radioactive and chemical constituents) and small-scale mixing between the stratosphere and troposphere.

Figure 10.10a shows an example of an upper-level front based on observations by conventional rawinsondes, instrumented aircraft, and satellites. A jet with maximum wind speed greater than $75 \, \text{m s}^{-1}$ is located on the eastern half of the trough, located downstream from the confluence of polar and subtropical jet streams. A composite of the aircraft data with synoptic soundings is made along the cross section AA′ (Fig. 10.10b), which connects Denver, Colorado and Longview, Texas. A front containing vertical wind shear of $30 \, \text{m s}^{-1}$ per 100 mb, horizontal potential temperature gradient of 14 K per 100 km, and horizontal wind shear of $35 \, \text{m s}^{-1}$ per 100 km is present. Two jet cores are found within the cross section: a polar jet near 300 mb and a subtropical jet at 230 mb. The *dynamic tropopause*, as denoted by 1 PVU ($= 10^{-6} \, \text{Km}^2 \, \text{s}^{-1} \, \text{kg}^{-1}$ of potential vorticity), is folded downward (*tropopause folding*) on the cyclonic shear side of the two jets. This figure indicates that the frontal layer beneath a jet stream core is bounded by discontinuities in the horizontal and vertical gradients of potential temperature and wind velocity. The spatial discontinuities of potential vorticity can be used to establish the dynamical relationship between tropopause surfaces and upper-level front wind and temperature discontinuities, and the approximate boundary of air having stratospheric versus tropospheric origin.

The significance of tropopause folding in upper-level frontal zones extends to the cyclogenesis problem (e.g., Whitaker *et al.* 1988; Lackmann *et al.* 1999). Studies of rapidly intensifying cyclones over the western North Atlantic have revealed upstream tropopause folding, which is associated with an intensification of upper-level short-wave troughs and strong dynamical forcing for ascent and vorticity stretching once the upper disturbance moves offshore. Specifically, cold advection in the along-front direction can serve to shift the frontal circulation towards the warm side of the

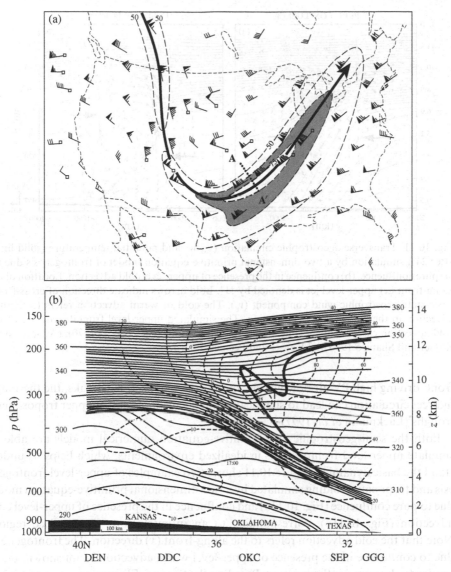

Fig. 10.10 Structure of an upper-level front. (a) 250 mb wind speed (dashed lines) at 0000 UTC April 5, 1981. Denoted are rawinsonde wind velocities (flag: $25\,\mathrm{m\,s^{-1}}$; barb: $5\,\mathrm{m\,s^{-1}}$; half barb: $2.5\,\mathrm{m\,s^{-1}}$) from rawinsonde, commercial airline, and satellite observations. The jet stream axis is denoted by a heavy solid arrow with area of $62.5\text{--}75\,\mathrm{m\,s^{-1}}$ wind speed shaded. The dashed line outside the shaded area represents wind speed of $50\,\mathrm{m\,s^{-1}}$. (b) Cross-section analysis of wind speed ($\mathrm{m\,s^{-1}}$, dashed lines) and θ (K, thin solid lines) along line AA' in (a) for 1900 UTC April 5, 1981. Tropopause is depicted by the heavy solid line with 1 PVU ($= 10^{-6}\,\mathrm{K\,m^2\,s^{-1}\,kg^{-1}}$ of potential vorticity). Rawinsonde soundings are from Denver, CO (DEN), Dodge City, KS (DDC), Oklahoma City, OK (OKC), Stephenville, TX (SEP), and Longview, TX (GGG). (Adapted after Shapiro 1983.)

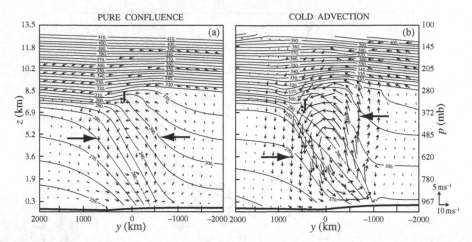

Fig. 10.11 Transverse ageostrophic circulation (v_a, w) and potential temperature (solid lines) after 24 h simulation by a two-dimensional primitive equation model of frontogenesis due to: (a) pure confluence, (b) confluence in the presence of upper-level cold advection. Location of the along-front (x) upper-level jet is denoted by "J". Bold arrows indicate direction of cross-front, basic-state geostrophic wind component (v_g). The cold or warm advection refers to thermal advection in the along-front (x) direction. The simulated upper-level frontal structure of the cold advection case is realistic and similar to observations (e.g., Bosart 1970). (Adapted after Keyser and Shapiro 1986 and Keyser and Pecnick 1985.)

front, driving high PV air downwards. In the exit regions of jet streaks, the thermally indirect circulation can lead to significant enhancement of mobile upper tropospheric troughs (Lackmann *et al.* 1997).

Both the semi-geostrophic and primitive-equation numerical models are able to simulate upper-level frontogenesis in idealized environments, which helps us understand the basic dynamics. Figure 10.11 shows two examples of upper-level frontogenesis and tropopause folding simulated by a two-dimensional primitive-equation model due to pure confluence (Fig. 10.11a) and confluence in the presence of upper-level cold advection (Fig. 10.11b), corresponding to an infinitely long jet entrance region. Note that the cold advection refers to the along-front (x) direction. The frontogenesis due to confluence in the presence of upper-level warm advection is not shown, but is similar to the pure confluence case. Based on the Sawyer–Eliassen equation (10.2.22), the cross-front, transverse circulation induced by external geostrophic forcing is thermally direct with the cold air descending and warm air ascending ($\partial w/\partial y < 0$) because $\partial v_g/\partial y < 0$, $\partial b/\partial y < 0$, and $\partial u_g/\partial y = 0$ in both cases. Feedbacks in such a two-dimensional model formulation are restricted to modifications to the cross-front potential temperature gradient ($\partial b/\partial y$) and the relative vorticity ($-\partial u_g/\partial y$) because $\partial v_g/\partial y$ and $\partial b/\partial x$ are specified externally.

The time evolution of $\partial b/\partial y$ following the motion can be described by rewriting (10.2.18) in the absence of diabatic forcing:

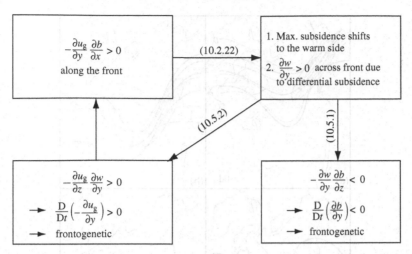

Fig. 10.12 A flowchart depicting the mechanism responsible for upper-level frontogenesis and tropopause folding for the case with confluence in the presence of upper-level cold advection (Fig. 10.11b). A positive feedback loop is established among upper boxes and the lower left box through the increase in cyclonic vorticity leading to horizontal shear forcing and frontogenetic subsidence (Shapiro (1981).

$$\frac{D}{Dt}\left(\frac{\partial b}{\partial y}\right) = -\left(\frac{\partial u_g}{\partial y}\frac{\partial b}{\partial x} + \frac{\partial v_g}{\partial y}\frac{\partial b}{\partial y} + \frac{\partial v_a}{\partial y}\frac{\partial b}{\partial y}\right) - \frac{\partial w}{\partial y}\frac{\partial b}{\partial z}, \qquad (10.5.1)$$

and the time evolution of $-\partial u_g/\partial y$ following the motion can be derived by taking the y-derivative of (10.2.3):

$$\frac{D}{Dt}\left(-\frac{\partial u_g}{\partial y}\right) = \left(f - \frac{\partial u_g}{\partial y}\right)\frac{\partial v_a}{\partial y} + \frac{\partial u_g}{\partial z}\frac{\partial w}{\partial y}. \qquad (10.5.2)$$

The bracket terms on the right-hand side of (10.5.1) and (10.5.2) contain the effects of frontogenesis due to horizontal motion. In the pure confluence case, these horizontal motion terms are frontogenetical in the upper- and mid-tropospheric portion of the frontal zone and dominate the tilting terms (terms including $\partial w/\partial y$) which are frontogenetical. The tilting terms are frontolytical because $\partial u_g/\partial z > 0$, $\partial b/\partial z > 0$, and $\partial w/\partial y < 0$ due to external geostrophic forcing, according to (10.5.1) and (10.5.2). The situation is totally reversed in the cold advection case. That is, the horizontal motions are frontolytical and the tilting terms are frontogenetical and tend to dominate. The tilting terms are frontogenetical because $\partial w/\partial y > 0$ across the upper-level front due to the shifting of circulation toward the warm (right) side. The lateral shifting of the downward motion toward the warm side of the upper-level front in the cold advection case (Fig. 10.11) is attributed to the combined effects of $\partial b/\partial x$ and $-\partial u_g/\partial y$ or shearing deformation. Figure 10.12 shows a flowchart depicting the mechanism responsible for upper-level frontogenesis and tropopause folding for this cold advection case

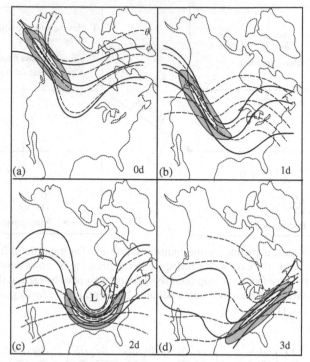

Fig. 10.13 Shapiro's model of the formation and evolution of an upper-level jet-front system through a midlatitude baroclinic wave: (a) equivalent-barotropic stage; (b) cold-advection stage; (c) closed-low stage; and (d) warm-advection stage. See text for details. Geopotential height (ϕ) contours are denoted by solid lines; isentropes (θ) are denoted by dashed lines; and isotachs are shaded. (After Shapiro 1981.)

(Fig. 10.11b). A positive feedback loop is established among upper boxes and the lower left box through the increase in cyclonic vorticity leading to horizontal shear forcing and frontogenetic subsidence. The simulated upper-level frontal structure of the cold advection case is realistic and similar to observations.

An upper-level front is often associated with the upper-tropospheric *jet-front system* which forms and moves through a midlatitude baroclinic wave. This process includes the following four stages (Fig. 10.13; Shapiro 1983). The first stage is the *equivalent-barotropic stage*, in which an upper-level jet-front forms within a confluent northwesterly flow and the isentropes are nearly parallel to height contours (isohypes). Approximately one day later, the jet-front evolves to its second stage, i.e., the *cold-advection stage*. At this stage, the trough in the isentropes lags the trough in the height contours, implying geostrophic cold advection along the length of the front. Approximately two days after the jet-front formation, it evolves to the third stage, i.e. the *closed-low stage*. At this stage, the trough deepens and develops into a closed, nearly symmetric low, with leading warm advection and trailing cold advection. About three days after the jet-front formation, it enters into its last stage, i.e. the *warm-advection stage*. At this stage, the closed low opens up in confluent southwesterly flow and geostrophic warm advection along the length of the front is implied.

Shapiro's conceptual model is able to describe many common features of the migration of an upper-level jet-front system through a baroclinic wave; thus, it has also been applied to upper-level frontogenesis. The formation and evolution stages of upper-level jet streaks progressing through a synoptic-scale baroclinc wave have also been verified in diagnostic studies. Based on vector frontogenesis analysis and the climatology of upper-level fronts associated with landfalling cyclones on the west coast of North America, it is found that in the earlier stage, such as the 12–24 h period, Shapiro's conceptual model can be revised to have two flow regimes: southwesterly and northwesterly (Schultz and Doswell 1999). Two major differences of this revised conceptual model from Shapiro's conceptual model are: (a) the upper-level frontogenesis occurs in a southwesterly flow, instead of in a northwesterly flow; and (b) the onset of cold advection is due to the migration of an upstream vorticity maximum toward the baroclinic wave.

In the southwesterly flow regime, the isotherms and the geopotential height contours are relatively parallel initially. The advance of an upstream vorticity maximum towards the baroclinic zone leads to the rotation of the isentropes relative to the height contours, resulting in the onset of cold advection. In the northwesterly flow regime, the initial cold advection along the length of the front becomes concentrated in the base of the thermal trough in conjunction with an intensifying and tightening vorticity maximum. Warm advection occurs upstream of the thermal trough, indicating a substantial along-front variation in thermal advection, unlike the cold advection along the length of the front in Shapiro's model. Note that although examples of strong frontogenesis in southwesterly flows are often observed, the most intense upper-level frontogenesis tends to occur in northwesterly flows. At a later stage, such as 2 d, the jet-front system enters the "closed-low stage." At this stage, the trough in the geopotential-height field closes off and becomes more symmetric, with leading warm advection and trailing cold advection (Fig. 10.13c). Finally, at 3 d, the jet-front system enters the "warm-advection stage," in which the closed low opens up in confluent southwesterly flow and geostrophic warm advection along the length of the front is implied (Fig. 10.13d).

Upper-level frontogenesis may occur when a *cold front aloft* (CFA) forms. In the lee of the Rocky Mountains, the CFA formation is associated with *cyclogenesis*. Many cyclones that form in the lee of the Rocky Mountains in the winter, spring, and even in early summer are produced by synoptic-scale waves moving eastward over the mountains and interacting with warm, moist subtropical air from the Gulf of Mexico and cold, dry air from the Canadian Arctic (Hobbs *et al.* 1996). This type of cyclogenesis differs from that which occurs in Europe. Thus the cyclones in the central United States possess structural differences from the conventional polar front theory as briefly discussed in Section 10.1. When a surface-based cold front passes over a mountain range, such as the Rockies in North America, the baroclinic zones near the surface may be eroded by the *dry trough*. The dry trough is formed by the *lee trough* produced by the adiabatic warming associated with the low-level downslope wind and a *dryline* (to be discussed in Subsection 10.5.2). The upper-level baroclinic zone continues to move over the mountain, cuts off from the lower-level baroclinic zone, and forms the CFA over the plains downstream of the mountain range. Upper-level frontogenesis is induced by this forward advancement of the CFA, which leads to upward motion in the upper- and mid-troposphere (Fig. 10.14a). In the meantime, the forward movement of the CFA also produces the upward motion beneath its forward edge. The upward motion can lift the potentially unstable lower-level air to produce the *CFA rainband* (Fig. 10.14b). A prefrontal surface trough may occur owing to the CFA.

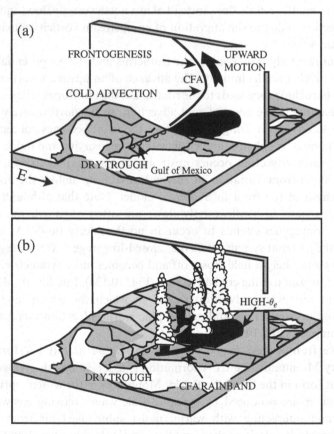

Fig. 10.14 (a) Upper-level frontogenesis associated with a cold front aloft (CFA) and (b) formation of the CFA rainband by the lower-level upward motion beneath the forward edge of the CFA, induced by the forward movement of the CFA. (Adapted after Hobbs *et al.* 1996.)

10.5.2 Drylines

A *dryline* is a low-level mesoscale boundary or narrow transition zone that separates dry and moist air masses. The length of a dryline is dictated by large-scale terrain and weather systems, while its width and movement are controlled by relevant mesoscale processes. Drylines differ from synoptic fronts since their scales are mesoscale and there is little density contrast. Drylines are found all over the world (Schaefer 1986). Over the USA, drylines are typically found over the Southern Plains to the east of the Rocky Mountains. The drylines align with the terrain in the spring and early summer, and separate warm, moist air from the Gulf of Mexico from hot, drier air from the southwestern states or the Mexican Plateau. The formation and movement of the dryline dictate the key dynamics required in forecasting the formation of severe thunderstorms and supercell thunderstorms and often determine the preferred location for convective initiation.

Figure 10.15 shows the synoptic situations for a dryline near the western boundary of Texas, the Oklahoma panhandle and the state boundary between Colorado and Kansas observed at 1500 UTC (10 am local time) May 26 and 0000 UTC May 27, 1991. At mid-morning, the dryline was located far west of its position later in the day

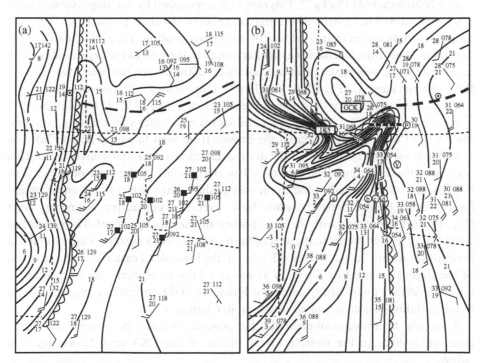

Fig. 10.15 Synoptic situations for a dryline observed at (a) 1500 UTC May 26 and (b) 0000 UTC May 27, 1991. The dryline is denoted by a scalloped line and the outflow boundary by a bold dashed line. Station symbols use conventional format with sea level pressure (mb and 10ths above 1000) at upper right. (Adapted after Hane *et al.* 1997)

(Fig. 10.15a). At mid day, the dryline was forced to move eastward in west Texas, due to the downward transport of westerly momentum caused by strong surface heating and vertical mixing (e.g., Schaefer 1986). The lack of eastward movement of the northern portion of the dryline is likely due to retardation of heating associated with land use (i.e., irrigation). The southern portion of the dryline continued to advance to the east by 0000 UTC May 27 (7 pm May 26 local time) as a pronounced *dryline bulge* advances into southwestern Kansas. Deep convection was initiated along the east edge of the bulge, but was in general short-lived. The dryline bulge may help explain why some thunderstorms initiate at particular locations. The relatively moist area east of the dryline in the northern Texas panhandle is still evident in Figure 10.15b. At night, the dryline retreats back to the west due to rapid surface cooling west of the dryline and strong easterly wind persisting to the east. Figure 10.16 shows the temperature and dew point profiles in the dry and moist air across the dryline in the evening of May 26 local time. In the dry air (Fig. 10.16a), west of the dryline, the lapse rate is approximately dry adiabatic from the surface to 580 mb due to vertical mixing through the layer. The wind speed and direction are quasi-uniform below 600 mb. In contrast, in the moist air (Fig. 10.16b) east of the dryline, the mean mixing ratio in the lowest 100 mb exceeds $15 \, g \, kg^{-1}$. This moist layer is capped by a strong inversion from 820 to 850 mb and has a strongly veering wind profile, consistent with the inhibition of deep convection except in areas of significant upward motion. The air to the east of the dryline and above the inversion was similar to that to the west of the dryline, indicating that the air in the well-mixed boundary layer west of the dryline has been advected over the moist air. Large CAPE, $2500–3000 \, J \, kg^{-1}$, is associated with this sounding.

The vertical structure of the dryline observed from 2158 to 2323 UTC May 26, 1991 is depicted in Fig. 10.17. There was a rapid increase in water vapor mixing ratio and the depth of the moist layer across the dryline from west to east (Fig. 10.17a). The elevated moist layer to the east of the dryline served as a potential source of moisture for convective initiation. The lower-level westerly wind to the west of the dryline converged with the easterly wind to the east of the dryline, while the stronger upper-level (above 1.5 km) westerly wind blew over the convective boundary layer (Fig. 10.17b), coinciding with the large horizontal moisture gradient. The convective boundary layer was deeper to the west of the dryline, where a westerly wind exhibited a forward shear with height. On the other hand, the lower-level easterly wind to the east of the dryline reversed its direction at about 1.2 km to westerly, thus exhibiting a critical level for a quasi-steady state flow. Existence of the critical level might also help in the initiation of convection, as discussed in Chapter 3.

Based on the above example and other observed events, the general characteristics of drylines in the western Great Plains of the USA may be summarized below.

(a) A typical dryline has an east–west moisture gradient of about $10 \, g \, kg^{-1}/100 \, km$, but with a negligible density gradient.

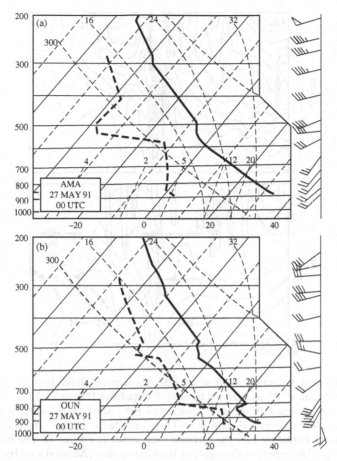

Fig. 10.16 Temperature and dew point profiles in (a) the dry air west of the dryline at Amarillo, Texas, and (b) the moist air east of the dryline at Norman, Oklahoma of the dryline at 0000 UTC May 27, 1991. (From Hane *et al.* 1997.)

(b) A dryline is nearly vertical in the lowest 1 km and then tilts eastward with height over the moist air.

(c) A veering wind shift is often associated with the passage of a dryline, but is not necessary for dryline existence.

(d) With weak synoptic forcing, the dryline moves eastward during the daytime and retrogrades at night, although different sections of the dryline may move at different speeds and even move in different directions.

(e) Drylines often form in late spring and early summer and are oriented in a north–south direction in the Texas Panhandle and the western portions of Oklahoma and Kansas.

(f) A dryline is a favored zone for cumulus convection. Severe thunderstorms and supercells, capable of producing tornadoes, often form along or just to the east of drylines.

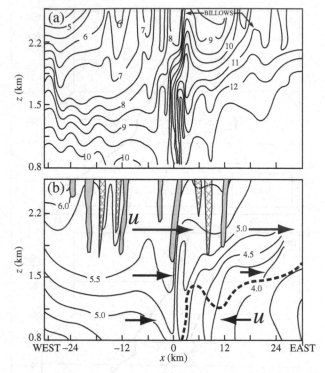

Fig. 10.17 Mesoscale vertical structure across a dryline across the bold-dashed line of Fig. 10.15b (to the west of location "P") as analyzed by aircraft data from 2158 to 2323 UTC May 26, 1991. (a) Water vapor mixing ratio (g kg^{-1}, solid); and (b) virtual potential temperature (excess above 310 K, solid) and vertical motion ($> 1\,\mathrm{m\,s}^{-1}$ shaded; $< -1\,\mathrm{m\,s}^{-1}$ hatched). The heavy dashed line is the boundary between east and west winds ($u = 0\,\mathrm{m\,s}^{-1}$). The horizontal wind direction and speed are represented roughly by the wind vectors. (Adapted after Hane *et al.* 1997.)

(g) A dryline is often located near a surface pressure trough, such as a lee trough or dry trough (described in Subsection 10.5.1), or a wind shift line. Neither feature is necessary for the existence of a dryline; nor does a dryline have to be coincident with the trough. The process does, however, set up a condition favorable for convective initiation by upper-level disturbances propagating eastward from the Rockies.

The formation of drylines involves both large-scale and mesoscale processes: the large-scale process helps build up a strong moisture gradient between moist and dry air masses and creates the lid inversion over the moist layer, while the mesoscale process helps strengthen the moisture gradient and the actual formation of the dryline. The large-scale process is associated with *differential moisture advection*. The moist, warm air over the Gulf of Mexico is advected northward in response to the trough situated in the lee of the Rockies and the anticyclone to the east, while the dry, hot air is advected eastward from the Mexican Plateau and the southwestern states through

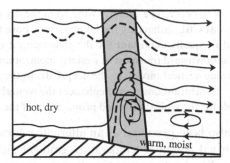

hot, dry

warm, moist

Fig. 10.18 A conceptual model of the formation of a dryline and initiation of convection along the dryline during afternoon and early evening. The solid curves denote airflow streamlines. The lower heavy dashed curve denotes the extent of the moist convective boundary layer, while the upper heavy dashed curve denotes the dry convective boundary layer which has a height of about 4 km along the left domain. The cross-dryline low-level jet points into the page and is denoted by "J". The dryline zone is shaded, which has a width of 1 to 10 km. (Adapted after Sun and Ogura 1979; Ziegler and Rasmussen 1998.)

the same mechanism. In the meantime, the elevated mixed layer of dry, hot air is advected over the moist, warm layer to form the lid inversion.

The mesoscale process involves the convergence and *solenoidal circulation* produced by *differential heating* across the maximum moisture gradient zone that is caused by the terrain slope and differential soil moisture. This process is depicted in the conceptual model of Fig. 10.18 and involves the following mesoscale circulations.

(a) During the day, the convective boundary layer (CBL) develops faster and deeper on the dry, hot (west) side of the maximum moisture gradient zone, compared to that on the moist, warm side. Due to a deeper stable layer and larger soil moisture on the moist side, more net amount of surface heating is needed. The elevated terrain on the west side further assists the development of a deeper CBL there.

(b) The easterly wind in the moist CBL is strengthened by the westward pressure gradient force enhanced by the differential heating and converges with the westerly wind at the maximum moisture gradient zone, which in turn strengthens the moisture gradient. A solenoidal circulation develops as a result and the dryline forms at the zone of maximum convergence. The formation mechanism of the solenoidal circulation is similar to that of sea-breeze circulation (Fig. 6.21) and mountain–plains solenoidal circulation (Fig. 6.26), as discussed in Chapter 6. Therefore, the vertical circulation across the dryline has also been called the "*inland sea-breeze circulation*" (Sun and Ogura 1979).

(c) A low-level southerly jet forms in the moist CBL near the maximum moisture gradient zone due to the rightward turning of the strengthening easterly wind by the Coriolis force. The low level jet can further enhance the northward moisture transport from the Gulf of Mexico.

(d) The enhanced convergence between the westerly and easterly winds produces an upward motion on the dry side of the maximum moisture gradient zone over the sloping terrain. The upward motion creates a zone of *moisture bulge* (i.e., *dryline bulge*), where convective initiation is most likely to occur. The westerly wind from the dry CBL flows continuously over the top of the moist CBL, enhancing the lid inversion above the moist CBL.

(e) Another process that is equally important for the enhancement of low-level convergence along the dryline is the downward transport of westerly momentum from the upper levels to the surface by the strong vertical mixing in the deep CBL to the west of the dryline. The transport of westerly momentum downward enhances the westerly winds on the west side and is also responsible for the typical eastward propagation of the dryline in the afternoon.

Frontogenesis has also been proposed as an alternative formation mechanism for drylines. Lack of a virtual potential temperature gradient and the positive relative vorticity across the dryline in the general dryline environment indicates that this mechanism may work only for some special environments, such as in the vicinity of a dryline–front interaction in which a density gradient (along the front) is present and frontogenesis can be a factor (Schaefer 1986). When a front overtakes a dryline, the wind shift line becomes more closely aligned with the line of maximum moisture gradient and a strong vorticity field is generated, which significantly enhances frontogenesis. The mesoscale processes associated with the creation of solenoidal circulations and the frontogenesis process at the later stage of the dryline formation are not clearly distinguishable, because the former may form a sharp moisture gradient zone in a way similar to the formation of the sea-breeze front during the day. Thus, calculations of the frontogenesis function, such as (10.1.11), from numerical modeling simulations may indicate whether both the differential heating term (term 1) and the stretching or confluence term (term 2) are significantly large during the later stage of dryline formation.

Once formed, a dryline moves eastward during the day and retreats to the west at night. The movement of drylines may be explained as follows (e.g., Sun and Wu 1992): During the day, the dry CBL to the west of the dryline develops faster and deeper than the moist CBL. The slope-related differential heating over the slope erodes the western portion of the moist boundary layer and forces the dryline to move eastward, but not necessarily in a continuous fashion. The westerly wind on the dry side is maintained by downward transport of westerly momentum due to strong vertical mixing. After sunset, the vertical mixing weakens and eventually vanishes due to the rapid cooling on the dry side of the dryline. The westerly wind decreases in speed and eventually reverses to easterly in response to the pressure gradient force associated with the cooling-induced high pressure on the dry air side and the lee trough to the west of the dryline. In the meantime, a stronger easterly wind persists on the moist side of the dryline and forces the dryline to retreat to the west at night. The westward movement decreases with the development of a drainage flow over the sloping terrain due to larger radiative cooling on the dry side.

Once the dryline forms, an upward bulge of thermal, moisture, and across-dryline wind component fields, similar to the head of density current, forms and a downward motion forms on the moist side of this upward bulge, as depicted in Fig. 10.18. Convective clouds tend to be initiated at the upward bulge due to conditional instability. Gravity waves, the release of potential instability, and an upper-level disturbance (such as a short-wave trough) may also help to initiate deep moist convection.

Although a dryline line defines the favorite location for convective initiation, in the real world, the initial initiation of deep convection often occurs at isolated locations along and east of the dryline position. The interaction of the smaller-scale *horizontal convective rolls* (HCRs), and/or cellular convection that develop within the CBL, with the dryline convergence zone has been proposed to be a possible cause for such preferred initiation. Based on observational data, it was shown (Atkins *et al.* 1998) that dryline variability forms in the along-line direction as a result of dryline interaction with the CBL HCRs. The rolls intersect with the dryline at periodic locations producing radar reflectivity and vertical velocity maxima, and initiating clouds at these intersection points.

Based on a detailed numerical study, a conceptual model (Fig. 10.19) of dryline convective initiation as related to the interaction of HCRs with the primary dryline convergence boundary (PDCB) was proposed (Xue and Martin 2006). In this conceptual model, HCRs develop on both sides of the dryline in the afternoon due to surface heating over sloping terrain. Close to the PDCB, the HCRs are often aligned at an angle with the dryline. The HCRs on the west side are more intense and deeper and their updraft speed can reach several meters per second. The HCRs west of PDCB are initially quasi-two-dimensional (Fig. 10.19a) but become more cellular with time (Fig. 10.19b) as the boundary layer becomes more unstable. The low-level convergence bands associated with the HCR updrafts are shown by gray shading in the figure and are enhanced at the dryline location. When the HCRs are fully developed, the convergence boundary (thick solid line in Fig. 10.19b) is distorted into a wavy pattern by the intersecting HCRs.

As the dryline strengthens in the afternoon, the easterly component of moist flow is increased. In the absence of other complications, convective initiation is preferred close to the central portion (marked by circles) of the leading HCR convergence bands (gray shaded bands) at the PDCB, where surface convergence is maximized due to opposing winds on each side of the bands as well as along-band flow convergence found at these locations. The low-level environment has been preconditioned for easy triggering of convection along this zone because of sustained mesoscale lifting at the PDCB, as discussed earlier, therefore convection tends to be triggered at these preferred locations. Localized vorticity centers or *misocyclones* (marked by **V** in Fig. 10.19b) often form along the dryline, due to the concentration of positive background vertical vorticity and the tilting of horizontal vorticity by the HCR circulation. A *misocyclone* is a horizontal vortex with a horizontal scale of between 40 m and 4 km. Maximum vertical lifting usually exists to the north or south of the misocyclones, where the low-level convergence can be enhanced by the vortex circulations. Within the misocyclones, downward pressure gradient force usually exists because the negative pressure perturbations induced by centrifugal force is the largest near the surface,

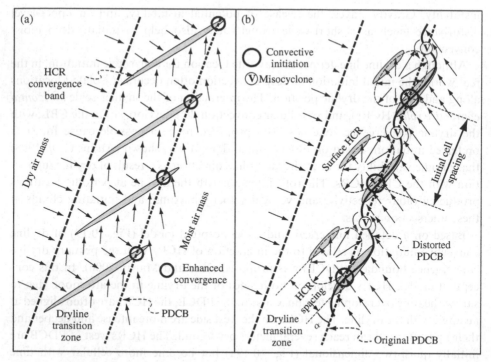

Fig. 10.19 A conceptual model of dryline convective initiation due to the interaction of the primary dryline convergence boundary (PDCB) with the evolving horizontal convective rolls (HCRs) that originate at and on the west side of the PDCB and are aligned at an acute angle, α, with the dryline. (a) The earlier stage of HCR development. (b) The low-level flow at the mature stage of HCR development. See text for details. The thin circles enclosing 'V' indicate locations of vorticity maxima (or misocyclones) along the PDCB. (From Xue and Martin 2006.)

and this pressure gradient force inhibits convective initiation at the misocyclone locations. Misocyclones usually do not co-locate with maximum surface convergence but their circulation can enhance convergence to their south and north, and non-supercell tornadoes can develop when their vertical vorticity is stretched by cumulus congestus clouds that move over them.

10.6 Jet streak dynamics

10.6.1 Upper-level jet streaks

As briefly mentioned at the beginning of this chapter, a *jet streak* is a region of local wind maximum embedded within a polar-front or subtropical jet stream. Cyclogenesis often occurs underneath the left side of the exit region of a jet streak, in association with positive vorticity advection downstream of an upper-level trough diffluent flow downstream of the axis of a large-amplitude, upper-level trough, as has been well documented. The jet streak associated with this type of development is often located within a jet stream constituting the main zone of the westerly. The left-exit region of a jet streak is also the region favorable for severe storm development due to the upward motion associated with the positive vertical gradient of horizontal vorticity advection. The strong vertical and horizontal shear in the vicinity of a jet streak also makes it a region conducive to the generation of turbulence. Thus, understanding the formation, propagation, and structure of upper-level jet streaks is essential to the prediction of rapid cyclogenesis, severe storm outbreaks, heavy rainfall, mesoscale gravity waves, and clear air turbulence, since these weather phenomena are intimately related to the location, horizontal and vertical structure, vertical circulation, and propagation of jet streaks.

a. General characteristics of a jet streak

A *polar jet streak* possesses the following general characteristics:

(1) It is located near 300 mb or above at midlatitudes.
(2) There is an abrupt increase with height in static stability and Ertel's PV across the jet streak.
(3) There is a strong wind shear across the jet streak in both horizontal and vertical directions.
(4) It is embedded in a westerly, southwesterly, or northwesterly jet stream.
(5) The maximum wind speed of a jet streak may reach as high as $75 \, \mathrm{m \, s^{-1}}$, comparable to the tangential wind of very strong tropical cyclones.
(6) Since it is embedded in the polar jet stream, its location is determined by the strong thermal gradient at the surface or the polar front at the surface through the thermal-wind relation.

A *subtropical jet streak* is similar to a polar jet streak except that:

(1) It is located higher than the polar jet streak, between 20° and 35° latitude.
(2) It is embedded in a mainly westerly subtropical jet stream which is larger in scale, more persistent and continuous around the globe than its polar front jet stream counterpart.
(3) It mainly occurs in winter.
(4) A band of cirrus clouds is often seen on the anticyclonic side of the subtropical jet streak.

A jet streak is often referred to as a balanced or *geostrophic jet streak* if the magnitude of the acceleration or deceleration of an air parcel associated with the wind speed maximum is much smaller than the acceleration induced by the Coriolis force (i.e., $R_o \ll 1$); otherwise, it is referred to as an unbalanced or *nongeostrophic jet streak*.

A jet streak propagates along and within a jet stream that serves as the basic flow, but at a slower speed. Therefore, an air parcel accelerates when it enters the entrance region of a jet streak and decelerates when it leaves through the exit region. The fluid surrounding the jet streak may be divided into left-entrance (left-rear), right-entrance (right-rear), left-exit (left-front), and right-exit (right-front) quadrants. The names of these quadrants refer to an observer who is located at the center and faces downstream of the jet streak. The horizontal and vertical flow circulation around a straight jet streak is depicted in Fig. 10.20. The ageostrophic flow in the entrance and exit regions

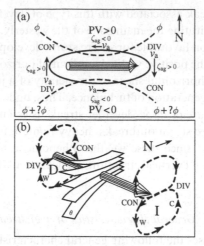

Fig. 10.20 A conceptual model of a steady-state, straight jet streak. (a) Horizontal wind structure: transverse ageostrophic wind component (v_a) and divergence field (CON and DIV denote convergence and divergence, respectively). Left side of the arrow is the entrance and the right side is the exit region. PV stands for potential vorticity. Signs of induced ageostrophic relative vorticity (ζ_{ag}) are also denoted. The dashed lines are representative geopotential contours (ϕ and $\phi + \Delta\phi$); (b) Transverse ageostrophic circulations in the entrance and exit regions of the jet streak, along with schematic isentropes (thin solid lines) and location of jet core (J). Symbols W, C, outlined I and D represent warm region, cold region, thermally indirect and direct circulations, respectively. The divergence fields at the jet streak level are also depicted. (Adapted after Uccellini and Kocin 1987, Cunningham and Keyser 2004, and Mattocks and Bleck 1986.)

of a jet streak can be derived from the inviscid, semi-geostrophic equation of motion, (10.2.13), which gives

$$V_a = \frac{1}{f} k \times \frac{DV_g}{Dt}. \tag{10.6.1}$$

Therefore, as a fluid particle enters a jet streak, its geostrophic flow speed increases and induces a southerly ageostrophic flow ($v_a > 0$), as shown in Fig. 10.20a. Similarly, a northerly ageostrophic flow is induced in the exit region. Note that the above equation may be rewritten as

$$V_a = \frac{1}{f} k \times \left[\underbrace{\frac{\partial}{\partial t} \left(\frac{1}{f} k \times \nabla\phi \right)}_{A} + \underbrace{(V_g \cdot \nabla) V_g}_{B} + \underbrace{w \frac{\partial V_g}{\partial z}}_{C} \right]. \tag{10.6.2}$$

Thus, the ageostrophic wind gets contributions from the isallobaric wind (term A), horizontally advective wind (term B), and vertically advective wind (term C). The isallobaric wind blows from regions of rising pressures to those of falling pressures. Term B may influence the convergence–divergence pattern, such as that associated with curved jet streaks. Term C may play a significant role if the vertical motion is induced by diabatic forcing associated with convection. In a forward shear geostrophic flow ($\partial u_g/\partial z > 0$), the upward motion tends to induce a northward ageostrophic wind and therefore eastward wind acceleration.

For a straight jet streak, the ageostrophic winds are associated with the cyclonic ageostrophic relative vorticity dipole in the entrance and exit regions, and the anticyclonic ageostrophic relative vorticity dipole to the north and south of the jet streak core (Fig. 10.20b, Cunningham and Keyser 2004). The ageostrophic flows induce divergence in the right-entrance and left-exit quadrants, and convergence in the left-entrance and right-exit quadrants. A four-cell pattern of vertical motion is then produced in the troposphere: upward motion in the left-exit and right-entrance quadrants and downward motion in the right-exit and left-entrance quadrants. The vertical motion is opposite above the jet streak level in the stratosphere. The compensated vertical motions are determined by mass continuity, which requires upward (downward) motion below the divergence (convergence) area. An alternative way to determine the vertical motion is to calculate the Q-vector convergence and divergence based on the semi-geostrophic or quasi-geostrophic omega equation, (10.3.21) or (10.3.23) respectively. Although earlier conceptual models of straight jet streaks, as represented in Fig. 10.20, are based on heuristic kinematic or parcel arguments, basic structures are verified theoretically by homogeneous and stratified quasi-geostrophic models with imposed, balanced monopolar or dipolar vortices of mesoscale dimension.

Figure 10.21 shows an example of a polar jet streak and its associated ageostrophic wind and vertical motion observed at 1200 UTC December 2, 1991. The jet streak belongs to a trough-over-ridge pattern with strong confluence upstream and diffluence downstream from the jet core. The 300-hPa height field (Fig. 10.21a) shows that the jet streak is approximately straight, with stronger curvature on the cyclonic-shear side. The 300-hPa ageostrophic wind is consistent with that depicted in the conceptual model of a straight jet streak (Fig. 10.20), directed toward lower (higher) geopotential heights in the jet entrance (exit) region. Figure 10.21b shows the ω vertical motion field at 500 hPa, analyzed by integrating the incompressible continuity equation vertically (i.e., the *kinematic method*), is much more complicated than the four-cell pattern depicted in the conceptual model of a straight jet streak. In fact, the ω field indicates a two-cell pattern, with downward motion in the entrance region and upward motion in the exit region. Variation of the four-cell pattern of vertical motion may be due to various reasons, such as nonlinear, curvature and thermal advection effects, as will be discussed later. The transverse circulations shown in the entrance region and the exit region (Figs. 10.21c–d) are consistent with the conceptual model of a straight jet streak.

The four-cell pattern of vertical motion and convergence/divergence, in turn, induces transverse, secondary circulations in both the entrance and exit regions (Fig. 10.20b). In the troposphere, the transverse circulation in the exit region forces the warm air to descend and cold air to ascend, often referred to as a *thermally indirect*

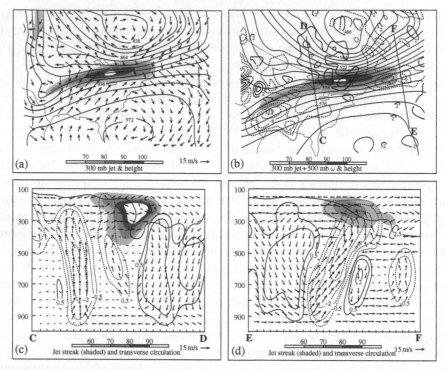

Fig. 10.21 An example of observed jet streak. (a) The 300-hPa total horizontal wind speed (shaded for values greater than 70 m s^{-1}), geopotential height (thin solid lines, contour interval, 21 dam) and ageostrophic wind vector fields at 1200 UTC Dec 2, 1991; (b) 500 hPa ω vertical motion (positive thick solid and negative dotted; contour interval, 0.2 Pa s^{-1}) and geopotential height (thin solid lines); (c) transverse circulation represented by wind vectors on the cross section CD of (b), oriented normal to the jet streak entrance region. The ω vertical motion is displayed (positive solid and negative dotted; contour interval, 0.1 Pa s^{-1}). Along-jet horizontal wind speed (perpendicular to the cross section) is shaded for values greater than 60 m s^{-1}. The 1.5 PVU (PV unit) contour is denoted by a bold line; and (d) same as (c) except for the cross section EF of (b), across the jet streak exit region. (Adapted after Pyle *et al.* 2004.)

circulation. The secondary circulation in the stratosphere is opposite in direction, but is still a thermally indirect circulation since the cross-stream thermal gradient is reversed from that in the troposphere. In the right-exit quadrant, the warm air in the troposphere warms adiabatically due to the forced descent, while in the left-exit quadrant, the cold air cools adiabatically due to the forced ascent. The thermally indirect circulation in the jet streak's exit region results in an increase of the horizontal temperature gradient in the troposphere. Such a thermally indirect circulation in the exit region is also frontogenetical. The averaged potential energy of the air column increases due to the lifting of cold air and the kinetic energy decreases.

On the other hand, the transverse circulation in the entrance region is thermally direct and is frontolytic because the horizontal temperature gradient decreases due to adiabatic warming and cooling associated with this transverse circulation. The potential energy of the air column decreases in the entrance region, which converts into kinetic energy and increases the wind speed at that location. As the air parcel passes the core of the jet streak, part of the kinetic energy converts into potential energy associated with the thermally indirect circulation in the exit region. Note that for a jet streak with an along-stream length smaller than the Rossby radius of deformation, $L_R = NL_z/f$ (4.2.12), the mass (temperature) field tends to adjust to the wind field; otherwise, the wind field tends to adjust to the mass field. The geostrophic adjustment process also occurs when the jet streak is propagating downstream. Cirrus clouds may form in the right-entrance region due to the upward motion if there is abundant supply of moisture, which are then advected horizontally at the jet streak level to the exit region. The cirrus clouds tend to last for a longer period in the region of upward motion at and above the jet streak level, but tend to dissipate in the region of downward motion.

b. Vertical motion associated with a jet streak

The conceptual model of a steady-state, straight jet streak (Fig. 10.20) and its associated ageostrophic circulations, as described above, is based on either quasi- or semigeostrophic dynamics which neglect the divergence tendencies necessary for the excitation of inertia-gravity waves. Thus, the four-cell pattern of vertical motion is not always observed in the atmosphere. For example, the nonlinear advection effect may make the quasi-geostrophically balanced four-cell pattern more localized and subsequently evolve toward a three-cell pattern as the Rossby number (R_o) increases from 0.26 to 0.39. It then evolves toward a two-cell pattern as R_o increases to 0.52 (Fig. 10.22). The numerical experiment with $R_o = 0.26$ shows some similarity to the classical four-cell pattern of the vertical motion, but the left-exit rising center is displaced toward the jet core (Fig. 10.22a). The distortion is stronger with $R_o = 0.39$, where the rising motion begins to dominate at a location upstream of the jet core (Fig. 10.22b). In fact, there are now three dominant cells aligned roughly along the jet axis. As the R_o increases to 0.52, the downward motion in the exit region increases significantly and, along with the upward motion near the jet streak core and entrance region, dominate the vertical motion field. With a stronger jet, the divergence in the entrance region and convergence in the exit region along the jet exis (i.e., the nonlinear advection effect) dominates the divergence fields produced by the y-component ageostrophic wind (v_a) field, and have thus generated upward and downward motion, respectively, below the jet streak level.

The vertical motion field may also be influenced by the curvature of a jet streak. For cyclonically curved jet streaks, previous research has proposed that the divergence–convergence pair on the cyclonic side of the jet is enhanced, while the pair on the anticyclonic side is weakened (Beebe and Bates 1955). The conceptual model for curved jet streaks has been verified by primitive equation models. Figure 10.23a shows the vertical motion associated with an initially balanced,

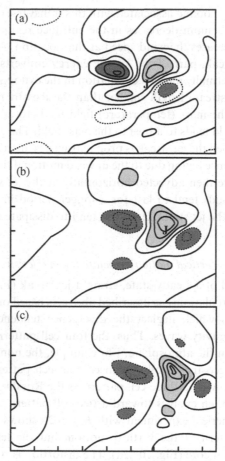

Fig. 10.22 Nonlinear effect on the vertical motion associated with a jet streak: 600 mb ω vertical velocity after 12 h simulations for jet streak forcing with the 400-mb Rossby number equal to (a) 0.26, (b) 0.39, and (c) 0.52. The numerical simulations are performed by a two-layer, nonlinear primitive equation model. The four-cell pattern subsequently evolves toward a three-cell and then two-cell pattern as the Rossby number (R_o) increases from 0.26 to 0.52. Upward (downward) motion is light (dark) shaded for (a) $\omega < -1 \times 10^{-4}\,\mathrm{mb\,s^{-1}}$ ($\omega > 1 \times 10^{-4}\,\mathrm{mb\,s^{-1}}$), (b) $\omega < -2 \times 10^{-4}\,\mathrm{mb\,s^{-1}}$ ($\omega > 2 \times 10^{-4}\,\mathrm{mb\,s^{-1}}$), and (c) $\omega < -5 \times 10^{-4}\,\mathrm{mb\,s^{-1}}$ ($\omega > 5 \times 10^{-4}\,\mathrm{mb\,s^{-1}}$). The jet streak center is denoted by J. (After Van Tuyl and Young 1982.)

cyclonically curved jet streak after a 12-h simulation with a primitive equation model. A two-cell pattern of the vertical motion, downward in the entrance region and upward in the exit region, is maintained and enhanced. The two-cell pattern can be explained conceptually by considering the superposition of a two-cell pattern of divergence field associated with a uniform jet stream (Fig. 10.23c) on a classical four-cell pattern of divergence field associated with a jet streak (Fig. 10.20b). At the jet streak level, the pattern results in areas of convergence and divergence in the entrance and

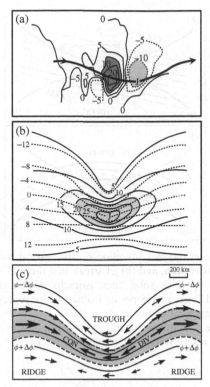

Fig. 10.23 Curvature effect on the vertical motion associated with a jet streak. (a) The 600-mb ω vertical velocity (10^{-4} mb s^{-1}; upward (downward) motion light (dark) shaded) field at 12 h, simulated by a primitive equation model. (b) The isotach (solid) and streamfunction (dotted) fields of a cyclonically curved jet streak used for the simulation of (a) at the initial time. The quasi-geostrophic vertical motion of (b) at the initial time is a pair of downward and upward motions in the entrance and exit regions, respectively. (After Moore and VanKnowe 1992.) (c) Convergence and divergence fields associated with a uniform jet stream within a stationary synoptic-scale trough. (After Shapiro and Kennedy 1981.)

exit regions, respectively, which subsequently generated downward and upward motions in the respective regions. Note that the convergence–divergence pattern of Fig. 10.23c may be determined qualitatively from the horizontal-advective ageostrophic wind (term B of (10.6.2)).

Changes of along-jet thermal advection pattern may also affect the pattern of vertical motion associated with a jet streak. Figure 10.24 displays two flow configurations of the potential temperature field, relative to a straight upper-tropospheric jet streak, which cause the departure of the vertical motion pattern from the classical four-cell pattern. Figure 10.24a depicts a case with straight isentropes rotated at an angle of $\alpha = 15°$ from the jet axis, resulting in cold advection in the along-jet direction. In addition to the stretching deformation responsible for the classical four-cell

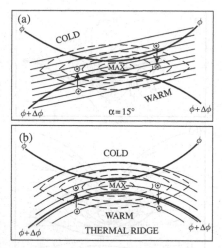

Fig. 10.24 Effect of along-jet thermal advection on the vertical motion associated with a jet streak: (a) along-jet cold advection, and (b) jet streak in a thermal ridge. Geopotential height contours are represented by thick solid lines; isotachs of the along-jet wind component are represented by dashed lines; isentropes or isotherms are represented by thin solid lines; cross-jet ageostrophic component of the winds at the jet streak level are represented by bold solid arrows; and the sense of midtropospheric upward and downward motion are denoted by arrow head and tail signs, respectively. (After Shapiro 1983.)

pattern, shear deformation (see discussion related to (10.2.22)) comes into play in this case. The transverse, ageostrophic circulation can be determined by applying either the semi- or quasi-geostrophic Sawyer–Eliassen equation ((10.2.27) or (10.2.31)) or by determining the convergence or divergence of the Q vectors ((10.3.21) or (10.3.23)). The shifting of the thermally direct circulation to the anticyclonic side of the jet streak in the entrance region (Fig. 10.24a) can be discerned because the cold air is shifted further to the south, compared to the case of straight jet streak with straight isentropes parallel to the jet axis. Similarly, the thermally indirect circulation in the exit region shifts to the cyclonic side. The shifting of ageostrophic circulation for the case for a straight jet streak with a thermal ridge (Fig. 10.24b) can be explained in the same fashion.

c. Formation and propagation of a jet streak

Jet streams have been hypothesized to form due to the confluence of a cyclonically rotating cold air mass from the north and an anticyclonically rotating warm air mass from the south (Namias and Clapp 1949), referred to as the *confluence mechanism*. The confluence is often the result of a warm midtropospheric airstream from the south curving anticyclonically, gradually flowing beside a cold cyclonically curved stream from the north. The type of upper-level flow deformation aforementioned may be generated by the differential propagation of waves in the westerlies at different latitudes. Similarly, a jet streak may form as a result of confluence–diffluence dipole forcing. Observations also indicate that jet streak formation is closely related to the upper-tropospheric frontogenesis and tropopause distortions (Shapiro 1983; Carlson 1998). As discussed earlier in Subsection 10.5.1, the jet-front system often propagates or moves through a midlatitude baroclinic wave. The formation and propagation of jet-front systems is depicted in and discussed along with Fig. 10.13.

The formation of jet streak can be understood by considering the following equation governing baroclinic perturbations in a uniform, continuously stratified, hydrostatic, Boussinesq flow on an f-plane (Weglarz and Lin 1997):

$$\left[\left(\frac{\partial^2}{\partial t^2}+f^2\right)\frac{\partial^2}{\partial z^2}+N^2\nabla_H^2\right]\nabla^2 u' =$$

$$-N^2\frac{\partial}{\partial y}\left(\nabla^2 q'_{pv}\right)+\frac{\partial^2}{\partial z^2}\left[\frac{\partial^2}{\partial t^2}\nabla^2 F_z+f\nabla^2 F_y\right]-\frac{g}{c_pT_o}\frac{\partial^2}{\partial z\partial x}\left(\nabla^2 \dot{q}\right). \quad (10.6.3)$$

$$A \qquad\qquad\qquad B \qquad\qquad\qquad C$$

In the above equation, the nonlinear terms have been moved to the right-hand side as momentum forcing terms. The above equation is similar to (4.2.20) except that: (1) it is expressed in terms of u'; (2) no basic winds are present ($U=V=0$); (3) nonlinear momentum forcing is included; and (4) a three-dimensional Laplacian operator has been performed. In (10.6.3), F_y and F_z are meridional and vertical momentum forcing, which are taken to represent the parameterized or prescribed effects of nonlinear inertial advection ($F_y=-V'\cdot\nabla v'$, $F_z=-V'\cdot\nabla w'$ where $V'=u'i+v'j+w'k$) in the interior of the stratified flow, and q'_{pv} is the perturbation PV as defined in (4.2.19) and repeated here:

$$q'_{pv}=\zeta'+\frac{f}{\rho_o N^2}\frac{\partial^2 p'}{\partial z^2}. \quad (10.6.4)$$

The strengthening of a localized wind maximum or jet streak ($\nabla^2 u' < 0$) may be caused by the three forcing terms on the right-hand side of (10.6.3). Term A approximately indicates that if a flow configuration with $(\partial/\partial y)(\nabla^2 q'_{pv})<0$ is present, then it tends to significantly increase the local wind maximum with time $[(\partial^2/\partial t^2)\partial^2(\nabla^2 u')/\partial z^2>0]$. For example, a PV dipole, with positive PV to the north and negative PV to the south, will provide the confluence–diffluence dipole forcing required for the jet streak formation, similar to the confluence mechanism hypothesis for jet stream formation. In addition to the differential propagation or movement of regions of the PV anomaly, as discussed above, a deepening trough or building ridge may also provide the confluence–diffluence dipole forcing. Since $\partial q'_{pv}/\partial y$ is proportional to $\partial(\partial\theta'/\partial y)/\partial z$, according to (10.6.4) and the hydrostatic equation, the PV dipole is supported by the thermal wind balance. The jet streak will eventually reach a balance according to

$$\left(f^2 \frac{\partial^2}{\partial z^2} + N^2 \nabla_H^2\right) \nabla^2 u' = -N^2 \frac{\partial}{\partial y} (\nabla^2 q'_{\mathrm{pv}}).$$
(10.6.5)

For a large-scale jet streak ($fL/NL_z \gg 1$), the first term involving Coriolis forcing dominates, which leads to the steady-state, quasi-geostrophically balanced straight jet streak in a continuously stratified atmosphere (Cunningham and Keyser 2004). As discussed in Chapter 4, gravity waves will be generated and will radiate out before the flow geostrophically adjusts to a steady-state balance for an initially unbalanced PV perturbation (Figs. 4.10 and 4.11). Term B of (10.6.3) contains the nonlinear terms which provide the momentum forcing for the formation of a jet streak and the subsequent flow patterns. Term C of (10.6.3) represents the diabatic forcing, in particular the latent heat released by cumulus clouds or mesoscale convective systems. Qualitatively speaking, cumulus heating in a quiescent atmosphere will produce two opposing low-level inflow jet streaks and two upper-level opposing outflow jets above the cloud top (e.g., Hamilton *et al.* 1998).

Following the derivation of the frontogenesis (frontogenetical) function, a jetogenesis (jetogenetical) function may be derived (Bluestein 1993):

$$J_s = \frac{\mathrm{D}}{\mathrm{D}t} \left(-\nabla^2 u\right) = -f \nabla^2 v_a - \nabla^2 F_{rx}$$
(10.6.6)

where F_{rx} is the component of friction in the direction of the jet streak. Thus, the jetogenesis function is positive whenever there is a local wind maximum in the cross-jet streak component of the ageostrophic wind.

For a straight jet streak, its propagation is associated with a pattern of marked upper-level convergence and divergence in its exit and entrance regions (Uccellini and Johnson 1979). Thus, the propagation speed and direction of a jet streak can be determined from the geopotential height tendencies since a jet streak is often associated with the juncture of a trough and a ridge. The geopotential height tendency equation may be derived based on quasi-geostrophic theory or semi-geostrophic theory. For a jet streak coupled with an upper-level front, the propagation is strongly influenced by the jet-front system, such as that shown in Shapiro's conceptual model (Fig. 10.13) and the northwesterly and southwesterly regimes in the earlier stage (Schultz and Doswell 1999), and is thus more challenging to predict. The jetogenesis function, (10.6.6), or (10.6.3) may be used for this purpose.

A kinematic approach (e.g., Cohen and Schultz 2005), may be constructed to represent the jet streak in terms of the growth rate and rotation rate of the *air parcel separation* by divergence, relative vorticity, and deformation. To improve the prediction of jet streak formation, propagation and evolution, a more fundamental understanding of the jet streak dynamics is needed. For example, it is found that some jet streaks coincide with large horizontal gradients of pressure and potential temperature on the dynamic tropopause (see Section 10.5), and of PV on tropopause-intersecting isentropic surfaces (Pyle *et al.* 2004). In addition, the jet streaks under a trough-over-ridge

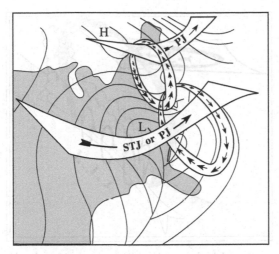

Fig. 10.25 A conceptual model of upper-level jet streak interaction through their transverse circulations. The upward motion in the right-entrance region of the northern polar jet streak (PJ) is in phase with and strengthened by that in the left-exit region of the southern subtropical or polar jet streak (STJ or PJ, respectively). This jet streak interaction is found to be a common feature for the majority of the US East Coast snowstorms. (Adapted after Uccellini and Kocin 1987.)

and southwesterly flow in their study are associated with a coherent tropopause disturbance which tends to enhance the pressure and potential temperature gradients over a mesoscale region, and these features may be used to help predict the evolution of jet streaks.

d. Interaction of jet streak circulations

An upper-level jet streak may interact with an adjacent upper-level jet streak through their transverse circulations. Figure 10.25 shows a conceptual model of this type of jet streak interaction. The upward motion in the right-entrance region of the northern (polar) jet streak is in phase with, and strengthened by the upward motion associated with the transverse circulation in the left-exit region of the southern (polar or sub-tropical) jet streak. The interaction is found to be a common feature of cyclogenetic events which produce heavy snow along the East Coast of the United States (Uccellini and Kocin 1987). The interaction of jet streak transverse circulations contributes to vertical motions and differential moisture and temperature advections necessary to produce heavy snowfall along the coast. The environment established by this jet streak interaction is conducive to mesoscale boundary layer processes, such as cold-air damming, coastal cyclogenesis, and low-level jet formation, that further contribute to the cyclogenesis. Note that an unbalanced mesoscale jetlet may develop due to the extraordinary pressure gradient forces produced by the superpositioning of different air masses and the diabatic perturbation associated with polar and subtropical upper-level jet streaks interactions (Kaplan *et al.* 1998).

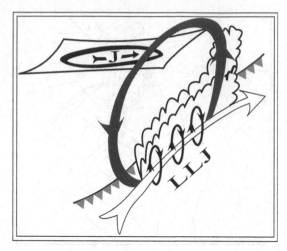

Fig. 10.26 A conceptual model of upper-level jet streak and low-level jet interaction ahead of a surface cold front. The upper-level jet streak and low-level jet are denoted by J and LLJ, respectively. The upper-level jet streak may also be associated with an upper-level jet-front system. The arrowed large circle denotes the transverse circulation associated with the upper-level jet streak in the exit region, while the arrowed small circles denote the transverse circulations associated with the LLJ. The prefrontal rainbands are produced by the convective clouds ahead of the surface front. (Adapted after Sortais *et al.* 1993 and Shapiro 1983.)

Besides interacting with an upper-level adjacent jet streak, an upper-level jet streak may also interact with a low-level jet through their transverse circulations. Figure 10.26 shows a conceptual model of an upper-level jet streak and low-level jet (LLJ) interaction ahead of a surface cold front. The conceptual model in Fig. 10.26 summarizes the coupling between an upper-level and a low-level jet-front system that occurred on November 11, 1987 during the Mesoscale Frontal Dynamic Project FRONTS 87 experiment, as revealed from data analysis and numerical simulations (Sortais *et al.* 1993). The conceptual model is consistent with that proposed by Shapiro (1983). The prefrontal rainbands ahead of the surface cold front on the left hand side of the LLJ are produced by convective clouds generated by the upward branch of the transverse circulation associated with the LLJ. The downward branch of the transverse circulation associated with the upper-level jet streak in the right exit region helped to intensify the LLJ, while the upward branch coupled with the upward motion associated with the LLJ's transverse circulation served to intensify the convection in the left-exit region of the upper-level jet streak. The upper-level jet streak may also be associated with an upper-level jet-front system. The upward advection term of (10.6.2) (Term C) helps to identify the strong upward motion enhanced by the coupling of transverse circulations of an upper-level jet streak and LLJ. Latent heat released due to this motion will feedback to strengthen the ageostrophic wind.

10.6.2 Low-level jets

A *low-level jet* (LLJ) is a fast moving, narrow, near-surface airstream with its maximum wind situated within the boundary layer, normally in the lowest 2–3 km from the surface. The LLJ has drawn significant attention from meteorologists since it plays important roles in the transport of moisture, generation of strong convergence at the head of the jet, organization of convective systems, and its interaction with orography and other weather systems. For example, LLJs which occur in the warm season over the Great Plains to the east of the US Rocky Mountains play important roles in transporting the heat and moisture from the Gulf of Mexico into the Great Plains to produce convective storms and heavy rainfall (Fig. 10.27a). Another

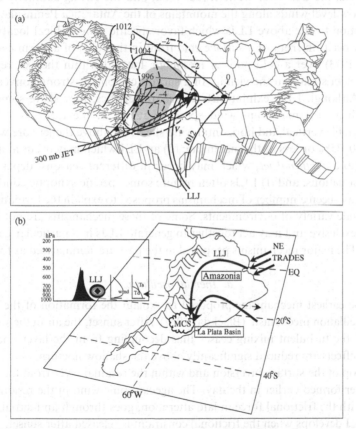

Fig. 10.27 (a) Pattern of surface isobars (solid curves in mb), surface pressure tendencies (dashed curves in mb/3h), low-level jet (LLJ), surface low (L), center of maximum pressure deepening (outlined L), ageostrophic wind at the surface (V_a) and upper-level jet over the southern Great Plains east of the US Rockies. Area of concentrated LLJ observations from January 1959 through December 1960 is shaded. (Adapted after Carlson 1998 and Bonner 1968.) (b) A conceptual model depicting the South American LLJ (SALLJ) east of Andes. (After Marengo *et al.* 2004.)

example is the South American LLJ (SALLJ), characterized as a narrow airstream that channels the near-surface flow between the tropics and midlatitudes east of the Andes (Fig. 10.27b). The SALLJ transports moisture from the Amazon region into the fertile lands of southern Brazil and northern Argentina, producing convection and rainfall at the exit region of the low-level jet in southeastern South America.

LLJs are known to occur in other parts of the world (e.g. Stensrud 1996), such as the *Somali jet* which occurs off the coast of Somalia and Oman during the Asian summer monsoon season, the *African easterly jet* which occurs to the south of the Sahara Desert, the northeasterly LLJ parallel to the Appalachian Mountains during cold-air damming events, the *California coastal jet* which occurs in summer months, the southwesterly LLJ in front of the Mei-Yu (also known as Baiu and Changma) front in East Asia, the LLJ over western Australia, and the strong southerly and south-westerly low-level winds along the mountains of the Antarctica Peninsula.

In addition to the above LLJs which favor specific geographical locations, other LLJs may occur in front of a surface cold front in the form of a warm *conveyor belt* (Harrold 1973) over a cold pool or mesohigh, located within the exit region of an upper-level jet streak, in the vicinity of a surface cyclone or strong convective latent heating. Although LLJs exhibit different characteristics, they share some common ones, such as: (a) maximum wind speeds that are in the range of 12–30 m s^{-1}; (b) appreciable vertical and horizontal wind shear; (c) above the jet core wind speeds that are 50–75% or less of the maximum; (d) an LLJ which may occur at any time of the day; (e) a *nocturnal jet*, which may occur in different seasons, depending upon synoptic situations; and (f) LLJs often require some specific synoptic conditions.

There have been a number of mechanisms proposed to explain the formation of LLJs under a wide variety of environments. Some of these mechanisms are not necessarily mutually exclusive and may act together to generate LLJs in a particular geographical location. The major mechanisms proposed in the past are summarized as follows.

a. Inertial oscillation

One of the earliest mechanisms proposed to explain the formation of the LLJ is the inertial oscillation mechanism (Blackdar 1957). After sunset, the air in the surface layer cools and the turbulent mixing ceases thus decoupling from the layer above it. The frictional effects are reduced significantly above the shallow nocturnal boundary layer near the top of the surface inversion and within the residual layer from the convective mixed layer formed earlier in the day. The ageostrophic wind in the residual layer, in balance with the frictional force in late afternoon, goes through an inertial oscillation and an LLJ develops when the frictional constraint is released after sunset.

Neglecting the frictional force and assuming the geostrophic winds are independent of time, the horizontal momentum equations reduce to

$$\frac{\partial u_a}{\partial t} = f v_a; \quad \frac{\partial v_a}{\partial t} = -f u_a, \tag{10.6.7}$$

where $u_a = \bar{u} - u_g$ and $v_a = \bar{v} - v_g$ are the ageostrophic wind velocities in x and y directions, respectively, and $\bar{u}$ and $\bar{v}$ are the mean wind velocities. The above two equations can be combined as

$$\frac{\partial^2 u_a}{\partial t^2} + f^2 u_a = 0; \quad \frac{\partial^2 v_a}{\partial t^2} + f^2 v_a = 0. \qquad (10.6.8)$$

The above equations have solution of the form

$$u_a = A\sin(ft + \alpha); \quad v_a = A\cos(ft + \alpha), \qquad (10.6.9)$$

where A is the initial wind speed and α is the phase angle determined by the initial wind direction. The above solution represents a clockwise oscillation and is referred to as *inertial oscillation*. The inertial oscillation has a period of $2\pi/f$ or half pendulum day. At $40°$, the half pendulum day is about 18.6 h. Thus, the LLJ that forms through this mechanism would generate a maximum wind speed about 9.3 h after sunset.

Although this mechanism may help explain the formation of some southerly LLJs observed in the Great Plains to the east of the US Rocky Mountains, the actual boundary layer processes are more complicated due to the presence of sloping terrain and synoptic forcing.

b. Baroclinic effects

Radiative heating over a flat terrain with different surface characteristics may produce strong low-level baroclinicity. LLJs can form along isotherms leaving the cool air to the left of the thermal wind in the Northern Hemisphere. Similar effects exist over the sloping terrain. During the daytime, air near the mountain surface is warmer than its surrounding air at the same level due to radiative heating, thus producing a strong temperature gradient toward the mountain range or strong baroclinicity within the planetary boundary layer. At night, the differences in land surface characteristics allows for a reversal in the temperature gradient orientation. An LLJ can then be produced over the sloping terrain by the geostrophic forcing associated with the baroclincity (Blecker and Andre 1951; Holton 1967), based on the thermal wind relations,

$$\frac{\partial u_g}{\partial z} = -\frac{g}{f\bar{T}}\frac{\partial \bar{T}}{\partial y}; \quad \frac{\partial v_g}{\partial z} = \frac{g}{f\bar{T}}\frac{\partial \bar{T}}{\partial x}. \qquad (10.6.10)$$

Over the Great Plains to the east of the US Rockies, this will produce southerly LLJs at night. Northerly LLJs do exist during daytime, but are less pronounced due to strong vertical mixing in the planetary boundary layer.

A low-level easterly jet is produced to the south of the Sahara Desert of North Africa in the summer months due to the baroclinicity created by sensible heating on surfaces with different characteristics.

c. Orographic deflection

An LLJ flowing upstream along a mountain range may be a manifestation of a *barrier jet* flowing parallel to the mountain barrier along the windward slope. The barrier jet forms when a low-level stable synoptic flow approaches the mountain barrier and is blocked for a half pendulum day $(2\pi/f)$ or longer. As discussed in Chapter 5, high pressure is generated by adiabatic cooling associated with the ascending air over the mountain. Thus, a pressure gradient force (PGF) is produced and directed away from the mountain and against the incoming basic flow. For example, an easterly basic flow will experience an additional PGF pointing eastward. In response to this eastward PGF, the incoming easterly flow speed is reduced and accelerated to the left (southward) in the Northern Hemisphere, due to the imbalance of the original (and larger) southward PGF and the reduced northward Coriolis force, in order to support the basic easterly flow. Barrier jets can extend laterally 100 km or more upstream of the barrier, and their maximum wind speeds may reach 15–30 m s^{-1}. Barrier jets have been observed upstream of the Sierra Nevada Mountains in California, to the north of the Brooks Range in Alaska, and in Antarctica along the Antarctic Peninsula and the Transantarctic Mountains (Glickman 2000). Another example is the northerly LLJ associated with cold-air damming along the eastern slopes of the Appalachian Mountains.

An LLJ may also be generated by the *down-valley wind*, which is associated with thermally induced mountain and valley winds, such as those that occur in the mouth of Austria's Inn Valley, and a *gap wind* when the air is channeling and accelerating through a mountain gap, as discussed in Subsection 5.6.4.

d. Synoptic forcing

Observations indicate that the organized, coherent LLJs in the Great Plains are frequently associated with leeside troughing, lee cyclogenesis, or a frontal passage with a cyclone located farther to the north, as shown in Fig. 10.27a. The lee cyclone tends to induce a southerly jet flowing from the Gulf of Mexico and to the east of the cyclone and the leeside troughing tends to induce *isallobaric wind* (term A of (10.6.2)) toward the center of maximum pressure deepening. During a frontal passage, a warm conveyor belt may also act as an LLJ transporting warm and moist air to the area ahead of a cold front.

Two patterns of upper-tropospheric flow have been proposed to prevail during the occurrence of an LLJ in the Great Plains: (a) a trough over the Rockies and a ridge located in the eastern third of the USA with a polar and a subtropical upper-tropospheric jet streaks propagating toward the Great Plains and (b) a strong ridge located over the front range of the Rockies with weak upper-tropospheric flow over the northern Texas/western Oklahoma-Kansas region (Uccellini 1980). The first type occurs more frequently than the second type. The development of a leeside cyclone or trough occurs with a particular type of strong upper-level flow. In this type of flow, the

LLJs (i) tend to be located within the exit region of the upper level jet streak, (ii) are directed toward the cyclonic side, (iii) are well-defined, coherent and more persistent even in the afternoon and extend beyond the planetary boundary layer, and (iv) tend to evolve with the propagation of an upper-level jet streak and weakening lee cyclone. Interactions between an upper-level jet streak and latent heat release may also influence the development of an LLJ.

The forcing associated with an upper-level jet streak may combine with the low-level isallobaric forcing to form an LLJ, as is the case over the Carolinas. Synoptic forcing may also combine with orographic forcing to produce an LLJ, such as the coastally trapped disturbance (CTD) or coastal jet that occurs along the California coast as discussed in Subsection 5.6.2.

e. Diabatic processes

Several previous studies have documented the enhancement of the LLJ due to diabatic alteration of the pressure field (e.g., Lackmann 2002). The alteration of the pressure field due to diabatic processes is perhaps most readily understood from a PV perspective: latent heat release accompanying a cold front or within a squall line produces a lower-tropospheric PV maximum. Piecewise PV inversion of this feature has quantified the diabatic contribution to the LLJ as up to 50% or more for some systems (Mahoney and Lackmann 2006). The argument is consistent with interpretations unrelated to PV; lower-tropospheric height falls, and an enhanced geostrophic LLJ flow, are expected to accompany organized regions of latent heat release.

References

Atkins, N. T., R. M. Wakimoto, and C. L. Ziegler, 1998. Observations of the finescale structure of a dryline during VORTEX 95. *Mon. Wea. Rev.*, **126**, 525–50.

Beebe, R. G. and F. C. Bates, 1955. A mechanism for assisting in the release of convective instability. *Mon. Wea. Rev.*, **83**, 1–10.

Bénard, P., J.-L. Redelsperger, and J.-P. Lafore, 1992. Nonhydrostatic simulation of frontogenesis in a moist atmosphere. Part I: General description and narrow rainbands. *J. Atmos. Sci.*, **49**, 2200–17.

Bergeron, T., 1937. On the physics of fronts. *Bull. Amer. Meteor. Soc.*, **18**, 265–75.

Bjerknes, J., 1919. On the structure of moving cyclones. *Geophys. Publ.*, **1**, 1–8.

Blackdar, A. K., 1957. Boundary layer wind maxima and their significance for the growth of nocturnal inversion. *Bull. Amer. Meteor. Soc.*, **38**, 283–90.

Blecker, W. and M. J. Andre 1951. On the diurnal variation of precipitation, particularly over central U.S.A., and its relation to large-scale orographic circulation systems. *Quart. J. Roy. Meteor. Soc.*, **77**, 260–71.

Bluestein, H. B., 1993. *Synoptic-Dynamic Meteorology in Midlatitudes. Vol. II: Observations and Theory of Weather Systems.* Oxford University Press.

Bonner, W., 1968. Climatology of the low-level jet. *Mon. Wea. Rev.*, **96**, 833–50.

Bosart, L. F., 1970. Mid-tropospheric frontogenesis. *Quart. J. Roy. Meteor. Soc.*, **96**, 442–71.

Carlson, T. N., 1998. Mid-Latitude Weather Systems. *Amer. Meteor. Soc.*

Charney, J. G., 1947. The dynamics of long waves in a baroclinic westerly current. *J. Meteor.*, **4**, 135–63.

Cohen, R. A. and D. M. Schultz, 2005. Contraction rate and its relationship to frontogenesis, the Lyapunov exponent, fluid trapping, and airstream boundaries. *Mon. Wea. Rev.*, **133**, 1353–69.

Cunningham, P. and D. Keyser, 2004. Dynamics of jet streaks in a stratified quasi-geostrophic atmosphere: Steady-state representations. *Quart. J. Roy. Meteor. Soc.*, **130**, 1579–1609.

Eady, E. T., 1949. Long waves and cyclone waves. *Tellus*, **1** (3), 33–52.

Eliassen, A., 1962. On the vertical circulation in frontal zones. *Geofys. Publ.*, **24**, 147–60.

Emanuel, K. A., M. Fantini, and A. J. Thorpe, 1987. Baroclinic instability in an environment of small stability to slantwise moist convection. Part I: Two-dimensional models. *J. Atmos. Sci.*, **44**, 1559–73.

Glickman, T. S., 2000. *Glossary of Meteorology*. 2nd edn., Amer. Meteor. Soc.

Hamilton, D. W., Y.-L. Lin, R. P. Weglarz, and M. L. Kaplan, 1998. Jetlet formation from diabatic forcing with applications to the 1994 Palm Sunday tornado outbreak. *Mon. Wea., Rev.*, **126**, 2061–89.

Hane, C. E., H. B. Bluestein, T. M. Crawford, M. E. Baldwin, and R. M. Rabin, 1997. Severe thunderstorm development in relation to along-dryline variability: A case study. *Mon. Wea. Rev.*, **125**, 231–51.

Harrold, T., 1973. Mechanisms influencing the distribution of precipitation within baroclinic disturbances. *Quart. J. Roy. Meteor. Soc.*, **99**, 232–51.

Hobbs, P. V., J. D. Locatelli, and J. E. Martin, 1996. A new conceptual model for cyclones generated in the lee of the Rocky Mountains. *Bull. Amer. Meteor. Soc.*, **77**, 1169–78.

Holton, J. R., 1967. The diurnal boundary layer oscillation above sloping terrain. *Tellus*, **19**, 199–205.

Hoskins, B. J., 1975. The geostrophic momentum approximation and the semi-geostrophic equations. *J. Atmos. Sci.*, **32**, 233–42.

Hoskins, B. J. and F. P. Bretherton, 1972. Atmospheric frontogenesis models: Mathematical formulation and solution. *J. Atmos. Sci.*, **29**, 11–37.

Hoskins, B. J. and I. Draghici, 1977. The forcing of ageostrophic motion according to the semi-geostrophic equations and in an isentropic coordinate model. *J. Atmos. Sci.*, **34**, 1859–67.

Hoskins, B. J., M. E. McIntyre, and A. W. Robertson, 1985. On the use and significance of isentropic potential vorticity maps. *Quart. J. Roy. Meteor. Soc.*, **111**, 877–946.

Kaplan, M. L., Y.-L. Lin, D. W. Hamilton, and R. A. Rozumalski, 1998. The numerical simulation of an unbalanced jetlet and its role in the Palm Sunday 1994 tornado outbreak in Alabama and Georgia. *Mon. Wea. Rev.*, **126**, 2133–65.

Keyser, D., 1986. Atmospheric fronts: An observational perspective. *Mesoscale Meteorology and Forecasting*, P. S. Ray (ed.), Amer. Meteor. Soc., 216–58.

Keyser, D. and R. A. Anthes, 1982. The influence of planetary boundary layer physics on frontal structure in the Hoskins-Bretherton horizontal shear model. *J. Atmos. Sci.*, **39**, 1783–1802.

Keyser, D. and M. J. Pecnick, 1985. A two-dimensional primitive equation model of frontogenesis forced by confluence and horizontal shear. *J. Atmos. Sci.*, **42**, 1259–82.

Keyser, D. and M. A. Shapiro, 1986. A review of the structure and dynamics of upper-level frontal zones. *Mon. Wea. Rev.*, **114**, 452–99.

Keyser, D., M. J. Reeder, and R. J. Reed, 1988. A generalization of Petterssen's frontogenesis function and its relation to the forcing of vertical motion. *Mon. Wea. Rev.*, **116**, 762–81.

Kuo, Y.-H. and S. Low-Nam, 1994. Effects of surface friction on the thermal structure of an extratropical cyclones. In *Proc. Int. Symp. On the Life Cycles of Extratropical Cyclones*, Vol. II, Bergen, Norway, University of Bergen, 129–134.

Lackmann, G. M. 2002. Potential vorticity redistribution, the low-level jet, and moisture transport in extratropical cyclones. *Mon. Wea. Rev.*, **130**, 59–74.

Lackmann, G. M., D. Keyser, and L. F. Bosart, 1997. A characteristic evolution of upper-tropospheric cyclogenetic precursors during the Experiment on Rapidly Intensifying Cyclones over the Atlantic (ERICA). *Mon. Wea. Rev.*, **125**, 2529–56.

Lackmann, G. M., D. Keyser, and L. F. Bosart 1999. Energetics of an intensifying midtropospheric jet streak during the Experiment on Rapidly Intensifying Cyclones over the Atlantic (ERICA). *Mon. Wea. Rev.*, **127**, 2777–95.

Mahoney, K. M. and G. M. Lackmann, 2006. The effect of upstream convection on downstream precipitation. *Wea. Forecasting*, **21**, 465–88.

Marengo, J. A., W. R. Soares, C. Saulo, and M. Nicolini, 2004. Climatology of the low-level jet east of the Andes as derived from the NCEP-NCAR reanalysis: Characteristics and temporal variability. *J. Climate*, **17**, 2261–80.

Mattocks, C. and R. Bleck, 1986. Jet streak dynamics and geostrophic adjustment processes during the initial stages of lee cyclogenesis. *Mon. Wea. Rev.*, **114**, 2033–56.

Miller, J. E., 1948. On the concept of frontogenesis. *J. Meteor.*, **5**, 169–71.

Moore, J. T. and G. E. VanKnowe, 1992. The effect of jet-streak curvature on kinematic fields. *Mon. Wea. Rev.*, **120**, 2429–41.

Namias, J. and P. F. Clapp, 1949. Confluence theory of the high tropospheric jet stream. *J. Meteor.*, **6**, 330–6.

Ogura, Y. and D. Portis, 1982. Structure of the cold front observed in SESAME-AVE III and its comparison with the Hoskins-Bretherton frontogenesis model. *J. Atmos. Sci.*, **39**, 2773–92.

Palmén, E. and C. W. Newton, 1948. A study of the mean wind and temperature distribution in the vicinity of the polar front in winter. *J. Meteor.*, **5**, 220–6.

Palmén, E. and C. W. Newton, 1969. *Atmospheric Circulation Systems: Their Structure and Physical Interpretations*. Academic Press.

Petterssen, S., 1956. *Weather Analysis and Forecasting, Vol. I, Motion and Motion Systems*, 2nd edn., McGraw-Hill Co.

Pyle, M. E., D. Keyser, and L. F. Bosart, 2004. A diagnostic study of jet streaks: Kinematic signatures and relationship to coherent tropopause disturbances. *Mon. Wea. Rev.*, **132**, 297–319.

Reed, R. J., 1955. A study of a characteristic type of upper-level frontogenesis. *J. Meteor.*, **12**, 226–37.

Rotunno, R., W. C. Skamarock, and C. Snyder, 1998. Effects of surface drag on fronts within numerically simulated baroclinic waves. *J. Atmos. Sci.*, **55**, 2119–29.

Sawyer, J. S., 1956. The vertical circulation at meteorological fronts and its relation to frontogenesis. *Proc. Roy. Soc. London, Ser. A*, **234**, 346–62.

Schaefer, J. T., 1986. The dryline. *Mesoscale Meteorology and Forecasting*. P. S. Ray (ed.), Amer. Meteor. Soc., 549–72.

Schultz, D. M. and C. A. Doswell III, 1999. Conceptual models of upper-level frontogenesis in southwesterly and northwesterly flow. *Quart. J. Roy. Meteor. Soc.*, **125**, 2535–62.

Schultz, D. M. and P. N. Schumacher, 1999. The use and misuse of conditional symmetric instability. *Mon. Wea. Rev.*, **127**, 2709–32.

Shapiro, M. A., 1981. Frontogenesis and geostrophically forced secondary circulation in the vicinity of jet stream-frontal zone system. *J. Atmos. Sci.*, **38**, 954–73.

Shapiro, M. A., 1983. Mesoscale weather systems of the central United States. *The National Storm Program: Scientific and Technical Bases and Major Objectives*, R. A. Anthes (ed.), University Corporation for Atmospheric Research, 3.1–3.41.

Shapiro, M. A. and P. J. Kennedy, 1981. Research aircraft measurements of jet stream geostrophic and ageostrophic winds. *J. Atmos. Sci.*, **38**, 2642–52.

Snyder, C., W. C. Skamarock, and R. Rotunno, 1991. A comparison of primitive-equation and semi-geostrophic simulation of baroclinic waves. *J. Atmos. Sci.*, **48**, 2179–94.

Sortais, J.-L., J.-P. Cammas, X. D. Yu, E. Richard, and R. Rosset, 1993. A case study of coupling between low- and upper-level jet-front systems: Investigation of dynamical and diabatic processes. *Mon. Wea. Rev.*, **121**, 2239–53.

Stensrud, D. J., 1996. Importance of low-level jets to climate: A review. *J. Climate*, **9**, 1698–1711.

Sun, W.-Y. and Y. Ogura, 1979. Boundary-layer forcing as a possible trigger to a squall-line formation. *J. Atmos. Sci.*, **36**, 235–54.

Sun, W.-Y. and C.-C. Wu, 1992. Formation and diurnal variation of the dryline. *J. Atmos. Sci.*, **49**, 1606–19.

Thorpe, A. J. and K. A. Emanuel, 1985. Frontogenesis in the presence of small stability to slantwise convection. *J. Atmos. Sci.*, **42**, 1809–24.

Uccellini, L. W., 1980. On the role of upper-tropospheric jet streaks and leeside cyclogenesis in the development of low-level jets in the Great Plains. *Mon. Wea. Rev.*, **108**, 1689–96.

Uccellini, L. W. and D. R. Johnson, 1979. The coupling of upper and lower jet streaks and implications for the development of severe convective storms. *Mon. Wea. Rev.*, **107**, 682–703.

Uccellini, L. W. and P. J. Kocin, 1987. The interaction of jet streak circulation during heavy snow events along the east coast of the United States. *Wea. Forecasting*, **2**, 289–308.

Van Tuyl, A. H. and J. A. Young, 1982. Numerical simulation of nonlinear jet streak adjustment. *Mon. Wea. Rev.*, **110**, 2038–54.

Weglarz, R. P. and Y.-L. Lin, 1997. A linear theory for jet streak formation due to zonal momentum forcing in a stably stratified atmosphere. *J. Atmos. Sci.*, **54**, 908–32.

Whitaker, J. S., L. W. Uccellini, and K. F. Brin, 1988. A model-based diagnostic study of the rapid development phase of the President's Day cyclone. *Mon. Wea. Rev.*, **116**, 2337–65.

Xu, Q., 1992. Formation and evolution of frontal rainbands and geostrophic potential vorticity anomalies. *J. Atmos. Sci.*, **49**, 629–48.

Xue, M. and W. J. Martin, 2006. A high-resolution modeling study of the 24 May 2002 case during IHOP. Part II: Horizontal convective rolls and convective initiation. *Mon. Wea. Rev.*, **134**, 172–91.

Ziegler, C. L. and E. N. Rasmussen 1998. The initiation of moist convection at the dryline: Forecasting issues from a case study perspective. *Wea. Forecasting*, **13**, 1106–31.

Problems

10.1 (a) Derive the thermal wind relation, $u_T = u_{g2} - u_{g1} = (-R/f)(\partial \bar{T}/\partial y) \ln(p_1/p_2)$, from the geostrophic wind relation, $u_g = (-1/f)\partial \phi/\partial p$, where ϕ is the geopotential, and given the hydrostatic relation, $\partial \phi/\partial p = -\alpha$; (b) Based on the above thermal wind relation and Fig. 10.1, find u_g at 275 hPa at 45° N by estimating u_g at 900 hPa and the mean layer temperature between 900 and 275 hPa. How does your calculated value compare with that in the figure?

10.2 Based on Fig. 10.2, derive (10.1.1).

10.3 Derive (10.1.13) from (10.1.12) and (10.1.14).

10.4 Derive (10.2.17), (10.2.18) and (10.2.20).

10.5 Consider an initial synoptic situation identical to that depicted in Fig. 10.3a except in the absence of the background stretching deformation and assume that the isentropic surfaces are tilted toward the cold air. (a) Sketch the initial flow structure (isentropic surfaces, v_g, cold and warm regions) on an *y–z* plane. (b) Determine the sign of $-(\partial u_g/\partial y)(\partial b/\partial x)$, and whether the streamfunction (ψ) is maximum or minimum near $u_g = 0$. (c) Based on (10.2.27), sketch the cross-frontal, ageostrophic secondary circulation.

10.6 Prove, based on the geostrophic coordinate transformation (10.3.2), that: (a) the X and Y velocities are geostrophic, i.e., (10.3.3): $DX/Dt = u_s$ and $DY/Dt = v_s$; and (b) the Jacobian of the semi-geostrophic transformation is as shown in (10.3.4).

10.7 Derive the *semi-geostrophic omega equation*, (10.3.21), from (10.3.16)–(10.3.18) with thermal wind relations and (10.3.19).

10.8 Based on the Q vector argument, show graphically that there is an upward motion at A in Fig. 10.4b, and a downward motion near the center between H and L in the left half of the system.

10.9 Based on the Sawyer–Eliassen equation (10.2.22), determine the signs of Q_2 and the transverse ageostrophic circulation induced by geostrophic flow in Figs. 10.11a–b.

10.10 Show that the air acceleration (deceleration) in the entrance (exit) region of the jet streak may be explained by the following equation,

$$\frac{D}{Dt}\left(\frac{V^2}{2}\right) = -gV \cdot \nabla Z.$$

Derive the above height equation from the inviscid equation of motion.

11

Dynamics of orographic precipitation

11.1 Orographic influence on climatological distribution of precipitation

Orographic influence on the formation of clouds and its associated precipitation amount and distribution is dramatic. The influence of orography was well recognized very early in human history and documented in numerous meteorological and climatological literatures. When a moist airflow impinges on a mountain, the dynamical and cloud microphysical characteristics of the airflow are modified by orographic lifting and blocking which may modify and/or trigger cloud and precipitation systems in the vicinity of the mountain. Figure 11.1 shows the mean annual precipitation for the period 1971–1990 over Western Europe. Areas of heavy precipitation are concentrated on the Alpine mountains. Note that precipitation over the Alps is produced by weather systems coming from different directions, in particular, from the northern and southern sides.

Over a meso-α/β or large-scale mountain range, precipitation is triggered or enhanced on the windward slope of a prevailing wind due to orographic lifting on the upwind slope. A larger advection time is required for an air parcel to pass over the windward slope compared to the formation time of orographic clouds and precipitation. On the lee side of the mountain, there is little or no rain due to the depletion of moisture over the upwind slope and the adiabatic warming associated with the descending air, known as *rain shadow*. In short, *the overall influence of mountains on climatological precipitation is orographic precipitation enhancement and suppression in the windward and lee sides of mountains, respectively.*

One revealing example of the orographic influence on precipitation distribution is shown on Fig. 11.2a. The figure reveals that the annual rainfall distribution over the Andes in South America has rainfall maxima located on the eastern slope in the northern Andes (to the north of 30° S) and on the western slope in the southern Andes under the prevailing easterly and westerly winds, respectively. Another example is shown in Fig. 11.2b, which indicates the total rainfall distribution accumulated from May 10 to June 27, 1987 during the Taiwan Area of Mesoscale Experiment (TAMEX). The rainfall was produced mainly from the lifting and blocking of the prevailing wind from the southwesterly monsoonal flow and the modification on mesoscale convective systems and Mei-Yu fronts moving from the west toward the

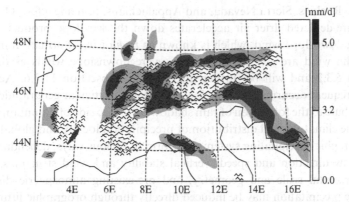

Fig. 11.1 Mean annual precipitation for the period 1971–1990 over Western Europe. The dark (light) shaded areas denote averaged daily precipitation higher than 5.0 mm d^{-1} (in between 3.2 and 5.0 mm d^{-1}). Terrain higher than 800 m is approximately denoted by mountain symbol. (Adapted after Frei and Schär 1998.)

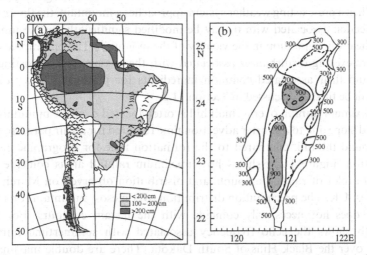

Fig. 11.2 (a) Annually averaged rainfall distribution over the land of the Andes in South America. (Adapted after Espenshade *et al.* 1990.) The dark (light) shaded areas denote annual precipitation higher than 200 cm (in between 100–200 cm). The terrain higher than 900 m is approximately denoted by the mountain symbol. (b) Total rainfall distribution during the TAMEX (10 May–27 June 1987) from 808 daily rainfall stations in Taiwan. The prevailing wind is the monsoonal flow from the southwest. Rainfall contours (thick) start at 300 mm with 200-mm intervals. The dashed contour denotes the terrain height of 1 km. (Adapted after Yeh and Chen 1998.)

western slopes of Taiwan's Central Mountain Range (CMR). The orographic control of global and local climatological precipitation distribution has been observed over many other meso-α/β and large-scale mountain ranges, such as the Himalayas, Western Ghats in India, New Zealand Alps, Scandinavian mountains, Pyrenees,

Apennines, Rockies, Sierra Nevada, and Appalachians, to name a few. Occasionally, the moisture-depleted drier air accelerates down the lee slope producing very dry and strong downslope wind, which is known as *foehn wind*. Formation mechanisms of the foehn wind are related to that of severe downslope winds as discussed in Subsection 5.3.2 and will be further discussed in Subsection 11.3.6. Accumulated effects of frequent occurrences of the foehn wind in a particular geographical location help contribute to the formation of rain shadow on the lee side of mountain ranges and to shape the climatological distribution of precipitation locally and globally.

An orographic precipitating system may form in various ways, depending upon the ambient flow direction and speed, vertical stability and wind structures, mountain height, horizontal scale and geometry, and pre-existing atmospheric disturbances. Orographic precipitation may be induced directly through orographic lifting and sur-face sensible heating, or indirectly through orographically modified local circulations and the release of moist instabilities of the impinging airflow. When an airflow impinges on a mountain, it may be associated with a prevailing wind, such as midlatitute westerlies and monsoonal flow, or weather systems, such as troughs, cyclones, and fronts. When a pre-existing weather system approaches a mountain, the flow circulation and convection associated with it may be modified significantly by the mountain to produce heavy precipitation in the vicinity of the mountain. The orographic modifica-tion of pre-existing disturbances associated with the passage of pre-existing weather systems and their associated common ingredients for producing heavy orographic rainfall will be further discussed in Section 11.2.

The maximum rainfall over a small hill is often located over the mountain peak or downwind slope due to a shorter advection time required for an air parcel to pass over or go around the hill, compared to the formation time for orographic clouds and precipitation. Figure 11.3a shows how maximum rainfall areas coincide with the terrain in a plot of rainfall amounts and distribution observed on March 28, 1953 over Wales, UK. The precipitation distribution over meso-γ mountains or small hills, however, does not necessarily coincide with the terrain contours. For example, Fig. 11.3b shows the echo frequency associated with summertime rain-showers observed over the Black Hills of South Dakota. There are double maxima in echo frequency: one located over the northeastern and the other over the southeastern section of the Black Hills. The double summertime rain-shower maxima are produced mainly by leeside convergence associated with prevailing southwesterly winds. The formation mechanisms of upslope and leeside orographic precipitation will be further discussed in Section 11.3.

The intensity, duration and distribution of orographic precipitation are mainly determined by the following factors (Sawyer 1956): (1) *large-scale environment* which determines the characteristics of the airflow upstream of the mountain, such as wind speed and direction, stability, and relative humidity; (2) *dynamics of the air motion* over and around the mountain, which determines the location and layer of orographic clouds and precipitation; and (3) *cloud microphysical processes* which

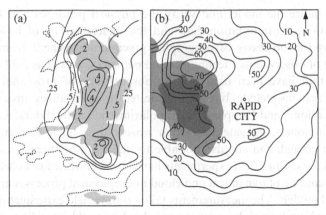

Fig. 11.3 Rainfall maxima located over the peak or leeside of small hills. (a) Rainfall observed over Wales, UK on March 28, 1953. Rainfall contours are in inches. Shaded areas denote the terrain higher than 150 m. (Adapted after Sawyer 1956.) (b) Radar echo frequency of rainshowers for all year in the southwesterly wind cases over four summers of 1967–70 in Black Hills, South Dakota. Shaded areas denote the terrain height contours (1372 and 1676 m). (Adapted after Kuo and Orville 1973.)

determine the characteristics of the cloud and precipitation systems and formation time in the vicinity of the mountain. Under the orographically modified upstream airflow associated with the large-scale environment, the interaction between the dynamical processes and cloud microphysical processes is extremely complicated, thus making quantitative forecasting of orographic precipitation very difficult and one of the most challenging problems in numerical weather prediction.

Improvements on numerical model techniques and computing powers have allowed for a continuous development of mesoscale models. These models have been used extensively as research tools to help understand the dynamics and cloud microphysics, and to forecast orographic precipitation. Despite the significant progress that has been made in cloud model development and in bulk microphysical parameterization schemes (see Chapter 14), progress in quantitative forecasting of orographic precipitation has been relatively modest. In particular, as model resolution has increased, model simulations of orographic cloud and precipitation systems have become more sensitive to model details, such as grid resolution and cloud microphysical parameterization schemes (e.g., Colle and Zeng 2004; Chiao *et al.* 2004). In general, errors in model-simulated results may come from various sources, such as insufficient model grid resolution, lack of initial data, deficiencies in initialization schemes, numerical errors, and problems with representations of cloud and precipitation processes, boundary layer, and radiative transfer.

In order to improve orographic precipitation forecasting, in addition to the improvements of the representation of moist processes, comprehensive data is also needed to initialize the model and verify the physical processes and hydrometeor fields. Collection of precipitation data and the understanding of basic dynamical and microphysical processes of orographic precipitation benefit from continuous improvement of observational techniques and networks, as well as field experiments. Instruments that have been used for the observations of orographic cloud and precipitation systems include, but are not limited to, rain gauges, instrumented airplanes, conventional and Doppler radars, polarimetric radars, lidars, wind profilers, dropsondes, mesonets, and satellites. Often, observations made by regular networks are not dense enough and are also not able to make *in situ* observation of elevated orographic clouds. Thus, field experiments have been conducted to collect data to aid in the understanding of dynamical and cloud microphysical processes and for verifying numerical models. The measurements taken during field experiments, along with more advanced analysis techniques, have helped our understanding of orographic precipitation, especially orographically induced convective systems.

This chapter places emphasis on the dynamics of orographic precipitation associated with extreme events that lead to flooding and form in unstable air. In addition to those extreme precipitation events, one should also pay attention to climatological patterns of orographic enhancement, which are the results of many mild to average events that do not individually cause floods. Current modeling studies have primarily focused on idealized cases and on a few spectacular events that lead to floods which are not representative of day-to-day average events. The dynamics of these average events, as well as extremely heavy orographic precipitation events, are thus considered unresolved and are topics of current research.

11.2 Orographic modification of preexisting disturbances

As briefly mentioned in Section 11.1, the impinging airflow on a mountain may be associated with prevailing winds or preexisting atmospheric disturbances. Prevailing winds are often associated with general circulations, such as the polar easterlies, midlatitude westerlies, equatorial easterly winds and trade winds, and monsoonal flows. The preexisting disturbances include, but are not limited to, troughs, fronts, as well as extratropical and tropical cyclones. In other words, *mountains usually modify, and often amplify, rather than create precipitation* (Smith 2006).

A prevailing wind or preexisting weather system influences the orographic precipitation by: (1) providing larger-scale upstream conditions, such as wind speed and direction, humidity, and stability; (2) generating additional forcing through dynamical processes due to its interaction with the topography. The additional forcing includes pressure gradient force associated with orographically modified flow circulations, convergence or divergence, and the latent heat release; and (3) affecting the microphysical processes in orographic cloud formation by providing the characteristics of hydrometeors and stability, and through its interaction with the topography. Orographic precipitation induced by orographic lifting and blocking is better understood when compared to precipitation events associated with a pre-existing disturbance. Most of the theoretical studies dealing with this subject are based on idealized upstream conditions which often assume simple structures in wind, stability, and humidity. Orographic precipitation induced by a prevailing or straight flow will be further discussed in Sections 11.3 and 11.4, while orographic precipitation associated with the passage of pre-existing disturbances will be discussed in this section.

11.2.1 Passage of troughs

Based on the observations of several flash floods that occurred in the USA, such as the 1972 Rapid City, South Dakota flood, 1977 Big Thompson Canyon, Colorado flood, the 1997 Fort Collins, Colorado flood, and the 1995 Madison County, Virginia flood, the following common synoptic and mesoscale environments were found to favor the slow-moving storms (Pontrelli *et al.* 1999; Fig. 11.4): (1) a mid- and upper-tropospheric shortwave trough just upstream from the flood location, which is propagating towards the high-pressure ridge axis; (2) a massive middle-upper tropospheric westward-tilted ridge of high pressure; (3) a very strong upslope, low-level jet (LLJ) from the east or southeast and a southward-propagating shallow

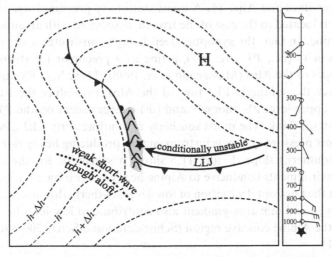

Fig. 11.4 Schematic of the synoptic and mesoscale conditions conducive to heavy orographic rainfall, based on the observations of the 1972 Rapid City, South Dakota flood, 1977 Big Thompson Canyon, Colorado flood, the 1997 Fort Collins, Colorado flood, and the 1995 Madison County, Virginia flood. Displayed are the surface synoptic features (fronts and high), threat region (star), mountains (shaded), low-level jet (LLJ), 500-hPa height field (dashed lines) and short-wave trough (heavy dotted curve), and typical wind profile above the threat region (right panel). (Adapted after Pontrelli *et al.* 1999.)

cold front; (4) a very high equivalent potential temperature associated with the LLJ; (5) conditional instability associated with the airflow upstream of the mountain; (6) weak mid- and upper-tropospheric steering currents. Note that a high CAPE has also been observed in some of these heavy orographic precipitation events, but is not consistently observed for others, such as in the Fort Collins storm. CAPE is of mesoscale nature and can be highly dependent on time and location. In addition, in the Fort Collins storm, an accelerated easterly LLJ impinged on the mountains and induced a quasi-stationary convective system causing the heaviest rainfall directly over Fort Collins.

Similar flash floods resulting from heavy orographic precipitation also occur over the Alps when a baroclinic trough moves eastward toward the Alps. During the fall, a warm and moist southerly LLJ impinges on the southern slopes of the Alps from the Mediterranean Sea, which may produce heavy rainfall and flash floods. Based on investigations of heavy orographic rain events over the Alps, the following flow features are found to play important roles in allowing the persistence of strong upward motion and moisture convergence (Buzzi and Foschini 2000). (1) A deep upper-tropospheric trough approaches the Alps from the west. (2) The flow at low levels remains southerly over the western Mediterranean, with pronounced confluence over the western Alps, between the post-frontal southwesterly flow and the pre-frontal southeasterly flow located more to the east. (3) A pre-frontal LLJ serves as a warm conveyor belt (Chapter 10) bringing in warm and moist air towards the Ligurian Apennines and Alps. (4) A quasi-stationary pressure ridge in the upper-troposphere is located to the east of the trough, associated with an anticyclone over Eastern Europe. In fact, the synoptic, deep trough constitutes a streamer of high potential vorticity (i.e., *PV streamer*), acting as a precursor for storms along the southern slopes of the Alps (Massacand *et al.* 1998). These high PV streamers tend to: (i) enhance the southerly LLJ toward the Alps, (ii) reduce the static stability beneath the upper-level PV anomaly and (iii) trigger ascent on the PV streamer's forward (eastern) flank. The moist southerly to southwesterly LLJ associated with the trough/low pressure system is instrumental in producing heavy rainfall over the windward (southern) slope. Figure 11.5 shows a schematic for the synoptic and mesoscale environments conducive to Alpine heavy orographic rain. In addition, it is found that the horizontal variation of low-level southerly flow associated with the synoptic trough and moisture gradient also contribute to low-level horizontal convergence at the Alpine concave region (Schneidereit and Schär 2000; Rotunno and Ferretti 2001).

Other common synoptic and mesoscale features associated with heavy orographic rain over the Alps include the elevated dry, Sahelian mixed layer and the easterly jet along the southern edge of the east-west oriented Alpine main ridge, conditionally unstable low-level flow, and concave geometry of the mountains. The superposition of the elevated dry, Sahelian mixed layer on top of the warm, moist LLJ provides an environment conducive to the potential (convective) instability by

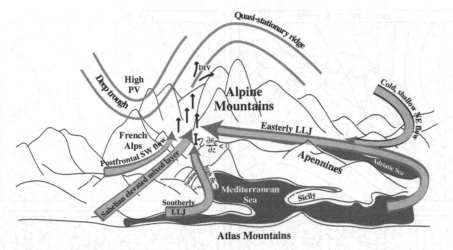

Fig. 11.5 A schematic for the synoptic and mesoscale environments conducive to heavy orographic rain in the Alps. (Adapted after Lin 2005.)

layer lifting. The easterly jet is either associated with a synoptic high pressure system located to the north of the Alps or the barrier jet induced by the southeasterly flow impinging on the eastern portion of the Alps, or a combination of these two low-level flows. Other observations indicate that the far upstream airflow often has a high CAPE that may lead to the release of conditional instability. For example, during Mesoscale Alpine Experiment (MAP) IOP-2, the CAPE at Cagliari (near the southern tip of Sardinia) at 0000 UTC September 20, 1999 reached 2793 J kg^{-1} (Fig. 11.14).

During heavy orographic rain events, convection often starts in the concave region south of Alps, such as the Lago Maggiore in Italy, and the Ticino region in Switzerland (e.g., Bougeault *et al.* 2001; Rotunno and Houze 2007). The convergence in the vicinity of the concave region tends to enhance the low-level upward motion induced by the upslope, which in turn triggers the convection. Figures 11.6a and b show the 850-hPa wind and 500-hPa height fields at 1800 UTC September 19 and 1200 UTC September 20, 1999, respectively, associated with MAP IOP-2B. During this period, the low-level flow was mainly from the south or southwest, as depicted in the schematic of Fig. 11.5. The low-level wind increased when the trough moves into northern Italy. The southerly LLJ has a maximum value of about 12.5 m s^{-1} at Milan, Italy. Figure 11.6c and 11.6d show the numerically simulated 6-h accumulated rainfall ending at 1800 UTC September 19 and 1200 UTC September 20, 1999, respectively. The maximum rainfall started at the concave area, i.e. the Lago Maggiore area (Fig. 11.6c), and then moved toward the east following the trough (Fig. 11.6d). The heavy rainfall area was associated with the impinging deep trough (Figs. 11.6a–b), which brought in air of high moisture

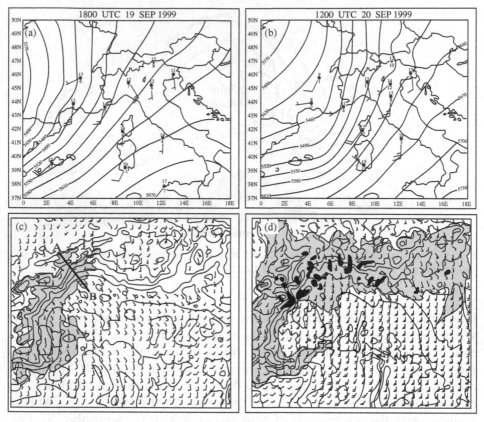

Fig. 11.6 Synoptic-scale flow and rainfall distributions during MAP IOP-2B. Geopotential height at 500 hPa (in m) and wind barbs (1 full bar = 5 m s^{-1}) at 850 hPa are shown in (a) 1800 UTC September 19 and (b) 1200 UTC September 20, 1999. Analysis is based on the ECMWF 0.5° reanalysis data. The 6-h accumulated rainfall ending at (c) 1800 UTC September 19 (shaded for greater than 10 mm) and (d) 0000 UTC September 20 (shaded for greater than 10 and 100 mm) 1999 are numerically simulated by a mesoscale model (MM5) with grid resolution of 1.67 km. The concave area, i.e., Lago Maggiore area, is near the southern end of the line AB. (Adapted after Chiao et al. 2004.)

content from the Ligurian Sea. Similar to the previously discussed historical events, the upper-level pressure ridge located to the east of the approaching deep trough is quasi-stationary. During the next 12-h period ending at 1200 UTC September 20, the heavy rainfall encompassed the entire southern slopes of the Alps (Fig. 11.6d). Similar phenomena also occur in heavy orographic rain events in other parts of the world, such as over the southwest concave region of the CMR of Taiwan (Fig. 11.11). Thus, evidence suggests that heavy orographic rain events occurring in the US mountains and in the Alps share some common synoptic and mesoscale environments.

11.2.2 Passage of midlatitude cyclones and fronts

When a midlatitude cyclone or a frontal system approaches a mesoscale mountain, the upstream flow conditions, such as wind speed and direction, stability, and relative humidity, keep changing due to the movement of the weather system and the orographic modification of the flow circulation associated with the weather system. In addition, microphysical processes may be altered and moist instabilities, such as conditional instability and/or potential instability, may be triggered due to orographic lifting of the lower layer of the incoming air.

Figure 11.7 shows a schematic for the passage of a typical storm over the San Juan Mountains of southern Colorado. During the passage of the storm over the mountains, a split front with the *warm conveyor belt* undergoes forward-sloping ascent. Based on the data analysis, the storm goes through four stages: (1) stable stage, (2) neutral stage, (3) unstable stage, and (4) dissipation stage during its passage over the mountains (Cooper and Marwitz 1980). During the stable stage, the low-level winds are not toward the mountain barrier, which result in little vertical displacement of the airflow over the mountain, and liquid water contents remain low. High ice crystal concentrations result from the high cloud tops and the long upwind extent of the clouds.

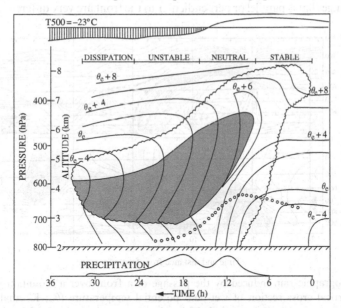

Fig. 11.7 Schematic of the four stages on the time-height cross-section for a typical storm over the San Juan Mountains of southern Colorado. The solid lines are lines of constant θ_e and the region conducive to potential (convective) instability ($\partial\theta_e/\partial z < 0$) is shaded. The typical 500 hPa temperature is shown at the top, which has a reference temperature of $-23\,°C$. The level below which the wind directions are $<160\,°$ is indicated by circles and is considered as blocked flow. The four stages of the storm during its passage over the mountains are indicated. (After Cooper and Marwitz 1980.)

Generally speaking, the cloud is completely glaciated except where gravity waves provide short-lived regions of low liquid water content. During the neutral stage, significant supercooled cloud water is present in the region above the upslope. Liquid water contents in this region are less than in the later unstable stage, but the ice crystals mentioned above do precipitate through this area. During the unstable stage, a convective region develops over the upslope of the mountains in association with a surface zone of horizontal convergence. Convection is triggered by the release of potential (convective) instability associated with layer lifting, as detected by $\partial\theta_e/\partial z < 0$. Due to convection, liquid water contents are highest during this later stage of the storm. The convective cloud system then dissipates following the unstable stage. Figure 11.7 shows the importance of orographic lifting of very moist air associated with the warm conveyor belt in triggering convective instability, which leads to heavy orographic precipitation. The storm's associated rainfall can be enhanced accordingly due to the dynamical and microphysical structure modification by the orography, such as wind direction shift, wind speed change, release of moist instabilities associated with the change of vertical stability structure, and changes in microphysical processes.

Figures 5.30, 5.38 and 5.39 provide examples for the horizontal and vertical distortion of a front when it passes over a mesoscale mountain. The behavior of a front passing over a mountain range that is parallel or perpendicular to the front are very different. Figure 11.8

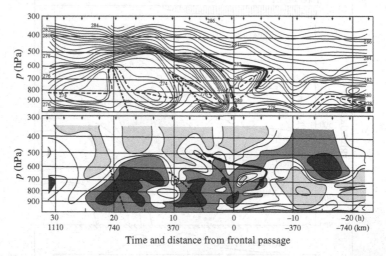

Time and distance from frontal passage

Fig. 11.8 Orographic rain induced by the passage of a front over a mountain range. (a) A west–east vertical cross section of wet-bulb potential temperature (θ_w; K) analyzed by the rawinsonde measurements for an occluded frontal system passing over the Cascade Mountains in western Washington State of the USA during March 15–16, 1973. The front is parallel to the north–south oriented Cascades. Region enclosed beneath dashed line is potentially unstable. The occluded front, the pre-frontal surge, and axes of post-frontal, high θ_w are denoted by the frontal symbols, bold line, and dotted lines, respectively. (b) Same as (a) except for relative humidity (RH) (with respect to liquid water). Regions of RH > 90%, 90% > RH > 70%, and RH < 30% are denoted by three levels of shading. (Adapted after Hobbs *et al.* 1975.)

shows an example of orographic rain induced by the passage of an occluded front passing over the Cascade Mountains in western Washington State of the USA during March 15–16, 1973. The front is parallel to the north–south oriented Cascades. The occluded front is located at the leading edge of a strong baroclinic zone which extends from upper troposphere downward into the surface trough. The air just below the front is characterized by cold advection and the air just above is characterized by nearly neutral or warm advection. From Fig. 11.8a, the cold air mass behind the occluded front is bounded by a zone of a strong wet-bulb potential temperature (θ_w) gradient. In Fig. 11.8b the occluded frontal surface coincides with a maximum of relative humidity. One important feature of the flow associated with the frontal passage over the mountain is that there is a region of low-θ_w air located aloft immediately ahead of the occluded front. The layer below this region of low-θ_w air is potentially (convectively) unstable since $\partial\theta_w/\partial z < 0$. Note that $\partial\theta_w/\partial z$ is approximately proportional to $\partial\theta_e/\partial z$, and thus both θ_w and θ_e have been alternatively used to identify the potential instability of a moist air. With sufficient moisture content (Fig. 11.8b), potential instability might occur. The air in a pre-frontal surge has been proposed to originate within the cold air mass and breaks away during a frontal evolutionary process (Hobbs *et al.* 1975). The aforementioned process may be caused by the differential advection of air behind the cold air and is essential in producing heavy orographic rainfall associated with the passage of a front over a mesoscale mountain (Smith 1982), as to be further discussed in Subsection 11.3.2.

Figure 11.9 shows a conceptual model of orographic rain triggered by the LLJ associated with a cold front passing over a perpendicular mountain range. It is based on observations over the California Coastal Range and Sierra Nevada mountains, including the mean characteristics of the vertical profiles of wind speed, stratification, and horizontal water vapor transport far upstream of the orography. The prefrontal LLJ is narrow, warm and almost moist neutral, which is very effective in transporting the water vapor toward the California coast. The narrow, almost moist neutral region of strong water vapor transport is also referred to as *atmospheric river*. The moist neutral LLJ often produces heavy rainfall on the upslope of the Coastal Range and Sierra Nevada.

11.2.3 Passage of tropical cyclones

When a tropical cyclone approaches a mountain range, precipitation may be enhanced by the orographic forcing on the mesoscale convective system embedded in the cyclone. In addition, the circulation associated with the cyclone may also interact with the mountain and induce extra precipitation. One example is the heavy rainfall event which occurred on August 29–30, 1974 over southeastern Kyushu, Japan (Figs. 11.10a and 11.10b). The heavy rainfall was induced by the Kyushu mountain range affecting a strong southeasterly LLJ associated with an approaching tropical depression. The major mountain range of Kyushu (the southernmost island of Japan), i.e. the Kyushu Sanchi, is oriented approximately parallel to the coast, roughly from

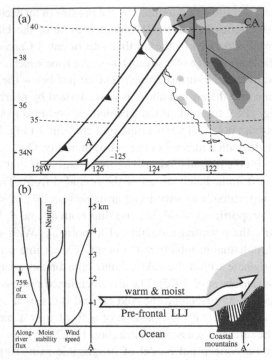

Fig. 11.9 A conceptual model of orographic rain triggered by the LLJ associated with a cold front and a land-falling extratropical cyclone over a perpendicular mesoscale mountain range. This is based on the observations over the California Coastal Range and Sierra Nevada mountains (shaded with 3 height levels). (a) Plan view schematic showing the relative positions of an LLJ and trailing polar cold front. (b) Schematic along the pre-frontal LLJ (i.e., along AA′ in (a)). The vertical structures of wind speed, moist static stability, and along-atmospheric-river moisture flux far upstream of the mountain ranges are depicted. Schematics of orographic clouds and precipitation are also shown. (Adapted after Ralph *et al.* 2005.)

north-northeast to south-southwest (Fig. 11.10a). The highest peak of the mountain range is higher than 1500 m. At 1200 UTC August 29, 1974, a tropical depression approached Kyushu from the southeast. The major low-level synoptic features at this time are shown on the 850 hPa map (Fig. 11.10b). An extratropical cyclone was located to the south of the Korean Peninsula. A cold front extended roughly southward from the cyclone, while a warm front extended east-northeastward from this cyclone over the Japan (East) Sea. The cyclone-front system was quasi-stationary during the heavy rainfall period, and was important in producing a long-lasting precipitation over the Kyushu mountain range. There were no other significant synoptic features in the upper troposphere at the time.

At 0000 UTC August 30 a cluster of small clouds started to form along the Pacific coast of the Japanese Islands even though the tropical depression was still located several hundred kilometers off the coast. The circulation of the tropical depression

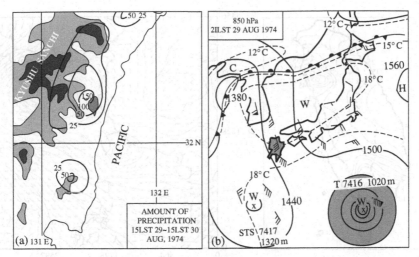

Fig. 11.10 Orographic rain associated with the passage of a tropical depression over the mountains in Kyushu island (both the island and the tropical depression are shaded in (b)). (a) Daily amount of precipitation (in mm) in the vicinity of the Kyushu Sanchi during the event (terrain higher than 500 m and 1000 m are light and dark shaded, respectively). (b) Synoptic situation at 1200 UTC (21LST) August 29, 1974 depicted by 850 hPa map around Japan Islands (bold curves). The solid and dashed lines represent geopotential height and temperature contours, respectively. (After Sakakibara 1979.)

helped produce heavy orographic rainfall on the upslope of the Kyushu Mountains. The echoes observed by the Tanegashima radar from 0400 UTC August 29 to 0600 UTC August 30, 1974 were restricted from the windward (southeastern) slope of the mountain range out into the ocean, approximately 150 km off the coast of Kyushu Island. The maximum intensity was located over the windward slope and the intensity of echoes sharply decreased near the crest of the mountain range. The heavy rainfall was induced by the mountains at this particular time and was not associated with the convection that had already existed within the tropical depression. The maximum daily rainfall reached an amount greater than 150 mm. The upslope convection was triggered by orographic lifting of the LLJ impinging on the Kyushu mountain range. The southeasterly low-level wind was enhanced by the cyclonic circulation associated with the major tropical depression, which approached Kyushu Island, and resulted in a LLJ upstream of the Kyushu mountains (Fig. 11.10b). Note that this LLJ was oriented almost perpendicular to the Kyushu mountains, which helped produce strong upward motion on the windward slope. The upstream air of the Kyushu mountains had high CAPE values as well as a high water vapor mixing ratio, with measured CAPE at an upstream station at 1200 UTC August 28 was 1149 J kg^{-1}.

Another example of enhanced orographic precipitation due to a tropical system is the approach of Tropical Storm (TS) Rachel to Taiwan at 1200 UTC August 6, 1999

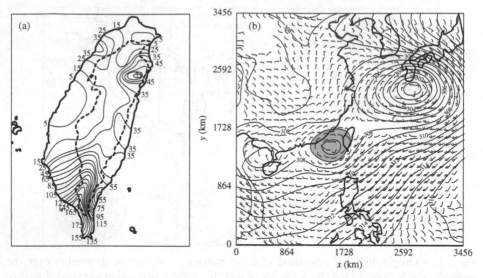

Fig. 11.11 Heavy orographic rainfall produced by the passage of a tropical storm (shaded in (b)) over the Central Mountain Range (CMR), Taiwan. The dashed contour denotes the terrain height of 1 km. (a) Accumulated rainfall (in mm) over Taiwan from 0000 UTC August 6 to 0000 UTC August 7, 1999. (b) NCEP reanalysis data over East Asia valid at 1200 UTC August 6, 1999 for vector wind (1 full bar = 5 m s^{-1}) and geopotential height at 700 hPa. (Adapted after Lin *et al.* 2001.)

from the southwest passing over the Taiwan Strait, where it produced heavy rainfall on the southwest coast of the country during the early morning of August 7 (Fig. 11.11). The CMR of Taiwan has a length of about 300 km, a width of about 100 km, an average height of about 2.5 km, orientation in a north-northeast to south-southwest direction, and, with a highest peak about 4 km. The CMR provides an almost ideal environment for studying orographic influence on precipitation caused by the passage of tropical cyclones due to the isolated nature of the mountains, since they are located in an island surrounded by open ocean and in the path of many typhoons.

The maximum accumulated rainfall in southwest Taiwan reached 347 mm for a two-day period starting from 0000 UTC August 6. At 1200 UTC August 6, the mesoscale convective system associated with the tropical storm covered a circular area with a diameter comparable to the meridional (north–south) extent of Taiwan (about 400 km). Some scattered convective clouds had already erupted before 0600 UTC August 6 over the CMR and also to the southwest of Taiwan. Since TS Rachel was located so far away at this time and the convection associated with it was strongest on its southwest side, the convection was induced by the orographic lifting of the southwesterly conditionally unstable airstream associated with Rachel's circulation. The low-level southwesterly flow was strengthened by the approaching tropical storm. In other words, a significant amount of heavy rainfall was not directly associated with the convection embedded within the tropical storm, but was instead triggered by the interaction of the storm's circulation and the mountains. At 1800 UTC August 6, the convection located over the southwest coast of Taiwan increased significantly and produced rainfall in that region. Note that the tropical storm was still about 500 km to the southwest of Taiwan at this time. By 0000 UTC August 7, the whole island of Taiwan was covered by this convective system and the 24-h accumulated maximum rainfall had reached about 160 mm (Fig. 11.11a).

At 1200 UTC August 6, 1999, the low-level south-southwesterly flow turned more southwesterly, was strengthened by TS Rachel, and impinged more perpendicularly on the CMR, which may have helped to induce conditional instability by lifting the air to its LFC or potential instability by layer lifting. The moist (conditional or potential) instability, in turn, might have induced convection over the southwestern coast of the island, resulting in heavy rainfall. The effect of orographic lifting of the low-level air is evident in the upstream sounding in southwest Taiwan, which gives a LCL of 0.2 km and a LFC lower than 0.6 km. Near the threat area, the CMR has a height of about 2 km. With an LLJ of about $10 \, \mathrm{m \, s^{-1}}$, the low-level flow should have enough kinetic energy to ascend to the LFC, which is a necessary condition for the conditional instability to occur. In addition to the orographic lifting, the concave geometry of the CMR near the southwest of Taiwan (see Fig. 11.2b) appears to also generate confluent flow and enhance the upward motion, as is often observed in other cases of heavy orographic rain. The TS Rachel case is similar to the convection that often starts in the concave region south of Alps, such as the Lago Maggiore in Italy, and the Ticino region in Switzerland (Section 11.2.1). The upstream flow has a very high CAPE ($2099 \, \mathrm{J \, kg^{-1}}$) and is potentially unstable ($\partial \theta_e / \partial z < 0$). Thus, it is evident that the tropical storm played a role in enhancing the southwesterly LLJ by transporting the conditionally and/or potentially unstable air to the threat area, which allowed the mountains to force the air upward and possibly release the instabilities. In addition to the orographic forcing, Typhoon Paul, which was located to the south of Japan (Fig. 11.11b), had slowed down the movement of TS Rachel. As will be discussed in Subsection 11.2.4, this helped produce more rainfall.

A very similar situation occurred in the evening of August 7, 1959 when a tropical depression approached Taiwan from the southwest, resulting in severe orographic rainfall and flash flooding over southern and central Taiwan. Concurrently, Typhoon Ellen was located to the north-northeast of Taiwan and south of Korea, thus mimicking the TS Rachel–Typhoon Paul situation described above. Similar synoptic and meso-scale features have also been observed for heavy orographic rain occurring over northern New Zealand when Tropical Cyclone Bola moved from north toward Northern Island on March 6–7, 1988 (Sinclair 1993). At approximately 0000 UTC March 7, Bola's southward movement was slowed down by an intensifying high-pressure ridge over the Southern Island. Thus, the high-pressure ridge over the southern island is seen to play the same role as the quasi-stationary front in Japan's case and the quasi-stationary tropical cyclones in Taiwan's cases.

11.2.4 Common ingredients of orographic precipitation

Based on the above discussion, the common synoptic and mesoscale features for producing heavy orographic rainfall associated with the passage of a weather system, such as a steep mountain, a LLJ, high θ_e, high CAPE, and a quasi-stationary synoptic system (e.g. a typhoon, a high pressure ridge, or a stationary front). Existence of these environments may be explained by taking a common ingredient approach as follows.

The total precipitation (P) is determined by the average rainfall rate (R) and the duration (D),

$$P = RD. \tag{11.2.1}$$

The rainfall rate can then be determined by the *precipitation efficiency*, E, and the *vertical moisture flux*, wq_v,

$$R = E(\rho/\rho_w)(wq_v), \tag{11.2.2}$$

where ρ_w is the liquid water density, ρ and q_v are the lower moist layer average air density and the water vapor mixing ratio. Substituting (11.2.2) into (11.2.1) leads to

$$P = E(\rho/\rho_w)(wq_v)D. \tag{11.2.3}$$

The duration of heavy rainfall may be estimated by

$$D = L_s/c_s, \tag{11.2.4}$$

where L_s and c_s are the horizontal scale of the convective system and its propagation speed in the direction of the system movement, respectively. Combining (11.2.3) and (11.2.4) leads to

$$P = E(\rho/\rho_w)(wq_v)(L_s/c_s). \tag{11.2.5}$$

Equation (11.2.5) implies that when the value of any combined factors on the right-hand side is large, there is a potential for heavy rainfall. Based on (11.2.5), a set of essential ingredients for flash floods over a flat surface have been proposed (Doswell *et al.* 1998). For flow over a mountain range, the low-level upward vertical motion is induced by either orography or the environment (e.g., upper-tropospheric divergence),

$$w = w_{oro} + w_{env} \tag{11.2.6}$$

The orographically forced upward vertical motion may be roughly estimated by the lower boundary condition for flow over mountains, which is done by extending (5.1.10) to three dimensions:

$$w_{oro} = \frac{Dh}{Dt} \approx V_H \cdot \nabla h, \tag{11.2.7}$$

where $h(x, y)$ is the mountain height as a function of x and y, and V_H is the low-level horizontal wind velocity. Similar forms of (11.2.7) have been proposed in different studies (e.g., Alpert 1986; Sinclair 1994; Lin *et al.* 2001). The environmentally forced upward vertical motion (w_{env}) can be determined by the transient synoptic setting, such as the divergence associated with an approaching trough or conditional instability. Combining (11.2.5)–(11.2.7) gives

$$P = E(\rho/\rho_w)(V_H \cdot \nabla h + w_{env})q_v(L_s/c_s). \tag{11.2.8}$$

Besides the ingredients included in (11.2.8), as discussed in Subsection 11.2.3, a conditionally or potentially unstable airstream may also help trigger deep convection by parcel or layer lifting forced by orography. Based on these arguments, the extremely heavy orographic rainfall requires significant contributions from any combination of the following common ingredients:

(1) high precipitation efficiency of the incoming airstream (large E),
(2) a low-level jet (large $|V_H|$),
(3) steep orography (large ∇h),
(4) favorable (e.g., concave) mountain geometry and a confluent flow field (large $V_H \cdot \nabla h$),
(5) strong synoptically forced upward vertical motion (large w_{env}),
(6) a moist unstable low-level flow (large w_{env}),
(7) a high moisture flow upstream (large q_v),
(8) presence of a large, preexisting convective system (large L_s), and
(9) slow or impeded movement of the convective system (small c_s).

The precipitation efficiency may be controlled by several different factors, such as the advection of hydrometeors, evaporation associated with rainfall below the cloud base, entrainment rate of environmental air into the cloud, and environment wind shear (e.g., Fankhauser 1988). The precipitation efficiency has been noted to be typically an unimportant issue unless there is reason to believe that it will be unusually low, as might be the case for high-altitude convection over much of the interior US Rockies (Doswell *et al.* 1996). However, in addition to the total rainfall amount, the location of the maximum rainfall is also highly related to the precipitation efficiency and timescale of hydrometeor generation, and to the downstream advection of those hydrometeors (Sawyer 1956). The moisture content may be represented either by the water vapor mixing ratio or by the amount of *precipitable water* in a vertical air column. The second ingredient, i.e. the existence of an LLJ, has been observed in many heavy orographic precipitation events around the world. The above common ingredients appear to be consistent with the factors responsible for enhancing precipitation in the seeder–feeder mechanism, to be discussed in Section 11.3.

Large CAPE is not consistently observed for the US and Alpine historical events of extremely heavy orographic precipitation. This may be due to either that the instability of the upstream airstream belonged to other types, such as the potential instability, or to the highly inherent temporal and spatial variations typically associated with the data used. For example, when a sounding farther upstream of the Alpine southern slopes, such as at Cagliari on Sardinia is used (Fig. 11.14), the CAPE is much larger than those observed near the Alps, such as at Milan. As discussed earlier, the Alpine and Taiwan heavy orographic rainfall events often start in a concave region since the strong vertical motion can be produced if the incoming flow is confluent as indicated by the fourth ingredient. Similarly, the common synoptic and mesoscale conditions found in the Taiwan and Japan cases can be explained by the common ingredient argument. Note that the nearby typhoons or fronts play the role of quasi-stationary synoptic systems in the Taiwan and Japan cases, respectively, to slow the movement of convective systems over the mountains. In both cases, the incoming airstream was maritime in nature, and able to obtain a high content of moisture and high values of CAPE. The moist and unstable incoming airstream noted in these cases may also explain why normally no mid-tropospheric trough is needed to enhance the low-level upward vertical motion, such as that observed in the US cases.

The above common ingredients argument has also been applied to explain events that occurred in other regions around the world, such as that on the South Island in New Zealand on December 27, 1989, the Sichuan flood in China from 11 to 15 July 1981, and monsoon rainfall on the Western Ghats of India and the Eastern Arabian Sea from 23 to 25 June 1979 (e.g., Lin *et al.* 2001). In the Indian heavy rainfall case, in addition to the importance of the interaction between latent heating and the basic flow, it is found that the presence of a strong low-level westerly jet, sheared environment, and latent and sensible heat fluxes from the ocean are essential (Ogura and Yoshizaki 1988). For this particular case, the wind reversal level or critical level appears to force the convective system to become stationary. The stationary nature of the convective system helps maximize the vertical motion generated by the latent heating in the vicinity of the critical level and to anchor the convective system to the upstream side of the Western Ghats.

Note that the vertical velocity is normally not a linear superposition of environmentally and orographically forced vertical motions, since they interact with each other nonlinearly. For example, w_{env} associated with the release of conditional instability requires a finite-amplitude orographically induced upward motion (w_{oro}). However, the partitioning of w into w_{oro} and w_{env} helps recognize the dynamics and the prediction of the precipitation distribution directly initiated by the orographic lifting, instead of the prediction of the amount of orographic precipitation.

Part of the two-dimensional form of (11.2.8), $U(\partial h/\partial x)q$ (where V_H reduces to U), can be used as an index for the prediction of upslope heavy orographic precipitation if the negative values are neglected (Alpert 1986). The index has been applied to some of the historical events of heavy orographic precipitation discussed earlier, and can help to provide additionally valuable information for the prediction of the occurrence of upstream heavy orographic precipitation events. In fact, the three-dimensional form, $(V_H \cdot \nabla h)q_v$, representing the *orographic moisture flux*, can be used to help predict the precipitation distribution (e.g., Witcraft *et al.* 2005), which is especially useful for climatological and hydrological applications. If $V_H \cdot \nabla h$ is replaced by the actual w, such as simulated or predicted w from a numerical model, then it represents the *general moisture flux*. Note that the above common ingredients are for extremely heavy orographic precipitation events only. For example, the combination of small hills (50–300 m) and differential friction of air across a coastline can combine to produce 30–40% enhancement of precipitation in a pre-existing extratropical cyclone (Colle and Yuter 2007).

11.3 Formation and enhancement mechanisms

Formation and enhancement mechanisms of orographic precipitation are extremely complicated and are highly dependent on the dynamical and thermodynamic conditions of the upstream airflow, interactions between different layers of clouds, interactions between dynamical and cloud microphysical processes, and mountain geometry. In the past, several major formation and enhancement mechanisms have been proposed: (1) stable ascent, (2) release of moist instabilities, (3) effects of mountain geometry, (4) combined thermal and orographic forcing, (5) seeder–feeder mechanism, and (6) interaction of dynamical and microphysical processes. These mechanisms are sketched in Fig. 11.12 and will be discussed in the following.

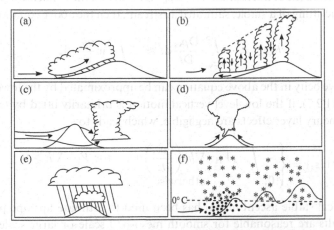

Fig. 11.12 Conceptual models of the formation and enhancement mechanisms of orographic precipitation: (a) stable ascent, (b) release of moist instabilities, (c) effects of mountain geometry, (d) combined thermal and orographic forcing, (e) seeder–feeder mechanism, and (f) interaction of dynamical and thermodynamic processes. (Adapted after Smith 1979; Banta 1990; Houze 1993; Medina and Houze 2003.)

11.3.1 Stable ascent mechanism

The earliest and simplest mechanism proposed to explain the formation of orographic rain is the *stable ascent mechanism*, which assumes the precipitation is produced by the forced ascent of a stable, moist airstream along the upslope of a mountain. Examples have been given in introductory textbooks and the mechanism has been applied to discussions of more complex precipitation systems produced by the passage of frontal cyclones in earlier literature which may not necessarily be correct. One approach to estimate the precipitation produced by the stable ascent mechanism is to adopt the ingredient approach discussed in Subsection 11.2.4 by slightly modifying (11.2.8) to

$$R = E(\rho/\rho_{\rm w})(V_{\rm H} \cdot \nabla h)q_{\rm v}. \tag{11.3.1}$$

One of the major problems of this type of model is that the vertical distribution of the water vapor mixing ratio is not taken into account. To improve the model described by (11.3.1), one may estimate the rainfall rate by vertically integrating the production rates of water vapor and liquid (cloud) water from the surface ($z_{\rm s}$) to the top of the atmosphere (Smith 1979):

$$R(z_{\rm s}) = - \int_{z_{\rm s}}^{\infty} \left(\frac{{\rm D}\rho_{\rm vs}}{{\rm D}t} + \frac{{\rm D}\rho_{\rm c}}{{\rm D}t} \right) {\rm d}z, \tag{11.3.2}$$

where $\rho_{\rm vs}$ is the saturation water vapor density and $\rho_{\rm c}$ is the cloud water density, assuming that raindrops fall directly to the ground without horizontal drifting or advection. The rainfall rate is measured in $\text{kg m}^{-2}\text{s}^{-1}$. Assuming raindrops form immediately from the condensation (i.e., ${\rm D}\rho_{\rm c}/{\rm D}t = 0$) and the condensation is triggered by orographic lifting of a stable, saturated airstream, then the above equation reduces to

$$R(z_{\rm s}) = - \int_{z_{\rm s}}^{\infty} \frac{{\rm D}\rho_{\rm vs}}{{\rm D}t} {\rm d}z \approx - \int_{z_{\rm s}}^{\infty} w \frac{\partial \rho_{\rm vs}}{\partial z} {\rm d}z. \tag{11.3.3}$$

The vertical velocity in the above equation can be approximated by the lower boundary condition, (11.2.7), if the low-level vertical motion is primarily lifted by the mountain and the boundary layer effects are negligible, which leads to:

$$R(x, y, z_{\rm s}) = \begin{cases} - \int_{z_{\rm s}}^{\infty} (V_{\rm H} \cdot \nabla h) \left(\frac{\partial \rho_{\rm vs}}{\partial z} \right) {\rm d}z, & \text{for } V_{\rm H} \cdot \nabla h \geq 0 \\ 0, & \text{otherwise} \end{cases} \tag{11.3.4}$$

The theoretical stable ascent model has been used to estimate upslope precipitation and the results are reasonable for smooth meso-α/β scale or large-scale mountains (Fig. 11.12a) and less satisfying for meso-γ scale hilly terrain.

One major deficiency of the model described by (11.3.4) is that the effect of wind drift of the raindrops is not included. The wind drift or advection of raindrops over a hill is influenced by two factors: (a) the horizontal drift of a raindrop of radius r

falling at a terminal velocity V_r through a cloud of depth h_c by a horizontal wind of speed U, $L_{\text{drift}} \approx U h_c / V_r$, and (b) the scale of horizontal variation of liquid-water content, $L_c \approx q_c / (\partial q_c / \partial x) \approx a$, where q_c is the liquid-water content, a is the half-width of the hill (Carruthers and Choularton 1983). If $a \leq L_{\text{drift}}$, such as over a narrow hill, wind drift will modify the distribution of orographic precipitation and shift the maximum to the downslope.

An alternative way to describe these effects is to compare different time scales involved in the cloud and precipitation formation processes, such as the advection time scale (τ_{advec}), raindrop falling time scale (τ_f), and cloud growth time scale (τ_c) (Jiang and Smith 2003; Chen and Lin 2005a). The advection time scale can be approximately estimated by a/U and the raindrop falling time can be estimated by h_c / V_r, where h_c is the depth of the cloud layer, and V_r is the averaged terminal velocity of raindrops of radius r. However, the cloud growth scale is controlled by the micro-physical process, which is rather difficult to estimate, unless using a more complicated numerical model with cloud microphysical processes explicitly simulated at the grid scale or parameterized. In a moist unstable atmosphere, the additional time scale related to the vertical motion forced by instability may come into play and make the estimate of orographic precipitation more complicated. The second major problem of the model of (11.3.4) is the lack of precipitation over the downslope. If no downslope evaporation is allowed, the model-estimated precipitation may exceed the incoming moisture flux when the moist airstream experiences multiple uplifts in flowing over a hilly terrain. The downslope evaporation can be incorporated approximately into the model by truncating negative precipitation values (Smith and Barstad 2004).

Another major deficiency of the above model is that the wave motion generated by the mountain is not included. The stable ascent model assumes the vertical motion generated by the mountain is independent of height, which is apparently not true, based on the orographic forced flow discussed in Chapter 5. One way to incorporate the orographically forced motion into the model is to represent the moist effects in the dry mountain wave theory, as partly discussed in Chapter 6. The most straightforward approach to treat the moisture effects is to assume that the diabatic heating is every-where proportional to the vertical velocity in a linear flow. The assumption implies that latent heat is released wherever there is upward motion and evaporative cooling is realized wherever there is downward motion; in other words, there is no precipitation. The assumption can be understood by considering the forced Taylor–Goldstein equa-tion for a two-dimensional, small-amplitude, inviscid, nonhydrostatic, nonrotating, Boussinesq fluid flow, similar to (6.2.1):

$$\left(\frac{\partial}{\partial t} + U \frac{\partial}{\partial x} \right)^2 \left(\frac{\partial^2 w'}{\partial x^2} + \frac{\partial^2 w'}{\partial z^2} \right) + N^2 \frac{\partial^2 w'}{\partial x^2} = \frac{\partial^2 \dot{Q}}{\partial x^2}, \tag{11.3.5}$$

where $\dot{Q} = (g/c_p T_0) \dot{q}$. If N is uniform far upstream and the latent heating is propor-tional to w' everywhere, $\dot{Q}$ may be written as

$$\dot{Q} = \varepsilon N^2 w'. \tag{11.3.6}$$

Substituting the above equation into (11.3.5) leads to a homogeneous differential equation (Barcilon *et al.* 1979):

$$\left(\frac{\partial}{\partial t} + U\frac{\partial}{\partial x}\right)^2 \left(\frac{\partial^2 w'}{\partial x^2} + \frac{\partial^2 w'}{\partial z^2}\right) + (1 - \varepsilon)N^2\frac{\partial^2 w'}{\partial x^2} = 0. \tag{11.3.7}$$

In (11.3.7), when the air parcel is outside the cloud, ε is 0; otherwise $\varepsilon < 1$ in a stable atmosphere and $\varepsilon > 1$ in an unstable atmosphere. Inside the cloud, the Brunt–Vaisala frequency becomes $N_m^2 = (1 - \varepsilon)N^2$. The mathematical problem found in determining the Brunt–Vaisala frequency inside the cloud (N_m) and in clear regions (N) of an orographic cloud system from (11.3.7) is highly nonlinear. Equation (11.3.7) implies that the heating effect is simply to modify the buoyancy frequency for a stable, nonprecipitating atmosphere. The particular treatment of $\dot{Q}$ in terms of w' requires that the heat added to an air parcel lag the vertical displacement by a quarter cycle (Smith and Lin 1982). Therefore, it follows that the buoyancy forces can do no work and therefore the heat can only modify, but not generate, internal gravity waves.

A similar approach, but assuming that the middle (cloud) layer of a steady-state, three-layer atmosphere is moist adiabatic and the basic wind has vertical shear, has been proposed and solved analytically for moist airflow over the Western Ghats of India. The rainfall intensity and distribution estimated by this dynamical model is in good agreement with the observed rainfall on the windward side, assuming precipitation rate is equal to the rate of condensation (Sarker 1966). However, they do not compare well on the lee side. A similar study found that the release of latent heat lowers the stability in saturated regions of the flow, and that the dynamical effects of latent heat are significant in some cases, but are generally secondary to the barrier effect of the terrain (Fraser *et al.* 1973). Adopting a similar approach, but in a nonlinear numerical model for simulating mountain lee waves, it is found that resonant waves in an absolutely stable environment are distorted and untrapped when the moisture is increased (Durran and Klemp 1982). Assuming the total water content is conserved, a nonlinear mathematical problem can be formulated, in which the cloud and dry regions are identified analytically by iteration method (Barcilon and Fitzjarrald 1985). The switching between cloud and dry regions is highly nonlinear mathematically, and it is found that the effect of moisture is to reduce the drag on the mountain. The precipitation is allowed to occur in this type of dynamical model by cutting off the evaporative cooling in regions of downward motion.

Some important dynamical processes have been developed by the above-mentioned theoretical studies, which have provided significant improvements to the parameterizations and representations of physical processes in numerical models. However, almost all theoretical models have some constraints, such as small-amplitude perturbations, overly simplified representations of moist processes, and simple upstream environments. Most deficiencies of the above-mentioned dynamical models can be overcome by performing numerical model simulations. However, further improvements in numerical

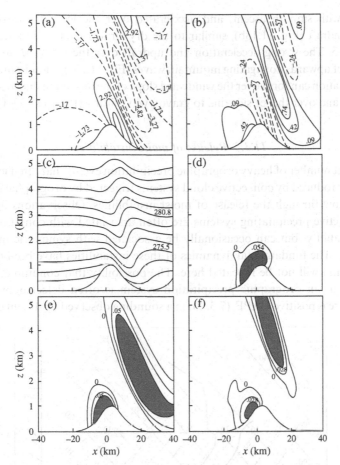

Fig. 11.13 An idealized, three-dimensional stratified moist flow over a Gaussian mountain, $h(x, y) = h_m \exp[-(x^2 + y^2)/2]$, as simulated by a mesoscale model (ARPS). The incoming flow is nearly saturated ($RH = 95\%$) and the simulation time is 5 h. Displayed are the vertical cross-sections across the center line $y = 0$ km. The upstream atmospheric conditions and orographic parameters are: $N = 0.011 \text{ s}^{-1}$, $U = 10 \text{ m s}^{-1}$, $T_o = 270 \text{ K}$, $a = 10$ km, and $h_m = 1$ km. Only part of the model domain is plotted. (a) Streamwise component of perturbation velocity (u', m s^{-1}); (b) vertical velocity (w', m s^{-1}); (c) potential temperature (K); (d) cloud water mixing ratio (g kg^{-1}); (e) snow mixing ratio (g kg^{-1}); and (f) cloud ice mixing ratio (g kg^{-1}) are plotted respectively. (Adapted after Jiang 2003.)

models are still required. In particular, the cumulus parameterizations with relatively coarser grid resolution and grid-explicit microphysical parameterizations for relatively finer grid resolution should be considered (Chapter 14). Figure 11.13 shows an example of an idealized, three-dimensional stratified moist flow over a Gaussian mountain, simulated by a mesoscale model. The incoming flow is nearly saturated with 95% relative humidity. The flow is decelerated over the windward (left) slope (Fig. 11.13a) associated with smooth ascent, and accelerated over the lee slope associated with strong downdraft (Fig. 11.13b), similar to the dry airflow over a mountain discussed in Chapter 5. The strong deceleration and updraft over the lee slope are due to the steepening of upward propagating mountain waves (Fig. 11.13c). Cloud water, cloud ice, and precipitation can form over the windward slope of the mountain due to stable ascent mechanism and over the lee side due to gravity wave lifting (Figs. 11.13d–f).

11.3.2 Release of moist instabilities

A significant number of heavy orographic precipitation events that often leads to flash flooding is produced by convective cloud systems triggered by orographic lifting on an incoming flow through the release of moist instabilities. Observations indicate that these convective precipitating systems are often associated with conditional or convective instability, but can occasionally be associated with symmetric instability or wave-CISK. The fundamental dynamics of these instabilities have been discussed in Chapter 7 and will not be repeated here. The possibility that conditional instability may occur in an orographic precipitation system is often detected by inspecting whether there is positive $CAPE_i$ (7.3.26) in a sounding observed upstream of the threat

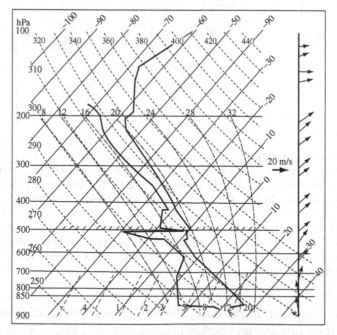

Fig. 11.14 Temperature and dewpoint profiles of a sounding observed at Cagliari, Sardinia, at 0000 UTC September 20, 1999 during MAP IOP-2B. The maximum CAPE calculated from the surface is $2692 \, \mathrm{J \, kg^{-1}}$ and the CIN is $119 \, \mathrm{J \, kg^{-1}}$. This conditionally unstable atmosphere far upstream of Alps is considered to be responsible for triggering conditional instability and leading to heavy precipitation over the southern slopes of the Alps. (Adapted after Asencio *et al.* 2003.)

area. The existence of conditional instability can also be detected by examining $\partial \theta_e^* / \partial z < 0$ (7.3.22), provided the air parcel can be lifted to its LFC. The former method is highly dependent on the level the air parcel is assumed to be lifted to. Note that these conditions are only necessary conditions, instead of sufficient conditions, for the conditional instability to occur. In other words, neither $\text{CAPE}_i > 0$ nor $\partial \theta_e^* / \partial z < 0$ can guarantee conditional instability to occur. As discussed in Section 11.2, high CAPE values have been observed in upstream soundings of extremely heavy orographic rainfall events, and may thus serve as a common ingredient.

Figure 11.14 shows an example of observed sounding far upstream of the Alps at Cagliari, Sardinia, Italy, at 0000 UTC September 20, 1999 during the MAP IOP-2B. The air associated with this sounding has a large CAPE of $2692 \, \text{J} \, \text{kg}^{-1}$, thus making the conditional instability able to trigger convective cloud systems and produce heavy rainfall on the southern slopes of the Alps during September 19–20, 1999 (Asencio *et al.* 2003). The upstream air has a relatively large CIN of $119 \, \text{J} \, \text{kg}^{-1}$, which will require strong lifting to overcome. Due to the strong low-level southerly flow (15 to $20 \, \text{m} \, \text{s}^{-1}$ in the 900 to 700 hPa layer) from the Mediterranean Sea over the steep slope of the Alps, the orographic lifting is considered to be able to overcome the CIN and lift the near-surface air parcel to its LFC and release the conditional instability. As discussed earlier in the ingredient argument (Subsection 11.2.4), some upstream soundings of historical extremely heavy orographic precipitation events in western Europe and the USA do not possess high CAPE. However, as shown in Fig. 11.14, CAPE is much larger if the sounding of a station is located further upstream.

Observations indicate that $\partial \theta_e / \partial z < 0$, a necessary condition for convective (potential) instability, exists in many upstream soundings of heavy precipitation events. Figure 11.15 shows an example of the possible development of convective cloud systems by potential (convective) instability for a heavy orographic precipitation event. The numerically simulated equivalent potential temperature field at 0000 UTC September 20, 1999 during MAP IOP-2B indicates a region of $\partial \theta_e / \partial z < 0$ just above the upslope of the first crest of the Alps along the cross section (Fig. 11.15a). The region of decreasing equivalent potential temperature with height is collocated with the region of heavy precipitation and large model-derived radar echoes, indicating that the potential instability might be working and has triggered the deep convection (Fig. 11.15b). In the case of orographic rain associated with the passage of a tropical cyclone over a mountain range, the orography should also be able to provide continuous layer lifting for potential instability.

On the other hand, Fig. 11.14 also indicates that the upstream air is conditionally unstable, thus making it difficult to articulate which instability is the actual cause of the moist instability responsible for the development of the convective cloud systems over the southern slopes of the Alps. A similar situation with both regions of $\partial \theta_e^* / \partial z < 0$ and $\partial \theta_e / \partial z < 0$ may occur within the outer circulation associated with tropical cyclones, which may induce heavy orographic rainfall as the tropical cyclone impinges on a mesoscale mountain range (e.g., Witcraft *et al.* 2005). As pointed out in Subsection 7.3.3, a layer that is conditionally unstable need not be potentially

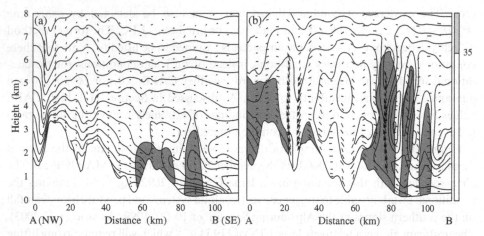

Fig. 11.15 Possible development of convective cloud systems by potential (convective) instability for a heavy orographic precipitation event across the Alps. Numerically simulated θ_e (solid every 1 K), model-derived reflectivity (shaded), and the winds along the cross section AB of Fig. 11.6c are shown in (a) 1800 UTC September 19 and (b) 1200 UTC September 20, 1999. The basic wind is from the southeast (SE). The grid resolution used by the mesoscale model (MM5) is 1.67 km. (Adapted after Chiao *et al.* 2004.)

unstable; nor is a potentially unstable layer necessarily conditionally unstable. The embedded convective cells in Fig. 9.5a appear to be the result of layer lifting (potential instability), while the single convective cell in Fig. 9.5b appears to be the result of parcel lifting (conditional instability). A similar situation may occur if the convergence is generated over the upwind slope of a mountain. Due to this, an exact relationship between the conditional instability and potential instability requires further investigation. Note that when the air is saturated, the vertical gradients of $\bar{\theta}_e$ and $\bar{\theta}_e^*$ are equivalent, and therefore conditional instability and potential instability are equivalent.

A moist instability may be detected by analyzing the saturated moist Brunt–Vaisala frequency (N_m) rather than the vertical profile of θ_e, where N_m is defined in (7.3.12). As discussed in Chapter 7, the cloudy air is statically or buoyant stable if $N_m^2 > 0$ and statically or buoyant unstable or *moist absolutely unstable* to an infinitesimal vertical displacement if $N_m^2 < 0$. The criterion of $N_m^2 > 0$ is equivalent to $\partial\theta_e/\partial z > 0$ for the environmental air when the total water mixing ratio (q_w) is uniform with height. A moist buoyant unstable airstream over a mountain may induce cellular convection (Kirshbaum and Durran 2004). Moist buoyant instability may also be released through differential advection of baroclinic airflow by orographic blocking, such as a front passing over a mountain (Smith 1982).

Occasionally extremely heavy orographic precipitation is produced by slantwise convection associated with symmetrically unstable airstream in an area of strong vertical shear (Buzzi and Alberoni 1992). In this situation, the slantwise air motion might be helped by the upslope orographic lifting. Figure 11.16 shows an example of

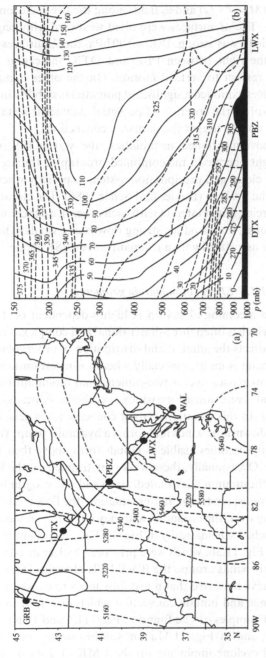

Fig. 11.16 Analysis of conditions for potential symmetric instability (PSI), which might trigger slantwise convection, associated with a cold front crossing over the central Appalachians. (a) The 500-hPa geopotential height contours (m) and locations of the five radiosonde launching stations used for the cross section of (b); (b) Vertical cross section of absolute momentum ($M = v + fx$) (solid, m s^{-1}) and θ_e (dashed, K). Symbols of DTX, PBZ, and LWX denote Detroit, Michigan (DTX), Pittsburgh, Pennsylvania (PBZ), and Sterling, Virginia (LWX), respectively. (After Barros and Kuligowski 1998.)

slantwise convection based on 1200 UTC January 19, 1996 orographic rain and flooding event associated with a cold front passage over the central Appalachians. The absolute momentum ($M = v + fx$) and θ_e fields along the five radiosonde stations are depicted in Fig. 11.16. The M surfaces were tilted westward with height west of LWX, where the baroclinicity was strong. Below 800 hPa, the θ_e surfaces were tilted slightly upward and to the right between PBZ and DTX, indicating only slight potential instability with respect to vertical motion. On the other hand, slantwise motions along the M surfaces exhibited a significant potential instability in this region below 800 hPa. In other words, the criterion of potential symmetric instability (PSI) was met and was possibly able to trigger the slantwise convection.

In addition to the above-mentioned instabilities, the wave-CISK mechanism (see Subsection 4.3.3) might operate in the combined orographic forcing and latent heating in the orographic cloud and precipitation system, especially when both forcings are in phase along the upslope (Davies and Schär 1986). The elevated thermal forcing and orographic forcing might interact with each other in a coherent manner to enhance the upslope response, and produce strong downslope winds. The enhanced upslope response might be associated with a resonant wave-CISK mode.

11.3.3 Effects of mountain geometry

Orographic forcing on an incoming airstream is highly dependent on the detailed geometry of a mountain, which is often three-dimensional and complex. The geometry of the mountain, in turn, affects the amount and distribution of orographic precipitation if the incoming airstream is moist, especially when it is moist unstable. When a low-Froude number airflow passes over a two-dimensional mountain ridge, it may produce wave breaking and severe downslope winds over the lee slope, as discussed in Chapter 5 (e.g., Fig. 5.10). In this situation, strong convergence is generated at the leading edge of the severe downslope winds and forms a hydraulic jump. When the air is moist, the hydraulic jump becomes visible as clouds form above their lifting condensation levels (Fig. 3.5). Occasionally, these waves are trapped on the lee side and rotor clouds may form. If the mountain is isolated, steep and tall, orographic blocking is strong on the impinging moist airstream. In this situation, the flow tends to split and go around, instead of going over, the mountain and may converge on the lee side and trigger convection, as sketched schematically in Fig. 11.12c. One example is the rain-shower maxima shown in Fig. 11.3b, which were produced by leeside convergence of the prevailing southwesterly winds around the Black Hills of South Dakota.

Previous observations have indicated that moist flow tends to converge in concave regions of a mountain range and initiate convection which may lead to heavy orographic rain or flooding. Examples can be found in Figs. 11.5 and 11.6 for southerly moist flow over the Alps and in Fig. 11.11a for southwesterly outer circulations associated with a tropical cyclone impinging on the CMR of Taiwan. In the Alps case, the moist flow could go through regime transition from flow-around the higher

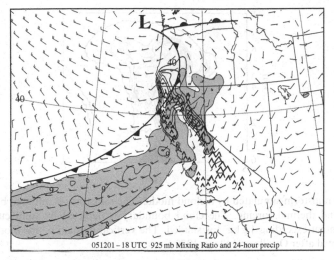

Fig. 11.17 Effects of mountain shape on orographic precipitation over the Coastal Range and Sierra Nevada in California. The average 24-h precipitation (in mm) is distributed along the mountains, leaving a rain shadow in the California basin. The storm tracks rainfall distributions are composited by 12 storms based on the North American Regional Reanalysis (NARR) dataset. The shaded area represents the moist tongue associated with the atmospheric river of water vapor mixing ratio higher than $8\,\mathrm{g\,kg^{-1}}$. (After Smith *et al.* 2006.)

east–west flank to flow-over the lower north–south flank near the concave region. Figure 11.17 shows effects of mountain shape on orographic precipitation over the Coastal Range and Sierra Nevada, California, based on the composites of 12 storm cases. The rainfall is distributed along the mountain ranges, leaving a rain shadow in the California basin.

Complex terrain is often composed by numerous irregular peaks and valleys. In order to make an accurate prediction, grid intervals of a numerical model are required to resolve the detailed topography, such as demonstrated in Fig. 11.15. The increase in model grid intervals demands not only computing power, but also more precise representation of the moist processes. Figure 11.15 also shows a fine-scale, compli-cated distribution of orographic precipitation. Although scales of individual patches of the orographic precipitation vary, they still more or less reflect the scales of the individual mountain peaks and valleys. Examples of small-scale terrain features can be found in the numerical simulations of orographic rainfall that occurred during the MAP IOP-2A (e.g., Richard *et al.* 2003; Gheusi and Stein 2003).

11.3.4 Combined thermal and orographic forcing

During the day, a mountain serves as an elevated heat source due to the sensible heat released by the mountain surface. In a quiescent atmosphere, this may induce upslope winds, which in turn may initiate cumuli or thunderstorms over the mountain peak

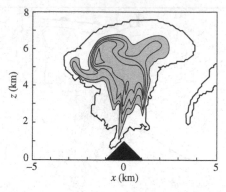

Fig. 11.18 A rainstorm simulated by a cloud model. The rainstorm is initiated by a combined orographic and thermal forcing. The cloud boundary is depicted by solid curves and the rainwater content exceeding $1 \, \text{g kg}^{-1}$ is shaded. The vertical domain of the model extends to 10 km. (Adapted after Orville and Sloan 1970.)

and produce orographic precipitation (Fig. 11.12d). The cumulus could be advected to the lee side of the mountain and produce precipitation over the lee slope if the ambient wind is strong enough. This type of orographic cumuli, as well as that induced by orographic lifting alone, has been observed by photogrammetry (following the stereo-photographic technique), instrumented airplanes in the 1950s, and by radars since 1960s. Understanding of the dynamics of these thermally and orographically forced cumuli and their associated precipitation has been advanced significantly by numerical modeling simulations since 1970s. Figure 11.18 shows an example of an orographic rainstorm simulated by an early cloud model in the 1960s.

Besides sensible heating, cooling over the mountain slope at night may influence the airflow and the formation of clouds and precipitation. For example, it was observed that a northerly down-valley flow originated within the deep valleys and over the lower slopes of the Alps, extended to the Mediterranean Sea and undercut a deep layer of southeasterly flow during MAP IOP-8 (Bousquet and Smull 2003; Steiner *et al.* 2003). As shown in Fig. 11.19, this layer of cold, stable air extends to the upstream (south) of the Ligurian Apennines and serves as an *effective mountain* in triggering convection over the Ligurian Sea (Lin *et al.* 2005; Reeves and Lin 2006). The cellular convection is present over the cold dome and is attributed to potential (convective) instability.

11.3.5 Seeder–feeder mechanism

Over small hills, there is a remarkably strong dependence of rainfall with height for stratus rain, although topography does not seem to have much effect on convective showers. The stable ascent mechanism could not explain the heavy orographic rainfall observed over such small hills, and the phenomenon was then explained by the *seeder–feeder mechanism* (Fig. 11.12e; Bergeron 1949). Based on this mechanism,

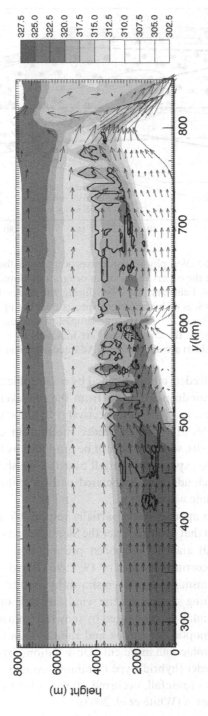

Fig. 11.19 An idealized simulation for a southerly flow over the Ligurian Apennines and Alps during MAP IOP-8. Shown are the equivalent potential temperature (shaded as in legend), wind vectors, and convective cloud areas ($N_m^2 < 0 \text{ s}^{-2}$; enclosed by bold contours) for a simulation with strong surface cooling after 15 hours. The cold, stable layer over the Po Valley and upstream (south) of Apennines helps block the flow and trigger convection at the leading edge of the cold dome over the Ligurian Sea, which is consistent with real-case simulations of Lin et al. (2005). Note that cellular convection is present over the cold dome, which is attributed to potential (convective) instability. (After Reeves and Lin 2006.)

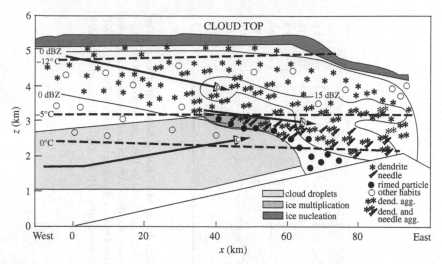

Fig. 11.20 A conceptual model of the microphysical processes occurring in the orographic cloud and precipitation system, based on the observations of an orographic cloud and precipitation system over the Sierra Nevada on February 12, 1986 during the Sierra Cooperative Pilot Project (SCPP). Two intersection zones of hydrometeor species over the upslope can be seen: $(x, z) = (40$–60 km, 3 km) and $(70$–90 km, 2–3 km). Cloud droplet (C), needle (N), and dentritic particle (D) trajectories in the plane of the cross section are indicated by arrows. Radar-echo contours (solid) and isotherms (dashed) are also shown. (Adapted after Rauber 1992.)

cloud droplets in the low-level (feeder) clouds formed by orographic lifting over small hills could be washed out by raindrops or snow from the midlevel (seeder) clouds through coalescence or aggregation process, respectively. In some cases, the feeder clouds may contain ice particles. The seeder–feeder mechanism can significantly enhance the rainfall over small hills, which could not be produced by the feeder clouds alone, which has been applied to explain the rainfall over the Welsh Hills (Browning *et al.* 1974), where the seeder clouds were associated with potentially unstable air during the passage of a wintertime warm sector.

The seeder–feeder mechanism appears to be partially responsible for the enhancement of orographic precipitation that occurred over the Sierra Nevada on February 12, 1986 (Rauber 1992; Fig. 11.20) and the enhanced precipitation just upstream of the Appalachian Mountains occurring on January 19, 1996 (Barros and Kuligowski 1998). The seeder–feeder mechanism has also been shown by vertically pointing radar data of two separated precipitating systems on the windward side of New Zealand's Southern Alps, where shallow rainfall was enhanced by snowfall from aloft (Purdy *et al.* 2005), where analysis of the synoptic conditions and rainfall data showed that this interaction did result in orographic enhancement. Observations by radar and rain gauges indicate that a seeder–feeder (hybrid) type of rainfall, composed by the brightband and non-brightband types of rainfall, occurred often for land-falling jet streaks over the coastal ranges of California (White *et al.* 2003).

Note that the "classic" seeder–feeder mechanism of Bergeron (1949) yields rain over cloud-covered hills that would not otherwise exist without the presence of the seeder particles. However, many cases of orographic precipitation represent enhancements of preexisting disturbances that produce rainfall prior to their interaction with topography. The seeder–feeder process may also combine with other processes that independently yield raindrops at low levels (Colle and Yuter 2007).

11.3.6 Dynamical–microphysical interaction mechanism

Microphysical processes may interact with dynamical processes induced by orography to enhance precipitation in various ways, such as (a) increase of vertical transport of water vapor and hydrometeors; (b) enhancing the accretion processes, such as coalescence between water (rain or cloud water) drops, aggregation between ice particles and riming between ice particles and water droplets, through the reduction of the lifting condensation level over the mountain surface; (c) changes in microphysical pathway in the production of precipitation; (d) interception of hydrometeors by a mountain surface at a higher level than that without a mountain; (e) changes in the horizontal advection time scale of air motion compared to the time scale for microphysical process may change the precipitation distribution over the mountain surface; (f) changes in precipitation efficiency; (g) enhanced evaporation over the downslope with the addition of adiabatic warming; and (h) unsteadiness of small-scale orographic precipitation via turbulent motion. In fact, the seeder–feeder mechanism may also be viewed as a special case of this dynamical–microphysical interaction mechanism.

Figure 11.20 shows a conceptual model of microphysical processes based on observations of an orographic cloud and precipitation system over the Sierra Nevada on February 12, 1987 during the Sierra Cooperative Pilot Project (SCPP). In this example, two intersection zones of hydrometeor species over the upslope during this event can be seen. One region is located over the middle upslope (about $x = 40$ to $60 \, km$) at about $z = 3 \, km$, where large cloud droplets rise from the cloud base and interact with the snow dendrites descending from the cloud top. The cloud droplet and snow dendrite interaction leads to secondary ice production from rime splintering and the formation of graupel. The other region is located near the mountain top (about $x = 70$ to $90 \, km$) in the layer of $z = 2$ to $3 \, km$. In this region, there is enhanced dendritic aggregate production as the ice crystals created by secondary ice production interact with the falling snow crystals. The conceptual model illustrated here demonstrates how the cloud microphysical processes are influenced by the orography.

The interactions between dynamical and microphysical processes become more complicated when they are associated with the passage of a front over orography, as shown in Fig. 11.8. Riming is found to play important roles in the growth of precipitation particles and the enhancement of precipitation amounts in these cases. The riming at low altitudes leads to precipitation hydrometeors falling out more quickly on the windward slope (Hobbs *et al.* 1973) and increased precipitation efficiency. Figure 11.12f shows a conceptual model of an unstable, unblocked air-stream impinging on a mesoscale mountain range which contain several mountain peaks, such as the Alps, in a north–south cross section. In this situation, the basic stratiform structure is enhanced because the lower layer of upstream flow rose up the mountain as a result of its high Froude number, and further because cellular convection is embedded in the stratiform background precipitation, such as happened during the MAP IOP-2 (Medina and Houze 2003). On the other hand, the microphysical–dynamical interactions are very different for a stable, blocked flow, such as the wide-spread stratiform precipitation consisting of dry snow aloft growing by deposition, melting and then falling out as rain, as happened during MAP IOP-8. The conceptual model for this stable, blocked flow is similar to Fig. 11.12f except with an area of blocked air upstream of the mountain.

Orographic precipitation can also be enhanced by turbulence. Turbulent motion can be generated in various ways, such as from cellular overturning embedded in a deep orographic convection (Smith *et al.* 2003; e.g., Fig. 11.15b) or in a shallow shear layer (Houze and Medina 2005). Based on data collected by vertically pointing S-band radar during MAP IOP-2B, rates of collection (below $0\,°C$) and riming (above $0\,°C$) are comparable. Each process makes significant contributions to precipitation growth within small-scale embedded updrafts and both collection and riming are needed to obtain high precipitation efficiencies in orographic precipitation (Yuter and Houze 2003).

In addition to the impacts of dynamical processes on the microphysical processes, some microphysical processes may also feedback and influence dynamical processes. For example, evaporative cooling associated with falling raindrops below the cloud base in a conditionally unstable atmosphere may produce a density current which then generates new convective cells and dictates the propagation of orographic precipitation system, as will be discussed in Section 11.4. Detailed interactions of dynamical and microphysical processes within orographic precipitation systems are complicated, but can be explored through field observations with advanced techniques and systematically designed numerical experiments.

Another example of the feedback of microphysical processes on dynamical processes is the formation of a deep foehn over the Alps. As briefly mentioned in Chapter 5, *foehn* is a special type of severe downslope wind occurring in many different mesoscale mountain ranges in addition to the Alps. When applying severe downslope wind theories to the Alpine foehn winds, one should keep in mind that foehn is usually restricted to north–south oriented valleys originating from a pass in the main Alpine range (e.g., Seibert 1990). Thus, the Alpine foehn winds are normally a byproduct of gap flows, which are driven by a pressure difference that is maintained between the two sides of a mountain range and a downslope wind over a ridge along the valley. Figure 11.21a shows a conceptual model of a *deep foehn* formed by the interactions among upslope clouds via latent heat release, severe downslope wind, down-valley gap flow from the adjacent mountain slopes, and the cold pool on the lee side (Steinacker 2006). The deep foehn often occurs in the Wipp Valley, Austria (e.g., Weissmann *et al.*

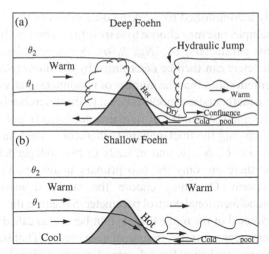

Fig. 11.21 Two types of foehn winds in the Alps: (a) deep foehn and (b) shallow foehn. The deep foehn is formed by the interactions among upslope clouds, severe downslope wind, gap flow, and cold pool. The shallow foehn is formed by the interactions among gravity waves, severe downslope wind, gap flow, and cold pool. ((a) Adapted after Steinacker 2006 and Weissmann *et al.* 2004; (b) Adapted after Steinacker 2006.)

2004). Note that not all the foehn winds require the presence of upslope clouds, such as the schematic shown in Fig. 11.21b. The *shallow foehn* is formed by the interactions among gravity waves, severe downslope winds, gap flows, and cold pools, and this type of foehn wind often occurs in the Rhine Valley, Switzerland (e.g., Zängle *et al.* 2004).

11.4 Control parameters and moist flow regimes

11.4.1 Control parameters

Based on the above discussion of synoptic and mesoscale environments and common ingredients for heavy orographic precipitation, the formation and propagation of an orographic precipitation system in a two-dimensional, approximately moist stable flow appear to be controlled by a number of flow and orographic parameters, such as the basic wind speed (U), saturated moist Brunt–Vaisala frequency (N_m, as defined in (7.3.12)), mountain height (h), mountain width scale (a), Earth rotation (f), and vertical wind shear (U_z). In order to thoroughly understand the complicated dynamics of orographic precipitation, the theoretical approach and/or a systematic approach for performing idealized nonlinear numerical simulations are required. Ideally, it is desirable to choose a set of independent, nondimensional control parameters to avoid varying too many dimensional control parameters. The use of control parameters will also make the theoretical results more general.

Practically, it is very difficult to take into account all of the dimensional parameters controlling the orographic precipitation formation and propagation. However, this can still be partially accomplished by working on a subset of the dimensional control parameters. For example, one may choose to start with a subset of the above-mentioned dimensional control parameters, (U, N_m, h, a). A set of independent, nondimensional control parameters can then be determined by the *Buckingham-Π theorem*. The Buckingham-Π theorem states that the number of nondimensional combinations (parameters) involved in a mathematical model is equal to the difference between the number of original quantities and the number of fundamental dimensional units, such as time, mass, and length. Applying the Buckingham-Π theorem to the four dimensional parameters listed above, i.e. U, N_m, h, and a, leads to two independent nondimensional parameters because there are only the two primary quantities, i.e. length and time, involved in the system. One may choose the saturated *moist Froude number*, $F_m = U/N_m h$, as a nondimensional control parameter. Similar to the dry Froude number ($F = U/Nh$), the reciprocal of the moist Froude number is also called the *nondimensional mountain height*, which represents the strength of nonlinearity. That is, larger (smaller) F_m represents a more linear (nonlinear) flow. A second nondimensional control parameter is the mountain slope (height–width aspect ratio), defined as h/a, or the nondimensional mountain width, $N_m a/U$. Both h/a and $N_m a/U$ measure the nonhydrostatic effects. The flow becomes more nonhydrostatic when h/a increases. Similar to dry airflow over mountains (Chapter 5), for $N_m a/U \gg 1$, the flow reaches hydrostatic balance. On the other hand, the nonhydrostatic effects cannot be ignored if this criterion is not met. Analogous to dry flow, among F_m, h/a, and $N_m a/U$, only two of them can be chosen for the nondimensional control parameter space since they are not independent of each other.

For a conditionally unstable flow over mountains, one may define the moist Froude number as $F_w = U/N_w h$ (Chen and Lin 2005a), where N_w is the unsaturated Brunt–Vaisala frequency defined as:

$$N_w^2 = \frac{g}{\theta_v} \frac{\partial \theta_v}{\partial z}. \tag{11.4.1}$$

In the above equation, θ_v is the virtual potential temperature for moist, but unsaturated air. In this situation, CAPE also becomes a factor in determining the flow responses. For a potentially unstable flow, the criterion $\partial \theta_e / \partial z < 0$ needs to be considered. When the mountain geometry is considered, the horizontal width aspect ratio a/b is applied, where a and b are the horizontal scales in the direction parallel and perpendicular to the basic wind, respectively. For a flow over a meso-α scale or larger scale mountain, then the rotational effects need to be considered. In this case, the Rossby number, U/fa, is a natural choice for an additional nondimensional control parameter.

11.4.2 Moist flow regimes

(a) Two-dimensional flow regimes

Figure 11.22 demonstrates the dependency of the flow on the moist Froude number with $F_m = 0.67$, 1.33, and 2.00. In the numerical experiments, the flow is designed to

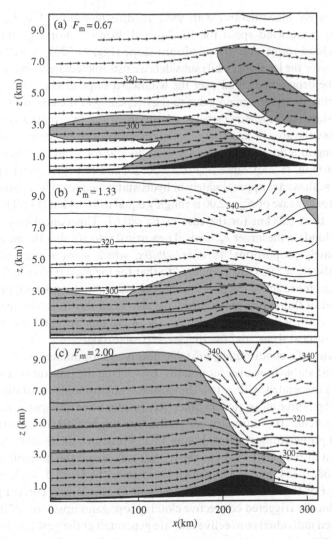

Fig. 11.22 Three moist flow regimes for a two-dimensional, nearly moist-neutral flow over a bell-shaped mountain with $F_m =$: (a) 0.67, (b) 1.33, and (c) 2.00. F_m is the saturated moist Froude number defined as $U/N_m h$. The flow and precipitation are averaged from numerically simulated fields for 6 to 12 h. The orographic and flow parameters are: $h = 1.5\,$km, $a = 50\,$km (half-width of the mountain), $N_m = 0.01\,$s^{-1}, and $U = 10$, 20, and 30 m s^{-1}. The corresponding $N_m a/U$ are 50, 25, and 16.67, respectively. Displayed are the potential temperature (solid), wind vectors, and total precipitation hydrometer (rain, snow and graupel) mixing ratio higher than 0.04 g kg^{-1} (shaded). (Adapted after Colle 2004.)

be approximately moist-neutral with 98% relative humidity. The F_m is varied by changing the basic wind speed for 10, 20, and 30 m s^{-1}. For $F_m = 0.67$, a shallow orographic cloud forms over the windward slope (Fig. 11.22a). A well-defined wave cloud forms over the lee slope, while subsidence exists immediately above the crest. For $F_m = 1.33$, the orographic cloud over the windward slope extends over the mountain peak and farther upstream, mainly due to the deepening of the upward motion to 3 km over the upslope and the downstream advection of the hydrometeors (Fig. 11.22b). For $F_m = 2.00$, the area of ascending motion deepens over the lower upslope to above 9 km, resulting in a deep orographic snow cloud and more precipitation upstream of the mountain (Fig. 11.22c). Since $N_m a / U$ ($= 50$, 25, and 16.67) is much larger than 1 for all cases, these flows are approximately in hydrostatic balance. The hydrostatic vertical wavelength for the case of $F_m = 2.00$ is roughly equal to 18.84 km (Fig. 11.22c), which is much larger than 6.28 km for the case of $F_m = 0.67$. This contributes to the deeper orographic clouds in the larger F_m cases. The vertical velocity near the mountain surface is approximately proportional to $w' \approx U \partial h / \partial x$, which gives a much higher vertical velocity for the $F_m = 2.00$ case, compared to that for the $F_m = 0.67$ case. Thus, a higher F_m may trigger more active microphysical processes and more precipitation. The moist stable or neutral flows over mountains have also been shown to be sensitive to other dimensional control parameters such as the mountain heights, mountain widths, strengths of vertical shear, and freezing levels (Colle 2004).

The behavior of a conditionally unstable flow over mountains is very different from that of a moist stable flow over mountains. By varying the unsaturated moist Froude number (F_w) associated with a two-dimensional conditionally unstable flow over a mountain ridge, three moist flow regimes may be identified: (I) upstream propagating convective system, (II) stationary convective system, and (III) both stationary and downstream propagating systems. Figure 11.23 illustrates the time evolution of rain-water content at the height of 1 km above the surface and the horizontal wind speed for $F_w = 0.208$, 0.354, and 0.833, which correspond to $U = 2.5$, 4.25, and 10 m s^{-1}, respectively. For the case of $F_w = 0.208$ (Regime I), the density current generated by the orographically triggered convective clouds propagates upstream of the mountain. The embedded individual convective cells are generated at the gust front of the density current and then propagate downstream once they form. However, the convective system propagates upstream along with the density current (Fig. 11.23a). For the case of $F_w = 0.354$ (Regime II), the density current, and thus the convective system, becomes quasi-stationary in the vicinity of the mountain peak (Fig. 11.23b). Both propagating and growing modes of convective cells found in multicell storms (Chapter 8) are present in this flow regime. The convective system is maintained by a balance between the orographic forcing and the density current forcing. For the case of $F_w = 0.833$ (Regime III), the density current propagates downstream and two convective systems exist: the quasi-stationary system and the downstream propagating system (Fig. 11.23c). For the stationary convective system, the cell generation is similar to that of regime II, but weaker in strength.

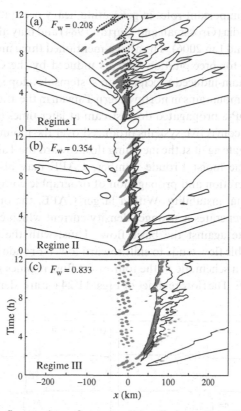

Fig. 11.23 Three moist flow regimes for a conditionally unstable airflow over a mesoscale mountain, based on the unsaturated, moist Froude number $F_w =$: (a) 0.208 (Regime I), (b) 0.354 (Regime II), and (c) 0.833 (Regime III). F_w is defined as $U/N_w h$. The corresponding basic flow speeds are $U = 2.5$, 4.25, and $10 \, \mathrm{m \, s^{-1}}$, respectively. Displayed are the time evolutions of surface wind (dotted, thin-solid, and heavy-solid contours for -5, 0, and $5 \, \mathrm{m \, s^{-1}}$, respectively) and rainwater (shaded for $q_r \geq 0.5 \, \mathrm{g \, kg^{-1}}$) at 1 km height above the surface, as simulated by a mesoscale model (ARPS, see Xue *et al.* 2001). (Adapted after Chu and Lin 2000.)

Observational and numerical studies of the MAP IOP-2B and IOP-8 precipitation systems provide evidence of the control of Froude number on the orographic pre-cipitation. During IOP-2B, much stronger convective precipitating systems were generated over the upslopes of the Alps due to the less stable flow upstream of the Alps, while more of the precipitation was widespread stratiform over the Po Valley and Alps due to the northward advection of the remnants of deep convective systems in the Gulf of Genoa (Rotunno and Houze 2007). Thus, airflow associated with IOP-8 upstream of the southern Alpine slopes is characterized by a larger Brunt–Vaisala frequency, generating a smaller F_w flow, while, on the other hand, IOP-2B as char-acterized by a smaller Brunt–Vaisala frequency, generating a larger F_w flow (Rotunno and Ferretti 2003; Medina and Houze 2003). Upstream propagating density currents appear also to play important roles in generating heavy precipitation upstream of the Western Ghats in India (Grossman and Durran 1984) and may also act like an effective mountain (Reeves and Lin 2006). The above-mentioned three moist flow regimes have also been observed for precipitation systems induced by the CMR of Taiwan. For example, the mountain-induced precipitation systems developed continuously on the western slope of the mountains in northwestern Taiwan in the afternoon of June 7, 1987 during TAMEX IOP-8 propagated downstream at later times (Chen *et al.* 1991). On the other hand, the convective systems produced over the mountain ridge in northern Taiwan may propagate against the incoming flow toward the Taipei Valley (Jou 1994).

In addition to the moist Froude number, CAPE may also come into play in controlling the generation and propagation of orographic convective systems via the release of conditional instability. With a larger CAPE, the orographically induced convective system generates a stronger density current which contains larger kinetic energy to propagate against the basic flow. Thus, with the increase in CAPE, a conditionally unstable flow tends to shift toward lower-Froude number flow regimes. Figure 11.24 shows a schematic of the four moist flow regimes generated by different sets of F_w and CAPE. The flow regimes of Figs. 11.24a–c are identical to Regimes I–III

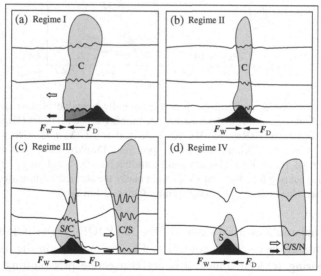

Fig. 11.24 Schematic of the four moist flow regimes for a conditionally unstable airflow over a mesoscale mountain for $(F_w, \text{CAPE}) =$: (a) (small, large), (b) (small, small), (c) (large, large), and (d) (large, small). Flow regimes in (a)–(c) are identical to regimes I–III shown in Figs. 11.23a–c. Flow regime IV is characterized by an orographic stratiform precipitation system over the mountain and a downstream-propagating cloud system. F_w and F_D denote the forcing associated with the basic flow and density current/cold pool, respectively. Isentropes and cloud boundaries are denoted by solid and dashed curves, respectively. Symbols C, S, and N denote convective, stratiform, and no-cloud types, respectively. Outline (filled) arrow denotes the propagation direction of the precipitation system (density current). (Adapted after Chen and Lin 2005a.)

shown in Figs. 11.23a–c, except that the cloud system in the vicinity of the mountain peak is a mixture of convective clouds and stratiform clouds (Fig. 11.23c). Flow regime IV is characterized by an orographic stratiform precipitation system over the mountain and a downstream-propagating cloud system. The long-lasting orographic stratiform precipitation system is due to the shorter advection time compared with cloud growth time. Strong upward propagating gravity waves are generated by the mountain and the downstream propagating cloud system, as revealed by the isentropes. The formation and propagation of different flow regimes may be interpreted from the competition between the forcing associated with the basic flow and the forcing associated with the density current or cold pool. The latter is strongly linked to the strength of the convective system and thus related to the magnitude of the CAPE.

An extreme case in the (F_w, CAPE) parameter space is a moist neutral flow over a mesoscale mountain ridge. Three flow subregimes are found: (1) for lower mountains, a saturated flow can be maintained everywhere given sufficient initial cloud water; (2) for higher mountains, the upstream atmosphere is maintained in a saturated state and transitions to an unsaturated downslope flow on the lee side, which has characteristics associated with downslope windstorms; and (3) for mountains of intermediate height, an upstream-propagating disturbance develops, which tends to de-saturate the atmosphere above the mountain (Miglietta and Rotunno 2005).

(b) Three-dimensional flow regimes

For flow over a three-dimensional isolated mountain, as expected, flow behaves differently from that over a two-dimensional mountain ridge. Numerical studies have demonstrated that the reduction of effective static stability of a southerly moist flow by latent heating may go through the regime transition from flow-around a higher mountain ridge to flow-over lower mountain ridge (Buzzi *et al.* 1998; Ferretti *et al.* 2000), which is consistent with the idea of F_m as a moist flow control parameter. For a nearly moist-neutral (RH = 98%) flow over a three-dimensional isolated mountain, three flow regimes, similar to those discussed in the two-dimensional moist flow regimes (Subsection 11.4.2(b)) are also identified (Miglietta and Buzzi 2001). The Froude number also controls the propagation of nocturnal cloud bands formed by the downslope drainage flow and the basic easterly trade wind in Hawaii (e.g., Smolarkiewicz *et al.* 1988).

Regimes I–III for a conditionally unstable flow over a two-dimensional mountain, as described above, are also found in flow over a three-dimensional, isolated mountain (Chen and Lin 2005b). In addition, a relatively stronger basic flow, such as a low-level jet, is able to produce a quasi-stationary mesoscale convective system and maximum rainfall on the windward slope (i.e., upslope precipitation), instead of on the mountain peak or on the lee slope, which is more consistent with observations. The change in preferred upslope precipitation accumulation is due to the fact that the three-dimensional isolated mountain allows the upstream airstream to split and flow around it. Regimes I and II discussed above have been reproduced by simulating a conditionally unstable flow represented by an upstream condition of MAP IOP-2B over the Alpine mountains (Stein 2004).

References

Alpert, P., 1986. Mesoscale indexing of the distribution of orographic precipitation over high mountains. *J. Atmos. Sci.*, **25**, 532–45.

Asencio, N., J. Stein, M. Chong, and F. Gheusi, 2003. Analysis and simulation of local regional conditions for the rainfall over the Lago Maggiore Target Area during MAP IOP 2b. *Quart. J. Roy. Meteor. Soc.*, **129**, 565–86.

Banta, R. M., 1990. The role of mountain flows in making clouds. *Atmospheric Processes over Complex Terrain*, W. Blumen (ed.), Meteor. Monogr., 45, Amer. Meteor. Soc., 229–283.

Barcilon, A. and D. Fitzjarrald, 1985. A nonlinear steady model for moist hydrostatic mountain waves. *J. Atmos. Sci.*, **42**, 58–67.

Barcilon, A., J. C. Jusem, and P. G. Drazin, 1979. On the two-dimensional, hydrostatic flow of a stream of moist air over a mountain ridge. *Geophys. Astrophys. Fluid Dyn.*, **12**, 1–16.

Barros, A. P. and R. Kuligowski, 1998. Orographic effects during a severe wintertime rainstorm in the Appalachian Mountains. *Mon. Wea. Rev.*, **126**, 2648–72.

Bergeron, T., 1949. The problem of artificial control of rainfall on the globe. II. The coastal orographic maxima of precipitation in autumn and winter. *Tellus*, **1**, 15–32.

Bougeault, P., P. Binder, A. Buzzi, R. Dirk, R. Houze, J. Kuettner, R. B. Smith, R. Steinacker, and H. Volkert, 2001. The MAP special observing period. *Bull. Amer. Meteor. Soc.*, **82**, 433–62.

Bousquet, O. and B. F. Smull, 2003. Observations and impacts of upstream blocking during a widespread orographic precipitation event. *Quart. J. Roy. Meteor. Soc.*, **129**, 391–409.

Browning, K. A., F. F. Hill, and C. W. Pardoe, 1974. Structure and mechanism of precipitation and the effect of orography in a wintertime warm sector. *Quart. J. Roy. Meteor. Soc.*, **100**, 309–30.

Buzzi, A. and P. P. Alberoni, 1992. Analysis and numerical modeling of a frontal passage associated with thunderstorm development over the Po Valley and the Adriatic Sea. *Meteor. Atmos. Phys.*, **48**, 205–24.

Buzzi, A. and L. Foschini, 2000. Mesoscale meteorological features associated with heavy precipitation in the Southern Alpine region. *Meteor. Atmos. Phys.*, **72**, 131–46.

Buzzi, A., N. Tartaglione, and P. Malguzzi, 1998. Numerical simulations of the 1994 Piemont flood: Role of orography and moist processes. *Mon. Wea. Rev.*, **126**, 2369–83.

Carruthers, D. J. and W. T. Choularton, 1983. A model of the feeder-seeder mechanism of orographic rain including stratification and wind-drift effects. *Quart. J. Roy. Meteor. Soc.*, **109**, 575–88.

Chen, C.-S., W.-S. Chen, and Z. Deng, 1991. A study of a mountain-generated precipitation system in northern Taiwan during TAMEX IOP 8. *Mon. Wea. Rev.*, **119**, 2574–606.

Chen, S.-H. and Y.-L. Lin, 2005a. Effects of moist Froude number and CAPE on a conditionally unstable flow over a mesoscale mountain ridge. *J. Atmos. Sci.*, **62**, 331–50.

Chen, S.-H. and Y.-L. Lin, 2005b. Orographic effects on a conditionally unstable flow over an idealized three-dimensional mesoscale mountain. *Meteor. Atmos. Phys.*, **88**, 1–21.

Chiao, S., Y.-L. Lin, and M. L. Kaplan, 2004. Numerical study of the orographic forcing of heavy precipitation during MAP IOP-2B. *Mon. Wea. Rev.*, **132**, 2184–203.

Chu, C.-M. and Y.-L. Lin, 2000. Effects of orography on the generation and propagation of mesoscale convective systems in a two-dimensional conditionally unstable flow. *J. Atmos. Sci.*, **57**, 3817–37.

Colle, B. A., 2004. Sensitivity of orographic precipitation to changing ambient conditions and terrain geometries: An idealized modeling perspective. *J. Atmos. Sci.*, **61**, 588–606.

Colle, B. A. and S. E. Yuter, 2007. The impact of coastal boundaries and small hills on the precipitation distribution across southern Connecticut. *Mon. Wea. Rev.*, **135**, 933–54.

Colle, B. A. and Y. Zeng, 2004. Bulk microphysical sensitivities within the MM5 for orographic precipitation. Part I: The Sierra 1986 event. *Mon. Wea. Rev.*, **132**, 2780–801.

Cooper, W. A. and J. D. Marwitz, 1980. Winter storms over the San Juan Mountains. Part III: Seeding potential. *J. Appl. Meteor.*, **19**, 942–9.

Davies, H. C. and C. Schär, 1986. Diabatic modification of airflow over a mesoscale orographic ridge: A model study of the coupled response. *Quart. J. Roy. Meteor. Soc.*, **112**, 711–30.

Doswell, C. A., III, H. Brooks, and R. Maddox, 1996. Flash flood forecasting: An ingredient-based methodology. *Wea. Forecasting*, **11**, 560–81.

Doswell, C. A., III, R. Romero, and S. Alonso, 1998. A diagnostic study of three heavy precipitation episodes in the western Mediterranean region. *Wea. Forecasting*, **13**, 102–24.

Durran, D. R. and J. B. Klemp, 1982. The effects of moisture on trapped mountain lee waves. *J. Atmos. Sci.*, **39**, 2490–506.

Espenshade, E. B. Jr., J. C. Hudson, and J. L. Morrison, 1990. *Goode's World Atlas*. 19th edn., Rand McNally.

Fankhauser, J. C., 1988. Estimates of thunderstorm precipitation efficiency from field measurements in CCOPE. *Mon. Wea. Rev.*, **116**, 663–84.

Ferretti, R., S. Low-Nam, and R. Rotunno, 2000: Numerical simulations of the 1994 Piedmont flood of 4–6 November. *Tellus*, **52A**, 162–80.

Fraser, A. B., R. C. Easter, and P. V. Hobbs, 1973: A theoretical study of air and fallout of solid precipitation over mountainous terrain: Part I. *J. Atmos. Sci.*, **30**, 801–12.

Frei, C. and C. Schär, 1998. A precipitation climatology of the Alps from high-resolution rain-gauge observations. *Int'l J. Clim.*, **18**, 873–900.

Gheusi, F. and J. Stein, 2003. Small-scale rainfall mechanisms for an idealized convective southerly flow. *Quart. J. Roy. Meteor. Soc.*, **129**, 1819–40.

Grossman, R. L. and D. R. Durran, 1984. Interaction of low-level flow with the Western Ghats Mountains and offshore convection in the summer monsoon. *Mon. Wea. Rev.*, **112**, 652–72.

Hobbs, P. V., R. C. Easter, and A. B. Fraser, 1973. A theoretical study of the flow of air and fallout of solid precipitation over mountainous terrain: Part II. *J. Atmos. Sci.*, **30**, 813–23.

Hobbs, P. V., R. Houze, Jr., and T. Matejka, 1975. The dynamical and microphysical structure of an occluded frontal system and its modification by orography. *J. Atmos. Sci.*, **32**, 1542–62.

Houze, R. A., Jr. 1993. *Cloud Dynamics.* Academic Press.

Houze, R. A. and S. Medina, 2005. Turbulence as a mechanism for orographic precipitation enhancement. *J. Atmos. Sci.,* **62,** 3599–623.

Jiang, Q., 2003. Moist dynamics and orographic precipitation. *Tellus,* **55A,** 301–16.

Jiang, Q. and R. B. Smith, 2003. Cloud timescales and orographic precipitation. *J. Atmos. Sci.,* **60,** 1543–59.

Jou, B. J.-D., 1994. Mountain-originated mesoscale precipitation system in northern Taiwan: A case study 21 June 1991. *Terr. Atmos. Ocean,* **5,** 169–97.

Kirshbaum, D. J. and D. R. Durran, 2004. Factors governing cellular convection in orographic precipitation. *J. Atmos. Sci.,* **61,** 682–98.

Kuo, J.-T. and H. D. Orville, 1973. A radar climatology of summertime convective clouds in the Black Hills. *J. Appl. Meteor.,* **12,** 359–68.

Lin, Y.-L., 2005. Dynamics of orographic precipitation. *2005 Yearbook of Science & Technology,* McGraw Hill, 248–250.

Lin, Y.-L., S. Chiao, T.-A. Wang, M. L. Kaplan, and R. P. Weglarz, 2001. Some common ingredients for heavy orographic rainfall. *Wea. Forecasting,* **16,** 633–60.

Lin, Y.-L., H. D. Reeves, S.-Y. Chen, and S. Chiao, 2005. Formation mechanisms for convection over the Ligurian Sea during MAP IOP-8. *Mon. Wea. Rev.,* **133,** 2227–45.

Massacand, A. C., H. Wernli, and H. C. Davies, 1998. Heavy precipitation on the Alpine southside: An upper-level precursor. *Geophys. Res. Lett.,* **25,** 1435–38.

Medina, S. and R. A. Houze, 2003. Air motions and precipitation growth in Alpine storms. *Quart. J. Roy. Meteor. Soc.,* **129,** 345–72.

Miglietta, M. M. and A. Buzzi, 2001. A numerical study of moist stratified flows over isolated topography. *Tellus,* **53A,** 481–99.

Miglietta, M. M. and R. Rotunno, 2005. Numerical simulations of moist nearly neutral flow. *J. Atmos. Sci.,* **62,** 1410–27.

Ogura, Y. and M. Yoshizaki, 1988. Numerical study of orographic-convective precipitation over the Eastern Arabian Sea and the Ghats Mountains during the summer monsoon. *J. Atmos. Sci.,* **45,** 2097–122.

Orville, H. D. and L. J. Sloan, 1970. A numerical simulation of the life history of a rainstorm. *J. Atmos. Sci.,* **27,** 1148–59.

Pontrelli, M. D., G. Bryan, and J. M. Fritsch, 1999. The Madison County, Virginia, flash flood of 27 June 1995. *Wea. Forecasting,* **14,** 384–404.

Purdy, J. C., G. L. Austin, A. W. Seed, and I. D. Cluckie, 2005. Radar evidence of orographic enhancement due to the seeder feeder mechanism. *Meteor. Appl.,* **12,** 199–206.

Ralph, F. M., P. J. Neiman, and R. Rotunno, 2005. Dropsonde observations in low-level jets over the northeastern Pacific Ocean from CALJET-1998 and PACJET-2001: Mean vertical-profile and atmospheric-river characteristics. *Mon. Wea. Rev.,* **133,** 889–910.

Rauber, R. M., 1992. Microphysical structure and evolution of a Central Sierra Nevada orographic cloud system. *J. Appl. Meteor.,* **31,** 3–24.

Reeves, H. D. and Y.-L. Lin, 2006. Effect of stable layer formation over the Po Valley on the development of convection during MAP IOP-8. *J. Atmos. Sci.,* **63,** 2567–84.

Richard, E., S. Cosma, P. Tabary, J.-P. Pinty, and M. Hagen, 2003. High-resolution numerical simulations of the convective system observed in the Lago Maggiore area on 17 September 1999 (MAP IOP 2a). *Quart. J. Roy. Meteor. Soc.,* **129,** 543–63.

Rotunno, R. and R. Ferretti, 2001. Mechanisms of intense Alpine rainfall. *J. Atmos. Sci.*, **58**, 1732–49.

Rotunno, R. and R. Ferretti, 2003. Orographic effects on rainfall in MAP cases IOP2B and IOP8. *Quart. J. Roy. Meteor. Soc.*, **129**, 373–90.

Rotunno, R. and R. A. Houze, Jr., 2007. Lessons on orographic precipitation from MAP. *Quart. J. Roy. Meteor. Soc.*, **133**, 811–30.

Sakakibara, H., 1979. Cumulus development on the windward side of a mountain range in convectively unstable air mass. *J. Meteor. Soc. Japan*, **57**, 341–8.

Sarker, R. P., 1966. A dynamical model of orographic rainfall. *Mon. Wea. Rev.*, **94**, 555–72.

Sawyer, J. S., 1956. The physical and dynamical problems of orographic rain. *Weather*, **11**, 375–81.

Schneidereit, M., and C. Schär, 2000. Idealised numerical experiments of Alpine flow regimes and southside precipitation events. *Meteor. Atmos. Phys.*, **72**, 233–50.

Seibert P., 1990. South foehn studies since the ALPEX experiment. *Meteor. Atmos. Phys.*, **43**, 91–103.

Sinclair, M. R., 1993. A diagnostic study of the extratropical precipitation resulting from Tropical Cyclone Bola. *Mon. Wea. Rev.*, **121**, 2690–707.

Sinclair, M. R., 1994. A diagnostic model for estimating orographic precipitation. *J. Appl. Meteor.*, **33**, 1163–75.

Smith, B., II, Y.-L. Lin, and H. D. Reeves, 2006. Effects of cyclone track on precipitation distribution along the California Coastal Range and Sierra Nevada. *12th Conf. Mount. Meteor.*, Amer. Meteor. Soc., Aug. 28–Sept. 1, Santa Fe, New Mexico.

Smith, R. B., 1979. The influence of mountains on the atmosphere. *Adv. in Geophys.*, **21**, 87–230.

Smith, R. B., 1982. A differential advection model of orographic rain. *Mon. Wea. Rev.*, **110**, 306–9.

Smith, R. B., 2006. Progress on the theory of orographic precipitation. In *Tectonics, Climate and Landscape Evolution*: Geol. Soc. Amer. Special Paper 398, S. D. Willett *et al.* (eds.), Penrose Conf. Series, 1–16, doi: 10.1130/2006.2398(01).

Smith, R. B. and I. Barstad, 2004. A linear theory of orographic precipitation. *J. Atmos. Sci.*, **61**, 1377–91.

Smith, R. B. and Y.-L. Lin, 1982. The addition of heat to a stratified airstream with application to the dynamics of orographic rain. *Quart. J. Roy. Meteor. Soc.*, **108**, 353–78.

Smith, R. B., Q. Jiang, M. G. Fearon, P. Tabary, M. Dorninger, J. D. Doyle, and R. Benoit, 2003. Orographic precipitation and air mass transformation: An Alpine example. *Quart. J. Roy. Meteor. Soc.*, **129**, 433–54.

Smolarkiewicz, P. K., R. M. Rasmussen, and T. L. Clark, 1988. On the dynamics of Hawaiian Cloud Bands: Island Forcing. *J. Atmos. Sci.*, **45**, 1872–905.

Stein, J., 2004. Exploration of some convective regimes over the Alpine orography. *Quart. J. Roy. Meteor. Soc.*, **130**, 481–502.

Steinacker, R. 2006. Alpiner Föhn – eine neue Strophe zu einem alten Liedin: Atmosphäre und Gebirge – Anregung von ausgeprägten Empfindlichkeiten Promet, **32**, 3–10, Deutscher Wetterdienst.

Steiner, M., O. Bousquet, R. A. Houze, Jr., B. F. Smull, and M. Mancini, 2003. Airflow within major Alpine river valleys under heavy rainfall. *Quart. J. Roy. Meteor. Soc.*, **129**, 411–32.

Weissmann, M. D., G. J. Mayr, R. M. Banta, and A. Gohm, 2004. Observations of the temporal evolution and spatial structure of the gap flow in the Wipp Valley on 2 and 3 October 1999. *Mon. Wea. Rev.*, **132**, 2684–97.

White, A. B., P. J. Neiman, F. M. Ralph, and D. E. Kingsmill, 2003. Coastal orographic rainfall processes observed by radar during the California land-falling jets experiment. *J. Hydrometeor.*, **4**, 264–82.

Witcraft, N. C., Y.-L. Lin, and Y.-H. Kuo, 2005. Dynamics of orographic rain associated with the passage of a tropical cyclone over a mesoscale mountain. *Terr. Atmos. Ocean*, **16**, 1133–61.

Xue, M., K. K. Droegemeier, V. Wong, A. Shapiro, K. Brewster, F. Carr, D. Weber, Y. Liu, and D.-H. Wang, 2001. The Advanced Regional Prediction System (ARPS) – a multi-scale nonhydrostatic atmospheric simulation and prediction tool. Part II: Model physics and applications. *Meteor. Atmos. Phys.*, **76**, 143–65.

Yeh, H.-C. and Y.-L. Chen, 1998. Characteristics of rainfall distributions over Taiwan during the Taiwan Area Mesoscale Experiment (TAMEX). *Mon. Wea. Rev.*, **37**, 1457–69.

Yuter, S. E., and R. A. Houze, Jr., 2003. Microphysical modes of precipitation growth determining by S-band vertically pointing radar in orographic precipitation during MAP. *Quart. J. Roy. Meteor. Soc.*, **129**, 455–76.

Zängle, G., B. Chimani, and C. Häberli, 2004. Numerical simulations of the foehn in the Rhine Valley on 24 October 1999. *Mon. Wea. Rev.*, **132**, 368–89.

12

Basic numerical methods

12.1 Introduction

In Chapter 2, we derived a set of nonlinear partial differential equations governing mesoscale atmospheric motions. One way to study the dynamics associated with these equations is to make the small-amplitude approximations and solve the linearized equations analytically, as demonstrated in earlier chapters. However, this approach limits us to study only mesoscale systems with small-amplitude perturbations. In addition, the number of available analytical methods at hand to solve these complicated equations is limited. As mentioned in earlier chapters, an alternative solution is to use numerical methods where the equations are discretized and solved numerically in space and time. The advantage of applying the numerical methods is that they are able to solve completely the nonlinear set of equations. Numerical methods also provide a powerful framework for sensitivity tests or experiments with forcing or physical processes. In these experiments, physical parameterizations or external forcing can be easily altered or completely deactivated. Examples of the parameterizations include those for planetary boundary layer processes, moist processes, and radiative processes, while external forcing can come from orography. In this regard, numerical simulations are more flexible than physical experiments, such as experiments conducted in a water tank, gas chamber or wind tunnel, and field experiments conducted in the real atmosphere.

When numerical methods are adopted to solve mathematically intractable governing equations, one needs to address the following important questions: (1) Does the solution of the approximate equations converge to that of the original differential equations when the time and grid intervals approach zero? (2) Is the numerical solution well-behaved in time, or more precisely, is the numerical scheme stable? (3) If the numerical scheme is stable, how well do the amplitudes and phases of the approximated waves or disturbances represent those of the exact solution? We will try to answer these questions in this chapter.

The major numerical methods that are used to solve partial differential equations can be categorized as (1) *finite difference methods*, (2) *Galerkin methods*, and (3) *Lagrangian methods*. Combinations of these methods for solving a set of time-dependent equations have also been developed. In the finite difference methods, dependent variables are defined at specific grid points in space and time, and the derivatives in the equations are approximated by Taylor series expansion or other approaches. For mesoscale numerical weather prediction models, the governing equations are solved in a finite region of the atmosphere. Thus, mesoscale models are often referred to as *limited-area* or *regional models*. In order to integrate the governing equations numerically, boundary conditions of the variables at the boundary of the integration domain are required. In addition, initial conditions are also required for integrating time-dependent partial differential equations so as to arrive at a future solution, which we call *prediction*. Finite difference methods are the most popular numerical methods adopted for mesoscale numerical weather prediction models.

In *Galerkin methods*, dependent variables are represented by a sum of functions that have prescribed spatial structures. The coefficient associated with each function is normally a function of time for a time-dependent problem, which transforms a partial differential equation into a set of ordinary differential equations (in time) for the coefficients. These equations are usually solved using finite difference approximations in time. The Galerkin methods can be divided into two major categories: the *spectral method* and the *finite element method*. In the spectral method, dependent variables are represented by orthogonal, global basis functions, such as a sinusoidal function. The spectral method is less popular with mesoscale models due to the difficulties posed by nonperiodic lateral boundary conditions with limited area models. The spectral method is, however, much more convenient with global models due to the periodic nature of their zonal boundary conditions. Techniques for treating nonperiodic boundary conditions have been developed in the last two decades; thus the spectral method has also been used in mesoscale models. The finite element method is similar to the spectral method except that it uses local instead of global (in terms of the integration domain) basis functions. The local basis functions include chapeau and tent functions. There is a growing interest in adopting the finite element methods for mesoscale models due to their accuracy and flexibility in treating the irregular geometry of the internal or external boundary. One of the disadvantages of finite element methods is that they usually require significantly more computing time to invert a normally large matrix every time step. Although finite element methods are more accurate compared to finite difference methods with the same order of accuracy, finite difference methods can achieve similar solution accuracy by using a higher-order scheme with less computing time. In addition, the advantage of finite element methods in treating irregular lower boundaries is significantly diminished when finite difference numerical models employ the so-called *terrain-following coordinates*, in which the irregular lower boundary becomes 'flat' or regular within the transformed computational domain.

In *Lagrangian methods*, the equations governing the fluid motion are solved by following a fixed set of particles throughout the period of integration. The advantage of a Lagrangian method is that it treats the total derivative at once, instead of treating the local rate of change and advection terms individually. However, in general, a set of fluid particles, which are initially distributed regularly, will soon become greatly deformed, and are thus rendered unsuitable for numerical integration. In order to avoid this problem, the *semi-Lagrangian method* is employed. Thus, the fluid variables at the predicted time step can be defined at the regular grid points and those at the previous time steps (which are often not located at the regular grid points) are interpolated from the known values at the regular grid points from the previous time step. The semi-Lagrangian method has become popular in recent years, especially with large-scale models, since a relatively large time interval for integration can be used due to its unconditional stability characteristics.

Other methods, such as the *upstream interpolation* (e.g., Pielke 2002) and *finite-volume* (e.g., Durran 1998) methods, have also been used in mesoscale numerical models. In an *interpolation method*, dependent variables at grid points are used to derive interpolation formulas that are then used to calculate spatial derivatives. Unlike finite difference methods, *finite volume methods* generate approximations to the grid-interval or grid-cell average. In a finite-volume method, the grid-point value f_i represents the average of a function, $f(x)$, over the interval (or grid cell) $[(i - \frac{1}{2})\Delta x, (i + \frac{1}{2})\Delta x]$, taking a one-dimensional problem as an example. Finite volume methods are very useful for approximating solutions that contain discontinuities (e.g., Colella and Woodward 1984).

12.2 Finite difference approximations of derivatives

Before discussing various finite difference methods, let us consider a simple example of a finite difference approximation to help us understand the approximation of derivatives of a given function. For example, one may use the Taylor series to approximate $f(x)$ at $x + \Delta x$,

$$f(x + \Delta x) = f(x) + f'(x)\Delta x + f''(x)\frac{\Delta x^2}{2!} + f'''(x)\frac{\Delta x^3}{3!} + \cdots, \qquad (12.2.1)$$

where Δx is the *grid interval*. For convenience, Δx is assumed to be greater than 0 in the following discussions. The derivative of $f(x)$ can be calculated from,

$$f'(x) = \frac{f(x + \Delta x) - f(x)}{\Delta x} + R(x, \Delta x), \qquad (12.2.2)$$

where

$$R(x, \Delta x) \equiv \Delta x \left\{ -\frac{f''(x)}{2!} - \frac{f'''(x)\Delta x}{3!} - \cdots \right\}$$

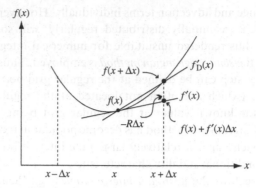

Fig. 12.1 A sketch of the forward finite difference scheme, based on (12.2.3).

is called the *remainder*, which has a magnitude of $O(\Delta x)$. If the remainder term is much smaller than the first term on the right side of (12.2.2), then the above equation can be approximated by

$$f'_{\mathrm{D}}(x) = \frac{f(x + \Delta x) - f(x)}{\Delta x}. \tag{12.2.3}$$

A differential equation becomes a *finite difference equation* when the derivatives are approximated by their finite difference forms, known as a *forward difference scheme* for (12.2.3), which has a first order of accuracy. The actual derivative, $f'(x)$, is approximated by the slope $f'_{\mathrm{D}}(x)$ (Fig. 12.1). The distance between $f(x) + f'(x)\Delta x$ and $f(x + \Delta x)$ is $-R\Delta x$. When Δx is reduced, the approximated derivative $f'(x)$, i.e. the approximated slope of $f(x)$, is closer to the real derivative. As seen from (12.2.2) and Fig. 12.1, there are two ways to reduce the truncation errors: (a) reducing the space interval (Δx) and (b) using a higher order approximation.

Similarly, the Taylor series expansion can also be expanded in a backward manner,

$$f(x - \Delta x) = f(x) - f'(x)\Delta x + f''(x)\frac{\Delta x^2}{2!} - f'''(x)\frac{\Delta x^3}{3!} + \cdots, \tag{12.2.4}$$

which can be rearranged in the following form:

$$f'(x) = \frac{f(x) - f(x - \Delta x)}{\Delta x} + \Delta x \left\{ \frac{f''(x)}{2!} - \frac{f'''(x)\Delta x}{3!} + \cdots \right\}. \tag{12.2.5}$$

Again, if the remainder term of the above equation is much smaller than the first term on the right side, then (12.2.5) can be approximated using the *backward difference scheme*,

$$f'_{\mathrm{D}}(x) = \frac{f(x) - f(x - \Delta x)}{\Delta x}. \tag{12.2.6}$$

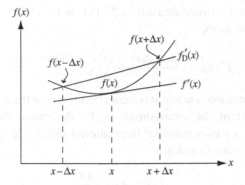

Fig. 12.2 A sketch of the relationship of $f'(x)$ and its centered difference approximation, $f'_D(x)$.

The meaning of (12.2.6) can be easily understood by replacing $x + \Delta x$ and x with x and $x - \Delta x$, respectively, in Fig. 12.1, where the approximated slope, $f'(x)$, is replaced by the straight line connecting $f(x - \Delta x)$ and $f(x)$. Like the forward difference scheme, the backward scheme has a first order of accuracy.

An alternative way to approximate the derivative is to subtract (12.2.4) from (12.2.1):

$$f(x + \Delta x) - f(x - \Delta x) = 2f'(x)\Delta x + 2f'''(x)\frac{\Delta x^3}{3!} + \cdots \qquad (12.2.7)$$

The derivative, $f'(x)$, can then be calculated from

$$f'(x) = \frac{f(x + \Delta x) - f(x - \Delta x)}{2\Delta x} + R, \qquad (12.2.8)$$

where the remainder term is defined as the following,

$$R = \Delta x^2 \left\{ -\frac{f'''(x)}{3!} - \frac{f^{(5)}(x)}{5!}\Delta x - \cdots \right\}. \qquad (12.2.9)$$

Neglecting the remainder term leads to the *centered difference scheme*,

$$f'_D(x) = \frac{f(x + \Delta x) - f(x - \Delta x)}{2\Delta x}. \qquad (12.2.10)$$

Based on (12.2.9), the centered difference scheme has an accuracy on the order of Δx^2, which is a second order of accuracy. The mathematical meaning of the centered difference scheme is depicted in Fig. 12.2. When compared with Fig. 12.1, it is apparent that this scheme is more accurate than the forward finite difference scheme.

Sometimes an *approximation of the second-order derivative* is needed in solving the governing equations, such as that appearing in the diffusion terms. One way to approximate the second-order derivative, $f''(x)$, is to add (12.2.4) to (12.2.1) and neglect the remainder term,

$$f''(x) = \frac{f(x+\Delta x) - 2f(x) + f(x-\Delta x)}{\Delta x^2}. \qquad (12.2.11)$$

When an approximation is made, it is important to know the *accuracy* of the scheme obtained as a result of the approximation. To determine the accuracy of finite difference methods, we may consider the centered difference approximation to the first derivative of the sine function,

$$f(x) = A \sin \frac{2\pi x}{L}. \qquad (12.2.12)$$

The first-order derivative can be easily obtained through analytical methods,

$$f'(x) = \frac{2\pi A}{L} \cos \frac{2\pi x}{L}. \qquad (12.2.13)$$

Now, we can apply the centered difference scheme, (12.2.10), to $f'(x)$,

$$f'_D(x) = \frac{A \sin(2\pi(x+\Delta x)/L) - A \sin(2\pi(x-\Delta x)/L)}{2\Delta x}, \qquad (12.2.14)$$

which can be rearranged as

$$f'_D(x) = \frac{A \cos(2\pi x/L) \sin(2\pi \Delta x/L)}{\Delta x}. \qquad (12.2.15)$$

Dividing the above approximation by $f'(x)$ yields

$$\frac{f'_D(x)}{f'(x)} = \frac{\sin(2\pi \Delta x/L)}{2\pi \Delta x/L}. \qquad (12.2.16)$$

The relationship between $f'_D(x)$ and $f'(x)$ is also sketched in Fig. 12.2. From the above expression, we obtain

$$\frac{f'_D(x)}{f'(x)} \to 1 \quad \text{as} \quad \frac{2\pi \Delta x}{L} \to 0 \qquad (12.2.17)$$

because $\sin \theta \to \theta$ when $\theta \to 0$. In other words, the truncation error of the centered difference scheme approaches 0 when $\Delta x \ll L$. In order to obtain a good approximation, the grid interval chosen should therefore be much smaller than the wavelength.

Now let us consider a special wave with $L = 2\Delta x$, which implies that one wavelength is exactly equal to two grid intervals. Such a wave is often called a $2\Delta x$ *wave*. Substituting $L = 2\Delta x$ on the right side of (12.2.16) leads to

$$\frac{f'_D(x)}{f'(x)} = \frac{\sin \pi}{\pi} = 0. \tag{12.2.18}$$

The above equation implies that the centered difference scheme fails to represent accurately a $2\Delta x$ wave. For a fixed grid interval, shorter waves are much more poorly represented by difference schemes than longer waves are. In fact, long waves can be very accurately represented.

Examples of finite difference approximations of derivatives with various orders of accuracy can be found in Appendix 12.1.

12.3 Finite difference approximations of the advection equation

One of the simplest finite difference time integration equations is the one-dimensional advection equation with a constant advection velocity (c), which composes of only one dependent variable, one time derivative and one spatial derivative,

$$\frac{\partial u}{\partial t} + c \frac{\partial u}{\partial x} = 0, \tag{12.3.1}$$

where u represents a quantity, such as horizontal velocity or temperature, being advected in the x direction at a speed c. If the advection speed c is replaced by u, and u is the advective velocity in the above equation, it leads to one of the simplest nonlinear equations, the inviscid *Burger equation*. An analytical solution of (12.3.1) is

$$u(x, t) = f(x - ct), \tag{12.3.2}$$

where f is an arbitrary function, whose functional form is determined by u at $t = 0$, or the *initial condition*, $f^0(x)$. For example, if

$$f^0(x) = \frac{u_o a^2}{x^2 + a^2}, \tag{12.3.3}$$

then

$$u(x, t) = f^0(x - ct) = \frac{u_o a^2}{(x - ct)^2 + a^2}. \tag{12.3.4}$$

If the advection velocity is positive (negative), then the wave propagates to the right (left). Note that (12.3.3), as mentioned in earlier chapters, is called the *bell-shaped function*, which has an amplitude u_o and a half-width a. The physical meaning of the above solution is that $u(x, t)$ constantly maintains its initial shape along the *phase line*, i.e. $x - ct = $ constant. This type of wave is also called nondispersive. Such a wave or disturbance propagation is illustrated in Fig. 12.3.

In the following, we will discuss the characteristics of several popular numerical approximations of the advection equation that have been adopted in mesoscale

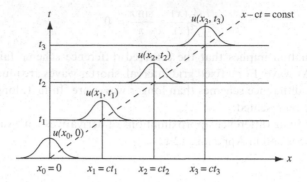

Fig. 12.3 A sketch of the propagation of $u(x,t)$ along a constant phase line, $x - ct = \text{constant} = 0$.

numerical models. Based on the number of time levels involved, these methods can be categorized as *two-time-level schemes* and *three-time-level schemes*.

12.3.1 Two-time-level schemes

The finite difference schemes used to approximate $f'(x)$, as discussed in Section 12.2, can also be applied to the time derivative. One can choose to adopt the forward, backward, or centered difference approximations. The methods of forward or backward finite difference in time belong to the so-called *two-time-level schemes* because only two time levels are involved in each step of time integration. On the other hand, using the second-order centered-difference-in-time (leapfrog) scheme would be to use the so-called *three-time-level scheme* since there are three times involved at each time step of integration.

(a) Forward-in-time and centered-in-space scheme

We use this scheme to demonstrate that not every finite difference method can be used to obtain a usable numerical solution and also to prove that *not every numerical method is numerically stable*. A natural choice in approximating the advection equation, (12.3.1), is by a combination of the forward difference of the time derivative and center difference of the spatial derivative on a time–spatial grid system shown in Fig. 12.4:

$$\frac{u_i^{\tau+1} - u_i^{\tau}}{\Delta t} + c\left(\frac{u_{i+1}^{\tau} - u_{i-1}^{\tau}}{2\Delta x}\right) = 0, \tag{12.3.5}$$

where the superscript τ and subscript i denote the *time step* and the *grid point* in space, respectively. Of particular interest is predicting u at time step $\tau + 1$ and grid point i, i.e. $u_i^{\tau+1}$, which can be obtained from the above equation,

$$u_i^{\tau+1} = u_i^{\tau} - \left(\frac{c\Delta t}{2\Delta x}\right)(u_{i+1}^{\tau} - u_{i-1}^{\tau}). \tag{12.3.6}$$

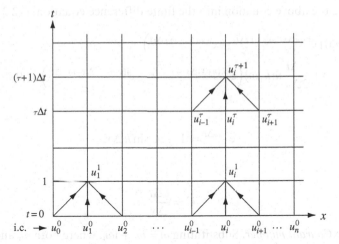

Fig. 12.4 The grid system and algorithm for the forward-in-time and centered-in-space finite difference scheme of the advection equation. The values of u at $t = 0$ are provided by the initial condition (i.c.), and the values at the left and right boundaries are determined by the boundary conditions (b.c.).

Equation (12.3.6) is called the difference equation of the advection equation obtained using the *forward-in-time and centered-in-space scheme*. The scheme's algorithm is sketched in Fig. 12.4. Initial conditions are needed in order to obtain values at time step 2, while boundary conditions are needed at both the left and right boundary points, $i = 1$ and $i = n$. The interior points u_i, $i = 2, 3, \ldots, n - 1$ at time step $\tau + 1$ are predicted by (12.3.6), using the values of u_{i-1}, u_i, u_{i+1} at time step τ. The values at the boundary points, i.e. $u_1^{\tau+1}$ and $u_n^{\tau+1}$, are determined by the boundary conditions.

Although many schemes to approximate a differential equation exist, there is no guarantee that every numerical solution is well-behaved or stable. A finite difference scheme is stable only if the solution at a fixed time $t = \tau \Delta t$ remains bounded as $\Delta t \to 0$. When this occurs, the scheme is *numerically stable*. Otherwise, it is *numerically unstable*. To examine the *numerical stability* of the forward-in-time and centered-in-space scheme, we consider the following sinusoidal wave in both time t and space x:

$$u(x, t) = \hat{u}(k, \omega)e^{i(kx - \omega t)}, \tag{12.3.7}$$

where $\hat{u}$ is the wave amplitude, k the wave number and ω the wave frequency. All three variables, $\hat{u}$, k, ω, are complex numbers. Both x and t can be represented by the grid and time intervals, respectively, $x = n\Delta x$, $t = \tau \Delta t$, where n and τ represent the grid and time intervals, respectively, from the origin $(x, t) = (0, 0)$. Using these expressions, (12.3.7) can be rewritten as

$$u(x, t) = u(n\Delta x, \tau \Delta t) = \hat{u}(k, \omega)e^{i(kn\Delta x - \omega \tau \Delta t)}. \tag{12.3.8}$$

Substituting the above equation into the finite difference equation, (12.3.5), yields

$$\hat{u}(k, \omega) \left(e^{i(kn\Delta x - \omega(\tau+1)\Delta t)} - e^{i(kn\Delta x - \omega\tau\Delta t)} \right)$$
$$+ \frac{c\Delta t}{2\Delta x} \hat{u}(k, \omega) \left(e^{i(k(n+1)\Delta x - \omega\tau\Delta t)} - e^{i(k(n-1)\Delta x - \omega\tau\Delta t)} \right) = 0, \tag{12.3.9}$$

or

$$e^{-i\omega\Delta t} = 1 - iC \sin k\Delta x, \tag{12.3.10}$$

where

$$C = \frac{c\Delta t}{\Delta x}, \tag{12.3.11}$$

is called the *Courant number*. Substituting $\omega = \omega_r + i\omega_i$, where both ω_r and ω_i are real numbers, on the left side of (12.3.10) yields

$$e^{-i\omega\Delta t} = e^{\omega_i\Delta t} e^{-i\omega_r\Delta t}. \tag{12.3.12}$$

The first term on the right side of the above equation represents the wave amplitude change in one time step Δt, while the second term represents the phase change per time step since the first term is a real number and the second term is an imaginary number. Let $\lambda = e^{\omega_i\Delta t}$, (12.3.10) becomes

$$\lambda e^{-i\omega_r\Delta t} = 1 - i C \sin k\Delta x. \tag{12.3.13}$$

Equating the real and imaginary parts yields

$$\lambda \cos \omega_r\Delta t = 1, \text{ and}$$
$$\lambda \sin \omega_r\Delta t = C \sin k\Delta x. \tag{12.3.14}$$

Summing the squares of the above two equations gives

$$\lambda = \pm\sqrt{1 + C^2 \sin^2 k\Delta x}. \tag{12.3.15}$$

Combining (12.3.8), (12.3.12) and (12.3.13) leads to

$$u(x, t) = \hat{u}(k, \omega) \left(e^{ikn\Delta x} e^{-i\omega_r\tau\Delta t} \right) \lambda^\tau. \tag{12.3.16}$$

In the above equation, only the last term on the right side, λ^τ, may change the amplitude as time proceeds. The other two terms inside the bracket, $e^{ikn\Delta x}$ and $e^{-i\omega_r\tau\Delta t}$, can only change the phase of these waves. In order for the *numerical stability* to occur for (12.3.16), $|\lambda| < 1$ is required. However, (12.3.15) implies that the absolute value of λ is always greater than 1. Thus, the amplitude will grow with time and the *scheme of forward-in-time and centered-in-space is unconditionally unstable* since any small perturbations will grow indefinitely given enough time, and this type of *stability analysis* should be conducted prior to adopting a numerical scheme to approximate a differential equation.

(b) Forward-in-time and upstream-in-space scheme

Another two-time-level scheme that has been adopted in mesoscale models is the *forward-in-time and upstream-in-space scheme*. Under this scheme, the advection equation, (12.3.1), is approximated by

$$\frac{u_i^{\tau+1} - u_i^{\tau}}{\Delta t} = \begin{cases} -c\dfrac{u_i^{\tau} - u_{i-1}^{\tau}}{\Delta x}, & \text{if } c > 0 \qquad (12.3.17a) \\[2ex] -c\dfrac{u_{i+1}^{\tau} - u_i^{\tau}}{\Delta x}, & \text{if } c \leq 0 \qquad (12.3.17b) \end{cases}$$

To check the stability of this scheme, we consider a positive constant advection velocity, $c > 0$, without loss of generality and substituting (12.3.8) into (12.3.17a)

$$e^{-i(\omega_r + i\omega_i)\Delta t} = 1 - C\left(1 - e^{-ik\Delta x}\right), \qquad (12.3.18)$$

or

$$\lambda \cos \omega_r \Delta t = 1 - C(1 - \cos k\Delta x), \qquad (12.3.19a)$$

$$\lambda \sin \omega_r \Delta t = C \sin k\Delta x. \qquad (12.3.19b)$$

Summing the squares of the above equations yields

$$\lambda = \pm\sqrt{1 + 2C(\cos k\Delta x - 1)(1 - C)}. \qquad (12.3.20)$$

To ensure the numerical stability ($|\lambda| < 1$), the above equation requires

$$2C(\cos k\Delta x - 1)(1 - C) \leq 0. \qquad (12.3.21)$$

It holds if $C \leq 1$ since ($\cos k\Delta x - 1$) is always negative. Thus, (12.3.21) requires $\Delta t \leq \Delta x / c$ for the scheme to be stable, called the *CFL* (Courant–Friedrichs–Lewy) *stability criterion*. Thus, *the numerical scheme of forward-in-time and upstream-in-space is conditionally stable*.

The phase of a wave can also be altered by the application of numerical methods to solve differential equations. We can investigate the phase characteristics of the forward-in-time and upstream-in-space scheme by dividing (12.3.19b) by (12.3.19a) to obtain

$$\tan \omega_r \Delta t = \frac{C \sin k\Delta x}{1 + C(\cos k\Delta x - 1)}. \qquad (12.3.22)$$

Based on the above equation, the *numerical phase speed* is

$$\tilde{c}_p = \frac{\omega_r}{k} = \frac{1}{k\Delta t}\tan^{-1}\left[\frac{C\sin k\Delta x}{1 + C(\cos k\Delta x - 1)}\right]. \tag{12.3.23}$$

Equation (12.3.23) indicates that the finite difference scheme of forward-in-time and upstream-in-space is dispersive because the numerical phase speed is a function of the wave number. Similar to what occurs in the physical dispersion, here waves with different wavelengths propagate at different speeds. The wave therefore cannot preserve its original wave pattern, making it a *dispersive wave* (Chapter 3). Interestingly enough, wave dispersion can be induced numerically i.e., *numerical dispersion*, as well as physically.

Based on the advection equation, (12.3.1), and (12.3.7), the physical phase speed for the advection equation can be obtained,

$$c_p = \frac{\omega}{k} = c. \tag{12.3.24}$$

According to the above equation, the wave is physically nondispersive since its physical phase speed is independent of the wave number. However, the numerical method applied here does introduce a numerical wave mode and makes the wave dispersive artificially. The ratio of the *numerical phase speed* ($\tilde{c}_p$) to the *physical phase speed* (c_p) is

$$\frac{\tilde{c}_p}{c_p} = \frac{1}{kc\Delta t}\tan^{-1}\left[\frac{C\sin k\Delta x}{1 + C(\cos k\Delta x - 1)}\right], \tag{12.3.25}$$

which indicates that

$$\begin{aligned}\tilde{c}_p > c_p \quad &\text{when } 0.5 < C < 1.0 \text{ and,}\\ \tilde{c}_p < c_p \quad &\text{when } 0 < C < 0.5.\end{aligned} \tag{12.3.26}$$

A numerical method may introduce, in addition to numerical instability and numerical dispersion, *numerical damping*. For example, when $C = 0$, 1, or $k \approx 0$ (very long waves), we have $\lambda = 1$. The aforementioned case means that the amplitude will be kept the same, which indicates that no damping exists under these special conditions. However, this scheme tends to damp waves in general, especially at $C = 0.5$. To demonstrate the damping characteristics, we use a truncated Taylor series approximation to the advection equation, (12.3.1),

$$u_i^{\tau+1} \cong u_i^{\tau} + \frac{\partial u}{\partial t}\Delta t + \frac{1}{2!}\frac{\partial^2 u}{\partial t^2}\Delta t^2, \text{ and} \tag{12.3.27}$$

$$u_{i-1}^{\tau} \cong u_i^{\tau} - \frac{\partial u}{\partial x}\Delta x + \frac{1}{2!}\frac{\partial^2 u}{\partial x^2}\Delta x^2. \tag{12.3.28}$$

Again, we may assume $c > 0$ without loss of generality and substitute the above approximations into (12.3.17a),

$$\frac{\left(u_i^\tau + \frac{\partial u}{\partial t}\Delta t + \frac{1}{2!}\frac{\partial^2 u}{\partial t^2}\Delta t^2 \right) - u_i^\tau}{\Delta t} = -c\left[\frac{u_i^\tau - \left(u_i^\tau - \frac{\partial u}{\partial x}\Delta x + \frac{1}{2!}\frac{\partial^2 u}{\partial x^2}\Delta x^2 \right)}{\Delta x} \right],$$

which can be rearranged to be

$$\frac{\partial u}{\partial t} + c\frac{\partial u}{\partial x} + \frac{1}{2}\frac{\partial^2 u}{\partial t^2}\Delta t - \frac{1}{2}\frac{\partial^2 u}{\partial x^2}c\Delta x = 0. \tag{12.3.29}$$

In addition to the original advection equation, this particular numerical scheme artificially introduces two additional terms that are caused by truncation errors. If both Δt and Δx approach 0, then the above equation reduces to the original differential equation. Under normal conditions, however, they have nonzero values that introduce errors to the numerical solution.

One can prove that the last two terms are related to each other because

$$\frac{\partial^2 u}{\partial t^2} = c^2\frac{\partial^2 u}{\partial x^2}. \tag{12.3.30}$$

Substituting the above equation into (12.3.29) leads to

$$\frac{\partial u}{\partial t} + c\frac{\partial u}{\partial x} = \nu_c\frac{\partial^2 u}{\partial x^2}, \quad \nu_c \equiv \frac{1}{2}c\Delta x(1 - C), \quad C = \frac{c\Delta t}{\Delta x}. \tag{12.3.31}$$

The solution obtained from the finite difference equation is similar to that of the above differential equation, which contains a spurious diffusion term. The solution obtained, therefore, tends to be damped. The term ν_c is called the *numerical diffusion coefficient*. In the early days of numerical model development, the forward-in-time and upstream-in-space scheme was used extensively due to its two-time-level simplicity and its low memory storage requirement. However, its strong numerical damping characteristics and failure to preserve the proper phase have generated serious criticism. The technique is acceptable if advection or wave propagation is not dominant for a particular mesoscale phenomenon. However, if the subgrid mixing is important, ν_c must be smaller than the corresponding physically relevant turbulent exchange coefficient in order to avoid excess damping. Thus, the development of increasingly accurate three-time-level schemes and the advancement of computing facility make this scheme less attractive to mesoscale modelers.

(c) Lax–Wendroff scheme

This scheme was originally proposed by Lax and Wendroff (1960). The procedure for computation is based on the grid stencil shown in Fig. 12.5 and is described as follows. First, provisional values of u at provisional time step $\tau + 1/2$ and grid points $i - 1/2$ and $i + 1/2$ are calculated at the points denoted by the cross symbol by applying the forward-in-time and centered-in-space scheme:

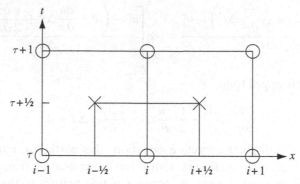

Fig. 12.5 Grid stencil for the Lax–Wendroff scheme.

$$\frac{u_{i+1/2}^{\tau+1/2} - (u_{i+1}^{\tau} + u_i^{\tau})/2}{\Delta t/2} = -c\frac{u_{i+1}^{\tau} - u_i^{\tau}}{\Delta x}, \tag{12.3.32a}$$

$$\frac{u_{i-1/2}^{\tau+1/2} - (u_i^{\tau} + u_{i-1}^{\tau})/2}{\Delta t/2} = -c\frac{u_i^{\tau} - u_{i-1}^{\tau}}{\Delta x}. \tag{12.3.32b}$$

Then, applying the second-order centered difference scheme in both time and space to values at grid points $u_{i+1/2}^{\tau+1/2}, u_{i-1/2}^{\tau+1/2}$, and u_i^{τ} gives,

$$\frac{u_i^{\tau+1} - u_i^{\tau}}{\Delta t} = -c\frac{u_{i+1/2}^{\tau+1/2} - u_{i-1/2}^{\tau+1/2}}{\Delta x}. \tag{12.3.33}$$

Finally, substituting the above provisional values of $u_{i+1/2}^{\tau+1/2}$ and $u_{i-1/2}^{\tau+1/2}$ from (12.3.32) into (12.3.33) leads to

$$u_i^{\tau+1} = u_i^{\tau} - \frac{C}{2}(u_{i+1}^{\tau} - u_{i-1}^{\tau}) + \frac{C^2}{2}(u_{i+1}^{\tau} - 2u_i^{\tau} + u_{i-1}^{\tau}). \tag{12.3.34}$$

From a computational point of view, $u_{i+1/2}^{\tau+1/2}$ and $u_{i-1/2}^{\tau+1/2}$ are provisional since they do not show up in (12.3.34); if (12.3.34) is used directly, there is no need to calculate or store them at all. The scheme as given by (12.3.34) is a two-time-level scheme, just like the forward-in-time upstream-in-space scheme discussed earlier, and has therefore the same low storage requirement.

The Lax–Wendroff scheme has a truncation error of $O[\Delta x^2] + O[\Delta t^2]$, meaning it has second-order accuracy in space and time, which is an improvement over the previous two-time-level scheme. It can be derived that

$$|\lambda| = \left[1 - 4C^2(1 - C^2)\sin^4\frac{k\Delta x}{2}\right]^{1/2}. \tag{12.3.35}$$

Therefore, the Lax–Wendroff scheme is stable if

$$C^2 \le 1 \quad \text{or} \quad \frac{|c|\Delta t}{\Delta x} \le 1. \tag{12.3.36}$$

That is, (12.3.36) satisfies the CFL stability criterion, and it can be proven that the last term of (12.3.34) serves as a damping term. In fact, the Lax–Wendroff scheme can be viewed as a modification of the forward-in-time and centered-in-space scheme with damping. For the shortest resolvable wavelength $2\Delta x$, we have $k = \pi/\Delta x$. Substituting k into (12.3.35) yields

$$|\lambda| = |1 - 2C^2|. \tag{12.3.37}$$

For the $4\Delta x$ wave, we have

$$|\lambda| = \left(1 - C^2 + C^4\right)^{1/2}. \tag{12.3.38}$$

Thus, the amount of damping is quite large for shorter waves.

The phase error, $1 - \tilde{c}_p/c$, can also be calculated from

$$\frac{\tilde{c}_p}{c} = \frac{\tan^{-1}\left\{-C\sin k\Delta x/\left(1 - C^2(1 - \cos k\Delta x)\right)\right\}}{-Ck\Delta x}. \tag{12.3.39}$$

Since $\tilde{c}_p$ is a function of wave number (k), *the Lax–Wendroff scheme* is *numerically dispersive*. The scheme has a predominantly lagging phase error except in cases of large wave numbers where $\sqrt{0.5} < C < 1$.

The Lax–Wendroff scheme has been modified by the following formula and is also known as the *Crowley scheme* (1968):

$$\begin{aligned} u_i^{\tau+1} &= u_i^\tau - \frac{C}{2}\left(u_{i+1}^\tau - u_{i-1}^\tau\right) + \frac{C^2}{2}\left(u_{i+1}^\tau - 2u_i^\tau + u_{i-1}^\tau\right) \\ &\quad + \frac{C}{12}\left(1 - C^2\right)\left(u_{i+2}^\tau - 2u_{i+1}^\tau + 2u_{i-1}^\tau - u_{i-2}^\tau\right) \end{aligned} \tag{12.3.40}$$

The last term in the above equation is the third-order space correction term.

(d) Multi-stage schemes

The advection equation (12.3.1) can be generalized through the following form:

$$\frac{\partial u}{\partial t} = F(u) \tag{12.3.41}$$

where $F(u)$ is the forcing term that includes the advection term of (12.3.1). To improve the accuracy of two-time-level schemes, the *multi-stage scheme* can be used to approximate (12.3.41),

$$\tilde{u}^{\tau+\alpha} = u^{\tau} + \alpha \Delta t F(u^{\tau}),$$
$$u^{\tau+1} = u^{\tau} + \Delta t [\beta F(\tilde{u}^{\tau+\alpha}) + (1-\beta)F(u^{\tau})]. \tag{12.3.42}$$

The above method reduces to the second-order *Runge–Kutta schemes* for any combinations of α and β leading to $\alpha\beta = 1/2$, which leads to a special scheme called the *Heun scheme* when $\alpha = 1$, $\beta = 1/2$. An example of a non-Runge–Kutta scheme is the *forward-backward (Matsuno) scheme* for which $\alpha = \beta = 1$ (Matsuno 1966).

12.3.2 Three-time-level schemes

(a) Adams–Bashforth scheme

Under the Adams–Bashforth scheme, (12.3.41) is approximated by

$$u^{\tau+1} = u^{\tau} + \Delta t \left(\frac{3}{2} F(u^{\tau}) - \frac{1}{2} F(u^{\tau-1}) \right). \tag{12.3.43}$$

The advantage of this scheme is that it generates neither the time splitting produced by the leapfrog scheme nor the numerical diffusion produced by the upstream difference (e.g., Lilly 1965; Durran 1998). The nonlinear advection terms and energy components may generate large errors with this scheme.

(b) Leapfrog-in-time and centered-in-space schemes

The advection equation can also be approximated by the leapfrog (second-order centered) in time and second-order centered difference in space scheme,

$$\frac{u_i^{\tau+1} - u_i^{\tau-1}}{2\Delta t} = -c \frac{u_{i+1}^{\tau} - u_{i-1}^{\tau}}{2\Delta x}. \tag{12.3.44}$$

Again, in order to examine the stability of this scheme, we substitute the wave solution (12.3.8) into (12.3.44). This yields

$$\left(\lambda e^{-i\omega_r \Delta t} - \frac{1}{\lambda e^{-i\omega_r \Delta t}} \right) = -2iC \sin k\Delta x, \ \lambda \equiv e^{\omega_i \Delta t}; \ C \equiv \frac{c\Delta t}{\Delta x}. \tag{12.3.45}$$

The above equation can be rearranged to obtain

$$\lambda^2 e^{-2i\omega_r \Delta t} + 2i\alpha \lambda e^{-i\omega_r \Delta t} - 1 = 0, \tag{12.3.46}$$

where $\alpha = C \sin k\Delta x$ is a temporary parameter. Regarding $\lambda e^{-i\omega_r \Delta t}$ as the unknown in the above equation, we obtain

$$\lambda e^{-i\omega_r \Delta t} = -i\alpha \pm \sqrt{1 - \alpha^2}. \tag{12.3.47}$$

Separating the real and imaginary parts of the above equation gives two possible cases, namely, (1) $\alpha^2 \leq 1$ and (2) $\alpha^2 > 1$. In case 1, we have

$$\lambda \cos \omega_r \Delta t = \pm \sqrt{1 - \alpha^2}, \tag{12.3.48a}$$

$$\lambda \sin \omega_r \Delta t = \alpha. \tag{12.3.48b}$$

Summing the squares of the above two equations yields

$$|\lambda| = 1.$$

Therefore, the *leapfrog-in-time and centered-in-space scheme is neutral when* $\alpha^2 \le 1$. In case 2, we have

$$\lambda \cos \omega_r \Delta t = 0, \tag{12.3.49a}$$

$$\lambda \sin \omega_r \Delta t = \alpha \mp \sqrt{\alpha^2 - 1}. \tag{12.3.49b}$$

Summing the squares of the above two equations leads to

$$\lambda^2 = \left(\alpha \mp \sqrt{\alpha^2 - 1} \right)^2 \quad \text{if} \quad \alpha^2 > 1. \tag{12.3.50}$$

We may claim that this scheme is unstable when $\alpha^2 > 1$. To prove that this scheme is unstable when $\alpha^2 > 1$, we only need to find one counter example for which $|\lambda| > 1$. We can assume $\alpha = 1 + \varepsilon$, where ε is a small positive number. Substituting α into the positive root of (12.3.50) gives

$$|\lambda| = 1 + \varepsilon \mp \sqrt{2\varepsilon + \varepsilon^2}. \tag{12.3.51}$$

Since either root is possible, we look at the solution with the positive root,

$$|\lambda| = 1 + \varepsilon + \sqrt{2\varepsilon + \varepsilon^2}. \tag{12.3.52}$$

The above equation gives $|\lambda| > 1$. Therefore, *the leapfrog-in-time and centered-in-space scheme is unstable when* $\alpha^2 > 1$. The stability is thus retained only when $\alpha^2 \le 1$. Based on the definition of α for (12.3.46), it requires

$$C^2 \sin^2 k\Delta x \le 1. \tag{12.3.53}$$

Since the maximum value of the sine square function is 1, the above equation is satisfied when

$$|C| \le 1. \tag{12.3.54}$$

In fact, the CFL criterion is not only a necessary condition but also a sufficient condition for the numerical stability of the leapfrog-in-time and centered-in-space scheme.

To demonstrate the phase characteristics associated with the leapfrog-in-time and second-order centered-in-space scheme, we divide (12.3.48b) by (12.3.48a):

$$\omega_r \Delta t = \tan^{-1}\left(\frac{\pm\alpha}{\sqrt{1-\alpha^2}}\right), \tag{12.3.55}$$

which gives us the numerical phase speeds

$$\tilde{c}_p = \frac{\omega_r}{k} = \frac{\pm 1}{k\Delta t}\tan^{-1}\left(\frac{\alpha}{\sqrt{1-\alpha^2}}\right). \tag{12.3.56}$$

The phase error can be obtained by comparing the numerical phase speed and the physical phase speed,

$$\frac{\tilde{c}_p}{c} = \frac{\pm 1}{kc\Delta t}\tan^{-1}\left(\frac{\alpha}{\sqrt{1-\alpha^2}}\right). \tag{12.3.57}$$

For $c > 0$, (12.3.56) has two solutions, one propagating to the right ($\tilde{c}_p > 0$), and the other propagating to the left ($\tilde{c}_p < 0$). The first solution represents the *physical mode* because it approximates the solution to the original advection equation. The second solution represents the *computational mode*, which is purely generated by the numerical scheme. If the computational mode is not damped, it slowly amplifies and eventually becomes unstable while it propagates in an opposite direction (left) to the physical mode during the simulation of wave propagation. The behavior of the computational (numerical) mode is a phenomenon known as *time-splitting*, and this scheme also induces numerical dispersion since the numerical phase speed is a function of wave number.

In summary, the leapfrog-in-time and second-order centered-in-space scheme preserves the amplitude when $\alpha^2 \leq 1$ (where $\alpha = C\sin k\Delta x$) but can generate phase errors. Figure 12.6 compares the numerical solution of the leapfrog-in-time and second-order centered-in-space scheme to that of the forward-in-time and upstream-in-space scheme and the analytical solution. The numerical solutions are initialized by a rectangular wave centered at $x = 0$. The figure indicates that this scheme preserves the amplitude of the initial rectangular wave much better than the forward-in-time and upstream-in-space scheme. However, it produces more severe numerical dispersion than the forward-in-time and upstream-in-space scheme (e.g., Haltiner and Williams 1980).

In addition to the second-order centered schemes for spatial difference, a scheme with higher-order accuracy can be derived. For example, consider the following Taylor series expansions for $f(x+\Delta x)$ and $f(x-\Delta x)$,

$$f(x+\Delta x) = f(x) + f'(x)\Delta x + f''(x)\frac{\Delta x^2}{2!} + f'''(x)\frac{\Delta x^3}{3!} + \cdots \tag{12.3.58}$$

$$f(x-\Delta x) = f(x) - f'(x)\Delta x + f''(x)\frac{\Delta x^2}{2!} - f'''(x)\frac{\Delta x^3}{3!} + \cdots \tag{12.3.59}$$

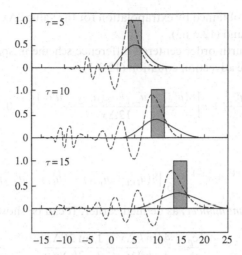

Fig. 12.6 An example of numerical damping and dispersion. Comparisons between the analytical solution (solid curve with shading) and numerical solutions of applying the leapfrog in time and second-order centered in space scheme (dashed curve) and the forward-in-time and upstream-in-space scheme (heavy solid curve) to the advection equation with an initial rectangular wave centered at $x = 0$. Three nondimensional times are shown. (Adapted after Wurtele 1961.)

Subtracting (12.3.59) from (12.3.58) leads to

$$f(x + \Delta x) - f(x - \Delta x) = 2f'(x)\Delta x + \frac{1}{3}f'''(x)\Delta x^3 + \cdots \qquad (12.3.60)$$

Now, consider the following Taylor series expansions for $f(x + 2\Delta x)$ and $f(x - 2\Delta x)$,

$$f(x + 2\Delta x) = f(x) + 2f'(x)\Delta x + f''(x)\frac{4\Delta x^2}{2!} + f'''(x)\frac{8\Delta x^3}{3!} + \ldots \qquad (12.3.61)$$

$$f(x - 2\Delta x) = f(x) - 2f'(x)\Delta x + f''(x)\frac{4\Delta x^2}{2!} - f'''(x)\frac{8\Delta x^3}{3!} + \ldots \qquad (12.3.62)$$

Subtracting (12.3.62) from (12.3.61) leads to

$$f(x + 2\Delta x) - f(x - 2\Delta x) = 4f'(x)\Delta x + \frac{8}{3}f'''(x)\Delta x^3 + \ldots \qquad (12.3.63)$$

Eliminating $f'''(x)$ terms from (12.3.60)–(12.3.63) yields

$$f'(x) = \frac{8[f(x + \Delta x) - f(x - \Delta x)] - [f(x + 2\Delta x) - f(x - 2\Delta x)]}{12\Delta x} + O(\Delta x^4).$$

$$(12.3.64)$$

Equation (12.3.64) gives a *fourth-order centered difference* scheme for $f'(x)$. Note that the boundary points are approximated by adjacent interior points. It can be shown that (12.3.64) can be obtained by extrapolation for the value $2\Delta x/3$ of the quotients of $f'(x)$ from (12.3.62) and (12.3.63).

Now apply the fourth-order centered difference scheme in space and the leapfrog scheme in time to the advection equation (12.3.1),

$$\frac{u_i^{\tau+1} - u_i^{\tau-1}}{2\Delta t} + c\left[\frac{8(u_{i+1}^{\tau} - u_{i-1}^{\tau}) - (u_{i+2}^{\tau} - u_{i-2}^{\tau})}{12\Delta x}\right] = 0. \tag{12.3.65}$$

Solve for $u_i^{\tau+1}$,

$$u_i^{\tau+1} = u_i^{\tau-1} - \frac{C}{6}\left[8(u_{i+1}^{\tau} - u_{i-1}^{\tau}) - (u_{i+2}^{\tau} - u_{i-2}^{\tau})\right], \tag{12.3.66}$$

where C is the *Courant number*, as defined earlier. It can be shown that

$$\frac{\tilde{c}_p}{c} = \frac{4}{3}\frac{\sin k\Delta x}{k\Delta x} - \frac{1}{3}\frac{\sin 2k\Delta x}{2k\Delta x} \tag{12.3.67}$$

for the fourth-order centered difference scheme. Compared with that of the second-order centered in space scheme,

$$\frac{\tilde{c}_p}{c} = \frac{\sin k\Delta x}{k\Delta x}, \tag{12.3.68}$$

we have

$$\tilde{c}_p = c\left[1 - \frac{4}{5!}(k\Delta x)^2 + \cdots\right] \quad \text{for the fourth-order scheme, and} \tag{12.3.69}$$

$$\tilde{c}_p = c\left[1 - \frac{1}{3!}(k\Delta x)^2 + \cdots\right] \quad \text{for the second-order scheme.} \tag{12.3.70}$$

Both schemes are therefore numerically dispersive. However, using the fourth-order scheme greatly increases in accuracy of the phase speed for longer waves (smaller k) (Fig. 12.7). In addition, for shorter waves there is more numerical dispersion associated with the fourth-order scheme since the slope of c_4 is larger than c_2.

12.4 Implicit schemes

With the above finite difference schemes, the advection term is evaluated at time step τ, thus the variables at time step $\tau + 1$ can be predicted explicitly by those at time step τ and/or τ-1. These schemes are referred to as *explicit schemes*. However, all explicit advection schemes are, at most, conditionally stable numerically, and the CFL condition type stability criterion imposes a severe restriction on the time interval with a resultant increase in computational time. This restriction can be relaxed by evaluating

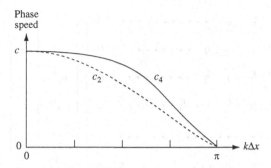

Fig. 12.7 The numerical phase speeds of the second-order (c_2) and fourth-order (c_4) centered diffcrence schemes. The phase speed of the linear advection equation (12.3.1) is c. (Adapted after Mesinger and Arakawa 1976.)

the advection term at time step $\tau + 1$. For example, the spatial differencing of (12.3.5), i.e. the forward-in-time and second-order-centered in space scheme, can be applied at time step $\tau + 1$,

$$\frac{u_i^{\tau+1} - u_i^{\tau}}{\Delta t} + c\left(\frac{u_{i+1}^{\tau+1} - u_{i-1}^{\tau+1}}{2\Delta x}\right) = 0, \qquad (12.4.1)$$

and this method is called the *Euler implicit method* (e.g., Tannehill *et al.* 1997). In order to solve for u at time step $\tau + 1$, we move all of them to the left side

$$-\frac{C}{2}u_{i-1}^{\tau+1} + u_i^{\tau+1} + \frac{C}{2}u_{i+1}^{\tau+1} = u_i^{\tau}, \quad i = 1, 2, 3, \ldots N - 1, \qquad (12.4.2)$$

where N is the total grid number. Thus, one cannot solve the equation for a general point, $u_i^{\tau+1}$, alone. Instead, we have to solve the system of algebraic equations, as shown in Fig. 12.8. In Fig. 12.8, we have assumed that boundary conditions are as follows: $u(0, t) = u_0^{\tau+1} = 0$ and $u(N\Delta x, t) = u_N^{\tau+1} = 0$. If a zero-gradient boundary condition $\partial u/\partial x = 0$ is imposed, then the coefficients at the upper-left and lower-right corners of the matrix in Fig. 12.8 become $1 - C/2$ and $1 + C/2$, respectively. In general, one can introduce a weighting factor α and replace (12.4.2) by

$$-\frac{\alpha C}{2}u_{i-1}^{\tau+1} + u_i^{\tau+1} + \frac{\alpha C}{2}u_{i+1}^{\tau+1} = \frac{(1-\alpha)C}{2}u_{i-1}^{\tau} + u_i^{\tau}$$
$$-\frac{(1-\alpha)C}{2}u_{i+1}^{\tau}, \quad i = 1, 2, 3, \ldots N - 1. \qquad (12.4.3)$$

If $\alpha = 0$, the above formula reduces to the completely explicit scheme, i.e. the forward-in-time and centered-in-space scheme, (12.3.6). If $\alpha = 1$, the Euler implicit scheme, (12.4.2) is recovered. To find out the numerical stability, we substitute (12.3.8) into (12.4.2) to obtain

$$\begin{pmatrix} 1 & \frac{C}{2} & 0 & \cdots\cdots\cdots\cdots & 0 \\ -\frac{C}{2} & 1 & \frac{C}{2} & 0 & \cdots\cdots\cdots & 0 \\ 0 & -\frac{C}{2} & 1 & \frac{C}{2} & \cdots\cdots\cdots & 0 \\ 0 & 0 & -\frac{C}{2} & 1 & \frac{C}{2}\cdots\cdots & 0 \\ \cdots\cdots\cdots\cdots\cdots\cdots\cdots\cdots \\ 0 & 0 & 0 & \cdots & -\frac{C}{2} & 1 & \frac{C}{2} & 0 \\ 0 & 0 & 0 & \cdots\cdots & -\frac{C}{2} & 1 & \frac{C}{2} \\ 0 & 0 & 0 & \cdots\cdots\cdots & -\frac{C}{2} & 1 \end{pmatrix} \begin{pmatrix} u_1^{\tau+1} \\ u_2^{\tau+1} \\ u_3^{\tau+1} \\ u_4^{\tau+1} \\ \vdots \\ u_{N-3}^{\tau+1} \\ u_{N-2}^{\tau+1} \\ u_{N-1}^{\tau+1} \end{pmatrix} = \begin{pmatrix} u_1^{\tau} \\ u_2^{\tau} \\ u_3^{\tau} \\ u_4^{\tau} \\ \vdots \\ u_{N-3}^{\tau} \\ u_{N-2}^{\tau} \\ u_{N-1}^{\tau} \end{pmatrix}$$

Fig. 12.8 The system of algebraic equations of (12.4.2) for the Euler implicit method with boundary conditions, $u(0, t) = u_0^{\tau+1} = 0$ and $u(N\Delta x, t) = u_N^{\tau+1} = 0$, where N is the total number of grid intervals.

$$|\lambda| = \frac{1}{\sqrt{1 + C^2 \sin^2 k\Delta x}}. \tag{12.4.4}$$

The above equation implies that *the Euler implicit scheme is unconditionally stable* because the right-hand side of (12.4.4) is always less than 1. In general, the use of an implicit scheme permits larger time steps than the explicit form without causing numerical instability. To invert the matrix, either direct (e.g., Gaussian elimination, LU decomposition) or iterative (e.g., Jacobi, Gauss–Seidel, relaxation) methods can be applied. Discussions of these methods can be found in numerical analysis textbooks.

To lessen the computational burden, *semi-implicit schemes* have been developed. In a semi-implicit scheme, terms which are primarily responsible for the propagation of faster waves (e.g., gravity waves) are treated implicitly, while other terms are treated explicitly. For example, the linear shallow water x-momentum equation, (3.4.7), can be shown by the *trapezoidal semi-implicit scheme* (with primes dropped) as

$$\left(\frac{u_i^{\tau+1} - u_i^{\tau}}{\Delta t}\right) + U\left(\frac{u_{i+1}^{\tau} - u_{i-1}^{\tau}}{2\Delta x}\right) + \frac{g}{2}\left[\left(\frac{h_{i+1}^{\tau} - h_{i-1}^{\tau}}{2\Delta x}\right) + \left(\frac{h_{i+1}^{\tau+1} - h_{i-1}^{\tau+1}}{2\Delta x}\right)\right] = 0. \tag{12.4.5}$$

Note that the advection term is treated in explicit manner and the spatial derivative is centered at the $\tau + 1/2$ time step by averaging values at time steps τ and $\tau + 1$, and it can be shown that *the trapezoidal semi-implicit scheme is unconditionally stable* (Mesinger and Arakawa, 1976). Although theoretically a very large Courant number (or say, time interval) can be used with implicit schemes, practically there is a limit in its use. For example, the trapezoidal semi-implicit scheme has a serious phase error when the Courant number is large (Haltiner and Williams, 1980).

12.5 Semi-Lagrangian methods

Ideally, one should be able to integrate the advection equation by following the fluid particles in a Lagrangian manner, so that the local rate of change and advection terms do not have to be considered separately. In fact, taking a *Lagrangian approach*, a graphical method has been developed to solve the barotropic vorticity equation using a single time step of 24 h by following a set of fluid particles (Fjortoft 1952). However, in general a set of fluid particles, which are initially distributed regularly, will soon become greatly deformed and are thus rendered unsuitable for numerical integration (Welander 1955). To avoid this difficulty, the *semi-Lagrangian method* (occasionally referred to as *quasi-Lagrangian method*) whereby a set of particles that arrive at a regular set of grid points are traced backward over a single time step to their departure points was proposed (Wiin-Nielsen 1959). The values of the dynamical quantities at the departure points are obtained by interpolating known values at neighboring grid points. Note that in a semi-Lagrangian method, the set of fluid particles in question changes at each time step, which is different from the pure Lagrangian method. In addition, a combination of these schemes, i.e. *semi-Lagrangian semi-implicit scheme*, has been proposed (Robert 1982; Staniforth and Côté 1991).

To examine the stability property of the semi-Lagrangian method, we consider the one-dimensional nonlinear advection equation in the form of total derivative,

$$\frac{D\psi}{Dt} = 0, \qquad (12.5.1)$$

where $D/Dt \equiv \partial/\partial t + u \partial/\partial x$ and ψ is any variable under consideration. By integrating over the trajectory of a fluid particle that arrives at a grid point $i\Delta x$, denoted as P in Fig. 12.9, and at time $(\tau + 1)\Delta t$, we have

$$\psi_i^{\tau+1} = \psi_*^\tau, \qquad (12.5.2)$$

where ψ_*^τ is the value of ψ at the departure point of the particle at time $\tau\Delta t$. The value ψ_*^τ is obtained by polynomial interpolation from the neighboring grid points. The stability and accuracy of the scheme depends on the interpolation method used. For example, we may consider the linear interpolation from the surrounding grid points $(i - p)$ and $(i - p - 1)$ for ψ_*^τ,

$$\frac{\psi_{i-p}^\tau - \psi_*^\tau}{u\Delta t - p\Delta x} = \frac{\psi_{i-p}^\tau - \psi_{i-p-1}^\tau}{\Delta x}, \qquad (12.5.3)$$

where u is the advection velocity as represented in (12.5.1). The above equation may be rearranged as

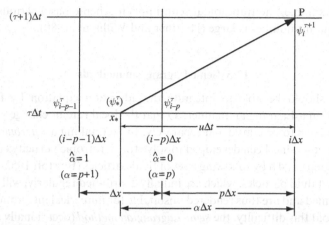

Fig. 12.9 A schematic of the semi-Lagrangian method. A fluid particle that arrives at a grid point $i\Delta x$ and at time $(\tau + 1)\Delta t$ is denoted as P, which is located at x_* and at time $\tau\Delta t$. The value of the variable at this time and location (ψ_*^τ) is obtained by polynomial interpolation from the neighboring grid points, ψ_{i-p-1}^τ and ψ_{i-p}^τ, as expressed in (12.5.4) or (12.5.5).

$$\psi_*^\tau = \psi_{i-p}^\tau - \left(\frac{u\Delta t}{\Delta x} - p\right)\left(\psi_{i-p}^\tau - \psi_{i-p-1}^\tau\right). \tag{12.5.4}$$

or

$$\psi_*^\tau = \psi_{i-p}^\tau - \hat{\alpha}(\psi_{i-p}^\tau - \psi_{i-p-1}^\tau), \tag{12.5.5}$$

where

$$\hat{\alpha} = \alpha - p, \quad \alpha = u\Delta t/\Delta x. \tag{12.5.6}$$

Therefore, from (12.5.2) we have

$$\psi_i^{\tau+1} = \psi_{i-p}^\tau - \hat{\alpha}(\psi_{i-p}^\tau - \psi_{i-p-1}^\tau). \tag{12.5.7}$$

According to (12.5.6) and Fig. 12.9, $\hat{\alpha}$ is the fractional part, and p is the integral part after advection of a non-dimensional distance $u\Delta t/\Delta x$.

To examine whether the semi-Lagrangian method is numerically stable or not, we may again assume a wave-like solution,

$$\psi_i^\tau = \hat{\psi}e^{-i\omega_r\tau\Delta t}e^{ikn\Delta x}\lambda^\tau. \tag{12.5.8}$$

Substituting (12.5.8) into (12.5.7) yields

$$\lambda^2 = 1 - 2\hat{\alpha}(1 - \hat{\alpha})(1 - \cos k\Delta x), \quad \lambda \equiv e^{\omega_i\Delta t}. \tag{12.5.9}$$

Thus, in order to have a numerically stable solution ($|\lambda| \leq 1$), we require

$$0 \leq \hat{\alpha} \leq 1. \tag{12.5.10}$$

That is, the departure points must lie within the interpolation interval $(i - p - 1, i - p)$: however, this is just the choice of points used for interpolation. Therefore, *the semi-Lagrangian scheme is unconditionally stable.*

The *semi-implicit method* may be incorporated into the integration by treating the other terms, such as the pressure gradient force term in the momentum equation, as time averages along the trajectory, while the total time derivative is evaluated by either leapfrog, forward or other time difference schemes. To elucidate this, let us consider the following Boussinesq, horizontal momentum equation:

$$\frac{Du}{Dt} - fv + \frac{1}{\rho_0}\frac{\partial p}{\partial x} = 0. \tag{12.5.11}$$

The total derivative of the above equation may be approximated by the forward-in-time scheme,

$$\frac{Du}{Dt} = \frac{u(x, t + \Delta t) - u(x - a, t)}{\Delta t}, \tag{12.5.12}$$

where

$$u = u(x - a, t)\Delta t \tag{12.5.13}$$

Equation (12.5.12) can be solved by using an iterative method to obtain the upstream displacement or the departure point, a. We can apply the semi-implicit approximation to the other terms on the left-hand side of (12.5.11)

$$\psi_{av}^t = \frac{\psi(x, t + \Delta t) + \psi(x - a, t)}{2}, \tag{12.5.14}$$

where the subscript av denotes the time average. Then the horizontal momentum equation can be approximated by the semi-implicit semi-Lagrangian scheme,

$$\frac{u(x, t + \Delta t) - u(x - a, t)}{\Delta t} - fv_{av}^t + \left(\frac{1}{\rho_0}\frac{\partial p}{\partial x}\right)_{av}^t = 0. \tag{12.5.15}$$

The above equation can also be rewritten as

$$\left([u]_x^{t+\Delta t} - [u]_{x-a}^t\right) - \frac{f\Delta t}{2}\left([v]_x^{t+\Delta t} + [v]_{x-a}^t\right) + \frac{\Delta t}{2\rho_0}\left(\left[\frac{\partial p}{\partial x}\right]_x^{t+\Delta t} + \left[\frac{\partial p}{\partial x}\right]_{x-a}^t\right) = 0. \tag{12.5.16}$$

Moving all terms at time $t + \Delta t$ to the left-hand side gives the following equation of the semi-implicit semi-Lagrangian equation:

$$[u]_x^{t+\Delta t} - \frac{f\Delta t}{2}[v]_x^{t+\Delta t} + \frac{\Delta t}{2\rho_0}\left[\frac{\partial p}{\partial x}\right]_x^{t+\Delta t} = [u]_{x-a}^t + \frac{f\Delta t}{2}[v]_{x-a}^t - \frac{\Delta t}{2\rho_0}\left[\frac{\partial p}{\partial x}\right]_{x-a}^t. \tag{12.5.17}$$

This forms a set of linear algebraic equations, which can be written in matrix form. Thus, a method for inverting the matrix can be applied to obtain the solution for time step $t + \Delta t$. Note that the advantage of the semi-implicit semi-Lagrangian scheme is that it is unconditionally stable so that a relatively large time step can be used. The disadvantage of this scheme is that the iterative method for finding the departure points and the method for inverting the matrix is computationally expensive.

Appendix 12.1:

Formulas for Finite Difference Approximations of Derivatives (Adapted after Gerald and Wheatley 2003)

Formulas for the first derivative:

$$f'(x) = \frac{f(x + \Delta x) - f(x)}{\Delta x} + O(\Delta x), \qquad \text{(forward difference)}$$

$$f'(x) = \frac{f(x + \Delta x) - f(x - \Delta x)}{2\Delta x} + O(\Delta x^2), \qquad \text{(second-order centered difference)}$$

$$f'(x) = \frac{-f(x + 2\Delta x) + 4f(x + \Delta x) - 3f(x)}{2\Delta x} + O(\Delta x^2),$$

(as above except for a left boundary point)

$$f'(x) = \frac{-f(x + 2\Delta x) + 8f(x + \Delta x) - 8f(x - \Delta x) + f(x - 2\Delta x)}{12\Delta x} + O(\Delta x^4).$$

(fourth-order centered difference)

Formulas for the second derivative:

$$f''(x) = \frac{f(x + 2\Delta x) - 2f(x + \Delta x) + f(x)}{\Delta x^2} + O(\Delta x), \qquad \text{(forward difference)}$$

$$f''(x) = \frac{f(x + \Delta x) - 2f(x) + f(x - \Delta x)}{\Delta x^2} + O(\Delta x^2),$$

(second-order centered difference)

$$f''(x) = \frac{-f(x + 2\Delta x) + 16f(x + \Delta x) - 30f(x) + 16f(x - \Delta x) - f(x - 2\Delta x)}{12\Delta x^2}$$
$$+ O(\Delta x^4). \qquad \text{(fourth-order centered difference)}$$

$$f''(x) = \frac{-f(x + 3\Delta x) + 4f(x + 2\Delta x) - 5f(x + \Delta x) + 2f(x)}{\Delta x^2} + O(\Delta x^2),$$

(as above except for a left boundary point)

Formulas for the third derivative:

$$f'''(x) = \frac{f(x + 2\Delta x) - 2f(x + \Delta x) + 2f(x - \Delta x) - f(x - 2\Delta x)}{2\Delta x^3} + O(\Delta x^2).$$

(centered difference)

$$f'''(x) = \frac{f(x + 3\Delta x) - 3f(x + 2\Delta x) + 3f(x + \Delta x) - f(x)}{\Delta x^3} + O(\Delta x),$$

(for a left boundary point)

Formulas for the fourth derivative:

$$f^{iv}(x) = \frac{f(x + 2\Delta x) - 4f(x + \Delta x) + 6f(x) - 4f(x - \Delta x) + f(x - 2\Delta x)}{\Delta x^4} + O(\Delta x^2).$$

(centered difference)

$$f^{iv}(x) = \frac{f(x + 4\Delta x) - 4f(x + 3\Delta x) + 6f(x + 2\Delta x) - 4f(x + \Delta x) + f(x)}{\Delta x^4} + O(\Delta x),$$

(for a left boundary point)

References

Colella, P. and P. R. Woodward, 1984. The piecewise parabolic method (PPM) for gas dynamical simulations. *J. Comput. Phys.*, **54**, 174–201.

Crowley, W. P., 1968. Numerical advection experiments. *Mon. Wea. Rev.*, **96**, 1–11.

Durran, D. R., 1998. *Numerical Methods for Wave Equations in Geophysical Fluid Dynamics.* Springer-Verlag.

Fjortoft, 1952. On a numerical method of integrating the barotropic vorticity equation. *Tellus*, **4**, 179–94.

Gerald, C. F. and P. O. Wheatley, 2003. *Applied Numerical Analysis.* 7th edn., Addison-Wesley.

Haltiner, G. J. and R. T. Williams, 1980. *Numerical Prediction and Dynamic Meteorology.* 2nd edn., John Wiley & Sons.

Lax, P. and B. Wendroff, 1960. Systems of conservation laws. *Comm. Pure Appl. Math.*, **13**, 217–37.

Lilly, D. K., 1965. On the computational stability of numerical solutions of time-dependent non-linear geophysical fluid dynamics problems. *Mon. Wea. Rev.*, **93**, 11–26.

Matsuno, T., 1966. Numerical integrations of the primitive equations by a simulated backward difference method. *J. Meteor. Soc. Japan*, Ser. 2, **44**, 76–84.

Mesinger, F. and A. Arakawa, 1976. Numerical methods used in atmospheric models. Vol. 1, GARP Publ. Ser., 17, WMO-ICSU.

Pielke, R. A., 2002. *Mesoscale Meteorological Modeling.* 2nd edn., Academic Press.

Robert, A., 1982. A semi-Lagrangian and semi-implicit numerical integration for the primitive meteorological equations. *J. Meteor. Soc. Japan*, **60**, 319–24.

Staniforth, A. and J. Côté, 1991. Semi-Lagrangian integration schemes for atmospheric models – A review, *Mon. Wea. Rev.*, **119**, 2206–23.

Tannehill, J. C., D. A. Anderson, and R. H. Pletcher, 1997. *Computational Fluid Mechanics and Heat Transfer*. 2nd edn., Taylor & Francis.
Welander, P., 1955. Studies on the general development of motion in a two-dimensional, ideal fluid. *Tellus*, **7**, 141–56.
Wiin-Nielsen, A., 1959. On the application of trajectory methods in numerical forecasting. *Tellus*, **11**, 180–96.
Wurtele, M. G., 1961. On the problem of truncation error. *Tellus*, **13**, 379–91.

Problems

12.1 Make a sketch similar to Fig. 12.1 except for the backward difference scheme of (12.2.6).

12.2 Replace the sine wave by a cosine wave in (12.2.12), and show that you can obtain the same conclusion as (12.2.17).

12.3 Show that if $C = 0.5$, then $2\Delta x$ waves are completely eliminated by the forward-in-time and upstream-in-space scheme. Note that the $2\Delta x$ wave can be represented by a function with two constants, a and b: $u_i^\tau = a + (-1)_b^i$, where i is an integer.

12.4 Prove that (12.3.26) by plotting $\tilde{c}_p/c_p$ as a function of the Courant number (C), according to (12.3.25).

12.5 Prove (12.3.34).

12.6 Derive (12.3.35).

12.7 Prove (12.3.67): $\dfrac{\tilde{c}_p}{c} = \dfrac{4}{3}\dfrac{\sin k\Delta x}{k\Delta x} - \dfrac{1}{3}\dfrac{\sin 2k\Delta x}{2k\Delta x}$.

12.8 Prove that the coefficients at the upper-left and lower-right corners of the matrix in Fig. 12.8 for the Euler implicit scheme become $1 - C/2$ and $1 + C/2$, respectively.

Modeling projects

Project A

12.A1 An advection model (**advec1.f**), which is written in Fortran, solves a one-dimensional advection equation numerically. This model may be obtained from the website: http://www.cambridge.org/9780521808750. There is no need to be concerned with the numerical details of the model at this moment. Run the program to generate a data set and modify the plotting program (**sample_plot.f**) to plot the curves (not contours). You can adjust the flag "NPR" to write out the data for one or more time steps, and then plot these curves. Explain your results.

12.A2 Modify the program to have two different initial fields (e.g., the shape and amplitude of the initial field U1), plot and then explain the results.

12.A3 Repeat project 12.A2 for a nonlinear case by setting NL = 1. Describe your results.

Project B

12.B1 Make some sensitivity tests on the Advection Model (Project A) to find out the maximum time interval (Δt) that gives a well-behaved solution. Construct a table for the cases you have performed that shows the maximum amplitude of u' versus the time interval.

12.B2 Based on the Advection Model, develop it into a Tank Model with the one-layer shallow water equations

$$\frac{\partial u'}{\partial t} + (U + u')\frac{\partial u'}{\partial x} + g\frac{\partial h'}{\partial x} = 0,$$

$$\frac{\partial h'}{\partial t} + (U + u')\frac{\partial h'}{\partial x} + (H + h')\frac{\partial u'}{\partial x} = 0,$$

The above equation set has been discussed in Section 3.4. You may follow the following procedure for the model development:

(i) Define a new variable for h' with the same dimension as u' adding a finite difference approximation of $g\partial h'/\partial x$ to the advection model by mimicking the approximation of $\partial u'/\partial x$.

(ii) Once the horizontal momentum equation, i.e. the first equation, is finished, then simply formulate the continuity equation, i.e. the second equation, by replacing u' by h' and g by $(H + h')$.

(iii) When you implement the second equation of h' into your model, you need to adopt a lateral boundary condition analogous to that used in u' by simply replacing u' by h'.

12.B3 (a) Run the Tank Model with $U = 0$ by giving an initial field, at your choice, in either u' or h',

(b) Repeat (a), but use a basic wind, i.e. $U \neq 0$.
   Plot and explain your results. Perform both linear and nonlinear simulations by resetting the flag NL.

12.B4 Extend your tank model to a two-layer tank model with a bottom topography, i.e.

$$\frac{\partial u'}{\partial t} + (U + u')\frac{\partial u'}{\partial x} + g'\frac{\partial h'}{\partial x} = 0,$$

$$\frac{\partial h'}{\partial t} + (U + u')\frac{\partial h'}{\partial x} + (H + h' - h_s)\frac{\partial u'}{\partial x} = (U + u')\frac{\partial h_s}{\partial x}.$$

Then simulate a flow with $U = 10\,\text{m s}^{-1}$ over a bell-shaped bump, i.e. $h_s = h_m a^2/(x^2 + a^2)$. Describe and explain your results. Again, run for both linear and nonlinear cases. Perform several experiments to obtain both solutions with supercritical and subcritical flow regimes (see Section 3.4).

13

Numerical modeling of geophysical fluid systems

In Chapter 12, we discussed various numerical approximations of the advection equation. However, to simulate a geophysical fluid system, such as the atmosphere and ocean, within a finite region, we need to choose the domain size, grid size, time interval, total integration time, and consider other factors, such as the initial condition and boundary conditions. In addition, when we deal with a real fluid system, the governing equations are much more complicated than the one-dimensional, linear advection equation, as considered in Chapter 12. For example, we have to integrate three-dimensional nonlinear governing equations with several dependent variables, instead of a one-dimensional advection equation with only one variable. When a nonlinear equation is being approximated by numerical methods, one may face new problems, such as nonlinear computational instability and nonlinear aliasing. Special numerical techniques are required to avoid these types of problems. Once optimal approximate forms of the equations are selected, it is still necessary to define the domain and grid structure over which the partial differential equations will be approximated. In this chapter, we will also briefly describe on how to build a basic numerical model based on a set of partial differential equations governing a shallow water system, and a hydrostatic or nonhydrostatic continuously stratified fluid system.

13.1 Grid systems and vertical coordinates

The first step in developing a mesoscale numerical model is to determine the appropriate domain size, grid intervals, time interval, and total integration time of the model. Selection of these numerical parameters in a mesoscale model is usually based on both physical and numerical factors, such as: (1) spatial scales and dimensionality of the forcing and physical processes, (2) time scales of the forcing and the fluid responses to the forcing, (3) stability criterion of the adopted numerical scheme, (4) limitations of predictability of the atmospheric phenomena, and (5) availability of computer resources. To represent mesoscale atmospheric systems properly, it is required that: (a) the meteorologically significant variations in the dependent variables caused by the mesoscale forcing and fluid responses be contained within the model domain, and that (b) the averaging volume used to define the model grid

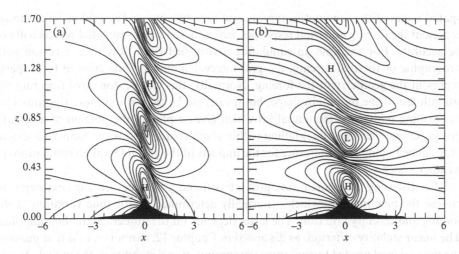

Fig. 13.1 Sensitivity of flow responses to the numerical domain size for a hydrostatic flow over a bell-shaped mountain. Displayed are the horizontal wind fields for different domain sizes: (a) 22.4a and (b) 12a, where a is the half-width of the mountain. The abscissa and ordinate are non-dimensionalized by a and $2\pi U/N$ (vertical wavelength), respectively. The Froude number is 1.2 for both cases.

spacing be small enough for the mesoscale forcing and responses to be accurately represented.

In making mesoscale numerical weather prediction (NWP), one needs to accurately represent multiscale processes in a finite domain. On one hand, it is important to capture the smaller-scale weather systems. Thus, the grid and time intervals should be fine enough to resolve small-scale processes. On the other hand, it is equally important to capture the larger-scale environment, which is responsible for the formation and modification of the smaller-scale weather systems. Therefore, the model domain should be large enough to contain the essential large-scale environment. Problems arise if a domain is inappropriately selected. For example, Fig. 13.1 shows two identical simulations for a hydrostatic flow over a bell-shaped mountain except with different domain sizes. With a smaller domain (Fig. 13.1b), the simulated horizontal wind field is more horizontally oriented and the wave weakens more rapidly with height. The difference in the horizontal orientation is due to that the upstream flow conditions at the left boundary, such as the wind speed and Brunt–Vaisala frequency of the basic flow, have been affected by the upstream propagating waves generated by the mountain. The flow field in the interior of the domain has thus been affected artificially by the wave reflection from the lateral boundaries, which means part of the waves cannot propagate upward, as the hydrostatic waves are supposed to do.

Once the domain interval is chosen, the next natural step is to choose an appropriate grid size (interval). The choice of grid interval used in a numerical model depends on spatial scales of the dominative forcing and flow responses. The grid interval should represent the forcing and fluid responses well; otherwise the simulated results will not be accurate. For example, in simulating a stratified fluid flow over a region with orographic or thermal forcing, the grid interval should be fine enough to properly represent the geometry of the forcing. Normally, this can be done by inspecting the smoothness of the mountain shape, but it can also be carried out more rigorously by performing power spectrum analysis which gives a plot of the portion of a signal's power (energy per unit time) falling within given frequency range. Numerous examples can be found in the literature for the impacts of the grid interval on simulations of weather systems.

After choosing the domain size, grid interval and grid structure, the next step is to choose the time interval, which is normally determined by the time variation of the forcing and stability criterion. The latter depends on the numerical method adopted. The linear stability criterion, as discussed in Chapter 12, can serve, as a first guess of the time interval needed to guarantee the computational stability of the model. As will be discussed later, nonlinear equations have a stricter criterion on computational stability, which is related to the time scale and the predictability of the weather phenomenon interested, as well as the computing resources available.

13.1.1 Grid systems

Different types of grid meshes have been adopted in mesoscale NWP models, such as (a) uniform grid mesh, (b) stretched grid mesh, (c) nested grid mesh, (d) movable grid mesh, (e) adaptive grid mesh, and (f) staggered grid mesh. Some of these grid meshes can be combined in a model.

For the *uniform grid mesh*, grid intervals are set to be equal in horizontal or in vertical. The advantages of this type of grid mesh are that it is relatively easy to code onto the computer, and it is relatively simple to input geographic features into the model. The disadvantages of the constant grid mesh include that it is difficult to properly incorporate both large and small features within the same model domain. For example, if one uses the same grid interval in vertical, it will be difficult to properly resolve the boundary layer circulation, while it is more than enough to resolve circulation in the free atmosphere, especially in the upper troposphere.

In order to improve the computing efficiency, *stretched grid mesh* is used, in which the grid intervals vary in space. The advantage of the stretched grid mesh is that with the same computing time, a much larger domain than the uniform grid mesh with the same number of grid points can be adopted for numerical simulations. Adopting a grid mesh stretched from a finer resolution near the surface to a relatively coarse resolution in the upper layer allows a model to resolve some smaller-scale disturbances present in the planetary boundary layer. If the grid intervals are stretched too abruptly, internal reflection may occur. In many mesoscale numerical simulations, the grid mesh is often stretched. The vertical turbulent mixing provides a measure of the needed grid resolution. The vertical stretch of the grid intervals may be defined by a known function, such as a logarithmic function, or specified by particular heights or horizontal distances. A two-way stretching may also be used. For example, in order to better represent the waves near a critical level, one may need to specify the grid mesh with very fine resolution near the critical level and stretches to coarser resolution both upward and downward.

An alternative approach to the stretched grid mesh is to insert a fine-mesh grid mesh, i.e. a *nested grid mesh*, within a coarse grid, which allows one to simulate smaller-scale features in the nested grid meshes. The smaller-scale features are not resolvable by the coarse grid mesh. During the simulation, the coarse mesh provides the boundary conditions for the fine mesh. There are two types of grid nesting techniques: (i) one-way nesting which only permits disturbances in the outer grid mesh to enter the finer nested grid mesh, but not the other way around, and (ii) two-way nesting in which the boundary values of the inner grid mesh are passed back to the outer grid mesh. Note that under the nested grid mesh system there is a discontinuity across the boundaries of the fine and coarse grid meshes. Nested-grid mesh techniques have been used in many research and operational models. The above-mentioned nested grid mesh can also be designed to move with a weather system, such as a thunderstorm, mesoscale convective system, front, midlatitude cyclone or tropical

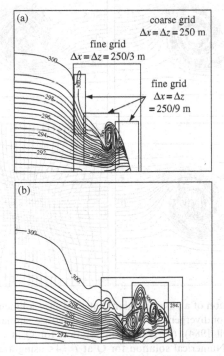

Fig. 13.2 Potential temperature fields at (a) 450 s and (b) 900 s in a cold pool collapse simulation using adaptive grid meshes. (After Skamarock 1989.)

cyclone. The advantage of a *moving grid mesh* is that it can follow the weather system and always provides a finer resolution needed in the vicinity of it.

To improve the nested and moving grid mesh techniques, *adaptive grid mesh* techniques have been developed. Grid points can be added in a structured manner through the placement of multiple and perhaps overlapping finer-scale grids in the domain. Figure 13.2 shows one example of using four adaptive grid meshes to simulate a cold pool collapse. Regions of strong potential temperature gradients along the gust front and Kelvin–Helmholtz billows are well simulated by using the adaptive grid mesh. In the adaptive grid mesh, a fixed number of grid points or collocation points may also be redistributed in a predetermined manner to provide locally increased resolution and thus an improved solution in certain regions of the domain. Figure 13.3 shows a simulation of kinematic frontogenesis, which is similar to a smoothed Rankine vortex being advected by a steady, nondivergent field, using this type of structured *continuous dynamic grid adaptation* (CDGA). With both 31×31 grid points, the simulated kinematic frontogenesis being advected by a steady, nondivergent flow field with

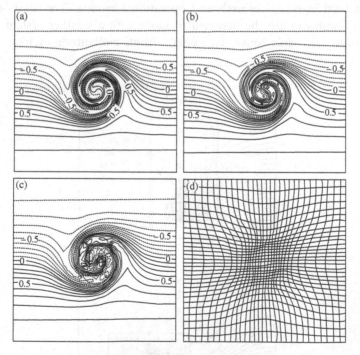

Fig. 13.3 (a) Exact solution of a passive scalar Q at $t = 4$ s for a kinematic frontogenesis being advected by a steady, nondivergent flow field with structure similar to that of a smoothed Rankine vortex (Doswell 1984). (b) Numerical solution for Q at $t = 4$ s on a fixed 31×31 uniform grid mesh. (c) Numerical solution for Q at $t = 4$ s using a continuous dynamic grid adaptation (CDGA). (d) Gridpoint distribution at $t = 4$ s with default parameters. (Adapted after Dietachmayer and Droegemeier 1992.)

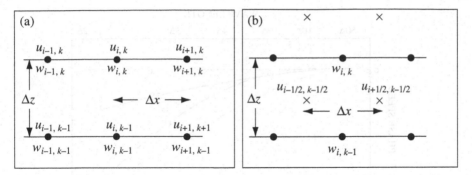

Fig. 13.4 Schematics of (a) an unstaggered grid mesh and (b) a staggered grid mesh for the computation of u and w with the two-dimensional incompressible continuity equation.

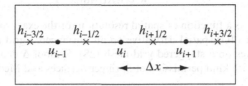

Fig. 13.5 A schematic of a staggered grid mesh for the shallow water system.

structure similar to that of a smoothed Rankine vortex (Fig. 13.3c) by using CDGA (Fig. 13.3d) is much better than that of uniform grid mesh (Fig. 13.3b), when comparing with the exact solution (Fig. 13.3a). One limitation of this type of structured adaptive grid mesh is that it is not suitable for dynamic grid adaptation because the grid generation requires a high degree of user interaction and user expertise. Thus, it is not an easy task to apply this type of method for real-case simulations. In order to resolve the problem, the *unstructured adaptive grid mesh* has been proposed in dealing with both large- and small-scale features without having to use a nested grid mesh (e.g., Bacon *et al.* 2000).

In order to preserve some conservation laws, a *staggered grid mesh* is used. In a staggered grid mesh, variables of a system of differential equations are defined at different grid points which are staggered with respect to each other. To elucidate the formulation of a staggered grid mesh, let us consider the two-dimensional incompressible continuity equation:

$$\frac{\partial u}{\partial x} + \frac{\partial w}{\partial z} = 0. \tag{13.1.1}$$

For an unstaggered grid mesh (Fig. 13.4a), a simple finite difference form can be written

$$w_{i,k} = w_{i,k-1} - \frac{u_{i+1,k-1/2} - u_{i-1,k-1/2}}{2\Delta x} \Delta z, \tag{13.1.2}$$

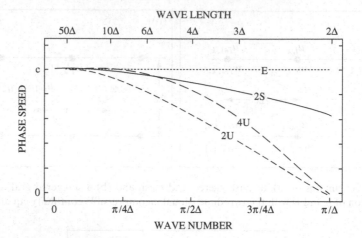

WAVE LENGTH

Fig. 13.6 Phase speed as a function of spatial resolution for the exact solution (E), for second-(2U) and fourth-order (4U) spatial derivative on an unstaggered grid mesh, and for second-order spatial derivatives on a staggered grid mesh (2S). Symbol Δ denotes one grid interval. (After Durran 1998, with kind permission of Springer Sciences and Media.)

where

$$u_{i+1,k-1/2} = (u_{i+1,k} + u_{i+1,k-1})/2; \ u_{i-1,k-1/2} = (u_{i-1,k} + u_{i-1,k-1})/2. \quad (13.1.3)$$

For a staggered grid mesh (Fig. 13.4b), a simple finite difference form can be written

$$w_{i,k} = w_{i,k-1} - \frac{u_{i+1/2,k-1/2} - u_{i-1/2,k-1/2}}{\Delta x} \Delta z, \quad (13.1.4)$$

where u is defined at half-way between the grid points at which w is defined. Thus, staggering the dependent variables as given by (13.1.4) increases the effective resolution by a factor of two, since derivatives are defined over an increment Δx, for instance, rather than $2\Delta x$, yet without requiring averaging as in (13.1.2).

To examine the computational stability and phase velocity associated with a staggered grid mesh, we can consider applying the leapfrog-in-time and second-order centered difference scheme to the two-dimensional, linear shallow water equations, (3.4.7) and (3.4.9), with $U = 0$ on a staggered grid mesh as shown in Fig. 13.5:

$$\frac{u_i^{\tau+1} - u_i^{\tau-1}}{2\Delta t} + g\frac{h_{i+1/2}^{\tau} - h_{i-1/2}^{\tau}}{\Delta x} = 0, \quad (13.1.5)$$

$$\frac{h_{i+1/2}^{\tau+1} - h_{i+1/2}^{\tau-1}}{2\Delta t} + H\frac{u_{i+1}^{\tau} - u_i^{\tau}}{\Delta x} = 0. \quad (13.1.6)$$

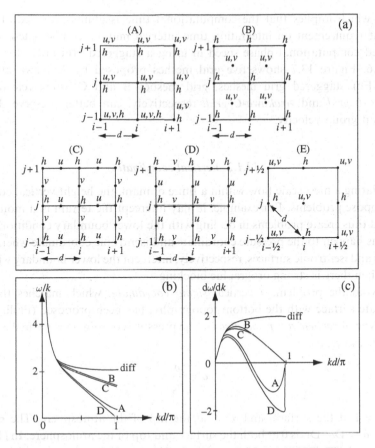

Fig. 13.7 (a) Five grid meshes proposed by Arakawa and Lamb (1977). The computational phase velocity ($c_p = \omega/k$) and the group velocity ($c_g = \partial\omega/\partial k$) analyzed as functions of kd/π for the four grids as shown in panels (b) and (c), respectively (Schoenstadt 1978). The differential equation solution is also included. These results use the following values: $\sqrt{gH} = 10^2 ms^{-1}$, $f = 10^{-4} s^{-1}$, and $d = 500\,m$, where d is the grid interval depicted in panel (a). (Adapted after Arakawa and Lamb 1977.)

The numerical dispersion relationship for the above system is

$$\sin \omega\Delta t = \pm \frac{2c\Delta t}{\Delta x} \sin\frac{k\Delta x}{2}, \qquad (13.1.7)$$

where $c = \pm\sqrt{gH}$ are the shallow-water phase speeds. Since the solutions of the two-dimensional shallow water wave system are neutral, a real ω is required, which implies that a stable solution requires $|C| = |c\Delta t/\Delta x| \leq 1/2$. The stability criterion for an unstaggered grid mesh can be derived to be $|C| \leq 1$. Therefore, the maximum time interval for a staggered grid system is half of the corresponding unstaggered mesh system, which implies that the computational time is almost doubled. The more stringent requirement on integration time interval may be compensated for by an improved computational phase speed in using a staggered grid mesh, as shown in Fig. 13.6. Figure 13.7 shows five grid meshes proposed by Arakawa and Lamb (1977). For staggered grid meshes, grid meshes B and C, often referred to as *Arakawa-B grid* and *Arakawa-C grid*, respectively, can better preserve the phase speed and group velocity.

13.1.2 Vertical coordinates

In simulating a mesoscale flow within a finite domain, the height vertical coordinate may propose problems. For example, it may intercept the terrain in a mountainous area and thus create problems in dealing with the lower boundary condition. Similar problems happen to the pressure coordinate and isentropic coordinate when isobaric surfaces and isentropic surfaces, respectively, intercept the lower boundary which may occur when there is strong orographic blocking.

To avoid the problem, a vertical *sigma coordinate*, which matches the lowest coordinate surface with the bottom topography, has been proposed (Phillips 1957). In this type of *vertical $\sigma - p$ coordinates*, the pressure coordinate is normalized by the surface pressure, p_S,

$$\sigma = \frac{p}{p_S}. \tag{13.1.8}$$

Thus, $\sigma = 1$ at the surface, and $\sigma = 0$ at the top of the atmosphere. The σ vertical velocity, $\dot{\sigma} = D\sigma/Dt$, is 0 at both the surface and top of the atmosphere. In (13.1.8), σ can also be defined as $\sigma = (p - p_T)/(p_S - p_T)$, where p_T is the pressure at the top of the numerical domain. The same concept may also be applied to the *isentropic coordinates*, in which σ can be defined as

$$\sigma = \frac{\theta - \theta_S}{\theta_T - \theta_S}. \tag{13.1.9}$$

The coordinate system in (13.1.9) is called *$\sigma - \theta$ coordinates*. One of the advantages of the vertical isentropic coordinate is that it can better resolve the vertical structure of weather systems, such as tropopause folding and upper-level frontogenesis. When the sigma coordinate is applied to the height coordinates, it is called *$\sigma - z$* or the *terrain-following coordinates* in which the σ can be defined as

$$\sigma = \frac{z_T(z - z_S)}{z_T - z_S}, \tag{13.1.10}$$

where z_S is the height of the lower surface in the $\sigma - z$ coordinates, which is independent of time, and z_T is a constant domain height or the constant height of the terrain-following part of the domain. In the general σ coordinates, the pressure, p, can be written as

$$p(x, y, z, t) = p[x, y, \sigma(x, y, z, t), t].\tag{13.1.11}$$

The pressure gradient in the x direction in the z coordinates can be obtained by performing the chain-rule,

$$\left(\frac{\partial p}{\partial x}\right)_z = \left(\frac{\partial p}{\partial x}\right)_\sigma + \frac{\partial p}{\partial \sigma}\left(\frac{\partial \sigma}{\partial x}\right)_z.\tag{13.1.12}$$

The pressure gradient $(\partial p/\partial x)$ in the σ coordinates can be obtained by deriving $(\partial \sigma/\partial x)_z$ from (13.1.10) and substituting it into (13.1.12):

$$\left(\frac{\partial p}{\partial x}\right)_\sigma = \left(\frac{\partial p}{\partial x}\right)_z - \left[\frac{\sigma - z_T}{z_T - z_S}\frac{\partial z_S}{\partial x}\right]\frac{\partial p}{\partial \sigma}.\tag{13.1.13}$$

The sigma coordinate transformation may also be applied to the *mass (hydrostatic pressure) coordinates* (Skamarock *et al.* 2005), in addition to the $\sigma - z$ coordinates. If one replaces p by a general variable A, then the above transformation may be used to derive the gradient of A in the x direction.

One problem of the sigma vertical coordinate systems is that errors in two terms of the pressure gradient force do not cancel out (Smagorinski *et al.* 1967). To avoid this problem, the *step-mountain* or *eta coordinates* have been proposed (Mesinger *et al.* 1988). In that model, *eta* (η) is defined as

$$\eta = \frac{p - p_T}{p_S - p_T}\eta_S\tag{13.1.14}$$

with

$$\eta_S = \frac{p_r(z_S) - p_T}{p_r(0) - p_T}.\tag{13.1.15}$$

In the above equations, the subscripts T and S denote the top and surface values of the model atmosphere; z is geometric height, and $p_r(z)$ is a suitably defined reference pressure as a function of z. The boundary conditions assumed are $p = $ constant at the top boundary $(\eta = 0)$, and $\dot{\eta} = 0$ at $\eta = 0$ and at the horizontal parts of the ground surface $(\eta = \eta_S)$. The advantage of this approach is that it does not transform the governing equations into complicated forms. However, the disadvantage of this coordinate is that it is of the first order of accuracy in representing the terrain, which is less accurate compared with the terrain-following coordinates, which is of the second order of accuracy. A third method is to adopt a *finite-element scheme*, which approximates the mountain surface by one side of the finite elements. It has the same advantage as not having to transform the governing equations into complicated forms as well as having a higher-order accuracy compared to the step-mountain coordinates.

So far, we have discussed about how to set the domain size, vertical and horizontal grid intervals, grid mesh and vertical coordinate. However, we still need to find (a) the temporal values of variables at the beginning of the integration, and (b) the boundary values of a finite domain. These are required by the fact that we are solving the initial-boundary value problems mathematically.

13.2 Boundary conditions

To make the mathematical problem well-posed, appropriate boundary conditions need to be specified in any limited-area models, such as mesoscale NWP models. If the model domain represents only part of the atmosphere in every direction, then boundary conditions are needed at the top, lateral, and lower boundaries of the model domain. The number of boundary conditions depends on the order of the differential equations in a particular direction involved.

13.2.1 Lateral boundary conditions

Basically, there are five types of lateral boundary conditions adopted in mesoscale NWP models: (1) closed boundary condition, (2) periodic boundary condition, (3) time-dependent boundary condition, (4) sponge layer boundary condition, and (5) open (radiation) boundary condition.

In a *closed boundary condition*, variables at the lateral boundaries are specified as constant values. Under such a situation, the waves or disturbances generated within the domain cannot propagate out of the domain; instead they are reflected back into the domain once they reach the lateral boundaries. The reflection of waves and disturbances back into the domain gives the name "closed boundary condition," since the model domain retains such disturbances. The use of the closed lateral boundary conditions works if the actual physical condition is closed, such as a solid wall. In addition, the boundary conditions will also work if the lateral boundaries are located far away so that the generated disturbances or waves do not reach the lateral boundaries at the desired integration time.

A *periodic boundary condition* assumes all the variables at the right boundary are equal to the left boundary, i.e. $\phi(x_N) = \phi(x_0)$, where ϕ is a field variable, or vice versa. Periodic boundary condition is an appropriate choice for a fluid flow system which does repeat itself at the lateral boundaries, such as sinusoidal waves propagating around a latitudinal circle of the Earth. A *time-dependent boundary condition* is often adopted when numerical integrations are performed concurrently at both the inner and outer domains, the lateral boundary values of the inner domain need to be specified by the updated values predicted by the outer domain. In this way, the weather systems, waves, or disturbances are able to propagate into the inner domain and produce the desired weather systems or fluid motion. Otherwise, the simulations of the inner domain cannot reflect the larger-scale environmental changes with time, which is called *one-way nesting*. If the lateral boundary values of the inner domain are passed back to the outer domain, then it is called *two-way nesting*.

The *sponge* or wave-*absorbing layer boundary condition* uses an enhanced filtering near the lateral boundaries to damp the waves or disturbances generated within the domain out of the lateral boundaries. For example, a sponge region can be formulated as follows (Perkey and Kreitzberg 1976):

$$\frac{\partial \phi_i}{\partial t} = W_i \left(\frac{\partial \phi_i}{\partial t}\right)_{\mathrm{m}} + (1 - W_i) \left(\frac{\partial \phi_i}{\partial t}\right)_{\mathrm{ls}}, \tag{13.2.1}$$

where m denotes the model calculated tendency of variable ϕ, ls denotes the larger scale specified tendency and W_i is a weighting factor which is given as follows:.

$$\begin{aligned} W_i &= 0.0 \text{ for the boundaries for } i = 0 \text{ or } N \\ &= 0.4 \text{ for the boundaries for } i = 1 \text{ or } N - 1 \\ &= 0.7 \text{ for the boundaries for } i = 2 \text{ or } N - 2 \\ &= 0.9 \text{ for the boundaries for } i = 3 \text{ or } N - 3 \\ &= 1.0 \text{ for all the interior points of } 4 \leq i \leq N - 4. \end{aligned} \tag{13.2.2}$$

Even though the wave-absorbing layer oversimplifies the boundary conditions, it has considerable practical applications (Davies 1983).

For pure gravity waves, the horizontal phase velocity is directed in the same sense as the horizontal group velocity, such as those shown in (3.5.14) and (3.5.15): thus it is possible to use the advection equation to advect the wave energy out of the lateral boundaries. Based on this concept, a *open (radiation) boundary condition* for a hyperbolic flow in a numerical model was proposed (Orlanski 1976). For the outflow boundary, the radiation boundary condition may be written

$$\frac{\partial \phi}{\partial t} + (U + c_o^*) \frac{\partial \phi}{\partial x} = 0, \qquad \text{at } x = L, \tag{13.2.3}$$

where $U + c_o^*$ is the propagation speed at the outflow boundary ($x = L$), which is yet to be determined. The leapfrog finite difference representation for time step $\tau - 1$ of the above equation may be written

$$\frac{\phi_{b-1}^\tau - \phi_{b-1}^{\tau-2}}{2\Delta t} = \frac{-(U + c_o^*)}{\Delta x} \left(\phi_{b-1}^{\tau-1} - \phi_{b-2}^{\tau-1}\right). \tag{13.2.4}$$

Based on the above approximation, the phase speed can be estimated by

$$U + c_o^* = -\frac{\Delta x}{2\Delta t}\left(\frac{\phi_{b-1}^\tau - \phi_{b-1}^{\tau-2}}{\phi_{b-1}^{\tau-1} - \phi_{b-2}^{\tau-1}}\right) \quad \text{at } b - 1, \tag{13.2.5}$$

$$U + c_o^* = 0 \qquad \text{if RHS of (13.2.5)} < 0,$$

$$U + c_o^* = \frac{\Delta x}{\Delta t} \qquad \text{if RHS of (13.2.5)} > \Delta x/\Delta t,$$

where subscript b denotes the boundary point and RHS means the right-hand side. For a hydrostatic and incompressible fluid system, since w is coupled with u, one can use the estimated phase speed of u for w. The phase speed estimation can also be applied to the coupled variables of potential temperature and pressure. Once the phase speed is estimated, then the boundary value at time step $\tau + 1$ can be determined

$$\phi_b^{\tau+1} = \phi_b^{\tau-1} - \frac{2\Delta t}{\Delta x}(U + c_o^*)(\phi_b^\tau - \phi_{b-1}^\tau), \tag{13.2.6}$$

where $U + c_o^*$ is estimated by (13.2.5). A similar formula can be formed for the inflow boundary:

$$\phi_b^{\tau+1} = \phi_b^{\tau-1} - \frac{2\Delta t}{\Delta x}(U - c_i^*)(\phi_{b+1}^\tau - \phi_b^\tau). \tag{13.2.7}$$

Note that the specification of ϕ at both boundaries will lead to an overdetermined problem for the first-order advection equation. In fact, this renders the problem ill-posed (Oliger and Sundström 1978).

In practice, the *constant gradient lateral boundary condition* has also been adopted in mesoscale NWP models, which specifies a constant gradient, such as zero gradient (e.g., $\partial\phi/\partial x = 0$, ϕ is the variable concerned), at the lateral boundaries. The effectiveness of the constant gradient lateral boundary condition to propagate the waves out of the domain depends on how close the specified constant gradient is to the advection speed of the physical waves. The constant gradient lateral boundary condition can be viewed as a special case of the radiation boundary condition. For example, a *zero gradient lateral boundary condition*, such as $\phi_N^{\tau+1} = \phi_{N-1}^\tau$, assumes the wave propagating out of the right boundary at a speed of $c = \Delta x/\Delta t$. If the real physical wave speed is very different from this numerical phase speed, then a large reflection from the boundary may occur.

13.2.2 Upper boundary conditions

The upper boundary of a mesoscale NWP model should be placed high enough above the region with active mesoscale waves and disturbances. Ideally, it should be placed at the top of the atmosphere, i.e. $p = 0$. However, practically, this is almost impossible due to the restriction of computing resources. Depending on the weather phenomena simulated by the model, it may require the top boundary of a numerical model domain be placed at deep within the stratosphere, at the tropopause, or within the stable layer of the troposphere. For example, a sea-breeze circulation in a stable boundary layer normally does not penetrate to a high altitude, thus allowing a top boundary of a mesoscale model to be placed at the mid-troposphere. On the other hand, in simulating a flow over a mesoscale mountain, the mountain waves often can propagate to a very high altitude, therefore a much higher vertical domain is required. No matter how high the model domain extends in vertical, an appropriate upper boundary condition is still needed. For example, the disturbances or waves generated by a mesoscale mountain in the lower stratosphere may look like very weak, but the energy per unit area associated with them could be very large because it is proportional to N (e.g. see (13.2.31)) and N is much larger in the stratosphere compared to that in the troposphere.

Upper boundary conditions that have been adopted in mesoscale NWP models can be categorized as (1) rigid lid, (2) sponge layer (Klemp and Lilly 1978; Anthes and Warner 1978; Mahrer and Pielke 1978), and (3) radiation condition (Klemp and Durran 1983; Bougeault 1983). A *rigid lid* upper boundary condition can be implemented by simply setting the desired variables to be constants. The sponge layer and radiation conditions are found to be much more effective in radiating the wave energy out from the interior of the numerical domain. Basically, these two approaches are taken to numerically approximate the *Sommerfeld* (1949) *radiation boundary condition*, which allows the energy associated with disturbances generated in the interior of a physical system to propagate out of the domain.

The addition of a *sponge (wave-absorbing) layer* to the top of the physical domain is a simple way to mimic the Sommerfeld (1949) radiation boundary condition in a numerical model. The sponge layer is designed to damp out disturbances generated in the physical layer out of the upper boundary. To elucidate the formulation of a sponge layer, we consider the following two-dimensional, steady-state, linear, hydrostatic, nonrotating, Boussinesq flow with Rayleigh friction and Newtonian cooling added to the momentum and thermodynamic equations, respectively:

$$U\frac{\partial u'}{\partial x} + \frac{1}{\rho_0}\frac{\partial p'}{\partial x} = -\nu u', \tag{13.2.8}$$

$$\frac{\partial p'}{\partial z} - \left(\frac{g\rho_0}{\theta_0}\right)\theta' = 0, \tag{13.2.9}$$

$$\frac{\partial u'}{\partial x} + \frac{\partial w'}{\partial z} = 0, \tag{13.2.10}$$

$$U\frac{\partial \theta'}{\partial x} + \left(\frac{N^2\theta_0}{g}\right)w' = -\nu\theta'. \tag{13.2.11}$$

The above equations may be obtained from (3.5.1)–(3.5.4). To minimize reflections caused by rapid increases in viscosity, one may consider the following function, which gradually increase ν from 0 at z_1 to ν_T at z_T,

$$\nu(z) = \nu_T \sin^2\left(\frac{\pi}{2} \frac{z - z_1}{z_T - z_1}\right), \quad z_1 \le z \le z_T. \tag{13.2.12}$$

To investigate the properties of wave reflection from the wave-absorbing layer, we assume a wave-like solution in the x direction,

$$(u', w', p', \theta') = (\hat{u}, \hat{w}, \hat{p}, \hat{\theta})e^{ikx}. \tag{13.2.13}$$

Substituting the above equation into (13.2.8)–(13.2.11) yields

$$\frac{\partial^2 \hat{w}}{\partial z^2} + \frac{N^2}{U^2}\hat{w} = 0, \quad z \le z_1, \tag{13.2.14a}$$

$$\frac{\partial^2 \hat{w}}{\partial z^2} + \frac{N^2}{U^2(1 - i\nu/kU)^2}\hat{w} = 0, \quad z_1 < z \le z_T. \tag{13.2.14b}$$

The general solutions of the above equations may be written

$$\hat{w} = c_1 e^{il_1 z} + c_2 e^{-il_1 z}, \quad \text{for } z \le z_1, \tag{13.2.15a}$$

$$\hat{w} = d_1 e^{il_2 z} + d_2 e^{-il_2 z}, \quad \text{for } z_1 \le z \le z_T, \tag{13.2.15b}$$

where

$$l_1 = \frac{N}{U} \quad \text{and} \quad l_2 = \frac{N}{U\sqrt{1 - i\nu/kU}} \tag{13.2.16}$$

are the Scorer parameters for uniform basic flow (U) in the physical and sponge layers, respectively. The four coefficients in (13.2.15) are determined by the upper boundary condition, lower boundary condition, and two interface conditions at z_1. According to the *Eliassen and Palm theorem* (Section 4.4), the term of c_1 (c_2) represents the upward (downward) propagation of the wave energy. Thus, the ratio

$$r = \left|\frac{c_2}{c_1}\right|, \tag{13.2.17}$$

represents the reflectivity produced by the upper viscous layer. Note that r can be obtained by applying the interface conditions at $z = z_1$ and the boundary condition $\hat{w} = 0$ at $z = z_T$. To minimize the reflection from the upper boundary, it is suggested that the depth of the sponge layer should be greater than the hydrostatic vertical wavelength ($\lambda = 2\pi U/N$) of the mesoscale disturbance (Klemp and Lilly 1978).

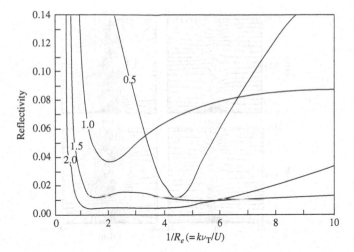

Fig. 13.8 Reflectivity, from the sponge layer as a function of a nondimensional inverse Reynolds number, $1/R_e$ for several nondimensional sponge layer depths ($d = 0.5$, 1.0, 1.5, and 2.0). A value of $r = 1.0$ corresponds to complete reflection from the top boundary of the computational domain. The nondimensional numbers are defined as $R_e = U/k\nu_T$ and $d = (z_T - z_1)/\lambda$, where $\lambda = 2\pi U/N$. The viscosity coefficient is defined as $\nu = \nu_T \sin^2[\pi \ln(\bar{\theta}/\theta_1)/2\ln(\theta_T/\theta_1)]$, where θ_T and θ_1 are the potential temperatures at the top and bottom of the sponge layer, respectively. (Adapted after Klemp and Lilly 1978.)

Figure 13.8 shows the reflectivity from the sponge layer as a function of the non-dimensional inverse Reynolds number, $1/R_e = k\nu_T / U$, where k is the horizontal wave number. In practice, $\nu_T/\nu_1 < 6$ is a better choice to avoid the reflection due to the rapid increase of the coefficient of viscosity (Klemp and Lilly 1978). If the physical layer is assumed to be inviscid ($\nu_1 = 0$), one may choose $2 \le |\nu_T/kU| \le 5$, where k is the horizontal wave number. For example, we may choose $\nu_T = 0.002\,\mathrm{s}^{-1}$ for a basic flow with $U = 10\,\mathrm{m\,s}^{-1}$ over a bell-shaped mountain with $a = 20\,\mathrm{km}$. Figure 13.9 shows the results from a hydrostatic numerical model (panels a and b) using a sponge layer for flow over a bell-shaped mountain and compared with those calculated from Long's (1953) nonlinear theory. A vertical domain of 3.4λ is used, in which the upper half is the sponge layer, and thus, the vertically propagating hydrostatic waves are effectively absorbed by the sponge layer.

Since the addition of a sponge layer increases the computational time significantly, a direct application of the Sommerfeld (1949) radiation condition has been proposed. To elucidate the numerical radiation boundary condition, we may consider the two-dimensional, linear, hydrostatic, Boussinesq equations for a uniform basic state in the absence of Coriolis force:

$$\frac{\partial u'}{\partial t} + U\frac{\partial u'}{\partial x} + \frac{1}{\rho_0}\frac{\partial p'}{\partial x} = 0, \qquad (13.2.18)$$

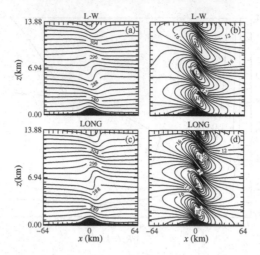

Fig. 13.9 Potential temperature ((a) and (c)) and total horizontal velocity fields ((b) and (d)) for a two-dimensional, continuously stratified, uniform flow over a bell-shaped mountain predicted at the nondimensional time $Ut/a = 100$ by a hydrostatic numerical model [(a) and (b)] and calculated by Long's steady state hydrostatic solution [(c) and (d)]. The Froude number ($F = U/Nh$) and hydrostatic parameter (Na/U) associated with the basic flow are 1.3 and 7.7, respectively. The dimensional flow and orographic parameters are $U = 13\,\mathrm{m\,s^{-1}}$, $N = 0.01\,\mathrm{s^{-1}}$, $h = 1\,\mathrm{km}$, and $a = 10\,\mathrm{km}$. The vertical coordinate is nondimensionalized by the hydrostatic wavelength $\lambda = 2\pi U/N$. (After Lin and Wang 1996.)

$$\frac{\partial p'}{\partial z} - \left(\frac{g\rho_0}{\theta_0}\right)\theta' = 0, \tag{13.2.19}$$

$$\frac{\partial u'}{\partial x} + \frac{\partial w'}{\partial z} = 0, \tag{13.2.20}$$

$$\frac{\partial \theta'}{\partial t} + U\frac{\partial \theta'}{\partial x} + \left(\frac{N^2\theta_0}{g}\right)w' = 0. \tag{13.2.21}$$

We assume a wave-like solution,

$$(u', w', p', \theta') = (u_0, w_0, p_0, \theta_0)e^{i(kx+mz-\omega t)}, \tag{13.2.22}$$

and substitute it into (13.2.18)–(13.2.21) to yield the dispersion relation,

$$m^2(\omega - kU)^2 = N^2k^2. \tag{13.2.23}$$

From the above equation, the horizontal phase speed and the horizontal group velocity characterize the horizontal propagation of hydrostatic gravity waves:

$$c_{px} = \frac{\omega}{k} = U \pm \frac{N}{m},$$ (13.2.24)

$$c_{gx} = \frac{\partial \omega}{\partial k} = U \pm \frac{N}{m}.$$ (13.2.25)

Thus, for each wavenumber pair (k, m), the horizontal propagation speeds of the phase lines and energy are identical. Consequently, the outward propagating wave energy can be transmitted through a lateral boundary by numerically advecting disturbances out of the boundary based on their horizontal phase speed, as discussed in Section 13.2.2. For nonhydrostatic waves, c_{px} and c_{gx} are not identical: however, they are propagating in the same direction. Thus, the radiation (open) lateral boundary condition is still able to advect the energy out by a simple advection equation.

In the vertical direction, the situation is completely different. For simplicity, we may assume $U = 0$. The phase speed and group velocity can be derived,

$$c_{pz} = \frac{\omega}{m} = \pm \frac{Nk}{m^2},$$ (13.2.26)

$$c_{gz} = \frac{\partial \omega}{\partial m} = \mp \frac{Nk}{m^2}.$$ (13.2.27)

Note that c_{pz} and c_{gz} have opposite signs, which implies that a positive c_{pz} corresponds to downward energy propagation. Thus, the advection equation is unable to advect wave energy generated within the domain out of the upper boundary, as adopted for the radiation (open) lateral boundary condition.

To identify wave modes with upward energy propagation, we consider a wave-like solution of the form

$$\phi(x, z, t) = \hat{\phi}(k, z, \omega) e^{i(kx - \omega t)},$$ (13.2.28)

where ϕ may represent any dependent variables, u', w', p', or θ'. Substituting the above equation into the governing equation, (13.2.18)–(13.2.21), yields

$$\frac{\partial^2 \hat{w}}{\partial z^2} + \frac{N^2}{(U - \omega/k)^2} \hat{w} = 0.$$ (13.2.29)

By assuming a positive k, the above equation has the following general solution:

$$\hat{w} = A e^{iNz/(U-\omega/k)} + B e^{-iNz/(U-\omega/k)}.$$ (13.2.30)

A similar argument may also be made easily for a negative k. The horizontally averaged vertical energy flux can then be obtained,

$$\overline{p'w'} = \frac{\rho_0 N}{2k}\left(|A|^2 - |B|^2\right),\tag{13.2.31}$$

where terms A and B represent the upward and downward propagation of wave energy, respectively. Thus, to avoid the wave reflection from the top boundary, we require $B = 0$. For upward propagating waves, we choose

$$\hat{w} = A\mathrm{e}^{\mathrm{i}Nz/(U-\omega/k)},\tag{13.2.32}$$

which implies

$$\frac{\partial \hat{w}}{\partial z} = \frac{\mathrm{i}N}{U-\omega/k}\hat{w}.\tag{13.2.33}$$

From the continuity equation and momentum equation, we have

$$\frac{\partial \hat{w}}{\partial z} = -\mathrm{i}k\hat{u} = \frac{\mathrm{i}k/\rho_0}{U-\omega/k}\hat{p}.\tag{13.2.34}$$

If both positive and negative k are taken into account, then the above two equations lead to

$$\hat{p} = \frac{\rho_0 N}{|k|}\hat{w}\quad\text{at}\quad z = z_{\mathrm{T}}.\tag{13.2.35}$$

Since the above equation has no frequency dependence, we can write the above upper radiation condition in the wave number or Fourier space,

$$\tilde{p}(z,t) = \frac{\rho_0 N}{|k|}\tilde{w}(z,t),\tag{13.2.36}$$

where $\tilde{p}$ and $\tilde{w}$ are defined as

$$(p,w) = (\tilde{p},\tilde{w})\mathrm{e}^{\mathrm{i}kx} = (\hat{p},\hat{w})\mathrm{e}^{\mathrm{i}(kx-\omega t)}.\tag{13.2.37}$$

The numerical implementation of the upper radiation boundary condition to the geophysical fluid system of (13.2.18)–(13.2.21) is sketched in Fig. 13.10:

(1) Integrate $w^{\tau+1}$ upward to z_{T} based on the continuity equation, (13.2.20).
(2) Make the Fourier transform of $w^{\tau+1}(z_{\mathrm{T}})$ to obtain $\tilde{w}^{\tau+1}(z_{\mathrm{T}})$. A Fast Fourier Transform (FFT) numerical software may accelerate the computation.
(3) Apply (13.2.36) to obtain $\tilde{p}^{\tau+1}(z_{\mathrm{T}})$.
(4) Make the inverse Fourier transform of $\tilde{p}^{\tau+1}(z_{\mathrm{T}})$ to obtain $p^{\tau+1}(z_{\mathrm{T}})$.
(5) Integrate the hydrostatic equation downward based on the upper boundary condition of $p^{\tau+1}(z_{\mathrm{T}})$ to obtain $p^{\tau+1}(z)$ at every height level in the domain.

Although the numerical radiation boundary condition is based more solidly on gravity wave theory, other factors, such as nonlinearity, can play roles in a more complicated

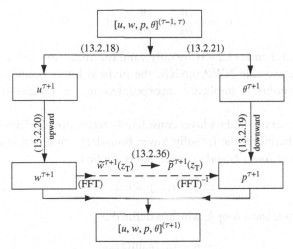

Fig. 13.10 A flow chart for modeling the fluid flow system of (13.2.18)–(13.2.21) and implementing the numerical upper radiation boundary condition at the top of computational domain.

fluid flow system (Klemp and Durran 1983). In addition, the flow response is sensitive to the domain height when a numerical upper radiation boundary condition is implemented.

13.2.3 Lower boundary conditions

The lower boundary condition for an inviscid flow over a flat surface is that the flow near the surface is allowed to flow over it freely, which is often referred to as the *free-slip lower boundary condition*. Since the normal velocity is required to be zero at a rigid surface, the inviscid flow is always tangential to the surface. For an inviscid flow over a mountainous terrain, the free-slip lower boundary condition requires that the flow be parallel to the surface. For a two-dimensional flow, this requires

$$\frac{w}{u} = \frac{dh}{dx} \quad \text{at} \quad z = h(x),\tag{13.2.38}$$

where $h(x)$ is the mountain profile. Thus, the *linear lower boundary condition* can be derived by making a linear approximation of the above equation,

$$w' = (U + u')\frac{dh}{dx} \approx U\frac{dh}{dx} \quad \text{at} \quad z = 0.\tag{13.2.39}$$

The above condition is often adopted in mountain wave theories. In deriving the above equation, two nonlinearities have been neglected by assuming: (1) $u' \ll U$, and (2) the lower boundary condition is applied at $z = 0$, instead of $z = h(x)$. Similarly, the linear lower boundary condition for a three-dimensional flow can be derived,

$$w' = U\frac{\partial h}{\partial x} + V\frac{\partial h}{\partial y}, \quad \text{at} \quad z = 0. \tag{13.2.40}$$

Equations (13.2.39) and (13.2.40) are only valid for linear flows over small-amplitude mountains. In mesoscale NWP models, the inviscid lower boundary condition for flow over mountains is implicitly incorporated in the terrain-following (sigma) coordinates.

With the planetary boundary layer considered, the frictional effects have to be taken into account. Therefore, the free-slip lower boundary condition is no longer valid. Instead, the *no-slip lower boundary condition* is applied,

$$u(z_0) = v(z_0) = w(z_0) = 0, \tag{13.2.41}$$

where z_0 is the *roughness length*, which is defined as

$$u = (u_*/k)\ln(z/z_0). \tag{13.2.42}$$

In the above equation, k is a universal constant called the *von Kármán constant*, which has a value of about 0.4 based on measurements, and u_* is the *friction velocity*, which can be obtained from the vertical momentum fluxes at surface,

$$u_*^2 = \sqrt{\left(\overline{u'w'}\right)^2 + \left(\overline{v'w'}\right)^2}, \tag{13.2.43}$$

where $\overline{u'w'}$ and $\overline{v'w'}$ are the turbulent momentum fluxes. Measurements indicate that the magnitude of the surface momentum flux is on the order of $0.1\,\mathrm{m^2\,s^{-2}}$. Thus, the friction velocity is typically on the order of $0.3\,\mathrm{m\,s^{-1}}$. In addition to the specifications of the velocities, we also have to specify pressure and potential temperature. The surface pressure can be approximately specified based on hydrostatic balance, while the potential temperature can be prescribed as a periodic heating function,

$$\theta(z_0) = \theta_0(z_0) + \Delta\theta_{max}\sin(2\pi t/24h), \tag{13.2.44}$$

where t is the time in hours after sunrise, $\theta_0(z_0)$ is the potential temperature at z_0 at sunrise, and $\Delta\theta_{max}$ the maximum temperature attained during the day. To permit interactions between ground and the atmosphere, calculations of surface heat energy budget are needed.

For flow over water surfaces, the *air–sea interaction* processes need to be considered. Basically, the water affects the atmospheric circulation through sensible and latent heat fluxes directly related to its sea surface temperature (SST), including its time and space variability, while the atmosphere feeds back to the water or ocean through the wind stress to produce a deepening of the *ocean mixed layer*, induces water or ocean currents, and alters the *upwelling-downwelling* pattern.

13.3 Initial conditions and data assimilation

Mathematically, NWP can be viewed as solving an initial-boundary value problem in which the governing equations of a geophysical fluid system are integrated forward in time in a finite region. Therefore, in addition to the boundary conditions as discussed in the Section 13.2, we must provide suitable initial conditions for the model. For idealized numerical simulations, the initial conditions can be prescribed by known functions or values. If the Coriolis force is included in the model, then the initial basic state should be in geostrophic balance. If shear is included in a rotating atmosphere, then the initial basic state should be in thermal wind balance. Otherwise, the initial state will be adjusted to reach a new balanced state by the model, which may not be desirable. For real data mesoscale NWP, the observational data must be modified to be dynamically consistent with the governing equations of the model. The process in producing initial conditions includes the following four components: (i) *quality control*, (ii) *objective analysis*, (iii) *initialization*, and (iv) *initial guess from a short-range forecast by an NWP model*. These components have been taken to form a continuous cycle of *data assimilation*, often called *four-dimensional data assimilation* (4DDA). Many of the methods described below are still current topics of research, thus only a brief review is appropriate. Although some of the techniques might be rarely used in today's NWP models, they still provide useful information in helping understand NWP technique development.

The necessity of performing *quality control* on meteorological data was recognized long ago, which is especially important when the data are used to initialize an NWP model because the errors associated with the data may be misrepresented (*nonlinear aliasing*) and amplified by the model. To reduce the errors in the sounding data, the following steps of quality control have been taken in NWP: (a) *plausibility check*, (b) *contradiction check*, (c) *gross check*, and (d) *buddy check* (Gandin 1988). In *plausibility check*, data values that cannot possibly occur in the real atmosphere or extremely exceed the climatological mean are rejected. For example, positive temperatures in Celsius at 300 hPa are rejected. In *contradiction check*, data values of two or more parameters at the same location contradicting each other are removed. For example, the occurrence of rain in the absence of clouds is removed. In *gross check*, observations with large deviations from the first guess field forecast by an operational model are removed. In *buddy check*, observations not agreeing with neighboring observations are removed.

Observational data are often not regularly spaced, and thus are not ready for use as initial fields for a mesoscale NWP model because they do not match the model grid mesh. In some areas, such as over ocean, observational data are sparse. Therefore, in order to use the observational data as initial fields for a mesoscale NWP model, one needs to interpolate or extrapolate the data to fit into the grid mesh of the model. This procedure is called *objective analysis*, and one example is the *Barnes objective analysis* (Barnes 1964). In an objective analysis, it is desirable to do the following: (1) filter out scales of motion that cannot be resolved by the mesoscale model, (2) use a first guess field or background field provided by an earlier forecast from the same model, which will help avoid the extrapolation of observation data in data sparse areas, and introduce dynamical consistency, and (3) make use of our knowledge of the probable errors associated with each observation, which can be weighted based on past records of accuracy. When the maximum information from data sources, including the observations, climatological records, space correlation among the meteorological variables, etc., are extracted statistically, the approach is called *optimal interpolation* (e.g., Kalnay 2003). The optimal interpolation often requires knowledge of the statistical structure of the fields of the variables. The variables may be analyzed separately or simultaneously, which is referred to as *univariate analysis* or *multivariate analysis*, respectively.

The objective analysis procedure generally does not provide fields of mass and motion that are consistent with model dynamics to initiate a forecast. Thus, the use of such objectively analyzed data to initialize an NWP model may generate large, spurious inertial-gravity wave modes. Theoretically, these inertial-gravity wave modes will be dispersed, dissipated or propagate out of the domain, after sufficient time due to redistribution of mass and wind fields. The inertia gravity-wave modes or noise, as often referred to by NWP modelers, however, cannot be dissipated locally because of the relatively low resolution in NWP models. Therefore, an additional procedure, called *initialization*, is required to force the data after objective analysis to be dynamically consistent with the model dynamics, and to allow the model to integrate forward in time with a minimum of noise and the maximum accuracy of the forecasts. To improve the NWP, a number of initialization techniques have been developed in the past, such as: (a) *damping method*, (b) *static initialization*, (c) *variational method*, (d) *normal mode initialization*, and (e) *dynamic initialization*.

A simple and straightforward way to reducing the *gravity wave mode* is to dampen or filter the *inertia-gravity wave 'noise'* by adding a divergence damping term to the horizontal momentum equation (Talagrand 1972). In this approach, the local rate of change of the divergence is diffused according to

$$\frac{\partial D}{\partial t} = -V \cdot \nabla D - \frac{1}{\rho}\nabla^2 p + \nu\nabla^2 D + f\left(\frac{\partial v}{\partial x} - \frac{\partial u}{\partial y}\right) - \left(\frac{\partial u}{\partial x}\right)^2 - \left(\frac{\partial v}{\partial y}\right)^2 - 2\frac{\partial v}{\partial x}\frac{\partial u}{\partial y}$$
$$- \left(\frac{\partial w}{\partial x}\frac{\partial u}{\partial z} + \frac{\partial w}{\partial y}\frac{\partial v}{\partial z}\right) + \frac{1}{\rho^2}\left(\frac{\partial p}{\partial x}\frac{\partial \rho}{\partial x} + \frac{\partial p}{\partial y}\frac{\partial \rho}{\partial y}\right).$$

(13.3.1)

The approach identified in (13.3.1) in initializing the data is called *damping method*. Another way to adjust the data at a single time level, usually to conform to some dynamical constraints in order to reduce or eliminate the generation of inertia-gravity wave "noise", is the *static initialization*. For example, in an isobaric model, one may (i) estimate the geopotential field (ϕ) from the pressure–height data and the geostrophic wind relations; (ii) calculate the streamfunction (ψ) from analyzed ϕ fields on the isobaric surfaces; and then (iii) compute the rotational wind component from the following relationship,

$$V_\psi = k \times \nabla\psi. \tag{13.3.2}$$

The above equation can be written as an elliptic function of ϕ, which may become hyperbolic in some areas. In order to ensure ellipticity in these areas so that the numerical method for the elliptic equations will apply, the geopotential fields must be altered (Haltiner and Williams 1980). In addition to this difficulty, the gravitational modes still exist even using the balance equation to determine a rotational wind for initialization. Another approach to initializing the data is to adopt the *variational method*, in which one or more of the conservation relations are applied to minimize the variance of the difference between the observations and the objectively analyzed fields. In performing the variational method, the principles of variational calculus is applied (Sasaki 1970). For example, the difference can be minimized in a least-square sense subjected to one or more dynamical constraints, such as the balance equation, hydrostatic relation, and steady-state momentum equation.

The *static initialization* described above is based on the distinction between gravity wave modes with relatively high divergence and other meteorological modes of the quasi-geostrophic type with small divergence and relatively high vorticity. However, in reality the separation is far less clear cut in some instances. Thus, it has been proposed to keep some normal modes, if they can be represented by the model grid resolution. Retaining these gravity wave modes are important since some severe weather has been found to be induced by gravity waves. Unlike applying the balance equation constraint, the *normal mode initialization* produces a divergent component as well. The normal mode initialization makes an optimal use of the observed data by adjusting both mass and motion fields while achieving dynamical consistency through appropriate constraints. In the *linear normal mode initialization* the original objectively analyzed fields are adjusted to the linearized versions of the model equations, and the undesirable gravity wave modes are removed. However, the disadvantage of this type of method is that nonlinear terms tend to regenerate the high-frequency wave modes, and also the curvature in the flow is neglected so that the fit with the original data may suffer, which is overcome by taking the *nonlinear normal mode initialization* technique. In the nonlinear normal mode initialization, the tendency of the undesirable wave modes, instead of the amplitude, is set to zero. The nonlinear normal mode initialization may also be applied to the vertical direction, too. Figure 13.11 shows the time evolution of a height field after applying two iterations of the implicit nonlinear normal mode initialization scheme. In comparison with the time evolution of the same height field with no initialization, the implicit nonlinear normal mode initialization appears to be able to remove high-frequency oscillations.

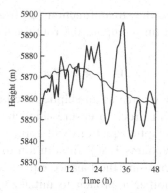

Fig. 13.11 Time evolution of height field after two iterations of the implicit nonlinear normal mode initialization scheme (thin curve) and with no initialization (bold curve). (After Temperton 1988.)

Since the normal mode initialization is performed separately right after the objective analysis, the initialized fields may no longer fit the observations as closely as possible. Therefore, the *dynamic initialization* was proposed (Miyakoda and Moyer 1968). The basic idea of dynamic initialization is to let the NWP model do the job by itself because any primitive equation models are supposed to inherently possess the mechanism for the geostrophic adjustment process. Indeed, the mass and velocity fields do mutually adjust to one another toward a quasi-geostrophic state when they are executed in an NWP model. In this way, observations are inserted intermittently or continuously over a period of time. In this type of initialization, the model is integrated forward and backward around the initial time, and this lets the model adjust itself before starting the forecast. During this process, it is desirable to use an integration scheme with selective damping technique. For example, the following iterative scheme consists of a forward step, then a backward step and finally an averaging:

$$u_*^{\tau+1} = u^\tau + \Delta t (\partial u/\partial t)^\tau,$$
$$u_{**}^\tau = u_*^{\tau+1} - \Delta t (\partial u_*/\partial t)^{\tau+1}, \tag{13.3.3}$$
$$\bar{u}^\tau = 3u^\tau - 2u_{**}^\tau.$$

Figure 13.12 shows an example of gravity wave activity after normal mode initialization and dynamic initialization for five vertical modes of a baroclinic model, as compared to that with no initialization. Gravity wave modes with large equivalent depths are dramatically reduced by the dynamic initialization. The disadvantages of the dynamic initialization scheme are that: (i) each iteration requires the equivalent of two prognostic steps, thus taking considerable computer time, (ii) it is unable to distinguish between large-scale gravity wave modes and small-scale Rossby modes, and (iii) backward integration may not be applicable to some irreversible physical processes.

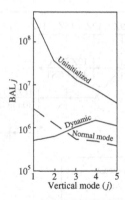

Fig. 13.12 Gravity wave activity after normal mode initialization, dynamic initialization for five vertical modes of a baroclinic model, and compared to that with no initialization. Note that the gravity wave modes of large equivalent depths (small j) are dramatically reduced by dynamic initialization. (After Sugi 1986.)

In order to incorporate the invaluable asynoptic data, such as Next Generation Weather Radar Doppler radar (NEXRAD), wind profilers, acoustic sounders, high-resolution dropsondes, satellite and aircraft, observed at nonstandard time (i.e., not at 0000 and 1200 UTC) into an NWP model system, the quality control, objective analysis, initialization, and initial guess forecast from the same model are combined into a 4DDA cycle. This 4DDA cycle can be carried out in an intermittent or continuous fashion. In the *intermittent* 4DDA, the data are assimilated intermittently at specified time intervals. The background or first guess fields forecast by the model plays a very important role, especially in data sparse regions. In data rich regions, usually the analysis is dominated by the information contained in the observations. The boundary conditions of regional NWP are provided by global model forecast. The intermittent 4DDA technique, such as three-dimensional variational data assimilation (3DVAR), is used in many global and regional operational NWP systems due to its computational efficiency. Figure 13.13 illustrates a 32-km data assimilation cycle adopted by NCEP's ETA model by using 3DVAR technique. Data types used in the 3DVAR of ETA model are rawinsonde mass and wind, pibal winds, dropwindsondes, wind profilers, surface land temperature and moisture, oceanic surface data (ships and buoys), aircraft winds, satellite cloud-drift winds, oceanic TOVS thickness retrievals, GOES and SSM/I precipitable water retrievals, ACARS temperature data, surface winds over land, VAD winds from NEXRAD, SSM/I oceanic surface winds, and tropical cyclone bogus data (see Roger *et al.* 1998 for detailed information). Note that more advanced techniques have been developed, including the use of the adjoint model in intermittent data assimilation systems (e.g. Huang 1999).

The intermittent updating process is appropriate as long as most available data are taken at a fixed time period, which may vary from 3 to 12 h in practice. However, in order to take advantage of the asynoptic data, which comes in much more frequently

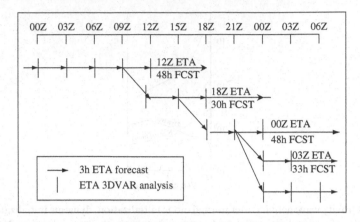

Fig. 13.13 An example of data assimilation cycle adopted by NCEP's ETA model by using 3DVAR technique. (Adapted after Rogers *et al*. 1998.)

than the synoptic data, methods of *continuous* or *dynamic 4DDA* are desired. In these methods, the observational data are essentially introduced into the assimilation system at each time step of the model integration during the assimilation time period. Examples of this type of continuous 4DDA are: (1) *nudging* or *Newtonian relaxation* (Hoke and Anthes 1976), (2) *variational assimilation* (also called *4DVAR*; Stephens 1970), and (3) *Kalman–Bucy filtering* (Kalman and Bucy 1961).

In the *nudging* or *Newtonian relaxation method*, there is a pre-forecast integration period during which the model variables are driven toward the observations by adding extra forcing terms in the equations. When the actual initial time is reached, the extra terms are dropped from the model equations, and the forecast proceeds without any forcing. For example, a forcing term is added to the *x*-momentum equation,

$$\frac{\partial u}{\partial t} = -\boldsymbol{V} \cdot \nabla u + fv - \frac{1}{\rho}\frac{\partial p}{\partial x} + \frac{u - u_{\text{obs}}}{\tau}. \tag{13.3.4}$$

The time scale for the relaxation, τ depends on the variable and is chosen to slowly increase (decrease) prior to (after) the time of the observation to prevent any shocks to the model during the assimilation time period. Nudging has been tested for use with the new generation of observing systems, such as dropsondes, wind profilers, and surface data. Compared to the variational assimilation and Kalman–Bucy filtering techniques, to be discussed below, the nudging or Newtonian relaxation technique is less elegant mathematically but very practical. In the four-dimensional *variational data assimilation (4DVAR)*, one tries to create the best possible fit between the model and the observational data such that the adjusted initial conditions are optimal for use in subsequent model forecasts. For example, Fig. 13.14 shows that the value produced by the first analysis is A. Although it fits the data well at $T - 3$ h, it leads to a forecast that does not match the observations well by T = 0 h. Note that even data collected at

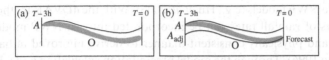

Fig. 13.14 A sketch of four-dimensional variational data assimilation (4DVAR). (a) The value produced by the first analysis is A, which fits the data well at $T - 3$ h, but leads to a forecast that does not match the observations well by $T = 0$ h. The shaded band is the observations. Note that even data collected at the same time do not necessarily agree with each other. (b) An iterative approach, i.e. the adjoint method, is taken to adjust the initial analysis so that it is optimal for prediction. (Courtesy of F. H. Carr.)

the same time do not necessarily agree with each other. One way to resolve the problem is to take the *adjoint method*, in which an iterative approach is used to adjust the initial analysis so that it is optimal for prediction, as one type of the 4DVAR. In other words, the adjusted analysis, A_{adj}, leads to a model trajectory (lower curve) that produces a better 3-h forecast for $T = 0$ h, even though it may not be the best fit at $T - 3$ h. In the *Kalman–Bucy filtering* technique, the data sequentially adjusts the assimilated fields as the model is integrated forward in time. The Kalman–Bucy filter minimizes the analysis error variance not only at every time step, but also over the entire assimilation period in which data are provided. In addition, the filter is able to extract all useful information from the observational increment or residual at each time step, thus allowing observations to be discarded as soon as they are assimilated.

In order to involve the standard or nonstandard data to reduce or eliminate the spin-up error caused by the lack, at the initial time, of the fully developed vertical circulation required to support regions of large rainfall rates, one may adopt the *diabatic* or *physical normal mode initialization*, which may improve quantitative precipitation forecasts, especially early in the forecast. Two key issues can be raised here: (1) choice of technique and (2) sources of hydrologic or hydrometeor data. Diabatic heating information in nonlinear normal mode initialization can be either from the model estimates or from observed rainfall data. Various methods have been developed to infer three-dimensional fields of latent heating, moisture and divergence from two-dimensional rainfall data, such as static methods, dynamic methods, adjusting the convective parameterization scheme to match the observed rainfall, and latent heat nudging. A major problem in all of the techniques presented is the need for accurate distribution of the heating and moistening rates. For example, (i) surface rain gauge data are not available on hourly basis, (ii) rawinsonde data are sparse horizontally (may be overcome by combining infrared and microwave satellite estimates), or (iii) cloud and hydrometeor data to be deduced from the network of Doppler radars are not complete (retrieval techniques are critical). To utilize the radar reflectivity measurements and satellite cloud observations, as well as surface-based cloud observations, complex cloud analysis packages have been developed that define the three-dimensional cloud and hydrometeor fields as well as the associated temperature perturbations, and such analyses have been successfully applied to the initialization of storm-scale NWP models (e.g., Hu *et al.* 2006). With the 4DVAR method, observed measurements of rainfall rates can be incorporated into an NWP model in a more dynamically and physically consistent manner to derive improved initial conditions (Zou and Kuo 1996). Through the model precipitation calculation and adjoint model, information from the space of the model variables, such as wind, temperature, and humidity, can be projected to that of the measured variables (i.e. rainfall rates), and back, in a consistent manner.

A numerical forecast model and a mesoscale or local analysis system can be combined as an operational atmospheric prediction system to initialize the model, such as the *Rapid Update Cycle* (*RUC*; Benjamin *et al.* 2004). The *Mesoscale Analysis and Prediction System* (*MAPS*) is the research counterpart to the RUC. The RUC has been developed to serve users needing short-range weather forecasts, including those in the US aviation community. In MAPS, a mesoscale model is employed to make a 3 h data assimilation in σ–θ coordinates. Advances in remote sensing from Earth- and space-borne systems, expanded *in situ* observation network, and increased low-cost computer capability allow an initialization for meso- and convective scale models. *Local Analysis and Prediction System* (*LAPS*; Shaw *et al.* 2004) uses data from local *mesonetwork* (*mesonet*) of surface observing systems, Doppler radars, satellites, wind and temperature profilers, as well as aircraft, which are incorporated every hour into a three-dimensional grid covering approximately a 1000 km × 1000 km horizontal area. The prediction component of the LAPS is configured using a model chosen from a suite of mesoscale NWP models. Any or all of these models, usually being initialized with LAPS analyses, is run to provide 0–18 h forecasts. Another example of this type of operational atmospheric prediction system is the ARPS Data Analysis System (ADAS) and its 3DVAR system which are often used in the intermittent assimilation cycle mode, at up to 5 minute assimilation intervals (e.g., Xue and Martin 2006). Surface mesonet data available as frequently as every 5 minutes are also routinely assimilated into the model system.

Many of the data assimilation methods developed for larger-scale models cannot be applied to the storm-scale models. For example, storm-scale phenomena are highly ageostrophic and divergent, so that the constraints between the mass and momentum field applied at the larger scale (geostrophic and thermal wind balances) cannot be applied to the storm scale. Furthermore, the mass field typically has to be inferred from the reflectivity and radial velocity measured by Doppler radars, instead of being measured directly. Thus, the retrieval techniques become very critical. The adjoint-based 4DVAR method can improve the accuracy in retrieving the thermodynamic fields, compared to the more conventional method that retrieves the thermodynamic fields from the retrieved wind fields (Sun and Crook 1996). Fig 13.15 shows an example of a vertical velocity field retrieved by an adjoint method.

Ensemble-based data assimilation, a collection of flexible state-estimation techniques that use short-term ensemble forecasts to estimate the flow-dependent background error covariance, has recently been implemented in various atmospheric

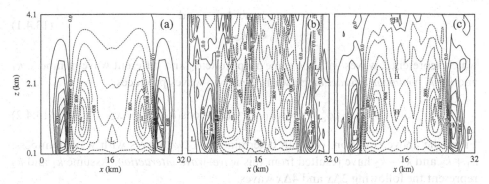

Fig. 13.15 Vertical velocity field from (a) control run, (b) vertical integration of the continuity equation, and (c) adjoint retrieval. (After Sun and Crook 1996.)

models. These experimental studies demonstrated the feasibility and effectiveness of the ensemble-based techniques for different scales and flows of interest (e.g., Houtekamer *et al.* 2005; Zhang *et al.* 2006a, b) and for parameter estimation, which offers hope to the treatment of different sources of parametric model error (e.g., Anderson 2001; Aksoy *et al.* 2006). The best-known form of ensemble-based assimilation is the ensemble Kalman filter (EnKF) (e.g., Evensen 2003; Snyder and Zhang 2003; Hamill 2006; Xue *et al.* 2006). Major advantages of using the EnKF over existing data assimilation schemes include its use of flow-dependent error covariance, its simplicity in implementation and maintenance, and its automatic generation of ensembles consistent with the analysis error covariance for the subsequent ensemble forecasts. The EnKF method is also capable of better handling nonlinear physical processes and/or nonlinear observational operators, such as that for the radar reflectivity (Tong and Xue 2005).

13.4 Nonlinear aliasing and instability

In discussing numerical instabilities in Chapter 12, we have neglected the nonlinear effects. However, in the real atmosphere, kinetic energy generated at the large scale or mesoscale tends to transfer to smaller scales. When it is transferred to the *inertial subrange*, the kinetic energy is neither produced nor dissipated, but handed down to smaller and smaller scales. The inertial subrange is an intermediate range of turbulent scales or wavelengths such that the kinetic energy is independent of original forcing of the motion and molecular dissipation. When the kinetic energy is transferred to an even smaller scale on the order of O(cm), which is called the *dissipation range*, the kinetic energy is converted into internal energy by molecular interaction.

In a numerical mesoscale model, this cascade of energy to smaller scales cannot occur because the smallest feature that can be resolved has a wavelength of 2Δx. For example, let us consider

$$\phi_1 = \phi_0 \cos k_1 \Delta x, \quad \text{and}$$
$$\phi_2 = \phi_0 \cos k_2 \Delta x, \tag{13.4.1}$$

which represent 2 waves with the same amplitude ϕ_0 and different wave numbers, k_1 and k_2. A *nonlinear interaction* between these two waves produces

$$\phi_1 \phi_2 = (\phi_0^2/2)[\cos(k_1 + k_2)\Delta x + \cos(k_1 - k_2)\Delta x]. \tag{13.4.2}$$

From the above equation, one can determine that two waves with wave numbers, $k_1 + k_2$ and $k_1 - k_2$ have resulted from this *wave–wave interaction*. Assume k_1 and k_2 represent the following $2\Delta x$ and $4\Delta x$ waves:

$$k_1 = 2\pi/(2\Delta x), \text{ and } k_2 = 2\pi/(4\Delta x), \tag{13.4.3}$$

then we have

$$\phi_1 \phi_2 = \left(\frac{\phi_0^2}{2}\right)\left\{\cos\left[2\pi\Delta x\left(\frac{3}{4\Delta x}\right)\right] + \cos\left[2\pi\Delta x\left(\frac{1}{4\Delta x}\right)\right]\right\}. \tag{13.4.4}$$

The second cosine term of the above equation is a $4\Delta x$ wave, which can be appropriately represented by the grid mesh. However, the first term is a $1.33\Delta x$ wave, which cannot be resolved by the grid mesh. The wave will be fictitiously represented by a $4\Delta x$ wave because the first integer multiple of $4\Delta x/3$ is $4\Delta x$. This phenomenon is called *nonlinear aliasing*. Figure 13.16 shows a schematic that illustrates how a physical solution with a wavelength of $4\Delta x/3$, caused by the nonlinear interaction of $2\Delta x$ and $4\Delta x$ waves, is seen as a computational $4\Delta x$ wave in the numerical grid mesh. In the real world, we have the large-scale disturbance generated by forcing, which then cascades to a mesoscale disturbance, small-scale disturbance, and then dissipates at an even smaller scale. However, it does not seem to happen in the same way in the numerical model, in which waves with wavelength shorter than $2\Delta x$ will be represented as larger-scale waves. Therefore, even if a numerical method is linearly stable, the results can degrade into computational noise. The erroneous accumulation of energy can cause the model dependent variables to increase in magnitude abruptly without bound, which is called *nonlinear instability*.

Two methods can be applied to avoid the nonlinear instability: (1) proper parameterization of the subgrid-scale correction terms, such as $\overline{u'w'}$, $\overline{v'w'}$, and $\overline{\theta'w'}$, so that energy is extracted from the averaged equations, or (2) the use of a spatial *numerical smoother* or *filter* to remove the shorter waves, which leaves the longer waves relatively unaffected. The first approach is better than the second one because it is based on physical principle. However, it requires a good knowledge about the *subgrid-scale correlation* terms. The second approach can be accomplished in a relatively easier manner (e.g, Shapiro 1975). To understand *numerical smoothing*, we may consider a simple one-dimensional, *three-point smoother*,

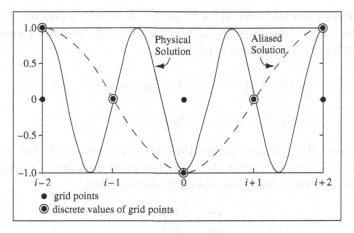

Fig. 13.16 Schematic illustration of nonlinear aliasing. A physical solution with a wavelength of $1.33\Delta x$, caused by the nonlinear interaction of waves of $2\Delta x$ and $4\Delta x$, is seen as a $4\Delta x$ wave in a numerical grid mesh. (Adapted after Pielke 2002, reproduced with permission from Elsevier.)

$$\bar{\phi}_j = (1 - s)\phi_j + (s/2)(\phi_{j-1} + \phi_{j+1}), \tag{13.4.5}$$

where $x = j\Delta x$ and s is a constant that can be negative. If this operator is applied to a sinusoidal wave

$$\phi = A\,e^{ikx}, \tag{13.4.6}$$

where $k = 2\pi/L$ is the wave number, and A is the wave amplitude that may be a complex number, then the result may be written as

$$\bar{\phi} = R\phi, \quad R = 1 - s(1 - \cos k\Delta x) = 1 - 2s\sin^2(\pi\Delta x/L). \tag{13.4.7}$$

In the above equation, R is referred to as the *response function*. If $R \geq 0$, then the wave number and phase are not affected but only the wave amplitude. If $|R| > 1$, then the wave is amplified by the operator. On the other hand, if $|R| < 1$, then the wave is damped by the operator. If $R < 0$, then the phase of the wave is shifted by $180°$, which is undesirable. With $s = 1/2$, we obtain the second-order *three-point smoother*:

$$\bar{\phi}_j = \frac{\phi_{j-1} + 2\phi_j + \phi_{j+1}}{4}, \tag{13.4.8}$$

and

$$R(1/2) = 1 - \frac{1 - \cos k\Delta x}{2} = \cos^2(\pi\Delta x/L). \tag{13.4.9}$$

From above, if $L = 2\Delta x$, then $R = 0$. Hence, for a $2\Delta x$ wave, the smoother will eliminate it immediately.

Since a three-point smoother, (13.4.8), damps the shorter waves too strongly, it is less desirable. A *five-point smoother* can be obtained by applying two successive three-point smoothers with $s = 1/2$ and $-1/2$:

$$
\begin{aligned}
\overline{\overline{\phi}}_j &= \frac{1}{16}\left[10\phi_j + 4(\phi_{j-1} + \phi_{j+1}) - (\phi_{j-2} + \phi_{j+2})\right] \\
&= \phi_j - \frac{1}{16}\left[6\phi_j - 4(\phi_{j-1} + \phi_{j+1}) + (\phi_{j-2} + \phi_{j+2})\right].
\end{aligned}
\tag{13.4.10}
$$

The above smoother will also remove the $2\Delta x$ wave immediately but will preserve more of the longer waves. In fact, the above five-point smoother is analogous to the finite difference form of the *fourth-order diffusion equation*,

$$
\frac{\partial \phi}{\partial t} + c\frac{\partial^4 \phi}{\partial x^4} = 0,
\tag{13.4.11}
$$

which has a finite difference form,

$$
\phi_j^{\tau+1} = \phi_j^{\tau} - \gamma_1[6\phi_j^{\tau} - 4(\phi_{j-1}^{\tau} + \phi_{j+1}^{\tau}) + (\phi_{j-2}^{\tau} + \phi_{j+2}^{\tau})],
\tag{13.4.12}
$$

where $\gamma_1 = c\Delta t/\Delta x^4$. If we choose $\gamma_1 = 1/16$, then the above equation is analogous to (13.4.10). Thus, in applying the five-point smoother, it has a similar effect as the fourth-order diffusion, which is why *numerical smoothing* has also been referred to as *numerical diffusion*. In order to retain the amplitude of longer waves, the coefficient $1/16$ in (13.4.10) or γ_1 is often reduced. Testing is needed to find out the most appropriate coefficient of the numerical smoothing or diffusion. In practice, smoothing is not applied to the boundary points. For the grid points adjacent to the boundaries, we may need to apply the three-point smoother or second-order diffusion

$$
\overline{\phi}_j = \phi_j - \gamma_2[2\phi_j - (\phi_{j-1} + \phi_{j+1})].
\tag{13.4.13}
$$

In order to make (13.4.13) consistent with (13.4.12), we require

$$
\gamma_2 = 4\gamma_1.
\tag{13.4.14}
$$

Notice that the leapfrog scheme also produces a computational mode with a $2\Delta t$ wave. To suppress this, we may apply the time smoother (Asselin, 1972)

$$
\phi^{\tau+1} = \bar{\phi}^{\tau-1} + 2\Delta t(\partial\phi/\partial t)^{\tau},
\tag{13.4.15}
$$

where

$$
\bar{\phi}^{\tau-1} = \phi^{\tau-1} + \gamma(\phi^{\tau} - 2\phi^{\tau-1} + \bar{\phi}^{\tau-2}).
\tag{13.4.16}
$$

Based on numerical testing, a choice of $\gamma < 0.25$ has been recommended.

In general, high-order filters or smoothers are more *scale selective*, i.e., they filter shorter waves at or close to two-grid interval wavelength most effectively while leaving longer waves less affected. High-order filters can have the effect of generating undesirable overshooting and undershooting in the filtered field, however. The erroneous effects in the filtered field can be avoided by a simple treatment that ensures that the diffusion flux is always down-gradient (Xue 2000).

13.5 Modeling a stratified fluid system

To elucidate how to model a stratified fluid flow system, we consider the nonlinear, hydrostatic, incompressible fluid system similar to that governed by (2.2.14)–(2.2.18) with $V = 0$,

$$\frac{\partial u'}{\partial t} + (U + u')\frac{\partial u'}{\partial x} + v'\frac{\partial u'}{\partial y} + w'\left(U_z + \frac{\partial u'}{\partial z}\right) - fv' + \frac{1}{\bar{\rho}}\frac{\partial p'}{\partial x} = \nu\nabla^2 u', \qquad (13.5.1)$$

$$\frac{\partial v'}{\partial t} + (U + u')\frac{\partial v'}{\partial x} + v'\frac{\partial v'}{\partial y} + w'\frac{\partial v'}{\partial z} + fu' + \frac{1}{\bar{\rho}}\frac{\partial p'}{\partial y} = \nu\nabla^2 v', \qquad (13.5.2)$$

$$\frac{1}{\bar{\rho}}\frac{\partial p'}{\partial z} = g\frac{\theta'}{\bar{\theta}}, \qquad (13.5.3)$$

$$\frac{\partial u'}{\partial x} + \frac{\partial v'}{\partial y} + \frac{\partial w'}{\partial z} = 0, \qquad (13.5.4)$$

$$\frac{\partial \theta'}{\partial t} + (U + u')\frac{\partial \theta'}{\partial x} + v'\frac{\partial \theta'}{\partial y} + w'\left(\frac{N^2\bar{\theta}}{g} + \frac{\partial \theta'}{\partial z}\right) = \frac{\theta_o}{c_p T_o}q' + \kappa\nabla^2\theta', \qquad (13.5.5)$$

where q' is the diabatic heating rate in J kg^{-1}s^{-1}, which may represent surface sensible and/or elevated heating, ν the *eddy viscosity*, κ *eddy thermal diffusivity*, and the Brunt–Vaisala frequency N is defined as $N^2 = (g/\bar{\theta})(d\bar{\theta}/dz)$. Other symbols are defined in Chapter 2. The basic state is assumed to be in geostrophic and hydrostatic balances. The basic and perturbation quantities have been separated in the above system, which allows one to examine the nonlinear effects by comparing with the corresponding linear simulation.

To elucidate how to model the nonlinear system of (13.5.1)–(13.5.5), we consider relatively simple and straightforward schemes, i.e. the leapfrog-in-time and second-order centered-in-space to the prognostic equations. The variables $u^{\tau+1}$, $v^{\tau+1}$ and $\theta^{\tau+1}$ at time step $\tau + 1$ are obtained from other variables at time steps τ and $\tau - 1$, based on (13.5.1), (13.5.2) and (13.5.5), respectively. The vertical velocity $w^{\tau+1}$ can then be obtained by integrating the continuity equation, (13.5.4), upward. The upper boundary condition can be approximated by a sponge layer with ν increasing from the top of the physical domain to the top of the model domain, or following the flow chart of Fig. 13.10 to apply the upper radiation boundary condition numerically. Following Fig. 13.10, we can obtain $p^{\tau+1}$ by integrating the hydrostatic equation (13.5.3) downward. In this approach, all variables at time step $\tau+1$ are then obtained numerically.

In some mesoscale models, the hydrostatic assumption is relaxed to more properly simulating *deep convection* and effects of steep topography. A set of fully compressible fluid system can be written as,

$$\frac{\mathrm{D}u}{\mathrm{D}t} = fv - \frac{1}{\rho}\frac{\partial p}{\partial x} + \nu\nabla^2 u, \tag{13.5.6}$$

$$\frac{\mathrm{D}v}{\mathrm{D}t} = -fu - \frac{1}{\rho}\frac{\partial p}{\partial y} + \nu\nabla^2 v, \tag{13.5.7}$$

$$\frac{\mathrm{D}w}{\mathrm{D}t} = -\frac{1}{\rho}\frac{\partial p}{\partial z} - g + \nu\nabla^2 w, \tag{13.5.8}$$

$$\frac{\mathrm{D}\theta_v}{\mathrm{D}t} = S_\theta + \kappa\nabla^2\theta \tag{13.5.9}$$

$$\frac{\mathrm{D}\phi}{\mathrm{D}t} = S_\phi + \kappa\nabla^2\phi, \quad \phi = q_v, q_c, q_i, q_r, q_s, q_g, \tag{13.5.10}$$

$$\frac{\mathrm{D}\rho}{\mathrm{D}t} + \rho\nabla\cdot V = 0, \tag{13.5.11}$$

$$p = \rho RT \tag{13.5.12}$$

$$\theta_v = T_v\left(\frac{p_s}{p}\right)^{R_d/c_p}, \tag{13.5.13}$$

$$T_v = T(1 + 0.61q_v), \tag{13.5.14}$$

where T_v is the *virtual temperature*; θ_v is the *virtual potential temperature*; p_s is the basic state pressure at the ground, usually taken as 1000 hPa; S_θ is any source or sink of θ_v, such as surface long-wave radiation and elevated latent heating; S_ϕ is any source or sink of the hydrometeor ϕ, such as mixing ratios of *water vapor* (q_v), *cloud water* (q_c), *rain* (q_r), *cloud ice* (q_i), *snow* (q_s), and *graupel/hail* (q_g). The virtual (potential) temperature is the (potential) temperature that a dry air parcel would have if its pressure and density were equal to those of a given sample of moist air. The virtual temperature is a fictitious temperature of a moist air parcel that satisfies the equation of state for dry air. More realistic and sophisticated parameterizations of planetary boundary layer processes can be adopted.

The above equations may also be represented in terms of the *Exner function*, to be defined below. Including the moisture, the equation of state can be written in the form,

$$p = \rho R_d T_v. \tag{13.5.15}$$

In order to avoid an explicit treatment of the density, an *Exner function* has been adopted in some mesoscale and cloud-scale models,

$$\pi = c_p \left(\frac{p}{p_s} \right)^{R_d/c_p} = \frac{c_p T_v}{\theta_v}. \tag{13.5.16}$$

The pressure gradient force terms can then be approximately represented by

$$\frac{1}{\rho} \frac{\partial p}{\partial x_i} = \bar{\theta}_v \frac{\partial \pi}{\partial x_i}, \quad i = 1, 2, 3, \tag{13.5.17}$$

where $\bar{\theta}_v$ is the initial undisturbed state θ_v and a function of z only, defined as $\theta_v = \bar{\theta}_v + \theta_v'$. The Exner function can be partitioned into $\pi = \bar{\pi} + \pi'$, where $\bar{\pi}$, is the initial basic state, and π' is the perturbation from the initial state $\bar{\pi}$. The initial basic state is assumed to be in geostrophic balance in horizontal and hydrostatic balance in vertical,

$$\frac{\partial \bar{\pi}}{\partial x} = f v_g; \quad \frac{\partial \bar{\pi}}{\partial y} = -f u_g, \tag{13.5.18}$$

$$\frac{\partial \bar{\pi}}{\partial z} = -\frac{g}{\bar{\theta}_v}. \tag{13.5.19}$$

The advantages of using π, instead of p are that: (a) ρ is not treated explicitly in the governing equations, (b) π does not present in the buoyancy term even if the vertical scale of the motion L_z is equivalent to the scale height H; (c) there is no need to compute the density perturbation; and (d) less truncation error is introduced since $\partial \bar{\pi}/\partial z$ is much less than $\partial \bar{p}/\partial z$ (Pielke 2002). One disadvantage of using the Exner function is that the anelastic continuity equation for π is much more complicated and needs a Poisson equation solver, which is very tedious in the terrain-following coordinates (Huang 2000).

Using the Exner function and approximating θ_v by $\bar{\theta}_v$ in pressure gradient forces, the momentum equations become

$$\frac{Du}{Dt} = f(v - v_g) - \bar{\theta}_v \frac{\partial \pi'}{\partial x} + \nu \nabla^2 u, \tag{13.5.20}$$

$$\frac{\mathrm{D}v}{\mathrm{D}t} = -f(u - u_\mathrm{g}) - \bar{\theta}_\mathrm{v}\frac{\partial \pi'}{\partial y} + \nu\nabla^2 v, \tag{13.5.21}$$

$$\frac{\mathrm{D}w}{\mathrm{D}t} = -\bar{\theta}_\mathrm{v}\frac{\partial \pi'}{\partial z} + g\left[\frac{\theta'_\mathrm{v}}{\bar{\theta}_\mathrm{v}} - q_\mathrm{H}\right] + \nu\nabla^2 w, \quad q_\mathrm{H} = q_\mathrm{c} + q_\mathrm{i} + q_\mathrm{r} + q_\mathrm{s} + q_\mathrm{g} \tag{13.5.22}$$

where $\mathrm{D}/\mathrm{D}t \equiv \partial/\partial t + u\partial/\partial x + v\partial/\partial y + w\partial/\partial z$. The pressure equation can be written as

$$\begin{aligned}
\frac{\mathrm{D}\pi'}{\mathrm{D}t} = &-\frac{1}{\bar{\theta}_\mathrm{v}}(\boldsymbol{V}\cdot\nabla\bar{\pi}) - \left(\frac{R_\mathrm{d}\bar{\pi}}{c_\mathrm{v}\bar{\rho}\bar{\theta}_\mathrm{v}^2}\right)\nabla\cdot(\bar{\rho}\bar{\theta}_\mathrm{v}\boldsymbol{V}) - \left(\frac{R_\mathrm{d}\pi'}{c_\mathrm{v}}\right)\nabla\cdot\boldsymbol{V} \\
&+ \frac{R_\mathrm{d}(\bar{\pi} + \pi')}{c_\mathrm{v}\bar{\theta}_\mathrm{v}}\dot{\theta} + \kappa\nabla^2\pi',
\end{aligned} \tag{13.5.23}$$

where $\dot{\theta} \equiv \mathrm{D}\theta/\mathrm{D}t$ is the diabatic heating rate (J s^{-1}). The thermodynamic equation and the equations governing hydrometeors can be written as

$$\frac{\mathrm{D}\theta}{\mathrm{D}t} = S_\theta + \kappa\nabla^2\theta, \tag{13.5.24}$$

$$\frac{\mathrm{D}\phi}{\mathrm{D}t} = S_\phi + \kappa\nabla^2\phi, \quad \phi = q_\mathrm{v},\ q_\mathrm{c},\ q_\mathrm{i},\ q_\mathrm{r},\ q_\mathrm{s},\ \text{and}\ q_\mathrm{g}. \tag{13.5.25}$$

The advantage of using the fully compressible fluid system of (13.5.20)–(13.5.25) is that every equation can be integrated alone, numerically, to obtain its own value at the next time step without having to couple with other equations, such as the hydrostatic, incompressible fluid system of (13.5.1)–(13.5.5). However, this set of equations contains *sound waves*, which propagate at much higher speeds than the gravity waves and require a very small time step to ensure the numerical stability. In practice, it is almost impossible to adopt such a small time step, even at the research mode of numerical simulations. In order to improve numerical efficiency of the above compressible atmospheric system, a *time-splitting scheme* was proposed (Klemp and Wilhelmson 1978). In the time-splitting scheme, equations with no sound wave modes, i.e. (13.5.24) and (13.5.25), embedded are marched with a large time step, Δt, while equations with sound waves embedded, i.e. (13.5.20)–(13.5.23), are integrated with a small time step $\Delta \tau$ from time $t - \Delta t$ to $t + \Delta t$.

Scale analysis indicates that the only important term in (13.5.23) for representing convection is the second term on the right-hand side, which leads to the anelastic continuity equation (Ogura and Philips 1962)

$$\nabla\cdot(\bar{\rho}\boldsymbol{V}) = 0. \tag{13.5.26}$$

Taking the time derivative of the above equation and using the momentum equations yield an elliptic equation

$$\nabla^2 \bar{\theta}_v \pi' + \nabla \bar{\theta}_v \cdot \nabla \pi' + \frac{\partial}{\partial z}(\ln \bar{\rho})\bar{\theta}_v \frac{\partial \pi'}{\partial z} = \text{source terms}, \qquad (13.5.27)$$

where the source terms include acceleration terms. One of the disadvantages in adopting the anelastic approximation is that (13.5.27) becomes very complicated and computationally expensive when it is transformed into terrain-following coordinates.

13.6 Predictability and ensemble forecasting

One major challenge in NWP is whether the weather phenomena of concern are predictable or not. If they are intrinsically unpredictable, then the improvements in more accurate initial and boundary conditions, numerical methods, and subgrid-scale parameterizations of a NWP model will be useless. If they are predictable, it is important to know how long we can make numerical prediction with a "perfect" NWP model, if there were one. As discussed briefly in Chapter 1, in reality, most of the weather systems more or less have limited predictability. Thus, it leaves us some room to make improvements in the accuracy of NWP models, which is still a topic of current research, especially for mesoscale NWP. Thus we will only make a very brief summary of the *predictability* problem and *ensemble forecasting*.

In the early 1950s, some meteorologists started to apply statistical methods to weather prediction to cope with the uncertainties encountered in forecasting (Gleeson 1961). The weather forecasting problem has been viewed as evolving probabilities. Even with improved model techniques, the weather prediction has its own inherent limitations due to the inevitable model deficiencies and errors in the initial conditions, or the predictability problem. The atmosphere, like any other dynamical system with instabilities, has an inherit time limit of predictability (Lorenz 1963). Based on Saltzman's (1962) convective equations, Lorenz found that two completely different solutions were predicted by the same model with slightly different initial conditions. He later found that errors of different spatial scales grow at different rates (Lorenz 1969). On average, the fastest error growth occurs at smallest scales, which would have significant impacts on the mesoscale NWP.

Predictability can have two fundamentally different meanings. Intrinsic predictability can be defined as "the extent to which prediction is possible if an optimum procedure is used" in the presence of infinitesimal initial errors (Lorenz 1969). Practical predictability, on the other hand, can be specified as the ability to predict based on the procedures that are currently available. Practical predictability is limited by realistic uncertainties in both the initial states and the forecast models, which in general are not infinitesimally small (Lorenz 1996). Compared to typical synoptic-scale flows, recent studies showed that predictability of mesoscale weather systems, particularly the associated quantitative precipitation, which are of most concern to the public, can be very limited (e.g., Ehrendorfer 1997). The predictability of quantitative precipitation is different from earlier results which indicated that the mesoscale enjoyed enhanced predictability. Through high-resolution mesoscale simulations, recent studies demonstrated that mesoscale predictability depends strongly on the background flow regime and dynamics. The simulations also show that moist convection is a primary mechanism for forecast-error growth at sufficiently small scales, and that convective-scale errors contaminate the mesoscale within lead times of interest to NWP, thus effectively limiting the predictability of the mesoscale (e.g., Zhang *et al.* 2006). Understanding of the limit of mesoscale predictability and the associated error growth dynamics is essential for setting up expectations and priorities for advancing deterministic mesoscale forecasting, and for providing guidance on the design, implementation and application of short-range ensemble prediction systems (e.g., Tracton and Kalnay 1993; Stensrud *et al.* 1999).

Over the past decade, *ensemble forecasting* has emerged as a powerful tool for operational numerical weather prediction (Molteni *et al.* 1996). Ensemble forecasting is a collection of different forecasts all valid at the same forecast time. Ensemble

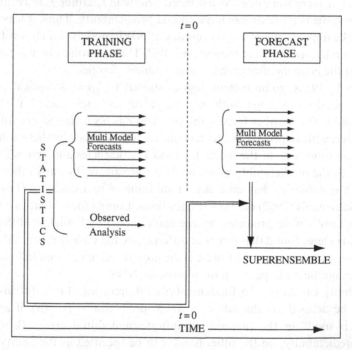

Fig. 13.17 A flow chart of multianalysis-multimodel superensemble forecasting. The vertical line in the center denotes the initial time ($t = 0$), and the area to the left denotes the training period where a large number of forecast experiments are carried out by the multianalysis-multimodel system. During the training period, the observed fields provide statistics that are then passed on to the period on the right, where $t > 0$. Here the multianalysis-multimodel forecasts along with the statistics provide the superensemble forecasts. (Adapted after Krishnamurti *et al.* 2001.)

forecasts start from different initial conditions, boundary conditions, parameter settings, or entirely independent NWP models. The various forecasts all represent possibilities given the uncertainties associated with forecasting. From these possibilities, one can estimate probabilities of various events as well as an averaged or consensus forecast. An ensemble of forecasts can be used to: (a) composite into a single forecast by means of a weighted average, (b) provide an excellent probabilistic alternative to explore the dynamics, predictability and background error covariance for mesoscale weather systems, (c) estimate the reliability of the composite forecast, (d) suggest where additional special observations might be targeted to improve forecast accuracy (Kalnay 2003; Hawblitzel *et al.* 2006), and (e) determine the sensitivity of forecasts to the model input parameters, including the initial and boundary conditions (Martin and Xue 2006).

In order to take advantage of forecasts from different models and analyses, the multianalysis-multimodel forecasts along with the aforementioned statistics are used to provide the superensemble forecasts. The superensemble has a higher accuracy compared to that of the ensemble mean because the superensemble is selective in assigning weights and the past history of performance of models from the past statistics. A real-time multianalysis-multimodel superensemble forecast can make a significant improvement in precipitation forecasts (Krishnamurti *et al.* 2001). As indicated in Fig. 13.17, during the training period, the observed fields provide statistics that are then passed on to the period on the right, where $t > 0$. Compared to medium-range ensemble forecasting with global models, limited-area short-range ensemble prediction is less widely used, which is at least partially due to our limited knowledge of mesoscale predictability (Stensrud *et al.* 1999). Still, limited studies on even high-resolution storm-scale ensemble predictions have been undertaken (Kong *et al.* 2006) and such research is expected to expand significantly in the future.

References

Aksoy A., F. Zhang, J. W. Nielsen-Gammon, 2006. Ensemble-based simultaneous state and parameter estimation with MM5. *Geophys. Res. Lett.*, **33**, L12801, doi:10.1029/2006GL026186, 2006.

Anderson, J. L., 2001. An ensemble adjustment Kalman filter for data assimilation. *Mon. Wea. Rev.*, **129**, 2884–903.

Anthes, R. A. and T. T. Warner, 1978. Development of hydrostatic models suitable for air pollution and other mesometeorological studies. *Mon. Wea. Rev.*, **106**, 1045–78.

Arakawa, A. and V. R. Lamb, 1977. Computational design of the basic dynamical processes of the UCLA general circulation model. In *Methods in Computational Physics*, **17**, Academic Press, 174–265.

Asselin, R. A., 1972. Frequency filter for time integration. *Mon. Wea. Rev.*, **100**, 487–90.

Bacon, D. P., N. N. Ahmad, Z. Boybeyi, T. J. Dunn, M. S. Hall, P. C. S. Lee, R. A. Sarma, M. D. Turner, K. T. Waight III, S. H. Young, and J. W. Zack, 2000.

A dynamically adapting weather and dispersion model: The operational multiscale environment model with grid adaptivity (OMEGA). *Mon. Wea. Rev.*, **128**, 2044–76.

Barnes, S. L., 1964. A technique for maximizing details in numerical weather map analysis. *J. Appl. Meteor.*, **3**, 396–409.

Benjamin, S. G., D. Dévényi, S. S. Weygandt, K. J. Brundage, J. M. Brown, G. A. Grell, D. Kim, B. E. Schwartz, T. G. Smirnova, T. L. Smith and G. S. Manikin, 2004. An hourly assimilation–forecast cycle: the RUC. *Mon. Wea. Rev.*, **132**, 495–518.

Bougeault, P., 1983. A non-reflective upper boundary condition for limited-height hydrostatic models. *Mon. Wea. Rev.*, **111**, 420–9.

Davies, H., 1983. Limitations of some common lateral boundary conditions used in regional NWP models. *Mon. Wea. Rev.*, **111**, 1002–12.

Dietachmayer, G. S. and K. K. Droegemeier, 1992. Application of continuous dynamic grid adaptation techniques to meteorological modeling. Part I: Basic formulation and accuracy. *Mon. Wea. Rev.*, **120**, 1675–1706.

Doswell, C. A. III, 1984. A kinematic analysis of frontogenesis associated with a nondivergent vortex. *J. Atmos. Sci.*, **41**, 1242–8.

Durran, D. R., 1998. *Numerical Methods for Wave Equations in Geophysical Fluid Mechanics*. Springer-Verlag.

Ehrendorfer, M., 1997. Predicting the uncertainty of numerical weather forecast: a review. *Meteor. Z.*, **6**, 147–83.

Evensen, G., 2003. The ensemble Kalman filter: theoretical formulation and practical implementation. *Ocean Dynamics*, **53**, 343–67.

Gandin, L. S., 1988. Complex quality control of meteorological observations. *Mon. Wea. Rev.*, **116**, 1137–56.

Gleeson, T. A., 1961. A statistical theory of meteorological measurements and predictions. *J. Meteor.*, **18**, 192–8.

Haltiner, G. J. and R. T. Williams, 1980. *Numerical Prediction and Dynamic Meteorology*. 2nd edn., John Wiley & Sons.

Hamill, T. M., 2006. Ensemble-based atmospheric data assimilation. In *Predictability of Weather and Climate*, T. Palmer (ed.), Cambridge University Press, 124–156.

Hoke, J. E. and R. A. Anthes, 1976. The initialization of numerical models by a dynamical initialization technique. *Mon. Wea. Rev.*, **104**, 1551–6.

Houtekamer, P. L., H. L. Mitchell, G. Pellerin, M. Buehner, M. Charron, L. Spacek, and B. Hansen, 2005. Atmospheric data assimilation with an ensemble Kalman filter: results with real observations. *Mon. Wea. Rev.*, **133**, 604–20.

Hu, M., M. Xue, J. Gao, and K. Brewster, 2006. 3DVAR and cloud analysis with WSR-88D level-II data for the prediction of Fort Worth tornadic thunderstorms. Part II: Impact of radial velocity analysis via 3DVAR. *Mon. Wea. Rev.*, **134**, 699–721.

Huang, C.-Y., 2000. A forward-in-time anelastic nonhydrostatic model in a terrain-following coordinate. *Mon. Wea. Rev.*, **128**, 2108–34.

Huang, X.-Y. 1999. A generalization of using an adjoint model in intermittent data assimilation systems. *Mon. Wea. Rev.*, **127**, 766–87.

Kalman, R. E. and R. S. Bucy, 1961. New results in linear filtering and prediction theory. Trans. ASME. *J. Basic Eng.*, **83D**, 95–108.

Kalnay. E., 2003. *Atmospheric Modeling, Data Assimilation and Prediction*. Cambridge University Press.

Klemp, J. B. and D. R. Durran, 1983. An upper boundary condition permitting internal gravity wave radiation in numerical mesoscale models. *Mon. Wea. Rev.*, **111**, 430–44.

Klemp, J. B. and D. K. Lilly, 1978: Numerical simulation of hydrostatic mountain waves. *J. Atmos. Sci.*, **35**, 78–107.

Klemp, J. B. and R. B. Wilhelmson, 1978. The simulation of three-dimensional convective storm dynamics. *J. Atmos. Sci.*, **35**, 1070–96.

Kong, F., K. K. Droegemeier, and N. L. Hickmon, 2006. Multi-resolution ensemble forecasts of an observed tornadic thunderstorm system, Part I: Comparison of coarse and fine-grid experiments. *Mon. Wea. Rev.*, **134**, 807–33.

Krishnamurti, T. N., S. Surendran, D. W. Shin, R. J. Correa-Torres, T. S. V. V. Kumar, E. Williford, C. Kummerow, R. F. Adler, J. Simpson, R. Kakar, W. S. Olson, and F. J. Turk, 2001. Real-time multianalysis-multimodel superensemble forecasts of precipitation using TRMM and SSM/I products. *Mon. Wea. Rev.*, **129**, 2861–83.

Lin, Y.-L. and T.-A. Wang, 1996. Flow regimes and transient dynamics of two-dimensional stratified flow over an isolated mountain ridge. *J. Atmos. Sci.*, **53**, 139–58.

Long, R. R., 1953. Some aspects of stratified fluids. I. A theoretical investigation. *Tellus*, **5**, 42–58.

Lorenz, E. N., 1963. Deterministic non-periodic flow. *J. Atmos. Sci.*, **20**, 130–41.

Lorenz, E. N. 1969. The predictability of a flow which possesses many scales of motion. *Tellus*, **21**, 289–307.

Lorenz, E. N., 1996. Predictability – A problem partly solved. *Seminar on Predictability, 1995 ECMWF Seminar Proceedings*, ECMWF. Volume **I**, 1–19.

Mahrer, Y. and R. A. Pielke, 1978. A test of an upstream spline interpolation technique for the advective terms in a numerical model. *Mon. Wea. Rev.*, **106**, 818–30.

Martin, W. J. and M. Xue, 2006. Initial condition sensitivity analysis of a mesoscale forecast using very-large ensembles. *Mon. Wea. Rev.*, **134**, 192–207.

Mesinger, F., Z. I. Janjić, S. Nickovic, D. Gavrilov, and D. G. Deaven, 1988. The step-mountain coordinate: model description and performance for cases of Alpine lee cyclogenesis and for a case of an Appalachian redevelopment. *Mon. Wea. Rev.*, **116**, 1493–518.

Miyakoda, K. and R. Moyer, 1968. A method of initialization for dynamical weather forecasting. *Tellus*, **20**, 115–128.

Molteni, F., R. Buizza, T. N. Palmer, and T. Petroliagis, 1996. The ECMWF ensemble prediction system. Methodology and validation. *Quart. J. Roy. Meteor. Soc.*, **122**, 73–120.

Ogura, Y. and N. A. Philips, 1962. Scale analysis of deep and shallow convection in the atmosphere. *J. Atmos. Sci.*, **19**, 173–9.

Oliger, J. and A. Sundström, 1978. Theoretical and practical aspects of some initial boundary value problems in fluid dynamics. *Int'l J. Numer. Methods Fluids*, **21**, 183–204.

Orlanski, I., 1976. A simple boundary condition for unbounded hyperbolic flows. *J. Compu. Phys.*, **21**, 251–269.

Perkey, D. J. and C. W. Kreitzberg, 1976. A time-dependent lateral boundary scheme for limited-area primitive equation models. *Mon. Wea. Rev.*, **104**, 744–55.

Pielke, R. A., 2002. *Mesoscale Meteorological Modeling.* 2nd edn., Academic Press.

Phillips, N. A., 1957. A coordinate system having some special advantages for numerical forecasting. *J. Meteor.*, **14**, 184–5.

Rogers, E., M. Baldwin, T. Black, K. Brill, F. Chen, C. DiMego, J. Gerrity, G. Manikin, F. Mesinger, K. Mitchell, D. Parrish, and Q. Zhao, 1998: *Changes to* the NCEP operational "early" Eta analysis/forecast system. NWS Tech. Proc. Bull., Ser. No. 447, NWS, NOAA. (http://www.nws.noaa.gov/om/tpb/447.htm)

Saltzman, B., 1962. Finite amplitude free convection as an initial value problem – I, *J. Atmos. Sci.*, **19**, 329–41.

Sasaki, Y., 1970. Some basic formulations in numerical variational analysis. *Mon. Wea. Rev.*, **98**, 875–83.

Schoenstadt, A. L., 1978. *A transfer function analysis of numerical schemes used to simulate geostrophic adjustment.* NPS Report. NPS-53-79-001.

Shapiro, R., 1975. Linear filtering. *Math. Compu.*, **29**, 1094–7.

Shaw, B. L., S. Albers, D. Birkenheuer, J. Brown, J. McGinley, P. Schultz, J. Smart, and E. Szoke, 2004. Application of the Local Analysis and Prediction System (LAPS) diabatic initialization of mesoscale numerical weather prediction models for the IHOP-2002 field experiment. *Preprints, 20th Conference on Weather Analysis and Forecasting and 16th Conference on Numerical Weather Prediction*, Amer. Meteor. Soc., Seattle, WA.

Skamarock, W. C., 1989. Truncation error estimates for refinement criteria in nested adaptive models. *Mon. Wea. Rev.*, **117**, 872–86.

Skamarock, W. C., J. B. Klemp, J. Dudhia, D. O. Giu, D. M. Barker, W. Wang and J. G. Powers, 2005. *A Description of the Advanced Research WRF Version 2.* NCAR Technical Note, NCAR/TN-468 + STR. [Available at http://wrf-model.org/wrfadmin/docs/arw_v2.pdf]

Smagorinski, J., J. L. Halloway, Jr., and G. D. Hembree, 1967. Prediction experiments with a general circulation model. Proceed. Int'l. Sympo. Dynamics Large Scale Atmospheric Processes, Nauka, Moscow, USSR, 70–134.

Snyder, C., and F. Zhang, 2003. Tests of an ensemble Kalman filter for convective-scale data assimilation. *Mon. Wea. Rev.*, **131**, 1663–77.

Sugi, M., 1986. Dynamic normal mode initialization. *J. Meteor. Soc. Japan*, **64**, 623–32.

Sommerfeld, A., 1949. *Partial Differential Equations in Physics.* Academic Press.

Stensrud, D. J., H. E. Brooks, J. Du, M. S. Tracton, and E. Rogers, 1999. Using ensembles for short-range forecasting. *Mon. Wea. Rev.*, **127**, 433–46.

Stephens, J., 1970. Variational initialization of the balance equation. *J. Appl. Meteor.*, **9**, 732–39.

Sun, J. and N. A. Crook, 1996. Comparison of thermodynamic retrieval by the adjoint method with the traditional retrieval method. *Mon. Wea. Rev.*, **124**, 308–24.

Talagrand, O., 1972. On the damping of high-frequency motions in four-dimensional assimilation of meteorological data. *J. Atmos. Sci.*, **29**, 1571–4.

Temperton, C., 1988. Implicit normal mode initialization. *Mon. Wea. Rev.*, **116**, 1013–1031.

Tong, M. and M. Xue, 2005. Ensemble Kalman filter assimilation of Doppler radar data with a compressible nonhydrostatic model: OSS Experiments. *Mon. Wea. Rev.*, **133**, 1789–1807.

Tracton, M. S. and E. Kalnay, 1993. Operational ensemble prediction at the National Meteorological Center: Practical aspects. *Weather and Forecasting*, **8**, 379–98.

Xue, M., 2000. High-order monotonic numerical diffusion and smoothing. *Mon. Wea. Rev.* **128**, 2853–64.

Xue, M. and W. J. Martin, 2006. A high-resolution modeling study of the 24 May 2002 case during IHOP. Part I: Numerical simulation and general evolution of the dryline and convection. *Mon. Wea. Rev.*, **134**, 149–71.

Xue, M., M. Tong, and K. K. Droegemeier, 2006. An OSSE framework based on the ensemble square-root Kalman filter for evaluating impact of data from radar networks on thunderstorm analysis and forecast. *J. Atmos. Ocean Tech.*, **23**, 46–66.

Zhang, F., Z. Meng, and A. Aksoy, 2006a. Tests of an ensemble Kalman filter for mesoscale and regional-scale data assimilation, Part I: Perfect model experiments. *Mon. Wea. Rev.*, **134**, 722–36.

Zhang, F., A. Odins, and J. W. Nielsen-Gammon, 2006b. Mesoscale predictability of an extreme warm-season rainfall event. *Weather and Forecasting*, **21**, 149–66.

Zou, X. and Y.-H. Kuo, 1996. Rainfall assimilation through an optimal control of initial and boundary conditions in a limited-area mesoscale model. *Mon. Wea. Rev.*, **124**, 2859–82.

Problems

13.1 From (13.1.5) and (13.1.6) derive the computational dispersion relationship, (13.1.17), of the two-dimensional shallow water system with $U = 0$.

13.2 Derive the computational dispersion relationship corresponding to (13.1.7) but for an unstaggered grid mesh.

13.3 Substituting $u(x, t) = \hat{u}(k, \omega) \exp[i(kx - \omega t)]$ into a particular finite difference form of the advection equation gives

$$e^{-2i\omega\Delta t} + i\beta e^{-i\omega\Delta t} - 1 = 0.$$

Find the stability criterion for this scheme if $0 \leq \beta \leq 2$.

13.4 Derive (13.1.13) from (13.1.12).

13.5 Applying the forward-in-time and backward-in-space scheme to the advection equation, (12.3.1), at the right boundary to show that the zero gradient boundary condition assumes the waves propagate out of the right boundary at a speed of $\Delta x / \Delta t$.

13.6 Derive a numerical radiation boundary condition at the inflow boundary, (13.2.7), and $U - c_i^*$.

13.7 Derive (13.2.23) from (13.2.18)–(13.2.21) by assuming a wave-like solution.

13.8 Derive the complete equation of (13.3.1).

13.9 Prove the response function (R) in (13.4.7) by applying the three-point smoother (13.4.5) to a sinusoidal wave (13.4.6).

13.10 Derive (13.4.10).

13.11 Prove $\gamma_2 = 4\gamma_1$ by considering a $2\Delta x$ wave, $\phi_i = (-1)^{i+1}$.

13.12 Derive the pressure gradient forces, (13.5.17), by using the Exner function.

Modeling projects

13.A1 Set the smoothing coefficient as 0 and different values in the advection model and run it. Discuss the results.

13.A2 Replace the five-point smoother of the shallow water model by a three-point smoother in the shallow water model and discuss the smoothing effects.

13.A3 Modify your 2D Tank Model to simulate a two-layer fluid system with $H = 1$ km and $\Delta\rho/\rho_0 = 0.1$. Rerun the model for a flow over a bell-shaped mountain with $U = 20$, 30, and 40 m s^{-1}. Describe and explain the results.

13.A4 (i) Extend the 2D Tank Model to a 3D Tank Model.

   (ii) Now run this 3D Tank Model with $U = 0\,\mathrm{m\,s^{-1}}$ and $h_\mathrm{m} = 0\,\mathrm{m}$ by giving a three-dimensional initial field in h'.

   (iii) Simulate a basic flow with $U = 20\,\mathrm{m\,s^{-1}}$ over a 3D bell-shaped mountain, $h_\mathrm{s} = h_\mathrm{m} / [1 + (x/a)^2 + (y/a)^2]^{3/2}$.

   Use the same H and $\Delta\rho/\rho_0$ as in project 13.A3 for (ii) and (iii). You may use *surface*, *contour* and *vector* plotting subroutines to plot u' and h' fields. Explain your results. The governing equations are expressed as (3.4.4)–(3.4.6).

13.A5 Change the current finite difference scheme (leapfrog, fourth-order central difference) to (i) forward-in-time and upstream-in-space scheme, and (ii) leapfrog-in-time and second-order central difference in space scheme.

   Using the above schemes to rerun project 13.A4 (ii) and compare the results.

13.A6 Model the three-dimensional, hydrostatic, Boussinesq system of (13.5.1)–(13.5.4).

14

Parameterizations of physical processes

14.1 Reynolds averaging

In the previous two chapters, we have described numerical methods for approximating
the primitive equations and the setup for numerical models, such as grid systems, and
initial and boundary conditions. As demonstrated in Section 13.5, a mesoscale model
may be developed to simulate a simple geophysical fluid system, such as a stratified,
inviscid flow over topography. However, in order to apply this type of geophysical
fluid dynamics model to simulate mesoscale atmospheric phenomena, some important
physical processes, such as boundary layer processes, moist processes, and radiative
transfer processes, need to be represented or parameterized in the model.

In order to numerically integrate the governing differential equations in a limited
area, a numerical method, such as a finite difference method, spectral method or finite
element method, must be used to approximately represent the atmospheric motion
and processes by the dependent variables at grid points or elements. The approximations limit the explicit representation of atmospheric motions and processes to a scale
smaller than that for the grid interval, truncated wavelength, or finite element. For
example, large-scale disturbances may cascade down to mesoscale, then further down to
the smallest turbulent eddies responsible for viscous dissipation in the atmosphere. If the
subgrid-scale disturbances are not appropriately represented by the grid point values,
they may cause nonlinear aliasing and nonlinear numerical instability, as discussed in
Chapter 13. An obvious way to resolve the problem is to provide sufficient resolution in
a numerical model, so that the model can explicitly simulate any significant small-scale
motions and processes. For example, *direct numerical simulation* (*DNS*) has been
developed to numerically simulate turbulent motions by fluid dynamicists, in which
the time-dependent Navier–Stokes equations with explicit terms for molecular diffusion
are integrated numerically to obtain the solution, without making any turbulence
parameterizations. The finest scales of the simulation are determined by the balance
between nonlinear advection and viscous diffusion, i.e. the *Reynolds number* ($R_e = UL/\nu$,
where U and L are the characteristic velocity and length scales, respectively, and ν the
kinematic viscosity coefficient) of the flow. A typical value of kinematic viscosity for the
air in the lower atmosphere is $1.5 \times 10^{-5} \mathrm{m}^2 \mathrm{s}^{-1}$. When $R_e \gg 1$, changes in motion by

advection are much more important than the dissipation due to molecular viscosity. In this type of turbulent flow, a *turbulent Reynolds number* is more appropriately used to describe the characteristic of the flow, in which the kinematic viscosity coefficient is replaced by the *turbulent exchange coefficient*. Boundary layers encountered in engineering practice have a fairly large Reynolds numbers ranging from 10^3 to 10^6, while the atmospheric boundary layers developing over most natural surfaces are characterized by even larger Reynolds numbers (10^6 to 10^9). The higher Reynolds number flows have also been observed in the free atmosphere, such as within cumulus cloud and in wave breaking regions.

DNS requires the whole range of spatial and temporal scales of the turbulence to be resolved by the grid interval (Δx), from the smallest dissipative scale (L_ε), where L_ε is approximately equal to $(\nu^3/\varepsilon)^{1/4}$ and ε is the kinetic energy dissipation. To satisfy these conditions, the number of grid intervals N in the grid direction must satisfy $N\Delta x > L$ and $\Delta x < L_\varepsilon$, where L is the integral scale. Since $\varepsilon \approx U^3/L$, a three-dimensional DNS will require a number of grid intervals $N^3 \geq R_e^{9/4}$. Thus, the computational cost for DNS is extremely high. With current computing power, it is unrealistic to apply DNS to mesoscale atmospheric modeling. On the other hand, even when the needed computing power is available; we still have to be careful in using the detailed information about small-scale turbulent motions and processes with sizes that cannot be resolved by available observational systems. Since these processes are not well understood at the present time, the governing equations of fluid motion cannot describe them accurately.

The second approach is to numerically integrate the Reynolds-averaged Navier–Stokes (*RANS*) equations of the mean motion. The ensemble properties of all time fluctuations in a turbulent flow are described by a turbulence closure. In this approach, the subgrid-scale motions and processes are parameterized. The parameterization approach gives a less detailed representation than the explicit representation (DNS), but it is more practical in terms of computing cost and may be sufficiently accurate for many mesoscale models since it considers grid interval and initial data, among other factors.

A third approach in numerically simulating turbulent flows is to simulate large turbulent eddies explicitly, while the unresolved subgrid scale motions associated with smaller turbulent eddies are either ignored or parameterized. In this type of *large-eddy simulations* (*LES*), the large turbulent eddies explicitly simulated by the numerical model fall in the range of the grid size to the domain size of the model. Although the LES of turbulent flows and neutral and unstable *planetary boundary layer* (PBL) flows have been demonstrated to be very encouraging, the simulations of the nocturnal boundary layer are less successful due to the fact that the characteristic large-eddy scale becomes too small, and that most of the energy transfer and other exchange processes are overly influenced or dominated by subgrid scale motions. Although the LES derive their credibility from the explicit resolution of large-scale turbulent eddies, they depend upon a small-scale *turbulence closure* and must, to some degree, inherit the many uncertainties associated with turbulence closure (Mason 1994). Most LES results obtained so far are very encouraging; however, there is still room for improvements to overcome certain limiations. Some improvements include (a) the quality of the simulation can depend sensitively on subgrid modeling, which is not fully developed; and (b) LES requires high numerical accuracy, and does not in particular tolerate numerical dissipation which is often adopted in mesoscale models. To take advantage of both the LES and RANS, a hybrid LES–RANS approach has been developed and applied, in particular, to engineering problems.

Unresolved turbulent eddies of various scales smaller than the grid interval often fluctuate rapidly in time, thus limiting the description of their behavior to statistical approaches. The use of statistical approaches requires the introduction of an averaging operator. Any averaging operator used in atmospheric modeling should be able to satisfy the following criteria (Cotton 1986): (a) The operator should provide a formal mechanism for distinguishing between resolvable and unresolvable eddies. (b) The operator should produce a set of equations more amenable to integration, either analytically or numerically, than the original system of equations. (c) The averaged set of atmospheric variables should be measurable by current or anticipated atmospheric sensing systems. Following the scheme originally developed by Reynolds (1895), each model variable is decomposed into a slow-varying mean field part and a rapid-varying turbulent part, such as $u = \bar{u} + u', v = \bar{v} + v', w = \bar{w} + w', \theta = \bar{\theta} + \theta', p = \bar{p} + p'$, and $\rho = \bar{\rho} + \rho'$. Some useful formulas for the *Reynolds averaging* can be derived, for example,

$$\overline{u + w} = \bar{u} + \bar{w}; \quad \overline{cw} = c\bar{w}; \quad \bar{\bar{w}} = \bar{w}; \quad \overline{w'} = 0;$$

$$\overline{w'\theta} = \overline{w'\bar{\theta}} = 0; \quad \overline{w\theta} = \overline{(\bar{w} + w')(\bar{\theta} + \theta')} = \bar{w}\bar{\theta} + \overline{w'\theta'}; \quad \overline{uw} = \bar{u}\bar{w} + \overline{u'w'}; \quad (14.1.1)$$

$$\frac{\overline{\partial w}}{\partial s} = \frac{\partial \bar{w}}{\partial s}; \quad \frac{\overline{\partial \bar{w}}}{\partial s} = \frac{\partial \bar{w}}{\partial s}; \quad \int \overline{w\,\mathrm{d}s} = \int \bar{w}\,\mathrm{d}s; \quad s = x, y, z, \text{ or } t,$$

where c is a constant, $\overline{u'w'}$ and $\overline{w'\theta'}$ are the *vertical turbulent flux of zonal momentum* and *vertical turbulent heat flux*, respectively. In statistical terms, these fluxes, as an average of the product of deviation components, are also called *covariances*. Figure 14.1 shows a sketch of subgrid-scale vertical velocity (w') and potential temperature (θ') and the subgrid scale covariance $\overline{w'\theta'}$. As can be seen from the figure, the vertical heat flux associated with the resolvable dependent variables is approximately zero, i.e. $\bar{w}\bar{\theta} \cong 0$ because $\bar{w} = 0$. However, the covariance or the vertical turbulent heat flux, $\overline{w'\theta'}$, is not zero. Both grid value averages are assumed to be constant over Δx.

If we apply the Reynolds averaging to a time interval and a grid volume of a numerical model, then the Reynolds-averaged value of a variable ϕ represents,

$$\bar{\phi} \equiv \frac{1}{\Delta x \,\Delta y \,\Delta z \,\Delta t} \int_t^{t+\Delta t} \int_x^{x+\Delta x} \int_y^{y+\Delta y} \int_z^{z+\Delta z} \phi \,\mathrm{d}z \,\mathrm{d}y \,\mathrm{d}x \,\mathrm{d}t. \qquad (14.1.2)$$

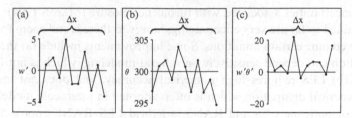

Fig. 14.1 Schematic illustration of subgrid scale values of vertical velocity w' (cm s^{-1}), potential temperature θ' (K), and the subgrid scale covariance $w'\theta'$ (K cm s^{-1}). In this example, the grid averaged value of vertical motion is required to be approximately 0 (i.e. $\bar{w} = 0$), and $\bar{\theta} = 299.5$ K is used. Both grid value averages are assumed to be constant over Δx. The grid-averaged subgrid-scale correlation $\overline{w'\theta'}$ is equal to 6.9 cm K s^{-1}. (Adapted after Pielke 2002, reproduced with permission from Elsevier.)

This is called *grid-volume averaging*. Thus, $\phi' = \phi - \bar{\phi}$ is the fluctuation or perturbation across the grid intervals, Δx, Δy, Δz, and time interval Δt from $\bar{\phi}$. Applying the Reynolds averaging to the grid volume of the mesoscale model system of (13.5.6)–(13.5.14) with Boussinesq approximation leads to

$$\frac{\overline{D}\bar{u}}{Dt} = f\bar{v} - \frac{1}{\rho_o}\frac{\partial \bar{p}}{\partial x} - \frac{1}{\rho_o}\left[\frac{\partial \left(\rho_o \overline{u'u'}\right)}{\partial x} + \frac{\partial \left(\rho_o \overline{u'v'}\right)}{\partial y} + \frac{\partial \left(\rho_o \overline{u'w'}\right)}{\partial z}\right] + \nu\nabla^2\bar{u}, \qquad (14.1.3)$$

$$\frac{\overline{D}\bar{v}}{Dt} = -f\bar{u} - \frac{1}{\rho_o}\frac{\partial \bar{p}}{\partial y} - \frac{1}{\rho_o}\left[\frac{\partial \left(\rho_o \overline{u'v'}\right)}{\partial x} + \frac{\partial \left(\rho_o \overline{v'v'}\right)}{\partial y} + \frac{\partial \left(\rho_o \overline{v'w'}\right)}{\partial z}\right] + \nu\nabla^2\bar{v}, \qquad (14.1.4)$$

$$\frac{\overline{D}\bar{w}}{Dt} = -\frac{1}{\rho_o}\frac{\partial p_1}{\partial z} - g\frac{\rho_1}{\rho_o} - \frac{1}{\rho_o}\left[\frac{\partial \left(\rho_o \overline{u'w'}\right)}{\partial x} + \frac{\partial \left(\rho_o \overline{v'w'}\right)}{\partial y} + \frac{\partial \left(\rho_o \overline{w'w'}\right)}{\partial z}\right] + \nu\nabla^2\bar{w}, \quad (14.1.5)$$

$$\frac{\overline{D}\bar{\theta}}{Dt} = \bar{S}_\theta - \frac{1}{\rho_o}\left[\frac{\partial \left(\rho_o \overline{u'\theta'}\right)}{\partial x} + \frac{\partial \left(\rho_o \overline{v'\theta'}\right)}{\partial y} + \frac{\partial \left(\rho_o \overline{w'\theta'}\right)}{\partial z}\right] + \kappa\nabla^2\bar{\theta}_v, \qquad (14.1.6)$$

$$\frac{\overline{D}\bar{\phi}}{Dt} = \bar{S}_\phi - \frac{1}{\rho_o}\left[\frac{\partial \left(\rho_o \overline{u'\phi'}\right)}{\partial x} + \frac{\partial \left(\rho_o \overline{v'\phi'}\right)}{\partial y} + \frac{\partial \left(\rho_o \overline{w'\phi'}\right)}{\partial z}\right] + \kappa\nabla^2\bar{\phi}, \qquad (14.1.7)$$

$$\phi = q_v, q_c, q_i, q_r, q_s, \text{ and } q_g,$$

$$\nabla \cdot \left(\rho_o \bar{V}\right) = 0, \quad \bar{V} = (\bar{u}, \bar{v}, \bar{w}), \qquad (14.1.8)$$

$$\bar{p} = \bar{\rho}R_d\bar{T}_v, \qquad (14.1.9)$$

$$\bar{\theta}_v = \overline{T}_v \left(\frac{p_s}{\bar{p}}\right)^{R_d/c_p}, \qquad (14.1.10)$$

$$\overline{T}_v = \overline{T}(1 + 0.61\bar{q}_v), \qquad (14.1.11)$$

$$\bar{p} = p_o + p_1; \quad \bar{\rho} = \rho_o + \rho_1; \quad \frac{\partial p_o}{\partial z} = -\rho_o g; \quad \bar{\phi} = \phi_o + \phi_1, \qquad (14.1.12)$$

where $p_s = 1000$ hPa, and

$$\frac{\overline{D}}{\overline{Dt}} = \frac{\partial}{\partial t} + \bar{u}\frac{\partial}{\partial x} + \bar{v}\frac{\partial}{\partial y} + \bar{w}\frac{\partial}{\partial z}, \qquad (14.1.13)$$

$$\phi_o \equiv \frac{1}{XY}\int_{-Y/2}^{Y/2}\int_{-X/2}^{X/2}\bar{\phi}\,dx\,dy. \qquad (14.1.14)$$

In the above equation, ϕ_o is the layer average over the domain (X,Y), which is assumed to be large enough compared with the mesoscale phenomena concerned, to ensure a hydrostatic balance. Since $\phi_1 = \bar{\phi} - \phi_o$, ϕ_1 is the nonhydrostatic part of $\bar{\phi}$ and/or the perturbation from large-scale ϕ_o. The terms $\overline{u'\theta'}$, $\overline{v'\theta'}$, and $\overline{w'\theta'}$ are the *turbulent heat fluxes*, $\overline{u'w'}$ and $\overline{v'w'}$ the *vertical turbulent fluxes* of horizontal momentum, and $\overline{u'v'}$ the *meridional turbulent flux* of zonal momentum. In deriving the above equations, we have assumed $|\rho_1/\rho_o| \ll 1$. We have also partitioned $\bar{p}$ and $\bar{\rho}$ into hydrostatic (with subscript "o") and nonhydrostatic parts (with subscript "1"). Above the boundary layer, these flux terms are very small compared with other terms and may thus be neglected. In the boundary layer, the turbulent flux divergence terms are of the same order as the other terms in (14.1.3)–(14.1.8). Therefore, they cannot be simply dropped from the equation system. Note that the equation set (14.1.3)–(14.1.11) is not a closed system mathematically since in addition to the unknown mean variables, other flux terms are also present. In order to make the equation set closed, we need to represent or parameterize the turbulent flux terms and the source and sink terms using the mean variables. The need for parameterizations poses a *closure problem*, which is a challenging task in parameterizing the PBL processes, as well as for moist and radiative transfer processes. The horizontal derivatives of the turbulent flux terms are normally associated with some horizontal inhomogeneities, such as cities and coastlines, which may be neglected over horizontally homogeneous regions.

In addition to the Reynolds averaging method, several different averaging methods may also be adopted. For example, for a data set measured discretely, an *ensemble averaging* may be adopted,

$$\bar{\phi}_e = \lim_{N \to \infty} \frac{1}{N}\sum_{k=1}^{N}\phi(x_o, y_o, z_o, t_o). \qquad (14.1.15)$$

If the turbulence is stationary and homogeneous, which is unlikely in the real world, then the above three averaging methods should give the same value. An alternative approach in averaging data set is to take the grid-volume averaging, as defined in (14.1.2). For cases where there are N data points to be averaged over a grid volume, one may take the *generalized ensemble averaging*,

$$\bar{\phi} = \frac{1}{TXYZ} \lim_{N\to\infty} \frac{1}{N} \sum_{k=1}^{N} \int_{-Z/2}^{Z/2} \int_{-Y/2}^{Y/2} \int_{-X/2}^{X/2} \int_{-T/2}^{T/2} \phi(x',y',z',t') dt' dx' dy' dz'. \quad (14.1.16)$$

In mesoscale models, physical processes such as the planetary boundary, moist and radiative transfer, land-surface interaction, and air–sea interaction processes need to be parameterized. In this chapter, we will limit our discussion to the introduction of basic principles and methods in parameterizing the first three processes. However, this does not mean that the importance of the parameterization of other physical processes will be neglected.

14.2 Parameterization of planetary boundary layer processes

The objective of the PBL parameterization is to use the grid-volume averaged mean variables to represent the turbulent flux terms associated with turbulent eddies, as well as the heat source and sink terms present in the PBL, such as those present in (14.1.3)–(14.1.8). In other words, we need to close the system mathematically. Thus, a suitable closure scheme is needed. Using the appropriate parameterizations of these terms, we are able to numerically integrate the Reynolds-averaged equations of the mean motion, such as (14.1.3)–(14.1.11).

One simple way to model the planetary boundary layer is to treat the whole layer as one slab and predict the vertically averaged properties of the PBL. In this approach, the details of the vertical structure of the PBL are ignored, which may work for cases where vertical gradients are small throughout much of the PBL, such as the convective boundary layer, or for use in models such as general circulation models (GCMs), which do not have enough vertical resolution. However, this approach is not appropriate for mesoscale models since the prediction of some detailed information of the PBL is expected. In mesoscale models, the PBL is divided into a number of layers, depending upon what the physical phenomenon concerned requires.

Physically, the PBL may be approximately divided into the *surface layer* and the layer above it. In the lowest part of the surface layer, i.e., the *viscous sublayer*, molecular motions dominate the transfer of dependent variables. The viscous sublayer forces the velocity to vanish (i.e., no-slip boundary condition) at the ground, which continuously leads to the development of turbulent eddies. Thus, the molecular viscosity and thermal diffusion terms in these equations are kept, but the turbulent flux terms in (14.1.3)–(14.1.7) are neglected. The viscous sublayer normally has a depth of O(1 cm), but can be as shallow as 0.001 cm over smooth ice. Over most

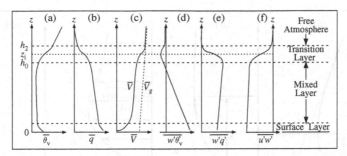

Fig. 14.2 Typical convective boundary layer profiles of (a) mean virtual potential temperature, (b) specific humidity, (c) wind speed ($\overline{V}$ and $\overline{V}_g$ denote mean wind speed and geostrophic wind speed respectively), (d) vertical heat flux, (e) vertical moisture flux, and (f) vertical momentum flux. (Adapted after Driedonks and Tennekes 1984.)

natural surfaces, a *roughness layer* or *canopy layer* forms above the viscous sublayer. The roughness layer may go up to 10 m over large buildings (Oke 1987), and the height is referred to as *aerodynamic roughness*, usually denoted as z_o. Due to the constraint of vertical resolution, the viscous sublayer and roughness layer of the surface layer are often neglected in mesoscale models.

Above the viscous sublayer or roughness layer exists the upper part of the surface layer, which usually occupies up to 10–100 m, about 10% of the entire PBL. The upper part of the surface layer, which is often referred to simply as the "surface layer," is mainly maintained by the vertical momentum transfer associated with turbulent eddies. Coriolis and pressure gradient forces do not play a major role in the surface layer. Therefore, for the purpose of modeling the surface layer, it is normally assumed that: (1) the subgrid scale fluxes are independent of z, (2) the Coriolis force is negligible, (3) the layer reaches a steady state, and (4) there is a horizontal homogeneity over a flat surface. Based on these assumptions, empirical formulas have been developed to specify the relationship between dependent variables and subgrid fluxes. The layer above the surface layer becomes the *mixed layer* under unstable and convective conditions and the *outer layer* under neutral and stable conditions.

The mixed layer and outer layer have very different characteristics in terms of wind, temperature, and humidity profiles. Figure 14.2 shows typical profiles of mean virtual potential temperature, specific humidity, wind speed, and heat, moisture, and momentum fluxes within a convective boundary layer. The virtual potential temperature and wind speed are quite uniform in the mixed layer due to mixing associated with turbulent eddies. On top of the mixed layer is the *transition layer*, which contains a temperature inversion and an increase in wind speed, and then the *free atmosphere*. On the other hand, a typical *stable boundary layer*, say induced by the nocturnal boundary layer, has an inversion layer above the surface (Fig. 14.3). Within the stable boundary layer, the atmosphere has its largest stability near the surface, decreasing smoothly toward neutral with height. A temperature inversion is often observed

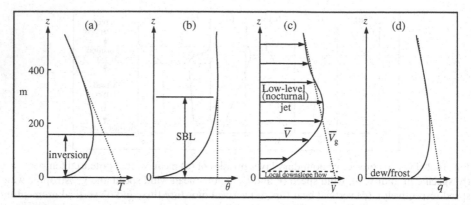

Fig. 14.3 Typical stable boundary layer (SBL) profiles of (a) mean temperature, (b) potential temperature, (c) wind speed, and (d) specific humidity. (Adapted after Stull 1988, with kind permission of Springer Sciences and Media.)

near the surface. Higher in the stable boundary layer, the wind speed may increase with height, reaching a maximum near the top of the stable layer, and thus becoming what is known as the *nocturnal low-level jet*. The whole depth of the stable boundary layer is on the order of several hundred meters, much shallower than the convective boundary layer.

14.2.1 Parameterization of the surface layer

Simulating the surface layer is important since momentum generated in the free atmosphere and the PBL tends to dissipate in this layer, while the heat and moisture are transported upward to the PBL from the ground through this layer. Analogous to molecular diffusion, the subgrid scale fluxes may be represented by

$$\overline{u'w'} = -K_{\mathrm{m}}\frac{\partial \bar{u}}{\partial z}; \quad \overline{v'w'} = -K_{\mathrm{m}}\frac{\partial \bar{v}}{\partial z}; \quad \overline{w'\theta'} = -K_{\mathrm{h}}\frac{\partial \bar{\theta}}{\partial z}; \quad \overline{w'q'} = -K_{\mathrm{q}}\frac{\partial \bar{q}}{\partial z}, \quad (14.2.1)$$

where K_{m} is called the *exchange coefficient of momentum* or simply *eddy viscosity*, and K_{h} and K_{q} are called the *exchange coefficients* or *eddy diffusivities of heat and water vapor*, respectively. In practice, K_{q} is often assigned as the same value as K_{h}. The exchange coefficients are often taken as constants, and are empirically related to height and stability as calculated from the NWP model output, and this approach of the parameterization of momentum, heat, and moisture fluxes is called *K theory*. The K theory is a *first-order closure* since the fluxes are parameterized proportional to the mean values, which have also been applied to the parameterization of the PBL above the surface layer. Since the subgrid fluxes are assumed to be independent of height in the surface layer, we may define a *friction velocity* u_*, *flux temperature* θ_*, and q_* as

$$u_*^2 = \tau_0/\bar{\rho} = \left(\overline{u'w'}^2 + \overline{v'w'}^2\right)^{1/2};$$

$$u_*^2 \cos\mu = -\overline{u'w'}; \quad u_*^2 \sin\mu = -\overline{v'w'}; \quad \mu = \tan^{-1}(\bar{v}/\bar{u}); \tag{14.2.2}$$

$$u_*\theta_* = -\overline{w'\theta'}; \quad u_*q_* = -\overline{w'q'},$$

where τ_0 is the shearing stress generated by the horizontal wind. If the x-axis is chosen such that $\bar{v} = \overline{u'v'} = 0$, then (14.2.2) reduces to $u_*^2 = -\overline{u'w'}$. From dimensional analysis, the wind shear $(\partial\bar{V}/\partial z)$ is proportional to the velocity scale (u_*) divided by the length scale (z),

$$\frac{\partial\bar{V}}{\partial z} = \frac{u_*}{kz}; \quad \bar{V} = \sqrt{\bar{u}^2 + \bar{v}^2}, \tag{14.2.3}$$

where k is the *von Kármán constant*, which has an empirical value of 0.4. Integrating (14.2.3) leads to the well-known logarithmic wind profile for a neutrally stratified, constant-flux surface layer,

$$\bar{V}(z) = \frac{u_*}{kz}\ln\left(\frac{z}{z_0}\right). \tag{14.2.4}$$

Based on the *Monin and Obukov similarity theory*, we have

$$\frac{\partial\bar{V}}{\partial z} = \frac{u_*}{kz}\phi_m(z/L), \tag{14.2.5}$$

where L is the *Monin–Obukov length* defined as $L = \bar{\theta}u_*^2/kg\theta_*$, $\theta_* = -\overline{w'\theta'}/u_*$, and ϕ_m is estimated by an empirical formula, such as (Businger 1973),

$$\phi_m \approx (1 - 15z/L)^{-1/4}, \quad z/L \leq 0$$

$$\approx 1 + 4.7z/L, \quad z/L > 0. \tag{14.2.6}$$

Integrating (14.2.5) from z_0, where $\bar{V} = 0$, to z gives

$$\bar{V}(z) = \frac{u_*}{k}\left[\ln\left(\frac{z}{z_0}\right) - \varphi_m\left(\frac{z}{L}\right)\right], \tag{14.2.7}$$

where

$$\varphi_m(z/L) = 2\ln\left((1 + \phi_m^{-1})/2\right) + \ln\left((1 + \phi_m^{-2})/2\right) - 2\tan^{-1}(1/\phi_m) + \pi/2, \quad z/L \leq 0$$

$$= -4.7z/L, \quad z/L > 0 \tag{14.2.8}$$

Similarly, the vertical profiles of $\bar{\theta}$ and $\bar{q}$ can be formulated as

$$\bar{\theta}(z) = \bar{\theta}_{z_0} + \frac{\theta_*}{\beta k}\left[\ln\left(\frac{z}{z_0}\right) - \varphi_{\rm h}\left(\frac{z}{L}\right)\right], \tag{14.2.9}$$

$$\bar{q}(z) = \bar{q}_{z_0} + \frac{q_*}{\beta k}\left[\ln\left(\frac{z}{z_0}\right) - \varphi_{\rm h}\left(\frac{z}{L}\right)\right], \tag{14.2.10}$$

where $\varphi_{\rm h}$ is an empirical formula. For example, the following $\varphi_{\rm h}$ has been proposed (Businger 1973),

$$\varphi_{\rm h}(z/L) = \begin{cases} 2\ln\left[(1+0.74\phi_{\rm h}^{-1})/2\right], & z/L \le 0 \\ -6.35z/L, & z/L > 0 \end{cases} \tag{14.2.11}$$

$$\phi_{\rm h} = \begin{cases} 0.74(1-9z/L)^{-1/2}, & z/L \le 0 \\ 0.74 + 4.7z/L, & z/L > 0 \end{cases} \tag{14.2.12}$$

The symbol β represents the characteristic *vertical mixing length* for $\bar{\theta}$ and $\bar{q}$. An empirical value of 1.35 has been used. Note that at the bottom of the surface layer (i.e. $z = z_0$) or the top of the viscous sublayer or roughness layer, the no free-slip conditions for the velocities are often assumed, i.e. $\bar{u} = \bar{v} = \bar{w} = 0$, and the potential temperature and specific humidity may be estimated by (Deardorff 1974)

$$\bar{\theta}_{z_0} = \bar{\theta}_{\rm G} + 0.0962(\theta_*/k)(u_* z_0/\nu)^{0.45} \quad \text{and} \tag{14.2.13}$$

$$\bar{q}_{z_0} = \bar{q}_{\rm G} + 0.0962(q_*/k)(u_* z_0/\nu)^{0.45}, \tag{14.2.14}$$

where ν is the kinematic viscosity coefficient of air, which has a value of about $1.5 \times 10^{-5} m^2 s^{-1}$.

14.2.2 Parameterization of the PBL

As mentioned above, the layer above the surface layer is called the mixed layer under unstable conditions, and the outer layer under neutral and stable conditions. The mixed layer extends from the top of the surface layer to 1–2 km or higher under unstable conditions, and several hundred meters under neutral and stable conditions. Three boundary layer flow regimes, based on different sets of force balances, have been proposed to help understand the dynamics of the PBL above the surface layer: (1) *Ekman layer*, which is supported by a balance among the pressure gradient force, Coriolis force, and frictional force, (2) *advective boundary layer*, which is supported by a balance among pressure gradient force, frictional force, and advective acceleration, and (3) *Stokes boundary layer*, which is supported by a balance between pressure gradient force and frictional force. In order to close the equation set (14.1.3)–(14.1.11), the subgrid scale fluxes need to be represented by the mean values averaged over grid and time intervals. From a numerical modeling point of view, the parameterizations of the PBL above the surface layer may be classified as: (a) bulk aerodynamic parameterization, (b) K-theory parameterization, (c) turbulent kinetic energy closure scheme, and (d) higher-order closure schemes.

a. Bulk aerodynamic parameterization

The *bulk aerodynamic parameterization* treats the boundary layer as a single slab and assumes the wind speed and potential temperature are independent of height, and the turbulence is horizontally homogeneous. Based on these assumptions, the horizontal turbulence flux divergence terms in (14.1.3)–(14.1.7) can be neglected, and the vertical subgrid turbulence fluxes are parameterized by

$$\overline{u'w'} = -C_d \overline{V}^2 \cos \mu; \quad \overline{v'w'} = -C_d \overline{V}^2 \sin \mu; \quad \overline{w'\theta'} = -C_h \overline{V}(\bar{\theta} - \bar{\theta}_{z_0}), \quad (14.2.15)$$

where C_d and C_h are nondimensional *drag* and *heat transfer coefficients*, respectively, $\overline{V} = (\bar{u}^2 + \bar{v}^2)^{1/2}$, $\mu = \tan^{-1}(\bar{v}/\bar{u})$, and z_0 is the roughness or top of the surface layer. The values of $\overline{V}, \bar{u}, \bar{v}$, and $\bar{\theta}$ are evaluated at the standard anemometer height, 10 m. The bulk aerodynamic parameterization has been adopted in some GCM and regional climate models. For a given reference height, C_d increases with increasing roughness, which ranges from 1.3×10^{-3} over ocean surface to 7×10^{-3} over rough land surface. From the formulas proposed in the parameterization of the surface layer, the expressions of C_d and C_h may be derived

$$C_d = k^2/(\ln(z/z_0) - \varphi_m(z/L))^2;$$
$$(14.2.16)$$
$$C_h = \beta k^2/(\ln(z/z_0) - \varphi_m(z/L))(\ln(z/z_0) - \varphi_h(z/L)),$$

where φ_m and φ_h are defined in (14.2.8) and (14.2.12), respectively, and β is an empirical value defined in (14.2.9). As mentioned earlier, an empirical value 1.35 has been proposed for β.

Due to the assumption of height-independent wind speed and potential temperature and horizontally homogeneous turbulence, the bulk aerodynamic parameterization is more suitable for representing a well-mixed boundary layer than the neutral and stable boundary layers. Based on these assumptions, further assuming a three-way balance among the Coriolis force, pressure gradient force, and the vertical gradient of the turbulent momentum flux from (14.1.3) and (14.1.4), and using the bulk parameterization, one may derive the following equations for $\bar{u}$ and $\bar{v}$,

$$\bar{u} = \bar{u}_g - \kappa_s \overline{V} \bar{v}; \quad \bar{v} = \kappa_s \overline{V} \bar{u}, \quad (14.2.17)$$

where $\kappa_s \equiv C_d/(fh)$, h is the mixed layer height, and $\bar{u}_g$ is the geostrophic wind speed at the bottom of the mixed layer. Equation (14.2.17) can also be rewritten as

$$fk \times \overline{V} = -\frac{1}{\rho_0}\nabla\overline{p} - \frac{C_d}{h}\overline{V}\,\overline{V}; \quad \overline{V} = (\overline{u}, \overline{v}), \tag{14.2.18}$$

which gives a three-way balance with the wind deflected toward the low pressure. In addition, the cross-isobar flow increases as the turbulent drag increases. Note that in a rotational frame of reference or in the presence of directional shear, the frictional force on a fluid element need not be parallel and opposite to the velocity vector (e.g., Fig. 6.4 of Arya 2001), as commonly depicted in many textbook schematics of the force balance in the frictional layer (e.g., Holton 2004).

b. K-theory parameterization

Although the bulk parameterization is simple and easy to implement in a numerical model, it cannot properly represent a neutrally and stably stratified boundary layer. The reason for this is that the wind speed and direction in this situation does vary significantly with height and the boundary layer above the surface cannot be treated as a single slab. In order to close the mathematical problem, the subgrid turbulent flux terms are assumed to be proportional to their corresponding local gradients of the mean values, analogous to molecular diffusion. In this approach, the turbulent flux terms in (14.1.3)–(14.1.7) are written as (14.2.1).

Similar to the bulk parameterization, the subgrid turbulent flux divergence terms are neglected. The simplest way to determine the exchange coefficients in the boundary layer is based on the *mixing length hypothesis*. Analogous to the mean free path of molecules, the mixing length hypothesis assumes that an air parcel that is displaced vertically will carry the mean properties of its original level for a characteristic length, i.e. the *mixing length* (l), before mixing with its environment. Since $lu' \approx \partial\overline{u}/\partial z$ and K_m is proportional to lu', based on dimensional argument, we then have

$$K_m = l^2\,\partial\overline{u}/\partial z. \tag{14.2.19}$$

The eddy and thermal diffusivity coefficients, K_m and K_h, respectively, are often taken as either constants or empirically related to height and stability as calculated from the NWP model output. As mentioned in the parameterization of the surface layer discussion, this approach of the parameterization of momentum, heat, and moisture fluxes is referred to as *K theory*. The K theory is a *first-order closure* because the fluxes are parameterized proportional to the mean values. If the exchange coefficients are taken as constants, then they are referred to as *local exchange coefficients*. For example, the local exchange coefficient may be expressed as (Blackadar 1979)

Stably stratified ($\partial\overline{\theta}/\partial z > 0$):

$$K_m = K_h = 1.1(R_{iC} - R_i)l^2(\partial\overline{V}/\partial z)/R_{iC}, R_i \le R_{iC}$$

$$= 0, \qquad\qquad\qquad\qquad R_i > R_{iC}, \tag{14.2.20a}$$

Unstably stratified ($\partial\bar{\theta}/\partial z \leq 0$):

$$K_m = l^2(\partial\bar{V}/\partial z)(1 - 21R_i)^{1/2}; \quad K_h = l^2(\partial\bar{V}/\partial z)(1 - 87R_i)^{1/2}, \quad (14.2.20b)$$

where $R_{iC} = 0.25$ is the critical Richardson number. Note that R_{iC} distinguishes whether the flow is shear stable or not. A value of $l = kz$ for $z < 200\,\text{m}$ (with $k = 0.35$) and $700\,\text{m}$ for $z \geq 200\,\text{m}$ in (14.2.20a) has been suggested (McNider and Pielke 1981).

This type of *local K-theory* approach has been adopted in a number of mesoscale models as an option. In addition to Blackadar's formulation, other formulations of local exchange coefficients have also been proposed. However, approaches such as the local K-theory scheme have been found to have some deficiencies. The most serious problem in this formulation is that the transport of mass and momentum in the PBL is mostly accomplished by the largest eddies and such eddies should be parameterized by the bulk properties of the PBL instead of the local properties (e.g., Wyngaard and Brost 1984; Holtslag and Moeng 1991). The discrepancy in eddy size makes the local K-theory problematic for unstable conditions, and its implementation could induce the appearance of countergradient fluxes. In order to resolve this problem, *non-local K-theory* has been proposed (e.g., Deardorff 1972; Troen and Mahrt 1986; Holtslag and Moeng 1991). For example, the turbulence diffusion equations for prognostic variables can be expressed by

$$\frac{\partial\phi}{\partial t} = \frac{\partial}{\partial z}\left[K_c\left(\frac{\partial\phi}{\partial z} - \gamma_c\right)\right], \quad \phi = u, v, w, \theta, \text{or } q \quad (14.2.21)$$

where $K_c = K_m$ or K_h and γ_c is a correction to the local gradient that incorporates the contribution of the large-scale eddies to the total flux. The eddy diffusivity coefficient can be formulated as

$$K_m = kw_s z\left(1 - \frac{z}{h}\right)^p, \quad (14.2.22)$$

where p is the profile shape exponent taken to be 2, k is the von Karman constant ($= 0.4$), z is the height from the surface, h is the height of PBL, and w_s is a mixed-layer velocity scale (e.g., Troen and Mahrt 1986; Hong and Pan 1996).

Assuming a three-way balance among the Coriolis force, pressure gradient force, and the vertical gradient of the turbulent momentum flux from (14.1.3) and (14.1.4), in addition to the use of the K-theory parameterization with constant K_m, one may derive the following *Ekman layer* relationships

$$K_m\frac{\partial^2\bar{u}}{\partial z^2} + f(\bar{v} - \bar{v}_g) = 0, \quad (14.2.23)$$

$$K_m\frac{\partial^2\bar{v}}{\partial z^2} - f(\bar{u} - \bar{u}_g) = 0. \quad (14.2.24)$$

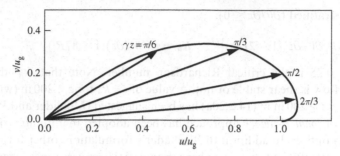

Fig. 14.4 A sketch of the wind vectors of the Ekman spiral (14.2.26). The arrows show the wind vectors at non-dimensional height $\gamma z = \pi/6$, $\pi/3$, $\pi/2$, $2\pi/3$, where γ is defined in (14.2.26). (Adapted after Batchelor 1967.)

The derivation of the above equations is similar to that of (14.2.17), except that the K-theory parameterization is adopted instead of the bulk parameterization. Introducing a new complex variable, $u + iv$, (14.2.23) and (14.2.24) can be combined into a single equation,

$$K_m \frac{\partial^2 (\bar{u} + i\bar{v})}{\partial z^2} - if(\bar{u} + i\bar{v}) = -if(\bar{u}_g + i\bar{v}_g).$$
(14.2.25)

The solution of (14.2.25) subjected to the no-slip boundary conditions at the ground, $\bar{u} = \bar{v} = 0$ at $z = 0$, and approaching geostrophic wind speeds far from the ground, i.e. $\bar{u} \rightarrow \bar{u}_g$ and $\bar{v} \rightarrow \bar{v}_g$ as $z \rightarrow \infty$ is

$$\bar{u} = \bar{u}_g(1 - e^{-\gamma z} \cos \gamma z); \quad \bar{v} = \bar{u}_g e^{-\gamma z} \sin \gamma z,$$
(14.2.26)

where $\gamma = (f/2K_m)^{1/2}$. The above solution is sketched in Fig. 14.4. The wind veers (i.e. turns clockwise) and increases with height to be slightly over the geostrophic value, and then reaches to be nearly the geostrophic value at $z = \pi/\gamma$, which may also be defined as the Ekman layer depth. The spiral wind profile is known as *Ekman spiral*.

c. Turbulent kinetic energy closure scheme

The first-order closure schemes, such as K-theory parameterization, may be improved by predicting one of the subgrid-scale variables, the turbulent kinetic energy (TKE) per unit mass $\left[e = (\overline{u'^2} + \overline{v'^2} + \overline{w'^2})/2\right]$, while the other subgrid scale turbulent flux terms are diagnosed and related to both the TKE and the grid-scale mean values. The prognostic prediction of TKE in the parameterization scheme is referred to as the *TKE* or *one-and-a-half-order closure*.

As the Reynolds number of a laminar flow increases, it may breakdown into a turbulent flow. A turbulent flow is characterized by high randomness, nonlinearity, diffusivity, vorticity, and dissipation. The breakdown is often associated with instability, such as shear instability or buoyant (static) instability. Shear and buoyancy are the two major sources of the production of TKE, which may be denoted as S and B, respectively. Once the turbulence is generated and fully developed into a steady state in terms of averaged flow properties, then the instability is no longer required to sustain the turbulent flow. In order to reach steady-state turbulence (statistically), certain mechanisms are required to remove and redistribute TKE. These mechanisms are often attributed to the dissipation (D) due to turbulent eddy viscosity and molecular viscosity, and the transport and redistribution (T_r) due to advection and pressure forces. Thus, the time evolution of TKE can be written as

$$De/Dt = S + B + T_r - D. \tag{14.2.27}$$

To derive the mathematical form of the TKE equation, we first substitute $u = \bar{u} + u'$, $v = \bar{v} + v'$, $w = \bar{w} + w'$, $p = \bar{p} + p' = p_o + p_1 + p'$, $\theta = \bar{\theta} + \theta' = \theta_o + \theta_1 + \theta'$, $\rho = \bar{\rho} + \rho' = \rho_o + \rho_1 + \rho'$ into (13.5.6)–(13.5.8) with f neglected, to obtain,

$$\frac{D(\bar{u} + u')}{Dt} = -\frac{1}{\rho_o}\frac{\partial(\bar{p} + p')}{\partial x} + \nu\nabla^2(\bar{u} + u'), \tag{14.2.28}$$

$$\frac{D(\bar{v} + v')}{Dt} = -\frac{1}{\rho_o}\frac{\partial(\bar{p} + p')}{\partial y} + \nu\nabla^2(\bar{v} + v'), \tag{14.2.29}$$

$$\frac{D(\bar{w} + w')}{Dt} = -\frac{1}{\rho_o}\frac{\partial(p_1 + p')}{\partial z} - \frac{g}{\rho_o}(\rho_1 + \rho') + \nu\nabla^2(\bar{w} + w'). \tag{14.2.30}$$

Unlike in Section 14.1, in deriving the above equation, we have used the partition of $\bar{p}$ and $\bar{\rho}$ into hydrostatic (p_o and ρ_o – large scale) and nonhydrostatic (p_1 and ρ_1 – mesoscale) parts, neglected ρ_1 and ρ' relative to ρ_o except in the buoyancy (associated with the gravity) term, and assumed an anelastic fluid. Now, multiplying (14.2.28)–(14.2.30) by u', v', w', respectively, and then taking the Reynolds averaging over a grid volume lead to the *TKE equation*,

$$\frac{\partial \bar{e}}{\partial t} = \underbrace{-\bar{V}\cdot\nabla\bar{e}}_{1} \underbrace{-\overline{V'\cdot\nabla e}}_{2} \underbrace{-\left(\frac{1}{\rho_o}\right)\left[\overline{(u'p')}_x + \overline{(v'p')}_y + \overline{(w'p')}_z\right]}_{3} \underbrace{-\left(\frac{g}{\rho_o}\right)\overline{\rho'w'}}_{4}$$

$$\underbrace{-\left[\left(\overline{u'u'}\bar{u}_x + \overline{u'v'}\bar{u}_y + \overline{u'w'}\bar{u}_z\right) + \left(\overline{u'v'}\bar{v}_x + \overline{v'v'}\bar{v}_y + \overline{v'w'}\bar{v}_z\right)\right.}_{5}$$

$$\underbrace{\left. + \left(\overline{u'w'}\bar{w}_x + \overline{v'w'}\bar{w}_y + \overline{w'w'}\bar{w}_z\right)\right]}_{} + \underbrace{\nu\nabla^2\bar{e}}_{6} \underbrace{- \nu\left(\overline{u_x'^2} + \overline{v_y'^2} + \overline{w_z'^2}\right)}_{7} \tag{14.2.31}$$

The left-hand side of (14.2.31) represents the local rate of change of the TKE. Term 1 is the advection of TKE by the grid-volume averaged velocity. Term 2 represents the grid-volume average of the advection of TKE by the subgrid-scale perturbation velocity. Term 3 represents the change in TKE by advection through the boundaries of the grid volume. Term 3 is difficult to measure and is thus often ignored in the closure problem. Term 4 represents the buoyancy production of the TKE, while Term 5 represents the shear production of the TKE. Term 6 represents the diffusion of turbulence by molecular diffusion. Term 7 is the sink of TKE by molecular diffusion. In mesoscale modeling, Terms 6 and 7 are often ignored.

d. Higher-order closure schemes

In fact, subgrid-scale perturbations such as u', v', w', and θ', can be predicted by subtracting the resolved flow equations from the full equations, similar to the derivation of TKE equation. The proposed *second-order closure scheme* will generate new unknown variables involving triple correlation of the perturbations, which must be represented by the mean variables and quadratic perturbation terms, in order to close the system mathematically. One can go further by deriving the prediction equations for the third moments and close the system on higher-order correlation terms (Mellor and Yamada 1974), commonly referred to as the *higher-order closures*. The higher-order closure schemes are capable of representing a well-mixed layer structure. Figure 14.5 shows a comparison of numerically simulated virtual potential temperature profiles in the boundary layer for Day 33 of the Wangara experiment by using a TKE closure scheme and a third-order closure scheme, and the observational data. The TKE closure scheme (Fig. 14.5a) is capable of capturing the observed major features (Fig. 14.5c) compared to the third-order closure scheme (Fig. 14.5b). The higher-order closure schemes are computationally expensive and do not necessarily make a significant improvement in accurately parameterizing the PBL compared to lower-order closure schemes, such as TKE.

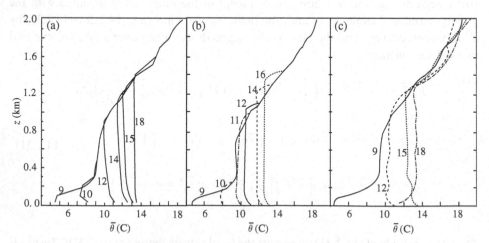

Fig. 14.5 Comparison of predictions of the virtual potential temperature profile using (a) TKE closure scheme (adapted after Sun and Chang 1986) and (b) third-order closure scheme (adapted after André *et al.* 1978) for Day 33 of the Wangara experiment against (c) observational data (adapted after André *et al.* 1978). The local times are denoted by the numbers adjacent to the curves.

14.3 Parameterization of moist processes

As discussed in Chapters 8 and 9, many severe mesoscale weather phenomena, such as thunderstorms, squall lines, mesoscale convective systems, rainbands, and frontal circulations, are associated with moist processes. In addition, the presence of water vapor and clouds in the atmosphere also play important roles in the reflection, absorption, and emission of both solar and terrestrial radiation. In most mesoscale NWP, the majority of clouds, especially convective clouds, cannot be resolved by the grid mesh system. Thus, the moist variables are parameterized by the grid-volume mean variables in a way analogous to the parameterization of turbulent eddies in the PBL. The situation is different in cloud models because normally the horizontal resolution is fine enough to roughly represent the clouds. The microphysical processes, however, still need to be parameterized or appropriately represented. Thus, the accurate representation of the moist processes in mesoscale NWP and cloud models has become one of the most challenging tasks in mesoscale modeling. Many details of the parameterization schemes are still topics of current research. Thus, in this section, it is only appropriate to make a brief summary of the representation or parameterization of these moist processes in mesoscale NWP models. Detailed discussions on individual parameterization schemes in microphysical and cumulus processes can be found in the relevant literature.

As discussed in Chapter 13, a dry atmospheric system can be described by the horizontal and vertical momentum equations, the continuity equation, the thermodynamic equation, and the equation of state, which is composed of six equations with six unknown variables. If potential temperature is adopted in the thermodynamic equation, then the Poisson equation is needed to close the system. For example, (13.5.6)–(13.5.9) and (13.5.11)–(13.5.13) with q_v set to zero describes this dry atmospheric system. When a moist atmospheric system is considered, however, effects of moist processes have to be represented in the heat source or sink term (S_θ) of the thermodynamic equation (13.5.9), and additional equations for the hydrometeors are needed to describe the moist processes. Equations (13.5.6)–(13.5.14) describe this type of moist atmospheric system. The derivation of equations for hydrometeors is similar to that of the continuity equation for dry air, based on the conservation of mass. Different types of clouds are described by a different number of equations for hydrometeor effects. For example, for warm clouds, three additional equations for the water vapor, cloud water, and rainwater are required, while for cold clouds we need to add equations for cloud ice, snow, and graupel or hail. Various source and sink terms portraying different hydrometeors are also represented by the S_ϕ term in (13.5.10). In addition, virtual temperature and virtual potential temperature, instead of temperature and potential temperature, are often used in the equation system because q_v is no longer zero.

The treatments of moist processes in mesoscale models may be divided into two categories: (1) *parameterization of microphysical processes*, and (2) *cumulus parameterization*. In the first category, the microphysical processes are represented by the continuity equations for each hydrometeor, such as those described in (13.5.10). The source terms on the right-hand side of (13.5.10) must be formulated in a way that all possible microphysical interactions among different categories of hydrometeors are included. Two approaches have been taken: (a) *explicit representation*, and (b) *bulk microphysics parameterization (BMP)*.

14.3.1 Parameterization of microphysical processes

a. Explicit representation

In the explicit representation of the microphysical processes, each category of the hydrometeors, such as water vapor, cloud water, cloud ice, rain, snow, and graupel/hail, is represented by a continuity equation, based on the conservation of mass. Each hydrometeor is further divided into different subcategories, based on the size. For example, the liquid water mixing ratio in a warm cloud may be approximated by

$$q_i = \frac{1}{\rho} \int_0^\infty mN(m)\mathrm{d}m \approx \frac{1}{\rho} \sum_{i=1}^k m_i N_i \Delta m_i, \qquad (14.3.1)$$

where ρ is the air density, m is the cloud water mass, $N(m)$ is the size distribution for cloud water, $mN(m)$ is the total number of cloud droplets in mass range m to $m + \mathrm{d}m$ per unit volume of air, and subscript i denotes a subcategory. The continuity equation for liquid water may then be written as

$$\frac{\mathrm{D}N_i}{\mathrm{D}t} = -N_i \nabla \cdot V + P_{\mathrm{AUTO}} + P_{\mathrm{DIFF}} + P_{\mathrm{ACCR}} + P_{\mathrm{BREK}} + P_{\mathrm{FALL}}, \qquad (14.3.2)$$

where the P terms represent microphysical processes responsible for production or reduction of cloud water in a warm cloud, which include condensation from water vapor (P_{AUTO}), vapor diffusion (condensation or evaporation) (P_{DIFF}), accretion (P_{ACCR}), drop breakup (P_{BREK}), and fallout (P_{FALL}).

The continuity equation for water vapor may be written as

$$\frac{\mathrm{D}q_v}{\mathrm{D}t} = -\frac{1}{\rho} \sum_{i=1}^k m_i (P_{\mathrm{AUTO}} + P_{\mathrm{DIFF}}) \Delta m_i + \kappa \nabla^2 q_v, \qquad (14.3.3)$$

Depending upon the number of size categories adopted, it may easily exceed 50 equations for cloud water alone (see (14.3.2)). In addition, the interactions among different size categories become very complicated and the calculations become very tedious and computationally expensive.

b. Bulk microphysics parameterization

Instead of explicit representation of microphysical processes, an alternative is to perform the *bulk microphysics parameterization* (BMP). In taking the bulk parameterization approach, each category of the hydrometeor is governed by its own continuity equation, based on the conservation of mass. In order to avoid calculations of complicated interactions among the different sizes of hydrometeor particles, the shape and size distributions are often assumed a priori and the basic microphysical processes are parameterized. The concept of bulk microphysical parameterization can be understood by considering a nonprecipitating cloud which contains only water vapor (q_v) and cloud water (q_c). Since the total water–substance mixing ratio, $q_T = q_v + q_c$, is conserved, the continuity equations for water vapor and cloud water are,

$$\frac{Dq_v}{Dt} = -c, \tag{14.3.4}$$

$$\frac{Dq_c}{Dt} = c, \tag{14.3.5}$$

where c represents the condensation rate of water vapor into cloud water when $c > 0$ and evaporation rate when $c < 0$. The rewritten contunity equations can be extended to the *warm clouds with precipitation*, which include water vapor, cloud water, and rain, such as the *warm-rain BMP* scheme proposed by Kessler (1969). In the warm-rain bulk parameterization, more source and sink terms associated with microphysical processes, such as evaporation of cloud water, evaporation of rainwater, autoconversion of cloud water to form rain, and accretion of cloud water by rainwater, need to be added to the right-hand side of the above equations. In addition, a continuity equation for rainwater with a fallout term should also be added to the system.

The warm-rain BMP has been extended to include ice phase components such as cloud ice, snow, and graupel/hail (e.g., Lin, Farley, and Orville 1983 – LFO scheme). In this BMP scheme, rain, snow, and graupel are assumed to have *terminal velocities* in order to precipitate downward, while the rest do not possess terminal velocities. Figure 14.6 illustrates the cloud microphysical processes in the LFO scheme. The symbols are explained in Table 14.1. The major microphysical processes include *autoconversion* (growth from only one category of hydrometeors, such as condensation and aggregation), *evaporation, sublimation, freezing, melting, accretion* (growth between different categories of hydrometeors), *Bergeron process* (growth of ice at the expense of cloud water in cold clouds because the saturation vapor pressure with respect to ice is less than that with respect to water), and *dry* and *wet growth* of graupel.

In the scheme sketched in Fig. 14.6, the shape of liquid water and ice particles are assumed to be spherical. The *size distributions* of precipitation particles, i.e., rain (q_r), snow (q_s), and graupel or hail (q_g), are hypothesized as

$$N_k(D) = N_{ok} \exp(-\lambda_k D_k), \quad k = r, s, \text{ or } g \tag{14.3.6}$$

where N_{ok} and λ_k are the *intercept* and *slope parameters* of the size distribution, respectively, and D_k is the diameter of the hydrometeor. Equation (14.3.6) is called the *Marshall–Palmer* (1948) *distribution*. The slope parameter λ_k is determined by

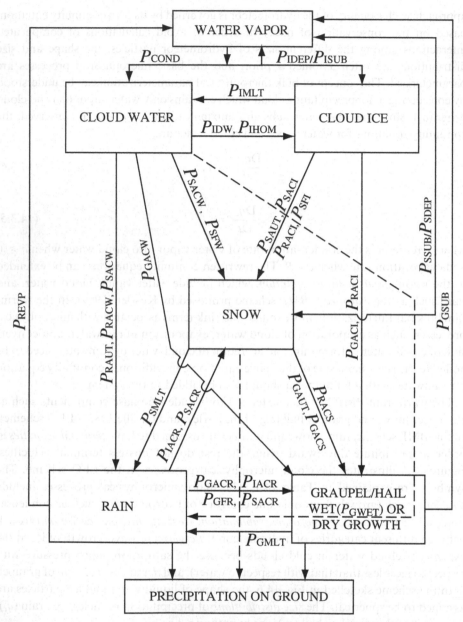

Fig. 14.6 A sketch of cloud microphysical processes in a bulk microphysics parameterization (LFO) scheme including ice phase. Meanings of the production terms (i.e., *P* terms) can be found in Table 14.1. (Adapted after Lin, Farley, and Orville 1983; Orville and Kopp 1977.)

Table 14.1 *Key to Fig. 14.6 (Adapted after Lin, Farley, and Orville 1983).*

Symbol	Meaning
P_{IMLT}	Melting of cloud ice to form cloud water, $T \geq T_o$, $T_o = 0^o C$.
P_{IDW}	Depositional growth of cloud ice at the expense of cloud water.
P_{IHOM}	Homogeneous freezing of cloud water to form cloud ice.
P_{IACR}	Accretion of rain by cloud ice; produces snow or graupel depending on the amount of rain.
P_{IDEP}	Generation of ice by depositional growth of ice.
P_{ISUB}	Sublimation of ice.
P_{COND}	Generation of cloud water by condensation.
P_{RACI}	Accretion of cloud ice by rain: produces snow or graupel depending on the amount of rain.
P_{RAUT}	Autoconversion of cloud water to form rain.
P_{RACW}	Accretion of cloud water by rain.
P_{REVP}	Evaporation of rain.
P_{RACS}	Accretion of snow by rain: produces graupel if rain or snow exceeds threshold and $T < T_o$.
P_{SACW}	Accretion of cloud water by snow: produces snow if $T < T_o$ or rain if $T \geq T_o$. Also enhances snow melting for $T \geq T_o$.
P_{SACR}	Accretion of rain by snow: For $T < T_o$, produces graupel if rain or snow exceeds threshold; if not, produces snow. For $T \geq T_o$, the accreted water enhances snow melting.
P_{SACI}	Accretion of cloud ice by snow.
P_{SAUT}	Autoconversion (aggregation) of cloud ice to form snow.
P_{SFW}	Bergeron process (deposition and riming) – transfer of cloud water to form snow.
P_{SFI}	Transfer rate of cloud ice to snow through growth of Bergeron process.
P_{SDEP}	Depositional growth of snow.
P_{SSUB}	Sublimation of snow.
P_{SMLT}	Melting of snow to form rain, $T \geq T_o$.
P_{GAUT}	Autoconversion (aggregation) of snow to form graupel.
P_{GFR}	Probabilistic freezing of rain to form graupel.
P_{GACW}	Accretion of cloud water by graupel.
P_{GACI}	Accretion of cloud ice by graupel.
P_{GACR}	Accretion of rain by graupel.
P_{GACS}	Accretion of snow by graupel.
P_{GSUB}	Sublimation of graupel.
P_{GMLT}	Melting of graupel to form rain, $T \geq T_o$. (In this regime, P_{GACW} is assumed to be shed as rain.)
P_{GWET}	Wet growth of graupel: may involve P_{GACS} and P_{GACI} and must include P_{GACW} and/or P_{GACR}. The amount of P_{GACW}, which is not able to freeze, is shed to rain.
P_{GDRY}	Dry growth of graupel: equal to $P_{GACS} + P_{GACI} + P_{GACW} + P_{GACR}$.

multiplying (14.3.6) by the particle mass and integrating over all diameters and equating the resulting quantities to the appropriate water contents, which leads to

$$\lambda_k = \left(\frac{\pi \rho_k N_{ok}}{\rho q_k}\right)^{0.25}. \tag{14.3.7}$$

Since N_{ok} is held constant and only λ_k is prognostic, this type of BMP scheme is called a *one-moment BMP scheme*.

The continuity equations for the water vapor and five categories of hydrometeors may be written as

$$\frac{\partial q_j}{\partial t} = -V \cdot \nabla q_j + \nabla \cdot K_h \nabla q_j + P_j, \quad j = v, c, \text{ or } i, \tag{14.3.8}$$

$$\frac{\partial q_k}{\partial t} = -V \cdot \nabla q_k + \nabla \cdot K_m \nabla q_k + P_k + \frac{1}{\rho}\frac{\partial}{\partial z}(\rho U_k q_k), \quad k = r, s, \text{ or } g, \tag{14.3.9}$$

where U_k represents the terminal velocities of precipitation hydrometeors (rain, snow, and graupel), K_m and K_h are the eddy viscosity and eddy thermal diffusivity, respectively, where the last term of (14.3.9) is the *fallout term*. The production (i.e. P terms, see Table 14.1) terms are sketched in Fig. 14.6. Note that the subgrid scale flux terms have been parameterized by the K-theory closure.

Figure 14.7 shows simulations of tropical cyclones using the LFO scheme and the importance of ice phase in the simulations. The warm-rain simulation shows an outward-sloping eyewall, subsidence inside the eyewall, and an area of mesoscale ascent extending 20–30 km out from the eyewall (Fig. 14.7a). The ice-phase simulation shows a similarly sloping eyewall below 5–6 km, but above this level the eyewall updrafts become more vertically oriented (Fig. 14.7c). An area of mesoscale ascent containing several convective updrafts is located radially outward from the convective ring at 60–70 km. The downdrafts in the ice-phase simulation are stronger and more coherent horizontally. They tend to originate near the melting level (dashed line). The tangential wind of the warm-rain simulation indicates maximum winds at $r = 33$–35 km in a deeper layer (Fig. 14.7b), while the ice-phase simulation shows maximum winds at $r = 17$–19 km in a shallow layer (Fig. 14.7d).

Different approaches have been taken to parameterize ice-phase microphysical processes, such as treating aggregates of ice crystals as a distinct snow species (e.g., Cotton *et al.* 1986), and to improve the microphysical parameterization schemes, such as the addition of collision between snow and cloud water (riming) (Rutledge and Hobbs 1984; Chen and Sun 2002; Lin *et al.* 2005). Other research has improved these schemes through the inclusion of four ice classes: small ice crystals, snow, graupel and frozen drops/hail (Ferrier 1994), ensuring that supersaturation (subsaturation) cannot exist at a grid point that is clear (cloudy) (Tao *et al.* 2003), and diagnosis of the cloud ice number concentration from its mixing ratio (Hong *et al.* 2004). In real-time NWP

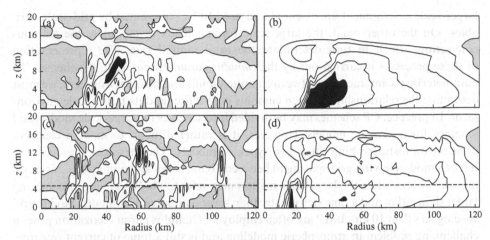

Fig. 14.7 Simulations of tropical cyclones using the LFO scheme (Fig. 14.6). Radius–height distributions of (a) vertical velocity (contours are 0, ± 1, 2.5, and $4\,\mathrm{m\,s^{-1}}$; areas greater than $4\,\mathrm{m\,s^{-1}}$ are dark-shaded, and areas of downward motion are light-shaded), and (b) tangential wind velocity (contours: 0, ± 5, 10, 15, 20 and $25\,\mathrm{m\,s^{-1}}$; dark-shaded for higher than $25\,\mathrm{m\,s^{-1}}$) for the warm-rain numerical simulation at 22 h. (c) and (d) are the same as (a) and (b), respectively, but for the ice-phase simulation at 36 h. The dashed lines in (c) and (d) denote the melting level. (Adapted after Lord *et al.* 1984.)

models, simplified BMP schemes have been developed to make numerical simulations more efficient computationally. The most significant improvement is the development of two- or multi-moment BMP schemes. In general, the size distribution (14.3.6) includes the shape factor and is written as

$$N_k(D) = N_{\mathrm{ok}} D_k^\alpha \exp(-\lambda_k D_k), \quad k = \mathrm{r, s, \, or \, g}, \tag{14.3.10}$$

where α is called the *shape parameter*. Thus, there are three parameters or moments, N_{ok}, λ_k, and α, to be determined. Following Kessler's (1969) warm-rain scheme, the LFO scheme ((14.3.6) and Fig. 14.6) assumes spherical precipitation particles ($\alpha = 0$) and that N_{ok} is a constant, which yields a one-moment scheme. If two of these parameters, such as N_{ok} and λ_k, are prognostic, and the third parameter (α) is held constant, the scheme is called a two-moment scheme (e.g., Ferrier 1994; Meyers *et al.* 1997; Reisner *et al.* 1998; Morrison and Pinto 2005; Seifert and Beheng 2006). If all of these three parameters are prognostic, then it is called a three-moment scheme (e.g., Milbrandt and Yau 2005).

14.3.2 Cumulus parameterization

Even though most individual cumulus clouds have horizontal scales smaller than the mesoscale model grid mesh, the collective effects of cumulus clouds, such as the convective condensation and transport of heat, moisture, and momentum, on the larger-scale environment are essential and need to be represented by grid-scale variables. On the other hand, the large-scale forcing tends to modulate the cumulus convection, which in turn determines the total rainfall rate. The representation of these processes is carried out by the *cumulus parameterization (CP)* schemes. To parameterize the interactions between cumulus clouds and their environment, we must determine the relationship between cumulus convection and its larger-scale environment. In practice, CP schemes may be divided into schemes for large-scale models and schemes for mesoscale models. However, it is rather difficult to make a clear-cut decision on distinguishing these two types of models. The mesoscale models may refer to models having grid spacing in between 10 to 50 km and a time step of several minutes or less, while the large-scale models may refer to models having grid spacing larger than 50 km and a time step greater than several minutes. For models having grid spacing less than 10 km, BMP are often employed. Cumulus parameterization poses a challenging problem in atmospheric modeling and is still a topic of current research. Nevertheless, due to the rapid advancement in computing power, higher grid resolution mesoscale NWP models will be able to adopt bulk microphysics parameterization schemes directly in the foreseeable future and reasonably resolve the mesoscale convective clouds and precipitating systems. However, many operational and research NWP models are initialized by global models which still rely heavily on CP schemes to represent the cumulus clouds, especially in areas with sparse data. Thus, some basic understanding of CP schemes is required for mesoscale modelers to interpret modeling results because some simulated features might have been implicitly inherited from the use of CP schemes in the large-scale or global models.

Existing parameterizations of cumulus convection for large-scale models may be divided into two groups: (a) equilibrium between mass or moisture supply and consumption is taken as the guiding principle, and (b) a balance between energy supply and consumption is postulated (Raymond 1994). The first group includes Kuo (1965) schemes and the second group includes convective adjustment schemes (e.g., Manabe *et al.* 1965, Kurihara 1973; Betts and Miller 1993) and Arakawa–Schubert (Arakawa and Schubert 1974) schemes. Schemes developed for mesoscale models include the Kain–Fritsch (Kain 2004) and Grell (1993) schemes, among others. In the following, we will briefly describe the convective adjustment scheme and Kuo schemes, which are presented in order to help understand the basic concepts of cumulus parameterization, as well as some of the schemes developed for mesoscale models.

a. Convective adjustment schemes

Convective adjustment refers to the concept that an unstable lapse rate cannot persist in the atmosphere and tends to be removed by either dry or moist convection. Thus, it is plausible to assume that it will do so by adjusting the vertical stratification toward a state that is approximately neutral for moist convection. If the time scale of convection is much smaller than that of circulations resolved in a numerical model, then an instantaneous adjustment to a neutral state can be applied as a first approximation.

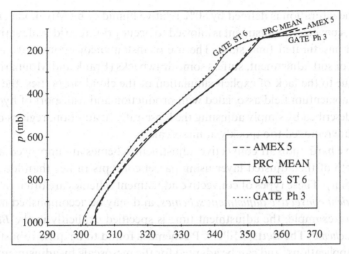

Fig. 14.8 Vertical profiles of virtual potential temperature (θ_v) for four tropical soundings. AMEX 5 (dashed line) is a composite sounding from the late decay stages of four cloud clusters, while PRC mean (solid line) is a six-week mean sounding from a ship during the Australian Monsoon Experiment (AMEX). GATE ST 6 (solid line) is a composite for the decay stages of eight GATE cloud clusters, and GATE ph 3 (dotted line) is the phase III mean for the GATE array. (Adapted after Frank and Molinari 1993.)

In addition, convective processes associated with cloud ensembles in nature are complex and are normally represented through a subgrid scale in mesoscale NWP models. Any attempts to simulate their integrated effects by defining the actual properties of subgrid-scale clouds require the use of many arbitrary parameters, whose values are poorly known in nature and that require enormous amounts of computing power. Based on these arguments, convective adjustment is a conceptually simple and straightforward approach in which the explicit convective processes do not need to be simulated. Figure 14.8 shows four vertical profiles of virtual potential temperature θ_v observed from four tropical regions of strong convection. The four soundings in Fig. 14.8 resemble moist adiabats to some degree.

The convective adjustment scheme may be further divided into the following groups: (a) *hard convective adjustment* schemes, (b) *soft convective adjustment* schemes, and (c) *time-dependent convective adjustment* schemes. In the hard convective adjustment scheme, the convective adjustment is involved only within layers that are saturated and convectively unstable (Manabe *et al.* 1965). In the hard convective adjustment scheme, an initial large-scale sounding in which $\partial \theta_e / \partial p > 0$, is adjusted so that θ_e, or equivalently, moist static energy h ($= c_p T + gz + Lq_v$), is set to be constant with height (Krishnamurti *et al.* 1980). In order to overcome the problem of rainfall rate over-prediction, the so-called *soft convective adjustment scheme* is proposed. In the soft convective adjustment scheme, saturation is assumed to occur only over a small fraction of the large-scale grid area, with the air between the clouds remaining unchanged. For example, the saturation is defined by 80% relative humidity by Miyakoda *et al.* (1969). In this way, convective adjustment is allowed to occur prior to grid-scale saturation and requires an unsaturated final state. The use of instantaneous convective adjustment, either hard or soft adjustment, suffers some drawbacks (Frank and Molinari 1993). For example, due to the lack of explicit simulation of the cloud properties, the convective effects on momentum field associated with production and transport of hydrometeors cannot be described by simply adjusting the lapse-rate. In addition, regions of potential instability are removed too quickly at mesoscale.

The above hard and soft convective adjustment schemes are improved by computing the depth of the adjusted layer using parcel concepts rather than localized grid-scale instability. These types of convective adjustment schemes are often referred to as *time-dependent convective adjustment schemes*, and may be accomplished in a number of ways. For example, the adjustment time is specified explicitly in the *Betts–Miller* (1986) *CP scheme*. The Betts–Miller CP scheme is found to be quite robust for a wide variety of applications, and can be adapted for the mesoscale by adjustment of several parameters. The drawbacks of the Betts–Miller CP scheme include the fact that the closure adopted appears to be less appropriate in cases of explosive deep convection and does not directly generate meso-β scale highs and lows (Seaman 1999).

b. Kuo schemes

Based on observations, Kuo (1965) proposed that cumulus convection occurs in deep layers of conditionally unstable stratification over areas of mean low-level convergence. The cloud base is taken to be the lifting condensation level (LCL) of the surface air, the vertical profiles of temperature distribution T_s and mixing ratio q_s follow a moist adiabat, and the cloud top is located at the level of neutral buoyancy (LNB). In addition, the cumulus clouds are assumed to dissolve immediately by mixing with the environmental air, imparting to it heat and moisture. The moisture cycle in an air column which contains convection in Kuo schemes is illustrated in Fig. 14.9. In the following, we will present the basic concept of the Kuo scheme.

The conservation equation of water vapor, (14.3.4), may be written in pressure coordinates, ignoring the detailed microphysical processes, and retaining only the vertical eddy flux of water vapor,

$$\frac{\partial q_v}{\partial t} + \nabla \cdot (q_v V) + \frac{\partial (q_v \omega)}{\partial p} = -(c - e) - \frac{\partial \overline{q_v' \omega'}}{\partial p}, \tag{14.3.11}$$

where c is the condensation rate per unit mass of air, e the evaporation rate, and ω the vertical motion in the pressure coordinates. Integrating (14.3.11) vertically from the surface (p_s) to the top of the atmosphere ($p = 0$) leads to

$$M_v + E = \frac{1}{g} \int_0^{p_s} (c - e) \mathrm{d}p + S_{qv}, \tag{14.3.12}$$

where M_v is the vertically integrated horizontal *moisture convergence*,

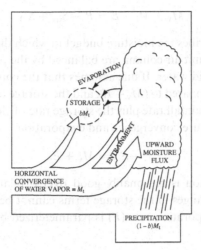

Fig. 14.9 A schematic for moisture cycle in a column which contains convection in Kuo schemes. See text for details. (Adapted from Anthes 1977.)

$$M_v = -\frac{1}{g}\int_0^{p_s} \nabla \cdot (q_v \boldsymbol{V})\mathrm{d}p. \tag{14.3.13}$$

In (14.3.12), E and S_{qv} are the surface evaporation rate and the storage rate of water vapor, respectively,

$$E = -\frac{1}{g}\left[\overline{q_v'\omega'}\right]_s, \tag{14.3.14}$$

$$S_{qv} = \frac{1}{g}\int_0^{p_s}\frac{\partial q_v}{\partial t}\mathrm{d}p. \tag{14.3.15}$$

The relationship between the integrated net condensation rate and the precipitation rate can be obtained by considering the conservation of cloud water q_c,

$$\frac{\partial q_c}{\partial t} + \nabla \cdot (q_c \boldsymbol{V}) + \frac{\partial(q_c\omega)}{\partial p} = (c - e) - P_{CR} - \frac{\partial \overline{q_c'\omega'}}{\partial p}, \tag{14.3.16}$$

where P_{CR} is the conversion rate of cloud water to precipitation. Integrating (14.3.16) with respect to pressure gives

$$\frac{1}{g}\int_0^{p_s}(c - e)\mathrm{d}p = P + S_{ql} - M_l, \tag{14.3.17}$$

where P is the precipitation rate, S_{ql} is the storage rate of liquid water and M_l is the vertically integrated horizontal convergence of cloud water. Substituting (14.3.17) into (14.3.12) yields

$$M_v + M_l + E = P + S_{qv} + S_{ql}. \tag{14.3.18}$$

Equation (14.3.18) describes the moisture budget in which the sources of water vapor and cloud water into a unit air column are balanced by the precipitation plus storage of water vapor and liquid water. If one assumes that the convergence of cloud water (M_l) is much smaller compared to ($M_v + E$), and the storage term of the water vapor is negligible, then the net rainfall rate plus the storage rate of cloud water is equal to the sum of large-scale moisture convergence and evaporation,

$$P + S_{ql} = M_v + E. \tag{14.3.19}$$

The above approximation is reasonably good over a relatively longer time scale, although substantial changes in the storage terms cannot be ignored at a short time scale. If the surface evaporation rate (E) is parameterized by the conventional bulk formula, then we have

$$M_t \equiv M_v + E = -\frac{1}{g} \int_0^{p_s} \nabla \cdot (q_v V) \mathrm{d}p + \rho_s C_d V_s (q_{ss} - q_s), \tag{14.3.20}$$

where M_t is the *moisture accession*, ρ_s is the surface air density, C_d the drag coefficient, V_s is the near surface wind speed, q_{ss} the saturation mixing ratio at the sea surface temperature and pressure, and q_s the near surface saturation mixing ratio. The cumulus convection in the Kuo schemes is driven primarily by the moisture convergence.

The large-scale equation of thermodynamics for potential temperature in pressure coordinates may be written as

$$\frac{\partial \theta}{\partial t} + \nabla \cdot (V\theta) + \frac{\partial(\omega\theta)}{\partial p} = \frac{1}{\pi}\left(L(c - e) - \frac{\partial \overline{\omega'\theta'}}{\partial p} + Q_r \right), \tag{14.3.21}$$

where L is the latent heat of condensation for water vapor, Q_r the radiative heating rate, and π the *Exner function* as defined in (13.5.16). The horizontal eddy flux of sensible heat is ignored in the above equation. We may define the net *cumulus heating* as

$$Q_c = L(c - e) - \frac{\partial \overline{\omega'\theta'}}{\partial p}, \tag{14.3.22}$$

which is part of the right-hand side of (14.3.21). Based on Kuo's (1965) original approach, taking the vertical integration of (14.3.22), neglecting the sensible heat flux term, and using (14.3.19) and (14.3.20) lead to

$$\int_0^{p_s} Q_c \mathrm{d}p = gLM_t. \tag{14.3.23}$$

The vertical structure of Q_c is assumed to be in the form of a relaxation toward a *moist adiabat* θ_{ma},

$$Q_c = \frac{\pi(\theta_{ma} - \theta)}{\tau}, \tag{14.3.24}$$

where τ is a relaxation time that is potentially a function of x, y, and t, but not of p. In regions where $M_t < 0$ and $\theta > \theta_{ma}$, Q_c is set to zero. From (14.3.24) and (14.3.23), the relaxation time may be estimated by the following equation,

$$\tau = \frac{1}{gLM_t} \int_0^{p_s} \pi(\theta_{ma} - \theta)dp. \tag{14.3.25}$$

Thus, the relaxation time is inversely proportional to the moisture accession.

The moisture convergence can be divided into bM_t, which increases the humidity of the air column, and $(1 - b)M_t$, which is condensed and precipitate as rain (Kuo 1974). Normally, b is much less than 1 and should depend on the mean relative humidity of the air column

$$
\begin{aligned}
b &= \left[\frac{1 - RH}{1 - RH_c}\right]^n \quad RH \geq RH_c \\
&= 1 \quad\quad\quad\quad\quad RH < RH_c,
\end{aligned}
\tag{14.3.26}
$$

where RH_c is a critical value of relative humidity and n is a positive exponent of order 1 which may be empirically determined (Anthes 1977). This modified form of the Kuo scheme is also known as the *Anthes–Kuo scheme*. Kuo and Anthes (1984) found that the best agreement between observed and diagnosed rainfall rates is when n is between 2 and 3 and RH_c is between 0.25 and 0.50. An alternative method for estimating b is (Krishnamurti *et al.* 1980)

$$b = -\frac{1}{gM_t} \int_0^{p_s} \nabla \cdot (q_v V)dp. \tag{14.3.27}$$

In the *Arakawa–Schubert* (1974) *scheme*, a spectrum of cloud types is considered and the scheme is coupled with a model of the mixed layer. The large-scale forcing function involves horizontal and vertical advection, radiation, and surface fluxes of heat and moisture.

c. Cumulus parameterization schemes for mesoscale models

Parameterization of cumulus clouds remains one of the most challenging problems in mesoscale modeling. The conceptual basis for cumulus parameterization requires, in principle, the existence of a spectral gap between the scales being parameterized and those being resolved on the grid points. The spectral gap ensures that all eddies have a time scale much smaller than the grid-scale motions, so that their integrated influence can be incorporated into a single time step. For mesoscale models with grid resolution of 10 to 50 km and time intervals on the order of several minutes, mesoscale circulations appear to be resolved reasonably well, but the models are still not fine enough to resolve cumulus convective clouds. In this situation, the convective scales and the resolvable scales are no longer as distinguishable as that assumed in the cumulus parameterization schemes adopted by large-scale models. Thus, the traditional cumulus parameterization schemes, such as Kuo schemes and the Arakawa–Schubert scheme, are not suitable for mesoscale models. The success of cumulus parameterization in numerical simulations of mature hurricanes using 10–20 km grid resolution is due to the fact that under strong rotation, the local *deformation radius* can shrink enough to produce a long-lasting, inertially stable disturbance. Thus, the time-scale separation requirement is met (Ooyama 1982).

Three approaches have been taken for the simulation of cumulus convection in mesoscale models (Molinari 1993): (1) the *traditional approach,* which utilizes cumulus parameterization, as those adopted by large-scale models, at convectively unstable grid points and explicit condensation at convectively stable grid points, (2) the *grid explicit approach,* which uses only explicit representations or bulk microphysics parameterization of microphysical processes regardless of stability, and (3) the *hybrid approach,* which parameterizes convectively unstable updrafts and downdrafts at convectively unstable grid points and also detrains a fraction of the parameterized cloud and precipitation particles to their respective grid-scale prediction equations. Due to the continuous advances in computer speed and memory, the grid explicit approach with microphysics parameterization schemes may become more applicable in resolving the cumulus parameterization problems in mesoscale models. In the following, we will briefly introduce some mesoscale cumulus parameterization schemes, such as the Kain–Fritsch scheme and the Grell scheme.

The *Kain–Fritsch (KF) scheme* (Kain and Fritsch 1993; Kain 2004) is a mass flux parameterization that uses the Lagrangian parcel method to estimate whether instability exists, whether any existing instability will become available for cloud development, and what the properties of any convective clouds might be. The scheme involves three parts: (1) the convective trigger function, (2) the mass flux formulation, and (3) the closure assumptions. The first part of the KF scheme is to identify potential source layers for convective clouds or *updraft source layers* (USLs). Beginning at the surface, vertically adjacent layers in the model are mixed until the depth of the mixed layer is at least 60 hPa. The combination of adjacent model layers composes the first potential USL which may be viewed as an "air parcel". The mean thermodynamics characteristics of this mixed layer are computed along with the temperature and height of this "air parcel" at its lifting condensation level (LCL). The potential of convective initiation is measured by $T_{\text{LCL}} - \overline{T}$ ($\overline{T}$ is the environmental temperature), which is typically negative indicating a negative buoyant air parcel. Observations suggest that convective initiation tends to be favored by background vertical motion (Fritsch and Chappell 1980). Thus, the parcel is assigned a temperature perturbation (δT_{vv}) linked to the magnitude of grid-resolved vertical motion, such as

$$\delta T_{\text{vv}} = k[w_{\text{g}} - c(z)]^{1/3}, \qquad (14.3.28)$$

where k is a unit number with dimensions $K \, cm^{-1/3} \, s^{1/3}$, w_g is an approximate running-mean grid-resolved vertical velocity ($cm \, s^{-1}$) at the LCL, and $c(z)$ is a threshold vertical velocity given by

$$c(z) = \begin{cases} w_o(z_{LCL}/2000), & z_{LCL} \leq 2000 \\ w_o, & z_{LCL} > 2000, \end{cases} \qquad (14.3.29)$$

where $w_o = 2 \, cm \, s^{-1}$ and z_{LCL} is the height of LCL above the ground in m. Equation (14.3.29) will effectively eliminate most air parcels (mixed layer) as candidates for deep convection. If $T_{LCL} + \delta T_{vv} < \overline{T}$, then this mixed layer is excluded for deep convection, the base of the USL is moved up one model level, and the above test is repeated for a new potential USL. Otherwise, the mixed layer or air parcel is allowed to proceed as a candidate for deep convection. At this stage, the parcel is released at its LCL with its original temperature and moisture content and a vertical velocity derived from the perturbation temperature, such as

$$w_{po} = 1 + 1.1\left[(z_{LCL} - z_{USL})\delta T_{vv}/\overline{T}\right]^{1/2}, \qquad (14.3.30)$$

where z_{USL} is the height at the base of the USL. The above equation yields initial vertical velocity for the air parcel up to several meters per second. Above the LCL, parcel vertical velocity is estimated at each model level using the Lagrangian parcel method (e.g., Perkey and Kreitzberg 1993), including the effects of entrainment, detrainment, and water loading (Bechtold *et al.* 2001). If the vertical velocity remains positive over a depth that exceeds a specified minimum cloud depth (typically 3–4 km), deep convection is activated using this USL. If deep convection is not activated, the base of the potential USL is moved up one model layer and the procedure is repeated until either the first suitable source layer is found or the sequential search has moved up above the lowest 300 hPa of the atmosphere. The set of criteria described here gives the *trigger function*. Note that the cloud depth is determined by the updraft model described in the mass flux formation which, in turn, determines whether the parameterization is activated.

The second part of the KF scheme is the *mass flux formation*. In this scheme, convective updrafts are represented using a steady-state entraining–detraining plume model, where both θ_e and q_v are entrained and detrained. Convective downdrafts are generated by evaporation of condensate that is produced within the updraft. A fraction of this total condensate is made available for evaporation within the downdraft, based on empirical formulas for precipitation efficiency as a function of vertical wind shear and cloud-base height. The downdraft is specified to start at the level of minimum θ_e^* in the cloud layer with a mixture of updraft and environmental air. It is moved downward in a Lagrangian sense, with a specified entrainment rate and a fixed relative humidity of 100% above the cloud base and 90% below the cloud base. When the downdraft is warmer than its environment, it is terminated and forced to detrain into the environment within and immediately above the termination level. The scheme also requires environmental mass fluxes to compensate for the upward and downward transports in updrafts and downdrafts so that there is no net convective mass flux at any level in the column. The third part of the KF scheme is the *closure assumptions*. The KF scheme rearranges mass in a column using the updraft, downdraft, and environmental mass fluxes until at least 90% of the CAPE is removed. CAPE is computed in the traditional way (see Chapter 7) and is removed by the combined effects of lowering θ_e in the USL and warming the environment aloft. The *convective time scale*, or *relaxation period* (τ_c), is based on the advective time scale in the cloud layer, which has upper and lower limits given as $0.5h \leq \tau_c \leq 1h$. The scheme feeds back to convective tendencies of temperature, water vapor mixing ratio, and cloud water mixing ratio.

Another widely adopted cumulus parameterization scheme developed mainly for mesoscale models is the *Grell scheme* (Grell 1993). The key features of the Grell scheme are (Seaman 1999): (a) deep convective clouds are all of one size; (b) the Arakawa–Schubert (1974) cloud work function was adopted for its closure, but this was later changed to use a CAPE closure, similar to that in Kain–Fritsch scheme; (c) no lateral mixing (i.e. no entrainment or detrainment) except at the levels of origin or termination of updrafts and downdrafts, thus making mass flux constant with height; and (d) it is not necessary to assume that the fractional area coverage of updrafts and downdrafts in the grid column is small since there is no lateral mixing. The absence of lateral mixing allows the scheme to operate relatively easier at finer scales, although some degree of scale separation is still important. The Grell scheme has been modified, based on some features developed in the Kain–Fritsch scheme. The advantages of the Grell scheme are that it includes effects of downdrafts and is well adapted for grids as fine as 10 to 12 km.

14.4 Parameterizations of radiative transfer processes

14.4.1 Introduction

Solar radiation is a major driving force of the atmospheric motion. The magnitude of radiative warming/cooling depends on many factors, including temperature, clouds, aerosols, water vapor, carbon dioxide, and ozone. Radiation is also the primary force for the soil model in terms of the surface energy budget. Figure 14.10 illustrates the radiative transfer processes in the Earth-atmosphere system, which include shortwave and longwave reflection, transmission, and absorption/emission. The radiative transfer processes are very complex and do not allow mesoscale modelers to make detailed adjustments due to limitations of computing time. Thus, similar to PBL and moist processes, the radiative transfer processes are parameterized in mesoscale models. The purpose of this section is to introduce the basic concepts of the parameterization of radiative transfer processes for mesoscale NWP models. Detailed

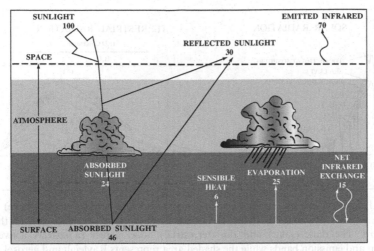

Fig. 14.10 Illustration of global energy balance through radiative transfer processes in the Earth-atmosphere system, based on data obtained from NASA Earth Observing System and various published model and empirical estimates. (Courtesy of Dr. S.-C. Tsay.)

discussions on radiative transfer processes and parameterizations can be found in advanced textbooks (e.g., Liou 1992, 2002).

The objective of parameterizing the atmospheric radiative transfer processes in a numerical model is to provide a simple, accurate and fast calculation of the total radiative flux profile within the atmosphere. The fast, simple calculation includes (i) the total radiative flux at the surface to calculate the surface energy balance, and (ii) the vertical radiative flux divergence to calculate the radiative warming and cooling rates of an atmospheric volume. The parameterization commonly includes the combined effects of absorption/emission and scattering by the radiatively active trace-gases of H_2O, CO_2, and O_3, together with cloud and haze particles.

Different levels of approximation have also been adopted, which depend on the desired accuracy for representing the type of interactions between radiation and dynamics. The factors often considered in selecting the parameterization include (Stephens 1984): (a) Radiation may simultaneously affect the dynamics in several different ways, and the accuracy required of the radiation computations depends on which process is important to the given dynamical problem. (b) The dynamics respond to the total heating fields, which include radiative, latent and sensible heating. The heating components are not always independent of each other and, as a result, radiation may influence the dynamics in a complex nonlinear manner that is often difficult to assess a priori. (c) The radiative warming and cooling may vary considerably in response to variations in temperature, which are caused by various dynamical factors.

Atmospheric radiation covers a broad spectrum of electromagnetic waves. As depicted in Fig. 14.11, the sun radiates approximately at a blackbody temperature

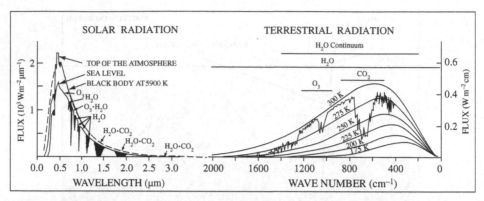

Fig. 14.11 Spectral energy curves of solar (shortwave) and terrestrial (longwave) radiation observed, as well as modeled, under cloud-free conditions at sea level and at the top of atmosphere. The dark (solar) and dip (terrestrial) areas depict radiatively active gaseous absorption and emission bands while the shaded area represents Rayleigh and aerosol scattering effects. (Adapted after Tsay *et al.* 1989.)

of 6000 K, which spans the entire spectrum of the electromagnetic waves, but the radiation outside the range 0.2–4.0 μm is negligible and referred to as *shortwave radiation*. Likewise, the Earth emits radiation at a blackbody temperature of about 250 K, which covers the entire spectrum of the the electromagnetic waves, but the radiation outside the range 4–200 μm is negligible, and is referred to as *longwave radiation*. The blackbody emission of electromagnetic radiation for a particular wavelength λ and temperature T may be derived,

$$B_\nu(T) = \frac{2h\nu^3 c^2}{e^{hc\nu/KT} - 1},$$ (14.4.1)

where B_ν is called the *Planck function*, $h = 6.6262 \times 10^{-34}$ J s is the *Planck constant*, $K = 1.380 \times 10^{-23}$ J K^{-1} is the *Boltzmann's constant*, ν is the wave number ($\nu = 1/\lambda$, λ is the wavelength), and c is the speed of light. The *total radiative flux, F,* emiting at a blackbody temperature T may be derived by integrating (14.4.1) over the entire spectral range and angles,

$$F = \sigma T^4,$$ (14.4.2)

where $\sigma = 5.67032 \times 10^{-8}$ Wm^{-2} K^{-4} is the *Stefan–Boltzmann constant*. The above equation is also known as the *Stefan–Boltzmann law*. However, the Earth has an atmosphere which contains molecules and particulates. Thus, it does not behave as a blackbody, which leads to

$$F = \varepsilon \sigma T^4,$$ (14.4.3)

where ε denotes the *emittance*. The emittance, ranging from 0 to 1, represents the ratio of the flux emitted by a graybody to that by a blackbody at the same temperature. Satisfying the condition of local thermal equilibrium, the emittance of a medium is equal to its absorptance, which is also known as *Kirchhoff's law of thermal radiation*.

The *net radiation heating* may be written as,

$$\frac{\partial \theta}{\partial t} = \frac{1}{\bar{\rho} c_p} \frac{dF_N}{dz}, \tag{14.4.4}$$

where $\partial \theta / \partial t$ has also been written as $\partial T / \partial t$ in the literature, F_N is the difference between downward ($F^\downarrow$) and upward ($F^\uparrow$) fluxes (in W m^{-2}) and dF_N/dz is the *vertical flux divergence*.

When electromagnetic radiation traverses a layer in the atmosphere, it can be transmitted, absorbed, or reflected. Based on the conservation of energy, it can be derived that

$$T_\nu + A_\nu + R_\nu = 1, \tag{14.4.5}$$

where T_ν, A_ν and R_ν are *transmissivity*, *absorptivity*, and *reflectivity*, respectively. A pencil of radiation traversing a medium may be weakened by extinction (i.e., *scattering + absorption*) of the material, strengthened by emission of the material, or undergo multiple scattering from all other directions into the pencil (Liou 2002). Thus, the change of the *intensity of radiation* traversing a medium may be expressed by

$$\left(\frac{1}{k_\nu \rho_a}\right) \frac{dI_\nu}{ds} = -I_\nu + B_\nu(T) + J_\nu, \tag{14.4.6}$$

where k_ν denotes the *extinction coefficient*; ρ_a, the *density* of media; I_ν, the *incident radiance*; $B_\nu(T)$, the *thermal emission*; and J_ν, the *source of radiation* from scattering into the line segment ds (Fig. 14.12). In the absence of emission and scattering, the *Beer–Bouger–Lambert law* can be derived, which states that the radiant intensity trasversing a homogeneous extinction medium decreases exponentially as $I_\nu = I_{\nu o}$ $\exp(-k_\nu u)$, where u is the *path length*.

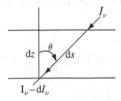

Fig. 14.12 A schematic illustration of radiant intensity attenuated by scattering and/or absorbing media between ds ($= \sec \theta \, dz$).

14.4.2 Longwave radiation

a. Clear atmosphere

In a clear air atmosphere, the scattering of longwave radiation (J_ν) may be neglected compared with the absorption and emission (Liou 2002). More specifically, the general problem of parameterizing the longwave radiative transfer in the clear air requires the suitable treatment of absorption and simultaneous emission by the ozone band (9.6 µm), the rotation and vibration bands of water vapor, the continuum absorption in the atmospheric window (between 8 µm and 14 µm) and the absorption by the carbon dioxide band (15 µm), which overlaps a portion of the rotation band (Stephens 1984).

By considering a monochromatic radiation of wavelength λ entering at angle θ (Fig. 14.12) from the vertical direction across a plane sheet of material of distance ds and vertical distance dz, (14.4.6) becomes

$$\left(\frac{\mu}{k_\nu \rho_a}\right)\frac{dI_\nu}{dz} = -\mu\frac{dI_\nu}{d\tau} = -I_\nu + B_\nu(T), \qquad (14.4.7)$$

where

$$\tau = \int_z^\infty k_\nu \rho_a dz, \qquad (14.4.8)$$

is the *optical thickness*, μ ($= \cos\theta$) is the cosine of the zenith angle, and dI_ν/dz is the intensity change in the vertical direction. In order to solve the first-order differential equation, (14.4.7), for both upward and downward components for an atmosphere with a total optical thickness of τ_*, two boundary conditions are required at the surface and the top of the atmosphere. For the atmospheric net radiation heating rate $\partial\theta/\partial t$ calculations, the required quantities are the upward and downward radiative fluxes ($F^\uparrow$ and $F^\downarrow$), as expressed in (14.4.4). The radiative fluxes at a particular wave frequency ($F_\nu^\uparrow$ and $F_\nu^\downarrow$) can be calculated by taking the integration of the intensities of radiation ($I_\nu^\uparrow$ and $I_\nu^\downarrow$) with respect to μ. The upward and downward radiative fluxes ($F^\uparrow$ and $F^\downarrow$) can then be obtained by taking integrations of $F_\nu^\uparrow$ and $F_\nu^\downarrow$, respectively, from $\mu = 0$ to 1.

Based on above equations with the absence of scattering, the *longwave radiative flux* may be derived to be

$$F^\uparrow(z) = \int_0^\infty \pi B_\nu(z=0)\tau_\nu^f(z, z=0)d\nu + \int_0^\infty \int_0^z \pi B_\nu(z')\frac{\partial}{\partial z'}\tau_\nu^f(z, z')dz'd\nu, \quad (14.4.9)$$

$$F^\downarrow(z) = \int_0^\infty \int_z^\infty \pi B_\nu(z')\frac{\partial}{\partial z'}\tau_\nu^f(z, z')dz'd\nu, \qquad (14.4.10)$$

where $F^\uparrow(z)$ and $F^\downarrow(z)$ are the upward and downward longwave radiative flux through level z, B_ν is the Planck function in terms of wave frequency and the *diffusion transmission function*, τ_ν^f, is defined by the hemispheric integral

$$\tau_\nu^f(z,z') = 2 \int_0^1 \tau_\nu(z,z',\mu)\mu d\mu, \qquad (14.4.11)$$

and

$$\tau_\nu(z,z',\mu) = \exp\left[-\frac{1}{\mu}\int_{u(z)}^{u(z')} k_\nu(p,T)du\right], \qquad (14.4.12)$$

where $k_\nu(p,T)$ is the *absorption coefficient* and u is the *absorption* or *optical path* from z to z'. Note that u should be the mass if the unit k is in the form of fraction/mass. The transmission function τ_ν^f is often referred to as the *slab transmission function*, as described in (14.4.11). The above equations may be combined and differentiated directly to obtain an equation of flux divergence, and thus of radiative cooling. However, for applications in a mesoscale or general circulation model, these fluxes may be calculated numerically at each model level, followed by the evaluation of the flux divergence for the layer between two levels. The evaluation of flux divergence between two levels automatically supplies fluxes at those levels where a radiation budget is required, such as at the Earth surface, the tropopause or at the model top.

There are three integrals involved in the calculations of *longwave flux*, which include (14.4.11), the inner integrations of (14.4.9) and (14.4.10) over all atmospheric layers (dz'), and the outer integrations of (14.4.9) and (14.4.10) over all spectral intervals ($d\nu$). The integration of (14.4.11) may be approximated by (Stephens 1984)

$$\tau_\nu^f(z,z') \propto \tau_\nu(z,z',1/\beta), \qquad (14.4.13)$$

where $\beta = 1.66$ and is known as the *diffusivity factor*. Basically, this approximation means the diffuse transmission can be approximated by intensity transmission with an angle of $\cos^{-1}(1/\beta)$. The integration of (14.4.9) and (14.4.10) over dz' may be approximated by a finite difference or other numerical method in the vertical direction. Thus, the objective of *parameterization of longwave radiation in a clear sky* is to find suitable approximations of integration over z' and ν in (14.4.9) and (14.4.10) and the integration over μ in (14.4.11).

The difficulties in the integration over optical path in (14.4.12) are due to the fact that the absorption coefficient k_ν is a function of both pressure and temperature, and most absorption data are collected in the laboratory at constant pressure and temperature, which are not necessarily applicable to the real atmosphere. Two commonly adopted approximations are (a) the *one-parameter scaling approximation* (Goody 1964a; Chou and Arking 1980) and (b) the *two-parameter scaling approximation* (Goody 1964b). The one-parameter scaling approximation method is able to provide a reasonable approximation to the problem of absorption along nonhomogeneous paths. Difficulties arise when one attempts to isolate the errors in the infrared cooling rate, which are likely to be larger than those of the two-parameter approximation method, especially in the upper atmosphere.

The problem inherent in simplifying the integration over frequency in (14.4.9) and (14.4.10) is more complicated than the rather simple and obvious task of averaging k_ν over some broad interval $\Delta\nu$. The finest frequency scale of absorption, that of an individual line, is described by a simple analytical function given by the Lorenz line absorption profile for an altitude below about 40 km (e.g., Liou 2002). Unfortunately a single absorbing line cannot be considered in isolation from neighboring lines, and it is not meaningful to average k_ν over a group of lines in a simple linear manner because k_ν is the sum of the contributions to the absorption coefficient at a given frequency from all lines. The properties of the single absorption line lead to the concept of a *band model*, which enables the averaging of the absorption properties for bands of lines, such as line strength, separation and position that are specified by well-defined statistical relationships.

An alternative approach is the *k-distribution method*, where k is the absorption coefficient which makes use of the fact that for a homogeneous atmosphere, transmission within a relatively wide spectral interval depends only on the fraction of the interval that is associated with a particular value of k. The k-distribution method has been demonstrated to be faster and more accurate than the band model (e.g., Arking and Grossman 1972; Chou and Arking 1980). In addition, treating molecular absorption and the scattering by cloud droplets in a self-consistent fashion in the k-distribution method is straightforward. Like other band-model methods, the k-distribution method was developed for homogeneous atmospheres. For nonhomogeneous atmospheres, the one-parameter scaling approximation is often adopted in the integration of (14.4.12) over optical path, which may lead to large deviation from results computed from line-by-line (LBL) methods (e.g. Rothman *et al.* 1987). In order to overcome this problem, the correlated k-distribution method has been proposed (e.g. Fu and Liou 1992; Mlawer *et al.* 1997). More details in recent developments of the k-distribution method can be found in Kratz *et al.* (1998) and Chou *et al.* (1999).

b. Cloudy atmosphere

The optical path length for infrared or longwave radiation is strongly influenced by the presence of clouds. Mesoscale models have often treated clouds as blackbodies in the longwave portion of the spectrum, where no infrared radiation is transmitted through the cloud. Although more advanced schemes have been developed, the parameterization for longwave radiation within a water cloud can be understood through the following scheme (Stephens 1978a):

$$F^\uparrow(z) = F^\uparrow(z_b)[1 - \varepsilon_b] + \varepsilon_b \sigma T_c^4, \text{ and} \qquad (14.4.14a)$$

$$F^{\downarrow}(z) = F^{\downarrow}(z_t)[1 - \varepsilon_t] + \varepsilon_t \sigma T_c^4, \tag{14.4.14b}$$

where $F^{\uparrow}(z_b)$ and $F^{\downarrow}(z_t)$ are the clear-air radiative flux at the cloud base (z_b) and cloud top (z_t), respectively, and T_c is the cloud temperature. The *cloud effective emissivity* $\varepsilon_b(z_b, z)$ and $\varepsilon_t(z_t, z)$ can be obtained by solving (14.4.14) using a detailed radiational model with eight cloud types in a US standard atmosphere to obtain $F^{\uparrow}(z_b)$ and $F^{\downarrow}(z_t)$ (Stephens 1978b). Liou and Ou (1981) also proposed a parameterization of longwave radiative transfer in the presence of a semitransparent cloud layer. They used a model with five broadband emissivity values to represent the five major absorption regions in the infrared spectrum. For application to a mesoscale model, (14.4.14) can be used when a grid volume is saturated with clouds (Pielke 2002).

14.4.3 Shortwave radiation

a. Clear atmosphere

The distribution of electromagnetic radiation emitted by the sun is approximately as blackbody radiation for a temperature of about 6000 K. The spectral distribution of solar or shortwave radiation received at sea level through a cloud free and haze free atmosphere is shown in Fig. 14.11. As can be seen in the figure, the primary absorptions in a clear atmosphere are by: (1) the ozone in the visible $(0.5\,\mu\text{m} \le \lambda \le 0.7\,\mu\text{m})$ and ultraviolet $(\lambda \le 0.3\,\mu\text{m})$ spectra; and (2) the water vapor in the near infrared (IR) spectrum $(0.7\,\mu\text{m} \le \lambda \le 4.0\,\mu\text{m})$. Thus, $\lambda = 0.7\,\mu\text{m}$ is a natural division of these two absorbers. The major absorption bands are shaded in Fig. 14.11. The absorptions by O_2 and CO_2 are substantially less than that of ozone and water vapor and their contributions can be ignored. The *solar irradiance* is composed by direct irradiance and diffuse irradiance. The diffuse irradiance is the irradiance observed at a point from directions other than the line of propagation, while the direct irradiance is the irradiance observed at a point without being absorbed or scattered from its line of propagation.

In the absence of scattering, the downward shortwave irradiance through level z for a collimated beam of solar irradiance may be derived,

$$F_{sw}^{\downarrow}(z, \mu_0) = \mu_0 \int_0^{\infty} S_{o\nu} T_{o\nu}(z, \mu_0)\,d\nu, \tag{14.4.15}$$

where $F_{sw}^{\downarrow}(z, \mu_0)$ is the downward irradiance through level z for a collimated beam of solar irradiance $(S_{o\nu})$ at the top of the atmosphere, inclined at a zenith angle θ_0 (or $\mu_0 = \cos\theta_0$). The monochromatic transmittance function can be calculated from

$$T_{o\nu}(z, \mu_0) = \exp\left(-\frac{1}{\mu_0} \int_z^{\infty} k_\nu\,du\right). \tag{14.4.16}$$

Calculation of the downward shortwave radiation (14.4.15) is less complicated than its longwave counterpart, (14.4.10), since it is not necessary to consider the complications imparted by simultaneous absorption and emission from layer to layer (Stephens 1984). The only variable that remains to be defined is the mean transmission function

$$\tau_{o\nu}(z, \mu_o) = \frac{1}{\Delta\nu} \int_{\Delta\nu} \exp\left(-m_r(\mu_o) \int_z^\infty k_\nu du\right) d\nu, \tag{14.4.17}$$

where $1/\mu_o$ is replaced by the mass factor $m_r(\mu_o)$ in (14.4.16), which is identical to $1/\mu_o$ except for larger solar zenith angle. Empirical formulas for $m_r(\mu_o)$ have been proposed. For example, Rodgers (1967) proposed that $m_r = 35\mu_o(1224\mu_o^2 + 1)^{-1/2}$ for ozone, otherwise $m_r = 1/\mu_o$. Thus, the clear-sky downward solar flux, (14.4.15), transmitted to level z along θ_o can be approximated by

$$F_{sw}^\downarrow = \mu_o \sum_{i=1}^N S_{oi}\tau_{\bar{\nu}i}(u). \tag{14.4.18}$$

The approach given in (14.4.18) is called the *discrete band approach*. The subscript i denotes the ith spectral interval. The upward solar flux received at level z by reflection from the Earth surface may be calculated in a similar fashion,

$$F_{sw}^\uparrow = \mu_o \sum_{i=1}^N \alpha_{gi}S_{oi}\tau_{\bar{\nu}i}(u^*), \tag{14.4.19}$$

where α_{gi} is the *surface albedo* for the ith spectral interval. The *path length* u^* is the effective total absorber amount traversed by the diffusely reflected radiation. Some useful empirical formulae have been proposed for estimating u^* (e.g., Lacis and Hansen 1974). Note that the mean transmittance is commonly defined as a convolution of the transmission function and the extraterrestrial flux $S_{o\nu}$. The mean transmittance function over the entire solar spectrum may be calculated by

$$\bar{\tau}_o(z, \mu_o) = \frac{1}{S_o} \int_0^\infty S_{o\nu}\tau_{o\nu}(z, \mu_o)d\nu, \tag{14.4.20}$$

in which case the downward solar flux at level z is defined by

$$S^-(z) = \mu_o S_o \bar{\tau}_o(z, \mu_o). \tag{14.4.21}$$

Using (14.4.20) and (14.4.21) to calculate the downward solar flux is also called the *broadband approach*, and has been used for calculating solar fluxes within the atmosphere. However, the broadband approach has hardly been used recently because Rayleigh scattering is important in clear atmospheres and is included in nearly all models. Furthermore, different values of land surface reflectivity are used for different spectral bands in most of the current models, which requires the division of the solar spectrum into multiple bands.

In the lower atmosphere, the absorption of solar radiation by water vapor is the major source of solar heating. As mentioned earlier, the absorption of water vapor is concentrated in the near IR spectrum, $0.7 \, \mu m \leq \lambda \leq 4.0 \, \mu m$ (see Fig. 14.11). In order to resolve the water vapor absorption and to apply Beer's law, the spectrum has to be divided into about a half million intervals (Chou 1992). The difficulty in parameterizing the water vapor absorption is due to the fact that: (1) the absorption fluctuates strongly within very narrow spectral intervals, (2) the absorption is complicated by the pressure and temperature dependencies, and (3) the absorption spectrum of water vapor overlaps with that of liquid water absorption. Ideally, line-by-line methods (e.g. Rothman *et al.* 1987) are required for achieving a high degree of accuracy in solar flux calculation, although it is computationally very expensive. In order to reduce the computational burden, there are numerous methods being proposed for parameterizing the absorption functions of the water vapor, such as the simple parameterization of the broadband absorption functions (e.g., Chou 1986; Lacis and Hansen 1974). In order to improve the accuracy, the k-distribution method has been proposed and applied in studies such as Chou (1986), where it was found that the solar near-infrared fluxes could be accurately computed with a maximum of nine values of k, instead of half a million spectral intervals, in each of the three near IR bands.

In the ultraviolet (UV) and visible spectrum ($\lambda < 0.7 \, \mu m$), the primary absorption of the solar radiation is due to ozone. The absorption spectrum of ozone is continuous in nature and requires less spectral intervals than the near-infrared spectrum for accurate calculations of the solar radiation. However, it is desirable to reduce the number of spectral intervals due to the relatively wide range of the absorption spectrum of ozone. Molecular scattering is significant at the absorption spectrum of ozone, but fortunately, they do not overlap much in high altitudes. The ozone absorption occurs in the higher atmosphere, while the molecular scattering occurs in the lower atmosphere. Again, many simple parameterizations of the broadband absorption functions have been proposed. Chou (1986) divided the spectrum between $0.175 \, \mu m$ and $0.7 \, \mu m$ into eight intervals and used a single mean value of k for each interval. When there are more than one absorber and scatterer in an atmospheric layer, the effective optical parameters are required for flux computation (Tsay *et al.* 1989).

b. Cloudy atmosphere

The interactions of clouds with solar radiation are extremely complicated since the attenuation of solar radiation includes scattering and absorption by wide spectrum of cloud droplets and ice crystals. Thus, in addition to making proper representation of solar radiative processes, one also has to represent or parameterize microphysical processes properly. In the absence of emission, the basic equation governing the *intensity of solar radiation* appropriate to a cloud medium may be written in the form

$$\mu \frac{dI(\tau,\mu)}{d\tau} = -I(\tau,\mu) + \frac{\tilde{\omega}_0}{4\pi} \int_{-1}^{1} \bar{p}(\tau,\mu,\mu') I(\tau,\mu') d\mu' + \frac{S_0}{4\pi} \bar{p}(\tau,\mu,\mu_0) e^{-\tau/\mu_0}, \quad (14.4.22)$$

where τ is the optical thickness, $\tilde{\omega}_o$ is the *single-scattering albedo*, $\bar{p}$ is the *scattering phase function*, $I(\tau,\mu)$ is the radiance along the angle $\mu\,(=\cos\theta)$ and S_o is the solar flux associated with a collimated beam incident on the cloud top. The optical parameters are functions of frequency, but for simplicity they are not shown in the equation. The difference between (14.4.22) and the flux equations of longwave radiation, such as (14.4.6), is as follows (Stephens 1984): First, the optical thickness now includes the contributions from scattering (τ_s) and absorption (τ_a) by cloud droplets, and by the intervening gas (τ_g),

$$\tau = \tau_s + \tau_a + \tau_g. \tag{14.4.23}$$

Second, the single-scattering albedo is included in (14.4.22), which is a ratio of the scattering optical thickness to the total optical thickness, i.e.,

$$\tilde{\omega}_o = \tau_s/\tau. \tag{14.4.24}$$

Thus, $\tilde{\omega}_o = 1$ for a nonabsorbing cloud and $\tilde{\omega}_o = 0$ if the scattering is negligible. Third, the *intensity of solar radiation* includes the scattering phase function $\bar{p}(\tau,\mu,\mu')$, which characterizes the angular distribution of the scattered radiation field. For spherical cloud droplets, this function exhibits an intense peak in the forward direction and produces rainbow and glory effects in the backward direction (Liou 2002). The most commonly adopted formula for the *scattering phase function* is (Henyey and Greenstein 1941),

$$\bar{p}(\tau,\mu,\mu') = \frac{1 - g^2}{1 + g^2 - 2g\mu\mu'}, \tag{14.4.25}$$

where g is the *asymmetry factor*, which is defined as

$$g = \frac{1}{2}\int_{-1}^{1}\bar{p}(\tau,\mu,\mu')\mu\,\mathrm{d}\mu. \tag{14.4.26}$$

Note that g varies from -1 for complete backscattering, to 0 for isotropic scattering, and to 1 for forward scattering.

In order to calculate the intensity of solar radiation with the presence of clouds, one needs to solve (14.4.22), which can be viewed in terms of three key properties: layer optical thickness (τ), layer single-scattering albedo ($\tilde{\omega}_o$), and asymmetric parameter (g). Due to the above complications, it is necessary to make some approximations for solving (14.4.22). After some manipulation, (14.4.22) may be approximated by two simultaneous differential equations and lead to the *Eddington approximation, two-stream approximation*, or to four first-order differential equations and lead to the *four-stream approximation* (see reviews in Liou 2002 and Stephens *et al.* 2001). Fu *et al.* (1997) have proposed a hybrid two-stream and four-stream method, which has been adopted by some mesoscale models.

References

André, J. C., G. DeMoor, P. Lacarcere, G. Therry, and R. du Vachat, 1978. Modeling the 24-hour evolution of the mean and turbulent structures of the planetary boundary layer. *J. Atmos. Sci.*, **35**, 1862–83.

Anthes, R. A., 1977. A cumulus parameterization scheme utilizing a one-dimensional cloud model. *Mon. Wea. Rev.*, **105**, 270–86.

Arakawa, A. and W. H. Schubert, 1974. Interaction of a cumulus cloud ensemble with the large scale environment. Part I. *J. Atmos. Sci.*, **31**, 674–701.

Arking, A. and K. Grossman, 1972. The influence of line shape and band structure on temperatures in planetary atmospheres. *J. Atmos. Sci.*, **29**, 937–49.

Arya, S. P., 2001. *Introduction to Micrometeorology*. 2nd edn., Academic Press.

Batchelor, G. K., 1967. *An Introduction to Fluid Dynamics*. Cambridge University Press.

Bechtold, P., E. Bazile, F. Guichard, P. Mascart, and E. Richard, 2001. A mass-flux convective scheme for regional and global models. *Quart. J. Roy. Meteor. Soc.*, **127**, 869–86.

Betts, A. K. and M. J. Miller, 1986. A new convective adjustment scheme. Part II. Single column tests using GATE wave, BOMEX, ATEX and arctic air-mass data sets. *Quart. J. Roy. Meteor. Soc.*, **112**, 693–709.

Betts, A. K. and M. J. Miller, 1993. The Betts-Miller Scheme. In *The Representation of Cumulus Convection in Numerical Models*, K. A. Emanuel and D. J. Raymond (eds.), Meteor. Monogr., No. 46, Amer. Meteor. Soc., 107–122.

Blackadar, A. K., 1979. High resolution models of the planetary boundary layer. *Adv. Environ. Sci. Eng.*, **1**, 50–85.

Businger, J. A., 1973. Turbulent transfer in the atmosphere surface layer. In *Workshop in Micrometeorology*, D. A. Hangen (ed.), Amer. Meteor. Soc., Chapter 2.

Chen, S.-H. and W.-Y. Sun, 2002. A one-dimensional time-dependent cloud model, *J. Meteor. Soc. Japan*, **80**, 99–118.

Chou, M.-D., 1986. Atmospheric solar heating rate in the water vapor bands. *J. Clim. Appl. Meteor.*, **25**, 1532–42.

Chou, M.-D., 1992. A solar radiation model for use in climate studies. *J. Atmos. Sci.*, **49**, 762–72.

Chou, M.-D. and A. Arking, 1980. Computation of infrared cooling rates in the water vapor bands. *J. Atmos. Sci.*, **37**, 855–67.

Chou, M.-D., K.-T. Lee, S.-C. Tsay, and Q. Fu, 1999. Parameterization for cloud longwave scattering for use in atmospheric models. *J. Climate*, **12**, 159–69.

Cotton, W. R., 1986. Averaging and the parameterization of physical processes in mesoscale models. In *Mesoscale Meteorology and Forecasting*, P. S. Ray (ed.), Amer. Meteor. Soc., 614–35.

Cotton, W. R., G. J. Tripoli, R. M. Rauber, and E. A. Mulvihill, 1986. Numerical simulation of the effects of varying ice crystal nucleation rates and aggregation processes on orographic snowfall. *J. Clim. Appl. Meteor.*, **25**, 1658–80.

Deardorff, J. W., 1972. Theoretical expression for the countergradient vertical heat flux. *J. Geophys. Res.*, **77**, 5900–4.

Deardorff, J. W., 1974. Three-dimensional numerical study of the height and mean structure of a heated planetary boundary layer. *Bound.-Layer Meteor.*, **7**, 81–106.

Driedonks, A. G. M. and H. Tennekes, 1984. Entrainment effects in the well-mixed atmospheric boundary layer. *Bound.-Layer Meteor.*, **30**, 75–105.

Ferrier, B. S., 1994. A double-moment multiple-phase four-class bulk ice scheme. Part I: Description. *J. Atmos. Sci.*, **51**, 249–80.

Frank, W. M. and J. Molinari, 1993. Convective adjustment. In *The Representation of Cumulus Convection in Numerical Models*, K. A. Emanuel and D. J. Raymond (eds.), Meteor. Monogr., No. 46, Amer. Meteor. Soc., 101–5.

Fritsch, J. M. and C. F. Chappell, 1980. Numerical prediction of convectively driven mesoscale pressure systems. Part I: Convective parameterization. *J. Atmos. Sci.*, **37**, 1722–33.

Fu, Q. and K. N. Liou, 1992. On the correlated k-distribution method for radiative transfer in nonhomogeneous atmospheres. *J. Atmos. Sci.*, **49**, 2139–56.

Fu, Q., K. N. Liou, M. C. Cribb, T. P. Charlock, and A. Grossman, 1997. Multiple scattering parameterization in thermal infrared radiative transfer. *J. Atmos. Sci.*, **54**, 2799–812.

Goody, R. M., 1964a. *Atmospheric Radiation I: Theoretical Basis*. Clarendon Press.

Goody, R. M., 1964b. The transmission of radiation through an inhomogeneous atmosphere. *J. Atmos. Sci.*, **21**, 575–81.

Grell, G. A., 1993. Prognostic evaluation of assumptions used by cumulus parameterization. *Mon. Wea. Rev.*, **121**, 764–87.

Henyey, L. G. and J. L. Greenstein, 1941. Diffuse radiation in the galaxy. *Astrophys. J.*, **93**, 70–83.

Holton, J. R., 2004. *Introduction to Dynamic Meteorology*. 4th edn., Elsevier Academic Press.

Holtslag, A. A. M. and C.-H. Moeng, 1991. Eddy diffusivity and countergradient transport in the convective atmospheric boundary layer. *J. Atmos. Sci.*, **48**, 1690–8.

Hong, S.-Y. and H.-L. Pan, 1996. Nonlocal boundary layer vertical diffusion in a medium-range forecast model. *Mon. Wea. Rev.*, **124**, 2322–39.

Hong, S.-Y., J. Dudhia, and S.-H. Chen, 2004. A revised approach to ice microphysical processes for the bulk parameterization of clouds and precipitation. *Mon. Wea. Rev.*, **132**, 103–20.

Kain, J. S., 2004. The Kain–Fritsch convective parameterization: An update. *J. Appl. Meteor.*, **43**, 170–81.

Kain, J. S. and J. M. Fritsch, 1993. Convective parameterization for mesoscale models: The Kain-Fritsch Scheme. In *The Representation of Cumulus Convection in Numerical Models*, K. A. Emanuel and D. J. Raymond (eds.), Meteor. Monogr., No. 46, Amer. Meteor. Soc., 165–70.

Kessler, E., 1969. *On the Distribution and Continuity of Water Substance in Atmospheric Circulations*. Meteor. Monogr., No. 32, Amer. Meteor., Soc.

Kratz, D. P., M.-D. Chou, M. M.-H. Yan, and C.-H. Ho, 1998. Minor trace gas radiative forcing calculations using the k-distribution method with one-parameter scaling. *J. Geophys. Res.*, **103**, 31 647–56.

Krishnamurti, T. N., Y. Ramanathan, H.-L. Pan, R. J. Pasch, and J. Molinari, 1980. Cumulus parameterization and rainfall rates I. *Mon. Wea. Rev.*, **108**, 465–72.

Kuo, H.-L., 1965. On formation and intensification of tropical cyclones through latent heat release by cumulus convection. *J. Atmos. Sci.*, **22**, 40–63.

Kuo, H.-L., 1974. Further studies of the parameterization of the influence of cumulus convection on large-scale flow. *J. Atmos. Sci.*, **31**, 1232–40.

Kuo, Y.-H. and R. A. Anthes, 1984. Semiprognostic tests of Kuo-type cumulus parameterization schemes in an extratropical convective system. *Mon. Wea. Rev.*, **112**, 1498–509.

Kurihara, Y., 1973. A scheme of moist convective adjustment. *Mon. Wea. Rev.*, **101**, 547–53.

Lacis, A. A. and J. E. Hansen, 1974. A parameterization for the absorption of solar radiation in the earth's atmosphere. *J. Atmos. Sci.*, **31**, 118–33.

Lin, H.-M., P.-K. Wang, and B. E. Schlesinger, 2005. Three-dimensional nonhydrostatic simulations of summer thunderstorms in the humid subtropics versus High Plains. *Atmos. Res.*, **78**, Issue 1, 103–45.

Lin, Y.-L., R. D. Farley, and H. D. Orville, 1983. Bulk parameterization of the snow field in a cloud model. *J. Clim. Appl. Meteor.*, **22**, 1065–92.

Liou, K.-N., 1992. *Radiation and Cloud Processes in the Atmosphere.* Oxford University Press.

Liou, K.-N., 2002. *An Introduction to Atmospheric Radiation.* 2nd edn., Academic Press, New York.

Liou, K.-N. and S. C. S. Ou, 1981. Parameterization of infrared radiative transfer in cloudy atmosphere. *J. Atmos. Sci.*, **38**, 2707–16.

Lord, S. J., H. E. Willoughby, and J. M. Piotrowicz, 1984. Role of a parameterized ice-phase microphysics in an axisymmetric, nonhydrostatic tropical cyclone model. *J. Atmos. Sci.*, **41**, 2836–48.

Manabe, S., J. Smogorinski, and R. F. Strickler, 1965. Simulated climatology of a general circulation model with a hydrological cycle. *Mon. Wea. Rev.*, **93**, 769–98.

Marshall, J. S. and W. M. Palmer, 1948. The distribution of raindrops with size. *J. Meteor.*, **5**, 165–6.

Mason, P. J., 1994. Large-eddy simulations: A critical review of the technique. *Quart. J. Roy. Meteor. Soc.*, **120**, 1–26.

McNider, R. T. and R. A. Pielke, 1981. Diurnal boundary-layer development over sloping terrain. *J. Atmos. Sci.*, **38**, 2198–212.

Mellor, G. L. and T. Yamada, 1974. A hierarchy of turbulence closure models for planetary boundary layers. *J. Atmos. Sci.*, **31**, 1791–806.

Meyers, M. P., R. L. Walko, J. Y. Harrington, and W. R. Cotton, 1997. New RAMS cloud microphysics parameterization. Part II: The two-moment scheme. *Atmos. Res.*, **45**, 3–39.

Milbrandt, J. A. and M. K. Yau, 2005. A multimoment bulk microphysics parameterization. Part I: Analysis of the role of the spectral shape parameter. *J. Atmos. Sci.*, **62**, 3051–64.

Miyakoda, K., J. Smagorinski, R. F. Strickler, and G. D. Hembree, 1969. Experimental extended predictions with a nine-level hemispheric model. *Mon. Wea. Rev.*, **97**, 1–76.

Mlawer, E. J., S. J. Taubman, P. D. Brown, M. J. Iacono, and S. A. Clough, 1997. Radiative transfer for inhomogeneous atmosphere: RRTM, a validated correlated-k model for the longwave. *J. Geophys. Res.*, **102**, D14, 16663–82.

Molinari, J., 1993. An overview of cumulus parameterization in mesoscale models. In *The Representation of Cumulus Convection in Numerical Models*, K. A. Emanuel and D. J. Raymond (eds.), Meteor. Monogr., No. 46, Amer. Meteor. Soc., 155–8.

Morrison, H. and J. O. Pinto, 2005. Mesoscale modeling of springtime arctic mixed-phase stratiform clouds using a new two-moment bulk microphysics scheme. *J. Atmos. Sci.*, **62**, 3683–704.

Oke, T. R., 1987. *Boundary Layer Climates.* 2nd edn., Methuen.

Ooyama, V. K., 1982. Conceptual evolution of the theory and modeling of the tropical cyclone. *J. Meteor. Soc. Japan*, **60**, 369–79.

Orville, H. D. and F. J. Kopp, 1977. Numerical simulation of the history of a hailstorm. *J. Atmos. Sci.*, **34**, 1596–618.

Perkey, D. J. and C. W. Kreitzberg, 1993. A method of parameterizing cumulus transports in a mesoscale primitive equation model: The sequential plume scheme. In *The Representation of Cumulus Convection in Numerical Models*, K. A. Emanuel and D. J. Raymond (eds.), Meteor. Monogr., No. 46, Amer. Meteor. Soc., 171–7.

Pielke, R. A., 2002. *Mesoscale Meteorological Modeling.* Academic Press.

Raymond, D. J., 1994. Cumulus convection and the Madden-Julian oscillation of the tropical troposphere. *Physica D*, **77**, 1–22.

Raymond, D. J. and K. A. Emanuel, 1993. The Kuo Cumulus Parameterization. In *The Representation of Cumulus Convection in Numerical Models*, K. A. Emanuel and D. J. Raymond (eds.), Meteor. Monogr., No. 46, Amer. Meteor. Soc., 145–7.

Reisner, J., R. M. Rasmussen, and R. T. Bruintjes, 1998. Explicit forecasting of supercooled liquid water in winter storms using the MM5 mesoscale model. *Quart. J. Roy. Meteor. Soc.*, **124**, 1071–107.

Reynolds, O., 1895. On the dynamical theory of incompressible viscous fluids and the determination of the criterion. *Phil. Trans. Roy. Soc. London*, **A174**, 935–82.

Rodgers, C. D., 1967. The use of emissivity in atmospheric radiation calculations. *Quart. J. Roy. Meteor. Soc.*, **93**, 43–54.

Rothman, L. S., R. R. Gamache, A. Goldman, L. R. Brown, R. A. Toth, H. M. Pickett, R. L. Poynter, J.-M. Flaud, C. Camy-Peyret, A. Barbe, N. Husson, C. P. Rinsland, and M. A. H. Smith, 1987. The HITRAN database: 1986 edition. *Appl. Opt.*, **26**, 4058–97.

Rutledge, S. A. and P. V. Hobbs, 1984. The mesoscale and microscale structure and organization of clouds and precipitation in midlatitude cyclones. XII: A diagnostic modeling study of precipitation development in narrow cold-frontal rainbands. *J. Atmos. Sci.*, **41**, 2949–72.

Seaman, N. L., 1999. Cumulus parameterization. COMET Faculty Course on Numerical Weather Prediction. June 7–11, NCAR, Boulder, CO.

Seifert, A. and K. D. Beheng, 2006. A two-moment cloud microphysics parameterization for mixed-phase clouds. Part I: Model description. *Meteor. Atmos. Phys.*, **92**, 45–66.

Stephens, G. L., 1978a and b. Radiative properties of extended water clouds. Parts I and II. *J. Atmos. Sci.*, **35**, 2111–32.

Stephens, G. L., 1984. The parameterization of radiation for numerical weather prediction and climate models. *Mon. Wea. Rev.*, **112**, 826–67.

Stephens, G. L., P. M. Gabriel, and P. T. Partain, 2001. Parameterization of atmospheric radiative transfer. Part I: Validity of simple models. *J. Atmos. Sci.*, **58**, 3391–409.

Stull, R. B., 1988. *An Introduction to Boundary Layer Meteorology.* Kluwer Academic Publishers.

Sun, W.-Y. and C.-Z. Chang, 1986. Diffusion model for a convective layer. Part I: Numerical simulation of convective boundary layer. *Clim. Appl. Meteor.*, **25**, 1445–53.

Tao, W.-K., J. Simpson, D. Baker, S. Braun, M.-D. Chou, B. Ferrier, D. Johnson, A. Khain, S. Lang, B. Lynn, C.-L. Shie, D. Starr, C.-H. Sui, Y. Wang, and P. Wetzel, 2003. Microphysics, radiation, and surface processes in the Goddard Cumulus Ensemble (GCE) model. *Meteor. Atmos. Phys.*, **82**, 97–137.

Troen, I. and L. Mahrt, 1986. A simple model of the atmospheric boundary layer: Sensitivity to surface evaporation. *Bound.-Layer Meteor.*, **37**, 129–48.

Tsay, S.-C., K. Stamnes, and K. Jayaweera, 1989. Radiative energy budget in the cloudy and hazy arctic. *J. Atmos. Sci.*, **46**, 1002–18.

Wyngaard, J. C. and R. A. Brost, 1984. Top-down and bottom-up diffusion of a scalar in the convective boundary layer. *J. Atmos. Sci.*, **41**, 102–12.

Problems

14.1 Derive the mean equations of (14.1.3)–(14.1.7) by applying Reynolds averaging.

14.2 Derive (14.2.16).

14.3 (a) Derive (14.2.17). (Hint: you need to integrate the reduced time-independent equation of (14.1.3)–(14.1.4) from surface at $z = 0$ to the top of the mixed layer, $z = h$) (b) Estimate $\overline{V}$ for a mixed layer of $\bar{u}_g = 10 \, \text{m s}^{-1}$, $C_d = 1.5 \times 10^{-3}$, $f = 10^{-4} \text{s}^{-1}$, and $h = 1 \, \text{km}$.

14.4 Obtain the solution (14.2.26) from (14.2.25) and the boundary conditions described in the text.

14.5 Derive (14.2.30).

14.6 Derive (14.3.18) by neglecting the sensible heat flux term in (14.3.16).

Appendices

610

F	(i) Froude number ($F = U/\sqrt{gH}$) for one-layer shallow water system; (ii) $F = U/Nh$ for stratified flow over mountains; (iii) $F = NH/c_0 = NH/\sqrt{g'H}$ for a two-layer fluid system; (iv) frictional force; (v) frontogenesis function; or (vi) radiative flux
F_r	thermal Froude number ($= U/Nd$)
$F^\uparrow, F^\downarrow$	upward, downward radiative flux
F_h	enthalpy flux
F_m	saturated moist Froude number
F_N	difference between downward ($F^\downarrow$) and upward ($F^\uparrow$) fluxes
F_{rx}, F_{ry}, F_{rz}	frictional (viscous) force in x, y, and z directions
F_w	unsaturated moist Froude number
g	gravitational acceleration; asymmetric factor
h	(i) scale for the depth of planetary boundary layer; (ii) fluid depth in a shallow water system; (iii) mountain height; (iv) Planck constant; or (v) enthalpy
h_m	mountain height
h'	vertical displacement in a shallow water system
h_s	bottom topography; height function of mountain geometry
h_0^*	saturated moist enthalpy
H	scale height of a stratified fluid ($H \equiv \gamma R\overline{T}/g$); depth of a homogeneous fluid layer; effective heating
H^*	wind reversal level
I_ν	incident radiance
J	total vorticity flux; Jacobian differential operator
J_ν	source of radiation
k	wave number in x direction; von Kármán constant
k_ν	extinction coefficient; absorption coefficient
$\mathbf{k}$	wave number vector ($= (k, l, m)$)
K	horizontal wave number ($= \sqrt{k^2 + l^2}$); kinetic energy; Boltzmann's constant (1.380×10^{-23} J K^{-1})
K_m	exchange coefficient of momentum (eddy viscosity)
K_h, K_q	exchange coefficients or eddy diffusivities of heat, water vapor
L	horizontal length scale or wavelength; latent heat of condensation; Monin–Obukov length
$L_x (L_y, L_z)$	horizontal length scale of the disturbance in x (y, z) direction
L_R	internal Rossby radius of deformation ($= NL_z/f$)
L_z	vertical length scale of a disturbance
m	wave numbers in z direction
m_v	molecular weight of water vapor
m_d	molecular weight of dry air
l	wave number in y direction; Scorer parameter ($l^2 = N^2/U^2 - U_{zz}/U$)
M	nondimensional mountain height ($= h_m/H$) in a shallow water system; absolute momentum
$\overline{M}$	geostrophic absolute momentum in x-direction
M_l	vertically integrated horizontal convergence of cloud water
M_g	geostrophic absolute momentum in y-direction
M_v	moisture convergence
MF	vertical transport of horizontal momentum (momentum flux)
MPV$_g^*$	saturated geostrophic potential vorticity

N	Brunt–Vaisala (buoyancy) frequency
N_{m}	saturated moist Brunt–Vaisala (buoyancy) frequency
N_{w}	unsaturated moist Brunt–Vaisala (buoyancy) frequency
p	pressure
p'	perturbation pressure
$\bar{p}$	basic-state pressure; scattering phase function
p'_{d}	dynamic pressure
p'_{b}	buoyancy pressure
p_{r}	reference pressure
p_{s}	constant reference pressure (1000 hPa)
q	potential vorticity
q_{c}	mixing ratio of cloud water
q_{g}	geostrophic potential vorticity
q^*_{g}	saturated geostrophic potential vorticity
q'_{pv}	perturbation potential vorticity
q_{r}	mixing ratio of rain water
q_{T}	total water–substance mixing ratio
q_{v}	water vapor mixing ratio
q_{vs}	saturation water vapor mixing ratio
q_{w}	total water mixing ratio
q^*	specific humidity
$\dot{q}$	diabatic heating rate (J kg^{-1} s^{-1})
$\dot{Q}$	diabatic heating rate per unit volume ($=(\theta/c_p T)\dot{q}$)
Q_{o}	maximum diabatic heating rate
$\mathbf{Q}$	Q vector
Q_{c}	cumulus heating
P	precipitation rate
r	radius; radial distance from the rotating axis of a cyclone
r_{c}	core radius
r_{o}	radius of updraft
R	radius of maximum wind scale; radius of circular motion; rainfall rate
R_{B}	bulk Richardson number (8.3.8)
R_{d}	gas constant for dry air
R_e	Reynolds number ($= UL/v$)
Re	real part of a complex number
R_i	gradient Richardson number ($= N^2/U_z^2$)
R_o	Rossby number ($= U/fL$ or U/fa); Lagrangian Rossby number ($= \omega/f = 2\pi/fT$)
R_ν	reflectivity for radiation
s	entropy ($ds = dq/T$)
s_{d}	dry entropy
s^*	moist (saturation) entropy
S	(i) initial perturbation PV distribution and forcing associated with momentum and diabatic sources; (ii) environmental wind shear vector; (iii) swirl ratio ($= v_{\mathrm{o}}/w_{\mathrm{o}}$); or (iii) vertical wind shear
S_{o}	solar flux
$S_{\mathrm{o}\nu}$	solar irradiance
S_{qv}	storage rate of water vapor
S_{ql}	storage rate of liquid water

S_θ	any source or sink of θ_v
S_ϕ	any source or sink of the hydrometeor ϕ
S^-	downward solar flux at level z
T	temperature; time scale; period of oscillation
T_o	constant reference temperature
T_s	sea surface temperature
T_{LCL}	temperature at the lifting condensation level
T_v	virtual temperature
T_ν	transmissivity
u	velocity in x direction
u'	perturbation velocity in x direction
u_g	geostrophic wind velocity in x direction
u_a	ageostrophic wind velocity in x direction
$u*$	path length
u_*	friction velocity
U	basic wind velocity in x direction
U_z	basic wind shear in x direction
U_k	terminal velocities of precipitating hydrometeors
U^*	scale for friction velocity
v	velocity in y direction
v'	perturbation velocity in y direction
v_a	ageostrophic velocity in y direction
v_g	geostrophic velocity in y direction
v_c	tangential velocity of the inflow at r_c (radius of updraft)
V	basic velocity in y direction
V_z	basic wind shear in y direction
V_a	ageostrophic wind velocity
V_g	geostrophic wind velocity
V_H	horizontal wind velocity
V_T	maximum tangential wind velocity or scale
V_ψ	rotational wind component
W	work
w	vertical velocity
w_o	amplitude of w
w'	perturbation w
w_{oro}	orographically-forced upward motion
z_c	critical level height
z_o	roughness length; aerodynamic roughness
z_S	height of the lower surface in the $\sigma - z$ coordinates
z_T	height of a numerical model domain
α	phase angle; specific volume
β	meridional gradient of the Coriolis parameter; diffusivity factor
ε	ratio of molecular weight of water vapor (m_v) to that of dry air (m_d); thermodynamic efficiency
Φ	geopotential function
ψ	streamfunction
ϕ	geopotential; latitude
ϕ'	perturbation geopotential
η	vertical displacement; y-component of the vorticity vector

Γ	environmental lapse rate
Γ_s	moist adiabatic lapse rate
Γ_d	dry adiabatic lapse rate
γ	c_p/c_v; environmental (actual) lapse rate ($\gamma \equiv -\partial T/\partial z$)
κ	eddy thermal diffusivity
λ_z	hydrostatic vertical wavelength ($= 2\pi U/N$)
λ_R	external Rossby radius of deformation ($= c/f$),
μ	dynamic viscosity coefficient ($= \rho\nu$)
ν	molecular kinematic viscosity coefficient; eddy viscosity
ν_c	numerical diffusion coefficient
Ω	Doppler-shifted or intrinsic frequency ($\equiv \omega - kU$); or Earth's rotation rate (7.272×10^{-5} s^{-1})
π	Exner function; kinematic pressure
π_b	buoyancy pressure (in Exner function)
π_d	dynamic pressure (in Exner function)
π'	perturbation pressure (Exner function); perturbation kinematic pressure ($= p'/\rho_o$)
ρ	air density
ρ_{av}	averaged air density of a layer
$\overline{\rho}$	basic-state air density
ρ_o	constant reference density
ρ_s	air density near surface
ρ'	perturbation density
Θ	maximum temperature anomaly of a warm region
θ	potential temperature
θ_o	constant reference potential temperature; zenith angle
θ'	perturbation potential temperature
$\overline{\theta}$	basic-state or environmental potential temperature
θ_e	equivalent potential temperature
θ_e^*	saturation equivalent potential temperature
$\overline{\theta}_e^*$	environmental equivalent potential temperature
$\overline{\theta}_e$	saturation equivalent potential temperature of the environmental air
θ_v	virtual potential temperature
θ_w	wet-bulb potential temperature
$\dot{\theta}$	diabatic heating rate
θ_*	flux temperature
θ_{ma}	moist adiabat
σ	isentropic coordinates; Stefan–Boltzmann constant (5.67032×10^{-8} Wm^{-2} K^{-4})
τ	period of oscillation; optical thickness; spin-up time for a tropical cyclone
τ_b	oscillation period due to buoyancy
τ_{advec}	advection time scale
τ_f	raindrop falling time scale
τ_c	cloud growth time scale
τ_*	optical thickness
τ_ν^f	diffusion transmission function or slab transmission function
ζ	vertical vorticity
ζ_a	absolute vertical vorticity ($= \zeta + f$)
ζ_g	geostrophic vorticity

ζ_s	environmental streamwise vorticity
ω	frequency; vertical velocity in the isobaric coordinates ($\equiv Dp/Dt$)
$\tilde{\omega}_o$	single-scattering albedo
$\boldsymbol{\omega}$	three-dimensional vorticity vector
$\boldsymbol{\omega}_a$	three-dimensional absolute vorticity vector
$\hat{}$ or $\tilde{}$	Fourier or Laplace transformed variables

Appendix B Nomenclature

Terminology	Meaning
ACARS	Aircraft Communication Addressing and Reporting System
ADAS	ARPS Data Analysis System
ALPEX	Alpine Experiment field project held in 1982
ARPS	Advanced Research Prediction System
BDO	Benjamin, Davis and Ono (for BDO or algebraic solitary waves)
BMP	bulk microphysics parameterization
BWER	bounded weak echo region
CBL	convective boundary layer
CAPE	convective available potential energy
CAT	clear air turbulence
CCOPE	Cooperative Convective Precipitation Experiment
CDGA	continuous dynamic grid adaptation
CFL	Courant–Friedrichs–Lewy stability criterion
CFA	cold front aloft
CIN	convective inhibition
CISK	Conditional Instability of the Second Kind
CMR	Central Mountain Range (Taiwan)
CP	cumulus parameterization
CSI	conditional symmetric instability
CTD	coastally trapped disturbance
DCAPE	downdraft convective available potential energy
DNS	direct numerical simulation
ECMWF	European Centre for Medium-Range Weather Forecasts
EH	Ethiopian Highlands
EnKF	ensemble Kalman filter
FFD	forward flank downdraft
FFLS	front-fed leading stratiform mesoscale convective system (MCS)
FFT	Fast Fourier Transform
GCM	general circulation model
GOES	Geostationary Operational Environmental Satellite
GFU	gust front updraft
HCR	horizontal convective rolls
HP	high precipitation supercell
IOP	Intensive Observation Period (in a field program)
IR	infrared radiation
ITCZ	Intertropical Convergence Zone
KdV	Korteweg and de Vries (for KdV or classical solitary waves)
KF	Kain–Fritsch cumulus parameterization scheme
Kessler scheme	Kessler (1969) warm rain scheme

K–H Instability	Kelvin–Helmholtz instability
LAPS	Local Analysis and Prediction System
LBL	line-by-line methods for integration over optical path
LCZ	leeside convergence zone
LCL	lifting condensation level
LES	large eddy simulation
LFC	level of free convection
LFO scheme	Lin–Farley–Orville (1983) microphysical parameterization scheme
LLJ	low-level jet
LLWM	low-level wind maximum
LNB	level of neutral buoyancy
LP	low precipitation supercell
LS MCS	leading stratiform precipitation mesoscale convective system
MAI	moist absolute instability
MAP	Mesoscale Alpine Program
MAPS	Mesoscale Analysis and Prediction System
MAUL	moist absolutely unstable layer
MBEs	meso-β scale convective elements
MCC	mesoscale convective complex
MCS	mesoscale convective system
MCV	mesoscale convective vortex
MJO	Madden–Julian oscillation
MM5	Penn State-NCAR Mesoscale Modeling System version 5
MPI	maximum potential intensity
MPS	mountain–plain solenoidal circulation
NA	negative area on a thermodynamic diagram
NCAR	National Center for Atmospheric Research
NCEP	National Centers for Environmental Prediction
NEXRAD	Next Generation Weather Radar
NWP	numerical weather prediction
OMEGA	Operational Mesoscale Environmental model with Grid Adaptivity
PA	positive area on a thermodynamic diagram
PBE	potential buoyant energy
PDCB	primary dryline convergence boundary
PE	primitive equations
PBL	planetary boundary layer
PS MCS	parallel stratiform precipitation mesoscale convective system
PSI	potential symmetric instability
PV	potential vorticity
QG	quasi-geostrophic approximation
RANS	Reynolds-averaged Navier–Stokes equations
RFD	rear flank downdraft
RFLS	rear-fed leading stratiform MCS
RH	relative humidity
RMW	radius of maximum wind
RKW theory	Rotunno–Klemp–Weisman (1988) theory for the longevity of a squall line
rms	root-mean-square

RUC	Rapid Update Cycle
SALLJ	South American Low Level Jet
SCAPE	slantwise convective available potential energy
SCIN	slant-wise convective inhibition
SCPP	Sierra Cooperative Pilot Project
SG	Semi-geostrophic model
SI	symmetric instability
SR	severe right storm or right mover
SREH	storm-relative environmental helicity
SRH	storm relative helicity
SSM/I	Special Sensor Microwave/Imager
SST	sea surface temperature
SSTA	sea surface temperature anomaly
TAMEX	Taiwan Area Mesoscale Experiment
TC	tropical cyclone
TKE	turbulent kinetic energy
TS	tropical storm
TS MCS	trailing stratiform precipitation mesoscale convective system
TUTT	tropical upper tropospheric trough
TVS	tornado vortex signature
USL	updraft source layer
UTC	Universal Time Coordinated
UV	ultraviolet radiation
VAD	velocity azimuth display
WADL	West African disturbance line
wave-CISK	wave-CISK mechanism
WER	weak echo region
WISHE	wind-induced surface-heat exchange
WKBJ	an asymptotic method named after G. Wentzel, H. A. Kramers, M. L. Brillouin, and Sir H. Jeffreys (e.g., Olver 1997)
WRF	Weather and Research Forecast model
3DVAR	three-dimensional variational data assimilation
4DVAR	four-dimensional variational data assimilation
4DDA	four-dimensional data assimilation

Index